AF540569

PLANT BIOSECURITY AND BIOSAFETY

NIPA® GENX ELECTRONIC RESOURCES & SOLUTIONS P. LTD.

New Delhi-110 034

About the Author

Dr. N.G. Ravichandra, Professor of Plant Pathology and Scheme Head, All India Co-ordinated Research Project (Nematodes), Department of Plant Pathology, University of Agricultural Sciences, GKVK, Bengaluru, Karnataka, India, is involved in teaching, research and extension activities. He has been a major advisor to the Post-Graduate students majoring in Plant Pathology for M.Sc., (Agri.) and Ph.D. Degree Programmes.

Dr. N.G. Ravichandra has authored ten text books in Plant Pathology & Nematology, has 269 scientific research papers published in National and International journals, contributed chapters to twelve text books, edited five books and also authored ten books in Kannada language.

Dr. N.G. Ravichandra has identified and reported a new species of root-knot nematode (*Meloidogyne enterolobii*) on Guava and a new host, Lac plant (*Flemingia semialalta* L.) for root-knot nematode (*Meloidogyne incognita*) from Karnataka. These are the new and first reports from India.

Dr. N.G. Ravichandra has been bestowed with three National awards for his research accomplishments and contributions. He has been recognized as the International Editor and Reviewer for three prestigious International Journals and National Editor for the eminent 'Indian Phytopathology' Journal. He has developed nine technologies for the management of root-knot and burrowing nematodes infecting tomato, brinjal, rice and banana, which have been included in the Package of Practices of University of Agricultural Sciences, Bengaluru, Karnataka, India.

Dr. N.G. Ravichandra has operated research projects funded by the Department of Bio-Technology & Indian Council of Agricultural Research, New Delhi, India as Principal Investigator and Co- Principal Investigator ; participated and presented original research papers on various aspects of plant diseases in National and International conferences; undergone advanced training programs sponsored by the Department of Biotechnology & Indian Council of Agricultural Research, New Delhi, India.

Dr. N.G. Ravichandra is an active life member of seven professional societies including Nematological Society of India, Indian Phytopathological Society, Society of Mycology & Plant Pathology, National Environmental Science Academy, Institute of Agricultural Technology, Glacier Journal Research Foundation and Institute of Scholars.

PLANT BIOSECURITY AND BIOSAFETY

N.G. Ravichandra
Professor of Plant Pathology & Scheme Head, AICRP (Nematodes)
Department of Plant Pathology
College of Agriculture
University of Agricultural Sciences, GKVK
Bengaluru-560 065, Karnataka, India

NIPA® GENX ELECTRONIC RESOURCES & SOLUTIONS P. LTD.
New Delhi-110 034

NIPA® GENX ELECTRONIC RESOURCES & SOLUTIONS P. LTD.

101,103, Vikas Surya Plaza, CU Block
L.S.C.Market, Pitam Pura, New Delhi-110 034
Ph : +91 11 27341616, 27341717, 27341718
E-mail: newindiapublishingagency@gmail.com
www: www.nipabooks.com
For customer assistance, please contact
Phone: + 91-11-27 34 17 17
Fax: + 91-11-27 34 16 16
E-Mail: feedbacks@nipabooks.com

ISBN: 978-93-95319-06-5

Composed and Designed by NIPA®.

To
The Department of Plant Pathology
UAS, GKVK, Bangalore

Acknowledgement

My respectful thanks to

Dr. S. Rajendra Prasad, Hon'ble Vice Chancellor, University of Agricultural Sciences, GKVK, Bangalore, Karnataka, India

Dr. K.B. Umesh, Director of Research, University of Agricultural Sciences, GKVK, Bangalore

Dr. Anil Sirohi, Project Coordinator, AICRP (Nematodes), Indian Agricultural Research Institute, New Delhi

My special thanks to Dr. Y.M. Somashekhara, Professor & Head, Dr. T. Narendrappa, Dr. N. Nagaraju, Dr. H.A. Prameela, Dr. H.K. Ramappa, Dr. T.R. Kavitha and Mr. B.R. Nagendra, Department of Plant Pathology, UAS, GKVK, Bangalore.

I would like to express my deep sense of gratitude to my parents Sri. N. Gurushankar and Smt. Parvathamma.

I am highly grateful to my in-laws Sri. M.S. Chandrashekhar and Smt. Kamala for their continued affection and encouragement.

I wish to record my appreciation and affection to my beloved wife, Ms. Deepthi and dear daughter, Ms. Anusha for their care, patience, keen interest and affection throughout the preparation of the manuscript.

My heartfelt thanks are due to my brother-in-law Mr. B.Sudesh Kumar and my beloved sister Ms. N.G. Tara, for their sustained moral support, affection and special care. I fondly acknowledge their keen interest, continued encouragement and blessings being showered on me.

My appreciation and thanks to M/s NIPA® GENX Electronic Resources & Solutions Pvt Ltd, New Delhi, India, who deserves commendation for their professional contribution.

Preface

Biosafety and biosecurity are complementary to each other for protection of human, animal, plant life and their health from harmful organisms and / or their toxins. The overall objective, however, is the protection of biodiversity. Under the present international scenario, the plant protection specialists have a major role to play not only in promoting and facilitating export and import in the interest of their respective countries, but also in protecting the environment from the onslaughts of invasive alien pests and the unforeseen ill effects of the introduction and trading of genetically modified organisms.

'Plant Biosecurity and Biosafety' is a core course prescribed by the Indian Council of Agricultural Research (ICAR), New Delhi for the Post Graduate students in Agricultural and Horticultural Universities. There is a dearth of a handy book that covers the entire syllabus prescribed by the ICAR for the course. The major objective of this book is to provide students and the faculty the recent developments and updated comprehensive information on various aspects of 'Plant Biosecurity and Biosafety'. This book fulfils the need for a comprehensive book on the fundamental and advanced aspects of 'Plant Biosecurity and Biosafety'.

The book explicates essential aspects related to 'Plant Biosecurity and Biosafety' and is conveniently divided into sixteen unique chapters, covering all the updated information and latest developments. The chapters covered include Introduction; Invasive Alien Species; Biowarfare, Bioterrorism and Bioethics; Early Warning and Forecasting System; Emerging Resurgence of Pests and Diseases; National Regulatory Mechanism and International Agreements Conventions; International Standards for Phytosanitary Measures; Pest Risk Analysis, Risk Management Models and Pest Information System; Global Positioning System and Geographic Information System for Plant Biosecurity; Pest, Disease and Epidemic Management; Agroterrorism and Biosecurity; Mitigation Planning and Integrated Approach for Biosecurity; Biosafety, History, Policies and Regulatory Mechanism ; Cartagena Protocol on Biosafety and Its Implications; Issues related to the Genetically Modified Crops; Operational Biosafety Practices and Procedures. In addition, 'Glossary of terms' is also furnished to present a list of frequently terms used in 'Plant Biosecurity and Biosafety'. Under each chapter, a bibliography has been included that provides the list of references cited for further reading. Diagrams of good quality, convincing tables, depictive graphs and appropriate illustrations have been included at suitable places for a meaningful reading.

The contents of this book, reflecting an extensive literature search, will be useful particularly for the students, teaching & research faculty and extension personnel in Agricultural and Horticultural Universities, the State Departments of Agriculture, Horticulture, Forestry, Sericulture & Fisheries, Plant Protection Organizations, Plant Quarantine Units, Administrators & Policy makers and all those who are interested and concerned with plant biosecurity and biosafety.

The author appreciates receiving suggestions and constructive criticism that would improve the quality of the book.

Bangalore

N.G. Ravichandra

Contents

1

Introduction to Plant Biosecurity

Biosecurity refers to the measures aimed at preventing the introduction and/or spread of harmful microorganisms to animals and plants in order to minimize the risk of transmission of infectious disease. Biosecurity is of growing interest as a result of developments at the international level, including globalization of the world economy, technological progress and the rapid increase in communications, transport and trade. Against this background, there is concern that the appropriate level of protection of human, animal and plant life and health is not being maintained as risks increase. Biosecurity is a strategic and integrated approach that encompasses the policy and regulatory frameworks for analyzing and managing relevant risks to human, animal and plant life and health, and associated risks to the environment. Biosecurity covers food safety, zoonoses, the introduction of animal and plant diseases and pests, the introduction and release of living modified organisms and their products (Genetically Modified Organisms) and the introduction and management of invasive alien species.

In agriculture, these measures are aimed at protecting food crops and livestock from pests, invasive species, and other organisms not conducive to the welfare of the human population. Biosecurity means putting in place procedures or measures designed to protect the population against harmful biological or biochemical substances. Biosecurity is designed to reduce the risk of transmission of infectious diseases in crops and livestock, quarantined pests, invasive alien species, and living modified organisms. These preventive measures include a combination of systems and practices put in place at bioscience laboratories, by border control, customs agents, agricultural and natural resource managers to prevent the spread of dangerous pathogens and toxins.

Biosecurity is a National security subject and should be a subset of CBRN (Chemical, Biological, Radiological and Nuclear) Security. The term includes biological threats to people, including those from pandemic diseases and bioterrorism. The definition has sometimes been broadened to embrace other concepts, and it is used for different purposes in different contexts. The COVID-19 pandemic is a recent example of a threat for which biosecurity measures have been needed in all countries of the world.

The term "biosecurity" has been defined differently by various disciplines. The term was first used by the agricultural and environmental communities to describe preventative measures against threats from naturally occurring diseases and pests, later expanded to introduced species. Australia and New Zealand, among other countries, had incorporated this definition within their legislation by 2010. New Zealand was the earliest adopter of a comprehensive approach with its Biosecurity Act 1993. In 2001, the US National Association of State Departments of Agriculture (NASDA) defined biosecurity as "the sum of risk management practices in defense against biological threats" and its main goal as "protection against the risk posed by disease and organisms".

1.1 Role of Biosecurity in Plant Health Management

Biosecurity is a set of precautions or an approach that encompasses the policy and regulatory frameworks to prevent the introduction and spread of harmful organism (insect pest, pathogen or invasive alien species). It analyses relevant risks to human, animal, plant life/health and to the environment. In general, bio-security is 'the safeguarding of resources from biological threats." Main focus of bio-security program is to restrict the entry of any threat responsible for hazardous effect on environment by collaboration with Central government (Sharma Preeti and Gaur Neeta, 2014). Plant quarantine, national and international stalk holders play a key role in plant bio security program. Pest Risk Analysis, Survey and Surveillance are the different step to detect the pest incidence and also provide a direction for the further process of investigation and control. Bio-security advocates a strategic and integrated approach, to meet the consumer expectations in relation to the food safety, preventing and controlling relevant risk and zoonotic aspects of public health, safeguarding the resources and protecting environment and biodiversity. Agricultural Bio-security Authority of India regulates the import and export of plants, animals and related products, preventing the introduction of quarantine pests from other countries and implementing post-entry quarantine measures. Biosecurity is an essential element in Plant health management by regulating different components require for plant growth and food production.

To prevent or manage the risks to plant health and to protect the environment, there is requirement for the development of a tool to manage this risk. In doing so, bio-security is an essential element to overcome this problem and direct relevance to sustainable agricultural development. In other words, bio security is a set of precautions or an approach that encompasses the policy and regulatory frameworks to prevent the introduction and spread of harmful organisms have relevant risks to human, animal, plant life/health and environment. The strategic and integrated approach of bio security provides a regulatory framework in association with higher authorities and policies to provide safety of food supply, prevent the zoonotic aspects of public health, ensure the sustainability of agriculture, and safeguard the terrestrial, freshwater and marine environments and biodiversity. Globalization in the world, ever increasing requirement of food production, adoption of new technologies, legal obligations for signatories of relevant international agreements, movement of people across borders increase the incidence of pest and disease across the world, dependence of some countries on food, import/ export are the important factors to meet the requirement of bio security at global level as well as for Plant Health Management. The increasing demand to gain more production by adopting new technologies and changing agricultural practices results in new hazards to health that are readily able to cross borders. Changing human ecology and behaviour also contribute to the greater incidence and spread of hazards. Due to which public awareness is increasing towards impact of adverse bio-security events and interventions.

Need of bio-security: In Agro ecosystem humans, plant, animal life and environment are associated to each other. If any harm taken at one stage then it will affect whole ecosystem. If homeostasis of environment disturb then it will affect all associated entities. This is the fundamental rationale for an integrated approach to bio security development (Sharma Preeti and Gaur Neeta, 2014). Bio security hazards of various types exist in each sector and have high potential to move and establish. In respect of food chains, hazards can be introduced anywhere from production to consumption level. If any pest enters though import/export or any other mean then it can affect trophic levels. At the national and international level, there are likely to be significant benefits in integrating bio security activities to overcome these problems. Plant biosecurity is a global issue that continues to grow in importance as the volume of trade between countries and the number of people travelling increases. Plant biosecurity is a continuum that draws together many different disciplines. It differs from plant protection in that it is risk management, being strategic for the future needs (McKirdy Simon *et.al.*, 2008).

The prevention and control of new pest and disease introductions is an agricultural challenge which is attracting growing public interest. This interest is in part driven by an impression that the threat is increasing, but there has been little analysis of the changing rates of biosecurity threat, and existing evidence is equivocal (Waage and Mumford, 2008). Traditional biosecurity systems for animals and plants differ substantially but are beginning to converge. Bio-economic modeling of risk will be a valuable tool in guiding the allocation of limited resources for biosecurity. The future of prevention and management systems will be strongly influenced by new technology and the growing role of the private sector. Overall, today's biosecurity systems are challenged by changing national priorities regarding trade, by new concerns about environmental effects of biological invasions and by the question 'who pays?'. Tomorrow's systems may need to be quite different to be effective. Three changes are suggested viz., an integration of plant and animal biosecurity around a common, proactive, risk-based approach; a greater focus on international cooperation to deal with threats at source; and a commitment to refocus biosecurity on building resilience to invasion into agroecosystems rather than building walls around them. Plant biosecurity has never been more important globally than it is today as plant pathogens, known and emerging, threaten food security and market access.

Areas of biosecurity: Major areas include human life and health (including food safety), animal life and health (including fish), plant life and health (including forests) apart from environmental protection. Biosecurity approach in a holistic manner used to manage biological risks associated with food and agriculture and also focus to enhance productivity, sustainability and profitability. It also covers food safety, zoonoses, the introduction of animal/plant diseases and pests, introduction/release of living modified organisms and their products (e.g. genetically modified organisms or GMOs), their introduction and management of invasive alien species (Sharma Preeti and Gaur Neeta, 2014). Thus, bio-security is a wide-ranging aspects to the public health, plant health and protection of the environment. The overarching goal of bio security is to prevent, control and/or manage risks to life and health. In accordance with, bio-security is an essential element

Process of biosecurity: Biosecurity is a strategic and integrated approach require collaboration, high level support across different agencies, their agreement on purpose, scope and process, profile the biosecurity context at the country level, assess existing bio-security capacity, describe the desired future situation of bio security, identify capacity needed to reach desired future and generate options to address identified needs are the essential components for bio security (Sharma Preeti and Gaur Neeta, 2014). Successful bio security program require conveying the information, needing to convince policy and decision-makers, to ensure transparency and best use of available resources, examine the context for bio-security at the national level. Bio-security issues, general needs, prevailing challenges and opportunities, context shape bio security goals, programs and activities for essentiality to establish bio-security as a national priority and ensure cross-sectoral collaboration and participation. Assess existing bio-security capacity by examine current situation of bio-security capacity and performance, ensure capacity building activities and provide broad framework for the successful development of bio-security system. Next steps for Bio-security program are to describe the desired future situation of bio-security (developing vision and goals across the bio security system, enhancement of the future outcome, effectiveness and maximized potential cross-sectoral gains), identify capacity needed to reach desired future and generate options (Bio security policy framework, legal and Institutional framework).

1.2 Biosecurity v/s Biosafety

Biosecurity is different from biosafety. According to WHO, biosafety is "the containment principles, technologies and practices that are implemented to prevent unintentional exposure to pathogens and toxins, or their accidental release". In other words, it is protecting yourself, others and the environment from bad bugs. The term "biosecurity" refers to the "institutional and personal security measures designed to prevent the loss, theft, misuse, diversion or intentional release of pathogens and toxins". In short, it can be said that biosecurity includes the means to protect bad bugs from bad people (Didier Breyer, 2021).

An illustration of the difference:

1. Use of PPE to prevent accidental exposure of research workers to pathogens/toxins is biosafety.
2. Preventing unauthorized access to facilities working on such pathogen/toxins is considered as biosecurity.

It is worth mentioning that, more or less 80% of what is called "biosecurity" is actually covered by "biosafety measures" as referred to an analysis done in Germany. Figure 1 shows a schematic representation of the complementarities and differences between biosafety and biosecurity with a common central goal: the protection of the Valuable Biological Materials (VBM) from misuse.

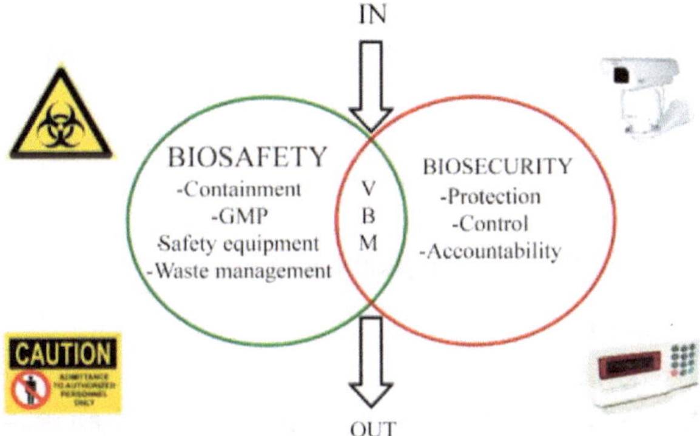

Fig. 1: GMP- Good Microbiological Practices, VBM: Valuable Biological Materials: IN and OUT refer to the complete traceability of all biological materials handled, stored and transported.

Why is biosecurity necessary?

- It holds direct relevance to sustainable agriculture, safe food production and protection of the environment and biodiversity.
- India is already susceptible to pest and weed attacks. In the last 15 years, there have been at least 10 major attacks. In recent times, India is facing invasion by locusts, fall armyworm and other pests.
- With the much-needed liberalization of the agricultural trade, the risk of introduction of non-native pests and weeds are on the rise. This poses a

serious threat to food security and economy– especially in a country where a significant portion of the population is still dependent on the agriculture sector.

- Climate change is expected to fuel threats from cross-boundary disease. Eg: Ug-99 wheat stem rust fungi and the avian influenza disease.
- Biosecurity is an important aspect of on-farm food safety programs to ensure the health of consumers.
- Along with health, biosecurity is of paramount importance to our national security.
- Extreme biological events like the ongoing COVID-19 pandemic shows the potential destruction that could ensue in the absence of biosecurity. The UN had termed it 'the biggest international crisis since World War II'.
- Not even the economically powerful nations are immune to the effects of mass contagion. The recent COVID-19 crisis has illustrated how such biological events can bring the global economy to its knees in a matter of months.
- The global concern about bioterrorism has been increasing over the years. India highlighted the threat posed by bioterrorism- terming it a 'contagious plague'- at the Shanghai Cooperation Organization's first Military Medicine Conference in September 2019.
- With the increasing advancement of technology, the creation of genetic chimaeras in the microbial world and manufacturing of increasingly hazardous compounds have become easier. There is a need to evolve the security plans to keep up with the fast-changing technologies.

1.3 Biosecurity-Global Scenario

Plant biosecurity is a global issue that continues to grow in importance as the volume of trade between countries and the number of people travelling increases (McKirdy Simon *et.al.*, 2008).Biosecurity is protecting the economy, environment and people's health from pests and disease. It includes trying to prevent new pests and diseases from arriving, and helping to control outbreaks when they do occur. While robust response arrangements are in place to combat outbreaks, preventing pest and disease incursions in the first place, remains a national priority." (Australian Government, Department of Agriculture, Fisheries and Forestry Plant biosecurity is a global issue that continues to grow in importance as the volume of trade between countries and the number of people travelling increases. Australia is free from many of the pests and diseases that affect plant industries and natural environments in other countries. This freedom provides a competitive advantage to Australia as a major agricultural exporter reliant on its international reputation as a producer of 'clean and green' agricultural and food products. Australia also places a high value on protecting our unique environment and lifestyle for future generations. Plant biosecurity is essential to protect these values. Plant biosecurity is focused on those pests (insects and plant pathogens) that are; not currently present in Australia, are present but not in all production regions of Australia and are being actively controlled, or those pests that represent a new threat as their biology has changed.

Plant biosecurity can impact on food safety, food security, trade, market access, market development, production costs and, ultimately, the profitability and sustainability of plant industries. Incursions of new pests directly threaten the economic viability of Australia's plant industries, which have an annual farm gate value of over $18 billion and annually contribute over $12 billion to export income. Even the perception that a pest is present in Australian produce can have a rapid and negative impact on Australia's reputation as a producer of safe, quality food products. The Australian lupin industry is threatened by several pests including *Sitona* spp. and *Uromyces lupinicola* (lupin rust). Both pests would significantly impact on lupin production in Australia should they be introduced.

The plant biosecurity is gaining significance with expansion of global trade of agricultural commodities and transboundary movement of plant genetic resources. The catastrophic impact through intrusion of pest and disease like historical Irish famine caused by potato late blight introduced from Central America and Golden nematode of potato introduced to India in 1960s from the UK furnished vivid evident that introduction and establishment of quarantine pest affects the crop production and economy of the country (Khetarpal and Gupta, 2007). Currently, International Plant Protection Convention (IPPC) requires all contracting parties to establish National Plant Protection Organization (NPPO) and develop international cooperation in prevention of introduction and spread of regulated pest.

In Australia, several areas of major biosecurity concerns are new race of wheat stem rust (Anonymous, 2020). Wheat stem rust epidemic in 1973 in Australia caused about $200-300 million and expected yield loss of more than 70 percent. Wheat stem rust, Ug99 is now only found in African countries and Middle East, although Ug99 is under control in Australia for more than 30 years, there is a potential threat to wheat industry as eradication and containment become difficult due to high mobility of the disease. The disease spread to larger distance by wind, movement of goods and people. Countries across the globe with strong biosecurity capabilities are also facing challenges from exotic pest and disease. Bhutan Agriculture and Food Regulatory Authority (BAFRA) as the NPPO has the mandate to protect health of plant, animal and human including the biodiversity and environment from exotic pest and disease. Currently, BAFRA is grappling with biosecurity issues with invasive species and pest such as outbreak of Giant African Land Snails (GALS) at Gyelposhing and widespread concern of Citrus greening disease in the country. Biosecurity catastrophic is yet to unveil should there be any complacency in phytosanitary measures taken and sound and improved biosecurity approach are adopted for risk management.

Today's agricultural biosecurity systems were built for the protection of national agriculture and food security. Their sectoral roots, in animal and plant protection, have strong historical features that make them different and explain their continuing segregation in most government services. Biosecurity systems have been able to exert substantial control over what comes into a country and to mount high-impact, country-wide eradication programmes. The potential trade conflicts created by different national government biosecurity systems have been substantially mitigated by international agreements which establish common standards and practices. They create a valuable platform for dialogue, and in a few circumstances they have fostered international cooperation in tackling common threats. Agricultural biosecurity remains largely as the business of governments (Waage and Mumford, 2008). Agricultural biosecurity has worked well for substantial threats where large, sustained investment has been made. Examples include the exclusion of several serious animal and plant diseases, and of fruit flies. Overall, however, biosecurity systems are 'leaky', and it is hard to resist the conclusion that in time, entropy will prevail and more pests and diseases will reach more parts of the world. Improving the 'tightness' of the existing system is certainly possible with sufficient government investment. Today, the perceived growth in the biosecurity threat is leading to calls for improvements in these systems.

Biosecurity is a term that has many interpretations. It is a key requirement for achieving the goals set out in the FAO Strategic Framework by promoting, developing and re-enforcing policy and regulatory frameworks for food, agriculture, fisheries, and forestry. Biosecurity has direct relevance to food safety, the conservation of the environment (including biodiversity), and sustainability of agriculture. Biosecurity encompasses all policy and regulatory frameworks (including instruments and activities) to manage risks associated with food and agriculture (including relevant environmental risks) including fisheries and forestry and constitute three sectors (namely food safety, plant life and health, and animal life and health) (Anonymous, 2021). These sectors include food production in relation to food safety the introduction of plant pests, animal pests and diseases and zoonoses the introduction and release of Genetically Modified Organisms (GMOs) and their products and the introduction and safe management of invasive alien species and genotypes. Growing interest in biosecurity is a result of major international developments such as globalization of the world economy, the rapid increase in volume of communications, transport and trade, technological progress and growing awareness of problems faced by biological diversity and environment. Members require effective efficient improved and updated international frameworks and standards to support appropriate national action. Members also require national frameworks to regulate manage and control biosecurity for food and agriculture including forestry and fisheries thus permitting practical implementation increasing cost effectiveness and improving consistency across sectors.

Recent developments related to biosecurity in food and agriculture includes the tendency toward integration of and cooperation across sectors. Internationally this tendency is demonstrated in the WTO Agreement on the Application of Sanitary and Phytosanitary Measures (SPS Agreement) and the Convention on Biological Diversity (CBD) and its Cartagena Protocol on Biosafety. It is further addressed in the FAO/WHO *Codex Alimentarius*, the International Plant Protection Convention (IPPC) and the FAO Code of Conduct for Responsible Fisheries. The IPPC makes a key contribution to biosecurity by reducing the risks of introduction of plant pests that may affect agriculture and the environment.

International plant biosecurity protocols have been overtaken by events – primarily global shifts in market structure and practice and by developments in scientific knowledge and are now outmoded, flawed, institutionalized, and too ineffectual. There is a need to be fully scientifically reviewed and appropriately overhauled, taking full account of the underlying scientific weaknesses and of the many other causes of security failure. Clearly, global initiatives and agreements, involving regulators and trade representatives from organizations such as the WTO, IPPC and EU, would be needed to service any such changes, The problem therefore has a marked political as well as scientific dimension (Brasie, 2008).

In relation to the EU, it should be noted that, in biosecurity terms, a considerable geographical advantage was possessed by the UK and Ireland up to the 1970s by virtue of their island status i.e. in being physically isolated. This geographical advantage was comparable to that of New Zealand and Australia; and both the latter countries have reinforced their geographic isolation by imposing what many would regard as among the world's toughest plant import regimes. It can be argued that much of the comparable advantage for the UK and Ireland was diluted in the interest of wider trade when the two countries joined the EU in the 1970s.24% of a European-wide sample of *Phytophthora ramorum* (Sudden oak death) isolates collected in 2004 were already resistant to the anti-Phytophthora chemical Metalaxyl Gold. These examples reflect the effectiveness of local science: Germany and Spain are not the member states referred to in the paragraph above. A vested interest in the status quo would not necessarily be confined to individuals within the plant trade. Internationally, plant health has an array of well meaning, highly professional inspectors, administrators, research scientists, risk analysts, and officers of plant protection organizations and of governmental and intergovernmental committees. I count myself among them. In essence, plant health has become an industry in itself. As in all professions, some may seek to stay within the comfort zone, becoming institutionalized and resistant to initiating change even when system failure is apparent. An exception is a Science Exchange hosted by the RHS at Reading University in November 2005.

Recently, there has been the beginning of an official awareness of the problem. In the USA, in the context of sudden oak death and other issues, a fundamental review of 'Plants for Planting' has been initiated by APHIS within the Department of Homeland Security. The IPPC has also recently set up a working group on 'Plants for Planting,' to consider the development of new standards; and the International Union of Forest Research Organizations has initiated a working group on 'Invasives' which has also identified this pathway as one needing scientific and regulatory attention. Hopefully the recognition process will accelerate, with the EU and UK playing a prominent role. One concern must be that the IPPC and other relevant organizations will take greater account of the growing scale of the international trade, which is quantifiable, and the significance of the trade lobby, which is large and influential, rather than the scale of the international threat and the environmental costs, which are difficult to define and quantify with any precision and lack a coherent institutional 'voice'. Another concern must be that, in consequence, any eventual proposals will amount to various prescriptions to enforce the status quo by tightening or improving existing inspection regimes (more certificates, more inspectors, improved on-site diagnostics, essentially 'more of the same'), rather than an aim to significantly reduce the scale of the risk by cutting and reversing import volumes. Since the main loopholes in current regulations are the access points for the large numbers of unknown threat organisms, an approach involving tightening of existing inspection and diagnosis alone will probably only delay the inevitable.

In 2010, the World Health Organization (WHO) provided an information note describing biosecurity as a strategic and integrated approach to analyzing and managing relevant risks to human, animal and plant life and health and associated risks for the environment. It is a strategic and integrated approach to analyzing and managing relevant risks to human, animal and plant life and health and associated risks for the environment. It is based on recognition of the critical linkages between sectors and the potential for hazards to move within and between sectors, with system-wide consequences. Reviewing national capacity provision for biosecurity as a whole helps identify any gaps in regulations and monitoring. Also, as technologies for the detection of pests and disease develop, it is likely that synergies will emerge between sectors in areas such as virology or detection of low levels of chemical contaminants. Ultimately the aim is to enhance national ability to protect human health, agricultural production systems, and the people and industries that depend on them. In another document, it describes the aim of biosecurity being "to enhance the ability to protect human health, agricultural production systems, and the people and industries that depend on them", with the overarching goal being "to prevent, control and/or manage risks to life and health as appropriate to the particular biosecurity sector" (Anonymous, 2020).

Measures taken to counter biosecurity risks typically include compulsory terms of quarantine, and are put in place to minimize the risk of invasive pests or diseases arriving at a specific location that could damage crops and livestock as well as the wider environment. In general, the term is today taken to include managing biological threats to people, industries or environment. These may be from foreign or endemic organisms, but they can also extend to pandemic diseases and the threat of bioterrorism, both of which pose threats to public health.

Plant biosecurity encompasses the policy and regulatory frameworks to analyse and manage risks in the sectors of plant life and health, and related environmental concerns. It covers the protection of a country from pests and diseases during trans-boundary movement of agricultural commodities as well as from emerging indigenous pests, introduction and release of GMOs (genetically modified organisms), and the threats from the use of pests/GMOs in biological warfare (Khetarpal and Kavita Gupta, 2008). The plant-quarantine legislation in India is aimed to secure protection from the ingress of exotic pests during import and export, while the Department of Biotechnology under the Ministry of Science and Technology, through its various committees takes care of the biosafety issues in dealing with GMOs, and the issue of biological warfare is presently dealt with under the Ministry of Home Affairs.

International trade poses a serious and growing threat to biosecurity through the introduction of invasive pests and disease: these have adverse impacts on plant and animal health and public goods such as biodiversity, as well as food production capacity. While international governmental bodies such as the World Trade Organization (WTO) recognize such threats, and permit governments to protect human, animal, and plant life or health, such measures must not be applied in a way that is restrictive to trade. This raises a fundamental (but little-examined) tension between effective biosecurity governance and the neoliberal priorities of international trade. A comparative approach enables close scrutiny of how trade liberalization and biosecurity are constituted as compatible objectives as well as the tensions and contradictions involved in making these domains a single governable problem (Damian Maye *et.al.*, 2012).

i. Australia: National biosecurity in Australia is governed and administered by two federal government departments, the Department of Health and the Department of Agriculture, Water and the Environment. The Biosecurity Act 2015 and related legislation is administered by the two departments and manages biosecurity risks at the national border. The Act aims to manage biosecurity risks to human health, agriculture, native flora and fauna and the environment. It also covers Australia's international rights and obligations, and lists specific diseases which are contagious and capable of causing severe harm to human health. Each state and territory has additional legislation and protocols to cover biosecurity in their jurisdiction (post-border). The Intergovernmental Agreement on Biosecurity (IGAB) created a framework for governments to coordinate and identify priority areas of reform and action to build a stronger and more effective national biosecurity system, and established the National Biosecurity Committee (NBC) in 2012. As of March 2020, the Department of Health has a page devoted to the COVID-19 pandemic, which is updated daily.

The Intergovernmental Agreement on Biosecurity (IGAB) was created in January 2012. It was an agreement between the federal, state and territory governments, with the exception of Tasmania, intended to "improve the national biosecurity system by identifying the roles and responsibilities of governments and outline the priority areas for collaboration to minimize the impact of pests and disease on Australia's economy, environment and the community". It was focused on controlling animal and plant pests rather than human biosecurity, as it was considered that this aspect was already covered by existing agreements, and set out to improve collaboration and understanding of shared responsibilities among all parties, including industry stakeholders.

The 2012 IGAB created a framework for governments to coordinate and identify priority areas of reform and action to build a stronger and more effective national biosecurity system. The agreement comprised two parts: the first part established the goal, objectives and principles of the system, as well as the purpose and scope of the agreement; the second part, the schedules, outlined the priority work areas for governments and their key decision-making committee, the NBC (National Biosecurity Committee). The work based on IGAB led to the development of significant and sound national policy principles and frameworks, including the National Environmental Biosecurity Response Agreement (NEBRA) (Craik Wendy *et.al.*, 2017).

ii. UK: The plant biosecurity strategy for Great Britain provides a high level overview of the activity that Defra and the devolved administrations in Scotland and Wales are undertaking to improve plant biosecurity. The strategy has been developed in consultation with stakeholders and details how this activity will meet the recommendations of the Independent Tree Health and Plant Biosecurity Expert Taskforce. The Taskforce was set up to review domestic and international risks presented from new and emerging tree and plant pathogens. The Tree Health and Plant Biosecurity Expert Taskforce was convened by Defra's Chief Scientific Adviser in November 2012 at the request of Defra's Secretary of State. The Taskforce is an independent, multi-disciplinary group of members of the academic community. Its remit was to provide advice on the current threats to tree health and plant biosecurity in the UK and make recommendations about how those threats could be mitigated. In UK, one of the main trigger to reinforce Biosecurity regulation was the Foot and Mouth Disease (FMD) outbreak at Pirbright in 2007 following the accidental release of FMD virus into the environment. For details, see the report ordered by the House of Commons. The oversight of laboratories involved in research activities with dual-use biological materials was already regulated under the counter-terrorism legal framework since 2001 (Anti-terrorism, Crime and Security Act).

Previously, plants could move around in the EU without restriction if they were grown in an EU country (including the UK), certified free from pest and diseases and for a traveler's own use or consumption. The UK has now left the EU and there are no longer exemptions for transporting plants for personal use. This means that, from the 1 January 2021, you must have a phytosanitary certificate for almost all plants and living parts of plants, including all seeds for planting and potatoes for human consumption, that you are bringing in in personal baggage. This applies to plant material from both the EU and any other country outside of the EU. A phytosanitary certificate is issued from the plant health authority of the country where the plant material originates from to guarantee that the material has been officially inspected, is free from quarantine pests and diseases, and meets the legal requirements for the material to enter GB.

Strategy Implementation: Governance for much of the activity described in the strategy falls within departments' existing portfolios and governance mechanisms. However, there are also commitments that can only be delivered if Government works together across departments. A new cross-departmental governance board will oversee these and any other new commitments. This governance board will report to the National Security Council (NSC), through the Security Minister. The Government Chief Scientific Adviser (GCSA) will maintain oversight of the strategy's outcomes (Anonymous, 2018).

iii. United States: Biosecurity in the United States is governed by the Bureau of Western Hemisphere Affairs, which is part of the US Department of State. It obtains guidance and advice on specific matters relating to biosecurity from various other government agencies. Biosecurity is set of measures aimed at preventing the introduction and/or spread of harmful organisms, in order to minimize the risk of transmission of infectious diseases to people, animals and plants caused by viruses, bacteria or other microorganisms. As well as protecting the agricultural economy and other industries of countries, it protects human health against biorisks caused by natural occurrences, accident, or deliberate acts of bioterrorism. The term also extends to dealing with epidemic and pandemic diseases, with the World Health Organization (WHO) playing an important role in the management of the latter. WHO has described biosecurity as a strategic and integrated approach to analyzing and managing relevant risks to human, animal and plant life and health and associated risks for the environment. Biosecurity protocols are also used in laboratories and research facilities to prevent dangerous biological materials from falling into the hands of malevolent parties, particularly where dual-use research is being undertaken, for both peaceful and military applications.

Plant biosecurity activities in the United States fall along a continuum ranging from offshore activities to the management of newly established exotic pests. For each step in the continuum, the roles, responsibilities, and information needs of the Animal and Plant Health Inspection Service are examined and other agencies involved in plant biosecurity. Both costs and information needs increase dramatically as a pest penetrates deeper into the continuum (Roger D. Magarey *et.al.*, 2009). To help meet these information needs, we propose a cyber infrastructure for plant biosecurity to link phytosanitary agencies, researchers, and stakeholders, including industry and the public. The cyber infrastructure should facilitate data collection, data integration, risk analysis, and reporting. The role of private industry in providing critical data for surveillance was emphasized.

iv. South Asia countries: The knowledge management in plant biosecurity will help in collating and harnessing inoformaitn for timely and accurate diagnosis of the pest and will be an indispensable tool for developing and implementing an early warning system and a rapid respsonse strategy in case epidemics break out (Khetrapal and Dashora, 2014). In addition, it will contribute to collate and harness the information on pests and will facilitate capturing the market for export. The South Asia countries need to come on one platform to strategize collectively in this aspect to harness the natural and knowledge resources in the region to ensure an effective plant biosecurity in the region. For this, the countries / organizations need to align their strategies with the knowledge base of the country by keeping pace with technology and using regional bodies as a platform to achieve the objectives.

A sectorial approach is presently being adopted to address the issues of biosecurity and this is true more so in South Asia countries. There is a lack of integrated approach to deal with its various components. The countries do not have their information on various aspects of plant protection that impinges upon plant biosecurity at one platform. To begin with, there is a need for a strong legislative mechanism, reorganization of infrastructural facilities and synergy of expertise from various organizations under the different concerned ministries including Agriculture, Environment, Commerce, Tourism etc., data bases of pests and diseases, research findings and the extension programmes all at one place.

There are number of Acts related to biosecutiy and biodiversity impringing upon plant health management. The Ministry of Agriculture in South Asia countries do have national programmes on quarantine, integrated pest management, registration of new molecules, locust and rodent control and capacity building. Besides, in some countries there are large numbers of research institutions and universities for a technical back-up to the ministers programme on extension and development. As an example, India has one of the largest agricultural scientific manpower in the wolrd spanned across 99 institutes on Indian Council of Agricultural Research and 70 State Agricultural Univeristies, many of which deal with plant health managmenet. In addition, there are large numbers of R & D on plant health management under various provate corporate of national and multinational origin. It has also been observed that South Asia is recently becoming increasing aware of integration of its policies, research, general plant protection measures and quarantine to develop an effective plant health management system. India has already ventured into a Naitonal Agricultural Biosecurity system, the recent outcome of which is the Integrated Agricultural biosecurity Bill, which is now tabled in the Parliament.

According to WHO, biosecurity is defined as "institutional and personal security measures designed to prevent the loss, theft, misuse, diversion or intentional release of pathogens and toxins".

This understanding includes various aspects like:

1. Access to facilities that deal with such biological agents
2. The storage of such materials and data related to them
3. Publication policies with respect to potentially harmful biological agents.
4. Technologies that deal with the resurrection of extinct viruses, construction of viruses that are guarded or viruses that are drug-resistant or have no current vaccine.

According to FAO, biosecurity is 'strategic and integrated approach' that includes policies and regulations that deal with the risks posed to food safety, animal and plant life and their health and associated environmental risks.

This understanding includes aspects like introduction of plant pests and diseases; introduction of animal pests and diseases; introduction of zoonoses; use and release of genetically modified organisms or their products and invasive species and their genotypes.

v. Belgium: To date there is no specific legislation in Belgium concerning biosecurity. Although all scientists receive an authorization (environmental permit delivered by the regional competent authority) for the contained use of genetically modified organisms and / or pathogens, the potential issues of laboratory security and biological weapons are not addressed (dual-use, Biosecurity risk assessment) (Didier Breyer, 2021). Some institutions decided a few years ago on their own initiative to adopt some biosecurity measures. In Belgium, the appointment of a biosafety officer, BSO ("Bioveiligheidscoördinator" or "Responsable de la biosécurité") and the setting of a biosafety committee became compulsory when Directive 98/81/EC was implemented in the 3 regional Decrees. This measure is required for every institution where genetically modified organisms and/or pathogens are contained. The role and the main tasks of the BSO and the biosafety committee are defined in the Decrees. It is worth mentioning that in the CWA 15793:2011 the role of the BSO covers both biosafety and biosecurity. In Belgium this interpretation is not in force.

vi. Denmark: Denmark adopted its law on biosecurity in 2008 followed by an executive order in 2009. The legislation regulates biological substances, related materials, as well as delivery systems and intangible technology which can be used in biological weapons. In Denmark, it is compulsory for the users (private and public sectors) to obtain a license from the Centre for Biosecurity and Biopreparedness (CBB) to work with or possess biological materials and technologies with dual-use potential. Awareness raising among scientists and students, working with dual-use materials, is an important factor to avoid misuse of biological materials. Therefore, biosecurity in Denmark is a dynamic process of minimizing the risk of a biological attack. This law requires the designation of biosecurity Officers and describes an inspection regime.

vii. The Netherlands: After a thorough risk assessment of the situation in 2010 and 2011 the Dutch authorities decided to implement a biosecurity regime without including biosecurity into the already present biosafety field. Biosecurity regime in the Netherlands encloses inter alia: overview of locations where biological agents are contained; overview of legal framework (rules and legislation); development of a toolkit for institutions/inspection bodies. A document entitled "Implementing biosecurity in scientific research in the Netherlands" is available on the internet (Policy Brief, Nov 2011). In 2013 the Royal Netherlands Academy of Arts and Sciences (Biosecurity Committee) published an advisory report entitled "Improving biosecurity assessment of dual-use research". The reflection was based on the "Fouchier case" showing that biosecurity is not limited to scientific considerations.

In collaboration with the Dutch platform of Biosafety professionals, an online self-assessment Biosecurity toolkit has been developed to enhance the anchorage of biorisk management inside life sciences, biotechnology and medical biology organizations. The biosecurity online toolkit consists of two separate parts. The first part is a self-check module for organizations to identify possible gaps in their security based on their current security status. The eight main aspects of Biosecurity will be addressed (the eight "biosecurity pillars"). The second part of the toolkit consists of measures to increase the institutional biosecurity regime with good practices such as the Code of Conduct (Sijnesael *et al.*, 2014). In a second stage a toolkit for inspections and supervision will be developed based on the toolkit for organisations handling hazardous biological agents.

viii. France: Since 2004, a legislation has been adopted in France regarding "Biosecurity" measures. Article L5139-2 explains on the production, preparation, transport, exportation, etc. and Article L. 5139-1 provides the use of the micro-organisms and toxins listed in . The 30 of April 2012 Decision entered into force in May 2012.An interesting dossier issued in 2012 by ANSES (Agence Nationale de Sécurité Sanitaire alimentation, environnement, travail) is available online : Euro Reference: Les cahiers de référence, cahier n°7 Spécial « Sécurité et sureté » (AFSSA, 2012).

ix. Germany: It worth mentioning the case of Germany that is, in some extent quite comparable to the situation in Belgium. Legislation on Pathogen Security in Germany did not introduce a biosecurity framework after the 2001 events. Germany relies on an extensive framework of 'biosafety' laws and regulations, which are designed to ensure the safe handling of dangerous pathogens by legitimate researchers and to minimize the risks to public health and the environment from research conducted for peaceful purposes ('biologische Sicherheit' is generally understood to mean 'biosafety'). Germany has extensive tracking and systems for dangerous pathogens and a well-developed infrastructure for inspection and training. The biosafety regulations are based on the inherent capacity of micro-organisms to cause illness and death in humans, animals, or plants. All known microbial agents are classified into four Risk Groups. Risk Group 4 includes pathogens such as the Ebola virus that are lethal and incurable, and hence demand highly stringent (Biosafety Level 4) containment measures, whereas Risk Group 1 agents require only minimal safety precautions. Because of this comprehensive approach, the total number of microbial and toxin agents covered by the German biosafety regulations is considerably larger than the US Select Agent List (comparable with Belgium or Switzerland).

x. China: China's Biosecurity Law is all set to come into force on April 15, 2021. It is the day China observes its National Security Education Day. The stated objective of the law is to strengthen regulations on the conduct of biotechnology research and development activities in China. However, the timing of the announcement of such a law at this juncture of time, when the world is still struggling to recover from the COVID-19 pandemic, is sufficient enough to raise doubts that whether this law is a by-product of the criticism China has been facing for being the origin of the virus. The law has already been approved by the Standing Committee of National People's Congress back in October 2020. Lingering its enactment for as long as 6 months post-approval, in contrast to swift affecting the New Defence Law within a week of its approval on December 26, it also seems deceptive. If China really needs this long gestation period for the enactment of its Biosecurity Law, it implies that either there was almost no biosecurity system in place beforehand or there are certainly other things to clean up.

Under the Biosecurity Law, lab animals have been prohibited to enter the market. This again raises the question of whether the law is intended to prevent the possibility of the spread of infections through these animals or a cover-up to a bio-disaster caused by some random sale of such lab animals in the markets of Wuhan. Another obvious objective of the Biosecurity Law is to diversify CCP's means of espionage (Manish Shukla, 2021). The Chinese government would be collecting biometric identifiers of people entering mainland China in the name of investigation. Through this, China aims at getting other countries' DNA sequencing for both civil and military applications.

On 17 October 2020, the Standing Committee of the PRC National People's Congress ("NPC") passed the PRC Biosecurity law, which will become effective on April 15, 2021 (the "Biosecurity Law"). Notably, the Biosecurity Law was passed by the NPC's Standing Committee in less than a year from its first review - this signifies the strengthening of regulations for the biotechnology and life sciences industry in the midst of the COVID-19 pandemic (Horace Lam *et.al.*, 2020). The Biosecurity Law establishes a comprehensive legislative framework for the somewhat piecemealed pre-existing regulations in the areas like Epidemic control of infectious diseases for humans, animals and plants; Research, development, and application of biology technology; Biosecurity management of pathogenic microbial laboratories; Security management of human genetic resources and biological resources; Countermeasures for microbial resistance; and Prevention of bioterrorism and defending threats of biological weapons.

xi. Russia: Biological threats have the potential to kill millions, cost billions in economic losses, and create political and economic instability, whether naturally occurring, accidental, or manmade. The risk of a catastrophic biological event is magnified by global travel, urbanization, terrorist interest in weapons of mass destruction as well as rapid advances in technology, including risks posed by newly developed or manipulated pathogens with pandemic potential. All of these factors taken together create an urgent need to strengthen biosecurity, reduce biological risks posed by advances in technology, create new approaches to improve infectious disease surveillance, and identify and fill gaps to measurably strengthen global health security capabilities. Despite these challenges, biosecurity remains an under-emphasized and under-financed global security priority (Anonymous, 2021).

Nuclear Threat Initiative (NTI) recognizes that threat reduction is a shared responsibility between governments and the private sector. We raise awareness, advocate for solutions, facilitate implementation of solutions, and foster new thinking about these challenges. Specifically, NTI seeks to reduce biological risks by:

- Defining concrete actions to reduce risks posed by advances in technology and engaging global stakeholders to implement them.
- Identifying gaps in the capacity of individual countries to mitigate high-consequence biological events and motivate governments and other stakeholders to fill them.
- Developing new approaches for curbing the catastrophic outcomes from a high-consequence biological event, including catalyzing progress toward real-time biosurveillance and pandemic forecasting.
- Increasing international biosecurity capabilities and raising the profile of biosecurity within the Global Health Security Agenda (GHSA), which is a group of over 60 countries and international organizations dedicated to preventing, detecting, and responding to infectious disease threats.

In March 2012, at a meeting convened by the recently reelected Russian president Vladimir Putin, Minister of Defense Serdyukov informed Mr. Putin that a plan was being prepared for "the development of weapons based on new physical principles: radiation, geophysical wave, genetic, psychophysical, etc." Subsequently, in response to concerns expressed both in Russia and abroad, the Russian government deleted the statement from the public transcript of the meeting. But the question remains: Is Russia developing an offensive biological warfare program?

xii. Japan: Management in terms of biosecurity can be made in three entities: material, human, and information. Japanese legislations intensely provide material controls. The pathogen control by the infectious diseases control law categorizes pathogens and like substances of high concern into following four groups: Possession of Group 1 is prohibited, only exempted under the designation by the Ministry of Health, Labour and Welfare (MHLW) (Makino Tomohiko, 2013). Possession of Group 2 needs prior permission by MHLW. Possession of Group 3 must be promptly reported to MHLW. Finally, possessors of Group 4 should comply with handling rules. Animal pathogens are also controlled by the Ministry of Agriculture, Forest and Fisheries. Conducting experiments that may handle genetic modifications is regulated by Cartagena Law. Although control of material are well managed as described above, control of human and information remains underdeveloped. There are no explicit legislations to clarify the security clearance of lab employees, nor investigational tracking of those who handle pathogens. Regarding informatic management, Japanese academic societies provoke DURC as the researchers' code of ethics on disclosure of their research. But neither publication nor grant funding processes necessarily explore potential malicious use of outcomes. This article analyzed the A(H5N1) mutation research case. This analysis revealed that Japanese regulations, although working generally well, would have a pinhole when the genetic modification could enhance the pathogenicity and transmissibility to the extent that would elevate the tier of pathogen control groups. Japanese regulations will face the challenge of best managing potential harmful use of research without hindering the benefit from the advance of science.

The 2004 Japanese 'Invasive Alien Species Act' was enacted to control invasive alien species and prevent the damage that they cause to ecosystems. The Act defines invasive alien species as those recognized as, or suspected of, causing damage to ecosystems, human safety, agriculture, forestry and fisheries (Goka, 2010). Invasive alien species are carefully regulated: raising, planting, keeping or transporting them is prohibited without the express permission of the relevant minister. The Act represents a revolutionary advance for biological conservation in Japan. However, enforcing the Act is problematic. Dealing with the European bumblebee (Bombus terrestris), for example, involved resolving a bitter dilemma between biological conservation and agricultural productivity. The Act also has a serious loophole; it does not include alien micro-organisms. The incursion of amphibian chytridiomycosis into Japan caused confusion for scientists and the Japanese Government because control of such an alien micro-organism was not anticipated in the Act. Japan faces particular difficulties in attempting to control alien species because of its reliance on imports.

1.4 Plant Biosecurity in India

Plant Biosecurity is of paramount importance to any country to safeguard food-security, sustainability of agricultural/horticultural production and also in protecting livelihood of people. Though incursion of alien pests into newer areas is not a new phenomenon, increased global trade has paved way for quicker entry of many exotic pests to hitherto unknown areas. Anthropogenic activities play a major role in such quick movement of alien pests either directly through trade or indirectly through illegal movement of seeds, planting material and agricultural commodities.

India is one among the 12 mega biodiversity centers in the world & is highly vulnerable to invasive pests. The far reaching consequences of entry, establishment, and spread of various kinds of pests are fearsome especially to a country like India due to its varied agro-climatic zones and due to its primary dependence on agriculture. As regards plants, according to the National Bureau of Plant Genetic Resources (NBPGR), several invasive alien species have been introduced into the country along with grain, seed and planting material imports. These introduced pests include bunchy top of banana, banana bract and streak viruses, vegetable/ pea leaf miner, spiraling white fly, American serpentine leaf miner, peanut stripe virus, cotton leaf curl, potato wart, sunflower downy mildew, coffee pod borer, apple San. Jose scale, Biotype B of white fly and invasive weeds like *Lantana camara* and Phalaris minor. Six of these were introduced in 1990s (Singh, 2008).

With the increasing intensification of agricultural production, productivity and trade, such invasive alien species will further threaten our crops. A new wheat stem rust pathotype Ug 99 is causing serious damage in Uganda, Kenya and a few other countries, and threatens to reach India. Wheat being our main pillar of national food security and rural economy, India must take proactive steps to prevent entry and establishment of this race in India.

Sounding the alarm on Global Stem Rust: Stem rust is a catastrophic disease because of its ability to cause complete annihilation of wheat crops over wide areas. The widespread use of PBW 343 wheat variety possessing 1BL.1RS translocation with Sr31 and its continuing stem rust protection over about 6 million ha in India alone had led to complacency throughout the wheat community. The discovery of race Ug99 with virulence for Sr31 and other important genes in Uganda in 1999, and possibly earlier in Kenya, was a reminder of the pathogen's ability to respond, but little seems to have happened in breeding programs until the emergence of current concerns following the continued incidence and spread of race Ug99 in Eastern Africa (Singh, 2008). The prospect of a stem rust epidemic in wheat in Africa, Asia and the Americas is real and must be stopped before it causes untold damage and human suffering. Fortunately, resistant sources against Ug 99 have been identified and the desirable agronomic bases are being used for developing resistant strains in collaboration with Kenya and CIMMYT.

Another disease of wheat that can be very important is Blast on wheat. This strain of blast was first found in Brazil and is now spread up to Bolivia. Very little is known about its likely effect to wheat crop in rice-wheat belt of India. If germplasm enhancement is initiated now (obviously selection will have to be done in Brazil or Bolivia), perhaps by the time diseases reaches India we may have resistant cultivars. An Expert panel on "Global Rust Initiative", 2005 recommended that diverse genetic resistance be identified in global wheat germplasm by testing in Kenya and Ethiopia. Because modern cultivars currently grown in Northern Africa and Asia are susceptible to race Ug99, a breeding strategy be implemented to incorporate diverse genetic resistance to Ug99 into such germplasm before the race migrates to those areas. DNA-marker assisted selection should be utilized where feasible (Singh, 2008).

The seed multiplication agencies and community-based organizations be encouraged to produce commercial seed of newly developed stem rust resistant varieties with stipulations that farmers and other stakeholders play a leading role; breeding programmes be supported in the maintenance and multiplication of Breeder's and Foundation seed; commercial seed be readily available to farmers; and on-farm demonstrations of elite varieties be conducted. The ex-ante and ex-post impact assessments should be undertaken, taking into account alternative crops and livelihood systems.

The pest incursions in the past caused serious economic damage, notable among them are; coffee berry borer (*Hypotehnemus hampei*), coconut eriophyid mite (*Aceria gurrreronis*), spiralling white fly (*Aleurodicus disperses*), cotton mealy bug (*Phenococcus solenopsis*), papaya mealy bug (*Paracoccus marginatus*), sunflower downy mildew (*Plasmopara halstedii*), bunchy top of banana etc. Most of these invasive alien species have established and invaded widespread areas in India, due to lack of awareness among stakeholders and due to inadequate surveillance, containment and eradication programmes.

Indian Initiatives: India has a robust legislative system to combat biosafety issues, refined and strengthened over the last four decades. National biosafety and bio-waste activities are governed by legislation through state health ministries and environment ministries (Pollution Control Boards). Under the Integrated Disease Surveillance Programme, a network of public laboratories with biosafety practices and infrastructure is being set up. Biosecurity has sadly remained a stepbrother of Biosafety (Col Ram Athavale, 2020).

FAO plays an important role in biosecurity through the International Plant Protection Convention (IPPC).

- Technical support is provided through the IPPC Secretariat for the national implementation of both the Convention itself and International Standards for Phytosanitary Measures
- ISPMs aim to prevent the introduction and spread of pests of plants and plant products and to promote phytosanitary measures for their control

Biosecurity laws in India: India has a plethora of laws which deal with Bio-security but it needs to be noted that they do not stem from an understanding of the term. The pieces of legislation have been enacted with differing objectives and public concerns in mind. Though disparate and scattered, these pieces of legislation serve an essential function in specifically addressing the sectoral concerns, and they carry forth the intent contained in the preambles. Likewise, the institutions, though numerous, have been established to serve the purposes of the original enactments (Sharma Preeti and Gaur Neeta, 2014).

Constitution of India: Though there is no specific reference or use of the term Bio security in the Constitution of India, a number of its provisions are of relevance to understanding the legal framework dealing with bio-security in the country. The Constitution is also the key to understanding how the general legal set-up works.

Fundamental rights: Part III of the Constitution of India contains the fundamental rights. Among these is the right to life, which is enshrined in article 21, and which has the most relevance for the legal framework for Bio-security. Since the late 1970s, the Supreme Court, which is the highest court of the country, has progressively widened the scope of the rights granted under this article. This has been achieved by giving an expansive interpretation of the term "life". As a result of judicial interpretation, the right to life has become a sort of repository of various human rights. Some of the pertinent rights thus included are the right to health ; the right to a healthy environment; the right to pollution-free water and air ; and protection against hazardous industries.

Federal scheme: Since India has a federal Constitution, it necessarily provides for a division of power and functions between the centre and the federal units (states). The Indian federal system leans slightly in favour of the centre while keeping a federal pattern and framework. The Constitution has created three functional areas regarding law-making by the two components of the federal system. These are, an exclusive area for the centre called the Union List ; an exclusive area for the states called the State List; and a common or concurrent area in which both the centre and the states may operate simultaneously, though with the centre having overall supremacy, called the Concurrent List.

Plant biosecurity programme and its need: "Plant biosecurity is a set of measures designed to protect a crop or a subgroup of crops from plant pests at national, regional and individual farm levels." Several control measures has been developed to control or manage exotic pests and diseases. Plant Biosecurity program develops quarantine policies that protect the plant health from exotic pests. Threat or quarantine pest identification and pest risk analysis are significant components of plant biosecurity (Sharma Preeti and Gaur Neeta, 2014). Plant Biosecurity policies are based on our national and international obligations under the World Trade Organization (WTO) and in particular the Sanitary and Phytosanitary (SPS) measures. Globalization and movement of people across world are increasing the chances of spreading the diseases and exotic pests among different forests and woodlands. Pests are mostly transported in soil or organic material, such as plant debris, that can be carried on footwear or by the wheels of vehicles and forest machinery. Diseases may also be spread via the equipment. Some pathogens are dispersed in water and so the risk of these being spread increases when conditions are wet. Several insect pests and diseases were introduced into country and now become a serious threats to our plant health and environment viz., *Agrobacterium tumefaciens, Corynebacterium michiganense,* bacterial pathogens, Ficus mosaic virus, Grape fan leaf virus, Puccinia helianthi from Europe, North America, North America, South Pasific Asia, Africa respectively. So there is essential to secure plant health by adopting Plant Biosecurity.

Survey and Surveillance in plant bio-security: To examine the situation of a region or incidence of insect pest or disease Survey and Surveillance of that region is essential. Survey is a procedure to determine the characteristics of a pest population or to determine which species occur in an area and in Surveillance collection and data on pest occurrence or absence are maintained for the future incidence of pest in that region. Current Status: Recently "The Agricultural Biosecurity Bill , 2013" as introduced in Lok Sabha, on March 11, 2013 by the Minister of Agriculture, Mr. Sharad Pawar. It contains different chapters related to agricultural bio- security authority, liabilities, powers and functions to regulate the Bio-security obligations under international agreements etc (Sharma Preeti and Gaur Neeta, 2014). The Bill aims to establish an integrated national bio-security system covering plant, animal and marine issues to combat threats of bioterrorism from pests and weeds.

Agricultural Biosecurity Authority of India, Faridabad: Central government specified the direction and policies of Agriculture Biosecurity Authority of India, regulating the import and export of plants, animals and related products, preventing the introduction of quarantine pests from another country and implementing post-entry quarantine measures. The Authority may prohibit import material whose use it considers risky. No person shall import any plant, animal, and plant or animal products in contravention of guidelines issued by the bill. In this case phytosanitary certificate will be required, issued by the Authority officers. Authority may notify any pest to be a quarantine pest and shall communicate the quarantine measures that the state government. If any biosecurity emergency declares by the central government in an area in case of an outbreak, distribution, or spreading of a pest or organism, which has the potential to cause a significant loss to biosecurity then centre may give direction to control the situation. The Authority shall act as the national organisation under various international conventions such as the International Plant Protection Convention and may award penalty to any person for loss or damage to non-infested plants, animals or related products incurred by him as a result of any sanitary or phytosanitary measures.

National Institute of Plant Health Management (NIPHM): NIPHM, Hyderabad (An autonomous Institution under DAC, Ministry of Agriculture) is a premier capacity building institution in the field of Plant Health Management, Bio-security and Incursion Management and Pesticide Management. The Institution also acts as a policy support centre to the Department of Agriculture and Cooperation. The Institute is the designated nodal centre at the National Level for capacity building in Plant Health Management, Biosecurity and Incursion Management, Rodent Pest Management and Pesticide Management. NIPHM organize different trainings related to Plant Bio-security and Incursion Management every year in to signify the Pest Risk Analysis, Pest Surveillance and diagnostics in bio-security management and threats of invasive alien species in the context of globalization are highlighted in the programme, besides exposure to SPS Measures and the relevant International standards.

Advantages: Enhanced International trade, Improved agriculture production, Improved public health and Protection of the environment.

Biosecurity from plant viruses in India: The liberalized trade of agricultural commodities under World Trade Organization has necessitated the strengthening of agricultural biosecurity and aligning the national efforts to the International Standards of Phytosanitary Measures (Chalam *et.al.*, 2017). In India an Agricultural Biosecurity Bill has been drafted and is under the consideration of the Government. The strategies for biosecurity for plant viruses include stringent quarantine measures for the imported material, domestic quarantine and use of certified disease-free seed and other planting material within the country. As far as the germplasm is concerned, National Bureau of Plant Genetic Resources has strengthened its quarantine measures for detection and interception of exotic plant viruses. During the last three decades, a number of viruses of serious economic concern have been intercepted including 16 viruses that are not known to exist in India. Besides, interceptions were made for 19 plant viruses that are not known to occur on specific host(s) in India. For facilitating export certification, exclusion of plant viruses from the fields is crucial. Strengthening the quality control of seeds with respect to seed health needs attention though an institutional mechanism for tissue culture-raised plants. Non availability of diagnostic reagents the key challenge in the detection of viruses in quarantine. Also strengthening of infrastructure, capabilities and methodologies for detection of viruses in bulk samples is essential. For virus detection, adoption of a right technique and strategy would help in ensuring biosecurity of Indian agriculture from transboundary movement of plant viruses. A need for National Plant Pests Diagnostics and Certification Network is proposed to meet the targets.

Nodal Ministries: Key ministries involved with biosecurity and biosafety are Ministry of Science and Technology (MoST), Ministry of Health and Family Welfare (MoHFW) and Ministry of Environment, Forests and Climate Change (MoEFCC). Many laws and regulations covering Biosecurity from WMD, laboratory biosafety and Biosecurity, vaccine development, disease research and DNA testing are managed by these ministries through the Indian Council of Medical Research (ICMR), Council of Scientific and Industrial Research (CISR) and Defence Research and Development Organization (DRDO).

The proposed Agricultural Biosecurity Bill 2013 and the National Biotechnology Regulatory Authority (NBRA) aim to establish an integrated national biosecurity system covering plant, animal and marine issues. The Bill (which is still pending approval) does not include epizootics/zoonoses like Corona virus and Avian Flu. Laboratory Biosafety and Biosecurity is managed by ICMR. In India, about 30 Bio-Safety Level (BSL) laboratories of the level of BSL3 or BSL2+ are currently under operation. The two BSL4 facilities in India are The National Institute of Virology (NIV) at Pune and The National Institute of High-Security Animal Diseases (NIHSAD) at Bhopal. Bio Laboratories are graded based on types of pathogen researched or analyzed, levels of containment and security/protection ranging from the lowest biosafety level 1 (BSL-1) to the highest at level 4 (BSL-4) (Col Ram Athavale, 2020).

International obligations and support: India has signed and ratified the Biological and Toxins Weapons Convention (BTWC) of 1972. It has also declared its support to all international efforts for strengthening norms against bioweapons and bioterrorism while maintaining a credible deterrent against the use or the threat of use of WMDs against India. India has expressed its support for UNSCR 1540. India is especially concerned about export controls on bioweapons. It has therefore been exercising control over the export of Special Chemicals, Organisms, Materials, Equipment and Technologies (SCOMET), which includes micro-organisms/toxins, including bacteria, fungi, parasites, viruses, rickettsiae, plant pathogens and genetically modified organisms.

The Indian subcontinent is still free from many devastating pests of Agriculture, Horticulture, Plantation & Forestry plants which are causing serious damage elsewhere. There is a need to protect the entry, establishment & spread of these exotic & invasive plant pests in India. Government of India has initiated action to establish an integrated National Agricultural Biosecurity system covering plant, animal and marine resources, to combat threats of bioinvasion of pests and build capacity to enhance SPS compliance & competitiveness to gain market access and enhance competitiveness in international trade. In order to strengthen Agricultural Biosecurity in the country NIPHM has been designated as the nodal centre by Department of Agriculture & Cooperation, Ministry of Agriculture, Government of India to develop capacity of all stake holders. The activities undertaken by the division are:

Capacity building programs: The division organizes a number of training programmes in Biosecurity & Incursion Management and special capacity building programmes to promote SPS compliance. In addition to regular training programmes, organizes Workshops, International programmes and exclusive programmes for the Officials of DPPQ&S.

Biosecurity and incursion management: Biosecurity and Incursion Management in transport, travel, tourism coupled with liberalization of trade pose increased risk of introduction of exotic and invasive pests into the country. Alien Plant Pests which gained entry into India are causing significant economic damage to agricultural production. The training programme of 21 days duration provides d exposure to participants in biosecurity concepts, impact of invasive alien species and exotic pest threats. The participants will able to understand various strategies that can be employed along the biosecurity continuum viz.,

i. **Pre border :** Off shore inspection, certification and treatments, PRA & quarantine regulations.

ii. **Border/ entry point :** Application of quarantine regulations, inspection, sampling, testing and treatment

iii. **Post border:** Regulations for growing in post entry quarantine facilities, pest surveillance, exotic pest monitoring, containment and/or eradication of pests.

The participants will be familiarized with the decision making tools such as Pest Risk Analysis to identify potential pests of concern to India, importance of pest listing and pest database, risk management measures, Diagnostic techniques and protocols for regulated pests. The significance of pest surveillance, and pest risk analysis in securing market access for Indian agricultural commodities will also be highlighted. The mock exercises and case study analysis in the programme enables the participants to improve their skills in understanding and carrying out Pest Risk Analysis, emergency response in the event of new pest reports, use of Pest Risk Analysis to analyze the cost-effectiveness of eradication / containment programme, delimitation of infested area and identification of buffer zones and the measures needed for initiating eradication.

Quarantine Insects- Detection & Identification: Insect pests cause great damage in field and storage and hence are considered as threat not only to biosecurity but also to the food security. Reliable detection methodologies and accurate & timely diagnosis is paramount in identification of insect pests of quarantine concern to prevent entry, establishment and spread of pests of concern to India. The participants learn about the concepts of quarantine, quarantine pests, regulated pests, economic impact of introduced pests in India, important insects of quarantine significance to India. The participants get exposure to the looming threats of insect pests to Agriculture, Horticulture, Plantation crops and Forest Trees, methods and protocols and online tools for identification of insect pests. The participants also get hands-on experience in collection, pinning and preservation of insect specimens and procedures to be followed for dispatching the specimens for identification to diagnostic centers.

Pest Surveillance: Pest surveillance provides insights into the health status of a country's agriculture and strengthens preparedness for preventive actions both in addressing the problems due to domestic pests of serious concern as well as in protection of native agricultural biodiversity from the incursion threats of exotic pests (Anonymous, 2020). Surveillance also provides vital information for development of robust quarantine policies and also facilitates early detection of invasive alien species which is very essential for their eradication. Establishment of pest free areas and areas of low pest prevalence to gain market access can be achieved only through well designed pest surveillance programmes. During the training programme the participants will learn various pest surveillance strategies such as detection, monitoring and delimiting surveys. The tools required for surveillance of target pests and the procedures for establishment of Pest Free Areas to gain Market Access, various lures and traps for carrying out fruit fly surveillance for monitoring as well as for area-wide control will be covered during the programme.

Stored Grain Pests-Detection and Identification & Phytosanitary Treatment (MBr & ALP): Increasing trade in stored grains and stored products is causing alarming biosecurity concerns. The international grain movement is increasingly becoming a contentious issue both for biosecurity protection and market access negotiations. The detection of such pests and precise identification is essential to determine the appropriate Phytosanitary treatment. Fumigation by and large is the most accepted Phytosanitary treatment. In India, fumigation with Phosphine and methyl bromide are accepted treatments. Modular approach is adopted in NIPHM training programmes, so as to facilitate stakeholders to opt for appropriate module or for entire course depending upon their role and need. Accordingly 21 day duration training programme having two built in sub-modules viz., Stored Grain Pests-Detection & Identification and Phytosanitary Treatment (MBr& ALP) are offered. The participants will learn inspection & sampling methods to detect insect pests & identification of different stored grain insect pests by using various identification keys and online tools. The participants will also learn the importance of specific pests which hamper the exportable commodities from India, with specific reference to Khapra beetle. In the second module, the participants will learn the use of approved fumigants for quarantine purposes i.e., Methyl bromide and Phosphine.

During the two weeks programme the participants will be familiarized with physical and chemical properties of Phosphine and Methyl bromide, safety precautions to be followed while handling fumigants, mode of action of fumigants, principles of fumigation, monitoring the fumigant concentration, appropriate use and maintenance of fumigants and safety equipments. The participants will be made to understand the guidelines laid in NSPM-11, 12 (MBr fumigation) and NSPM-22 (Phosphine fumigation) to conduct appropriate fumigation procedures as well as the accreditation procedure of fumigation operators prescribed by the

Directorate of PPQ&S. The participants will gain hands-on practical experience in creating gas-tight enclosure, laying gas supply and monitoring lines, use of vaporizer, fan, leak detector and gas concentration monitor.

Stored Grain Pest Management for FCI and CWCs : India is witnessing a record grain production in the recent years and the godowns are overflowing and warehouse management has become a tough task (Anonymous, 2020). The mere presence of insects, even dead, like that of Khapra beetle, results in serious setback in international trade of stored grains and products. The domestic management is also of serious concern especially in the context of food security requirement. An exclusive programme of 6 days duration for officers of Food Corporation of India (FCI), Central Warehousing Corporation (CWC) and State Warehousing Corporations is being is designed to create better skills and competency in managing the food grains from the ravages of pests at warehouses. The participants will learn the importance of stored grain pest management in the context of national food security and global grain trade. The trainees will also learn methods to detect and identify various stored grain insect pests by employing appropriate identification tools, use of pheromones and traps, the significance of Systems Approach in managing the food grains and scientific fumigation of systems approach in managing the food grains and Scientific Fumigation using Aluminium phosphide as per NSPM 22 will be emphasized.

Orientation for PEQ Inspection Authorities: Imported consignments of plants have the potential to introduce quarantine pests. In some cases, period of quarantine is necessary for a specific consignment because of the impossibility of verifying the presence of quarantine pests in that consignment at the port of entry. Post entry quarantine allows for testing for the presence of pests, time for the expression of signs or symptoms, and appropriate treatment if necessary. The imported planting materials such as live plants, cuttings, saplings, bulbs, tubers, rhizomes, seed sprouts, bud wood etc., are to be grown under Post Entry Quarantine facility (confinement area) for stipulated period under supervision by notified Inspection Authorities viz., Heads of the Department of Plant Pathology of State Agriculture / Horticulture University, Crop specific ICAR Institutes and IARI. In Six day training programme on orientation for PEQ Inspection Authorities the participants will learn the role and responsibilities of Inspection Authorities in safeguarding the nation from ingress of exotic pests, protocols to be followed in establishing and certification of open field or closed PEQ facilities. The programme covers the significance of PEQ inspections at required intervals and skills for detection and identification of quarantine pests, preservation of specimens, forwarding of specimens to nodal laboratories for authentication, applying appropriate mitigation measures in the event of detection of quarantine pests, reporting the clearance from PEQ and Import non-compliance (if any) for different types of planting materials as per the Standard Operating Procedures will be covered.

Sanitary and Phytosanitary (SPS) Measures: Globalization of trade has triggered large scale movement of plant materials and criss-crossing of the exotic pests across the continents through the commodities traded. To ward off entry and establishment of unwanted pests, Pest Risk Analysis helps to identify various Phytosanitary Measures that can be employed to mitigate the pest risk. Fumigation employing Methyl Bromide and Phosphine is most commonly employed Phytosanitary Treatment for Seeds, Grains, Fruits, Timber etc.; whereas, Forced Hot Air Treatment is being enforced for Solid Wood Packing Materials (SWPM) (Anonymous, 2020). Plant Biosecurity Division organizes programmes to build the capacity of stakeholders in Public & Private sector as detailed below.

1.5 Programmes for Public Sector

Plant Quarantine National Regulations and Procedures: As per Article IV of IPPC, each contracting party of IPPC shall make provision for an official National Plant Protection Organization to safeguard the country's agricultural economy and the biodiversity from the ravages of exotic pests. The main responsibilities of public authorities involved are; issuance of Phytosanitary certificates, pest surveillance, pest risk analysis, inspection, disinfestations, post entry quarantine etc. The stakeholders viz., plant quarantine officials, PSC issuing authorities and PEQ inspection authorities, need to acquire appropriate knowledge on plant quarantine regulations, procedures and documentation in order to safeguard Biosecurity and to facilitate safe trade. Further, adherence to procedures facilitates safe international trade and reduces non-compliances. The Six day duration training on 'Plant Quarantine National Regulations and Procedures' covers the important aspects such as importance of SPS Agreement, International conventions, National Regulations, SOPs on imports and exports. Further, through mock exercises & practical scenarios, the participants will learn the procedures for use of on-line PQIS software and procedures to be followed in import/export of seeds, plants, bulbs, grains, fruits, GMOs, Germplasm and bio-control agents.

Pest Risk Analysis: Pest risk analysis (PRA) is a science based tool to tackle the alien pests of concern to any nation while facilitating international trade. PRA is a process which helps to assess the risks of entry, establishment and spread potential of exotic pests. PRA helps to identify the options to prevent the entry and management options in the event of pest establishment. The international standards brought out by IPPC serve as guidance for carrying out PRA. The participants will learn the importance of International conventions & National regulations, SPS obligations for regulating trade based on pest risk analysis, concept of risk and risk analysis, PRA process for assessing the likelihood of pests being associated with the pathway, transport, its direct and indirect impact in the event of pest establishment, spread and the risk management options to minimize such event to happen. The participants will also learn the importance of PRA for market access for new commodities in the international trade through mock exercises.

Orientation for Phytosanitary Certification (PSC) Issuing Authorities: Phytosanitary certification is one of the basic measures employed by the IPPC member countries to prevent global movement of plant pests in traded agricultural commodities. Phytosanitary certificates are issued by the exporting NPPO as a plant health certificate after carrying out inspection, sampling, testing and treatment (if required) to promote safe trade. To promote export, D.A.C, MoA has notified more than 155 public Officers of Central/State Government, ICAR institutes and SAU's for carrying out phytosanitary certification. The participants will get acquainted on international regulations and obligations under IPPC to promote safe agricultural trade, the role and responsibilities of NPPO and PSC issuing authorities. They also learn procedures for use of on-line PQIS software skills for inspection & sampling, testing for pests of concern to importing countries, importing country's regulations. The importance of PFA/ ALPP in export promotion and role of phytosanitary treatments for mitigating the pest risks as per the Standard Operating Procedures for Phytosanitary Certification will be covered in the programme.

1.6 Programmes for Private Sector

Plant Quarantine Procedures for Imports and Exports: The liberalization of trade in the post WTO scenario has opened up new avenues for international trade in agricultural commodities (Anonymous, 2020). One of the main components in the international trade of agricultural commodities is application of Biosecurity and quarantine rules & procedures, by exporting and importing countries. The PQ rules, procedures and degree of implementation vary in specificity and protocols from country to country. Proper understanding and following the procedures reduces the chances of consignment rejections, penalties and non-compliances which are of major trade concerns. A five day training programme will give exposure to participants on SPS and Technical Barriers, International conventions, National Regulations, SOPs on imports and exports. Practical scenarios on procedures for on-line PQIS software use in import/export of seeds, plants, bulbs, grains, fruits, GMOs, germplasm and bio-control agents will be explained.

Phytosanitary treatments: Phytosanitary Treatments often serve as one stop solution at the end point of export. The increased trade in agricultural products is accompanied by the increased risk of entry of inadvertently transporting quarantine pests to countries or regions. Quarantine pests can seriously disrupt trade of fresh agricultural products not only between countries, but also between geographical areas within countries unless accepted post-harvest quarantine treatments are available. Phytosanitary treatments are helpful in safeguarding biosecurity and also in gaining market access. Following programmes on Phytosanitary Treatments were offered.

a. **Fumigation (MBr& ALP):** Among the phytosanitary treatments, Fumigation is most accepted treatment. Fumigation treatment providers form an important and indispensable part for the import/export of agricultural commodities in international trade and knowledge and skill sets possessed by them can make a great difference in the success of phytosanitary treatments. NIPHM is one of the notified Institutes under Insecticides Rules 1971 Chapter III -10, (3a) (iii) for imparting training for commercial pest control operators on fumigation using Methyl bromide and Phosphine. The participants will learn use of approved fumigants, their physical and chemical properties, safety precautions to be followed while handling fumigants, modes of action of fumigants, principles of fumigation, monitoring the fumigant concentration, appropriate use and maintenance of fumigants and safety equipments. The participants will be able to understand the guidelines laid in NSPM-11, 12 (MBr fumigation) and NSPM-22 (Phosphine fumigation) to conduct appropriate fumigation procedures as well as the accreditation procedure of fumigation operators prescribed by the Directorate of PPQ&S. The participants will be given hands-on practical experience in creating gas-tight enclosure, laying gas supply and monitoring lines, use of vaporizer, fan, leak detector and gas concentration monitor

b. **Forced Hot Air Treatment (FHAT):** The packaging material is one of the most threatening pathway for incursion of timber pests across the globe. Forced Hot Air Treatment (FHAT) is one of the approved treatments for packaging material under ISPM -15. National Standard for Forced Hot Air Treatment (NSPM-9) has been developed which prescribes treatment procedures and the steps to register the facility. It is essential to certify the FHAT facilities to ensure that wood packaging material is treated and marked in consistence with the provisions of ISPM -15. NIPHM is the only Institute in India to offer a specialized training programme on FHAT for industry stakeholders. The participants learn the critical requirements for establishing FHAT facilities, calibration of sensors, placement of sensor, identification of coolest point, safety precautions, conducting the treatments, use of appropriate mark and record keeping in accordance with ISPM – 15 and NSPM – 9.

iii. The Gaps and Thoughts on Restructuring: Many essential factors imperative to strengthen our national biosecurity paradigm need to be considered. While well ahead of the global pack and proactive in a larger perspective, India's stance to the ongoing Corona Biosecurity crisis has been largely responsive, and not proactive from a biosecurity perspective (Col Ram Athavale,, 2020).

1.7 National Security Issue and National Biosecurity Strategy

Until now, biosecurity has been a health matter (state subject) and the Ministry of Health has been issuing only guidelines. This has caused varied implementation and imbalanced response by states causing vulnerabilities in the National

Biosecurity framework. Biosecurity is a national security matter and needs to be viewed with appropriate seriousness and importance. It needs to be a Central, not a State subject. Highlighting its importance, in many countries like the US, UK and Japan, Biosecurity is under the Prime Minister's Office (PMO), Homeland Security or Ministry of Defence. Biosecurity should come under National Chemical, Biological, Radiological and Nuclear (CBRN) Security paradigm. Apropos, the Central Government should be issuing Directives and not Guidelines to all states for compliance. Directives legally compel action, guidelines do not. This will iron out the inequalities and help build a robust system across the Nation.

National Biosecurity Strategy: India needs an effective strategy to mitigate biosecurity risks of criminal, accidental or natural origin that requires a very high level of co-operation and co-ordination both between different national agencies as well as among international and regional organizations. Lack of harmonization of national preparedness and fragmentation of responsibilities (even between states as seen during the Corona crisis) can reduce the effectiveness of prevention strategies and cause a delay in response (Col Ram Athavale, 2020). A National Biosecurity Experts group should ideally be created under the PMO. Other Experts, scientists and agencies can provide advice on specialist issues. Nodal Ministries, like MoHFW or MoST, may be nominated to deal with various aspects, tasks and activities and to develop plans and SOPs.

Intelligence: There is a need to develop a sound bio-threat intelligence system integrated into existing intelligence networks. This would include bio-surveillance at ports, airports, cyber intelligence and monitoring local health and suspicious activities in areas of interest. Special training of intelligence analysts in biosecurity aspects is essential.

How is India dealing with this issue?

- The **key** ministries tackling biosecurity in India are the ministries of health and family welfare, science and technology and the environment ministry.
- The **various a**spects of biosecurity in India are managed by ICMR (Indian Council of Medical Research), CSIR (Council of Scientific and Industrial Research) and DRDO (Defence Research and Development Organization).
- Biosecurity is considered as a health and agriculture matter in India- hence it is largely dealt with by the states. The centre issues guidelines which the states modify to suit their local needs.
- In 2004, the National Farmers Commission headed by M. S.Swaminathan had recommended the establishment of a National Agricultural Biosecurity Program.
- In 2013, the Agricultural Biosecurity Bill sought to set up an 'Agricultural Biosecurity Authority'– a high powered body to cover 4 sectors: animal health, plant health, marine organisms and agriculturally important microbes. However, this is still waiting for approval.

- The import of invasive pests and weeds are curbed by the customs department. The Plant Quarantine Order of 2003 categorized species as restricted, prohibited, etc. with respect to their import into India. The categorization under CITES is also adhered to for controlling the species' introduction.
- The ICMR manages several bio-safety level (BSL) labs in India. There are 30 labs of BSL-3 and BSL-2+ that are operational. There are 2 BSL-4 (highest safety level) labs- one at Pune (National Institute of Virology) and another at Bhopal (National Institute of High-Security Animal Diseases).
- India is a signatory to the Biological and Toxins Weapons Convention (BTWC) of 1972– the first multilateral treaty to ban an entire class of weapons. It has also ratified the treaty. The convention makes use of 'confidence-building measures' like consultations among the parties, complaints to the UNSC, assistance to victims, etc.

1.8 The Agricultural Biosecurity Bill, 2013

The inflow of pests/diseases of plants and animals into countries through imports is considered one of the biggest threats to diversity, leading to huge economic losses. It is known that the weed *Parthenium hysterophorous*, is a highly prevalent invasive species in India. This species is originally a native to the American Tropics and it was introduced in India by the contaminated PL-480 wheat, which used to come to us from USA as food support once upon a time in 1950s and 1960s. Similar weeds are *Phalaris minor* (guli danda) and *Lanatana camara* have got established in the country via the same ways (Anonymous, 2021). Against this backdrop, On March 11, 2013, the Agricultural Biosecurity Bill, 2013 has been introduced in Lok Sabha. The Agricultural Biosecurity Bill aims bring all aspects of plant, animal and marine protection and quarantine under a high powered statutory body Agricultural Biosecurity Authority with adequate powers.

Objectives of the Bill:

- A better regime of quarantining and controlling pests and even "exotic species".
- To cover *four sectors of agricultural insecurity* viz. plant health, animal health, living aquatic resources (like fisheries) and agriculturally important micro-organisms.

Proposed Agricultural Biosecurity Authority of India

- To be established via the proposed act at Faridabad.
- To be headed by a Director General, appointed by the central government.
- Comprises experts in plant and animal pests and diseases, and representatives of various ministries and organizations.

Functions of the authority

- Regulation of the import and export of plants, animals and related products
- Prevention of the introduction of quarantine pests from outside India
- Implementation of the post-entry quarantine measures.
- It can issue directions to importers and exporters of such products for the discharge of its functions

Powers of the authority

- No person shall import any plant, animal, and plant or animal products in contravention of notifications or guidelines issued by the Authority.
- Exceptions shall be provided to those imports that are issued permits by the Authority, and imports with sanitary or phytosanitary (relating to the health of plants) certificates issued by the respective authority in the country of origin and the country of re-export.
- Exports of the above products shall not be allowed except in cases where sanitary or phytosanitary certificates have been issued by an officer of the authority. However, such certificates shall not be necessary if the country of destination does not require it.
- The Customs Act, 1962 or other laws in force, that prohibit the import of certain customs and goods shall also apply to those pests, plants and animals, which require permits or are prohibited by the Authority.
- If an officer has reason to believe that any product was imported into India in contravention of the Act, the officer may require the holder to remove the product from India within a period of thirty days. If the holder fails to do so, the officer may seize such a product from him and remove or destroy it.
- The Authority may award a reasonable compensation to a person for loss or damage to non-infested plants, animals or related products incurred by him as a result of any sanitary or phytosanitary measures.

Control of quarantine pest

- No person shall possess, move, grow, raise, culture, breed or produce any plant, animal and related products if he has reason to believe that such a product is or may be carrying a quarantine pest.
- A person shall be responsible for providing information immediately when he becomes aware of the existence of quarantine pests or plant or animal diseases in an area.
- The Authority may notify any pest to be a quarantine pest. It can also notify an area to be a controlled area if it suspects or determines that the area is infested or infected with a quarantine pest.

- When an area is notified to be controlled, the Authority shall communicate the quarantine measures that the state government shall implement. Such measures shall include the treatment or disposal of plants, animals, their prohibition or control and other measures.
- If the state government fails to take measures in a controlled area, then the Authority can take necessary steps for the eradication or containment of the quarantine pest. The state government shall reimburse the costs incurred for such purposes.

Biosecurity emergencies

- On the recommendation of the Authority, Central government can declare a biosecurity emergency in an area in case of an outbreak, distribution, or spreading of a pest or organism, which has the potential to cause a significant loss to biosecurity.
- Such declaration shall cease to have effect after *six months* unless it is revoked earlier.
- During the biosecurity emergency, the centre may give directions to the Authority for managing or eradicating the organism due to which the emergency has been declared.
- Authority may notify a scheme for the management or eradication of such an organism, with the prior approval of the central government.

Agricultural Biosecurity Fund: The bill proposes to establish a Agricultural Biosecurity Fund, in which the money obtained by the authority as for purposes specified in the Act will be credited and will be utilized there from. The authority can also borrow funds from any source through the issue of bonds and debentures to discharge its functions. However, to raise such funds, the authority needs prior consent of the central government.

National Organization for International Collaboration: The Authority shall act as the national organization to discharge obligations under various international conventions such as the International Plant Protection Convention. It shall provide information related to the import, export and technical requirements for plants, plant products and other objects, to similar international, regional or other organizations, free of charge or on a reciprocal basis. Thus, Bill aims to establish an authority to biosecure global trade in farm items. The integrated national biosecurity system thus established will be covering plant, animal and marine issues to combat threats of bio-terrorism from pests and weeds. Such a system would not only increase the national capacity to protect human health and agricultural production, it would also equip the country to meet obligations under several trade and sanitary agreements in food and agricultural products.

Highlights of the Bill

- The Bill establishes the Agricultural Biosecurity Authority of India (ABAI) to protect plants, animals and related products from pests and diseases to ensure agricultural biosecurity.
- ABAI will regulate imports and exports of plants and animals, as well as their inter-state movement. Imports and exports of plants and animals shall only be allowed if they are issued permits by the respective authority in the originating country or by ABAI, respectively.
- ABAI will also conduct surveillance of pests and diseases in the country, undertake pest risk analysis, and interact with research institutes and state governments on plant and animal protection.
- ABAI may notify quarantine pests. It may also declare an area as a controlled area if it is suspected of being infested with pests. It shall communicate the measures to be taken by the state government.
- If a state government fails to take the required measures, ABAI can take necessary steps to eradicate or contain a quarantine pest in a controlled area. The state government shall reimburse ABAI with the costs incurred for such purposes.
- The central government may declare a biosecurity emergency on the recommendation of ABAI in case of a pest or disease outbreak.
- ABAI shall discharge international obligations under various international trade, sanitary and phytosanitary agreements.

Key Issues and Analysis

- Currently, import, export, quarantine, and the inter-state spread of plant and animal diseases are regulated by various entities under different laws. These functions will be subsumed under the proposed ABAI.
- Other countries have established biosecurity systems similar to the one proposed under the Bill. Most of them have established national authorities that regulate imports, exports, quarantine and inter-state movement of plants and animals.
- The Standing Committee examining the Bill recommended a higher representation of states in ABAI. It also recommended removing the requirement for states to reimburse ABAI for measures taken by it to contain a quarantine pest or disease.

What are the challenges?

- The implementation of biosecurity measures in India is not uniform as it's under the purview of individual states.

- The fact that India is already susceptible to pest invasions implies that even detecting an act of agro-terrorism (bioterrorism directed towards agricultural sector), let alone tracing its origin is difficult (*Karishma,* 2020).
- The import of potentially invasive pests and biological agents is to be curbed by the customs officials, who have been criticized for lacking training in the area. Eg: identification of a potentially invasive species' seed among the baggage of incoming travellers.
- Quarantine officers have been rendered essentially toothless as **D**estructive Insects and Pests Act of 1914 and the Livestock Importation Act of 1898 are only subsidiaries to the Customs Act of 1962. This is one of the aspects the 2013 Biosecurity Bill sought to change.
- The biosecurity bill drafted in India has been pending approval since 2014. Also, it does not take zoonoses (like Coronaviruses) into account.
- Unlike many conventional threats to national security, novel biological agents like the SARS CoV 2 cannot be effectively anticipated.
- There is also a significant time lag in coming up with a potent treatment/ vaccine, which makes the issue even more dangerous.
- Biological agents like viruses show a higher level of mutations and there is also the issue of latency period- which impedes the disease detection and control measures.
- Such biological attacks (intentional/ accidental or natural) poses double jeopardy to the defence forces of the country- the armed forces may get affected and weakened by the biological agent and their capacity to handle the conventional threats of terrorist attacks and WMDs is diluted as resources are diverted for the domestic response- thus posing a security challenge.
- In light of the discussion and accusations about the role of Wuhan Institute of Virology in the COVID-19 crisis, the challenge of differentiating between offensive (or aggressive) and defensive (or peaceful) purposes of such biological agents have come into focus and posing a challenge.
- Even local mismanagement of a biosecurity threat has the potential to expand and cause an effect on an international scale. This calls for international level cooperation characterized by transparency, credibility and timely action.
- Lack of verification regime under the BTWC. Any nation with a developed enough pharmaceutical sector can potentially develop a biological WMD which makes the framing of a verification regime a difficult task.
- Development of such a verification regime is further complicated by the consideration that a non-state actor may develop a bioweapon.

- The ability to detect and conclude such non-compliances is affected by how quickly an international investigation team is launched (as fresh forensic evidence is vital) and full access to the affected area by the investigating team. Eg: the 1981 investigations into an accusation by the US of the Soviet's use of mycotoxins were inconclusive.

What is the way forward?

- There is a need for an integrated approach to ensure biosecurity in India- in line with the One Health approach (*Karishma,* 2020).
- The healthcare, economy and the general infrastructure must be made bio-attack and pandemic proof.
- There is a dire need for mainstreaming of biosecurity often under-represented as policy and financial priority.
- Biosecurity must be made a central subject as in the case of countries like Japan and the UK. This will ensure a more uniform implementation of measures.
- Authorities dealing with Phyto-sanitation and regulating the entry of invasive biological agents must be provided with adequate training.
- There is also a need for close coordination with the national public health system for all-round bio-security.
- Improving self-sufficiency in medical supply chains, technologies, essential goods and services, etc. is necessary for improving immunity to such attacks.
- For differentiating the peaceful and anti-social uses of the biological agents, India needs to reengineer and revamp the system of verification and certification.
- It is critical to make even the normal labs and medical facilities capable of seamlessly transforming into biosecurity infrastructure for expedited and reliable testing, vaccine development and deployment.
- Multi-utility biological agents and their related technologies must be subject to export control and non-proliferation measures.
- There is a need to revamp the ethical overview of the R&D works in areas like genetic engineering.
- Ensuring biosafety is an important part of ensuring biosecurity. Research facilities that work with potentially dangerous biological agents need to be categorized according to the safety levels (eg: BSL categorization) and the necessary precautions for that level must be taken. This will reduce the possibility of loss or theft of such hazardous agents from the facilities.

- In case of accidents or thefts at these facilities, personnel must have a well-oiled set of protocols to follow to ensure maximum mitigation of possible adverse effects and better chances of retrieval.
- It should also be mainstreamed into security, defence and counter-terrorism strategies. Experts have called for the establishment of a National Rapid Deployment Biosecurity Force composed of defence and police personnel and health responders. This Force can be used to undertake bio-defence activities and play related relief and response roles.
- There is a need for effective and credible bio-intelligence and bio-surveillance system at both- the national and international levels. Such a system can continuously look for the emergence of new infectious/ harmful agents and potential bioweapons and provide intelligence for deterrence.
- There is a need to incorporate medical intelligence, infectious disease risk assessment and pandemic predictions into the national defence intelligence.
- Like in the case of conventional counter-terrorism, cooperating with friendly biosecurity powers is essential.
- It is also imperative to deter the development of bioweapons– by both state and non-state actors.
- Instead of sacrificing WHO to geopolitical pressures, the countries need to empower the global health body to strengthen its health security mechanisms so that it can freely assess such health emergencies in ground zero countries.
- India can call for reinforcing the BTWC using a legally-binding verification protocol at the upcoming review conference in 2021.
- Comprehensive Convention on International Terrorism, which is being championed by India, needs to be updated with regards to biosecurity.
- Implementing 2019 UN General Assembly Resolution 1540 on disarmament measures to prevent the acquisition of WMD (weapons of mass destruction) by terrorists.

1.9 Components of Biosecurity

Biosecurity is one of the three components of biorisk management, which ensures the safe use and security of biological materials in laboratories. Biosecurity focuses on protecting biological agents from theft, loss, or misuse. Laboratory biosecurity refers to the protection, control of, and accountability for high-consequence biological agents and toxins, and critical relevant biological materials and information within laboratories to prevent unauthorized possession, loss, theft, misuse, diversion, and intentional release.

A biosecurity plan encompasses three major components of protection, viz., physical security, personnel reliability, and information security. A research facility should consider all three aspects of biosecurity to ensure the safety of their personnel and the security of the biological agents and toxins in use there (Anonymous, 2018). Physical security focuses on preventing unauthorized access to biological facilities and ensuring only the appropriate people within the facilities can access agents. Personnel reliability focuses on ensuring that all staff at a biological facility are responsible and are suitable to work with sensitive materials. Information security, which includes cyber security, focuses on ensuring all electronic information is safe from theft or misuse. However, components of a laboratory biosecurity program include physical security, personnel security, material control and accountability, transport security, information security, program management and biological security. The handling of biological materials, including transportation, is an element of biosecurity. These aspects are addressed by a variety of agencies focused on aspects such as export controls, import, and chemical safety. In addition, a biosafety program to address the risks posed by biological research to scientists, laboratory workers, the community, and the environment is a component of a biorisk management approach.

1.9.1. Physical Security

Physical Security is one of the aspects of biosecurity intended to prevent the misuse, loss, or theft of biological agents and toxins. Physical security encompasses measures to safeguard and prevent non-official access to these biological assets in the laboratory, building, or medical/research campus. A biosecurity risk assessment determines procedures and practices to ensure that biological materials remain secure. The risk assessment includes a thorough review of the building and premises, the laboratories, and biological material storage areas. Physical security elements should be implemented, as needed, based upon the risk assessment process.

The National Institutes of Health (NIH) Design Requirements Manual for Biomedical Laboratories and Animal Research Facilities (DRM), formerly called the NIH Design Policy and Guidelines, is the only detailed design requirements and guidance manual for biomedical research laboratories and animal research facilities in the U.S. Compliance with the DRM, which sets minimum performance design standards for NIH-owned and leased buildings, ensures that those facilities include physical security measures to protect against unauthorized access in all laboratories. In addition to the design requirements stipulated by the DRM, the Department of Defense Manual 6055.18-M: Safety Standards for Microbiological and Biomedical Laboratories lists BSL-3/4 commissioning criteria and inspection checklists.

A separate NIH Security Design Policy and Guideline has been developed that contains specific security requirements for the construction of new NIH facilities, as well as for modifications, renovations, and alterations of existing NIH facilities. The Security Design Policy and Guideline also contains specific security requirements for NIH-leased facilities. For security reasons, access to copies of the NIH Physical Security Guidelines is limited to key project personnel and is based on a project-specific need–to–know. The Whole Building Design Guide (WBDG) has also published information on how to safely and securely design laboratories. The section titled Security and Safety in Laboratories addresses these concerns and specific building code standards with respect to research facilities. All laboratories should adopt biosecurity practices to minimize opportunities for unauthorized entry into laboratories, animal and storage areas, as well as the unauthorized removal of toxins and infectious materials from their facility. Photo identification badges for employees and temporary badges for escorted visitors can also be used to identify individuals with clearance to enter restricted areas. Support for a physical security element for laboratories and the biological agents and toxins within those laboratories can be found in several key National Strategies which categorize these elements as critical infrastructure.

The National Strategy for the Physical Protection of Critical Infrastructures and Key Assets outlines the guiding principles to secure the infrastructures and assets vital to national security, governance, public health and safety, economy, and public confidence. Hazardous materials and public health are two of the key critical infrastructure assets that require protection, including physical security of facilities. The National Infrastructure Protection Plan outlines a strategy to build resilient and secure infrastructure by preventing, deterring, neutralizing, or mitigating the effects of deliberate efforts by terrorists to destroy, incapacitate, or exploit critical infrastructure. The National Critical Infrastructure Protection Research and Development (NCIP R&D) Plan addresses physical, cyber and human elements of the critical infrastructure sectors.

The physical security of laboratories and secure mechanisms for storing biological agents and toxins are key National objectives. For entities registered with the Federal Select Agent Program, the Federal Select Agent Program has published Security Guidance for Select Agent and Toxin Facilities. The NIH and CDC's Biosafety in Microbiological and Biomedical Laboratories (BMBL) publication is a key reference for laboratory biosecurity planning for microbiological laboratories. Section VI highlights the principles of biosecurity. The BMBL recommends that physical security should be included within a laboratory biosecurity program. It states: "An evaluation of the physical security measures should include a thorough review of the building and premises, the laboratories, and biological material storage areas. Many requirements for a biosecurity plan may already exist in a facility's overall security plan. Access should be limited to authorized and designated employees based on the need to enter sensitive areas. Methods for limiting access could be as simple as locking doors or having a card key system in place. Evaluations of the levels of access should consider all facets of the laboratory's operations and programs (e.g., laboratory entrance requirements, freezer access). The need for entry by visitors, laboratory workers, management officials, students, cleaning/maintenance staff, and emergency response personnel should be considered."

Facilities that work with BSAT have high requirements for personnel screening and monitoring to ensure the security of BSAT. In addition, facilities working with Tier 1 BSAT have enhanced personnel screening, training, and monitoring responsibilities. The Federal Select Agent Program has prepared "Security Guidance for Select Agent or Toxin Facilities" to assist in complying with the requirements of the select agents regulations.

1.9.2 Personnel Reliability

One of the fundamental aspects of a biosecurity plan is personnel management, which aims to keep biological agents and toxins out of the possession of individuals who might intend to misuse them. It is in the best interest of every facility working with biological agents and toxins to have a complete biosecurity plan, including policies and procedures to screen and evaluate individuals. Personnel reliability measures for biological facilities must consider both insider and outsider threats. These factors should be considered as part of a comprehensive biosecurity risk assessment process when developing a biosecurity plan. Insider threats occur when a person who has gained legitimate access to biological agents and toxins for research chooses to misuse his/her access for nefarious purposes. In many cases, the existing screening procedures and risk assessments may not suffice. Although it is important to address these biosecurity concerns, life sciences research should not be unduly hindered in the process. The potential benefits of enhanced personnel reliability measures must be carefully weighed against the potential negative consequences that such measures would likely have on the research community. It is equally important to establish a culture of responsibility within the life sciences research community.

Personnel are required to be screened prior to being granted access to biological select agents and toxins (BSAT). A Security Risk Assessment (SRA) is one method of evaluating personnel reliability and it is specifically required for those individuals who wish to possess, use, or transfer BSAT. This is completed by the Federal Bureau of Investigation (FBI) in coordination with the Federal Select Agent Program. For Tier 1 BSAT, personnel must also undergo ongoing suitability assessments and the entity has increased responsibilities for personnel monitoring, reporting and coordination with security and safety officials. The Federal Select Agent Program has also published Guidance for Suitability Assessment, which is intended for laboratories working with Tier 1 BSAT. Biosafety in Microbiological and Biomedical Laboratories (BMBL) provides some standard voluntary guidance for what should be included when a laboratory or other facility with any biological agents and toxins considers their personnel management plan, "Personnel management includes identifying the roles and responsibilities for employees who handle, use, store and transport dangerous pathogens and/or other important assets. The effectiveness of a biosecurity program against identified threats depends, first and foremost, on the integrity of those individuals who have access to pathogens, toxins, sensitive information and/or other assets. Employee screening policies and procedures are used to help evaluate these individuals. Policies should be developed for personnel and visitor identification, visitor management, access procedures, and reporting of security incidents."

The BMBL provides these recommendations, but there are no distinct requirements or prescriptive measures for personnel reliability. Background checks and security clearances may be required before employees are granted access to certain containment facilities. Procedures should be developed for approving and granting visitors access to controlled areas. In this capacity, the access to agents and toxins storage facilities can be limited to individuals having a legitimate need to access such areas. Biosecurity training should be provided to all personnel who are given access to laboratory facilities.

Personnel Reliability Programs: Certain research facilities, including those owned and/or operated by the federal government, have instituted formal Personnel Reliability Programs (PRPs) to provide additional measures to help ensure that individuals with access to Tier 1 BSAT meet additional standards of reliability. Current PRPs are modeled after those within the traditional reliability programs and may include extensive background investigations with interviews of character references, security clearances, medical evaluations that may include a review of complete medical records, psychological testing, drug and alcohol testing, polygraph examinations, credit checks, and a comprehensive review of service and employment records.

PRPs usually also involve formal mechanisms for ongoing monitoring that can include requirements for self-reporting, peer-reporting, ongoing monitoring by supervisors, and penalties for noncompliance. Individuals enrolled in a PRP typically undergo periodic reassessments including annual physical examinations, random drug tests, re-evaluation of medical records and medications, recurring psychological evaluations, and renewal of security clearances. Importantly, personnel reliability measures can help reduce but cannot eliminate the risk of an insider threat. In addition, strong institutional and laboratory leadership, clear articulation of priorities and expectations, and an institutional framework that provides relevant education, training, performance review, and employee support will facilitate responsible practices, personnel reliability, safety, and security, while allowing research on BSAT to flourish. Many of the principles of a culture of responsibility that underlie these recommendations can be applied to all biological facilities and other scientific endeavors.

1.9.3 Information Security

Information security, also known as cyber security or data security, is the next component of a laboratory biosecurity plan. The objective of an information security program is to protect information from unauthorized release and ensure that the appropriate level of confidentiality is preserved. It is critical to the security of laboratory equipment and materials. Loss of data and computer systems from sabotage, viruses, or other means can be devastating for a laboratory. Information, like physical property, is considered to be critical infrastructure and must be properly protected and secured because of its value to the Nation. The National Infrastructure Protection Plan considers a full range of physical, cyber, and human risk elements across sectors. Insider threats and a range of other pervasive cyber threats to critical infrastructure highlight the need for public, private, academic, and international entities to collaborate and enhance cyber security awareness and preparedness efforts. The National Critical Infrastructure Protection Research and Development (NCIP R&D) Plan also addresses physical, cyber, and human elements of the critical infrastructure sectors.

Biosafety in Microbiological and Biomedical Laboratories provides some standard voluntary guidance for what should be included when a laboratory or other facility possessing biological agents and toxins considers their information security plan. For the purposes of the BMBL, "sensitive information" is that which is related to the security of pathogens and toxins, or other critical infrastructure information. Sensitive information may also include personnel names, identifying information, and any Personally Identifiable Information on patients or the origin of samples held by the laboratory. Examples of sensitive information may include facility security plans, access control codes, agent inventories and storage locations. Information security in this context does not include US Government classified information which is governed by US law or unrestricted research-related information. The BMBL states,"Facilities should develop policies that govern the identification, marking and handling of sensitive information. The information security program should be tailored to meet the needs of the business environment, support the mission of the organization, and mitigate the identified threats. It is critical that access to sensitive information be controlled."

Over the years, several incidents of cyber security breaches have led to loss of sensitive information. A detailed description of a laboratory procedure may find its way into the public domain, creating a new resource for those with illicit intentions, or simply depriving the researchers of recognition for their work. Most institutions and firms have information security policies and procedures and information technology support staff who can help implement security systems. Laboratory managers and personnel should be familiar with and follow all appropriate protocols. People at all levels within an organization have a role in managing information security risks to the organization's missions and business functions and the information systems that support those missions/business functions. Managing risk is a comprehensive and complex process that involves many activities and functions of an organization its programs, investments, budgets, legal and safety issues, inventory and supply chain matters, and security. An integrated approach to managing risk brings together the best collective judgments of individuals and groups within the organization who are responsible for strategic planning, oversight, management, and day-to-day operations.

The policies developed for oversight of dual use research of concern (DURC) also provide a mechanism for protecting research information. If research is defined to be DURC, according to the US Government definition, the framework for oversight includes the development of a risk mitigation plan. Within this risk mitigation plan, there is a responsibility to communicate the research and research findings in a responsible manner throughout the research process, not only at the point of publication. These policies help to support the conduct of responsible life sciences research in a manner that protects information that could be utilized for harmful purposes. More detailed information on DURC. The National Institute of Standards and Technology (NIST) has developed several documents in support of information security and to provide assistance to fulfill the requirements of two key cyber security laws, Executive Order 13636 Improving Critical Infrastructure Cyber security and the Federal Information Security Management Act (FISMA) of 2002. The FISMA directs Federal government organizations to develop and implement programs to protect their information and information systems.

EO 13636 focuses on the importance of improving cyber security for critical infrastructure to protect the Nation against both inside and external threats. NIST Special Publication 800-39: Managing Information Security Risk: Organization, Mission, and Information System View is the flagship document in the series of information security standards and guidelines developed by NIST in response to FISMA. The purpose of Special Publication 800-39 is to provide guidance for an integrated, organization-wide program for managing information security risks to organizational operations, assets, individuals, and the Nation resulting from the operation and use of federal information systems. Special Publication 800-39 provides a structured, yet flexible approach for managing risk that is intentionally broad-based, with the specific details of assessing, responding to, and monitoring risk on an ongoing basis provided by other supporting NIST security standards and guidelines.

In support of this, NIST was directed to work with stakeholders to develop a voluntary framework for reducing cyber risks to critical infrastructure. The NIST Cyber security Framework consists of standards, guidelines, and best practices to promote the protection of critical infrastructure. The prioritized, flexible, repeatable, and cost-effective approach of the framework will help owners and operators of critical infrastructure manage cyber security-related risks while protecting business confidentiality, individual privacy, and civil liberties. The selection and specification of security controls for an information system is accomplished as part of an organization-wide information security program that involves the management of organizational risk. The management of organizational risk is a key element in the organization's information security program and provides an effective framework for selecting the appropriate security controls for an information system. NIST has developed a Risk Management Framework Overview which describes this approach in detail.

Risk assessments are an integral part of the risk management process. NIST SP 800-30 Revision 1, Guide for Conducting Risk Assessments was developed by the Joint Task Force Transformation Initiative Interagency Working Group to provide guidance for federal agencies in conducting risk assessments of information systems and organizations for each of the steps in the risk assessment process. The Working Group is collaborating on the development of a unified information security framework for the federal government to address the challenges of protecting federal information and information systems as well as the Nation's critical information infrastructure. A common foundation for information security will also provide a strong basis for reciprocal acceptance of security assessments and will facilitate information sharing.

Many of these guidelines are not specific to protecting information within biological laboratories, but many of the same tenets are true regardless of the type of information protected. Facilities registered with the Federal Select Agent Program have specific requirements they must follow as well. The Federal Select Agent Program has prepared an Information Systems Security Control Guidance Document to assist in complying with the requirements of the select agent regulations.

1.10 Levels of Biosecurity

Three major levels are Operational biosecurity, Conceptual biosecurity and Structural biosecurity.

1.10.1 Operational Biosecurity

Includes processes, protocols, management practices, and standard operating procedures , viz., keeping disease agents out, preventing disease spread, measures conducted on-premises and addresses personnel, vectors, animals, equipment, and other materials. In emergency management the three phases of action are planning – response – recovery. With proper planning, including a strong biosecurity program, farms will know what to do when diseases should hit and then progress smoothly to a recovery program to minimize loss and provide for business continuity (Gregory Martin, 2020). A stepwise approach in each of these phases is required so as to not miss a step in the fight against a large threat to the farm.

A good biosecurity program will entail one or more of three protocols / methods of preventing disease:

Physical: Barriers to disease from being carried on the farm or house are placed. Most common method deployed. This would include pest (IPM) controls.

Chemical: Agents used to help break down or destroy disease agents. Soaps, heat treatments (including steam), disinfectants, and other similar methods are part of this group.

Logical: Controls that by action help reduce the chance of diseases. Staying home from the duck pond would be a simple example of this type of method.

Biosecurity is more of an attitude than that of a practice. Anyone who is not fully vested in a good farm biosecurity plan may cut corners and expose the farm to undue risk of infection. All involved must do their part to keep the level of biosecurity high. Also, a good review of movement of all people on the farm is very important to stopping the spread of disease agents through normal farm activities. An in farm lane disinfection station within an insulated box at this line means all who pass should be hosed down to prevent any agents form coming onto the farm. As things escalate all normal traffic must stop, allowing only for essential traffic in planned routes on the farm to minimize contact with poultry housing. Placing a lockable drop box for UPS/FedEx/USPS deliveries keeping them off the farm may help. During AI control zone activity farms should negotiate an embargo of auditors, inspectors, meter readers, and salesmen from coming onto the farm. If need be a heavy cable across the farm lane with a sign with a cell number for access to the farm may be in order. Combo / pushbutton locks could be used so that feed trucks and other trusted sources may gain access (Gregory Martin, 2020). Each house on the farm should be compartmentalized so that contact by anyone is minimal during an outbreak. Multiple animal species enterprises on the farm should divide labor as much as possible to help prevent cross contamination

People and equipment movements should be reviewed and evaluated as to if this is a necessary practice. Workers who wear the same footwear everywhere are the most suspect. The practice of the Danish System of work wear donning/duffing with some modifications should help. Meeting outsiders to the farm should be done neutral locations. A change of clothing or disinfection protocol should take place to prevent fomite transfer of disease agents following a trip off the farm. Vehicles should be washable and have washable floor mats installed for protection. Sanitation during travel should be conducted to help reduce agent loads in the car. Washing is considered an important part of disease control. This includes cell phones and computers. Footwear should be washable, and a two-step wet system will be needed to C&D feet and hands during an outbreak. Washing to remove soils on the entire surface is needed to help reduce fomites. These systems require a higher level of maintenance to be effective

By taking time to review, revamp and retrain if needed all should know and understand the farm biosecurity plan. If done correctly a farm may be able to continue operation regardless of what is taking place around them.

1.10.2 Conceptual Biosecurity

It is best to build farm in an isolated area, at least three km away from nearest infested farm. It revolves around the location of animal facilities and their various components, i.e., physical isolation – away from public roads, limiting the use of common vehicles, limiting access by personnel not directly involved with the operation, and controlling vermin, wild animals, and wind.

1.10.3 Structural Biosecurity

It includes fencing of farm to prevent unwanted visitors; test water source for minerals, bacteria, chemical contamination and pathogen load (once in 2-3 months) ;suitable location for storage of bagged feed ;facilities for safe scientific disposal of dead birds and waste ;safe housing, with suitable wild birds and rodent proofing ;feed, litter and equipment should be stored in a section separated from live bird area to prevent contamination ;proper decontamination and disinfection of equipment, houses; concrete stage with suitable water and power supply for sanitation of vehicles ; all-weather roads within the farm to ease cleaning and to prevent spreading of microbes by vehicles and foot wear ; a three meter boundary of land around the building must be kept free of all vegetation to prevent rodent and wild life activity.

Factors Influencing Biosecurity: Various factors include high dependence of some countries on food imports ; Scarcity of technical and operational resources; Greater public attention to biodiversity, the environment and the impact of agriculture on both ; Advances in communications and global access to biosecurity information ; Increasing travel and movement of people across borders ′ Increased trade in food and agricultural products ; New agricultural production and food processing technologies and Globalization.

1.11 Agricultural Biosecurity-A Changing landscape

Today's agricultural biosecurity systems were built for the protection of national agriculture and food security. Their sectoral roots, in animal and plant protection, have strong historical features that make them different and explain their continuing segregation in most government services. Biosecurity systems have been able to exert substantial control over what comes into a country and to mount high-impact, country-wide eradication programmes. The potential trade conflicts created by different national government biosecurity systems have been substantially mitigated by international agreements which establish common standards and practices (Waage and Mumford, 2008). They create a valuable platform for dialogue, and in a few circumstances they have fostered international cooperation in tackling common threats. Agricultural biosecurity remains largely as the business of governments.

Agricultural biosecurity has worked well for substantial threats where large, sustained investment has been made. Examples include the exclusion of several serious animal and plant diseases, and of fruit flies. Overall, however, biosecurity systems are 'leaky', and it is hard to resist the conclusion that in time, entropy will prevail and more pests and diseases will reach more parts of the world. Improving the 'tightness' of the existing system is certainly possible with sufficient government investment. Today, the perceived growth in the biosecurity threat is leading to calls for improvements in these systems. But are these existing systems the ones we want to improve? In this section, we explore three trends which challenge today's biosecurity systems, and in the next we suggest some alternative approaches for the future.

Trade and consumers: The traditional status-awarded biosecurity as a vehicle for food security is changing today. Trade liberalization has contributed to this change in two ways. Firstly, it has made it easier and less expensive to import food, reducing concern about food security, and hence domestic food biosecurity. Secondly, it has created a new pro-trade political agenda, associated with the WTO. While the WTO has adopted IPPC and OIE biosecurity practices into its SPS Agreement, it has done so in a way that makes trade barriers a last resort, with the burden of justification falling on the barrier maker (Waage and Mumford, 2008). Regional trade and economic agreements, such as those created under the EU and in economic zones in Africa, South America and other regions, will also reduce national control of biosecurity risks.

Trade liberalization will also affect the price differential between domestic and imported products, with important potential implications for government policy on biosecurity. The Organization for Economic Development (OECD) and Consumer Support Estimates (CSE) suggest that, over recent years, EU agricultural prices have been significantly higher than world prices, costing consumers between €50.6 billion and €62.8 billion per annum. Dominating this calculation are milk products and the beef and veal industries, while the sugar industry also achieves a high percentage CSE due to high EU prices. In an economic model, Cook and Fraser (2002) have shown that, where the price of the imported product is less than the locally produced product, consumers' interests will be enhanced by a lower level of biosecurity. This is because biosecurity systems may tend to restrict import of cheaper goods. Add to this the effect of reducing CAP subsidy and support for local industry, and not only might this price differential grow, but local industry may also decrease, making the value to the UK economy of excluding new pests and disease even less.

For the foreseeable future, the interests of free trade systems and consumers will reduce the regulation of trade as a valuable biosecurity measure.

1.11.1 Environment and the New Biosecurity Agenda

Agricultural biosecurity finds itself today part of a much larger biosecurity agenda. In the 1990s, public and political awareness of the substantial environmental impacts of invasive alien species increased with the inclusion in the Convention on Biological Diversity (CBD) of Article 8h, which required signatory governments to 'prevent, eradicate or control those alien species which affect species, habitats or ecosystems'. This in turn was based on substantial research over recent decades into 'environmental invasives' including, for instance, predators like cats and rats that exterminated native species, plants which out-competed native flora and covered water bodies, diseases which decimated wildlife and a wide range of alien marine organisms that changed coastal ecosystems (Waage and Mumford, 2008).

While the impact of biological invasions on native species and extinction are important, perhaps of greater future significance is their broader effect on ecosystem processes, such as water and nutrient cycling, vegetational succession and food chains. Most environmental invasives are intentional introductions, and many are agricultural in origin, such as forage grasses, forestry trees, new commercial fish and shellfish, or species associated with agricultural introductions, like introduced diseases that move on to native flora or wildlife.

What this means is that agricultural biosecurity is becoming only part of a new national biosecurity agenda. New governmental structures give responsibility for biosecurity to inter-ministerial bodies which link environment, agriculture, trade and other agencies (e.g. the US National Invasive Species Council) or create new biosecurity agencies with this broad coverage, as in Australia and New Zealand. A similar approach has been proposed for the UK. As this happens, environmental biosecurity priorities may compete with agricultural ones. In the UK today, one of the greatest perceived plant health risks is *P. ramorum*, a fungal disease introduced on horticultural stock which threatens native non-commercial tree species (Defra, 2005). Internationally, the IPPC has recently re-interpreted its mandate to include the protection of plants in natural as well as agricultural systems.

The societal importance of environmental invasions relative to agricultural ones is difficult to assess—there have been far fewer economic evaluations of environmental invasions and their non-market effects make comparisons difficult (Mumford, 2001. But the growing value of local environmental goods to wealthier societies and the fact that, unlike agricultural goods, they are difficult to substitute suggest that their comparative value will not diminish (Waage, *et.al.*, 2005). Environmental issues are not the only factors broadening national biosecurity agendas. Concern for human health has dominated recent political discussion of avian influenza, even though it is still only an epidemic disease of poultry, and the high proportion of human diseases of animal origin have focused attention on this zoonotic threat. Growing environmental and health-related biosecurity agendas will compete with those for agricultural biosecurity, and existing infrastructure for agricultural biosecurity (e.g. inspection services) will be stretched further to cover these new threats.

1.11.2 Who Pays for Tomorrow's Biosecurity?

A recent series of economic assessments has had a substantial effect on public and political awareness of biosecurity issues. Studies in the USA estimated the national costs of alien species to be in the tens of billions per year (Waage and Mumford, 2008) and were a major impetus for a US Executive Order establishing an inter-Departmental National Invasive Species Council in 1996. Now, governments also have substantial, binding contractual commitments, under the CBD, OIE, IPPC and other conventions, to prevent, eradicate or control biosecurity risks (Waage and Mumford, 2008). If biosecurity threats are increasing, and we certainly know they are accumulating, the price tag for solving national problems and complying to international agreements is potentially large.

All of these factors have generated considerable interest today in who should pay for biosecurity or, particularly, in moving this burden from the government and tax payer to those responsible for creating the risk. While the government has a broad quarantine remit, investment is often quite focused on particular industries. In 2000, before the FMD outbreak, approximately 90% of operational funding for quarantine activities in the UK was directed at animal disease, and the small allocation to plant disease was directed largely at potato pests. Beyond quarantine, government usually picks up costs of eradication and sometimes compensation as for most losses of livestock due to introduced diseases in the UK. This is not generally true, however, for plant biosecurity; if an importer of nursery stock is shown to be responsible for introducing a pest or disease, they must pay themselves for the cost of destroying infested plants and there is no compensation for their loss. Thus, government investment in biosecurity is actually quite heterogeneous across agriculture.

Private sector participation in biosecurity is particularly likely in the food industry. Where food industries are organized around maintaining thriving export markets, like the Spanish clementine producers described earlier, or have an interest in keeping food chains moving efficiently, like the big multiples that command much of today's food retail market, industry will itself invest in biosecurity as good business. Often, it will do this offshore, by improving the standard of imported produce. However, food is only one pathway of pest and disease introduction, and often a minor one. Further, not all agricultural industries are well organized in this way.

There is a general view that sharing the burden of biosecurity must involve some element of 'polluter pays' and the cost of preventing pests and disease, and of cleaning up outbreaks, should be borne by those who benefit from the process by which they are introduced. This approach, however, has several problems. Firstly, while most kinds of industry-related pollution attract one-off 'clean up' costs, biological invasions by their nature grow and become exponentially worse, making the potential cost of a single error so great that the perpetrator cannot possibly pay it. Secondly, biosecurity breaches happen in a context of enormous uncertainty; events are generally so infrequent as to make difficult the identification of who should pay and how much, which makes paying for prevention or even contingency difficult. Finally, the pathways of biological invasion make it difficult to identify those responsible. If a plant disease were to enter the UK from China on a horticultural plant, who is the polluter, the retailer, the importer, the exporter, the producer or all four?

Much research is currently being directed at mechanisms which address this problem have reviewed a range of financial mechanisms that include import tariffs, which pay for inspection and the potential cost of cleanup, bonds or even tradable pollution rights. A levy-like system on livestock producers is under consideration by Defra as a means of financing management of future FMD outbreaks in the UK. The common feature of all of these economic models is to extract payments for 'biosecurity risky' activity in order to build up a fund to pay for prevention or control of future biological invasions. Whatever the mechanism, it would seem inevitable that agricultural biosecurity can no longer be just a game of governments and will need to be played by many more parties.

Who is involved in Biosecurity?

i. **National stakeholders:** the sector-specific government agencies have a primary interest in dealing with biosecurity threats, but industry, scientific research institutes, specialist interest groups, nongovernmental organizations (NGOs) and the general public all have a vital role to play. Even within government, bodies responsible for the sectors usually associated with biosecurity – food safety, public health, agriculture, forestry, fisheries and the environment – play the primary role in a contemporary integrated approach to biosecurity (Anonymous, 2010). However, other parts of government responsible for sectors such as trade, customs, transport, finance and tourism may also become involved depending on national circumstances.

ii. **International stakeholders:** International standard-setting organizations, international bodies and international legal instruments and agreements constitute the governance framework for biosecurity. International standard-setting organizations and bodies like the *Codex Alimentarius* Commission (CAC), the World Organization for Animal Health (OIE) and the Commission on Phytosanitary Measures (CPM) develop standards according to their mandates, which have become international reference points through the World Trade Organization (WTO) Agreement on the Application of Sanitary and Phytosanitary Measures (SPS Agreement), 1995. Other relevant agreements include the Cartagena Protocol on Biosafety, the *Codex Alimentarius*, the Convention on Biological Diversity (CBD), the Food and Agriculture Organization of the United Nations (FAO), General Agreement on Tariffs and Trade (GATT 1947), the International Health Regulations 2005 (IHR, 2005), the International Plant Protection Convention (IPPC), the International Maritime Organization (IMO), the Organization for Economic Cooperation and Development (OECD) and the World Health Organization (WHO) are the potentially most important and relevant global and regional agreements, soft-law instruments, international organizations and bodies that are associated with biosecurity.

1.12 Harmonization and Integration of Approaches to Biosecurity

A traditional sector-based approach to biosecurity is increasingly under challenge, and many countries are revising the relevant legal and regulatory systems, institutional responsibilities and resources available for essential infrastructure in response. The aim is to ensure a more integrated approach, and ensure a faster and more effective response to biosecurity threats. Some countries have even made major changes to institutions to join all relevant responsibilities 'under one roof' – an example would be New Zealand; others have combined some responsibilities e.g. for plant and animal health inspection; while a larger group have instituted formal communication mechanisms (such as a national biosecurity committee or task force) to ensure regular and effective dialogue between the different stakeholders.

Requirements for a harmonized and integrated approach to biosecurity: The successful implementation of a harmonized and integrated biosecurity approach requires a clear policy and legal framework, an institutional framework that defines the roles and responsibilities of relevant stakeholders, adequate technical and scientific capability, including use of risk analysis, a well-functioning infrastructure for testing and control, and a system for communication and information exchange (Anonymous, 2010). In a modern biosecurity environment, considerable importance is placed on a holistic approach. Countries are encouraged to base their controls, as far as possible, on international standards where they exist. At the national level and internationally, there are likely to be significant benefits in integrating biosecurity activities to the extent practical. Examples of enhanced outcomes of biosecurity include better risk analysis, ability to consider complete exposure pathways, integrated responses to new and emerging diseases, rationalization of controls, improved emergency preparedness and response, integrated surveillance or traceability systems and more efficient use of available resources.

Practical biosecurity approaches proposed by the FAO Toolkit: The FAO Toolkit examines critically the nature and performance of an existing biosecurity system, and describes first the needs for an integrated biosecurity approach:

- Why assess biosecurity capacity needs;
- What does biosecurity capacity encompass;
- An analytical framework to assess biosecurity capacity needs;
- A process to assess biosecurity capacity needs.

It also describes the seven steps to assess biosecurity capacity needs:

- Step 1: Obtain high-level support;
- Step 2: Agree on the purpose, scope and process;
- Step 3: The profile the biosecurity context at the country level;

- Step 4: Assess existing biosecurity capacity and performance;
- Step 5: Describe the desired future situation (goals and objectives) of biosecurity;
- Step 6: Identify capacity needed to reach the desired future situation;
- Step 7: Generate options to address the identified capacity needs.

Once these needs assessed, governments are better able to set priorities and organize their work, improve the use of available resources and raise additional resources for unmet needs.

In step 3, a process to assess biosecurity capacity is illustrated. It provides a systematic and analytical means to critically examine the nature and performance of the existing biosecurity system, pinpoint areas for improvement and identify options to address these needs. At the sectoral/organizational level, the guide examines the capability of relevant competent authorities (in terms of their mandate, structure, processes, resources, infrastructure, etc.) to deliver core normative and technical functions of biosecurity based on a risk analysis approach.

The 3rd and 4d step in the capacity assessment process ask: what is the current situation of biosecurity capacity and performance? Understanding existing biosecurity capacity is indeed essential to be able to identify capacity needs accurately and to ensure that the needs identified and any capacity building activities subsequently developed fully reflect local circumstances.

Step 5 in the capacity assessment process is developing on this basis a shared vision of desired future biosecurity which is crucial to identifying capacity needs and actions to effectively respond to these needs through discussions and brainstorming sessions involving competent authorities and bodies. Some countries may decide to involve other groups (such as industry, academic or scientific institutes) given their contribution to biosecurity, for instance through compliance with regulations or their creation and provision of scientific knowledge.

The report highlights in particular the case of Norway or the one of New Zealand which has a high performing, integrated system for managing biosecurity risks to the economy, environment and human health underlining that "New Zealanders understand and have confidence in their biosecurity system, that they are committed and playing their vital role, from pre-border through to pest management".

The steps 6 and 7 in the capacity assessment process focus on the diagnosis and analysis of needs and options to address them:

Step 6 focuses on the identification of capacity needs at the various interfaces between human, animal and plant life and health, and associated environmental protection, in terms of opportunities to take advantage of cross-sectoral synergies and/or to reduce overlaps. It will be important to differentiate between what is essential and what is simply desirable, and to prioritize the identified needs by focusing on the areas, resources and capabilities considered most important.

Step 7 seeks then to determine which actions and activities would be most effective to achieve the desired future situation in terms of expected biosecurity gains, costs and benefits, feasibility, affordability, legitimacy and timeliness.

1.13 History of Biosecurity

The term "biosecurity" has been defined differently by various disciplines. The term was first used by the agricultural and environmental communities to describe preventative measures against threats from naturally occurring diseases and pests, later expanded to introduced species. Australia and New Zealand, among other countries, had incorporated this definition within their legislation by 2010 (Koblentz Gregory, 2010). New Zealand was the earliest adopter of a comprehensive approach with its Biosecurity Act 1993. In 2001, the US National Association of State Departments of Agriculture (NASDA) defined biosecurity as "the sum of risk management practices in defense against biological threats", and its main goal as "protecting against the risk posed by disease and organisms".

In 2010, the World Health Organization (WHO) provided an information note describing biosecurity as a strategic and integrated approach to analyzing and managing relevant risks to human, animal and plant life and health and associated risks for the environment. In another document, it describes the aim of biosecurity being "to enhance the ability to protect human health, agricultural production systems, and the people and industries that depend on them", with the overarching goal being "to prevent, control and/or manage risks to life and health as appropriate to the particular biosecurity sector". Measures taken to counter biosecurity risks typically include compulsory terms of quarantine, and are put in place to minimize the risk of invasive pests or diseases arriving at a specific location that could damage crops and livestock as well as the wider environment (Fitt Gary, 2013). In general, the term is today taken to include managing biological threats to people, industries or environment. These may be from foreign or endemic organisms, but they can also extend to pandemic diseases and the threat of bioterrorism, both of which pose threats to public health.

The current focus on biosecurity evolved from a series of events that made the need for more focus on security around laboratories clear. In 1984, members of the Rajneeshee commune in The Dalles, Oregon, purchased a strain of Salmonella from a medical supply company in Seattle, Washington to contaminate ten local salad bars, sickening over 750 individuals (Anonymous, 2015). Although not immediately recognized as an attack, this incident was a clear indication of the potential impact of the misuse of biological agents.

In 2001, at least five envelopes containing *Bacillus anthracis* spores (the etiologic agent of the disease anthrax) were mailed to U.S. Senators and media organizations. At least 22 individuals contracted anthrax as a result of the mailings; five of the individuals died. After a nearly ten year investigation, known as Amerithrax, it was determined through genetic analysis that the spores in the letters were derived from a single spore-batch of *Bacillus anthracis* (Ames strain), isolated from a cow in Texas and distributed to a number of research laboratories, including the United States Army Medical Research Institute for Infectious Diseases (USAMRIID). On February 19, 2010, the Justice Department, the FBI, and the U.S. Postal Inspection Service formally concluded the investigation into the 2001 anthrax attacks, which determined that Dr. Bruce Ivins, a USAMRIID employee, acted along in planning and executing these attacks.

Events such as the 1984 Rajneeshee Salmonella incident and the 2001 anthrax mailings demonstrate vulnerabilities to acts of bioterrorism and the importance of biosecurity training and personnel reliability programs. Both incidents have led to policy changes intended to reduce the risk of future misuse of organisms intended for research and disease prevention. In October 2003, a report from the National Research Council of the National Academies entitled, Biotechnology Research in an Age of Terrorism, often referred to as the "Fink Report" was released. This report called for more oversight from and self-policing by scientists in life sciences and recommended that the Department of Health and Human Services (HHS) set up a national board to offer guidance to funding agencies. In 2005, HHS established the National Science Advisory Board for Biosecurity (NSABB) to advise the Federal government on biosecurity issues and options for addressing them.

In 2009, Executive Order (EO) 13486, Strengthening Laboratory Biosecurity in the United States, was signed by the President. EO 13486 established the Working Group on Strengthening the Biosecurity of the United States and states that its scope of activities pertains to the United States' policy which states: "...Facilities that possess biological select agents and toxins have appropriate security and personnel assurance practices to protect against theft, misuse, or diversion to unlawful activity of such agents and toxins." Although not the only biological incidents in U.S. history, these incidents have been very influential on the formation of U.S. policy for handling and storing biological agents. There have been many reports, committees, recommendations, and policies to address protecting biological agents and toxins from theft, loss, and misuse in the past 20 years. Each has made a unique contribution to the way the U.S. approaches protecting the biological sciences and future reports will continue to contribute to an ever evolving program.

1.13.1 Major Milestones

Biosecurity is a holistic term which encompasses policy and regulation to protect agriculture, food and the environment from biological risk (Kanwar Kumar *et.al.*, 2019).

i) Potato late blight, *Phytophthora infestans*, responsible for the Irish potato famine that led to the death or emigration of millions of rural poor in the mid-1800s, continues to evolve and spread new virulent forms

ii) Karnal bunt, *Tilletia indica*, an Asian fungal disease of wheat, was introduced into the US in 1996, where it is under containment in south-western states

iii) In East Africa, wheat stem rust, *Puccinia graminis*, has recently reemerged after decades of suppression with resistant varieties, creating a global biosecurity risk if it now spreads

iv) In the UK, a booming trade in horticultural plants has led to many new introductions (Independent 2003) one of which, the fungus *Phytophthora ramorum*, may threaten indigenous forest trees

v) With respect to livestock, an outbreak of classic Swine Fever in 1997 cost The Netherlands approximately £2.4bn

vi) The Foot and Mouth Disease (FMD) outbreak in 2001 cost the UK approximately £7bn

vii) Until 2003, Avian Influenza had been considered a relatively rare animal disease, but in that year a devastating outbreak occurred in The Netherlands and a series of outbreaks began in Asia which have continued and spread to other regions

viii) The explosive growth of aquaculture and mariculture worldwide has led to the spread of many serious diseases and parasites of fishes and shrimp

ix) The introduction of the New World Screwworm fly, *Cochliomyia hominivorax*, into Africa in 1988 led to a successful emergency eradication programme coordinated by the UN

Global need for biosecurity : Main international regulatory instruments and organizations that led FAO to adopt the concept and promote a specific work programme in relation to the Biosecurity approach are: I) World Trade Organization Agreement on the Application of Sanitary and Phytosanitary Measures (SPS Agreement) II) International Plant Protection Convention for plant health III) *Codex Alimentarius* Commission for food safety IV) Convention on Biological Diversity (CBD) and the Cartagena Protocol on Biosafety for living modified organisms V) Office international des épizooties (OIE, or World Organization for Animal Health) (Kanwar Kumar *et.al.*, 2019).

1.13.2 Recent Biosecurity Threats

Recent years have seen a range of biosecurity problems which are notable for their high costs and public profile. With respect to livestock, an outbreak of classic swine fever in 1997 cost The Netherlands approximately £2.4bn. The foot and mouth disease (FMD) outbreak in 2001 cost the UK approximately £7bn, while the appearance of bovine spongiform encephalopathy (BSE) in Canada and the USA in 2003 is estimated to have cost each of those countries $3–4bn in lost trade revenue. Until 2003, avian influenza had been considered a relatively rare animal disease, but in that year a devastating outbreak occurred in The Netherlands and a series of outbreaks began in Asia which have continued and spread to other regions (Waage and Mumford, 2008).

The explosive growth of aquaculture and mariculture worldwide has led to the spread of many serious diseases and parasites of fishes and shrimp. In Europe, the recent spread of *Gyrodactylus salaris*, a small, leech-like parasite of salmonids, threatens salmon fishing. Insect pests of animals have also been moving worldwide. The introduction of the New World screwworm fly, *Cochliomyia hominivorax*, into Africa in 1988 led to a successful emergency eradication programme coordinated by the UN. While accidental, human-assisted movement of pests and pathogens appears to be the major cause of animal biosecurity problems, other mechanisms of introduction are emerging. For instance, the spread of bluetongue disease of sheep in Europe appears to be related to range extension of its culicoid fly vectors, probably as a result of climate change. The African bont tick, *Amblyomma variegatum*, a potential vector of important cattle diseases, was accidentally introduced into the Caribbean in the 1800s, but its rapid spread into new countries in recent decades has been stimulated by another introduction, that of the cattle egret, which spreads the tick to new island countries.

The great diversity of crops and their rich and often cryptic insect and pathogen complexes guarantee a continuing and high level of new pest and disease introductions. In recent years, particularly worrying introductions have occurred on four crops: wheat, rice, maize and potatoes that constitute 50% of the world's food supply. Karnal bunt, *Tilletia indica*, an Asian fungal disease of wheat, was introduced into the US in 1996, where it is under containment in southwestern states. Potato late blight, *Phytophthora infestans*, responsible for the Irish potato famine that led to the death or emigration of millions of rural poor in the mid-1800s, continues to evolve and spread new virulent forms. In East Africa, wheat stem rust, *Puccinia graminis*, has recently re-emerged after decades of suppression with resistant varieties, creating a global biosecurity risk if it now spreads. In the UK, a booming trade in horticultural plants has led to many new introductions one of which, the fungus *Phytophthora ramorum*, may threaten indigenous forest trees. Forestry in general has seen a dramatic pattern of new disease and pest introduction, particularly through the recent opening of trade between East Asia and other regions.

This brief snapshot of agricultural biosecurity threats illustrates their diversity and considerable potential impact on agriculture. It also illustrates a policy-relevant phenomenon. Since biosecurity problems are usually one-off, distinctive, time-bound events, they tend to be presented as clusters of cases as above. Taken all together, we may get the impression of an immense and urgent crisis. Further, discovery of one new threat can lead to greater surveillance which in turn increases the likelihood of finding another, creating a rising spiral of new problems. In fact, at a national level, truly new introductions may be relatively few and far between, while chronic, indigenous or long-established pest or disease problems may be of greater economic significance than those posed by new biosecurity threats. One of the challenges of biosecurity research is to move beyond anecdote into rigorous analysis of the nature and impact of these diverse threats in the proper context.

Bibliography

Anonymous, 2010. International Food Safety Authorities Network (INFOSAN). INFOSAN Information Note No. 1/2010 – Biosecurity. WHO-FAO.

Anonymous, 2015. Biosecurity History. Science safety Security. U.S. Department of Health and Human Services, Washington, D.C. 20201

Anonymous, 2018. Science Safety Security, U.S. Department of Health and Human Services, Washington, D.C. 20201

Anonymous, 2018. UK Biological Security Strategy, 2 Marsham Street, London, SW1P 4DF

Anonymous, 2019. FAO Biosecurity Tool Kit-Part I.20 pp.

Anonymous, 2020. International Food Safety Authorities Network (INFOSAN) (3 March 2010). "Biosecurity: An integrated approach to manage risk to human, animal and plant life and health" (PDF). INFOSAN Information Note No. 1/2010 - Biosecurity. World Health Organization & Food and Agriculture Organization of the United Nations. Retrieved 23 May 2020.

Anonymous, 2020. National Institute of Plant Health Management (NIPHM), (An Organization of Department of Agriculture, Cooperation & Farmers Welfare, Ministry of Agriculture, Govt. of India), Rajendranagar, Hyderabad - 500 030 (Telangana) India.

Anonymous, 2020. Plant Biosecurity Report (2019-2020)- Bhutan Agriculture and Food Regulatory Authority (BAFRA) Ministry of Agriculture and Forests (MoAF) Thimphu.

Anonymous, 2021, Biosecurity in Food and Agriculture. The secretariat of the International Plant Protection Convention is located at the headquarters of the Food and Agriculture Organization (FAO) of the United Nations in Rome, Italy.

Anonymous, 2021. Nuclear Threat Initiative, Washington, DC 20006

Anonymous, 2021. The Agricultural Biosecurity Bill, 2013. Bills & Acts- Ministry: Agriculture. PRS Legeislative Research, New Delhi, India.

Brasie, C. M., 2008. The biosecurity threat to the UK and global environment from international trade in plants. *Pl.Pathol.*, 57: 792-808

Chalam, V.C., Parakh, D.B., Maurya, Khetarpal, R.K.,2017. Biosecurity from Plant Viruses in India. In: A Century of Plant Virology in India. Mandal B., Rao G., Baranwal V., Jain R. (Eds.), Springer, Singapore.

Col Ram Athavale,, 2020. Biosecurity in India: The Way Forward. L2P Research® Labs. Immuno-Oncology Studies, Tumor Xenograft, Focal Radiation,Spontaneous Tumor Studies

Cook, D.C. and Fraser, R.W., 2002. Exploring the regional implications of interstate quarantine policies in Western Australia. *Food Policy,* 27:143–157.

Craik Wendy, Palmer David and Sheldrake Richard,2017. Priorities for Australia's biosecurity system: An Independent Review of the Capacity of the National Biosecurity System and Its Underpinning Intergovernmental Agreement (PDF). Commonwealth of Australia. ISBN 978-1-76003-131-2. Retrieved 23 May 2020.

Damian Maye, Jacqui Dibden, Vaughan Higgins and Clive Potter, 2012. Governing Biosecurity in a Neoliberal World: Comparative Perspectives from Australia and the United Kingdom. Environment and Planning A: *Economy and Space,* 44: 150-168

Defra, 2005. Biosecurity guidance to prevent the spread of animal diseases. (http://www.defra.gov.uk/animalh/diseases/pdf/biosecurity_guidance.pdf)

Didier Breyer, 2021. Belgium Biosafety Server, Sciensano Service Biosafety and Biotechnology (SBB), Belgium

Fitt Gary, 2013. "Explainer: why Australia needs biosecurity". The Conversation. Retrieved 21 May2020

Goka, T., 2010. Biosecurity measures to prevent the incursion of invasive alien species into Japan and to mitigate their impact. *Rev. Sci Tech.,* 29:299-310

Gregory Martin, 2020. Operational Biosecurity – Continued Improvement. PAS; Extension Educator in Poultry Pennsylvania State Cooperative Extension Lancaster, PA

Hawkes, C. and Ruel, M., 2006. The links between agriculture and health: an intersectoral opportunity to improve the health and livelihoods of the poor. Bull. World Health Organ., 84 (http://www.who.int/bulletin/volumes/84/12/ 05-025650.pdf).

Horace Lam, William Fisher and Ting Xiao, 2020. China signs off on PRC Biosecurity Law: What this means for industry players in China. DLA Piper Life scinces Alert, Cybersecurity Law Alert

Kanwar Kumar, Jayant Yadav and Naveen Rao, 2019. History, Concept and Components of Biosecurity. In: Advances in Agriculture Sciences, Vol.19, Dr. R.K. Naresh (Ed.), AkiNik Publications, New Delhi

Karishma, 2020. Biosecurity in India – Need, Challenges, Solutions. IAS Express, Internal security / international security

Khetrapal, R.K. and Kavita Gupta, 2008. Agriculture for food security and rural growth. Teri Press, The Energy and Resources Institute New Delhi India. 247-266

Koblentz Gregory, D., 2010. "Biosecurity Reconsidered: Calibrating Biological Threats and Responses". *Intel. Security,* 34: 96–132

Khetarpal, R. K. and Dashora, K., 2014. Knowledge management for plant biosecurity in South Asia. *Scientific Res. & Essays,* 9: 997 - 1002.

Makino Tomohiko, 2013. Japanese Regulatory Space on Biosecurity and Dual Use Research of Concern. *J. Disaster Res.,* 8: 686-692.

Manish Shukla, 2021. What is China's Biosecurity Law? Know all details here. Zee Media source

Mankad, A., 2016. Psychological influences on biosecurity control and farmer decision-making. *A review. Agron. Sustain. Dev.,* 36: 40

Mankad, A.,2012. Decentralised water systems: emotional influences on resource decision making. *Environ. Int.,* 44:128–40.

McKirdy Simon, Shea Greg Hardie Darryl and Eagling David, 2008. Why plant biosecurity?. International Lupin Association. 0-86476.

Monroe, M.C.,2003. Two avenues for encouraging conservation behaviors. *Res. Hum. Ecol.,* 10:113–125

Mumford, J.D., 2001. Environmental risk evaluation in quarantine decision making. In: The economics of quarantine and the SPS agreement. Anderson K, McRae C, Wilson D, (Eds.). Centre for International Economic Studies/AFFA Biosecurity; Adelaide/ Canberra, Australia: 2001. pp. 353–383.

Roger D. Magarey, Manuel Colunga-Garcia and Daniel A. Fieselmann, 2009. Plant Biosecurity in the United States: Roles, Responsibilities, and Information Needs. *BioScience*, 59: 875-884

Sharma Preeti and Gaur Neeta, 2014. Role of Bio security in Plant Health Management. *Res. J. Agriculture and Forestry Sci*., 2: 14-19

Singh, R.B., 2008. Biosecurity for Food Security. Paper prepared for the National Commission on Farmers, 2006, for the South Asian Conference, 2008 and a Memorial Lecture, 2008 (Ex-ADG, FAO and Ex-Member, National Commission on Farmers)

Waage, J. K., Fraser, R. W., Mumford, J. D., Cook, D. C. and Wilby, A., 2005. *A new agenda for biosecurity* Horizon Scanning Programme, Department for Environment, Food and Rural Affairs, UK

Waage, J.K. and Mumford, J.D., 2008. Agricultural biosecurity, *Philos. Trans. R. Soc. Lond. B Biol. Sci.*, 363: 863–876

[illegible]

2

Invasive Alien Species (IAS)

An invasive species is an organism that causes ecological or economic harm in a new environment where it is not native. Invasive species are capable of causing extinctions of native plants and animals, reducing biodiversity, competing with native organisms for limited resources, and altering habitats. This can result in huge economic impacts and fundamental disruptions of coastal and Great Lakes ecosystems. Invasive species threaten and can alter our natural environment and habitats and disrupt essential ecosystem functions. Invasive plants specifically displace native vegetation through competition for water, nutrients, and space. Once established, invasive species can reduce soil productivity. The negative effects of invasive alien species on biodiversity can be intensified by climate change, habitat destruction and pollution. Isolated ecosystems such as islands are particularly affected. Loss of biodiversity will have major consequences on human well-being.

Because they are not native to a location, they can quickly dominate the wildlife. They may not have any predators to keep them in check. They basically destroy a food chain which can have a domino affect of killing off other native species that require certain food to survive. Invasive species are harmful to our natural resources (fish, wildlife, plants and overall ecosystem health) because they disrupt natural communities and ecological processes. The invasive species can outcompete the native species for food and habitats and sometimes even cause their extinction. Since invasive species are in a new environment, free from natural predators, parasites, or competitors, they often develop large population sizes very rapidly. These high populations can out-compete, displace or kill native species or can reduce wildlife food and habitat.

2.1 Characteristic Features of Invasive Species

Invasive species possess characteristic features like “pioneer species” in varied landscapes, tolerant of a wide range of soil and weather conditions, generalist in distribution, produces copious amounts of seed that disperse easily, grows aggressive root systems, short generation time, high dispersal rates, long flowering and fruiting periods, broad native range, abundant in native range. Preliminary data from one interesting study shows that invasive species are likely to have relatively small amounts of DNA in their cell nuclei (ENVIS Resource Partner on Biodiversity, 2020). Apparently, the cells in these plants are able to divide and multiply more quickly and consequently the entire plant can grow more rapidly than species with higher cellular DNA content. This gives them a leg up in disturbed sites.

According to World Conservation Monitoring Centre (WCMC), 1,604,000 species have been described at the global level. Thus India accounts for 8% of the global biodiversity existing in only 2.4% land area of the world. According to Nayar (1989) the number of flowering plant species endemic to the present political boundaries of this country is 4900 out of a total of 15000, i.e. 33%. Hajra & Mudgal (1997) reported 5400 endemics in 17000 angiospermous species of India, which comes to 31.76 %. India is an important center of agri-biodiversity having contributed 167 species to the world agriculture and homeland for 320 species of wild relatives of crops. This study focused on 173 species of invasive alien plants in India that include the most serious invasives, such as *Alternanthera philoxeroides*, *Cassia uniflora*, *Chromolaena odorata*, *Eichhornia crassipes*, *Lantana camara*, *Parthenium hysterophorus*, *Prosopis juliflora* and others.

2.2 Criteria Adopted for Designating an Alien Species as Invasive

Centre for Biodiversity Policy and Law (CEBPOL), National Biodiversity Authority (NBA) collect the readily available ecosystem-wise literature of invasive alien species and thoroughly analyze. During the consolidation, there will be lots of confusions, wrong citations, biased definitions and information in most of the published lists of invasive alien species. For instance, some of the lists declare the naturalized species as invasive, and conversely some lists declare the invasive species as naturalized alien species. Besides, the accepted name and the synonym of a species were simultaneously reported in the same list and mentioned as different species. In a worst-case scenario, the native species has also been reported as invasive alien species. CEBPOL, NBA realizes the need for avoiding this kind of ambiguity and at the same time feel the necessity for criteria to be adopted for declaring a species as invasive alien species (Sandilyan, 2020).

The compiled list is primarily screened to confirm the alien status and invasiveness of the species based on a simple methodology developed by CEBPOL. After the initial filtration/confirmation, the confirmed list is placed in the NBA's invasive species expert committee for scrutiny. The committee deliberates on the lists compiled by CEBPOL, NBA and suggests to include the invasive attributes on a graded scale for confirmation of the invasiveness of the species in India. After reviewing the available literature, the committee suggests to adopt the important invasive attributes viz., invasiveness, impacts, range of extension and others to designate the alien species as invasive in India (Table 2). Details of the Invasive Alien species reported in India are furnished in table 3 (Sandilyan, 2020).

Besides, the committee also takes into account the personal experiences of the researchers and their view in declaring a species as invasive if there is non-availability/inadequate literature. Based on the aforesaid criteria, the committee finalizes a list of 170 invasive alien species in different ecosystems. The committee also feels the list might further be expanded. For example, when some species are designated as invasive based upon the specific criteria, there may be many more invasive species which may satisfy the above criteria, but due to lack of adequate information of the concerned species it is not included in the present lists. Keeping this aspect in view, the committee requests the NBA to host the lists on its website for public access and comments. Once adequate information is available on the new invasive species in Indian provinces, it may be included in the lists in the near future after due consultation with the expert committee.

Table 2: Invasive attributes used to confirm the invasive status of the species reported in Indian ecosystems

S.No	Invasive attributes
Invasiveness	
1.	IE -Invasive Elsewhere
2.	RMS – Rapid Multiplication and Spread in different ecosystems
3.	MMR – Multiple Modes of Reproduction
4.	MMD – Multiple Modes of Dispersion
Impacts	
1.	B1 – Affecting ecosystem functions and services
2.	B2 – Biodiversity loss
3.	B3 – Economic loss and health hazard
Invasion areas (Continues spread	
	RE – Range Extension

Table 3: Details of the Invasive Alien species reported in India

S.No.	Details of the Species and Ecosystem	Total
1.	Terrestrial Ecosystem	54
	Total	54
2.	Aquatic Ecosystem	
1	Microorganism reported in freshwater and brackish water	15
2	Aquatic plants (inland)	8
3	Fishes	14
4	Marine invasive species	19
	Total	56
3.	Agriculture Ecosystem	
1	Fungi	16
2	Bacteria	5
3	Viruses	3
4	Nematode	1
5	Invasive Insects	22
	Total	47
4.	Major Island Ecosystem	
1	Insects	2

S.No.	Details of the Species and Ecosystem	Total
2	Cnidaria	1
3	Mollusca	1
4	Fishes	2
5	Amphibian	1
6	Reptile	1
7	Birds	2
8	Mammals	14
	Total	14
	Overall Indian IAS species:	173

2.3 Major IAS in India and Their Introduction Pathways

Acacia farnesiana (tree, shrub), *Achatina fulica* (mollusc), *Ambrosia artemisiifolia* (herb), *Aristichthys nobilis* (fish), *Bemisia tabaci* (insect), *Cabomba caroliniana* (aquatic plant), *Chromolaena odorata* (herb), *Chromolaena odorata* (herb), *Columba livia* (bird), Cryphonectria parasitica (fungus), *Cyprinus carpio* (fish), *Eichhornia crassipes* (aquatic plant), *Eugenia uniflora* (tree, shrub), *Gambusia affinis* (fish), *Gymnocoronis spilanthoides* (aquatic plant), *Hypophthalmichthys molitrix* (fish), *Lantana camara* (shrub), *Ludwigia peruviana* (aquatic plant), *Mikania micrantha* (vine, climber), *Mimosa diplotricha* (vine, climber, shrub), *Monomorium pharaonis* (insect), *Oncorhynchus mykiss* (fish), *Oreochromis mossambicus* (fish), *Oreochromis* spp. (fish), *Phalaris arundinacea* (grass), *Parthenium hysterophorus* (herb), *Prosopis* spp. (tree, shrub), *Ricinus communis* (tree, shrub), *Salmo trutta* (fish), *Salvelinus fontinalis* (fish), *Salvinia molesta* (aquatic plant, herb), *Solenopsis geminata* (insect), *Tinca tinca* (fish), *Vibrio cholerae* (micro-organism), *Zosterops japonicus* (bird) (Anonymous, 2009).

Introduction pathways: A pathway is broadly defined as the means (e.g. aircraft, vessel or person), purpose or activity (e.g. farming, shipping or pet trade), or a commodity (e.g. fisheries) by which an invasive alien species may be transported to a new location, either intentionally or unintentionally. Note that species can also expand their range through natural means. For example, birds can fly or be blown by storms to new locations. Some species or their propagules can be moved to new locations by wind, currents and in or on animals. This is referred to as natural dispersal and not an introduction. Natural dispersal may play a significant role in the subsequent spreading of an alien species once introduced to a new region or country.

2.4 World's Worst 100 IAS

It is very difficult to identify 100 invasive species from around the world that really are "worse" than any others. Species and their interactions with ecosystems are very complex. Some species may have invaded only a restricted region, but have a high probability of expanding and causing further great damage (e.g. Boiga irregularis: the brown tree snake). Other species may already be globally widespread and causing cumulative but less visible damage. Many biological families or genera contain large numbers of invasive species, often with similar impacts (Global Invasive Species Database, 2021). Species were selected for the list according to two criteria: their serious impact on biological diversity and/or human activities, and their illustration of important issues surrounding biological invasion. To ensure the inclusion of a wide variety of examples, only one species from each genus was selected. Absence from the list does not imply that a species poses a lesser threat

2013 Update: Rinderpest virus a species of morbillivirus causing cattle plague, a highly fatal viral disease of domestic cattle, buffaloes and yaks was listed as one of the '100 of the World's Worst Invasive Alien Species'. Rinderpest virus was declared eradicated in the wild in 2010.

A global survey was conducted in 2013 to nominate a replacement for the Rinderpest virus on the '100 of the World's Worst Invasive Alien Species' list. Over 650 invasion biologists participated in the survey. The floating aquatic fern *Salvinia molesta* gained the most votes and was selected to replace the Rinderpest virus. *S. molesta* thrives in slow-moving, nutrient-rich warm freshwater. A rapidly growing competitive plant, it is dispersed long distances within a waterbody (via water currents) and between waterbodies (via animals and contaminated equipment, boats or vehicles). *S. molesta* can form dense vegetation mats that reduce water-flow and lower light and oxygen levels in the water. This stagnant dark environment negatively affects, the biodiversity and abundance of freshwater species, including fish and submerged aquatic plants. *S. molesta* can alter wetland ecosystems and cause wetland habitat loss. *Salvinia* invasion also poses a severe threat to socio-economic activities that are dependent on open, flowing and/or high quality waterbodies, including hydro-electricity generation, fishing and boat transport.

1. *Acacia mearnsii*: *Acacia mearnsii* is a fast growing leguminous (nitrogen fixing) tree. Native to Australia, it is often used as a commercial source of tannin or a source of fire wood for local communities. It threatens native habitats by competing with indigenous vegetation, replacing grass communities, reducing native biodiversity and increasing water loss from riparian zones.

Common names: Acácia-negra, Australian acacia, Australische akazie, black wattle, swartwattel, uwatela

2. *Achatina fulica*: *Achatina fulica* feeds on a wide variety of crop plants and may present a threat to local flora. Populations of this pest often crash over time (20 to 60 years) and this should not be percieved as effectiveness of the rosy wolfsnail (*Euglandina rosea*) as a biocontrol agent. Natural chemicals from the fruit of Thevetia peruviana have activity against *A. fulica* and the cuttings of the alligator apple (*Annona glabra*) can be used as repellent hedges against *A. fulica.*

Common names: Afrikanische Riesenschnecke, giant African land snail, giant African snail.

***Acridotheres tristis*:** The common myna (*Acridotheres tristis*), also called the Indian myna, is a highly commensal Passerine that lives in close association with humans. It competes with small mammals and bird for nesting hollows and on some islands, such as Hawaii and Fiji, it preys on other birds eggs and chicks. It presents a threat to indigenous biota, particularly parrots and other birdlife, in Australia and elsewhere.

Common names: Calcutta myna, common myna, German Indischer mynah etc.,

3. *Aedes albopictus*: The Asian tiger mosquito is spread via the international tire trade (due to the rainwater retained in the tires when stored outside). In order to control its spread such trading routes must be highlighted for the introduction of sterilisation or quarantine measures. The tiger mosquito is associated with the transmission of many human diseases, including the viruses: Dengue, West Nile and Japanese Encephalitis.

Common names: Asian tiger mosquito, forest day mosquito, mosquito tigre, moustique tigre, tiger mosquito, tigermücke, zanzara tigre.

4. *Anopheles quadrimaculatus*: *Anopheles quadrimaculatus* a mosquito is the chief vector of malaria in North America. This species prefers habitats with well-developed beds of submergent, floating leaf or emergent aquatic vegetation. Larvae are typically found in sites with abundant rooted aquatic vegetation, such as rice fields and adjacent irrigation ditches, freshwater marshes and the vegetated margins of lakes, ponds and reservoirs.

Common names: Common malaria mosquito, Gabelmüche.

5. *Anoplolepis gracilipes*: *Anoplolepis gracilipes* (so called because of their frenetic movements) have invaded native ecosystems and caused environmental damage from Hawaii to the Seychelles and Zanzibar. On Christmas Island in the Indian Ocean, they have formed multi-queen supercolonies. They are also decimating the red land crab (*Gecarcoidea natalis*) populations. Crazy ants also prey on, or interfere in, the reproduction of a variety of arthropods, reptiles, birds and mammals on the forest floor and canopy. Their ability to form and protect sap-sucking scale insects which damage the forest canopy on Christmas Island, is one of their more surprising attributes. Although less than 5% of the rainforest on Christmas Island has been invaded so far, scientists are concerned that endangered birds such as the Abbott's booby (*Sula abbotti*), which nests nowhere else in the world, could eventually be driven to extinction through habitat alteration and direct attack by the ants.

Common names: Ashinaga-ki-ari, crazy ant, Gelbe Spinnerameise, gramang ant, long-legged ant, Maldive ant, yellow crazy ant.

6. *Anoplophora glabripennis*: The Asian longhorn beetle Anoplophora glabripennis is a large wood-boring beetle that is native to countries in Asia, such as Japan, Korea and China. The beetle spends most of its life within the inner wood of a variety of hardwood trees as larvae which tunnel and feed on the cambium layer, eventually killing the tree. It was first detected in New York 1996, although it is thought to have arrived in the 1980s in solid wood packing material from China. It has since been detected in Massachusetts, New Jersey, Illinois, California, Ontario (Canada) and parts of Europe. The Asian longhorn beetle threatens 30-35% of the trees in urban areas of eastern USA. The economic, ecological and aesthetic impacts on the United States would be devastating if the beetle continues to spread. Potential losses have been estimated in the tens to hundreds of billions of US dollars. Current control measures focus on rapidly delimiting new infestations, imposing quarantine and cutting down and burning of infected trees.

Common names: ALB, Asian longhorned beetle, Asiatischer Laubholzkäfer, longicorne Asiatique, starry sky beetle.

7. *Aphanomyces astaci* : *Aphanomyces astaci* commonly referred to as crayfish plague is an oomycete or water mould that infects only crayfish species. It is endemic of North America and is carried by North American crayfish species; signal crayfish *Pacifastacus leniusculus*, *Procambarus clarkii* and *Orconectes limosus*. *A. astaci* was introduced into Europe through imports of North American species of crayfish. Native European crayfish populations are not resistant to this oomycete. It has since devastated native crayfish stocks throughout the continent.

Common names: Crayfish plague, Wasserschimmel.

8. *Ardisia elliptica*: *Ardisia elliptica* is a shade tolerant evergreen tree whose fast growth and attractive fruit made it a popular ornamental plant in the past. It has escaped from private and public gardens to invade natural areas. Due to high reproductive output and high shade-tolerance, carpets of seedlings can form underneath adult trees. High seed viability (99%) and seed consumption by both avian and mammalian frugivores can lead to rapid spread across a landscape.

Common names: Ati popa'a, shoebutton ardisia.

9. *Arundo donax*: Giant reed (*Arundo donax*) invades riparian areas, altering the hydrology, nutrient cycling and fire regime and displacing native species. Long 'lag times' between introduction and development of negative impacts are documented in some invasive species; the development of giant reed as a serious problem in California may have taken more than 400 years. The opportunity to control this weed before it becomes a problem should be taken as once established it becomes difficult to control.

Common names: Arundo grass, bamboo reed, caña, caña común, caña de Castilla, caña de la reina, etc.,

10. *Asterias amurensis*: Originally found in far north Pacific waters and areas surrounding Japan, Russia, North China, and Korea, the northern Pacific seastar (*Asterias amurensis*) has successfully invaded the southern coasts of Australia and has the potential to move as far north as Sydney. The seastar will eat a wide range of prey and has the potential for ecological and economic harm in its introduced range. Because the seastar is well established and abundantly widespread, eradication is almost impossible. However, prevention and control measures are being implemented to stop the species from establishing in new waters.

Common names: Flatbottom seastar, Japanese seastar, Japanese starfish, Nordpazifischer Seestern, northern Pacific seastar, North Pacific seastar, purple-orange seastar.

11. Banana Bunchy Top Virus (BBTV): Banana bunchy top virus (BBTV) is a deadly pathogen which affects many areas of the world-wide banana industry. Infected banana plants produce increasingly smaller leaves on shorter petioles giving the plants a bunched appearance. Fruits may be distorted and plants become sterile before the whole mat (rhizome) eventually dies. The international spread of BBTV is primarily through infected planting materials.

Common names: Abaca bunchy top virus, banana bunchy top disease (BBTD), BBTV, bunchy top, bunchy top virus.

12. *Batrachochytrium dendrobatidis*: *Batrachochytrium dendrobatidis* is a non-hyphal parasitic chytrid fungus that has been associated with population declines in endemic amphibian species in upland montane rain forests in Australia and Panama. It causes cutaneous mycosis (fungal infection of the skin), or more specifically chytridiomycosis, in wild and captive amphibians. First described in 1998, the fungus is the only chytrid known to parasitise vertebrates. *B. dendrobatidis* can remain viable in the environment (especially aquatic environments) for weeks on its own, and may persist in latent infections. Common names: Chytrid frog fungi, chytridiomycosis, Chytrid-Pilz, frog chytrid fungus

13. *Bemisia tabaci* : *Bemisia tabaci* has been reported from all continents except Antarctica. Over 900 host plants have been recorded for *B. tabaci* and it reportedly transmits 111 virus species. It is believed that *B. tabaci* has been spread throughout the world through the transport of plant products that were infested with whiteflies. Once established, *B. tabaci* quickly spreads and through its feeding habits and the transmission of diseases, it causes destruction to crops around the world. *B. tabaci* is believed to be a species complex, with a number of recognised biotypes and two described extant cryptic species.

Common names: Cotton whitefly, mosca blanca, sweet potato whitefly, Weisse Fliege.

14. *Boiga irregularis* : Native island species are predisposed and vulnerable to local extinction by invaders. When the brown tree snake (*Boiga irregularis*) was accidentally introduced to Guam it caused the local extinction of most of the island's native bird and lizard species. It also caused cascading ecological effects by removing native pollinators, causing the subsequent decline of native plant species. The ecosystem fragility of other Pacific islands to which cargo flows from Guam has made the potential spread of the brown tree snake from Guam a major concern.

Common names: Braune Nachtbaumnatter, brown catsnake, brown tree snake, culepla, kulebla.

15. *Rhinella marina* : Cane toads were introduced to many countries as biological control agents for various insect pests of sugarcane and other crops. The cane toads have proved to be pests themselves. They will feed on almost any terrestrial animal and compete with native amphibians for food and breeding habitats. Their toxic secretions are known to cause illness and death in domestic animals that come into contact with them, such as dogs and cats, and wildlife, such as snakes and lizards. Human fatalities have been recorded following ingestion of the eggs or adults.

Common names: Aga-Kröte, bufo toad, bullfrog, cane toad, crapaud, giant American toad, giant toad etc.,

16. *Capra hircus* : The goat (*Capra hircus*) was domesticated 10,000 years ago in the highlands of western Iran. These herbivores have a highly varied diet and are able to ultilise a larger number of plant species than other livestock. Goats alter plant communities and forest structure and threaten vulnerable plant species. The reduction of vegetation reduces shelter options for native animals and overgrazing in native communitties leads to ecosystem degradation. Feral goats spread disease to native animals. Native fauna on islands are particularly susceptible.

Common names: Goat, Hausziege.

17. *Carcinus maenas* : *Carcinus maenas* is native to Europe and northern Africa and has been introduced to the North America, Australia, parts of South America and South Africa. It is a voracious food generalist and in some locations of its introduced range it has caused the decline of other crab and bivalve species. Its success with invasion has also caused numerous other problems that require management. Common names: European green crab, European shore crab, green crab, le crabe enragé, le crabe vert etc.

18. *Caulerpa taxifolia*: *Caulerpa taxifolia* is an invasive marine alga that is widely used as a decorative plant in aquaria. A cold-tolerant strain was inadvertently introduced into the Mediterranean Sea in wastewater from the Oceanographic Museum at Monaco, where it has now spread over more than 13,000 hectares of seabed. Caulerpa taxifolia forms dense monocultures that prevent the establishment of native seaweeds and excludes almost all marine life, affecting the livelihoods of local fishermen. Common names: Caulerpa, killer alga, lukay-lukay, Schlauchalge, sea weed.

19. *Cecropia peltata* : *Cecropia peltata* is a fast-growing, short-lived tree that grows in neotropical regions. It is light-demanding and rapidly invades disturbed areas, such as forest canopy gaps, roadsides, lava flows, agricultural sites, urban locations, and other disturbed areas. It naturally occurs in tropical Central and South America, as well as some Caribbean islands and has been introduced to Malaysia, Africa, and Pacific Islands. It may be replacing, or competing with, other native pioneer species in some locations.

Common names: Bois cannon, faux ricin, guarumo, papyrus géant, parasolier, pisse-roux, pop-a-gun, etc.,

20. *Cercopagis pengoi* : *Cercopagis pengoi* is a water flea native to the Ponto-Aralo-Caspian basin in South Eastern Europe, at the meeting point of the Middle East, Europe and Asia. It has spread from its native range and become invasive in some waterways of Eastern Europe and in the Baltic Sea. It has been introduced to the Great Lakes of North America, quickly becoming established and is now increasing its range and abundance. Cercopagis pengoi is a voracious predator and may compete with other planktivores. Through this competition, C. pengoi has the potential to affect the abundance and condition of zooplanktivorous fish and fish larvae. It also interferes with fisheries by clogging nets and fishing gear.

Common names: Cercopag, fishhook waterflea, Kaspischer Wasserfloh, petovesikirppu, rovvattenloppa, tserkopag.

21. *Cervus elaphus* : Red deer (*Cervus elaphus*) were introduced to several countries, including North and South America, New Zealand and Australia. In Argentina they have invaded several National parks, influencing native flora and fauna and possibly disrupting ecological processes. Of particular concern is possible competition with an endangered deer endemic to the southern parts of Chile and Argentina. They also compete with livestock.

Common names: Cerf elaphe, Ciervo colorado, deer, Edelhirsch, elk, European red deer, red deer, etc.,

22. *Chromolaena odorata* : *Chromolaena odorata* is a fast-growing perennial shrub, native to South America and Central America. It has been introduced into the tropical regions of Asia, Africa and the Pacific, where it is an invasive weed. Also known as Siam weed, it forms dense stands that prevent the establishment of other plant species. It is an aggressive competitor and may have allelopathic effects. It is also a nuisance weed in agricultural land and commercial plantations.

Common names: Agonoi, bitter bush, chromolaena, hagonoy, herbe du Laos, huluhagonoi, jack in the bush etc.,.

23. *Cinara cupressi* : *Cinara cupressi* is a brownish soft-bodied insect classified as an aphid. It has been discovered around the world feeding on various trees from the following genus : Cupressus, Juniperus, Thuja, Callitris, Widdringtonia, Chamaecyparis, Austrocedrus, and the hybrid Cupressocyparis. C. cupressi sucks the sap from twigs causing yellowing to browning of the foliage on the affected twig. The overall effect on the tree ranges from partial damage to eventual death of the entire tree. This aphid has seriously damaged commercial and ornamental plantings of trees around the globe.

Common names: Cypress aphid, Zypressen Blattlaus.

24. *Cinchona pubescens*: *Cinchona pubescens* is a widely cultivated tropical forest tree which invades a variety of forest and non-forest habitats. It spreads by wind-dispersed seeds and vegetatively via multiple suckers up to several metres away from original tree once it is established. *C. pubescens* replaces and outshades native vegetation.

Common names: Arbre à quinine, cascarilla, chinarindenbaum, hoja ahumada, hoja de zambo, quinine, etc.,

25. *Clarias batrachus* : *Clarias batrachus* is native to southeastern Asia and has been introduced into many places for fish farming. Walking catfish, as it is commonly known (named for their ability to move over land), is an opportunistic feeder and can go for months without food. During a drought large numbers of walking catfish may congregate in isolated pools and consume other species. They are known to have invaded aquaculture farms, entering ponds where they prey on fish stocks. *C. batrachus* has been described as a benthic, nocturnal, tactile omnivore that consumes detritus and opportunistically forages on large aquatic insects, tadpoles, and fish.

Common names: Alimudan, cá trèn trang, cá trê tráng, clarias catfish, climbing perch, freshwater catfish, Froschwels, etc.,

26. *Clidemia hirta* : *Clidemia hirta* is a problem in tropical forest understories in its introduced range, where it invades gaps in the forest, preventing native plant species from regenerating. The spread of *Clidemia hirta* has been linked to soil disturbances, particularly that caused by the wild pig, another invasive species. It has proven to negatively affect native ecosystems and is difficult to control in the Hawaiian archipelago. It is feared it will have a similar effect in other regions where it has been introduced such as in various Indian Ocean Islands (Seychelles), the Malaysian Peninsula and parts of Micronesia (Palau).

Common names: Clidemia, faux vatouk, Hirten-Schwarzmundgewaechs, kaurasiga, kauresinga, Koster's curse etc.,

27. *Coptotermes formosanus* : *Coptotermes formosanus* is a subterranean termite with an affinity for damp places. Wherever there is wood (cellulose) and moisture there is the possibility that this species can inhabit that location.

Common names: Formosan subterranean termite, Formosa termite.

28. *Cryphonectria parasitica* : *Cryphonectria parasitica* is a fungus that attacks primarily *Castanea* spp. but also has been known to cause damage to various *Quercus* spp. along with other species of hardwood trees. American chestnut, *C. dentata*, was a dominant overstorey species in United States forests, but now they have been completely replaced within the ecosystem. *C. dentata* still exists in the forests but only within the understorey as sprout shoots from the root system of chestnuts killed by the blight years ago. A virus that attacks this fungus appears to be the best hope for the future of *Castanea* spp., and current research is focused primarily on this virus and variants of it for biological control. Chestnut blight only infects the above-ground parts of trees, causing cankers that enlarge, girdle and kill branches and trunks.

Common names: Chestnut blight, Edelkastanienkrebs

29. ***Cyprinus carpio*** **:** The introduction of fish as a source of protein for human consumption into tropical and subtropical lake systems is continuing apace. The common carp (*Cyprinus carpio*) has been cultured for 2500 years and is also a popular angling and ornamental fish; is the third most frequently introduced species in the world. Its method of feeding churns up the sediments on the bottom of the water and uproots macrophytes, making it an keystone ecosystem engineer that altering habitats for native fish and other native aquatic species.

Common names: Cá Chép, carp, carpa, carpat, carpe, carpeau, carpe commune, carpo, cerpyn, ciortan, ciortanica, ciortocrap, ciuciulean, common carp, crapcean, cyprinos, escarpo, etc.,

30. ***Dreissena polymorpha*** **:** The zebra mussel (*Dreissena polymorpha*) is a small freshwater mussel. The species was originally native to the lakes of southern Russia and Ukraine, but has been accidentally introduced to numerous other areas and has become an invasive species in many countries worldwide. Since the 1980s, the species has invaded the Great Lakes, Hudson River, and Lake Travis. The species was first described in 1769 by German zoologist Peter Simon Pallas in the Ural, Volga, and Dnieper Rivers. Zebra mussels get their name from a striped pattern commonly seen on their shells, though it is not universally present.

Common names: Dreiecksmuschel, Dreikantmuschel, dreisena, Eurasian zebra mussel, moule zebra, etc.,

31. ***Eichhornia crassipes*** **:** Originally from South America, *Eichhornia crassipes* is one of the worst aquatic weeds in the world. Its beautiful, large purple and violet flowers make it a popular ornamental plant for ponds. It is now found in more than 50 countries on five continents. Water hyacinth is a very fast growing plant, with populations known to double in as little as 12 days. Infestations of this weed block waterways, limiting boat traffic, swimming and fishing. Water hyacinth also prevents sunlight and oxygen from reaching the water column and submerged plants. Its shading and crowding of native aquatic plants dramatically reduces biological diversity in aquatic ecosystems.

Common names: Aguapé, bekabe kairanga, bung el ralm, floating water hyacinth, jacinthe d'eau, jacinto-aquatico, etc.,

32. ***Eleutherodactylus coqui*****:** *Eleutherodactylus coqui* is a relatively small tree frog native to Puerto Rico. The frogs are quite adaptable to different ecological zones and elevations. Their loud call is the main reason they are considered a pest. *E. coqui* s mating call is its namesake, a high-pitched, two-note co-qui (ko-kee) which attains nearly 100 decibels at 0.5 metres. *E. coqui* have a voracious appetite and there is concern in Hawai'i, where it has been introduced, that E. coqui may put Hawai'i's endemic insect and spider species at risk and compete with endemic birds and other native fauna which rely on insects for food.

Common names: Caribbean tree frog, common coqui, coqui, Puerto Rican tree frog.

33. *Eriocheir sinensis* : *Eriocheir sinensis* (the Chinese mitten crab) is a migrating crab which has invaded Europe and North America from its native region of Asia. During its mass migrations it contributes to the temporary local extinction of native invertebrates. It modifies habitats by causing erosion due to its intensive burrowing activity and costs fisheries and aquaculture several hundreds of thousands of dollars per year by consuming bait and trapped fish as well as by damaging gear.

Common names: Chineesche Wolhandkrab, Chinese freshwater edible crab, Chinese mitten crab, etc.,.

34. *Euglandina rosea* : The carnivorous rosy wolfsnail *Euglandina rosea* was introduced to Indian and Pacific Ocean Islands from the 1950s onwards as a biological control agent for the giant African snail (*Achatina fulica*). E. rosea is not host specific meaning that native molluscs species are at risk of expatriation or even extinction if this mollusc-eating snail is introduced. Partulid tree snails of the French Polynesian Islands were particularly affected; having evolved separately from each other in isolated valleys, many Partulid tree snails have been lost and today almost all the survivors exist only in zoos.

Common names: Cannibal snail, Rosige Wolfsschnecke, rosy wolf snail.

35. *Euphorbia esula* : Native to Europe and temperate Asia, *Euphorbia esula* (leafy spurge) is found throughout the world, with the exception of Australia. This aggressive invader is one of the first plants to emerge in the spring and displaces native vegetation by shading and out-competing them for available water and nutrients. Leafy spurge contains a highly irritating substance called ingenol that, when consumed by livestock, is an irritant, emetic and purgative.

Common names: Esels-Wolfsmilch, euphorbe feuillue, euphorbia, euphorbia esule, faitours-grass , etc.,

36. *Polygonum cuspidatum* : *Polygonum cuspidatum* is an herbaceous perennial native to Japan. It has been introduced to Europe and North America as an ornamental and is also used to stabilise soil, especially in coastal areas. It requires full sun and is found primarily in moist habitats but also grows in waste places, along roadways and other disturbed areas. Once established, *P. cuspidatum* forms dense stands that shade and crowd out all other vegetation, displacing native flora and fauna, and the overwintering canes and leaves are slow to decompose.

Common names: Crimson beauty, donkey rhubarb, German sausage, huzhang , itadori , Japanese bamboo, etc.,

37. *Felis catus*: *Felis catus* was domesticated in the eastern Mediterranean c. 3000 years ago. Considering the extent to which cats are valued as pets, it is not surprising that they have since been translocated by humans to almost all parts of the world. Notable predators, cats threaten native birdlife and other fauna, especially on islands where native species have evolved in relative isolation from predators.

Common names: Cat, domestic cat, feral cat, Hauskatze, house cat, poti, pusiniveikau

38. *Gambusia affinis*: *Gambusia affinis* is a small fish native to the fresh waters of the eastern and southern United States. It has become a pest in many waterways around the world following initial introductions early last century as a biological control agent for mosquitoes. In general, it is considered to be no more effective than native predators of mosquitoes. The highly predatory mosquito fish eats the eggs of economically desirable fish and preys on and endangers rare indigenous fish and invertebrate species. Mosquito fish are difficult to eliminate once established, so the best way to reduce their effects is to control their further spread. One of the main avenues of spread is continued, intentional release by mosquito-control agencies. *G. affinis* is closely related to he eastern mosquito fish (*G. holbrooki*), which was formerly classed as a sub-species. Their appearance, behaviour and impacts are almost identical, and they can therefore be treated the same when it comes to management techniques. Records of G. affinis in Australia actually refer to *G. holbrooki*.

Common names: Barkaleci, Dai to ue, Gambusia, Gambusie, Gambusino, Gambuzia, Gambuzia pospolita, etc.,

39. *Hedychium gardnerianum* : *Hedychium gardnerianum* is a showy ornamental which grows over a metre tall in wet climates and grows from sea level to an altitude of 1700 metres. It displaces native plants, forms vast, dense colonies and chokes the understorey vegetation. It can also block stream edges, altering water flow. It is dispersed by birds over short distances and by man over long distances (in garden waste or via the horticultural industry). Even small root fragments will re-sprout, making it difficult to control.

Common names: Awapuhi kahili, cevuga dromodromo, conteira, Girlandenblume, Jin jiang hua, kahila garland-lily etc.,

40. *Herpestes javanicus*: The small Indian mongoose (*Herpestes auropunctatus*) has been introduced to many islands worldwide for control of rats and snakes, mainly in tropical areas, but also to islands in the Adriatic Sea. Moreover, it has been introduced successfully in two continental areas: the northeast coast of South America and a Croatian peninsula. Mongooses are diurnal generalist carnivores that thrive in human-altered habitats. Predation by mongoose has had severe impacts on native biodiversity leading to the decline and extirpation of native mammals, birds, reptiles, and amphibians. At least seven species of native vertebrates, including mammals, birds, reptiles, and amphibians, have almost disappeared on Amami-oshima Island since the introduction of the mongoose in 1979. In addition, mongoose carries human and animal diseases, including rabies and human Leptospira bacterium.

Common names: Beji, Kleiner Mungo, mangouste, mangus, mweyba, newla, small Indian mongoose.

41. *Hiptage benghalensis*: *Hiptage benghalensis* is a native of India, Southeast Asia and the Philippines. The genus name, Hiptage, is derived from the Greek hiptamai which means to fly and refers to its unique three-winged fruit known as samara . Due to the beautiful unique form of its flowers, it is often cultivated as a tropical ornamental in gardens. It has been recorded as being a weed in Australian rainforests and is extremely invasive on Mauritius and Réunion, where it thrives in dry lowland forests, forming impenetrable thickets and smothering native vegetation.

Common names: Adimurtte, adirganti, atimukta, benghalen-Liane, chandravalli, haldavel, hiptage, kampti, kamuka etc.,

42. *Imperata cylindrica*: Native to Asia, cogon grass (*Imperata cylindrica*) is common in the humid tropics and has spread to the warmer temperate zones worldwide. Cogon grass is considered to be one of the top ten worst weeds in the world. Its extensive rhizome system, adaptation to poor soils, drought tolerance, genetic plasticity and fire adaptability make it a formidable invasive grass. Increases in cogon grass concern ecologists and conservationists because of the fact that this species displaces native plant and animal species and alters fire regimes.

Common names: Alang-alang, blady grass, Blutgras, carrizo, cogon grass, gi, impérata cylindrique, etc.,

43. *Lantana camara*: *Lantana camara* is a significant weed of which there are some 650 varieties in over 60 countries. It is established and expanding in many regions of the world, often as a result of clearing of forest for timber or agriculture. It impacts severely on agriculture as well as on natural ecosystems. The plants can grow individually in clumps or as dense thickets, crowding out more desirable species. In disturbed native forests it can become the dominant understorey species, disrupting succession and decreasing biodiversity. At some sites, infestations have been so persistent that they have completely stalled the regeneration of rainforest for three decades. Its allelopathic qualities can reduce vigour of nearby plant species and reduce productivity in orchards. *Lantana camara* has been the focus of biological control attempts for a century, yet still poses major problems in many regions.

Common names: Ach man, angel lips, ayam, big sage, blacksage, bunga tayi, cambara de espinto, etc.,

44. *Lates niloticus* : The Nile perch (*Lates niloticus*) is a large freshwater fish. Also known as capitaine, mputa or sangara, it can grow up to 200kg and two metres in length. It was introduced to Lake Victoria in 1954 where it has contributed to the extinction of more than 200 endemic fish species through predation and competition for food.

Common names: Chengu, mbuta, nijlbaars, nilabborre, Nilbarsch, nile perch, perca di nilo, perche du nil, etc.,

45. *Leucaena leucocephala* : The fast-growing, nitrogen-fixing tree/shrub *Leucaena leucocephala*, is cultivated as a fodder plant, for green manure, as a windbreak, for reforestation, as a biofuel crop etc. Leucaena has been widely introduced due to its beneficial qualities; it has become an aggressive invader in disturbed areas in many tropical and sub-tropical locations and is listed as one of the '100 of the World's Worst Invasive Alien Species'. This thornless tree can form dense monospecific thickets and is difficult to eradicate once established. It renders extensive areas unusable and inaccessible and threatens native plants.

Common names: Acacia palida, aroma blanca, balori, bo chet, cassis, false koa, faux-acacia, faux mimosa etc.,,

46. *Ligustrum robustum* : *Ligustrum robustum* subsp.*walkeri* is a highly invasive weed in the Mascarene Achipelago in the Indian Ocean. It was introduced to Mauritius over a century ago and to La Réunion Island in the 1960s. On the oceanic islands that it has invaded, it disrupts primary forest regeneration and threatens native floral biodiversity. Its high fruit production, due to a lack of natural enemies in regions where it has invaded, has been cited as one reason for its high invasiveness.

Common names: Bora-bora, Ceylon privét, Sri Lankan privet, tree privet, troene.

47. *Linepithema humile*: *Linepithema humile* (the Argentine ant) invades sub-tropical and temperate regions and is established on six continents. Introduced populations exhibit a different genetic and social makeup that confers a higher level of invasiveness (due to an increase in co-operation between workers in the colony). This allows the formation of fast growing, high density colonies, which place huge pressures on native ecosystems. For example, *Linepithema humile* is the greatest threat to the survival of various endemic Hawaiian arthropods and displaces native ant species around the world (some of which may be important seed-dispersers or plant-pollinators) resulting in a decrease in ant biodiversity and the disruption of native ecosystems.

Common names: Argentine ant, Argentinische Ameise, formiga-Argentina.

48. *Lymantria dispar*: *Lymantria dispar* commonly known as the Asian gypsy moth, is one of the most destructive pests of shade, fruit and ornamental trees throughout the Northern hemisphere. It is also a major pest of hardwood forests. Asian gypsy moth caterpillars cause extensive defoliation, leading to reduced growth or even mortality of the host tree. Their presence can destroy the aesthetic beauty of an area by defoliating and killing the trees and covering the area with their waste products and silk. Scenic areas that were once beautiful have become spotted with dead standing trees where the Asian gypsy moth has invaded. Also, urticacious hairs on larvae and egg masses cause allergies in some people.

Common names: Asian gypsy moth, erdei gyapjaslepke, gubar, gypsy moth, lagarta peluda, limantria, etc.,

49. *Lythrum salicaria* : *Lythrum salicaria* is an erect perennial herb with a woody stem and whirled leaves. It has the ability to reproduce prolifically by both seed dispersal and vegetative propagation. Any sunny or partly shaded wetland is vulnerable to L. salicaria invasion, but disturbed areas with exposed soil accelerate the process by providing ideal conditions for seed germination.

Common names: Blutweiderich, purple loosestrife, rainbow weed, salicaire pourpre, spiked loosestrife.

50. *Macaca fascicularis* : Macaca fascicularis (crab-eating macaque) are native to south-east Asia and have been introduced into Mauritius, Palau (Angaur Island), Hong Kong and parts of Indonesia (Tinjil Island and Papua). They are considered to be invasive, or potentially invasive, throughout their introduced range and management may be needed to prevent them from becoming invasive in areas such as Papua and Tinjil. They are opportunistic mammals and reach higher densities in degraded forest areas, including habitats disturbed by humans. They have few natural predators in their introduced ranges. Macaca fascicularis impact native biodiversity by consuming native plants and competing with birds for fruit and seed resources. In addition, they facilitate the dispersal of seeds of exotic plants. *M. fascicularis* may also impact on the commercial sector through their consuming of agriculturally important plant species and damaging of crops.

Common names: Crab-eating macaque, long-tailed macaque.

51. *Melaleuca quinquenervia* : The broad-leaved paperbark tree or melaleuca (Melaleuca quinquenervia) can reach heights of 25 meters and hold up to 9 million viable seeds in a massive canopy-held seed bank. This fire-resistant wetland-invader aggressively displaces native sawgrass and pine communities in south Florida, alters soil chemistry and modifies Everglades ecosystem processes. Melaleuca is notoriously difficult to control, however, bio-control (integrated with herbicidal and other methods) holds a promising alternative to traditional control methods.

Common names: Aceite de cayeput, ahambo, balsamo de cayeput, belbowrie, bottle brush tree, broadleaf paperbark tree etc.,

52. *Miconia calvescens* : *Miconia calvescens* is a small tree native to rainforests of tropical America where it primarily invades treefall gaps and is uncommon. Miconia is now considered one of the most destructive invaders in insular tropical rain forest habitats in its introduced range. It has invaded relatively intact vegetation and displaces native plants on various islands even without habitat disturbance. Miconia has earned itself the descriptions "green cancer of Tahiti and "purple plague of Hawaii . More than half of Tahiti is heavily invaded by this plant. Miconia has a superficial root system which may make landslides more likely. It shades out the native forest understorey and threatens endemic species with extinction.

Common names: Bush currant, cancer vert, miconia, purple plague, velvet tree.

53. *Micropterus salmoides* : *Micropterus salmoides* has been widely introduced throughout the world due to its appeal as a sport fish and for its tasty flesh. In some places introduced *Micropterus salmoides* have affected populations of small native fish through predation, sometimes resulting in the decline or extinction. Its diet includes fish, crayfish, amphibians and insects.

Common names: Achigã, achigan, achigan à grande bouche, American black bass, bas dehanbozorg, bass etc.,

54. *Mikania micrantha* : *Mikania micrantha* is a perennial creeping climber known for its vigorous and rampant growth. It grows best where fertility, organic matter, soil moisture and humidity are all high. It damages or kills other plants by cutting out the light and smothering them. A native of Central and South America, M. micrantha was introduced to India after the Second World War to camouflage airfields and is now a major weed. It is also one of the most widespread and problematic weeds in the Pacific region. Its seeds are dispersed by wind and also on clothing or hair.

Common names: American rope, Chinese creeper, Chinesischer Sommerefeu, fue saina, liane americaine etc.,

55. *Mimosa pigra*: *Mimosa pigra* is invasive, especially in parts of South East Asia and Australia. It reproduces via buoyant seed pods that can be spread long distances in flood waters. *M. pigra* has the potential to spread through natural grassland floodplain ecosystems and pastures, converting them into unproductive scrubland which are only able to sustain lower levels of biodiversity. In Thailand M. pigra blocks irrigation systems that supply rice fields, reducing crop yield and harming farming livelihoods. In Vietnam it has invaded unique ecosystems in protected areas, threatening the biodiversity of seasonally inundated grasslands.

Common names: Bashful plant, catclaw, catclaw mimosa, chi yop, columbi-da-lagoa, eomrmidera, espino, giant sensitive plant etc.,.

56. *Mnemiopsis leidyi* : The ctenophore, Mnemiopsis ledyi, is a major carnivorous predator of edible zooplankton (including meroplankton), pelagic fish eggs and larvae and is associated with fishery crashes. Commonly called the comb jelly or sea walnut, it is indigenous to temperate, subtropical estuaries along the Atlantic coast of North and South America. In the early 1980s, it was accidentally introduced via the ballast water of ships to the Black Sea, where it had a catastrophic effect on the entire ecosystem. In the last two decades of the twentieth century, it has invaded the Azov, Marmara, Aegean Seas and recently it was introduced into the Caspian Sea via the ballast water of oil tankers.

Common names: American comb jelly, comb jelly, comb jellyfish, Rippenqualle, sea gooseberry, sea walnut, Venus' girdle, warty comb jelly.

57. *Mus musculus* : The house mouse (*Mus musculus*) probably has a world distribution more extensive than any mammal, apart from humans. Its geographic spread has been facilitated by its commensal relationship with humans which extends back at least 8,000 years. They cause considerable damage to human activities by destroying crops and consuming and/or contaminating food supplies intended for human consumption. They are prolific breeders, sometimes erupting and reaching plague proportions. They have also been implicated in the extinction of indigenous species in ecosystems they have invaded and colonised. An important factor in the success of *M. musculus* is its behavioural plasticity brought about by the decoupling of genetics and behaviour. This enables *M. musculus* to adapt quickly and to survive and prosper in new environments.

Common names: Biganuelo, field mouse, Hausmaus, house mouse, kiore-iti, raton casero, souris commune, wood mouse.

58. *Mustela erminea* : *Mustela erminea* (the stoat) is an intelligent, versatile predator specialising in small mammals and birds. It is fearless in attacking animals larger than itself and adapted to surviving periodic shortages by storage of surplus kills. In New Zealand it is responsible for a significant amount of damage to populations of native species.

Common names: Ermine, Grosswiesel, Hermelin, hermine, short-tailed weasel, stoat.

59. *Myocastor coypus*: *Myocastor coypus* (coypu) is a large semi-aquatic rodent which originated from South America. However, due to escapes and releases from fur farms there are now large feral populations in North America, Europe and Asia. Their burrows penetrate and damage river banks, dykes and irrigation facilities. Myocastor coypus feeding methods lead to the destruction of large areas of reed swamp. Habitat loss caused by coypus impacts plant, insect, bird and fish species.

Common names: Biberratte, coipù, coypu, nutria, ragondin, ratão-do-banhado, Sumpfbiber.

60. *Morella faya*: *Morella faya*, commonly called the fire tree, is a native to the Azores, Madeira Islands and the Canary Islands. It has been introduced to several places including Hawaii, New Zealand and Australia. This fast growing tree, whose dispersal is facilitated by introduced frugivorous birds, is capable of rapidly forming dense stands and has a negative effect on the recruitment and persistence of native plant species.

Common names: Candleberry myrtle, fayatree, Feuerbaum, firebush, fire tree.

61. *Mytilus galloprovincialis* : *Mytilus galloprovincialis* (blue mussel or the Mediterranean mussel) is native to the Mediterranean coast and the Black and Adriatic Seas. It has succeeded in establishing itself at widely distributed points around the globe, with nearly all introductions occurring in temperate regions and at localities where there are large shipping ports (Branch and Stephanni 2004). Ship hull fouling and transport of ballast water have been implicated in its spread and its impact on native communities and native mussels has been suggested by a number of studies and observations (Carlton 1992; Robinson and Griffiths 2002; Geller 1999).

Common names: Bay mussel, blue mussel, Mediterranean mussel, Mittelmeer-Miesmuschel.

62. *Oncorhynchus mykiss* : *Oncorhynchus mykiss* (rainbow trout) are one of the most widely introduced fish species in the world. Native to western North America, from Alaska to the Baja Peninsula, O. mykiss have been introduced to numerous countries for sport and commercial aquaculture. O. mykiss is highly valued as a sportfish, with regular stocking occurring in many locations where wild populations cannot support the pressure from anglers. Concerns have been raised about the effects of introduced trout in some areas, as they may affect native fish and invertebrates through predation and competition.

Common names: Alabalik, Alabalik türü, Amerikaniki Pestrofa, Aure, Baiser, Baja California rainbow trout, Brown trout, Coast angel trout, Coast rainbow trout etc.,

63. *Ophiostoma ulmi sensu lato* : Dutch elm disease (DED) is a wilt disease caused by a pathogenic fungus disseminated by specialised bark beetles. There have been two destructive pandemics of the disease in Europe and North America during the last century, caused by the successive introduction of two fungal pathogens,*Ophiostoma ulmi* and *O.novo-ulmi,* the latter much more aggressive. The vector is represented by bark beetles, various different species of scolyts living on elm trees. These beetles breed under the bark of dying elm trees. The young adults fly from the DED infected pupal chambers to feed on healthy elm trees. As a consequence, spores of the fungus carried on the bodies of these beetles are deposited in healthy plant tissue. *Ophiostoma ulmi sensu* lato can also spread via root grafts.

Common names: Dutch elm disease, Schlauchpilz.

64. *Opuntia stricta* : *Opuntia stricta* is a cactus that can grow up to 2 metres in height and originates in central America. This spiny shrub favours habitats such as rocky slopes, river banks and urban areas. *O. stricta* was considered to be Australia s worst ever weed. *O. stricta* is also invasive in South Africa, where biological options are currently being explored to control the problem.

Common names: Araluen pear, Australian pest pear, chumbera, common pest pear, common prickly pear, erect prickly pear etc.,.

65. *Oreochromis mossambicus* : *Oreochromis mossambicus* (*Mozambique tilapia*) has spread worldwide through introductions for aquaculture. Established populations of *O. mossambicus* in the wild are result of intentional release or escapes from fish farms. *O. mossambicus* is omnivorous and feeds on almost anything, from algae to insects.

Common names: Blou kurper, common tilapia, fai chau chak ue, Java tilapia, kawasuzume, kurper bream, malea etc.,

66. *Oryctolagus cuniculus*: Native to southern Europe and North Africa, the rabbit (*Oryctolagus cuniculus)* has been introduced to all continents, except Antarctica and Asia. In many countries, rabbits cause serious erosion of soils by overgrazing and burrowing, impacting on native species that depend on undamaged ecosystems.

Common names: Europäisches Wildkaninchen, kaninchen, lapin, rabbit

67. *Pheidole megacephala* : *Pheidole megacephala* is one of the world s worst invasive ant species. Believed to be native to southern Africa, it is now found throughout the temperate and tropical zones of the world. It is a serious threat to biodiversity through the displacement of native invertebrate fauna and is a pest of agriculture as it harvests seeds and harbours phytophagous insects that reduce crop productivity. P. megacephala are also known to chew on irrigation and telephone cabling as well as electrical wires.

Common names: Big-headed ant, brown house-ant, coastal brown-ant, Grosskopfameise, lion ant.

68. *Phytophthora cinnamomi* : The oomycete, Phytophthora cinnamomi, is a widespread soil-borne pathogen that infects woody plants causing root rot and cankering. It needs moist soil conditions and warm temperatures to thrive, and is particularly damaging to susceptible plants (e.g. drought stressed plants in the summer). *P. cinnamomi* poses a threat to forestry, ornamental and fruit industries, and infects over 900 woody perennial species. Diagnostic techniques are expensive and require expert identification. Prevention and chemical use are typically used to lessen the impact of *P. cinnamomi.*

Common names: Cinnamon fungus, Green fruit rot, Heart rot, Jarrah dieback, Phytophthora crown and root rot etc.,

69. *Pinus pinaster* : Pinus pinaster, originally from the Mediterranean Basin, has been planted in temperate regions within and outside its natural range for a wide range of reasons. It regenerates readily almost everywhere it is planted and in many places it invades natural shrubland, forest and grassland. Pinus pinaster forms dense thickets which supress native plants, changes fire regimes and hydrological properties and alters habitats for many animals.

Common names: Cluster pine, maritime pine.

70. *Plasmodium relictum* : The protozoon, Plasmodium relictum, is one of the causative parasites of avian malaria and may be lethal to species which have not evolved resistance to the disease (e.g. penguins). It may be devastating to highly susceptible avifauna that has evolved in the absence of this organism, such as native Hawaiian birds. The parasite cannot be transmitted directly from one bird to another, but requires a mosquito to move from one bird to another. In Hawaii, the mosquito that transmits *P. relictum* is the common house mosquito, Culex quinquefasciatus. Passerine birds are the most common victims of avian malaria. Common names: Avian malaria, paludisme des oiseaux, Vogelmalaria.

71. *Platydemus manokwari* : Worldwide land snail diversity is second only to that of arthropods. Tropical oceanic islands support unique land snail faunas with high endemism; biodiversity of land snails in Pacific islands is estimated to be around 5 000 species, most of which are endemic to single islands or archipelagos. Many are already under threat from the rosy wolfsnail (*Euglandina rosea*), an introduced predatory snail. They now face a newer but no less formidable threat, the introduced flatworm Platydemus manokwari (Platyhelminthes). Both biocontrol species continue to be dispersed to new areas in attempts to control *Achatina fulica*.

Common names: Flachwurm, flatworm, snail-eating flatworm.

72. *Pomacea canaliculata* : *Pomacea canaliculata* is a freshwater snail with a voracious appetite for water plants including lotus, water chestnut, taro and rice. Introduced widely from its native South America by the aquarium trade and as a source of human food, it is a major crop pest in south east Asia (primarily in rice) and Hawaii (taro) and poses a serious threat to many wetlands around the world through potential habitat modification and competition with native species.

Common names: Apple snail, channeled apple snail, Gelbe Apfelschnecke, golden apple snail, golden kuhol, miracle snail.

73. *Potamocorbula amurensis*: The suspension-feeding clam, *Potamocorbula amurensis* is native to Japan, China and Korea in tropical to cold temperate waters. Known as the Asian or Chinese clam, it has been designated as a major bilogical disturbance with significant ecological consequences in the San Francisco Bay area of California where large populations have become established.

Common names: Amur river clam, Amur river corbula, Asian bivalve, Asian clam, brackish-water corbula, etc.,

74. *Prosopis glandulosa*: *Prosopis glandulosa* (mesquite) is a perennial, woody, deciduous shrub or small tree. It forms impenetrable thickets that compete strongly with native species for available soil water, suppress grass growth and may reduce understory species diversity.

Common names: Honey mesquite, mesquite, Mesquite-Busch, Texas mesquite.

75. *Psidium cattleianum*: *Psidium cattleianum* is native to Brazil, but has been naturalised in Florida, Hawai i, tropical Polynesia, Norfolk Island and Mauritius for its edible fruit. It forms thickets and shades out native vegetation in tropical forests and woodlands. It has had a devastating effect on native habitats in Mauritius and is considered the worst plant pest in Hawai i, where it has invaded a variety of natural areas. It benefits from feral pigs (Sus scrofa) which, by feeding on its fruit, serve as a dispersal agent for its seeds. In turn, the guava provides favourable conditions for feral pigs, facilitating further habitat degradation.

Common names: Cattley guava, cherry guava, Chinese guava, Erdbeer-Guave, goyave de Chine, etc.,

76. *Pueraria montana* var. *lobata* : Kudzu (*Pueraria montana* var. *lobata*) roots can eventually comprise over 50% of the plant's biomass, serving as an organ for carbohydrate storage for recovery after disturbance and making it difficult to control with herbicides. Only in the eastern United States is kudzu considered a serious pest, although it is also established in Oregon in the northwestern USA, in Italy and Switzerland, and one infestation on the northern shore of Lake Erie in Canada. Kudzu is considered naturalized in the Ukraine, Caucasus, central Asia, southern Africa, Hawai, Hispaniola, and Panama. Impacts of kudzu in the southeastern USA include loss of productivity of forestry plantations (estimated at about 120 USD per hectare per year), smothering and killing of native plants and denying access to lands for hunting, hiking and bird watching.

Common names: Acha, aka, aka fala, akataha, fen ge, fen ke, foot-a-night vine, gan ge, gan ge teng, Japanese arrowroot etc.,.

77. *Pycnonotus cafer* : *Pycnonotus cafer* (red-vented bulbul) is a noisy, gregarious bird distinguished by a conspicuous crimson patch below the root of the tail. It is aggressive and chases off other bird species and may also help to spread the seeds of other invasive species. It is an agricultural pest, destroying fruit, flowers, beans, tomatoes and peas. It occurs naturally from Pakistan to southwest China and has been introduced to many Pacific Islands, where it has caused serious problems by eating fruit and vegetable crops, as well as nectar, seeds and buds.

Common names: Bulbul à ventre rouge, Bulbul cafre, red-vented bulbul, Rubülbül.

78. *Lithobates catesbeianus* : The American bullfrog (*Lithobates catesbeianus* (=Rana catesbeiana)) is native to North America. It has been introduced all over the world to over 40 countries and four continents. Many introductions have been intentional with the purpose of establishing new food sources for human consumption. Other populations have been established from unintentional escapes from bullfrog farms. Consequences of the introduction of non-native amphibians to native herpetofauna can be severe. The American bullfrog has been held responsible for outbreaks of the chytrid fungus found to be responsible for declining amphibian populations in Central America and elsewhere. They are also important predators and competitors of endangered native amphibians and fish. The control of this invasive in Europe partly relies upon increasing awareness, monitoring and education about the dangers of releasing pets into the wild. Strict laws are also in place to prevent further introductions. Eradication is achieved largely by physical means including shooting, spears/gigs, bow and arrow, nets and traps.

Common names: Bullfrog, grenouille taureau, North American bullfrog, Ochsenfrosch, rana toro, Stierkikker.

79. *Rattus rattus* : A native of the Indian sub-continent, the ship rat (*Rattus rattus*) has now spread throughout the world. It is widespread in forest and woodlands as well as being able to live in and around buildings. It will feed on and damage almost any edible thing. The ship rat is most frequently identified with catastrophic declines of birds on islands. It is very agile and often frequents tree tops searching for food and nesting there in bunches of leaves and twigs.

Common names: Black rat, blue rat, bush rat, European house rat, Hausratte, roof rat, ship rat.

80. *Rubus ellipticus* : *Rubus ellipticus* is a thorny shrub that originates from southern Asia. It has been introduced to several places, including Hawaii, Southern USA and the UK, and is grown in cultivation for its edible fruits. This plant has become a major pest in Hawai i, threatening its own native species of raspberry (*Rubus hawaiiensis*), and the ability of this plant to thrive in diverse habitat types makes it a particularly threatening invasive plant.

Common names: Asian wild raspberry, broadleafed bramble, Ceylon blackberry, eelkek, golden evergreen raspberry, Himalaya-Wildhimbeere etc.,

81. *Salmo trutta* : *Salmo trutta* has been introduced around the world for aquaculture and stocked for sport fisheries. It is blamed for reducing native fish populations, especially other salmonids, through predation, displacement and food competition. It is a popular angling fish.

Common names: An breac geal, aure, bachforelle, blacktail, breac geal, brook trout, brown trout, denizalabaligi, denizalasi etc.,

82. *Salvinia molesta* : *Salvinia molesta* is a floating aquatic fern that thrives in slow-moving, nutrient-rich, warm, freshwater. A rapidly growing competitive plant, it is dispersed long distances within a waterbody (via water currents) and between waterbodies (via animals and contaminated equipment, boats or vehicles). It is cultivated by aquarium and pond owners and it is sometimes released by flooding, or by intentional dumping. *S. molesta* can form dense vegetation mats that reduce water-flow and lower the light and oxygen levels in the water. This stagnant dark environment negatively affects the biodiversity and abundance of freshwater species, including fish and submerged aquatic plants.*Salvinia* invasions can alter wetland ecosystems and cause wetland habitat loss. *Salvinia* invasions also pose a severe threat to socio-economic activities dependent on open, flowing and/or high quality waterbodies, including hydro-electricity generation, fishing and boat transport. *S. molesta* in 2013 was elected as the one of the 100 of the World' s Worst Invasive Alien Species to replace the Rinderpest virus which was declared eradicated in the wild in 2010.

Common names: African payal, African pyle, aquarium watermoss, fougre deau, giant *salvinia* , kariba weed, koi kandy, *salvinia*, water fern , water spangles.

83. *Schinus terebinthifolius* : Native to Argentina, Paraguay and Brazil, Schinus terebinthifolius is a pioneer of disturbed sites, but is also successful in undisturbed natural environments. It is an aggressive evergreen shrub or small tree, 3-7 metres in height that grows in a variety of soil types and prefers partial sun. Schinus terebinthifolius produces shady habitats that repel other plant species and discourage colonisation by native fauna and alter the natural fire regime. Its fruit has a paralysing effect on birds and even grazing animals when ingested. *Schinus terebinthifolius* seeds are dispersed by birds and mammals and it readily escapes from garden environments. It is planted as both an ornamental and shade tree and has many uses.

Common names: Baie rose , Brazilian holly, Brazilian pepper, Brazilian pepper tree, Christmas berry, copal etc.,

84. *Sciurus carolinensis* : The grey squirrel (*Sciurus carolinensis*) is native to deciduous forests in the USA and has been introduced to the UK, Ireland, Italy and South Africa. In the introduced range grey squirrels damage trees by eating the bark and in Europe they cause the local extinction of red squirrel (*Sciurus vulgaris*) populations through competition and disease.

Common names: Grauhoernchen, gray squirrel, grey squirrel, scoiattolo grigio.

85. *Solenopsis invicta* : *Solenopsis invicta* is an aggressive generalist forager ant that occurs in high densities and can thus dominate most potential food sources. They breed and spread rapidly and, if disturbed, can relocate quickly so as to ensure survival of the colony. Their stinging ability allows them to subdue prey and repel even larger vertebrate competitors from resources.

Common names: Fourmi de feu, red imported fire ant (RIFA), rote importierte Feuerameise.

86. *Spartina anglica* : *Spartina anglica* is a perennial salt marsh grass which has been planted widely to stablize tidal mud flats. Its invasion and spread leads to the exclusion of native plant species and the reduction of suitable feeding habitat for wildfowl and waders.

Common names: Common cord grass, Englisches Schlickgras, rice grass, townsends grass.

87. *Spathodea campanulata*: The African tulip tree (*Spathodea campanulata*) is an evergreen tree native to West Africa. It has been introduced throughout the tropics, and, has naturalised in many parts of the Pacific. It favours moist habitats and will grow best in sheltered tropical areas. It is invasive in Hawaii, Fiji, Guam, Vanuatu, the Cook Islands and Samoa, and is a potential invader in several other tropical locations.

Common names: African tulip tree, Afrikanischer Tulpenbaum, amapola, apär, baton du sorcier, fa'apasī, fireball, flame of the forest, fountain tree, Indian Cedar, ko'i'i, mata kō'ī'ī, mimi, orsachel kui, patiti vai, pisse-pisse, pititi vai, rarningobchey, Santo Domingo Mahogany, taga mimi, tiulipe, tuhke dulip, tulipan africano, tulipier du Gabon.

88. *Sphagneticola trilobata* : Although *Sphagneticola trilobata* is the accepted name for this species, it is widely known as Wedelia trilobata. S. trilobata is native to the tropics of Central America and has naturalised in many wet tropical areas of the world. Cultivated as an ornamental, it readily escapes from gardens and forms a dense ground cover, crowding out or preventing regeneration of other species. In plantations, it will compete with crops for nutrients, light and water, and reduce crop yields.

Common names: Ate, atiat, creeping ox-eye, dihpw ongohng, Hasenfuss, ngesil ra ngebard, rosrangrang, Singapore daisy etc.,

89. *Sturnus vulgaris* : Native to Europe, Asia and North Africa, *Sturnus vulgaris* (the European starling) has been introduced globally, save in neotropic regions. The starling prefers lowland habitats and is an aggressive omnivore. *Sturnus vulgaris* cost hundreds of millions of dollars in agricultural damage each year and contribute to the decline of local native bird species through competition for resources and nesting spaces.

Common names: Blackbird, common starling, English starling, estornino pinto, etourneau sansonnet, étourneau sansonnet, Europäischer Star, European starling.

90. *Sus scrofa* : *Sus scrofa* (feral pigs) are escaped or released domestic animals which have been introduced to many parts of the world. They damage crops, stock and property, and transmit many diseases such as Leptospirosis and Foot and Mouth disease. Rooting pigs dig up large areas of native vegetation and spread weeds, disrupting ecological processes such as succession and species composition. *S. scrofa* are omnivorous and their diet can include juvenile land tortoises, sea turtles, sea birds, endemic reptiles and macro-invertebrates. Management of *S. scrofa* is complicated by the fact that complete eradication is often not acceptable to communities that value feral pigs for hunting and food.

Common names: Kuhukuhu, kune-kune, petapeta, pig, poretere, razorback, te poaka, Wildschwein.

91. *Tamarix ramosissima* : *Tamarix ramosissima* is a rampantly invasive shrub that has dominated riparian zones of arid climates. A massive invasion of *T. ramosissmia* in the western United States has dominated over a million acres. Typically found in conjunction with other Tamarix species and resultant hybrids, *T. ramosissima* displaces native plants, drastically alters habitat and food webs for animals, depletes water sources, increases erosion, flood damage, soil salinity, and fire potential.

Common names: Salt cedar, Sommertamariske, tamarisk, tamarix.

92. *Trachemysscripta elegans* : The red-eared slider (*Trachemysscripta elegans*) has been the most popular turtle in the pet trade with more than 52 million individuals exported from the United States to foreign markets between 1989 and 1997. Despite the vast worldwide occurrence of the sliders little is known of their impact on indigenous ecosystems, clearly research and education on the dangers of releasing pet turtles into the wild are needed. Their omnivorous diet and ability to adapt to various habitats, gives them great potential for impacting indigenous habitats.

Common names: Buchstaben-Schmuckschildkröte, Krasnoukhaya cherepakha, Nordamerikansk terrapin, punakorvakilpikonna etc.,

93. *Trichosurus vulpecula* : The brushtail possum (*Trichosurus vulpecula*) is a solitary, nocturnal, arboreal marsupial introduced from Australia. It damages native forests in New Zealand by selective feeding on foliage and fruits and also preys on bird nests and is a vector for bovine tuberculosis.

Common names: Brushtail possum, Fuchskusu.

94. *Trogoderma granarium* : *Trogoderma granarium* are considered a pest of considerable impact to stored foodstuffs. It maintains its presence in food storage in very low numbers and is able to survive long periods of time in an inactive state.

Common names: Escarabajo khapra, khapra beetle, khaprakfer, trogoderma (dermeste) du grain.

95. *Ulex europaeus* : *Ulex europaeus* is a spiny, perennial, evergreen shrub that grows in dense and impenetrable thickets which exclude grazing animals. It is common in disturbed areas, grasslands, shrublands, forest margins, coastal habitats and waste places. *Ulex europaeus* is a very successful and tenacious plant once it becomes established and is extremely competitive, displacing cultivated and native plants, and altering soil conditions by fixing nitrogen and acidifying the soil. It creates an extreme fire hazard due to abundant dead material and its oily, highly flammable foliage and seeds. Soil is often bare between individual plants, which increases erosion on steep slopes where *U.europaeus* has replaced grasses or forbs. Spiny and mostly unpalatable when mature, *Ulex europaeus* reduces pasture quality where it invades rangeland. *U. europaeus* understorey in cultivated forests interferes with operations; increasing pruning and thinning costs and can interfere with the growth of conifer seedlings.

Common names: Ajonc, bois jonc, chacay, furze, Gaspeldoorn, Ginestra spinosa, gorse, jonc marin, kolcolist zachodni, picapica, Stechginster, Tojo, vigneau, vIrish furze, whin.

96. *Undaria pinnatifida* : The kelp (*Undaria pinnatifida*) is native to Japan where it is cultivated for human consumption. It is an opportunistic weed which spreads mainly by fouling ship hulls. It forms dense underwater forests, resulting in competition for light and space which may lead to the exclusion or displacement of native plant and animal species.

Common names: Apron-ribbon vegetable, Asian kelp, haijiecai, Japanese kelp, miyeuk, qundaicai, wakame.

97. *Vespula vulgaris* : *Vespula vulgaris* (the common wasp) nest underground and in the cavities of trees and buildings. In addition to causing painful stings to humans, they compete with other insects and birds for insect prey and sugar sources. They will also eat fruit crops and scavenge around rubbish bins and picnic sites.

Common names: Common wasp, common yellowjacket, Gemeine Wespe.

98. *Vulpes vulpes* : The European red fox is probably responsible for declines of some small canids and ground-nesting birds in North America, and numerous small- and medium-sized rodents and marsupials in Australia. A programme to reduce predation pressure on native fauna within the critical weight range of 35 g to 5.5 kg in Western Australia has involved the use of 1080 fox baits.

Common names: Fuchs, lape, lis, raposa, red fox, renard, rev, Rotfuchs, silver, black or cross fox, volpe, vos, zorro.

99. *Wasmannia auropunctata* : *Wasmannia auropunctata* (the little fire ant) is blamed for reducing species diversity, reducing overall abundance of flying and tree-dwelling insects, and eliminating arachnid populations. It is also known for its painful stings. On the Galapagos, it eats the hatchlings of tortoises and attacks the eyes and cloacae of the adult tortoises. It is considered to be perhaps the greatest ant species threat in the Pacific region.

Common names: Albayalde, cocoa tree-ant, formi électrique, formiga pixixica, fourmi électrique, fourmi rouge, hormiga colorada, West Indian stinging ant etc.,

Invasive Alien Species of Microbes /strains (other than plants) in Agriculture Ecosystems

S.No	Species Name	Common Name
1. Fungi		
1	*Hemileia vastatrix*	Coffee rust
2	*Phytophthora infestans*	Late blight of potato
3	*Urocystis tritici*	Flag smut of wheat
4	*Puccinia carthami*	Rust of chrysanthemum
5	*Venturia inequalis*	Apple Scab
6	*Plasmopara viticola*	Downy mildew of grapes
7	*Sclerospora phillipinensis*	Downy mildew of maize
8	*Pyricularia grisea*	Blast of paddy
9	*Fusarium moniliforme*	Foot rot of Rice
10	*Phyllachora sorghi*	Leaf spot of sorghum
11	*Oidium heavea*	Powdery mildew of rubber
12	*Phytophthora nicotianae* var. *nicotianae*	Tobacco black shank
13	*Sphaeropsis* spp.	Canker of apple

S.No	Species Name	Common Name
14	*Synchytrium endobioticum*	Potato wart
15	*Fusariumoxysporum* f.sp *cubense* (TR4)	Fusarium wilt of Banana
16	*Plasmopara halstedii*	Downy mildew of sunflower
Bacteria		
1	*Xanthomonas campestris* p.v. *campestris*	Black rot of crucifers
2	*Agrobacterium tumefaciens*	Crown gall of apple/pear
3	*Agrobacterium rhizogenes*	Hairy root of apple/pear
4	*Erwinia amylovora*	Fire blight of pear
5	*Xanthomonas oryzae* p.v. *oryzae*	Bacterial leaf blight of paddy
Viruses		
1	Banana Bunchy Top Virus (Babu virus)	Banana bunchy top
2	Sunflower necrosis illar virus	Sunflower necrosis
3	Peanut stripe virus	Bud necrosis
Nematode		
1	*Globodera rostochiensis*	Potato golden cyst nematode

2.5 Favourbale effects of IAS

Invasive species have the potential to provide a suitable habitat or food source for other organisms. In areas where a native has become extinct or reached a point that it cannot be restored, non-native species can fill their role. An example of this is the Tamarisk, a non-native woody plant, and the Southwestern Willow Flycatcher, an endangered bird. 75% of Southwestern Willow Flycatchers were found to nest in these plants and their success was the same as the flycatchers that had nested in native plants. The removal of Tamarisk would be detrimental to Southwestern Willow Flycatcher as their native nesting sites are unable to be restored (Schlaepfer *et.al.*, 2011).

The California clapper rail (*Rallus longirostris obsoletus*), had grown partial to the new hybrid grass of Spartina alterniflora/Spartina foliosa (invasive). The new grass grew more densely than the local version and didn't die back during the winter, providing better cover and nesting habitat for the secretive bird. During the 1990s, as the hybrid spread, the rail population had soared.In addition since zebra mussels became established, the clarity of the once-murky water in Lake Erie has increased substantially, increasing visibility to thirty feet in some areas, compared to less than six inches at the middle of the 20th century. This has encouraged growth of some aquatic plants, which in turn have become nurseries for fish such as the yellow perch. The zebra mussel also constitutes a food source for fish species such as the smallmouth bass and the previously endangered lake sturgeon, with demonstrable effects on population sizes. Lake Erie is now reportedly the world's premier smallmouth bass fishery. Migrating ducks have also started to make use of the mussels as a food source.

The second way that non-native species can be beneficial is that they act as catalysts for restoration. This is because the presence of non-native species increases the heterogeneity and biodiversity in an ecosystem. This increase in heterogeneity can create microclimates in sparse and eroded ecosystems, which then promotes the growth and reestablishment of native species. In Kenya, guava has real potential as a tool in the restoration of tropical forest. Studies of isolated guava trees in farmland showed that they were extremely attractive to a wide range of fruit-eating birds. In the course of visiting them, birds dropped seeds beneath the guavas, many of them from trees in nearby fragments of rainforest, and many of these seeds germinated and grew into young trees. Surprisingly, distance to the nearest forest didn't seem to matter at all, trees up to 2 km away (the longest distance studied) were just as good as trees much nearer to forest fragments. Guavas establish easily on degraded land, and each tree is potentially the nucleus of a patch of regenerating rainforest. Of course, most seedlings that grow beneath guavas are just more guavas, but guava is an early-successional tree that soon dies out when overtopped by bigger trees, nor does it actively invade primary forest. Invasive alien trees can also be useful for restoring native forest. In Puerto Rico, native pioneer trees could cope with natural disturbances such as drought, hurricanes, floods and landslides, but are mostly unable to colonise land that has undergone deforestation, extended agricultural use and eventual abandonment. In these sites, low-diversity pioneer communities of invasive trees develop, but over time native trees invade. Alien pioneers may dominate for 30 to 40 years but the eventual outcome, after 60 to 80 years, is a diverse mixture of native and alien species, but with a majority of native species. In the absence of the initial alien colonists, abandoned agricultural land tends to become pasture and remain that way almost indefinitely.

The last benefit of non-native species is that they provided ecosystem services. Furthermore, non-native species can function as biocontrol agents to limit the effects of invasive species, such as the use of non-native species to control agricultural pests. Asian oysters, for example, filter water pollutants better than native oysters to Chesapeake Bay. A study by the Johns Hopkins School of Public Health found the Asian oyster could significantly benefit the bay's deteriorating water quality. Additionally, some species have invaded an area so long ago that they have found their own beneficial niche in the environment, a term referred to as naturalisation (Zayed Amro *et.al.*, 2007)

For example, the bee *L. leucozonium*, shown by population genetic analysis to be an invasive species in North America, has become an important pollinator of caneberry as well as cucurbit, apple trees, and blueberry bushes. The checkerspot butterfly had an advantage to any female that laid her eggs on ribwort plantain an invasive plant. The plantain leaves remained green long enough for the caterpillars to survive during dry summers, which seemed to be getting a little drier with the first signs of climate change. In contrast, the native plants they used to eat shriveled up and most of the caterpillars starved or desiccated. With this difference in survival, the butterflies started to evolve a liking for laying their eggs on plantains: the proportion of female butterflies content to lay their eggs on this plant rose from under a third in 1984 to three-quarters in 1987. A few years later, the switch was complete. The federally endangered Taylor's checkerspot Euphydryas editha taylori (a subspecies of Edith's checkerspot, whose historical habitats have been lost) is so reliant on it that conservationists are actively planting plantains out into the wild. To provide a supply of butterflies, prisoners at the Mission Creek Corrections Center for Women in Washington State breed checkerspots in a greenhouse so that they can be released into these new habitats. Odd as it might seem, actively encouraging an alien plant (increasing gains) is helping to conserve a much-loved native insect (reducing losses).

Some invasions offer potential commercial benefits. For instance, silver carp and common carp can be harvested for human food and exported to markets already familiar with the product, or processed into pet foods, or mink feed. Water hyacinth can be turned into fuel by methane digesters, and other invasive plants can also be harvested and utilized as a source of bioenergy (Van Meerbeek *et.al.*, 2015). But elsewhere, most of the time, the tens of thousands of introduced species usually either swiftly die out or settle down and become model eco-citizens, pollinating crops, spreading seeds, controlling predators, and providing food and habitat for native species. They rarely eliminate natives. Rather than reducing biodiversity, the novel new worlds that result are usually richer in species than what went before.

2.6 Invasivorism

Invasive species are flora and fauna whose introduction into a habitat disrupts the native eco-system. In response, Invasivorism is a movement that explores the idea of eating invasive species in order to control, reduce, or eliminate their populations. Chefs from around the world have begun seeking out and using invasive species as alternative ingredients (Jacobsen Rowan, 2019). In 2005 Chef Bun Lai of Miya's Sushi in New Haven, Connecticut created the first menu dedicated to the idea of using invasive species, during which time half the menus invasive species offerings were conceptual because invasive species were not yet commercially available. Today, Miya's offers a plethora of invasive species such as Chesapeake blue catfish, Florida lionfish, Kentucky silver carp, Georgia cannonball jellyfish, and invasive edible plants such as Japanese knotweed and Autumn olive.

Joe Roman, a Harvard and University of Vermont conservation biologist who is the recipient of the Rachel Carson Environmental award, is the editor and chief of Eat The Invaders, a website dedicated to encouraging people to eat invasive species as part of a solution to the problem. Skeptics point out that once a foreign species has entrenched itself in a new place such as the Indo-Pacific lionfish that has now virtually taken over the waters of the Western Atlantic, Caribbean and Gulf of Mexico eradication is almost impossible. Critics argue that encouraging consumption might have the unintended effect of spreading harmful species even more widely.Proponents of invasivorism argue that humans have the ability to eat away any species that it has an appetite for, pointing to the many animals which humans have been able to hunt to extinction such as the Caribbean monk seal, and the passenger pigeon. Proponents of invasivorism also point to the success that Jamaica has had in significantly decreasing the population of lionfish by encouraging the consumption of the fish. In recent years, organizations including Reef Environmental Educational Foundation and the Institute for Applied Ecology, among others, have published cookbooks and recipes that include invasive species as ingredients.

Species-based mechanisms: While all species compete to survive, invasive species appear to have specific traits or specific combinations of traits that allow them to outcompete native species. In some cases, the competition is about rates of growth and reproduction. In other cases, species interact with each other more directly (Kolar, 2001). Researchers disagree about the usefulness of traits as invasiveness markers. One study found that of a list of invasive and noninvasive species, 86% of the invasive species could be identified from the traits alone. Another study found invasive species tended to have only a small subset of the presumed traits and that many similar traits were found in noninvasive species, requiring other explanations. Common invasive species traits include the following:

- Fast growth
- Rapid reproduction
- High dispersal ability
- Phenotype plasticity (the ability to alter growth form to suit current conditions)
- Tolerance of a wide range of environmental conditions (Ecological competence)
- Ability to live off of a wide range of food types (generalist)
- Association with humans
- Prior successful invasions

An introduced species might become invasive if it can outcompete native species for resources such as nutrients, light, physical space, water, or food. If these species evolved under great competition or predation, then the new environment may host fewer able competitors, allowing the invader to proliferate quickly. Ecosystems which are being used to their fullest capacity by native species can be modeled as zero-sum systems in which any gain for the invader is a loss for the native. However, such unilateral competitive superiority (and extinction of native species with increased populations of the invader) is not the rule. Invasive species often coexist with native species for an extended time, and gradually, the superior competitive ability of an invasive species becomes apparent as its population grows larger and denser and it adapts to its new location.

An invasive species might be able to use resources that were previously unavailable to native species, such as deep water sources accessed by a long taproot, or an ability to live on previously uninhabited soil types. For example, barbed goatgrass (*Aegilops triuncialis*) was introduced to California on serpentine soils, which have low water-retention, low nutrient levels, a high magnesium/calcium ratio, and possible heavy metal toxicity. Plant populations on these soils tend to show low density, but goatgrass can form dense stands on these soils and crowd out native species that have adapted poorly to serpentine soils. Invasive species might alter their environment by releasing chemical compounds, modifying abiotic factors, or affecting the behaviour of herbivores, creating a positive or negative impact on other species. Some species, like *Kalanchoe daigremontana*, produce allelopathic compounds, that might have an inhibitory effect on competing species, and influence some soil processes like carbon and nitrogen mineralization. Other species like *Stapelia gigantea* facilitates the recruitment of seedlings of other species in arid environments by providing appropriate microclimatic conditions and preventing herbivory in early stages of development (Herrera Ileana *et.al.*, 2018).

Other examples are *Centaurea solstitialis* (yellow starthistle) and *Centaurea diffusa* (diffuse knapweed). These Eastern European noxious weeds have spread through the western and West Coast states. Experiments show that 8-hydroxyquinoline, a chemical produced at the root of *C. diffusa*, has a negative effect only on plants that have not co-evolved with it. Such co-evolved native plants have also evolved defenses. *C. diffusa* and *C. solstitialis* do not appear in their native habitats to be overwhelmingly successful competitors. Success or lack of success in one habitat does not necessarily imply success in others. Conversely, examining habitats in which a species is less successful can reveal novel weapons to defeat invasiveness. Changes in fire regimens are another form of facilitation. Bromus tectorum, originally from Eurasia, is highly fire-adapted. It not only spreads rapidly after burning but also increases the frequency and intensity (heat) of fires by providing large amounts of dry detritus during the fire season in western North America. In areas where it is widespread, it has altered the local fire regimen so much that native plants cannot survive the frequent fires, allowing B. tectorum to further extend and maintain dominance in its introduced range.

Ecological facilitation also occurs where one species physically modifies a habitat in ways that are advantageous to other species. For example, zebra mussels increase habitat complexity on lake floors, providing crevices in which invertebrates live. This increase in complexity, together with the nutrition provided by the waste products of mussel filter-feeding, increases the density and diversity of benthic invertebrate communities. Studies of invasive species have shown that introduced species have great potential for rapid adaptation. This explains how many introduced species are able to establish and become invasive in new environments. In addition, the rate at which an invasive species can spread can be difficult to ascertain by biologists since population growth occurs geometrically, rather than linearly. When bottlenecks and founder effects cause a great decrease in the population size and may constrict genetic variation, the individuals begin to show additive variance as opposed to epistatic variance (Prenti Peter, 2008).This conversion can actually lead to increased variance in the founding populations which then allows for rapid adaptive evolution. Following invasion events, selection may initially act on the capacity to disperse as well as physiological tolerance to the new stressors in the environment. Adaptation then proceeds to respond to the selective pressures of the new environment. These responses would most likely be due to temperature and climate change, or the presence of native species whether it be predator or prey.Adaptations include changes in morphology, physiology, phenology and plasticity.

Rapid adaptive evolution in these species leads to offspring that have higher fitness and are better suited for their environment. Intraspecific phenotypic plasticity, pre-adaptation and post-introduction evolution are all major factors in adaptive evolution. Plasticity in populations allows room for changes to better suit the individual in its environment. This is key in adaptive evolution because the main goal is how to best be suited to the ecosystem to which the species has been introduced. The ability to accomplish this as quickly as possible will lead to a population with a very high fitness. Pre-adaptations and evolution after the initial introduction also play a role in the success of the introduced species. If the species has adapted to a similar ecosystem or contains traits that happen to be well suited to the area where it is introduced, it is more likely to fare better in the new environment. This, in addition to evolution that takes place after introduction, all determine if the species will be able to become established in the new ecosystem and if it will reproduce and thrive. The enemy-release hypothesis states that the process of evolution has led to every ecosystem having an ecological balance. Any one species cannot occupy a majority of the ecosystem due to the presences of competitors, predators, and diseases. Introduced species moved to a novel habitat can become invasive when these controls competitors, predators, and diseases - do not exist in the new ecosystem. The absence of appropriate controls leads to rapid population growth.

Native species exported / introduced to non-native environments: About 28 species native to India have been found to be invasive to other biogeographical zones (Anonymous, 2008). *Acridotheres fuscus* (bird), *Albizia julibrissin* (tree), *Alternanthera sessilis* (herb), *Axis axis* (mammal), *Caesalpinia decapetala* (tree, shrub), *Casuarina equisetifolia* (tree), *Channa marulius* (fish), *Clarias batrachus* (fish), *Dalbergia sissoo* (tree), *Dioscorea bulbifera* (herb, vine, climber), *Dioscorea oppositifolia* (herb, vine, climber), *Hedychium flavescens* (herb), *Hiptage benghalensis* (vine, climber, shrub), *Hygrophila polysperma* (aquatic plant), *Hypericum perforatum* (herb), *Landoltia punctata* (aquatic plant),

Lepidium latifolium (herb), *Lutjanus kasmira* (fish), *Microstegium vimineum* (grass), *Paspalum scrobiculatum* (grass), *Pennisetum polystachion* (grass), *Rattus rattus* (mammal), *Streptopelia decaocto* (bird), *Suncus murinus* (mammal), *Syzygium cumini* (tree), *Ziziphus mauritiana* (tree, shrub).

2.7 Actions to Prevent, Detect and Manage IAS

Actions to prevent, detect and manage IAS is categorized into three themes, viz., biodiversity, human health and economic. Actions (such as projects, publications and programs) are classified according to the most obvious theme but may also fit into the dimensions of another (The Ministry of Environment & Forests, 2008).

Theme	Action
Biodiversity	National Biodiversity Action Plan: Regulation of introduction of invasive alien species and their management. As presently there is no exclusive legislation or policy in India to deal with the invasive alien species, actions include: • Develop a unified national system for regulation of all introductions and carrying out rigorous quarantine checks. • Strengthen domestic quarantine measures to contain the spread of invasive species to neighbouring areas. • Promote intersectoral linkages to check unintended introductions and contain and manage the spread of invasive alien species. • Develop a national database on invasive alien species reported in India. • Develop appropriate early warning and awareness system in response to new sightings of invasive alien species. • Provide priority funding to basic research on managing invasive species. • Support capacity building for managing invasive alien species at different levels with priority on local area activities. • Promote restorative measures of degraded ecosystems using preferably locally adapted native species for this purpose. • Promote regional cooperation in adoption of uniform quarantine measures and containment of invasive exotics. • At the central level, there are two relevant departments in the Ministry of Agriculture – the Department of Agriculture and Cooperation (DAC) and Department of Agricultural Research and Education (DARE) - which are concerned with plant protection outreach and research, respectively. Through Indian Council of Agricultural Research (ICAR), about 90 Institutes and more than 100 universities in the country have programmes on various invasive alien species. • The Guidelines on Quarantine and Strategic Plan for exotic introduction was published.1 (see Case study "The Plant Quarantine Order in India") • The threat of invasive pest species gaining entry into India (imported plant/ planting material) is addressed under The Plant Quarantine (Regulation of Import into India) Order, 2003, under the ICAR. However, the risk analysis for invasiveness of a plant species per se is not taken care of under this order.

- National workshop sponsored by The Ministry of Environment & Forests and organized by the Department of Botany, Banaras Hindu University was held in August 2004 to discuss various aspects relating to invasive species and biodiversity.
- India has assessed the risks posed to ecosystems, habitats and species by some invasives within their borders. However, most assessments were done at the local level. Some states such as West Bengal and Tamil Nadu have adopted legislative and administrative measures for eradicating and preventing further invasion of the most noxious weed species and exotic fish carnivores (such as the Big Head Carp) replacing native species.
- In terms of the restoration of degraded forest ecosystems, priority is given to regenerating native and locally adapted species groups. Furthermore, in freshwater ecosystems are given priority for clearing the effects of invasvies.
- Active cooperation among the concerned central and state government departments like agriculture, livestock, fisheries, forests, water resources, tourism, commerce, shipping, environment and rural development while involving lead institutions and NGOs are being developed on case-to-case basis. The Ministry of Environment & Forests supports the Ministry of Agriculture in the eradication and control of invasive species and the restoration of degraded ecosystems where the infrastructure and expertise is necessary to with deal with the issues.
- In 2002 a major training course was held at the National Police Academy of India involving wildlife law enforcement officials from 12 countries in Asia.
- Invasive growth of the grass Paspalum distichum has changed the ecological character of large areas of the Keoladeo National Park, reducing its suitability for certain waterbird species including the Siberian Crane. In the Kanjli Wetlands the water hyacinth which was introduced is now invasive; from time to time it is removed using mechanical means. At the Ropar wetlands invasive weeds are also a concern and management plans are under development.
- The Chilika Development Authority (CDA) received the Ramsar Wetland Conservation Award in 2002 for its work in restoring the Chilika Lake Ramsar site… increased the biodiversity, but also increased fish catches and other socio-economic benefits to the local population. Chilika Lake became degraded mainly through siltation and the choking of the seawater inlet channel, this resulted in the proliferation of invasive freshwater species, a decrease in fish productivity and an overall loss in biodiversity.
- In selected wetlands of national and international importance, changes in the ecology has been assessed and vulnerability to invasive species has been critically evaluated, leading to specific actions for mitigation of the impacts. Management Action Plans have been formulated for 30 out of 66 wetlands identified for conservation and sustainable use these MAPs have a focus on biodiversity conservation and restoration of ecosystem processes and functions one of the activities carried out in association with these plans is the control of alien invasive species.
- Legislation on Environmental Impact Assessments (EIA) applicable to wetlands has been put in place. It is mandatory to carry out EIAs for

developmental projects on wetlands by the proponent organizations to obtain environmental clearance (Environmental Impact Assessment Notification).

- Targets for control of IAS are the water hyacinth, *Salvinia* and Ipomea in 19 Ramsar sites... planned activities have included the control of the water hyacinth and other alien plant species with MAPs and other measures being developed to control the proliferation of these species.
- A Draft National Wetland Strategy has been developed with a clear focus on control of invasive species. Also, several initiatives have been undertaken under the CBD to control proliferation of invasive species in wetlands and other aquatic bodies.
- The Wildlife Institute of India periodically organizes training on wetland related issues including specific training modules and materials being developed to impart training to wetland managers in India and some selected participants from the South Asian region including the field of IAS

Human health

Livestock Importation Act, 1898 was amended by the Livestock Importation (Amendment) Ordinance, 2001. The Act now applies to livestock and livestock products to regulate the import of such products in a manner so that they do not adversely affect human and animal health within India. The Act applies to

- meat and meat products of all kinds
- egg and egg powder
- milk and milk products
- bovine, ovine and caprine embryos, ova or semen
- pet food products of animal origin

Imports require a valid sanitary import permit issued by the Department of Animal Husbandry and Dairying and are only allowed through airports and seaports which have Animal Quarantine and Certification Services Stations.

Economic

The National Bureau of Fish Genetic Resources has prepared a list of exotic/alien species under aquaculture, fisheries and aquarium trade. Their impacts of these invasives have been evaluated and a strategic plan for quarantine and exotic fish introductions has been prepared and published.

Bio-security and Sanitary-Phytosanitary Import Permit will be issued by the Directorate of Plant Protection Quarantine and Storage, Department of Agriculture and Cooperation, and the Government of India. National standards as set by the Plant Quaratine Organization of India, includes (standard operating procedures) SOP for Export Inspection & Phytosanitary Certification of plants / plant products and other regulated articles.

With regards to the prevention of introduction of invasives there are six agencies which are responsible for issuance of certificate of export/import of bioresources:

1. Plant Quarantine Division, NBPGR issues phytosanitary certificate for export of material and permits for import of germplasm, under the Plant Quarantine Order (PQO) 2003 of the Destructive Insects and Pests Act, 1914.
2. The Plant Protection Adviser issues permits for import of live insects and microbial cultures, plants and plant products, and phytosanitary certificates along with the organism for export under the PQO.

3. The Department of Animal Husbandry and Dairying deals with import of livestock issues health certificates of the livestock to be exported if required by the importing country under the Livestock Importation Act, 1898.
4. Directorate General of Foreign Trade issues licenses before export of any living organism or their product from the county under the Foreign Trade (Development & Regulation) Act, 1992.
5. The Ministry of Environment and Forests issues approval along with quarantine certificates for the export of wild animals and articles under the Wildlife (Protection) Act 1972.
6. The National Biodiversity Authority of The Ministry of Environment & Forests (MoEF) is empowered to issue approval for export of biological material from the country under the Biological Diversity Act 2002

The Destructive Insects and Pests Act 1914 aims to prevent introductions into India, and the transport from one province to another, of any fungus or other pest which is, or may be destructive to crops. 7 Furthermore, Customs Clearance Procedure for Food Items, Livestock Products, Plant and Plant Materials are in place under the Destructive Insects & Pests Act, 1914 to prevent introduction of exotic pests and diseases into the country.

Actions on IAS in cooperation with other countries (Convention on Biological Diversity, 2005)

Agreement/ Organization	Countries/ Member	Action
Asia-Pacific Forest Invasive Species Network	Australia, Bangladesh, Bhutan, Cambodia, China, Fiji, India, Indonesia, Japan, Republic of Korea, Laos, Malaysia, Maldives, Mongolia, Myanmar, Nepal, New Zealand, Pakistan, Papua New Guinea, Philippines, Samoa, Solomon Islands, Sri Lanka, Thailand, Timor-Leste, Tonga, US, Vanuatu, Vietnam, Tuvalu, Kiribati, France, and Russia	The APFISN has been established as a response to the immense costs and dangers posed by invasive species to the sustainable management of forests in the Asia-Pacific region. It is a cooperative alliance of 32 member countries of the Asia-Pacific Forestry Commission (APFC). The network operates under the umbrella of APFC which is a statutory body of the Food and Agricultural Organization of the United Nations. The APFISN focuses on inter-country cooperation that helps to detect, prevent, monitor, eradicate and/or control forest invasive species in the Asia-Pacific region
		1. Raises awareness of FIS throughout the Asia-Pacific region 2. Exchanges and shares information on FIS among member countries
		3. Facilitates access to technical expertise, research results and training and education opportunities

Agreement/ Organization	Countries/ Member	Action
		4. Strengthens capacities of member countries to conduct research, manage FIS and prevent new incursions 5. Develop strategies for regional cooperation and collaboration in combating threats posed by FIS
		The Ministry of Environment & Forests is implementing Asia-Pacific Forest Invasive Species Network Project of the FAO, and has completed a country report on IAS.
		Country Reports: Status of Forest Invasive Species in India
India and USA sign joint declaration on knowledge initiative in agriculture ICAR web portal USDA web portal	Indian Council of Agricultural Research and U.S. Department of Agriculture (USDA)	A Joint Declaration was signed between Ministry of Agriculture of India and the United States Department of Agriculture regarding support for India-United States Knowledge Initiative on Agricultural Education, Research, Service and Commercial Linkages at New Delhi on 12.11.2005. [The original declaration does not mention IAS or phytosanitary issues]
		Fifth US-India Agricultural Knowledge Initiative Board Meeting (Washington D.C. June 2007):
		- In view of the emerging socioeconomic volatility and global threats, biosecurity as realted to agriculture becomes an important area. It is required to be addressed as a new component under already identified focus areas of emerging biotechnology. Starting from minimizing the risk of introduction of alien invasive species up to averting the release of bioagents of mass destruction, a global partnership is necessary between the like-minded nations.
		- The Board would be reconstituted upon completion of 2 years with each side deciding on the composition, which is limited to 7 members on each side

Agreement/ Organization	Countries/ Member	Action
		U.S.–India Agricultural Knowledge Initiative: Board Members: Both countries created a board comprised of academia, government, and private sector representatives from the United States and India. The board agreed to a three-year work plan that supports the "Evergreen Revolution," which is based on environmentally sustainable, marketoriented agriculture. To jump start the Initiative, the United States secured funding of $8 million in fiscal year 2006, with a total of $24 million pledged through 2008.

Biosecurity frameworks for cross-border movement of invasive alien species: The policy background and legislative frameworks underlying the regulation of transboundary movement of potentially invasive alien species (IAS) was examined (Robert Black and Debbie M.F. Bartlett, 2020). The starting point is the examination of the fundamental regulatory concepts for IAS that are found in (1) the International Plant Protection Convention (IPPC) (in accordance with the World Trade Organisation's Agreement on the Application of Sanitary and Phytosanitary Measures, 'SPS Agreement') and (2) the Convention on Biological Diversity (CBD), together with a discussion on whether IAS are legally regarded as 'pests'. How these concepts are applied in different transboundary situations is then examined with examples from within federal jurisdictions (USA, Australia) and across external and internal boundaries in transnational jurisdictions (European Union and Eurasian Economic Union). Special attention is paid to IAS in aquatic environments and the question of naturalisation of once alien species.

2.8 The International Plant Protection Convention and IAS

The Convention on Biological Diversity (CBD), which was adopted in 1992, incorporates provisions regarding those alien species which threaten ecosystems, habitats or species. The International Plant Protection Convention (IPPC), which has existed since the 1950s, aims to prevent the introduction and spread of plant pests. National plant protection services and the governing body of the IPPC, the Interim Commission on Phytosanitary Measures (ICPM), recognized that the aim of the CBD to prevent the introduction of alien species corresponds in large measure to the aim of the IPPC. Since 1999, the ICPM has been actively engaged in clarifying its role in regard to invasive alien species that are plant pests. In 2001, it determined that such species should be considered quarantine pests and should be subjected to measures according to IPPC provisions. The ICPM also decided that IPPC standards should be reviewed to ensure that they adequately address environmental risks of plant pests(IPPC Secretariat, 2005). In 2003, the ICPM adopted supplements to two of the international standards for phytosanitary measures (namely Glossary of phytosanitary terms and Pest risk analysis for quarantine pests). These supplements elaborated on environmental considerations. To avoid conflicting developments within the IPPC and the CBD regarding invasive alien species and plant pests, the secretariats of the two conventions have established a Memorandum of Cooperation and developed a joint work plan as was called for by the Conference of Parties to the CBD at its seventh meeting. In 2002, the Conference of Parties to the CBD formalized a set of guiding principles on alien species that threaten ecosystems, habitats or species. A comparison between these guiding principles and provisions of the IPPC shows strong correspondence and considerable overlap. This paper discusses the need for closer cooperation between environmental and phytosanitary authorities on a national and international level as well as future activities in regard to invasive alien species.

2.8.1 Relationship Between IAS and Quarantine Pests

The parameters for analysing the relationship between invasive alien species and quarantine pests are the agreed or official definitions of both conventions. The CBD defines an alien species as "a species, subspecies or lower taxon, introduced outside its natural past or present distribution; includes any part, gametes, seeds, eggs, or propagules of such species that might survive and subsequently reproduce" and further prescribes that an invasive alien species is "an alien species whose introduction and/or spread threaten biological diversity" (annex footnote 57, CBD, 2002). The IPPC, on the other hand, defines a [plant] pest as "any species, strain or biotype of plant, animal or pathogenic agent injurious to plants or plant products" and a quarantine pest as "a pest of potential economic importance to the area endangered thereby and not yet present there, or present but not widely distributed and being officially controlled" ((IPPC Secretariat, 2005).The IPPC definition of a quarantine pest covers much, but not all, of what is considered as an invasive alien species under the CBD. Both definitions cover any organism that is injurious to plants and that has an environmental impact (threatens biological diversity). Both definitions prescribe in different words that the environmental impact results from the organism's introduction and/or spread. It can be argued that most quarantine pests are invasive alien species and that those invasive alien species which are directly or indirectly injurious to plants are quarantine pests.

One marked difference between the definitions is that quarantine pests do not necessarily threaten biological diversity. For example, they may affect only agricultural or horticultural plants that are alien species in their own right.

2.8.2 Overlapping Mandates of International and Regional Organizations

The relationship between the CBD and the IPPC in relation to invasive alien species not only results in a partial overlap of the mandates of both conventions, but also draws several other international organizations into the picture (Fig. 3) (IPPC Secretariat, 2005).

Regional plant protection organizations (RPPOs) may also be drawn into the international framework dealing with invasive alien species that are plant pests. According to Article IX of the IPPC, RPPOs function as the coordinating bodies in the geographical areas covered and participate in various activities to achieve the objectives of the IPPC. Consequently, RPPOs may be active in relation to invasive alien species. For example, the European and Mediterranean Plant Protection Organization (EPPO) has initiated an extensive work programme in this area.

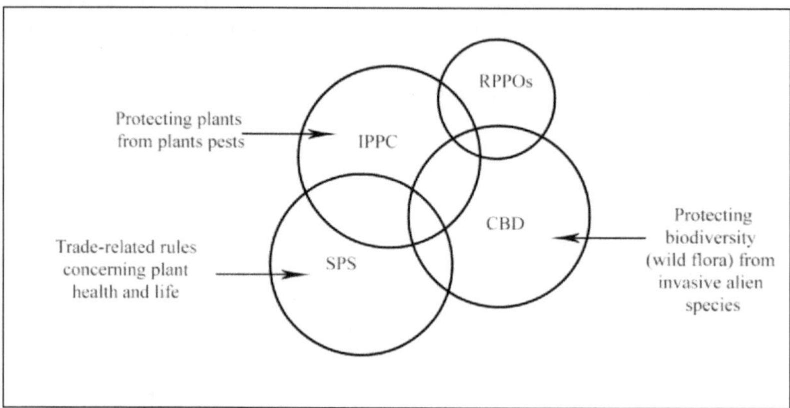

Fig. 3: Overlapping mandates of international and regional organizations

2.8.3 PPC Activities in Relation to IAS

The governing body of the IPPC, the Interim Commission on Phytosanitary Measures (ICPM), addressed the topic of biosafety at its second meeting (ICPM 2) in 1999 (IPPC Secretariat, 2005). It concluded that the concept of invasive alien species had considerable implications for the IPPC and that coordination between government authorities at a national and international level was necessary to avoid conflicting views in different fora (FAO, 1999). As a result of these discussions, ICPM 2 created an informal open-ended working group to consider issues regarding GMOs, biosafety and invasive alien species. This group met in 2000 and, based on its recommendations, ICPM 3 decided in 2001 that

- Species that may be invasive and that directly or indirectly affect plants or plant products should be assessed, monitored and managed, if necessary, according to IPPC provisions.
- Invasive alien species that are plant pests and that are absent from an area (or if present, are of limited distribution and under official control) should be considered quarantine pests and should be subjected to measures according to IPPC provisions.

- Implementation of the IPPC is directly relevant to implementation of Article 8(h) of the CBD.
- Many provisions and standards of the IPPC are directly relevant to, or overlap with, the (then interim) guiding principles of the CBD.
- Standards should be reviewed to ensure that they adequately address environmental risks of plant pests.
- A supplement to ISPM 11: Pest risk analysis for quarantine pests should be developed to address in detail the environmental risks of plant pests (FAO 2001)

ICPM 3 also decided that the IPPC Secretariat should cooperate with the CBD on matters related to invasive alien species. The IPPC Secretariat was asked to seek observer status for the IPPC with the CBD, to attend relevant meetings of the CBD and invite the CBD to attend relevant IPPC meetings. To avoid conflicting developments in different international organizations, ICPM 3 charged the IPPC Secretariat to cooperate with other standard-setting bodies to ensure that common areas of interest would be adequately covered (IPPC Secretariat, 2005). The decisions of ICPM 3 on invasive alien species have continued to occupy the activities of the organization.

In 2003, ICPM 5 adopted two ISPMs with direct relevance to invasive alien species and the protection of the environment:

A supplement to ISPM 5: Glossary of phytosanitary terms. Supplement no. 2: Guidelines on the understanding of potential economic importance and related terms including reference to environmental considerations (FAO, 2003)

A supplement to ISPM 11 [2001]: Pest risk analysis for quarantine pests. Supplement: Analysis of environmental risks (FAO, 2003).

In addition to these amendments to standards, the cooperation between the IPPC Secretariat and the CBD has developed. A Memorandum of Cooperation between FAO and the Secretariat of the CBD on cooperation between the secretariats of the IPPC and the CBD aims to promote synergy, to avoid overlaps and unnecessary duplication as well as to ensure effective cooperation in joint activities. A joint work plan has been developed as was called for by the seventh meeting of the Conference of Parties to the CBD (CBD, 2004).

2.8.4 The CBD's Guiding Principles on IAS and the IPPC

Comparing the aim and scope of the CBD and the IPPC reveals numerous overlapping elements. ICPM 3 noted that many provisions and standards of the IPPC are directly relevant to, or overlap with, the provisions of the (then interim) guiding principles of the CBD (Appendix XIII, FAO, 2001).

Relevant areas include providing legal and regulatory frameworks; building capacity and technical assistance for developing countries; assessing and managing potential plant pest risks; protecting areas that may be threatened by plant pests; applying measures to prevent unintentional introduction of plant pests; certifying that risk management procedures have been applied; assessing and managing the intentional introduction of organisms that may be pests of plants including claimed beneficial and biological control organisms; exchanging of scientific and regulatory information relevant to plant pests; cooperating between countries to minimize the impact of plant pests; and detecting, controlling, and eradicating pests in agricultural and wild flora.

Guiding principles : In 2002, the sixth Conference of the Parties (COP) to the CBD adopted the guiding principles as non-binding recommendations intended to provide all governments and organizations with guidance for developing effective strategies to minimize the spread and impact of invasive alien species (CBD, 2002). A detailed comparison between the recommendations made in the 15 guiding principles on invasive alien species and the IPPC provisions and its ISPMs may further clarify the relationship between invasive alien species and plant pests and the work of the IPPC and CBD (IPPC Secretariat, 2005).

Guiding principle 1: Precautionary approach-The precautionary approach in principle 1 refers to the Rio Declaration and the preamble of the CBD, which lays down that "... where there is a threat of significant reduction or loss of biological diversity, lack of full scientific certainty should not be used as a reason for postponing measures to avoid or minimize such a threat". The IPPC does not have a similar provision. Indeed, its Article VII.2a specifies that phytosanitary measures shall not be taken without technical justification, which is understood to be a pest risk analysis. This apparent contradiction between the two conventions, however, becomes less evident on closer inspection. The use of the terms "significant reduction" and "lack of full scientific certainty" in the CBD's precautionary approach seems to imply that certain knowledge should be available. On the other hand, it is also understood by the IPPC that the availability of full scientific evidence is not always possible. ISPM 11 makes provision for uncertainties in the PRA process.

Guiding principle 2: Three-stage hierarchical approach-A three-stage hierarchical approach to minimize the risk and spread of invasive alien species gives preference to the prevention of their introduction. In the event of an introduction, early detection and rapid action (e.g. eradication) are recommended. If this fails, the third stage is containment and control.The approach can be considered as the underlying principle in phytosanitary policy. For decades, governments have operated their phytosanitary policy on the principle of preventing the introduction of quarantine pests, and if this fails, eradicating or containing them. This is also reflected in many provisions of the IPPC, although a detailed hierarchical approach is not explicitly specified.

Guiding principle 3: Ecosystem approach-Principle 3 recommends that measures to deal with invasive alien species should, as appropriate, be based on the ecosystem approach. This approach may be described as a strategy to protect complex and dynamic plant, animal and micro-organism communities and their non-living environment, which together interact as functional units, through integrated management of land, water and living resources (CBD, 2000).

Relationship between the CBD's guiding principles and the IPPC and its standards.

Guiding principle, CBD	Corresponding IPPC provisions and/or ISPMs
1. Precautionary approach	Article VII.2a (potential contradiction)
	Relevant ISPMs: ISPM 1 (partial application)
2. Three-stage hierarchical approach	General aim of IPPC (full application)
3. Ecosystem approach	Relevant ISPMs: ISPM 11 supplement on environmental risks; ISPM 5 supplement on potential economic importance
4. The role of states	General aim of IPPC (full application), Article IV
	Relevant ISPMs: ISPM 3, 6, 17, 19
5. Research and monitoring	Article IV.2b and Article IV.3b
	Relevant ISPMs: ISPM 6, 8
6. Education and public awareness	not covered (Article IV.3a provides very limited coverage)
7. Border control and quarantine measures	especially Article V and Article VII
	Relevant ISPMs: ISPM 1, 7, 12, 13, 14
8. Exchange of information	especially Article VIII
	Relevant ISPMs: ISPM 1, 6, 8, 13, 17, 19
9. Cooperation, including capacity-building	especially Articles VIII, IX and XX
10. Intentional introduction	especially Article VII
	Relevant ISPMs: ISPM 1, 2, 3, 11
11. Unintentional introductions	especially Article VII
	Relevant ISPMs: ISPM 1, 2, 11, 14
12. Mitigation of impacts	Article I
13. Eradication	Article I
	Relevant ISPMs: ISPM 9
14. Containment	Article I
	Relevant ISPMs: ISPM 4, 9, 10
15. Control	Article I
	Relevant ISPMs: ISPM 14

Guiding principle 4: The role of states-This guiding principle calls for states to recognize that activities within their jurisdiction or under their control, such as intentional and unintentional introductions, may pose risks to other states. Guiding principle 4 stipulates that states should take actions to minimize the spread and impact of invasive alien species. This would include the identification of invasive alien species or species that could become invasive as well as providing information on such species to other states.The IPPC recognizes the role of states in combating plant pests. In fact, the IPPC goes further by requesting the establishment of adequate structures in states. Especially Article IV of the IPPC lays down general provisions for the organizational arrangements for national plant protection, such as the establishment of an official national plant protection organization (NPPO) and its functions in each state. Several ISPMs give guidance on the role of states in their actions to minimize risks. ISPM 3: Code of conduct for the import and release of exotic biological control agents is important in relation to the intentional introduction of organisms, and ISPM 6: Guidelines for surveillance and ISPM 19: Guidelines on lists of regulated pests to the identification, surveillance and listing of pests, while ISPM 17: Pest reporting describes responsibilities of states in reporting the occurrence, outbreak and spread of pests.

Guiding principle 5: Research and monitoring-This guiding principle states that it is important for states to conduct research on and monitoring of invasive alien species in order to increase the knowledge about such species and their status in the country.The IPPC also specifically addresses research and monitoring. Article IV.2b of the IPPC attributes the surveillance of cultivated plants and wild flora to the responsibilities of NPPOs. Article IV.3b prescribes that each state shall make provisions concerning research and investigations in plant protection. Practical guidance in relation to monitoring is provided in ISPM 6: Guidelines for surveillance and ISPM 8: Determination of pest status in an area.

Guiding principle 6: Education and public awareness-Guiding principle 6 attributes importance to public awareness in the management of invasive alien species. It recommends that states should promote education and public awareness of the causes of invasion and the risks associated with the introduction of alien species. In cases of mitigation measures, such as control or containment programmes, this should be done in a way to involve local communities and appropriate interest groups.The IPPC does not have similar provisions. It partly covers the provisions in Article IV.3a, which prescribes that states should make provisions for the distribution of information within their territory. Public awareness initiatives, however, are carried out by many NPPOs as a regular and useful tool in preventing the introduction of pests or in involving the public in specific surveillance programmes.

Guiding principle 7: Border control and quarantine measures-This guiding principle recommends that states should implement border controls and quarantine measures to minimize the risks of introduction of alien species that are or could become invasive. The quarantine measures should be based on risk assessment and existing appropriate government bodies should be strengthened as necessary to implement the measures.Provisions concerning quarantine measures and border control are the main substance of the IPPC. In particular, Articles V and VII deal with phytosanitary certification and requirements in relation to imports, respectively. Several ISPMs provide practical guidance in relation to border controls and quarantine measures. The most important are: ISPM 1: Principles of plant quarantine as related to international trade, ISPM 7: Export certification system, ISPM 12: Guidelines for phytosanitary certificates, ISPM 13: Guidelines for the notification of non-compliance and emergency action and ISPM 14: The use of integrated measures in a systems approach for pest risk management.

Guiding principle 8: Exchange of information-Provisions regarding information exchange on alien species are laid down in principle 8. It recommends the development of information systems in regard to relevant biological information on alien species as well as the dissemination of information. Information on import requirements for alien species should be made available to other states.Perhaps one of the most important differences between principle 8 and the IPPC provisions is that many of the IPPC provisions on information exchange are obligatory to contracting parties. Article VIII specifies that contracting parties shall cooperate in the exchange of information on plant pests and that they shall designate an official contact point through which the information exchange is facilitated. Several articles of the IPPC specify which type of information contracting parties are obliged to exchange and with whom. The IPPC Secretariat has put into operation the International Phytosanitary Portal, an Internet-based information system and network, which facilitates international information exchange. Several ISPMs also provide guidance on information exchange, especially ISPM 17: Pest reporting, which describes detailed reporting responsibilities. Other ISPMs dealing partially with information exchange are ISPM 1: Principles of plant quarantine as related to international trade, ISPM 6: Guidelines for surveillance, ISPM 8: Determination of pest status in an area, ISPM 13: Guidelines for the notification of non-compliance and emergency action and ISPM 19: Guidelines on lists of regulated pests.

Guiding principle 9: Cooperation, including capacity building-This guiding principle points out that a state's response to minimizing the spread and impact of invasive alien species not only may be applied internally within the country but also may require a bilateral or multilateral approach with other countries. Cooperative efforts may include the development of programmes to share information and the establishment of bilateral or multilateral agreements to regulate trade in certain alien species, as well as cooperation in research and its funding. Capacity-building programmes for states that lack expertise and resources are advocated. Such programmes may involve technology transfer and the development of training programmes.The IPPC also calls for cooperation and capacity building to achieve its aims. Multilateral cooperation is especially advocated in Articles VIII and IX of the IPPC. Capacity building has an especially high profile in the IPPC. In Article XX, contracting parties agree to promote technical assistance to facilitate implementation of the convention. This is reflected in the work of the IPPC Secretariat which devotes considerable resources to technical assistance. The development and practical application of the Phytosanitary Capacity Evaluation (PCE) module, a tool to evaluate the phytosanitary capacity of countries, has increased considerably the understanding of technical needs of developing countries.

Guiding principle 10: Intentional introduction-Guiding principle 10 provides recommendations regarding the intentional introduction of alien species into countries or into new ecological areas within a country. The principle recommends that such intentional introductions should take place only after they have been evaluated and authorized. A risk assessment should be part of the evaluation and the authorization should be based on the precautionary principle. Furthermore, principle 10 recommends that the burden of proof that a proposed introduction is unlikely to threaten biological diversity should be with the proponent of the introduction or be assigned as appropriate by the recipient state. The IPPC covers requirements in relation to imports especially in Article VII. Specifically, Article VII.5 addresses the intentional import of pests and regulated articles for research, education or specific use. In cases of intentional import, special and adequate safeguards should be established to prevent the "escape" of the pest. Particularly relevant to the intentional introduction of alien species, ISPM 3: Code of conduct for the import and release of biological control agents describes the responsibilities of authorities of governments, importers and exporters in relation to the importation of exotic biological control agents capable of multiplication. Other IPPC standards relevant to this guiding principle are ISPM 1: Principles of plant quarantine as related to international trade, ISPM 2: Pest risk analysis and ISPM 11 [2004]: Pest risk analysis for quarantine pests, including analysis of environmental risks and living modified organisms.

Guiding principle 11: Unintentional introduction-This guiding principle recommends that every state should have in place provisions to prevent unintentional introductions of invasive alien species. Such provisions could incorporate legislative measures and the establishment or strengthening of institutions. Guiding principle 11 also mentions the most common pathways for unintentional introductions, such as agriculture, forestry, shipping or tourism. It recommends that environmental impact assessments of such activities should consider unintentional introductions of invasive alien species and that risk assessments should be carried out for these pathways, where appropriate. Unintentional introductions are probably the most common problem in the phytosanitary field. The IPPC addresses in several of its articles the provisions recommended in the guiding principle 11. Most relevant is Article VII with its requirements in relation to imports. ISPMs of relevance to guiding principle 11 are ISPM 1: Principles of plant quarantine as related to international trade, ISPM 2: Pest risk analysis, ISPM 11 [2004]: Pest risk analysis for quarantine pests, including analysis of environmental risks and living modified organisms and ISPM 14: The use of integrated measures in a systems approach for pest risk management.

Guiding principle 12: Mitigation of impacts-Guiding principle 12 (as well as the three following guiding principles) deals with the mitigation of impacts once the establishment of an invasive alien species has been detected. Mitigation measures, which should be initiated at the earliest possible date, may be eradication, containment or control programmes that are safe to humans, the environment and agriculture, as well as ethically acceptable to stakeholders. Principle 12 also recommends that, consistent with national policy or legislation, an individual or entity responsible for the introduction of invasive alien species should bear the costs of control measures and biological diversity restoration where their failure to comply with the national laws and regulations is established.The IPPC does not go into detail of what should happen if a quarantine pest has been introduced into a country. Article I specifies that one of the purposes of the IPPC is to promote appropriate measures to control pests. It is, however, common practice for countries to implement mitigation measures once a pest has been detected. In relation to the safety of the mitigation measures and their ethical acceptability to stakeholders, the IPPC has no provisions equivalent to those of the CBD's guiding principles on invasive alien species. Also the "polluter pays" principle as laid down in principle 12 is not reflected in the IPPC or any of its standards.

Guiding principle 13: Eradication-This guiding principle recommends dealing with the introduction and establishment of invasive alien species by eradication, where feasible. Eradication is best carried out in the early stages of an invasion, and community support is often essential for the success of an eradication campaign.The eradication of pests is implied in Article I of the IPPC. More detailed provisions concerning the establishment of pest eradication programmes are provided in ISPM 9: Guidelines for pest eradication programmes, which specify administrative and technical components of such programmes.

Guiding principle 14: Containment-Guiding principle 14 suggests that containment, or limiting the spread, of alien invasive species may be an appropriate strategy where eradication is not feasible.As in the case of eradication, the containment of pests is implied in Article I of the IPPC. Several ISPMs provide partial guidance on limiting the spread of pests. These are ISPM 4: Requirements for the establishment of pest free areas, ISPM 9: Guidelines for pest eradication programmes and ISPM 10: Requirements for the establishment of pest free places of production and pest free production sites.

Guiding principle 15: Control-In cases where eradication and/or containment has failed, control measures are the last step in efforts to minimize the impact of alien invasive species. Guiding principle 15 recommends that control measures focus on reducing the damage caused by invasive alien species as well as reducing their number. This principle highlights the use of integrated management measures. Article I of the IPPC defines that promoting appropriate measures for the control of pests is one of the purposes of the convention. Practical guidance on the control of pests through integrated measures is given in ISPM 14: The use of integrated measures in a systems approach for pest risk management, although this standard is focused on the development of pest risk management options for the import of plants and plant products.

As presented above, there are considerable overlaps in the work of the CBD on invasive alien species and the IPPC activities in regard to plant pests (IPPC Secretariat, 2005). To avoid inconsistencies and duplication of effort and to achieve effective action, we should consider the following questions:

- How can plant health authorities assist environmental authorities in realizing their objectives?
- How should national and international environmental and phytosanitary bodies cooperate in future in dealing with invasive alien species?
- What are important future activities in the field of invasive alien species?

These issues are discussed below.

2.9 The Role of Phytosanitary Authorities in the Work on IAS

For almost 100 years, phytosanitary authorities worldwide have carried out the important task of preventing the introduction of quarantine pests. An efficient infrastructure (such as border controls, national surveillance programmes, technical and scientific institutions, as well as export-oriented certification programmes) has been established to achieve the tasks of phytosanitary authorities. The long experience of phytosanitary authorities in the assessment and management of biological risks related to the introduction of organisms provides these authorities with the knowledge of how to deal with risks posed by plant pests and invasive alien species that are plant pests.On a national level, this existing infrastructure and know-how should be utilized by environmental authorities in their efforts to implement the guiding principles of the CBD. Such utilization would have considerable advantages for governments since existing structures and know-how would be used without significant new investments and a duplication of activities would be prevented.

Cooperation between international bodies: The argument for close national cooperation between environmental and phytosanitary authorities also holds true on an international level. To avoid duplication of activities, contradictory approaches and a confusion of competences, the IPPC and the CBD should work closely together in relation to invasive alien species. Such a close cooperation need not be limited to the secretarial levels of both conventions, but may also include joint activities of the relevant governing bodies. This could be achieved either through a declaration by the CBD that the IPPC is a competent authority for the development of technical standards on invasive alien species that are pests of plants, including invasive plants, or through the establishment of a formal inter-organizational working group developing recommendations for invasive alien species (IPPC Secretariat, 2005).Improving the cooperation between international organizations should not be limited to the IPPC and the CBD. The SPS Agreement of the WTO should also be included in such a closer collaboration. Measures to prevent the introduction of invasive alien species may, by their nature, be very trade restrictive. A close cooperation between the CBD, the IPPC, the SPS Agreement and possibly other international organizations could certainly help to achieve the objectives of these instruments without restricting trade unnecessarily.

Closer cooperation between international organizations depends on the efforts of national governments because it is ultimately their responsibility to determine the policy of the international organizations to which they belong. A number of countries have not ratified or accepted the CBD and/or the IPPC. If countries wish to influence the policy of these organizations their primary objective should be to ratify them.CBD matters are administered in many countries by environmental authorities and IPPC-related activities by agricultural authorities. Hence, communication between these authorities in relation to invasive alien species is important. If governments are to address matters related to invasive alien species and plant pests in the CBD and the IPPC in a consistent way, it would be advisable for national coordination strategies to be developed.

Future activities: It is of importance to states that different international agreements partly covering the same substance do not provide contradictory approaches or room for conflicting interpretations. Situations where national governments by implementing the provision of one international agreement may violate the provisions of another should be avoided (IPPC Secretariat, 2005). Thus it would be advisable if joint activities could be conducted to analyse the guiding principles of the CBD and the IPPC and its standards in order to resolve possible conflicting approaches and to agree on a common understanding and terminology. The success of the CBD and the IPPC in protecting plants and the environment is very much dependent on the implementation of their provisions by national governments. The prevention of the spread of invasive alien species and plant pests is primarily an international approach in which countries must cooperate to prevent the natural or man-facilitated spread of such organisms. For many developing countries and especially for least-developed countries, however, the protection of the environment may not be located high on their list of national priorities. These countries may need to use their scarce resources to establish basic economic conditions taken for granted in the developed world. Thus the provision of technical assistance for developing countries should be seen as one of the priorities for the CBD and the IPPC to further the implementation of their provisions. A close cooperation between the CBD and the IPPC on technical assistance activities in relation to invasive alien species and plant pests would maximize the use of resources provided for this purpose and utilize FAO's experience in providing technical assistance.

2.10 Regulations and Management

i. **International agreements, legislation, and voluntary self-regulation:** Invasion biologists and policy-makers generally agree that efficient responses to biological invasions require prioritizing measures to prevent the arrival of potentially invasive alien species, the timely management of incursions, and effective management of those already established (Petr Pysek, 2020). Achieving these goals requires implementing mechanisms to regulate the intentional introduction of alien species and identifying pathways and mitigation methods for unintentional arrivals. It also demands enforcing preventive measures and ensuring the timely deployment of protocols for detection and rapid response to deal with new incursions.

ii. **National biosecurity programs:** The term 'biosecurity' refers to measures to prevent and manage biological invasions. A close correspondence exists between the various stages of the invasion process and different biosecurity activities. For example 'border biosecurity' refers to measures, such as inspection, quarantines (bans on imports), and sanitary treatments (e.g. fumigation) of imported goods at or near the border (Petr Pysek, 2020). These activities contrast with surveillance and eradication, which aim to locate and eliminate nascent invaders before they establish populations. Nearly every country operates biosecurity measures to protect natural resources and citizens from invasion-related impacts. For some nations, biosecurity has become a national priority (e.g. Australia and New Zealand), and in these countries there have been long-term successes such as eradication of rats and cats on increasingly large islands or biological control of weeds across continental areas. International trade creates important pathways for the accidental movement of alien species, and the trend of increasingly globalized economies has contributed to increased invasion rates.

iii. **Technological advances in management- From classical control to gene editing:** Established populations of IAS have long been managed to low densities or even eradicated, primarily by three methods – mechanical or physical control, chemical control, and biological control. Each method has recorded substantial successes as well as failures, but incremental technological advances have improved all three methods and lessened non-target impacts (Petr Pysek, 2020). Significant advances have occasionally allowed successful management or eradication of a much greater range of invasions. Although the majority of management projects for established invaders employ one or more of the above methods, other technologies have been applied in more limited domains and are being developed for a greater range of applications. For instance, invasive insects, especially lepidopterans, have long been managed with pheromones, especially through attract-and-kill or mating disruption. Two pheromones have now been isolated for the sea lamprey (*Petromyzon marinus*) with an eye towards control in the Laurentian Great Lakes. Similarly, the male-sterilization technique has been widely used to manage or eradicate invasive insect populations and is now being used against the sea lamprey.

iv. **Surveillance and Monitoring- the key role of Citizen Science:** The importance of early-warning and rapid-response initiatives, and concurrently the need for surveillance to inform such approaches, is widely recognized. Most countries do not implement integrated national invasive alien species surveillance programs. Also, many IAS that can affect biodiversity and ecosystems adversely do not fulfil the criteria for inclusion under government-funded schemes (Petr Pysek, 2020). Engaging volunteers in surveillance and monitoring is a low-cost, large-scale, and long-term option. There are many benefits of engaging the public in recording IAS; the collected data are valuable, and the process of raising awareness has important consequences for increasing acceptance of biosecurity. Citizen scientists with smartphones and appropriate apps such as iNaturalist and IveGot1 plus a program to record and evaluate images, such as EDDMaps, can greatly increase early detection ability and also aid in recording the spread and location of invasive alien species. The emergence of new tools and technologies to detect new invasions, including image recognition, use of machine learning, and remote sensing, will be influential in advancing citizen science for surveillance and monitoring of IAS. Progress has also been made on developing more cost-effective strategies for deploying surveillance networks, targeting surveillance in high-risk areas to increase efficiency.

2.11 Global Overview of the Management of IAS

2.11.1 Instruments Pertaining to the Management of IAS

The foremost international convention obliging its contracting parties to take action on invasive alien species is the Convention on Biological Diversity, which was adopted in 1992. The objectives of the convention are the conservation and sustainable use of biological diversity and the fair and equitable sharing of benefits arising out of the utilization of genetic resources. Article 8(h) of the CBD requires contracting parties to prevent the introduction of, control or eradicate those alien species which threaten ecosystems, habitats or species (IPPC Secretariat, 2005). Some 50 instruments or guidelines deal with particular aspects of the management of invasive alien species..

Major international instruments relating to invasive alien species:

- Convention on Biological Diversity
- International Plant Protection Convention
- Office International des Épizooties.

Instruments dealing with invasive alien species in the context of particular species groups include:

- Convention on the Conservation of Migratory Species of Wild Animals
- Agreement on the Conservation of African-Eurasian Migratory Waterbirds

- Convention on International Trade in Endangered Species of Wild Fauna and Flora.

Examples of instruments and programmes dealing with invasive alien species in the context of particular ecosystems are:

- United Nations Convention on the Law of the Sea
- United Nations Environment Programme Regional Seas Programme
- Ramsar Convention on Wetlands.

Instruments dealing with invasive alien species in the context of particular pathways for their introduction include:

- International Convention for the Control and Management of Ships' Ballast Water and Sediments
- FAO Code of Conduct for Responsible Fisheries.

Given the large number of such instruments, it is important for countries to be guided in how to develop national strategies on invasive alien species and understand how to put them into practice without running into conflict between one instrument and another. Measures to prevent the introduction of invasive alien species may, by their nature, be very trade restrictive. Therefore, members of the World Trade Organization must ensure also that any measures relating to invasive alien species are in accord with their obligations under international trade rules.

2.11.2 Recent Developments

In 2002, the CBD formalized a set of guiding principles on alien species that threaten ecosystems, habitats or species (refer footnote on page 7). These guiding principles comprise six general principles, three dealing with prevention, two with introduction of species and four with mitigation of impacts (IPPC Secretariat, 2005). The 15 subject areas are:

- Precautionary approach
- Three-stage hierarchical approach
- Ecosystem approach
- The role of states
- Research and monitoring
- Education and public awareness
- Border control and quarantine measures
- Exchange of information
- Cooperation, including capacity building
- Intentional introduction
- Unintentional introductions
- Mitigation of impacts

- Eradication
- Containment
- Control.

Many of the recommendations within these guiding principles correspond with provisions of the IPPC and its international standards for phytosanitary measures. Other aspects of the CBD's programme of work on invasive alien species include a toolkit of prevention and management practices, a guide to designing legal and institutional frameworks, an analysis of gaps and inconsistencies in the international regulatory framework and an analysis of the ecological and socio-economic impact on island ecosystems and on inland water ecosystems. All of these relate to invasive alien species.

Recent developments under the IPPC framework: Since 1999, the governing body of the IPPC has devoted considerable effort to clarifying the role of the convention in relation to invasive alien species. In 2001, it formally determined that implementation of the IPPC is directly relevant to implementation of Article 8(h) of the CBD. It also agreed that many provisions and standards of the IPPC are directly relevant to, or overlap with, the (then interim) guiding principles of the CBD on invasive alien species. In particular, ISPM 11 [2001]: Pest risk analysis for quarantine pests was to be developed to address in detail the environmental risks of plant pests. The revised standard, ISPM 11 rev. 1: Pest risk analysis for quarantine pests including analysis of environmental risks, was adopted in 2003 and further revised and supplemented in 2004.

Also in 2003, ISPM 5: Glossary of phytosanitary terms, was supplemented with Guidelines on the understanding of potential economic importance and related terms including reference to environmental considerations. This clarifies that the IPPC can account for environmental concerns in economic terms using monetary or non-monetary values; market impacts are not the sole indicator of pest consequence. Thus the scope of the IPPC covers the protection not only of cultivated plants but also of uncultivated/unmanaged plants, wild flora, habitats and ecosystems.

Collaborative international initiatives: Since 2001, the CBD and the IPPC have worked together in their efforts to help countries deal with the threat posed by invasive alien species. The aim is to avoid duplication of activities, contradictory approaches and a confusion of competences. A Memorandum of Cooperation has been established to formalize the arrangement between the two secretariats (IPPC Secretariat, 2005).The CBD has recommended that its contracting parties consider ratifying the IPPC. It has invited collaboration in development of standards (such as those of the IPPC) when they could incorporate elements relating to threats to biodiversity posed by alien species. There has been close collaboration between both conventions in the 2003 revision of ISPM 11. The CBD also works in partnership with the IPPC in information exchange endeavours such as the International Phytosanitary Portal and International Portal on Food Safety, Animal and Plant Health.

Range of adverse impacts: Cooperative international effort to foster the management of invasive alien species is particularly important because the problem is global. No single country can successfully control the spread and impacts of alien species when trade and travel promote the dispersal of such organisms and when their effects in a new territory may be unforeseen. The extent and range of adverse impacts makes cooperative action essential.Although many alien species offer great benefits to a country (e.g. in agriculture, forestry, aquaculture), those species which become invasive can have devastating impacts. The negative impacts may be environmental through loss of biodiversity, economic through a loss of production by affected species or the cost of control measures, health-related (e.g. when the invasive organism is a host or vector for disease) or political through effects on international trade, food security, water supply, regional stability, poverty, migration etc. (IPPC Secretariat, 2005).Invasive alien plants have a range of impacts on native biodiversity and ecosystems. They may compete with native taxa of flora, hybridize with genetically close species, alter the physical and chemical characteristics of soil, modify natural and semi-natural habitats, and propagate pests and diseases. The ecosystems most vulnerable to invasion are geographically and evolutionarily isolated ecosystems (islands, mountain ranges, lakes etc.) whose flora and fauna have evolved over millions of years.

Using the IPPC framework: National phytosanitary authorities have a long experience in the assessment and management of biological risks related to the introduction of organisms. They possess the knowledge of how to deal with risks posed by plant pests. This existing infrastructure and know-how could be utilized by environmental authorities in their efforts to implement the guiding principles of the CBD.The IPPC has been in effect for more than 50 years; over the past decade it has been developing a suite of international standards for the use of national plant protection organizations. IPPC-related activities are administered in many countries by agricultural authorities and CBD matters by environmental authorities. Communication between these authorities in relation to invasive alien species is important. It is suggested that, with its increased focus on environmental risks of plant pests, the IPPC provides a suitable framework to manage the introduction, control or eradication of invasive alien species that are plant pests. The infrastructure of national plant protection organizations is in place; the body of experience is well developed; relevant, user-friendly, international standards are in existence and more are being developed or revised; the necessary information exchange systems are established.

2.12 Gaps and Constraints in Mitigating the Effects of IAS

In addition to a lack of resources, common constraints at the national level include the lack of public, political and media awareness of environmental impacts of invasive alien species; fragmented or outdated legislation that does not cover the full range of agricultural, environment, marine and public health concerns and lack of a strategic approach and poor coordination between key departments and agencies.

The impacts of invasive alien species may extend to a group of neighbouring countries. Constraints on effective action at the regional level include lack of awareness, and poor coordination and cooperation between sectors and between neighbouring countries. Another problem arises with changes to regional free trade arrangements without consideration of the risks associated with liberalizing border controls and of management needs for invasive alien species already present within the free trade area (IPPC Secretariat, 2005). Many pathways and vectors are still not adequately covered by international rules, guidance or codes of best practice. For example, seeds, food and other commodities moving in the course of development assistance, humanitarian or military operations fall outside the regulatory framework for conventional trade pathways. In 2004, the Conference of the Parties to the CBD identified a number of cases in which invasive alien species are not covered by the international regulatory framework.

The approach to invasive alien species is usually a defensive one. Existing frameworks do little to support "prevention through avoidance". There are few deterrents to the continued use of environmentally harmful species in some sectors and few incentives to promote the use of native species as an alternative to introduced species. There are few preventive rules to restrict exports of high-risk invasive plants or animals to countries where they are likely to be problematic. There are virtually no international early warning systems for invasive alien species as they affect wild species and natural or semi-natural systems. Plants and animals that are invasive in their own right are covered under IPPC or OIE only if they qualify as plant pests or animal diseases. There is a lack of international standards for "environmental pests". Recent supplements to ISPMs do not extend the IPPC definition of "plant pest". The extension of some ISPMs to address environmental risks and costs will open up a new area for many plant health regulators. More advanced tools are needed to deal with risk assessment and environmental assessment. Outside mainstream agriculture and forestry sectors information is lacking and impacts, except for particularly invasive species, are largely unknown. Existing tools do not usually cover ways to deal with species already present in a country and to determine factors of vulnerability for receiving environments.Constraints on funding and gaps in institutional coordination are also areas that need addressing.

Recommended steps: Common goals for all levels (international, regional, national) should include making best use of existing regulatory frameworks, strengthening cooperation between key organizations and targeting existing tools and resources more effectively to encompass biodiversity-related impacts.

Recommendations for action at the national level include development of policy and regulatory frameworks and the engagement of relevant community groups and entities:

- Policy and legal frameworks need to support broad objectives of protection: human, plant and animal health and life; species, subspecies and races against contamination, hybridization or extinction; native biodiversity against impacts resulting from invasive alien species; and against biosecurity threats generally.
- Each country needs the legal basis, capacity and resources to support prevention, early detection and rapid response and eradication. Efficient use should be made of existing tools.
- Interagency cooperation and improved information-sharing between environment, agricultural and plant health sectors will facilitate the use of pest risk analysis for environmental risks, taking account of issues such as ecosystem dynamics, secondary pests and indirect impacts.
- The multiple roles of border and quarantine services in trade facilitation, food security, human health and environmental protection should be recognized and capacity strengthened where possible.
- Industry, retailers, users and other groups should be actively engaged to ensure that measures selected are seen as legitimate and proportionate to the level of risk. The development of codes of conduct and of best practice should be promoted.
- Appropriate education and communication strategies may be tailored to different target audiences and groups, including enforcement personnel.
- Positive strategies need to be developed to encourage the use of native plant species and seeds of known local provenance in landscaping, countryside management, revegetation, erosion control, protected area management and international assistance programmes.

2.13 Regional Management of IAS

2.13.1 Regional Standards for Phytosanitary Measures

The 1997 new revised text of the IPPC includes mention of regional standards as a component of the standard-setting objectives of the convention (see box 1). Over many years EPPO has developed a large body of regional standards, now exceeding 400, many relating to pest-specific phytosanitary measures (IPPC Secretariat, 2005).These standards provide support for EPPO members (not only European countries but also countries of North Africa and Near East) in dealing with quarantine pests and, more recently, with invasive alien species that are quarantine pests. Members are recommended to manage these pests through national phytosanitary regulations. The work of the regional plant protection organization and the regional standards helps promote a harmonized regional approach. This is very important in the European situation where there are numerous internal borders in a large continent.

Many of the regional standards refer to one particular general standard providing two lists of pests recommended for regulation as quarantine pests. In 2002, EPPO resolved that invasive alien species that have an effect on plants are quarantine pests under the IPPC. Quarantine pests may be:

- Pests of agriculture (the conventional target of pest regulation)
- Pests of forests (now including natural and seminatural communities as well as commercial silviculture)
- Pests of wild flora (at this stage, still a theoretical situation in the EPPO region)
- Indirect pests of plants (an example being bee parasites)
- Plants (as parasitic plants, as weeds of cultivated plants or as invasive alien plants in environments other than cultivated land).

National plant protection organizations are encouraged to consider their responsibilities for the management of such species. Where the pest is an invasive plant primarily affecting the environment, the NPPOs would be expected to consult with national environmental authorities and should also respect the CBD guiding principles. EPPO also recommends its members, in addition to targeting specific pests, take general background measures against alien pests; such general measures may be relevant for invasive alien plants. The existing regional standards include many that are relevant to invasive alien species. Invasive species could become a component of the lists of quarantine pests and could be covered by standards on specific measures or procedures, pest risk analysis or diagnostic protocols. The development of regional standards on commodity specific measures could cover pest plants. In particular, regional standards on national regulatory control systems may be developed to provide for the management of invasive alien species.

In a further role of support for the phytosanitary organizations of member countries, EPPO publishes an up-to-date “alert list” on its Web site, drawing attention to any pest incident that could be of phytosanitary significance. Several invasive plants are now included in this list. EPPO has also established an expert panel on invasive alien species, which is developing a list of invasive plants in the region and selecting a high-priority group for pest risk analysis.

2.13.2 The European Strategy on IAS

Europe needs a strategy on invasive alien species because,

- Invasive alien species from all taxonomic groups are a threat to European biodiversity.

- Europe has many countries, shared borders and growing freedom of trade and movement.
- European countries actively supported the adoption of guiding principles on invasive alien species under the CBD .
- Common approaches are needed to make the CBD guiding principles operational and to reach agreement on priorities, especially for transboundary problems.

As well as damaging biodiversity, invasive alien species have imposed huge losses on the European economy, affecting, for example, agriculture, forestry, fisheries and land stability. The European Community has identified proliferation of invasive alien species as an emerging issue (IPPC Secretariat, 2005).

Many European states face similar constraints in prevention and management efforts. Depending on the country, these may include:

- Low public awareness of invasive alien species, reluctance to increase the regulatory burden
- Shortage and/or inaccessibility of scientific information
- Absence of clear and agreed priorities for action
- Ease of introduction and movement (e.g. through the post), inadequate inspection and quarantine
- Inadequate monitoring capacity
- Lack of effective emergency response measures
- Outdated or inadequate legislation
- Poor coordination between government agencies, states and other stakeholders.

In 2000, the Bern Convention's expert group on invasive alien species began developing a strategy to address the above constraints. In December 2003, the Bern Convention Standing Committee adopted a recommendation urging contracting parties to develop and implement national strategies on invasive alien species taking into account the European strategy.The European Strategy on Invasive Alien Species contains an introductory section and eight substantive sections, cross-referenced to relevant CBD guiding principles. Each of these eight sections sets out a specific aim, key actions and practical indicators for additional actions.

The eight specific aims of the strategy are building awareness and support; collecting, managing and sharing information; strengthening policy, legal and institutional frameworks; regional cooperation and responsibility; prevention; early detection and rapid response; mitigation of impacts and restoration of native biodiversity.

Many aspects of implementing the strategy will be delivered through existing plant, animal and human health agencies with longstanding expertise in specific areas. The strategy encourages the active engagement of stakeholders involved in the movement, use and control of potential invasive alien species (industry and trade, transporters, retailers etc.) as well as competent non-governmental organizations and research institutes. Many of the proposed actions call for joint or complementary initiatives by private and public stakeholders. The strategy recognizes that contracting parties' existing legal obligations may constrain or influence the measures that can be taken, particularly as regards trade-related aspects.

2.13.3 The EU Phytosanitary Framework and IAS

The phytosanitary provisions and systems of the member states of the EU have been fully harmonized since 1993. Most of the measures have to be applied in the same way in all member states. In principle, the EU phytosanitary system covers most of the CBD's guiding principles on invasive alien species (IPPC Secretariat, 2005).Measures applied to imports and the internal movement of plants and plant products are based on directive 2000/29/EC of the Council of the European Union. All EU member states are obliged to prohibit the import and internal movement of specified quarantine organisms (listed in annexes of the directive) and of other alien organisms potentially harmful to plants. Traditionally, these organisms are plant pests directly harmful to plants or plant products; thus, the EU plant health provisions usually relate to preventing and controlling unintentional introductions. All main pathways for alien harmful organisms affecting plants are regulated and controlled by the EU phytosanitary system in order to minimize the probability of introduction of these organisms.

These pathways are mainly plants (and plant parts); plant products (including wood) ; wooden packaging and soil.

If not already prohibited from import, such items are subjected to inspection at the EU borders on entry. It implements elements of the CBD guiding principles 7 and 11, which call for border controls and measures to minimize unintentional introductions.The phytosanitary control of intentional import of species is much less developed than the measures against unintentional introduction. The EU provisions are focused only on those alien organisms that are directly harmful to plants and plant products, such as bacteria and insects. A 2002 revision of directive 2000/29/EC allows application of protective measures to organisms that are suspected of being harmful to plants or plant products but are not specified in the annexed lists. In the same revision, the directive adopts the IPPC definition of a plant "pest" as its definition of "harmful organism". Thus there is the legal basis to regulate on the EU level the intentional introduction of such invasive alien species as weeds and invasive alien plants. The current EU phytosanitary system partially fulfils the requirements of the CBD guiding principles 7 and 10 dealing with border controls and measures to minimize risk associated with intentional introductions of alien species that are or could become invasive.

If quarantine organisms listed in the annexes of directive 2000/29/EC are identified in an area in the EU where they have not been found before, the member state concerned has to take effective action against the outbreak with the aim to stop its spread and, if possible, to eradicate or suppress the population of the organism in the infested area. If outbreaks are identified for new harmful organisms, the member state concerned is obliged to take preliminary measures that at least limit the spread of the organism to other member states. The CBD guiding principles 12-15 provide for the mitigation of impacts of invasive alien species, their eradication or, if this is not possible, their control. The EU provisions on this are in principle a powerful tool ensuring that all member states take the required action individually or, if necessary, community wide.

Pest risk analysis: It is an essential component in the protection of habitats, ecosystems, plants and other organisms and features in seven of the CBD guiding principles. Pest risk analysis in the EU plant health system is based on IPPC provisions and standards. In principle, it covers all requirements of the guiding principles, including the consideration of the precautionary approach, the application of a cost-benefit analysis and research.The EU phytosanitary system partly covers the role of states (guiding principle 4). The CBD recommendation that states should identify, as far as possible, species that could become invasive and make such information available to other states may be interpreted to require pest risk analysis done by the exporting state for the importing state. This is not considered to be a realistic requirement. Pest risk analyses in the EU plant health system are done on the importing side. However, once informed of species that may be invasive to other countries, member states have to take this into account before exports of potentially problematic products can take place and the required plant health certificates are issued.The only guiding principle covered poorly in most member states is that on education and public awareness (guiding principle 6). Risk communication not only to the stakeholders but also to the public is crucial for a long-term acceptance of effective measures against invasive alien species that are plant pests. The EU population is not sufficiently aware of the threat posed by invasive alien species to agriculture, forestry and the uncultivated environment.

Regional cooperation and responsibility: Cooperation is the subject of CBD guiding principle 9, which points out that a state's response to minimizing the spread and impact of invasive alien species may require a bilateral or multilateral approach with other countries. Cooperation between contracting parties and the EU member states is a main prerequisite for the EU phytosanitary system in general and one of the main goals of EPPO. Cooperation between different stakeholders in the management of invasive alien species in Europe is already taking place, for example between EPPO (plant health) and the Bern Convention (nature conservation).

The European Strategy on Invasive Alien Species includes in its fourth aim:

- Cooperation between parties to the Bern Convention, where states recognize the risk that activities within their jurisdiction or control may pose to other states as a potential source of invasive alien species and take appropriate individual and cooperative actions to minimize that risk
- The role of the Bern Convention in strengthening cooperation with relevant regional and global institutions
- Subregional cooperation, where states sharing common problems in a subregion (including states not party to the Bern Convention) are encouraged to develop and participate in relevant programmes.

2.13.4 Information and Research Aspects of IAS Management in Europe

For regional initiatives in risk identification and management of invasive alien species to be successful, relevant information must be accurate, timely and accessible to all concerned (IPPC Secretariat, 2005).

Monitoring and early warning are crucial elements of risk identification. Examples in the European region include:

- EPPO's "alert list", published on the Internet, draws attention to any pest incident that could be of phytosanitary significance.
- The phytosanitary system of the EU stipulates that if quarantine pests or other potential invasive alien species that may pose a risk to plants in the EU are found in import inspections of consignments from non-EU countries, immediate notification must be made to the other responsible bodies in the member states and to the European Commission. More than 2 500 such notifications in a standardized format are circulated annually. A computerized information exchange system coming into operation will allow direct input of the relevant information into the system and its immediate availability to all services that have access to it.
- Any finding of a quarantine pest in a previously uninfested area inside the EU has to be notified. So, also, do outbreaks of harmful organisms that are new to the EU or that show unexpected characteristics of phytosanitary relevance.
- The sixth aim of the European Strategy on Invasive Alien Species involves early detection and rapid response. It recommends that parties have comprehensive and cost-effective surveillance procedures in place.

In a more general information role, the European Strategy on Invasive Alien Species includes in its second aim:

- Species inventories by parties, to help identify species that are invasive, set priorities for research, prevention, monitoring and mitigation and rapidly detect new arrivals not already present in the country or part of the country

- Research and monitoring for a better understanding of the ecology, distribution, patterns of spread and response to management of invasive alien species, as well as to provide a stronger scientific basis for decision-making and allocation of resource
- Regional exchange of information through ensuring effective systems are in place to share information relating to invasive alien species with neighbouring countries, trading partners and regions with similar ecosystems.

Some aspects are noteworthy on a more general basis for application in any region:

- A regional strategy is important, given that invasive alien species are a cross-border problem.
- Regional plant protection organizations provide a framework for developing regional standards to deal with invasive alien species that are pests of plants.
- Where a group of countries have harmonized regulations relating to phytosanitary measures, there is opportunity for a strong regional approach to the management of invasive alien species.
- Regional bodies are increasingly giving formal recognition of the threats posed by invasive alien species and encouraging the cooperative efforts of individual states to manage these risks.
- Cooperation is an essential element of regional initiatives.
- Systems for information collection, management and sharing, including monitoring and early warning, must be developed with attention given to ensuring the information is timely and relevant.

2.14 Directorate of Plant Protection, Quarantine & Storage (PPQS)

The Directorate of Plant Protection Quarantine & Storage was established in the year 1946 on the recommendation of 'Woodhead Commission' as an apex organization for advising the Government of India and state governments on all the matter related to Plant Protection. The Directorate is headed by Plant Protection adviser. Plant Protection strategy and activities have significant importance in the overall crop production programmes for sustainable agriculture. Plant protection activities encompasses activities aimed to minimizing crop losses due to pests through integrated pest management, plant quarantine, regulation of pesticides, locust warning & control and training in desert areas besides training and capacity building in plant protection. It is an attached Office of Ministry of Agriculture and Farmers Welfare. It has various Sub-Offices throughout India.

Mandate : To prevent the entry, establishment and spread of exotic pests in India as per the provisions of The Destructive Insects & Pests Act, 1914 and the notifications issued there under.

Objective : To provide an efficient and effective service, that fully satisfies our customers, such as importers, exporters, individuals and the Government.

Mission: To protect our plant life from ravages of destructive pests by preventing their entry, establishment and spread and thereby increasing agriculture productivity in order to improve the economy of our country ; To facilitate export certification of plants and plant products for safe global trade in agricultural commodities and thereby fulfilling our legal obligations under the international agreements and to adopt safe quarantine practices to protect our environment.

Vision: Dynamic quarantine programmes to protect our plant life and environment; Sound inspection systems to safe guard the interests of farming community and the consumers and overseas marketing agencies and Sustained quality export certification to allow our competitiveness in the global trade in agriculture.

2.14.1 Locust Swarm in India

Locusts are insects that travel in large swarms that can travel up to 150 kilometers in a day depending on the wind speed. Locust swarms devastate crops and cause major agricultural damage, which can lead to famine and starvation. Locusts devour leaves, flowers, fruits, seeds, bark and growing points, and also destroy plants by their sheer weight as they descend on them in massive numbers. A small swarm of the desert locust eats on an average as much food in one day as about 10 elephants, 25 camels, or 2,500 people. But swarms are not always small. In 1875, the US reported a swarm estimated to be 1,98,000 square miles or 5,12,817 square kilometers in size. Delhi-NCR is only 1,500 square kilometres, for comparison. A swarm the size of Delhi may consume the same amount of food in one day as every inhabitant in Rajasthan or Madhya Pradesh in one day. Locust swarms have been recorded in the Arabian Peninsula and some African countries since biblical times, but unusual weather patterns exacerbated by climate change have created ideal conditions for insect numbers to surge, scientists say. These warms have infested 23 countries across East Africa, the Middle East and South Asia in 2020, the biggest outbreak in 70 years, the World Bank said.

According to the Food and Agricultural Organization (FAO) desert locusts typically attack the western part of India and some parts of the state of Gujarat from June to November. However, the Ministry of Agriculture's Locust Warning Organization spotted them in India as early as April this year. A swarm of 40 million locusts can eat as much food as 35,000 humans, according to FAO estimates. The current swarm has destroyed seasonal crops in the states of Rajasthan and Madhya Pradesh. This will lead to lower production than usual and a rise in prices of foodstuff. An agrarian crisis and subsequent food inflation will severely impede India's response to the coronavirus pandemic. Thousands of migrant workers have died from hunger after India suddenly imposed a nationwide lockdown to contain the spread of the coronavirus, leaving workers penniless. An agrarian crisis because of a locust swarm will further hamper relief efforts of the government.

Heavy rains and cyclones in the Indian Ocean are being cited by experts as reasons for increased breeding of locusts this year. The attack is also spread over a wider geography in India. The FAO has warned that the locust infestation will increase next month, when locusts breeding in East Africa reach India.

The desert locust (*Schistocerca gregaria*) is a short-horned grasshopper. Innocuous when solitary, locusts undergo a behavioural change when their population builds up rapidly. They enter the 'gregarious phase' by forming huge swarms that can travel up to 150 km per day, eating up every bit of greenery on their way. These insects feed on a large variety of crops. If not controlled, locust swarms can threaten the food security of a country. At present countries in the Horn of Africa such as Ethiopia and Somalia are witnessing one of the worst locusts attacks in the last 25 years. In India, locusts are normally sighted during July- October along the Pakistan border. In 2019, parts of Western Rajasthan and Northern Gujarat reported swarms that caused damage to growing rabi crops. These were the first swarms reported in India since 1997. During 2020 the first sightings of small groups were reported early on April 11 by scientists of the Agriculture Ministry's Locust Warning Organization (LWO), from Sri Ganganagar and Jaisalmer districts of Rajasthan.

Other parts of the world affected by locusts: India isn't the only country attacked by a huge swarm of locusts this year. Pakistan, countries in East Africa, and Yemen have also faced the desert pests and their destruction. In February, Pakistan declared a national emergency because of locust attacks in the eastern part of the country. The pests damaged cotton, wheat, maize and other crops. Earlier this month of 2020, the FAO said that it had made a headway in dealing with the locust invasion by saving 720,000 tons of cereal in 10 countries.

2.14.2 Locust Swarm During 2020 in India

Locusts fly in swarms of millions, and quite rapidly. They travel tens of miles in a day and have tremendous endurance. They can remain in the air for a long time, covering huge distances. They are capable of doing massive crop damage. If not checked, they can clear out fields in a few hours. Each insect can eat as much as it weighs. Desert locusts originate in East Africa and Sudan and travel in swarms across Saudi Arabia and Iran to Pakistan and India. Then the bigger swarm breaks into small swarms, affecting different parts of the country.

Early arrival of locusts in 2020 can be traced back to the cyclonic storms Mekunu and Luban that had struck Oman and Yemen respectively in 2018. These turned large deserts tracts into lakes, facilitating locust breeding that continued through 2019. Swarms attacking crops in East Africa reached peak populations from November, and built up in southern Iran and Pakistan since the beginning of 2020, with heavy rains in East Africa in March-April enabling further breeding.

During 2020, swarms of desert locusts have been spreading through India from Rajasthan to Uttar Pradesh, destroying crops and pastures at a voracious pace.

Currently, 16 out of 33 Rajasthan districts are affected by locust swarms. The state's Kharif crop is at risk. The desert locusts entered Madhya Pradesh Chief Minister Shivraj Singh's constituency Budhni in Sehore earlier this week. The pests entered through Neemuch district in the state, subsequently travelled to parts of Malwa Nimar and were close to Bhopal. The Madhya Pradesh agriculture department has issued an advisory to the farmers in villages of the affected districts to keep continuous vigil over the desert locusts. They have been asked to keep the insects at bay by using loud sounds through drums, banging of utensils and shouting.

Experts have warned that if the swarms are not controlled soon enough, they can destroy the standing Moong cereal crop worth around Rs 8,000 crore. According to the Food and Agriculture Organisation (FAO), an agency of the United Nations, the desert locust is considered the most dangerous of all migratory pest species in the world. It threatens people's livelihoods, food security, the environment and economic development, it said. Experts say global warming can increase the chances of locust attacks. Rising temperature diminishes rain, which means more dry spells and more locust swarms.

Insects have reined the world earlier than mankind. They are omnipresent right from below the earth to hill top. Insects are very much associated with man's life. Some are useful and some are highly harmful to mankind, one of which is Desert locust, the most harmful insect in the world. They are scourge of mankind since time immemorial. Locusts are the short-horned grasshoppers with highly migratory habit, marked polymorphism and voracious feeding behavior. They are capable of forming swarms (adult's congregation) and hopper bands (nymphal congregation). They cause great devastation to natural and cultivated vegetation (Anonymous, 2020). They are indeed the sleeping giants that can flare up any time to inflict heavy damage to the crops leading to national emergency of food and fodder. Desert locust, Migratory locust , Bombay Locust and Tree locust are found in India. The desert locust is most important pest species in India as well as in intercontinental context.

History of locust invasion in India: Historically, the Desert Locust has always been a major threat to man's well-being. The Desert Locust is mentioned as curse to mankind in ancient writings viz. Old Testament-Bible and the Holy Koran. The magnitude of the damage and loss caused by the locusts is very gigantic beyond imagination as they have caused the starvation due to its being polyphagous feeder, and on an average small locust swarm eats as much food in one day as about 10 elephants, 25 camels or 2500 people. Locust do cause damage by devouring the leaves, flowers, fruits, seeds, bark and growing points and also by breaking down trees because of their weight when they settle down in masses.

Locust plagues and upsurges: The attack of the desert locust used to occur earlier in a phases of plague cycles (a period of more than two consecutive years of wide-spread breeding, swarm production and thereby damaging of crops is called a plague period) followed by a period of 1-8 years of very little locust activity called as the recession period again to be followed by another spell of plague. India witnessed several locust plague and locust upsurges and incursions during last two centuries. Small scale localized locust breeding have also been reported and controlled during the period 1998, 2002, 2005, 2007, 2010 and 2020. Since 2010 till 2012-13, situation remained calm and no large scale breeding and swarms have been reported. However, solitary phase of Desert locust has been reported from time to time at some locations in the State of Rajasthan and Gujarat as during 2020.

The invasion area of desert locust covers about 30 million sq km which includes whole or parts of nearly 64 countries. This includes countries like North West and East African countries, Arabian Peninsula, the Southern Republic of USSR, Iran, Afghanistan, the Indian sub-continent. During recession periods when locust occurs in low densities, it inhabits a broad belt of arid and semi-arid land which stretches from the Atlantic Ocean to North West India. Thus, it covers over 16 million sq kms in 30 countries (Anonymous, 2020).

Economic importance: In India, in spite of taking control measures, damage to crops caused by locusts during 1926-31 cycles, on a conservative estimate, was about Rupees 10 crore. During 1940-46 and 1949-55 locusts cycles the damage was estimated at Rs. 2.00 crore each and it was only Rs. 50.00 lakh during the last locust cycle (1959-62). Although no locust plague cycles have been observed after 1962, however, during 1978 and 1993, large scale upsurges were reported. Damage estimated was Rs. 2.00 lakh in 1978 and Rs. 7.18 lakh in 1993.Thereafter, insignificant damage by locust upsurges were reported largely due to the efforts of National, Regional and International Organizations established to prevent plague under the overall coordination of the Food and Agriculture Organization.

Objectives

i. To monitor, forewarn and control locust in Scheduled Desert Area (SDA) being International obligation and commitment.

ii. To conduct research on locust and grasshoppers.

iii. Liaison and coordination with National and International Organizations.

iv. Human resource development through training and demonstration for staff of Locust Warning Organization (LWO), State officials, BSF personnel and Farmers.

v. To maintain control potential to combat locust emergency by organizing locust control campaign.

However, the main objective of Locust Warning Organisation (LWO) is protection of standing crops and other green vegetation from the ravages of Desert locust which is one of the most dangerous pests occurring in desert areas throughout the world (Anonymous, 2020).

Functions

i. Keep constant vigil through field survey to prevent crop losses due to locust attack in approximately 2 lakh sq. kms. Scheduled Desert Area in the States of Rajasthan and Gujarat

ii. Avoid upsurge of locust population in SDA and entry of locust swarms into India through prompt control operations.

iii. Hold Indo-Pak Border meetings for exchange of locust situation information between two Countries to effectively monitor the situation and ensure preparedness to tackle the emerging locust threat, if any.

iv. Train the Farmers, State functionaries and locust staff on latest locust control technologies.

v. Advise state functionaries, BSF personnel, Panchayat Raj Institutions to inform the nearest LWO office if any locust activity was reported in their areas for needful action.

vi. Issuance of Desert Locust Situation Bulletin at fortnightly intervals to inform all concerned stakeholders about emerging locust situation in India.

vii. Conduct research at Field Station on Investigation on Locusts(FSIL) at Bikaner on bio-efficacy of pesticides and bio-pesticides for locust control.

Mandate of Field Station for Investigations on Locust (FSIL)

- To conduct research study on various aspects of biology, ecology and behavior of Locusts and Grasshoppers.
- To study and evaluate the different chemical and biological pesticides against Locusts and Grasshoppers in Laboratory and field conditions.
- To evaluate and evolve new techniques for the control and monitoring of Locusts and grasshoppers.

Mandate: Locust Warning Organisation (LWO), Directorate of Plant Protection Quarantine and Storage, Ministry of Agriculture, Department of Agriculture and Cooperation are responsible for monitoring, survey and control of Desert Locust in Scheduled Desert Areas mainly in the States of Rajasthan and Gujarat. Incursion of exotic locust swarms into India is prevented through organization of suitable control operation. LWO keeps itself abreast with the prevailing locust situation at National and International level through monthly Desert Locust Bulletins of FAO issued by the Desert Locust Information Service (DLIS), AGP Division Rome, Italy. Survey data are collected by the field functionaries from the fields which are transmitted to LWO circle offices, field HQ Jodhpur and Central HQ Faridabad where these are compiled and analyzed to forewarn the probability of locust outbreak and upsurges (Anonymous, 2020). The locust situation is appraised to the State Governments of Rajasthan and Gujarat with the advice to gear up their field functionaries to keep a constant vigil on locust situation in their areas and intimate the same to nearest LWO offices for taking necessary action at their end. Lot of innovations have been made in the field of locust survey and surveillance for quick transmission of locust survey data, their analysis, decision making, mapping of survey areas through computerization, adoption of new software like eLocust2/ eLocust3 and RAMSES.

Organizations of locust control campaign: With the onset of locust season an alert should be issued to the Agriculture authorities of Rajasthan, Gujarat, Haryana and Punjab States. Other stake holders like Ministry of Home Affairs, Defence, Science and Technology, Civil Aviation, Communication, Aircraft Companies and Pesticides Manufacturing Firms etc. may also be sounded for providing needful assistance, if required, during locust emergency. The role of different Stake holders is given as under: Role of Different Stake holders.

Ministry of Home affairs

1. To advise BSF authorities to extend help and to provide facilities in border surveys.
2. To advise BSF authorities to extend help in arranging Indo-Pak border meetings.
3. To grant permission for establishing direct wireless linkage between Jodhpur and Karachi.
4. To extent help in reporting of locust population/swarm through BSF staff

Ministry of Defence: To provide wireless sets (HF and VHF), trained manpower for wireless and vehicles during locust emergency. Also request Defence Ministry to coordinate in using the HF frequency allotted for establishing direct wireless link between Jodhpur (India) and Karachi (Pak) for exchange of locust information (Anonymous, 2020).

Ministry of Science and Technology: To provide meteorological data.

Ministry of Civil Aviation: To get permission from Air Traffic Control (ATC) for flying aircraft during locust control operation.

Ministry of Communication: To approach Ministry of Communication for timely renewal of wireless telegraph licence granted to operate the Locust Warning Organisation wireless communication network.

Government departments

1. To report locust information to LWO.
2. To provide assistance in form of vehicles and manpower during locust campaign.
3. To conduct survey and surveillance of locust in cropped areas.
4. To control locust in cropped areas.
5. To create awareness among public and farmers about locust.

6. To provide facilities to LWO staff during locust survey and control campaign.

Aircraft Companies: To provide aircrafts/helicopters on hire basis for locust control.

Pesticide Manufacturing Firms: To supply the pesticides on short notice during locust emergency.

2.14.3 Fall Armyworm (*Spodoptera frugiperda*) in India

The pest which arrived in India in 2018, after ravaging cornfields of sub-Saharan Africa, and has already been reported from maize farms in 20 states. Maize farmers in many parts of Karnataka were taken by surprise in July last year when an unknown caterpillar attacked their crop. It didn't take scientists long to identify the new pest. By the second week of July, researchers from the National Bureau of Agricultural Insect Resources (NBAIR), an institute under the Indian Council for Agricultural Research, said the new pest was the Fall Armyworm (FAW). Spotted in a maize field in Chikkaballapur, around 60 km from Karnataka state capital Bangalore, the appearance of FAW in India is a cause for serious concern. Native to tropical and subtropical regions of the Americas, the dreaded caterpillar appeared and spread rapidly in Africa in 2016, and has since then devastated millions of hectares of maize crop in all parts of sub-Saharan Africa (Manupriya, 2019). And sure enough, the worm spread very fast through the maize fields of India as well. In a matter of months, more than 14 states in the country reported the infestation last year, seriously compromising the corn harvest. The infestation has since spread even wider this year to 20 states, with the northeastern parts of the country the worst affected.

Native to the Americas, the pest is known to eat over 80 plant species, with a particular preference for maize, a main staple crop around the world. The fall armyworm was first officially reported in Nigeria in West Africa in 2016, and rapidly spread across 44 countries in sub-Saharan Africa. Sightings of damage to maize crops in India due to fall armyworm mark the first report of the pest in Asia.. The pest was first detected in China in the southwest province of Yunnan in January 2019. Through 2019, the pest infested a total of 26 provinces. The armyworm is expected in 2020 to hit China's Northeast wheat belt. A report issued by the Ministry of Agriculture and Rural Affairs rates the situation as "very grave". In February 2020 S. frugiperda was detected in Queensland, Australia. It was observed in traps baited with a male pheromone lure, firstly on Erub and Saibai islands in the Torres Strait, and subsequently on the mainland near Croyden. Within a week it was officially declared ineradicable. In April 2020, it was then detected in Papua New Guinea, spreading through the Torres Strait .

Fast spread: The spread of FAW through the Indian subcontinent has been particularly fast. In 2019, the pest has spread as far as Mizoram in the northeast,

Uttar Pradesh in the north, Gujarat in the west, Chhattisgarh in central India, and several states in the south. This year, the biggest victims so far have been farmers in the northeastern states, where a cumulative of 10,772 hectares of maize crop has been affected. The pestilence has been reported from 20 states in India. The northeastern states with their "high humidity and moderately high temperatures" are suitable for the spread of FAW. Its metabolic rate is well supported in these conditions, sometimes even leading to "intensification of infestation," said Chaudhry. It means that the pest is able to complete its lifecycle in a shorter period of time, resulting in more pests, more quickly. Farmers and scientists are now fighting to contain the infestation. Maize is India's third most important cereal crop after rice and wheat. In 2016, 25.9 million metric tons of maize was produced in India. In 2017, that number rose to 28.7 million tons. In 2018, however, production fell by 3.2% to 27.8 million tons. It is expected that the net production will decline further in 2019 due to the pest attack. It is not just the feed and starch industries that are feeling the heat. Maize is too facing additional challenges in continuing to grow the crop. They've had to endure crop losses and bear the additional cost of rescuing their crop from FAW and preventing further infestation.

The pest has the potential to spread quickly not only within India, but also to other neighboring countries in Asia, owing to suitable climatic conditions. As FAW continues its march across India and other Asian countries, the need for effective protective measures will only grow stronger. "Increasing monoculture of maize around the year and wrong pest management practices with excessive dependence on chemical pesticides, which increased the resistance in the insect to pesticides, have contributed to FAW becoming a serious pest, which works with smallholder farmers. Any pest is always a function of practices followed and local weather conditions. Therefore, a shift towards agro-ecological approaches like non-pesticidal management, organic or natural farming, and multiple cropping systems are the ways to manage such pest outbreaks. Other regions are at risk as well. Researchers have warned of the potential impacts if FAW spreads to Europe, where customs inspectors have already reported having discovered and destroyed the pest on quarantined crops imported from Africa on several occasions.

FAW spread from the Americas to other continents is probably due to shipments, trade, people displacements. Adults can also move over 100 km a night. In a world of climate change and increased global connectedness through trade and tourism, the frequency of invasive pest attacks is likely to only go up. It obviously calls for stepped-up phytosanitary and quarantine efforts to prevent the onset of transboundary pathogens and pests. But that apart, effective monitoring, surveillance and early warning systems, coupled with an IPM approach, are vital to respond to any new insect-pest threat, in order to safeguard the crops and protect the incomes of mostly smallholder farmers. In the case of pests such as FAW, India must simply "fail" its spread in the beginning itself, through IPM and synergistic inter-institutional and multi-disciplinary efforts. Fall armyworm has rapidly spread across maize growing States since it was detected in Southern India in late mid-2018. Most of countries in Asia including Bangladesh, China, Indonesia, Myanmar, Nepal, Sri Lanka, Taiwan and Vietnam have reported severe infestation of FAW in last one year".

2.14.4 *Tuta absoluta*- A New Invasive Pest Alert

Invasive species, alien species, exotic pests, or invasive alien species, are common names that categorize non-native animals, insects, microbes, diseases, or plants that are pests. These pests are not native in areas in which they cause problems and they are considered "invasive" because they invade and establish populations in new areas and the resulting uncontrolled population growth and spread causes economic or environmental problems. South American tomato pinworm, *Tuta absoluta* (Lepidoptera: Gelechiidae) also known as the tomato leaf miner is one of the destructive invasive pest observed for the first time infesting tomato crop in Maharashtra, India (Anoymous, 2017). This pest has been classified as the most serious threat for tomato production worldwide. The pest has spread from South America to several parts of Europe, entire Africa and has now spread to India. Plants are damaged by direct feeding on leaves, stems, buds, calyces, young fruit, or ripe fruit and by the invasion of secondary pathogens which enter through the wounds made by the pest. It can cause up to 90% loss of yield and fruit quality under greenhouses and field conditions.

The pest was initially observed in Pune on tomato plants grown in polyhouse and fields during October 2014. The specimens were collected, identified and deposited at National Pusa Collection (NPC), Division of Entomology, ICAR-IARI, New Delhi by P.R. Shashank and K. Chandrashekar, ICAR-IARI scientists. Subsequently the pest was observed in the farmer's fields in major tomato growing districts of Maharashtra viz., Pune, Ahmadnagar, Dhule, Jalgaon, Nashik, and Satara. Severe infestation (>50% plants affected) was observed in several tomato fields.

Following the reports of Maharashtra, recent surveys conducted by researchers of Network Project on Insect Biosystematics (NPIB), University of Agricultural Sciences, Bengaluru and ICAR-NBAIR, Bengaluru in January, 2015 observed the presence of this pest in Kolar and Bengaluru districts of Karnataka. The current report of T. absoluta from India is alarming because this pest is oligophagous and can attack several suitable solanaceous host plants. Present information is useful for adaptation of rapid response strategies against its invasion by educating farmers, extension entomologists and other stakeholders.

2.15 Policy for the Control of IAS in India and Related Policy Initiatives

The national integrated pest management (IPM) programme is considered to be the mechanism with which to prevent and control the threat posed by invasive alien species within the country. State governments, non-governmental organizations, private sector organizations, research institutions and farmer self-help groups are all increasingly involved in the surveillance and detection of pests and diseases. They are capable of taking environmentally friendly corrective action within the IPM scheme (IPPC Secretariat. 2005).

International cooperation has helped in dealing with migratory locust, a pest of great concern for the Asian region. India maintains active coordination with FAO and with neighbouring countries for surveillance, early detection and control measures for locust. There was no major incidence reported in the region in 2003. A peculiar cyclic problem, thankfully confined to a small hilly region of north-eastern India, relates to unexplained but sudden surges in rodent population and activity. The menace reaches peak proportions at the same time as the periodic gregarious flowering of bamboo. The problem has surfaced again and is likely to peak in 2006-2007 when the next mass flowering is predicted, causing crop losses. Research and preventive control measures under way include study of the rodent characteristics, damage capacity, pathways associated with the pest and an environmentally friendly control strategy. The traditional knowledge of the local agrarian community of the region is also utilized.

Of particular interest to India is research being conducted to study the impact of climate change on the threat posed by invasive alien species. The topic is of greater importance since a serious white woolly aphid infestation of the sugarcane crop in parts of peninsular India in 2002 caused substantial crop damage and losses. This pest had never before infested sugarcane in India. The task of research, future prevention and control measures for white woolly aphid is being handled by the Ministry of Agriculture in coordination with other central government departments, concerned state governments, the Indian Council of Agricultural Research, other research institutions and agriculture universities, private sector organizations and sugar factories. The severity of the white woolly aphid infestation, recorded in 2002 in over 200 000 ha of sugarcane, has subsequently reduced substantially. However, almost 75 000 ha of the crop was still infested in 2003 and the matter continues to be of concern.

2.16 Goals for Addressing the Problem of IAS

Goals include the following (Nirmalie Pallewatta *et.al.*, 2002).

Prevention: Keeping an IAS from being introduced into a new ecosystem. Ideally, this usually means keeping alien organisms from entering a new country.

Early detection: Locating IAS before they have a chance to establish and spread. This usually requires effective, site-based inventory and monitoring programmes.

Eradication: Killing the entire population of IAS. Typically, this can only be accomplished when the organisms are detected early.

Control: The process of long-term management of the IAS' population size and distribution when eradication is no longer feasible. Control and eradication methods can take one or more of three forms (see below). Integrated pest management (IPM) is their combined application:

Mechanical control: The physical removal of organisms – pulling weeds, for example. The process requires a long-term investment of human resources.

Chemical control: The use of chemicals to kill organisms – poisons for wildlife and herbicides for plants, for example. The process can be quite costly and typically requires repeat applications.

Biological control: The introduction of a highly specific predator, parasite or pathogen that will attack the IAS. This process is not likely to result in eradication of the organism but often can reduce the population of the IAS to tolerable levels. The initial costs associated with research and development may be high, but the long-term costs once applied are low and relatively little maintenance is required.

Restoration: The process of re-establishing natural populations and ecosystem functions. In theory, this increases the ecosystem's resistance to future invasions. These goals are best accomplished through a strategic, holistic approach incorporating Risk assessment and risk management ; Research ; Inventory and monitoring ; Policy and regulation ; Information management; Education and outreach and International cooperation and capacity building. International cooperation and capacity building are crucial, as IAS is an international problem by their very definition. However, these processes are probably the "weakest link" in any country's efforts to minimize the spread of IAS.

The Government of India in 2003 has approved the notification of a new plant quarantine order harmonizing India's regulatory framework with the International plant protection convention and internationally accepted standards and the tenets of the SPS agreement of the World Trade Organization. Community based approaches, for alien species management, can best be complemented with biological control.

Brief description of alien species and their present management strategies in India were discussed in this communication along with a note on strengthening the said strategy.

Eradication is difficult and expensive, but possible. Rapid response is crucial. Because immediate response is more cost-effective and more likely to succeed than action after a species has become established, we recommend an early warning system (EWS) for IAS. Containment action is often needed for a successful eradication program. Such a program must be science-based and have a reasonable chance of success. The involvement of all relevant stakeholders is essential. Public support and acceptance of eradication methods are also important. Monitoring and control after initial efforts are often necessary, and restoration of affected systems is an important consideration.

Containment, suppression, and control are second options, but often have more benefits than costs: Given the high complexity of the ecological characteristics of both IAS and the habitats and native species they affect, control measures must be developed and applied on the basis of the best current scientific understanding. Specific costbenefit analyses should be developed and applied for eradication and control programs for IAS. Selection of control methods must also be based on thorough scientific knowledge. For chemical control the possible problem of negative effects on non-target species and the potential development of resistant types and strains must be carefully addressed. For biological control the possibility of the control agent itself becoming invasive must be avoided. An integrated management approach to IAS involving a combination of mechanical, chemical and biological control measures is often most appropriate. Careful monitoring and co-ordination are needed. Because the cost and benefit factor influences decisions that results in risk analysis that are often very difficult politically, the criteria for such making such decisions should be clearly developed.

Comprehensive international and national action is required: Numerous global and regional policies are already addressing the problem of IAS. Coordination of implementation and practical co-operation among those responsible for these instruments however, are highly insufficient. Practical prevention, eradication and control measures are also inadequate. We therefore recommend a consolidated action plan. The Convention on Biodiversity (CBD) and the International Plant Protection Convention (IPPC) could take the lead, but trade, transport, travel, and other economic sectors must be closely involved. Other institutions, including the United Nations Environment Programme (UNEP), the World Trade Organization (WTO), Food and Agriculture Organization (FAO), and the International Maritime Organization (IMO) are key components at the international level (McNeely *et.al.*, 2001). These institutions are supported by international non-governmental organizations (NGOs) such as The World Conservation Union (IUCN), World Wildlife Fund (WWF), Wetlands International, Conservation International, and The Nature Conservancy (TNC).

Likewise, at the national level, consolidated and coordinated action is required. This could be part of a national biodiversity strategy and action plan, with close involvement of the economic sectors and identifying people responsible for operative actions involving potential IAS as a key prerequisite. Clear responsibilities for each relevant sector should be identified. Insurance mechanisms and liability regulations for the spread of IAS are almost non-existent, presenting a major deficiency for controlling the problem. Governments should cooperate with the insurance sector to find solutions, beginning with feasibility studies. Capacity and expertise to deal with IAS are highly insufficient in many countries. Capacity building and further research on the biology and control of IAS and biosecurity issues should therefore be given attention and priority. This also relates to financial institutions and other organizations responsible for environment and development co-operation, at national and international levels.

A global information system regarding the biology and control of IAS is urgently needed. Tools, mechanisms, best management practices, control techniques and resources should be provided and exchanged. The information system must be linked to the Clearing House Mechanism of the Convention on Biological Diversity. Awareness raising and education regarding IAS should be given high priority in action plans, and development of economic tools and incentives for prevention are urgently needed.

2.17 Strategies to Prevent and Manage the IAS

i. Set up a regional forum for the development of a regional strategy that emphasizes prevention and management of Invasive Alien Species (IAS) : Regional action needs a dual focus: a) limit the entry of IAS to the region and b) for IAS already present, limit their spread and share information on how to deal with the problem. See objectives below for related proposed actions that have relevance to the above (Anonymous, 2012).

ii. Greater integration of national and regional approaches:

- The region needs to formulate national, sub-regional and regional strategies and action plans that are co-ordinated and mutually supportive of one another, as well as to recognise the different values and perspectives of each country. Biological invasions are very complex issues where some invaders may be viewed as beneficial by some stakeholders. Such conflicting views need to be resolved as much as possible through consultation to ensure greater cooperation within and between countries.
- National strategies and action plans should take into account the insidious and pervasive threat posed by IAS, and the importance of biogeographical rather than political boundaries in the spread of IAS.

iii. Agreement on common terminology and lists of major IAS, and an understanding of IAS issues to provide a common basis for discussion and actions:

- The region needs to clarify and agree on common definitions of terms (specifically terms such as agricultural pests). There was agreement that terms should (as appropriate) cover both economic and environmental aspects. A glossary of key terms could be developed cooperatively. See also answers to question two below.
- Adopt a system of classification of alien biocontrol agents.

iv. Harmonisation of legal instruments at national, sub-regional and regional levels:

- National legal frameworks should undergo sectoral review for harmonisation and for filling gaps to cover the full range of IAS as pests with environmental impacts, such as threatening native biodiversity, and not only as agricultural, medical or veterinary pests.

- Take steps to harmonise national frameworks with regional concerns, such as through development of a master list (a "negative"/"black" list) of IAS at the regional or sub-regional level for national guidance (see also below objective on capacity building).
- Harmonisation of plant quarantine protocols within the region.

v. Effective sharing of information at national and regional levels:

- National clearing house mechanisms (CHMs) can play a key role in bringing together organisations that have different types of data collection systems and different kinds of data. They can also acquire more information on IAS with environmental impacts which are urgently needed for use at a regional level.
- A regional IAS information exchange mechanism to enable national information to be shared regionally and promote development of appropriate approaches for common problems. More efficient use could be made of neighbouring country's databases and regional/international technical expertise. Most countries suffer from IAS due to lack of information, but in some cases, it may not be the best use of resources and time to concentrate on primary research or build new national databases.
- An operational regional database that addresses environmental, agricultural and social aspects of IAS to be perhaps maintained by a global programme such as GISP.
- Strengthen or establish Asia wide networks that provide technical support from IAS management specialists and authorities in public health, quarantine, and other agencies; maximize the lessons learned and exchange experiences from national to regional levels and vice versa.

vi. Capacity building and greater use of available expertise: Asian countries are at very different stages of development and some will require more resources and capacity building than others.

- The taxonomic impediment is a serious issue that needs to be addressed on a priority basis. Regional networks of taxonomic specialists, such as ASEANET (as part of Bionet International), and programmes of Convention on Biological Diversity enhancing taxonomic capability could be enlisted in this issue.
- Awareness raising is critical to combat successfully the problem of IAS. The profile of IAS needs to be raised across stakeholders spanning different societies and across the region. Special attention should be paid to local communities that depend on subsistence agriculture.
- Enhanced ability to implement risk analysis procedures, early warning tools, and guidelines for the prevention and management of IAS.

Absence of clearing house mechanisms, technical specialists, weak or absent relevant data, and development of legislative frameworks, are addressed in actions set out under previous objectives.

Greater cooperation, especially for problems with a trans-boundary/regional dimension:

- Strengthen and establish structures and mechanisms for regional cooperation
- At present there are virtually no opportunities for trans-boundary assessment and management of IAS. Three major areas were identified as suitable for regional/trans-boundary cooperation.
- Measuring impacts - several countries noted that particularly for species with serious environmental impacts, there is often no available assessment/ information on the possible (future) impacts. Assessment of the economic costs of at least major IAS should be undertaken on a region wide basis
- Engaging regional support for research to search for solutions and to obtain greater understanding of the vulnerability of ecosystems to invasions (why and how questions)
- Implementing IAS management programmes.

2.18 Research Priorities and Conclusions

Despite some notable successes in preventing some invasions, reducing the impacts of others, and putting various measures in place to tackle invasions and their impacts more systematically, the magnitude of the challenges is extremely daunting. A major problem is that changes in the extent and impacts of invasions are occurring not just incrementally (through the increase in numbers of invaders and invaded area, and steady accumulation of impacts), but also through non-linearities and synergisms with other components of global change. Unlike some other components of global change, biological invasions can be effectively managed and mitigated. The following priorities may be suggested to ensure progress in dealing effectively with the many dimensions of biological invasions (Petr Pysek, 2020).

Bibliography

Anonymous, 2008. The Ministry of Environment & Forests. National Biodiversity Action Plan. Retrieved on 2 February 2009, from http://envfor.nic.in/divisions/csurv/Approved_NBAP.pdf

Anonymous, 2009. Global Invasive Species Program. "India." Retrieved on 2 February 2009, from http://www.issg.org/database/species/search.asp?sts=sss&st=sss&fr=1&x=0&y=0&sn=&rn=India&hci=-1&ei=-1& lang=EN

Anoymous, 2017. Tuta absoluta- A new invasive pest alert. Indian Council of Agricultural Research (Ministry of Agriculture and Farmers Welfare), New Delhi, India 110001

Cardenas, L., Leclerc, J.C., Brunning, P., Garrido, I., Detree, C., Figueroa, A., Astorga, M., Navarro, J. M., Johnson, L. E., Carlton, J. T. and Pardo, L.,2020. First mussel settlement observed in Antarctica reveals the potential for future invasions. *Sci. Rep.,* 10, 5552

Convention on Biological Diversity, 2005. "India's Third National Report." Retrieved 19 October 2008, from http://www.cbd.int/doc/world/in/in-nr-03-en.doc

ENVIS Resource Partner on Biodiversity, 2020. Botanical Survey of India, Kolkata, West Bengal, Ministry of Environment, Forest & Climate Change, Govt of India

Global Invasive Species Database, 2021. Downloaded from http://www.iucngisd.org/gisd/100_worst.php on 12-05-202

Herrera Ileana, Ferrer-Paris José R, ; Benzo Diana, Flores Saúl, García Belkis and Nassar Jafet, M.,2018. An Invasive Succulent Plant (Kalanchoe daigremontiana) Influences Soil Carbon and Nitrogen Mineralization in a Neotropical Semiarid Zone. *Pedosphere*. 28: 632–643

Hulme, P. E.,2020. Plant invasions in New Zealand: global lessons in prevention, eradication and control. *Biological Invasions*, 22: 1539– 1562.

IPPC Secretariat. 2005. Identification of risks and management of invasive alien species using the IPPC framework. Proceedings of the workshop on invasive alien species and the International Plant Protection Convention, Braunschweig, Germany, 22-26 September 2003. Rome, Italy, FAO. xii + 301 pp

Jacobsen Rowan, 2019. The Invasivore's Dilemma. Outside, Retrieved May 28, 2019.

Kolar, C.S.,2001. Progress in invasion biology: predicting invaders. *Trends in Ecology & Evolution*. 16: 199–204

Leonard, S.P., Powell, J.E., Perutka, J., Geng, P., Heckmann, L. C., Horak, R. D., Davies, B. W., Ellington, A. D., Barrick, J. E. and Moran, N. A.,2020. Engineered symbionts activate honey bee immunity and limit pathogens. *Science*, 367: 573– 576.

Nirmalie Pallewatta, Jamie K. Reaser, and Alexis T. Gutierrez, 2002. Prevention and Management of Invasive Alien Species: Proceedings of a Workshop on Forging Cooperation throughout South and Southeast Asia, 14-16 August 2002 Bangkok, Thailand.

Paap, T., de Beer, Z. W., Migliorini, D., Nel, W. J. and Wingfield, M. J.,2018. The polyphagous shot hole borer (PSHB) and its fungal symbiont Fusarium euwallaceae: a new invasion in South Africa. *Australasian Pl. Pathol.*, 47: 231– 237.

Peltzer, D. A., Bellingham, P. J., Dickie, I. A., Houliston, G., Hulme, P. E., Lyver, P. O. B., McGlone, M., Richardson, S. J. and Wood, J.,2019. Scale and complexity implications of making New Zealand predator-free by 2050. *J. Royal Soc. New Zealand*, 49: 412– 439

Petr Pyšek, Philip E. Hulme, Dan Simberloff , Sven Bacher, Tim M. Blackburn, James T. Carlton, Wayne Dawson, Franz Essl, Llewellyn C. Foxcroft, Piero Genovesi, Jonathan M. Jeschke, Ingolf Kühn , Andrew M. Liebhold, Nicholas E. Mandrak, Laura A. Meyerson, Aníbal Pauchard, Jan Pergl, Helen E. Roy, Hanno Seebens, Mark van Kleunen, Montserrat Vilà, Michael J. Wingfield and David M. Richardson, 2020. Scientists' warning on invasive alien species. *Biol. Rev.*, 95: 1511-1534

Prentis Peter,2008. Adaptive evolution in invasive species. Trends in Pl. Sci., 13: 288–294

Robert Black and Debbie M.F. Bartlett, 2020. Biosecurity frameworks for cross-border movement of invasive alien species. *Environ. Sci. & Policy*, 105: 113-119

Roy, H. E., Rabitsch, W., Scalera, R., Stewart, A., Gallardo, B., Genovesi, P., Essl, F., Adriaens, T., Bacher, S., Booy, O., Branquart, E., Brunel, S., Copp, G. H., Dean, H., D'hondt, B., Josefsson, M., Kenis, M., Kettunen, M., Linnamagi, M., Lucy, F., Martinou, A., Moore, N., Nentwig, W., Nieto, A., Pergl, J., Peyton, J., Roques, A., Schindler, S., Schönrogge, K., Solarz, W., Stebbing, P. D., Trichkova, T., Vanderhoeven, S., van Valkenburg, J. and Zenetos, A.,2018. Developing a framework of minimum standards for the risk assessment of alien species. *J. Appl.Ecol.*, 55: 526– 538

Sandilyan, S., 2020. Invasive Alien Species of India. Centre for Biodiversity Policy and Law (CEBPOL) National Biodiversity Authority, Ministry of Environment Forests and Climate Change, Government of India, Chennai – 600 113, India.

Sardain, A., Sardain, E. and Leung, B., 2019. Global forecasts of shipping traffic and biological invasions to 2050. *Nature Sustainability*, 2: 274– 282.

Schlaepfer Martin, A., Sax DOV, F. and Olden Julian, D., 2011. The Potential Conservation Value of Non-Native Species. *Conservation Biol.*, 25: 428–437

Shackleton, R. T., Foxcroft, L. C., Pyšek, P., Wood, L. E. and Richardson, D. M.,2020. Assessing biological invasions in protected areas after 30 years: revisiting nature reserves targeted by the 1980s SCOPE programme. *Biological Conservation*, 243: 108424.

Van Meerbeek Koenraad, Appels Lise, Dewil Raf, Calmeyn Annelies, Lemmens Pieter; Muys, Bart Hermy and Martin, 2015. Biomass of invasive plant species as a potential feedstock for bioenergy production. *Biofuels, Bioproducts and Biorefining*, 9: 273–282

Zavaleta, E., 2000. Valuing ecosystem services lost to Tamarix invasion in the United States. In Invasive Species in a Changing World (eds H. A. Mooney and R. J. Hobbs), pp. 261–300. Island Press, Washington.

Zayed Amro, Constantin Serban, A. and Packer Laurence,2007 .Successful Biological Invasion despite a Severe Genetic Load. *PLOS ONE*, 2(9): e868

3

Biowarfare, Bioterrorism and Bioethics

3.1 Biowarfare

Biowarfare, also known as Biological warfare or germ warfare, is the use of biological toxins or infectious agents such as bacteria, viruses, insects, and fungi with the intent to kill, harm or incapacitate humans, animals or plants as an act of war. Biological weapons (often termed "bio-weapons", "biological threat agents", or "bio-agents") are living organisms or replicating entities (like viruses, which are not universally considered "alive"). Entomological (insect) warfare is a subtype of biological warfare.

Offensive biological warfare is prohibited under customary international humanitarian law and several international treaties. In particular, the 1972 Biological Weapons Convention (BWC) bans the development, production, acquisition, transfer, stockpiling and use of biological weapons. Therefore, the use of biological agents in armed conflict is a war crime. In contrast, defensive biological research for prophylactic, protective or other peaceful purposes is not prohibited by the BWC. Biological warfare is distinct from warfare involving other types of weapons of mass destruction (WMD), including nuclear warfare, chemical warfare, and radiological warfare. None of these are considered conventional weapons, which are deployed primarily for their explosive, kinetic, or incendiary potential.

Biological warfare is the intentional use of microorganisms, and toxins, generally of microbial, plant or animal origin to produce disease and death in humans, livestock and crops. The attraction of bioweapons in war, and for use in terroristic attacks is attributed to easy access to a wide range of disease-producing biological agents, to their low production costs, to their non-detection by routine security systems, and to their easy transportation from one place to another. In addition, novel and accessible technologies give rise to proliferation of such weapons that have implications for regional and global security. In counteraction of such threats, and in securing the culture and defence of peace, the need for leadership and example in devising preventive and protective strategies has been emphasized through international consultation and cooperation (Edgar J. DaSilva, 1999). Adherence to the Biological and Toxin Weapons Convention reinforced by confidence building measures sustained by use of monitoring and verification protocols, is indeed, an important and necessary step in reducing and eliminating the threats of biological warfare and bioterrorism.

A biological attack could conceivably result in large numbers of civilian casualities and cause severe disruption to economic and societal infrastructure. A nation or group that can pose a credible threat of mass casualty has the ability to alter the terms under which other nations or groups interact with it. When indexed to weapon mass and cost of development and storage, biological weapons possess destructive potential and loss of life far in excess of nuclear, chemical or conventional weapons. Accordingly, biological agents are potentially useful as strategic deterrents, in addition to their utility as offensive weapons on the battlefield.

As a tactical weapon for military use, a significant problem with biological warfare is that it would take days to be effective, and therefore might not immediately stop an opposing force. Some biological agents (smallpox, pneumonic plague) have the capability of person-to-person transmission via aerosolized respiratory droplets. This feature can be undesirable, as the agent(s) may be transmitted by this mechanism to unintended populations, including neutral or even friendly forces. Worse still, such a weapon could "escape" the laboratory where it was developed, even if there was no intent to use it ,for example by infecting a researcher who then transmits it to the outside world before realizing that they were infected. Several cases are known of researchers becoming infected and dying of Ebola, which they had been working with in the lab (though nobody else was infected in those cases), while there is no evidence that their work was directed towards biological warfare, it demonstrates the potential for accidental infection even of careful researchers fully aware of the dangers. While containment of biological warfare is less of a concern for certain criminal or terrorist organizations, it remains a significant concern for the military and civilian populations of virtually all nations.

3.1.1 History of Biowarfare

Because of the increased threat of terrorism, the risk posed by various microorganisms as biological weapons needs to be evaluated and the historical development and use of biological agents better understood. Biological warfare agents may be more potent than conventional and chemical weapons (Riedel, 2004). During the past century, the progress made in biotechnology and biochemistry has simplified the development and production of such weapons. In addition, genetic engineering holds perhaps the most dangerous potential. Ease of production and the broad availability of biological agents and technical knowhow have led to a further spread of biological weapons and an increased desire among developing countries to have them. This article explains the concepts of biological warfare and its states of development, its utilization, and the attempts to control its proliferation throughout history. The threat of bioterrorism is real and significant; it is neither in the realm of science fiction nor confined to our nation.

Man has used poisons for assassination purposes ever since the dawn of civilization, not only against individual enemies but also occasionally against armies (Table 2) (Frischknecht, 2003). However, the foundation of microbiology by Louis Pasteur and Robert Koch offered new prospects for those interested in biological weapons because it allowed agents to be chosen and designed on a rational basis. These dangers were soon recognized, and resulted in two international declarations in 1874 in Brussels and in 1899 in The Hague that prohibited the use of poisoned weapons. However, although these, as well as later treaties, were all made in good faith, they contained no means of control, and so failed to prevent interested parties from developing and using biological weapons. During the past century, more than 500 million people died of infectious diseases. Several tens of thousands of these deaths were due to the deliberate release of pathogens or toxins, mostly by the Japanese during their attacks on China during the Second World War. Two international treaties outlawed biological weapons in 1925 and 1972, but they have largely failed to stop countries from conducting offensive weapons research and large-scale production of biological weapons. And as our knowledge of the biology of disease causing agents, viruses, bacteria and toxins increases, it is legitimate to fear that modified pathogens could constitute devastating agents for biological warfare.

3.1.2 Examples of Biological Warfare During the Past Millennium

Year	Event
600 BC	Solon uses the purgative herb hellebore during the siege of Krissa
1155	Emperor Barbarossa poisons water wells with human bodies, Tortona, Italy
1346	Mongols catapult bodies of plague victims over the city walls of Caffa, Crimean Peninsula
1495	Spanish mix wine with blood of leprosy patients to sell to their French foes, Naples, Italy
1650	Polish fire saliva from rabid dogs towards their enemies
1675	First deal between German and French forces not to use 'poison bullets'
1763	British distribute blankets from smallpox patients to native Americans
1797	Napoleon floods the plains around Mantua, Italy, to enhance the spread of malaria
1863	Confederates sell clothing from yellow fever and smallpox patients to Union troops, USA
World war I	German and French agents use glanders and anthrax
World war II	Japan uses plague, anthrax, and other diseases; several other countries experiment with and develop biological weapons programs
1980–1988	Iraq uses mustard gas, sarin, and tabun against Iran and ethnic groups inside Iraq during the Persian Gulf War
1995	Aum Shinrikyo uses sarin gas in the Tokyo subway system

During World War II, some of the mentioned countries began a rather ambitious biological warfare research program. Various allegations and countercharges clouded the events during and after World War II. Japan conducted biological weapons research from approximately 1932 until the end of World War II (Riedel, 2004).

3.1.3 Biological Warfare Programs During World War II

Nation	Numbers of workers (estimated)	Focus
Germany	100–200	Offense research forbidden
Canada	Small	Animal and crop diseases, rinderpest, anthrax
United Kingdom	40–50	Animal and crop diseases, anthrax, foot and mouth disease
Japan	Several thousand	Extensive; official information suppressed by a treaty with USA in which all charges for war crimes were dropped for exchange of information from experiments
Soviet Union	Several thousand	Typhus, plague
USA	1500–3000	Chemical herbicides, anthrax (started too late to be important)

3.1.4 Crucial biological agents (Centers for Disease Control, Atlanta, Georgia, USA)

Disease	Pathogen	Abused[1]
Category A (Major public health hazards)		
Anthrax	*Bacillus antracis* (B)	First World War
		Second World War
		Soviet Union, 1979
		Japan, 1995
		USA, 2001
Botulism	*Clostridium botulinum* (T)	–
Haemorrhagic fever	Marburg virus (V)	Soviet bioweapons programme
	Ebola virus (V)	–
	Arenaviruses (V)	–
Plague	*Yersinia pestis* (B)	Fourteenth-century Europe
		Second World War
Smallpox	*Variola major* (V)	Eighteenth-century N. America
Tularemia	*Francisella tularensis* (B)	Second World War
Category B (Public health hazards)		
Brucellosis	*Brucella* (B)	–
Cholera	*Vibrio cholerae* (B)	Second World War
Encephalitis	Alphaviruses (V)	Second World War
Food poisoning	*Salmonella, Shigella* (B)	Second World War
		USA, 1990s
Glanders	*Burkholderia mallei* (B)	First World War
		Second World War

Psittacosis	*Chlamydia psittaci* (B)	–
Q fever	*Coxiella burnetti* (B)	–
Typhus	*Rickettsia prowazekii* (B)	Second World War
Various toxic syndromes	Various bacteria	Second World War

Category C includes emerging pathogens and pathogens that are made more pathogenic by genetic engineering, including hantavirus, Nipah virus, tick-borne encephalitis and haemorrhagic fever viruses, yellow fever virus and multidrug-resistant bacteria.

[1]Does not include time and place of production, but only indicates where agents were applied and probably resulted in casualties, in war, in research or as a terror agent. B, bacterium; P, parasite; T, toxin; V, virus.

3.1.5 Antiquity and Middle Ages

Rudimentary forms of biological warfare have been practiced since antiquity (Mayor,2003). The earliest documented incident of the intention to use biological weapons is recorded in Hittite texts of 1500–1200 BCE, in which victims of tularemia were driven into enemy lands, causing an epidemic. Although the Assyrians knew of ergot, a parasitic fungus of rye which produces ergotism when ingested, there is no evidence that they poisoned enemy wells with the fungus, as has been claimed. Scythian archers dipped their arrows and Roman soldiers their swords into excrements and cadavers victims were commonly infected by tetanus as result. In 1346, the bodies of Mongol warriors of the Golden Horde who had died of plague were thrown over the walls of the besieged Crimean city of Kaffa. Specialists disagree about whether this operation was responsible for the spread of the Black Death into Europe, Near East and North Africa, resulting in the deaths of approximately 25 million Europeans (Barras and Greub, 2014). Biological were extensively used in many parts of Africa from the sixteenth century AD, most of the time in the form of poisoned arrows, or powder spread on the war front as well as poisoning of horses and water supply of the enemy forces. In Borgu, there were specific mixtures to kill, hypnotize, make the enemy bold, and to act as an antidote against the poison of the enemy as well. The creation of biologicals was reserved for a specific and professional class of medicine-men.

3.1.6 Modern History

The British Army attempted use of smallpox against Native Americans during the Siege of Fort Pitt in June 1763. A reported outbreak that began the spring before left as many as one hundred Native Americans dead in Ohio Country from 1763 to 1764. It is not clear, however, whether the smallpox was a result of the Fort Pitt incident or the virus was already present among the Delaware people as outbreaks happened on their own every dozen or so years and the delegates were met again later and seemingly had not contracted smallpox (King, 2016). It is likely that the British Marines used smallpox in New South Wales, Australia, in 1789. Carus (2015) states "Ultimately, we have a strong circumstantial case supporting the theory that someone deliberately introduced smallpox in the Aboriginal population."

During the final months of World War II, Japan planned to use plague as a biological weapon against U.S. civilians in San Diego, California, during Operation Cherry Blossoms at Night. The plan was set to launch on 22 September 1945, but it was not executed because of Japan's surrender on 15 August 1945. In Britain, the 1950s saw the weaponization of plague, brucellosis, tularemia and later equine encephalomyelitis and vaccinia viruses, but the programme was unilaterally cancelled in 1956. The United States Army Biological Warfare Laboratories weaponized anthrax, tularemia, brucellosis, Q-fever and others. In 1969, US President Richard Nixon decided to unilaterally terminate the offensive biological weapons program of the US, allowing only scientific research for defensive measures. This decision increased the momentum of the negotiations for a ban on biological warfare, which took place from 1969 to 1972 in the United Nation's Conference of the Committee on Disarmament in Geneva. These negotiations resulted in the Biological Weapons Convention, which was opened for signature on 10 April 1972 and entered into force on 26 March 1975 after the ratification by 22 states (Anonymous, 2021). Despite being a party and depositary to the BWC, the Soviet Union continued and expanded its massive offensive biological weapons program, under the leadership of the allegedly civilian institution Biopreparat. The Soviet Union attracted international suspicion after the 1979 Sverdlovsk anthrax leak killed approximately 65 to 100 people.

Their properties of invisibility and virtual weightlessness render detection and verification procedures ineffectual and make non-proliferation of such weapons impossibility. Consequently, national security decision-makers defence professionals and security personnel will increasingly be confronted by biological warfare as it unfolds in the battlefields of the future from the increasing number of countries that are engaged in the proliferation of such weapons i.e. from about four in the mid-1970s to about 17 today (Edgar J. DaSilva, 1999). A similar development has been observed with the proliferation of chemical weapons i.e. from about 4 countries in the recent past to some 20 countries in the mid-1990s. Other alarming issues are the contamination of the environment resulting from dump burial , the use of disease-producing micro-organisms in terroristic attacks on civilian populations; and noncompliance with the 1972 Biological and Toxins Weapons Convention (Table 4). The diverse roles of micro-organisms interacting with humans as "pathogens and pals" has been described with Leishmania infections, and with the presence of Bacteroides thetaiotaomicron in the intestines of humans and mice (Strauss, 1999). Also the development of "battle strains" of anthrax, bubonic plague, smallpox, Ebola virus, and of a microbe-based "double agent" has been reported.

Table 4: Chronological Summary of Conventions, Protocols and Resolutions curbing biological warfare

Year	Convention	Remarks
1899 Hague, Netherlands*	The Laws and Customs of War on Land (II)	Entering into force in 1900, the Convention in defining the rules, laws and customs of war, based on deliberation of the Brussels Peace Conference of 1874, prohibited the use of poison and poisoned weapons as well as the use of arms, projectiles and/or material calculated to cause unnecessary suffering
1907 Hague, Netherlands**	The Laws and Customs of War on Land (IV)	Entering into force in 1910, the Convention covers issues, and customs in more detail, relating to belligerents, prisoners of war, the sick and wounded, means of injuring the enemy, and bombardments, etc.
1925 Geneva, Switzerland	Prohibition of the Use in War of Asphyxiating, Poisonous or other Gases, and of Bacteriological Methods of Warfare	In force since 1928, the protocol prohibits the use in war of asphyxiating, poisonous or other gases, and of all analogous liquids, materials or devices, and, the use of bacteriological methods of warfare
1972 Geneva, Switzerland	Prohibition of the Development, Production and Stockpiling of Bacteriological (Biological) and Toxin Weapons and on Their Destruction	Entering into force in 1975, the Convention - prohibits the development, production, stockpiling, acquisition and retention of microbial or other biological agents or toxins that have no justification for prophylactic, protective or other peaceful purposes - their use as weapons, or in military equipment, missiles and other means of delivery for hostile use or in armed conflict - furthers development and application of scientific discoveries in the field of bacteriology (biology) for the prevention of disease, or for other peaceful purposes
1974 Paris, France	Prevention of Marine Pollution from Land-Based Sources	Amended by a protocol in march, 1986, the Convention covers - prevention of pollution of the sea inclusive of marine estuaries, by humankind either by direct or indirect means, through introduction of substances of energy resulting in deleterious effects as hazards to human health, living marine resources, marine ecosystems, and damage to amenities, or interference with other legitimate uses of the sea

Year	Convention	Remarks
1976 Geneva, U.N.	Prohibition of Military or Any Other Hostile Use of Environmental Modification Techniques	Adopted by the Resolution 31/72 of the U.N. General Assembly on 10 December, 1976, and open for signature in Geneva, 18 May, 1877, the Convention focuses on any technique that changes "through deliberate manipulation of natural processes-- the dynamics, the composition or structure of the Earth, including its biota, lithosphere, hydrosphere and atmosphere, or of outer space"
1981 Abidjan, Cote d'Ivoire	Co-operation in the Protection and Development of the Marine and Coastal Environment of the West and Central African Region	The Convention which entered into force in 1984 covers - the marine environment, coastal zones, and related inland waters within the jurisdiction of the States of the West and Central African Region - the introduction, directly or indirectly, of substances or energy into the marine environment, coastal zones, and related inland waters resulting in deleterious effects that harm living resources, endanger human health, obstruct marine activities (inclusive of fishing) and alters the quality and use of seawater and reduction of amenities. - promotes scientific and technological co-operation to monitor and assess direct and/or indirect pollution, and to engage in networking exchange of scientific data and technical information.
1983 Bonn, Germany	Co-operation in Dealing with Pollution of the North Sea by Oil and Other Harmful	Agreement, by the governments of Belgium, Denmark, France, Germany, the Netherlands, Norway, Sweden, the U.K., and the European Economic Community, based on an agreement reached in Bonn, 1969 covers - prevention of pollution of the sea by oil and other hazardous substances - development of mutual assistance and co-operation in combating marine pollution and destruction of marine bioresources

Year	Convention	Remarks
1989 Basle, Switzerland	Control of Transboundary Movements of Hazardous Wastes and Their Disposal	Known as the Basel Convention, it entered into force in 1992, and covers a variety of hazardous wastes resulting from wastes such as clinical wastes, household wastes, radioactive wastes, and toxic wastes resulting from the production of biologicals, medicines, the chemical industry, etc.***
1991 Bamako, Mali	Ban of the Import into Africa and the Control of Transboundary Movement and Management of Hazardous Wastes with Africa	Known as the Bamako Convention, and yet to enter into force, the Convention focuses on the - need to promote the development of clean production methods, including clean technologies, for the sound management of hazardous wastes produced in Africa, in particular, to avoid, minimise, and eliminate the generation of such wastes - protection, through strict control, the human health of the African population against the adverse effects which may result from the generation and movement of hazardous wastes within the African Continent.
1992 Bucharest, Romania	Protection of the Black Sea against Pollution	The Convention takes into account the - special hydrological and ecological characteristics of the Black Sea, and the susceptibility of its flora and fauna to pollutants and noxious wastes of biological and chemical origin resulting from disposal systems, and dumping by aircraft and seaborne craft - need to develop co-operative scientific monitoring systems to minimize and eliminate pollution of the Black Sea
1993 Geneva, Switzerland	Prohibition of the Production, Stockpiling, and Use of Chemical Weapons and on Their Destruction	Entering into force in 1997, the Convention prohibits the development, production, stockpiling, acquisition or retention of chemical weapons, their transfer, directly or indirectly to anyone, as well as their use in any military preparations or in missile delivery systems or weapons

*Year of the First International Peace Conference based on invitations from Czar Nicholas II of Russia and Queen Wilhelmina of the Netherlands

**Year of the Second International Peace Conference. The Third Conference scheduled for 1915 never took place due to outbreak of the First World War.

***The reader is referred to Annexes I – V appended to the Treaty and which covers the range, categories and characteristics of hazardous wastes and conditions concerning their transboundary movement and disposal.

3.2 Biological/Chemical Warfare Characteristics

Biological, chemical and nuclear weapons possess the common property of wreaking mass destruction. Though biological warfare is different from chemical warfare, there has always been the tendency to discuss one in terms of the other or both together. This wide practice probably arises from the fact that the victims of such warfare are biological in origin unlike that in the Kosovo War in which destruction of civic infrastructure, and large-scale disruption of routine facilities were the primary goals, e.g. the loss of electricity supplies through the use of graphite bombs. Another consideration is that several biological agents e.g., toxic metabolites produced by either micro-organisms, animals or plants are also produced through chemical synthesis (Edgar J. DaSilva, 1999). One of the main goals of biological warfare is the undermining and destruction of economic progress and stability. The emergence of bio-economic warfare as a weapon of mass destruction can be traced to the development and use of biological agents against economic targets such as crops, livestock and ecosystems. Furthermore, such warfare can always be carried out under the pretexts that such traumatic occurrences are the result of natural circumstances that lead to outbreaks of diseases and disasters of either endemic or epidemic proportions.

Biological and chemical warfare share several common features. A rather comprehensive study of the characteristics of chemical and biological weapons, the types of agents, their acquisition and delivery has been made. Formulae and recipes for experimenting and fabricating both types of weapons result from increasing academic proficiency in biology, chemistry, engineering and genetic manipulations. Both types of weapons, to date, have been used in bio- and chemoterroristic attacks against small groups of individuals. Again, defence measures, such as emergency responses to these types of terrorism, are unfamiliar and unknown. A general state of helplessness resulting from a total lack of preparedness and absence of decontaminating strategies further complicates the issue. The widespread ability and interest of non-military personnel to engage in developing chemical and biologically based weapons is linked directly to easy access to academic excellence world-wide. Another factor is the tempting misuse of freely available electronic data and knowledge concerning the production of antibiotics and vaccines, and of conventional weapons with their varying details of sophistication.

Several other factors make biological agents more attractive for weaponization, and use by terrorists in comparison to chemical agents (Table 5). Production of biological weapons has a higher cost efficiency index since financial investments are not as massive as those required for the manufacture of chemical and nuclear weapons. Again, lower casualty numbers are encountered with bigger payloads of chemical and nuclear weapons in contrast to the much higher numbers of the dead that result from the use of invisible and microgram payloads of biological agents. To a great extent, application or delivery systems for biological agents differ with those employed for chemical and nuclear weapons. With humans and animals, systems range from the use of live vectors such as insects, pests and rodents to aerosol sprays of dried spores and infective powders. In the case of plants, proliferation of plant disease is carried out through delivery systems that use propagative material such as contaminated seeds, plant and root tissue culture materials, organic carriers such as soil and compost dressing, and use of water from contaminated garden reservoirs.

Table 5: Biological and Chemical Warfare Characteristics

Biological	**Chemical**
Natural odourless occurrence ·	Obtained synthetically with characteristic odour
Invisible particles normally dispersed through aerosol spray	Normally volatile in nature and dispersed either through mists or aerosol sprays
Entry through inhalation or ingestion ·	Entry through inhalation or dermal absorption
Pre-exposure treatment confers or enhances immunity through use of toxoids, vaccines, antibacterial protective clothing, biosensors and smoke -detectors·	Pre-exposure treatment relies on use of gas-masks, antichemical protective clothing and use of chemosensors for toxic substances
Post-exposure treatment relies on antibiotics or in combination with vaccines	Post-exposure treatment relies on use of antidotes and neutralizing agents
Effects of biological agents and toxins are diverse resulting in incapacitation or death occurring after contraction of disease resulting from infection by a specific biological agent e.g. anthrax caused by *Bacillus anthracis* and plague caused by *Yersinia pestis* ·	Effects of chemical agents are either instantaneous or delayed for a few hours, with the onset of symptoms such as allergy, respiratory discomfort, intense irritation of mucous membranes, malfunctioning of physiological processes, resulting in dosedependent death or incapacitation
Can be weaponized into artillery rounds, cluster bombs, and missile warheads	Long history of use as poison bombs, in artillery rounds, and in missile warheads
Production methods are simple and cheap relying on non-sophisticated technology and easily obtainable knowledge in biology, genetics engineering, medicine and agriculture ·	Simple and complex production methods needing appropriate corresponding equipment and technology for simple and sophisticated chemical synthesis, purification and development of lethal doses
Not easily detected in export control and searches by routine detection systems, e.g. X-rays ·	Detection facilitated through odour escape, and packaging in inert metallic containers showing up on X-ray screens

In terms of lethality, the most lethal chemical warfare agents cannot compare with the killing power of the most lethal biological agents. Amongst all lethal weapons of mass destruction chemical, biological and nuclear, the ones most feared are bioweapons. Biological agents listed for use in weaponization and wars are many. Those commonly identified for prohibition by monitoring authorities are the causative agents of the bacterial diseases anthrax and brucellosis; the rickettsial disease Q fever; the viral disease Venezuela equine encephalitis (VEE), and several toxins such as enterotoxin and botulinum toxin. As a rule, microbiologists have pioneered research in the development of a bioarmoury comprised of powerful antibiotics, antisera, toxoids and vaccines to neutralise and eliminate a wide range of diseases. However, despite the use of biological agents in military campaigns and wars, it is only since the mid1980s that the attention of the military intelligence has been attracted by the spectacular breakthroughs in the life sciences. Military interest, in harnessing genetic engineering and DNA recombinant technology for updating and devising effective lethal bioweapons is spurred on by the easy availability of funding, even in times of economic regression, for contractual research leading to the development of the following.

- Vaccines against a wide variety of bacteria and viruses identified in core control and warning lists of biological agents used in biowarfare
- Rapid detection, identification and neutralisation of biological and chemical warfare agents
- Antidotes and antitoxins for use against venoms, microbial toxins, and aerosol sprays of toxic biological agents
- Development of genetically-modified organisms
- Development of bioweapons with either incapacitating or lethal characteristics
- Development of poisons e.g. ricin, and contagious elements e.g. viruses, bacteria
- Development of antianimal agents e.g. rabbit calcivirus disease (RCD) to curb overpopulation growth of rabbits in Australia and New Zealand
- Development of antiplant contagious agents e.g. causative agents of rust, smut, etc.

3.2.1 International Law

International restrictions on biological warfare began with the 1925 Geneva Protocol, which prohibits the use but not the possession or development of biological and chemical weapons. Upon ratification of the Geneva Protocol, several countries made reservations regarding its applicability and use in retaliation. Due to these reservations, it was in practice a "no-first-use" agreement only (Anonymous, 2021a). The 1972 Biological Weapons Convention (BWC) supplements the Geneva Protocol by prohibiting the development, production, acquisition, transfer, stockpiling and use of biological weapons. Having entered into force on 26 March 1975, the BWC was the first multilateral disarmament treaty to ban the production of an entire category of weapons of mass destruction. As of March 2021, 183 states have become party to the treaty. The BWC is considered to have established a strong global norm against biological weapons, which is reflected in the treaty's preamble, stating that the use of biological weapons would be "repugnant to the conscience of mankind". However, the BWC's effectiveness has been limited due to insufficient institutional support and the absence of any formal verification regime to monitor compliance.

In 1985, the Australia Group was established, a multilateral export control regime of 43 countries aiming to prevent the proliferation of chemical and biological weapons. In 2004, the United Nations Security Council passed Resolution 1540, which obligates all UN Member States to develop and enforce appropriate legal and regulatory measures against the proliferation of chemical, biological, radiological, and nuclear weapons and their means of delivery, in particular, to prevent the spread of weapons of mass destruction to non-state actors (Anonymous, 2021b).

3.2.2 Entomological Warfare (EW)

It is a type of biological warfare that uses insects to attack the enemy. The concept has existed for centuries and research and development have continued into the modern era. EW has been used in battle by Japan and several other nations have developed and been accused of using an entomological warfare program. EW may employ insects in a direct attack or as vectors to deliver a biological agent, such as plague. Essentially, EW exists in three varieties. One type of EW involves infecting insects with a pathogen and then dispersing the insects over target areas. The insects then act as a vector, infecting any person or animal they might bite. Another type of EW is a direct insect attack against crops; the insect may not be infected with any pathogen but instead represents a threat to agriculture. The final method uses uninfected insects, such as bees, wasps, etc., to directly attack the enemy (Lockwood,2008).

History: Entomological warfare is not a new concept; historians and writers have studied EW in connection to multiple historic events. A 14th century plague epidemic in Asia Minor that eventually became known as the Black Death (carried by fleas) is one such event that has drawn attention from historians as a possible early incident of entomological warfare. That plague's spread over Europe may have been the result of a biological attack on the Crimean city of Kaffa (Kirby Reid, 2008).

During the Second Parthian War, King Barsamia used scorpion-stuffed pots thrown at the enemy to defend the ancient Middle Eastern city of Hatra from the Romans (Ryan C. Gott, 2018). It's possible that these literal bug bombs also contained rove beetles in the genus Paederus. These small rove beetles' hemolymph contains the compound pederin. Pederin causes dermatitis and blistering when contacting skin, a likely scenario when panicked warriors began smashing beetles thrown onto them. King Mithridates VI of Pontus also enlisted arthropods in his wartime maneuvers but favored those of the hymenopteran persuasion. During the Third Mithridatic War, Mithridates ordered grayanotoxin-laden honey created by rhododendron-foraging honey bees to be left along roads for pursuing Roman invaders. Warriors eating this honey as part of their pillaged loot experienced intense sickness and hallucinations, giving it the name "mad honey." The incapacitated Romans were then easy targets for Mithridates' army. Mithridates also ordered the release of hornets and bees into sapper tunnels dug beneath battlefields. Clearly, applied entomology has a very long, if brutal, history.

- 1346: Genghis Khan catapulted plague ridden Mongol corpses over the castle walls of Kafa (now Feodosia, Crimea) Fleas dispersed and spread the disease to the enemy
- 1710 :Russia attacked Sweden by catapulting plague infected corpses over the city walls of Reval
- WWII: 1940 Japan's unit 731 led by Lt. General Shiró Ishii dispersed plague infected fleas and flies covered with cholera via low flying planes to infect the populations of China (Kyle Brinson,2019).
- WWII: July 1944 Battle of Saipan Japan intended on releasing plague infested fleas onto U.S. combatants; however, the Japanese submarine carrying fleas was intercepted and su WWII –March 26, 1945 "Operation Cherry Blossoms at Night" Japan finalized plans to spread plague fleas over Southern California scheduled for September 22, 1945. • Plan halted with Japanese surrender on August 15, 1945.nk by U.S. Submarine "Swordfish"
- Germany experimented with massproduction and dispersion of the Colorado Potato Beetle. Release of 54,000 beetles resulted in an infestation to their own country in 1944.
- Cold War :1954 U.S. operation "Big Itch" Dugway Proving Grounds, Utah. - Tested munitions loaded with uninfected fleas.
- 1955 U.S. Operation "Big Buzz" -Dropped 300,000 Aedes aegypti over Georgia to see survival rates, feeding results, and dispersion. Partially declassified in 1981 Cost/Death: 50% vector mortality Operation Drop Kick Operation May Day
- 1961 U.S. Operation "Bellwether II" -Released uninfected, starved, virgin female mosquitoes on U.S. soldiers to test varying vector to host ratios.
- 1965 U.S. Operation "Magic Sword" -Dropped Aedes aegypti over the SE coast to assess how well mosquitoes could find their way to land while battling strong oceanic winds.
- 1989 California Med Fly Attack "The Breeders" claimed responsibility for releasing Mediterranean fruit flies: Indirect Attack to damage crops, Financial retaliation for aerial spraying of Malathion and $60 million dollars in eradication efforts
- The Biological Weapons Convention of 1975 "prohibits the development, production, and stockpiling of biological agents as well as related equipment and delivery systems that are intended for hostile use."

- According to Jeffrey Lockwood, author of Six-Legged Soldiers (a book about EW), the earliest incident of entomological warfare was probably the use of bees by early humans. The bees or their nests were thrown into caves to force the enemy out and into the open. Lockwood theorizes that the Ark of the Covenant may have been deadly when opened because it contained deadly fleas (Baumann Peter, 2008).
- During the American Civil War the Confederacy accused the Union of purposely introducing the harlequin bug in the South. These accusations were never proven, and modern research has shown it more likely that the insect arrived by other means.[1] The world did not experience large-scale entomological warfare until World War II; Japanese attacks in China were the only verified instance of BW or EW during the war. During, and following, the war other nations began their own EW programs.

3.2.3 Genetically Engineered Insects

US intelligence officials have suggested that insects could be genetically engineered via technologies such as CRISPR to create GMO "killer mosquitoes" or plagues that wipe out staple crops. There is research ongoing to genetically modify mosquitoes to curb the spread of diseases, such as Zika, and the West Nile virus by using mosquitoes modified using CRISPR to no longer carry the pathogen. However this research also shows that it may also be possible to implant diseases or pathogens via genetic modification. It has been suggested by the Max Planck Institute for Evolutionary Biology that current US research into genetically modified insects for crop protection via infectious diseases which spread genetic modifications to crops en masse could lead to the creation of genetically modified insects for use in warfare (Reeves *et.al.*, 2019).

3.2.4 Bioweapons

Bioweapons are characterized by a dual-use dilemma. On a lower scale, a bioweapons production facility is a virtual routine run-of the-mill microbiological laboratory. Research with a microbial discovery in pathology and epidemiology, resulting in the development of a vaccine to combat and control the outbreak of disease could be intentionally used with the aid of genetic engineering techniques to produce vaccine-resistant strains for terroristic or warfare purposes. The best known example,reported by UNSCOM , is the masquerading of an anthrax-weapon production facility as a routine civil biotechnological laboratory at Al Hakam. In summary the dual-use dilemma is inherent in the inability to distinctively define between offence -and defence- oriented research and development work concerning infectious diseases and toxins. Whilst progress in immunology, medicine, and the conservation of human power resources are dependent on research on the very same agents of infectious diseases, bans and nonproliferation treaties are associated with the research and production of offensive bioweapons.

Genetic engineering and information are increasingly open to misuse in the development and improvement of infective agents as bioweapons (Edgar J. DaSilva, 1999). Such misuse could be envisaged in the development of antibiotic-resistant micro-organisms, and in the enhanced invasiveness and pathogenicity of commensals. Resistance to new and potent antibiotics constitutes a weak point in the biobased arsenal designed to protect urban and rural populations against lethal bioweapons. An attack with bioweapons using antibiotic-resistant strains could initiate the occurrence and spread of communicable diseases, such as anthrax and plague, on either an endemic or epidemic scale. The evolution of chemical and biological weapons is broadly categorized into four phases. World War I saw the introduction of the first phase, in which gaseous chemicals like chlorine and phosgene were used in Ypres. The second phase ushered in the era of the use of nerve agents e.g. tabun, a cholinesterase inhibitor, and the beginnings of the anthrax and the plague bombs in World War II. The Vietnam War in 1970 constituted the third phase which was characterized by the use of lethal chemical agents e.g. Agent Orange, a mix of herbicides stimulating hormonal function resulting in defoliation and crop destruction. This phase included also the use of the new group of Novichok and mid-spectrum agents that possess the characteristics of chemical and biological agents such as auxins, bioregulators, and physiologically active compounds. Concern has been expressed in regard to the handling and disposal of these mid-spectrum agents by "chemobio " experts rather than by biologists.

The fourth phase coincides with the era of the biotechnological revolution and the use of genetic engineering. Gene-designed organisms can be used to produce a wide variety of potential bioweapons such as organisms functioning as microscopic factories producing a toxin, venom or bioregulator; organisms with enhanced aerosol and environmental stability; organisms resistant to antibiotics, routine vaccines, and therapeutics; organisms with altered immunologic profiles that do not match known identification and diagnostic indices and organisms that escape detection by antibody-based sensor systems

Public attention and concerns, in recent times, have been focused on the dangers of nuclear, biological and chemical-based terrorist threats. This concern is valid given the significant differences between the speed at which an attack results in illness and in which a medical intervention is made, the distribution of affected persons, the nature of the first response, detection of the release site of the weapon used, decontamination of the environment, and post-care of patients and victims. Pollution and alteration of natural environments occurs with the passage of time, as a consequence of reliance on conventional processes such as dumping of chemical munitions in the oceans; disposal of chemical and biological weapons through open-pit burning; and in-depth burial in soil in concrete containers or metallic coffins. Incineration, seemingly the preferred method in the destruction and disposal of chemical weapons, is in the near future likely to be replaced by micro-organisms. Laboratory-scale experimentation has shown that blistering agents, such as mustard mixtures e.g. lewisite and adamsite, and nerve agents e.g.

tabun, sarin and saman are susceptible to the enzymatic action of *Pseudomonas* diminuta, Alteromonas haloplanktis, and Alcaligenes xylosoxidans. In disposing of the chemical weapon stockpile of diverse blister and nerve agents, research now focuses on several microbial processes that are environment-friendly and inexpensive in preference to costly conventional chemical processes in inactivating dangerous chemical agents, and degrading further their residues.

Chemical weapons are intended to kill, seriously injure or incapacitate living systems. Choking agents such as phosgene cause death; blood agents such as cyanide based compounds are more lethal than choking agents; and nerve agents such as sarin and tabun are still more lethal than blood agents. The use of bioweapons is dependent upon several stages. These involve research, development and demonstration programmes, large-scale production of the invasive agent, devising and testing of efficiency of appropriate delivery systems, and maintenance of lethal and pathogenic properties during delivery, storage and stockpiling. Projectile weapons in the form of a minuscule pellet containing ricin, a plant-derived toxin are ingenuously delivered through the spike of an umbrella. Well known examples of the use of such a delivery system are the targeted deaths of foreign nationals that occurred in London and Paris in the autumn of 1978.

Deliberately contaminated food containing herbicide, pesticide or heavy metal residues, and use of land for crops for production of luxurious ornamental plants and cut flowers, is another constituent of food insecurity. Again, new and emerging plant diseases affect food security and agricultural sustainability, which in turn aggravate malnutrition and render human beings more susceptible to re-emerging human diseases (DaSilva and Iaccarino, 1999). The deliberate release of harmful and pathogenic organisms, that kill cash crops and destroy the reserves of an enemy, constitute an awesome weapon of biological warfare and bioterrorism.

Anticrop warfare, involving biological agents and herbicides, results in debilitating famines, severe malnutrition, decimation of agriculture-based economies, and food insecurity. Several instances using late blight of potatoes, anthrax, yellow and black wheat rusts and insect infestations with the Colorado beetle, the rapeseed beetle, and the corn beetle in World Wars I and II have been documented. Defoliants in the Vietnam War have been widely used as agents of anticrop warfare. Cash crops that have been targeted in anticrop warfare are sweet potatoes, soybeans, sugar beets, cotton, wheat, and rice. The agents used to cause economic losses with the latter two foreign-exchange earnings were *Puccinia graminis* tritici and Piricularia oryzae respectively. Wheat smut, caused by the fungus Tilettia caries or T. foetida has been used as a biowarfare weapon. The use of such warfare focuses on the destruction of national economies benefiting from export earnings of wheat, an important cereal cash crop in the Gulf region. In addition, the personal health and safety of the harvesters is also endangered by the flammable trimethylamine gas produced by the pathogen. Species of the fungus Fusarium have been used as a source of the mycotoxin warfare in Southeast and Central Asia.

Food borne pathogens are estimated to be responsible for some 6.5 to 33 million cases on human illnesses and up to 9000 deaths in the USA per annum (Edgar J. DaSilva, 1999). The costs of human illnesses attributed to foodborne causes are between US$2.9 and 6.7 billion, and are attributed to six bacterial pathogens, Salmonella typhosa, *Campylobacter jejuni*, *Escherichia coli* 0157H:H7, Listeria monocytogenes, *Staphylococcus aureus* and *Clostridium perfringens* found in animal products. Consequently, there is the dangerous risk that such organisms could be used in biological warfare and bioterrorism given that *Salmonella*, *Campylobacter* and *Listeria* have been encountered in outbreaks of food borne infections, and that cases of food poisoning have been caused by *Clostridium*, Escherichia and Staphylococcus. Bacterial and fungal diseases are significant factors in economic losses of vegetable and fruit exports. Viral diseases, transmitted by the white fly *Bemisia tabaci* are responsible for severe economic losses resulting from damage to melons, potatoes, tomatoes and aubergines. The pest, first encountered in the mid-1970s in the English-speaking Caribbean region has contributed to estimated losses of US$50 million p.a in the Dominican Republic. Economic losses resulting from infestation of over 125 plant species, inclusive of food crops, fruits, vegetables and ornamental plants have been severe in St. Lucia, St. Kitts and Nevis, St. Vincent and the Grenadines, Trinidad and Tobago, and the Windward Islands. In Grenada, crop losses in the mid-1990s were estimated at UD$50 million following an attack by Maconnellicoccus hirsutus, the Hibsicus Mealy Bug. The existence of natural occurring or endemic agricultural pests or diseases and outbreaks permits an adversary to use biological warfare with plausible denial" and has drawn attention to several imaginative possibilities.

The interaction of biological warfare, genetic engineering and biodiversity is of crucial significance to the industrialized and non-industrialized societies. Developing countries that possess a rich biodiversity of cash crops have a better chance of weathering anti crop warfare. On the other hand, the food security of the industrialized societies, especially in the Northern Hemisphere, is imperiled by their reliance on one or two varieties of their major food crops. The use of genetic engineering, whilst enhancing crop yields and food security, could result in more effective anticrop weapons using gene-modified pathogens that are herbicide resistant, and non-susceptible to antibiotics. Threats to human health exist with the biocontrol and bioremediation agent *Burkholderia cepacia* during agricultural and aquacultural use (Holmes et al, 1998). Attention has also been drawn to the new and potential threats arising from the uncontrolled release of genetically modified organisms. Another aspect of biological warfare involves the corruption of the youth of tomorrow, the bastion of a nation's human power with cocaine, heroin and marijuana derived from drug and narcotic plantations reared by conventional and/or genetically engineered agriculture. On the other hand, the eradication of such drugs plant crops through infection with plant pathogens could prove counterproductive in yielding more knowledge and skills to wipe out food crops, and animal-based agriculture.

Techniques to enhance efficacy of bioweapon: Scientists and genetic engineers are considering several techniques to increase the efficacy of pathogens in warfare (Anonymous, 2013).

i. **Binary biological weapons:** This technique involves inserting plasmids, small bacterial DNA fragments, into the DNA of other bacteria in order to increase virulence or other pathogenic properties within the host bacteria .

ii. **Designer genes:** According to the European Bioinformatics Institute, as of December 2012, scientists had sequenced the genomes of 3139 viruses, 1016 plasmids, and 2167 bacteria, some of which are published on the internet and are therefore accessible to the public . With complete genomes available and the aforementioned advances in gene synthesis, scientists will soon be able to design pathogens by creating synthetic genes, synthetic viruses, and possibly entirely new organisms.

iii. **Gene therapy:** Gene therapy involves repairing or replacing a gene of an organism, permanently changing its genetic composition. By replacing existing genes with harmful genes, this technique can be used to manufacture bioweapons.

iv. **Stealth viruses:** Stealth viruses are viral infections that enter cells and remain dormant for an extended amount of time until triggered externally to cause disease. In the context of warfare, these viruses could be spread to a large population, and activation could either be delayed or used as a threat for blackmail.

v. **Host-swapping diseases:** Much like the naturally occurring West Nile and Ebola viruses, animal viruses could potentially be genetically modified and developed to infect humans as a potent biowarfare tactic .

vi. **Designer diseases:** Biotechnology may be used to manipulate cellular mechanisms to cause disease. For example, an agent could be designed to induce cells to multiply uncontrollably, as in cancer, or to initiate apoptosis, programmed cell death.

vii. **Personalized bioweapons:** In coming years it may be conceivable to design a pathogen that targets a specific person's genome. This agent may spread through populations showing minimal or no symptoms, yet it would be fatal to the intended target.

3.2.5 Biodefense

In addition to creating bioweapons, the emerging tools of genetic knowledge and biological technology may be used as a means of defense against these weapons (Ainscough, 2012).

i. **Human genome literacy:** As scientific research continues to reveal the functions of specific genes and how genetic components affect disease in humans, vaccines and drugs can be designed to combat particular pathogens based on analysis of their particular molecular effect on the human cell .

ii. **Immune system enhancement:** In addition to enabling more effective drug development, human genome literacy allows for a better understanding of the immune system. Thus, genetic engineering can be used to enhance human immune response to pathogens. As an example, Dr. Ken Alibek is conducting cellular research in pursuit of protection against the bioweapon anthrax.

iii. **Viral and bacterial genome literacy :** Decoding the genomes of viruses and bacteria will lead to molecular explanations behind virulence and drug resistance. With this information, bacteria can be engineered to produce bioregulators against pathogens. For example, Xoma Corporation has patented a bactericidal/permeability-increasing (BPI) protein, made from genes inserted into bacterial DNA, which reverses the resistance characteristic of particular bacteria against some popular antibiotics.

iv. **Efficient bio-agent detection and identification equipment :** Because the capability of comparing genomes using DNA assays has already been acquired, such technology may be developed to identify pathogens using information from bacterial and viral genomes. Such a detector could be used to identify the composition of bioweapons based on their genomes, reducing present-day delays in resultant treatment and/or preventive measures.

v. **New vaccines :** Current scientific research projects involve genetic manipulation of viruses to create vaccines that provide immunity against multiple diseases with a single treatment .

vi. **New antibiotics and antiviral drugs :** Currently, antibiotic drugs target DNA synthesis, protein synthesis, and cell-wall synthesis processes in bacterial cells. With an increased understanding of microbial genomes, other proteins essential to bacterial viability can be targeted to create new classes of antibiotics. Eventually, broad-spectrum, rather than protein-specific, anti-microbial drugs may be developed.

3.3 Future of Biowarfare

The revolution in molecular biology and biotechnology can be considered as a potential Revolution of Military Affairs (RMA). According to Andrew Krepinevich, who originally coined the term RMA, "technological advancement, incorporation of this new technology into military systems, military operational advancement, and organizational adaptation in a way that fundamentally alters the character and conduct of conflict" are the four components that make up an RMA. For instance, the Gulf War has been classified as the beginning of the space information warfare RMA. From the technological advances in biotechnology, biowarfare with genetically engineered pathogens may constitute a future such RMA. In addition, the exponential increase in computational power combined with the accessibility of genetic information and biological tools to the general public and lack of governmental regulation raise concerns about the threat of biowarfare arising from outside the military . The US government has cited the efforts of terrorist networks, such as al Qaida, to recruit scientists capable of creating bioweapons as a national security concern and "has urged countries to be more open about their efforts to clamp down on the threat of bioweapons" (Anonymous, 2012).

Despite these efforts, biological research that can potentially lead to bioweapon development is "far more international, far more spread out, and far more diverse than nuclear science [...] researchers communicate much more rapidly with one another by means that no government can control [...] this was not true in the nuclear era," according to David Kay, former chief U.S. weapons inspector in Iraq. Kay is "extraordinarily pessimistic that we [the United States] will take any of the necessary steps to avoid the threat of bioweapons absent their first actual use".

There are those who say 'the First World War was chemical; the Second World War was nuclear; and that the Third World War – God forbid – will be biological'" (Ainscough,2012).

3.4 Genetic Warfare and Next Generation Bioweapons

Rapid developments in biotechnology, genetics and genomics are undoubtedly creating a variety of environmental, ethical, political and social challenges for advanced societies. But they also have severe implications for international peace and security because they open up tremendous avenues for the creation of new biological weapons. The genetically engineered 'superbug' highly lethal and resistant to environmental influence or any medical treatment is only a small part of this story. Much more alarming, from an arms-control perspective, are the possibilities of developing completely novel weapons on the basis of knowledge provided by biomedical research developments that are already taking place. Such weapons, designed for new types of conflicts and warfare scenarios, secret operations or sabotage activities, are not mere science fiction, but are increasingly becoming a reality that we have to face. a systematic overview of the possible impact of biotechnology on the development of biological weapons was provided by (van Aken and Hammond , 2003).

More efficient classical biowarfare agents will probably have only a marginal role, even if the genetically engineered 'superbug' is still routinely featured in newspaper reports. More likely and more alarming are weapons for new types of conflicts and warfare scenarios, namely low-intensity warfare or secret operations, for economic warfare or for sabotage activities. To prevent the hostile exploitation of biology now and forever, a bundle of measures must be taken, from strengthening the Biological and Toxin Weapons Convention to building awareness in the scientific community about the possibilities and dangers of abuse. Any kind of biotechnological or biomedical research, development or production must be performed in an internationally transparent and controlled manner. In cases in which military abuse seems to be imminent and likely, alternative ways to pursue the same research goal have to be developed.

Genome sequencing has given rise to a new generation of genetically engineered bioweapons carrying the potential to change the nature of modern warfare and defense. Biological weapons are designed to spread disease among people, plants, and animals through the introduction of toxins and microorganisms such as viruses and bacteria. The method through which a biological weapon is deployed depends on the agent itself, its preparation, its durability, and the route of infection. Attackers may disperse these agents through aerosols or food and water supplies (Anonymous, 2013).

Although bioweapons have been used in war for many centuries, a recent surge in genetic understanding, as well as a rapid growth in computational power, has allowed genetic engineering to play a larger role in the development of new bioweapons. In the bioweapon industry, genetic engineering can be used to manipulate genes to create new pathogenic characteristics aimed at enhancing the efficacy of the weapon through increased survivability, infectivity, virulence, and drug resistance. While the positive societal implications of improved biotechnology are apparent, the "black biology" of bioweapon development may be "one of the gravest threats we will face".

Limits of past bioweapons and recent advances: Prior to recent advances in genetic engineering, bioweapons were exclusively natural pathogens. Agents must fulfill numerous prerequisites to be considered effective military bioweapons, and most naturally occurring pathogens are ill suited for this purpose . First, bioweapons must be produced in large quantities. A pathogen can be obtained from the natural environment if enough can be collected to allow purification and testing of its properties. Otherwise, pathogens could be produced in a microbiology laboratory or bank, a process which is limited by pathogen accessibility and the safety with which the pathogens can be handled in facilities. To replicate viruses and some bacteria, living cells are required. The growth of large quantities of an agent can be limited by equipment, space, and the health risks associated with the handling of hazardous germs. In addition to large-scale production, effective bioweapons must act quickly, be environmentally robust, and their effects must be treatable for those who are implementing the bioweapon (van Aken and E. Hammond).

Though biological weapons were internationally banned by the 1925 Geneva Convention, state biowarfare programs continued and in many cases expanded during World War II and the Cold War. In 1972, as evidence of these violations mounted, 103 nations signed a treaty known as the Biological Weapons

Convention (BWC). The treaty bans the creation of biological arsenals and outlaws offensive biological research, though defensive research is permissible. Each year, signatories are required to submit certain information about their biological research programs to the United Nations, and violations reported to the UN Security Council may result in an inspection.

As researchers continue to transition from the era of DNA sequencing into the era of DNA synthesis, it may soon become feasible to synthesize any virus whose DNA sequence is known (Anonymous, 2013). This was first demonstrated in 2001 when Dr. Eckard Wimmer re-created the poliovirus and again in 2005 when Dr. Jeffrey Taubenberger and Terrence Tumpey re-created the 1918 influenza virus). The progress of DNA synthesis technology will also allow for the creation of novel pathogens. According to biological warfare expert Dr. Steven Block, genetically engineered pathogens "could be made safer to handle, easier to distribute, capable of ethnic specificity, or be made to cause higher mortality rates". The growing accessibility of DNA synthesis capabilities, computational power, and information means that a growing number of people will have the capacity to produce bioweapons. Scientists have been able to transform the four letters of DNA , A (adenine), C (cytosine), G (guanine), and T (thymine) into the ones and zeroes of binary code. This transformation makes genetic engineering a matter of electronic manipulation, which decreases the cost of the technique (4). According to former Secretary of State Hillary Clinton, "the emerging gene synthesis industry is making genetic material more widely available. A crude but effective terrorist weapon can be made using a small sample of any number of widely available pathogens, inexpensive equipment, and college-level chemistry and biology." The international biological hazard symbol is provided in figure 4.

Fig. 4: The international biological hazard symbol

Agents considered for weaponization or known to be weaponized, include bacteria such as *Bacillus anthracis*, Brucella spp., Burkholderia *mallei*, Burkholderia pseudo*mallei*, Chlamydophila psittaci, Coxiella burnetii, Francisella tularensis, some of the Rickettsiaceae (especially Rickettsia prowazekii and Rickettsia rickettsii), Shigella spp., *Vibrio cholerae*, and *Yersinia pestis*. Many viral agents have been studied and/or weaponized, including some of the Bunyaviridae (especially Rift Valley fever virus), Ebolavirus, many of the Flaviviridae (especially

Japanese encephalitis virus), Machupo virus, Marburg virus, Variola virus, and yellow fever virus. Fungal agents that have been studied include Coccidioides spp. (Hassani *et.al.*, 2004).

3.5 Anti-agriculture

3.5.1 Anti-crop/Anti-vegetation/Anti-fisheries

The United States developed an anti-crop capability during the Cold War that used plant diseases (bioherbicides, or mycoherbicides) for destroying enemy agriculture. Biological weapons also target fisheries as well as water-based vegetation. It was believed that the destruction of enemy agriculture on a strategic scale could thwart Sino-Soviet aggression in a general war. Diseases such as wheat blast and rice blast were weaponized in aerial spray tanks and cluster bombs for delivery to enemy watersheds in agricultural regions to initiate epiphytotic (epidemics among plants). When the United States renounced its offensive biological warfare program in 1969 and 1970, the vast majority of its biological arsenal was composed of these plant diseases. Enterotoxins and Mycotoxins were not affected by Nixon's order (Franz, 2018). Though herbicides are chemicals, they are often grouped with biological warfare and chemical warfare because they may work in a similar manner as biotoxins or bioregulators. The Army Biological Laboratory tested each agent and the Army's Technical Escort Unit was responsible for the transport of all chemical, biological, radiological (nuclear) materials. Scorched earth tactics or destroying livestock and farmland were carried out in the Vietnam war (cf. Agent Orange)[J] and Eelam War in Sri Lanka. Biological warfare can also specifically target plants to destroy crops or defoliate vegetation. The United States and Britain discovered plant growth regulators (i.e., herbicides) during the Second World War, and initiated a herbicidal warfare program that was eventually used in Malaya and Vietnam in counterinsurgency operations.

3.5.2 Anti-livestock

During World War I, German saboteurs used anthrax and glanders to sicken cavalry horses in U.S. and France, sheep in Romania, and livestock in Argentina intended for the Entente forces. One of these German saboteurs was Anton Dilger. Also, Germany itself became a victim of similar attacks horses bound for Germany was infected with Burkholderia by French operatives in Switzerland (Anonymous, 2020a). During World War II, the U.S. and Canada secretly investigated the use of rinderpest, a highly lethal disease of cattle, as a bioweapon. In the 1980s Soviet Ministry of Agriculture had successfully developed variants of foot-and-mouth disease, and rinderpest against cows, African swine fever for pigs, and psittacosis to kill the chicken. These agents were prepared to spray them down from tanks attached to airplanes over hundreds of miles. The secret program was code-named "Ecology". During the Mau Mau Uprising in 1952, the poisonous latex of the African milk bush was used to kill cattle.

Common epidemiological clues that may signal biological attack: From most specific to least specific (Treadwell *et.al.*, 2003).

- Single cause of a certain disease caused by an uncommon agent, with lack of an epidemiological explanation.
- Unusual, rare, genetically engineered strain of an agent.
- High morbidity and mortality rates in regards to patients with the same or similar symptoms.
- Unusual presentation of the disease.
- Unusual geographic or seasonal distribution.
- Stable endemic disease, but with an unexplained increase in relevance.
- Rare transmission (aerosols, food, water).
- No illness presented in people who were/are not exposed to "common ventilation systems (have separate closed ventilation systems) when illness is seen in persons in close proximity who have a common ventilation system."
- Different and unexplained diseases coexisting in the same patient without any other explanation.
- Rare illness that affects a large, disparate population (respiratory disease might suggest the pathogen or agent was inhaled).
- Illness is unusual for a certain population or age-group in which it takes presence.
- Unusual trends of death and/or illness in animal populations, previous to or accompanying illness in humans.
- Many affected reaching out for treatment at the same time.
- Similar genetic makeup of agents in affected individuals.
- Simultaneous collections of similar illness in non-contiguous areas, domestic, or foreign.
- An abundance of cases of unexplained diseases and deaths.

3.5.3 Identification of Bioweapons

List of some biowarfare institutions, programmes and projects is furnished in table 6.

Researchers at Ben Gurion University in Israel are developing a different device called the BioPen, essentially a "Lab-in-a-Pen", which can detect known biological agents in under 20 minutes using an adaptation of the ELISA, a similar widely employed immunological technique, that in this case incorporates fiber optics (Genuth and Fresco-Cohen, 2006).

Table 6 : List of some biowarfare institutions, programmes and projects

United States:	**United Kingdom:**	**Soviet Union and Russia:**
• Fort Detrick, Maryland • U.S. Army Biological Warfare Laboratories (1943–69) • Building 470 • One-Million-Liter Test Sphere • Operation Whitecoat (1954–73) • U.S. entomological warfare program • Operation Big Itch • Operation Big Buzz • Operation Drop Kick • Operation May Day • Project Bacchus • Project Clear Vision • Project SHAD • Project 112 • Horn Island Testing Station • Fort Terry • Granite Peak Installation • Vigo Ordnance Plant	• Porton Down • Gruinard Island • Nancekuke • Operation Vegetarian (1942–1944) • Open-air field tests: • Operation Harness off Antigua, 1948–1950. • Operation Cauldron off Stornoway, 1952. • Operation Hesperus off Stornoway, 1953. • Operation Ozone off Nassau, 1954. • Operation Negation off Nassau, 1954-5.	• Biopreparat (18 labs and production centers) • Stepnagorsk Scientific and Technical Institute for Microbiology, Stepnogorsk, northern Kazakhstan • Institute of Ultra Pure Biochemical Preparations, Leningrad, a weaponized plague center • Vector State Research Center of Virology and Biotechnology (VECTOR), a weaponized smallpox center • Institute of Applied Biochemistry, Omutninsk • Kirov bioweapons production facility, Kirov, Kirov Oblast • Zagorsk smallpox production facility, Zagorsk o Berdsk bioweapons production facility, Berdsk o Bioweapons research facility, Obolensk o Sverdlovsk bioweapons production facility (Military Compound 19), Sverdlovsk, a weaponized anthrax center • Institute of Virus Preparations • Poison laboratory of the Soviet secret services • Vozrozhdeniya • Project Bonfire • Project Factor
Japan: • Unit 731 • Zhongma Fortress • Kaimingjie germ weapon attack • Khabarovsk War Crime Trials • Epidemic Prevention and Water Purification Department	Iraq: • Al Hakum • Salman Pak facility • Al Manal facility	South Africa: • Project Coast • Delta G Scientific Company • Roodeplaat Research Laboratories • Protechnik

Canada:	India :
• Grosse Isle, Quebec, site (1939–45) of research into anthrax and other agents • Experimental Station Suffield, Suffield, Alberta	• Development of Biogenic 3D nanoporous silica based sensor for enhanced sensitivity against phytofungi • Development of an electrochemical biosensor system for the rapid detection of biological warfare agents • Design and development of bio-potential signal analysis system for control of mobility assistive device • Design and development of on-evasive, multimodal neuromodulatory device studies on feasibility, safety and defence application • Usability of smartphone to evaluate the efficiency of polyniline based sensor for indication of freshness of fish fillet during chemical spoilage • Design and development of biosignal controlled hand exoskeleton • Cutomized bioactive porous titanium implants with improved tissue-integration and attenuated aseptic loosening for orthopardic applications

3.6 Characteristics of a Bioweapon

Intrinsic features of biological agents which influence their potential for use as weapons include infectivity; virulence; toxicity; pathogenicity; incubation period; transmissibility; lethality; and stability. Unique to many of these agents, and distinctive from their chemical counterparts, is the ability to multiply in the body over time and actually increase their effect.

i. **Infectivity:** The infectivity of an agent reflects the relative ease with which microorganisms establish themselves in a host species. Pathogens with high infectivity cause disease with relatively few organisms, while those with low infectivity require a larger number. High infectivity does not necessarily mean that the symptoms and signs of disease appear more quickly, nor that the illness is more severe.

ii. **Virulence:** The virulence of an agent reflects the relative severity of disease produced by that agent. Different microorganisms and different strains of the same microorganism may cause diseases of different severity.

iii. **Toxicity:** The toxicity of an agent reflects the relative severity of illness or incapacitation produced by a toxin.

iv. **Pathogenicity:** This reflects the capability of an infectious agent to cause disease in a susceptible host.

v. **Incubation Period:** A sufficient number of microorganisms or quantity of toxin must penetrate the body to initiate infection (the infective dose), or intoxication (the intoxicating dose). Infectious agents must then multiply (replicate) to produce disease. The time between exposure and the appearance of symptoms is known as the incubation period. This is dose;

virulence; route of entry; governed by many variables, including: the initial rate of replication; and host immunological factors.

vi. **Transmissibility:** Some biological agents can be transmitted from person-to-person directly. Indirect transmission (for example, via arthropod vectors) may be a significant means of spread as well. In the context of BW casualty management, the relative ease with which an agent is passed from person-to-person (that is, its transmissibility) constitutes the principal concern.

vii. **Lethality:** Lethality reflects the relative ease with which an agent causes death in a susceptible population.

viii. **Stability:** The viability of an agent is affected by various environmental factors, including temperature, relative humidity, atmospheric pollution, and sunlight. A quantitative measure of stability is an agent's decay rate (for example, "aerosol decay rate").

ix. **Additionally Factors:** Additional factors which may influence the suitability of a microorganism or toxin as a biological weapon include: ease of production; stability when stored or transported; and ease of dissemination.

3.7 Ethnic/Biogenetic Bioweapon

An ethnic bioweapon is a type of theoretical bioweapon which could only target or primarily target people of specific ethnicities or people with specific genotypes. One of the first modern fictional discussions of ethnic weapons is in Robert A. Heinlein's 1942 novel Sixth Column (republished as The Day After Tomorrow), in which a race-specific radiation weapon is used against a so-called "Pan-Asian" invader.

3.7.1 Genetic Weapons

- In 1997, U.S. Secretary of Defense William Cohen referred to the concept of an ethnic bioweapon as a possible risk. In 1998 some biological weapon experts considered such a "genetic weapon" plausible, and believed the former Soviet Union had undertaken some research on the influence of various substances on human genes (William Cohen, 2006).
- In its 2000 policy paper Rebuilding America's Defenses, think-tank Project for the New American Century (PNAC) described ethnic bioweapons as a "politically useful tool" that US adversaries could have incentive to develop and utilize.
- The possibility of a "genetic bomb" is presented in Vincent Sarich's and Frank Miele's book, Race: The Reality of Human Differences, published in 2004. These authors view such weapons as technically feasible but not very likely to be used. (page 248 of paperback edition.)

- In 2004, The Guardian reported that the British Medical Association (BMA) considered bioweapons designed to target certain ethnic groups as a possibility, and highlighted problems that advances in science for such things as "treatment to Alzheimer's and other debilitating diseases could also be used for malign purposes".
- In 2005, the official view of the International Committee of the Red Cross was "The potential to target a particular ethnic group with a biological agent is probably not far off. These scenarios are not the product of the ICRC's imagination but have either occurred or been identified by countless independent and governmental experts.
- In 2008, the US government held a congressional committee, 'Genetics and other human modification technologies: sensible international regulation or a new kind of arms race?', during which it was discussed how "we can anticipate a world where rogue (and even not-so-rogue) states and non-state actors attempt to manipulate human genetics in ways that will horrify us".
- In 2012, The Atlantic wrote that a specific virus that targets individuals with a specific DNA sequence is within possibility in the near future. The magazine put forward a hypothetical scenario of a virus which caused mild flu to the general population but deadly symptoms to the President of the United States. They cite advances in personalized gene therapy as evidence
- In 2016, Foreign Policy magazine suggested the possibility of a virus used as an ethnic bioweapon that could sterilize a "genetically-related ethnic population.

Israeli 'ethnobomb' controversy: In November 1998, The Sunday Times reported that Israel was attempting to build an "ethno-bomb" containing a biological agent that could specifically target genetic traits present amongst Arab populations. Wired News also reported the story as did Foreign Report (Uzi Mahnaimi and Marie Colvin, 1998).

Microbiologists and geneticists were skeptical towards the scientific plausibility of such a biological agent. The New York Post, describing the claims as "blood libel", reported that the likely source for the story was a work of science fiction by Israeli academic Doron Stanitsky. Stanitsky had sent his completely fictional work about such a weapon to Israeli newspapers two years before. The article also noted the views of genetic researchers who claimed the idea as "wholly fantastical", with others claiming that the weapon was theoretically possible.

Russian ban on export of biological samples: In May 2007, a Russian newspaper Kommersant reported that the Russian government banned all exports of human biosamples. The report claims that the reason for the ban was a secret FSB report about on-going development of "genetic bioweapons" targeting Russian population by Western institutions. The report mentions the Harvard School of Public Health, American International Health Alliance, Department of Medical Biotechnology of Jagiellonian University, United States Department of Justice Environment and Natural Resources Division, Institute of Genetics and Biotechnology Warsaw University, and United States Agency for International Development.

3.7.2 Control, Monitoring and Reporting Systems

Reporting of outbreaks of disease, often attributed to natural causes, should always be taken seriously since such outbreaks often result from non-compliance with the prohibitions embodied in international conventions in force. Potential nosocomial transmission of biological warfare agents occurs through blood or body fluids (e.g. haemorrhagic fever and hepatitis viruses); drainags and secretions (e.g. anthrax, plague, smallpox); and respiratory droplets (e.g. influenza plague, smallpox). The obligatory notification and reporting of outbreaks of diseases in humans, animals and plants helps to contain and neutralise the threats of biological warfare and bioterrorism (Edgar J. DaSilva, 1999). Such practice, in accordance with existing health codes and complementary reporting systems, helps to develop a reservoir of preparedness capacity.

The development of a response strategy and technology in monitoring the control of weapons is at the core of a state of preparedness in the USA. Current anti-bioterrorism measures involve the devising of unconventional effective countermeasures to combat misuse of pathogens encountered either naturally or in a genetically modified state. Such a strategic response involves:

- The use of bacterial RNA-based signatures and corresponding structural templates through which all pathogens can be potentially identified through appropriate trial and error testing, and verification
- Development of a data base of virtual pathogenic molecules responding to the bacterial signature templates
- Development, evaluation and use of effective antibacterial molecules that eliminate pathogens but do not harm humans nor animals

Guidelines and recommendations have been formulated for use by public health administrators and policy-makers, medical and para-clinical practitioners, and technology designers and engineers in developing civilian preparedness for terrorist attack (Edgar J. DaSilva, 1999). Areas covered deal with rapid detection of biological and chemical agents, pre-incident analysis of the targeted area, protective clothing, and use of vaccines and pharmaceuticals in treatment and decontamination of mass casualties. The lack of basic hygienic procedures accompanying the use of domestic and public health facilities in the discharge, and disposal of human wastes has contributed to a large extent of the state of unpreparedness in responding to obnoxious biological weapons. Furthermore, the indiscriminate use of chemotherapeutics, and the overuse of antibiotics, has contributed to a complacent sense of invincibility in confronting once easily eradicated causative agents of disease. Crucial elements of appropriate and timely responses are the renovation and modernization of the public health infrastructure, the necessary networking of the para-clinical and specialized medical forces involving nurses, general health practitioners, epidemiologists, quarantine specialists and experts in communicable diseases. In brief, an appropriate optimal response constitutes a co-ordinated management of medical capability and restorative efforts backed up by supporting extension services.

Several examples of scientific societies, and of national, regional and global initiatives addressing the global threats of emerging infections and disease have been documented (DaSilva and Iaccarino, 1999). The African biotechnological community is aware of the need of safety considerations and risk assessment in the development and use of bioengineering micro-organisms (Van der Meer et al, 1993). Activities in Uganda, Kenya, Zimbabwe, Tanzania, South Africa, and the Southern African Development Community (Angola, Botswana and Zimbabwe) constitute a revelation of regional academic capacity and competence in addressing issues formulating guidelines, and programming initiatives concerning food security, recombinant DNA biosafety guidelines, and environmental biosafety protocols.

Destruction and deterioration of the environment is usually preceded by the emergence and spread of infectious diseases. In Southern Africa, beset by warplagued conditions, migration of tribal populations and overnight development of nomadic villages, the loss of life and erosion of human resources results from the occurrence of AIDS, malaria, tuberculosis, meningitis and dysentery. Academic and affluent societies are often stricken by outbreaks of hamburger disease. The causative agent is a virulent commensal Escherichia coli. AIDS in South Africa is likely to become a notifiable disease as a consequence of governmental concern in containing the widespread occurrence of the disease. The Department of Industrial Health in Singapore, in fostering a favourable workplace environment, requires the reporting of an outbreak or occurrence of anthrax listed amongst 31 notifiable industrial diseases. The rare outbreak of encephalitis in Malaysia, more recently, reached alarming proportions of concern with severe economic and health implications for other Southeast Asian countries e.g. Laos and Vietnam, thus prompting the destruction of large numbers the porcine population suspected of harbouring the virus.

The role of chemical protective clothing in the performance of military personnel in combat and surveillance situations has been reviewed. The performance and output of military and auxiliary personnel is severely affected following exposure to chemical weapons using nerve agents and disabling chemicals. Interference with a loss of physiological functions such as loss of muscle control, paralysis of body movements, loss of memory, dermal discoloration, prolonged deterioration of vision, speech intelligibility, and the like result in loss of psychological confidence, and professional competence (Edgar J. DaSilva, 1999). The development of chemical protective clothing incorporating chemical and biochemical protectants, such as hypochlorites, phenolics, soap waxes, and antidotes, helps offset psychological stress and trauma, and combat anxiety. Anti-biowarfare and anti-bioterrorism research has led to the development of rub-on polymer creams and anti-germ warfare lotions that provide protection also against the influenza virus. Chemical protection in the form of rubberised hoods and tunics, gloves, boots, and gas masks helps guard against tear gas agents, nerve agents and chemical irritants delivered either by aerosols or liquid sprays. Recently, the incorporation of antibiotics in routine textiles as antiodour and anti-infection agents has been reported.

Weapons of mass destruction, be they nuclear, chemical or biological in nature, constitute a threat to national security, and to regional and international co-operation. Civilian and military vulnerability to biological weapons can be overcome by resorting to the development of biosensors, fast-reacting bio-detection agents, advanced medical diagnostics, and effective vaccination and immunization programmes. Bio -detection has been spurred on through the development of biorobots. Mechanized insects with computerized artificial systems mimic through microchips or biochips certain biological processes such as neural networks that gather and process neural impulses that influence behavioural sensitivities to stress and dangerous responses to substances of biological and chemical origin. These micro-gadgets can carry out in a single operation tasks such as DNA processing, screening of blood samples, scans for the presence and identifications of disease genes, and monitoring of genetic cell activity normally carried out by several laboratory technicians.

The ability to incorporate such dual-use cyberinsects and biorobots in the potential weaponnization of biological agents needs to be addressed and curbed. Biorobots of the household pestthe cockroach, Blaberus discoidalis, the desert ant Cataglyphis, and the cricket Gryllus bimaculatus are already the subject of in situ research. The cricket robot is being developed, in the USA, through academic research within the framework of the Defence Advanced Research Projects Agency (DARPA) robotics program. The main raison d'être of robobiology is the development of miniaturized models with biomechanical minds that could be used also in space biology exploration. Moreover, like humans and other living systems, their life span is not limited by the deleterious effects of toxic chemicals and wastes. To help the medical community save lives during and in the immediate aftermath of bioterrorist attack, DARPA has sponsored projects that rapidly identify pathogens for treatment either with a combination of antimicrobial substances or nannobombing with potent biosurfactant emulsions. The development of advanced biological and medical technologies aim at saving the 30 to 50 per cent of lives that are traditionally lost in frontline battlefield areas, and, reducing drastically the 90 per cent combat deaths that occur in close combat prior to medical intervention (Edgar J. DaSilva, 1999). Such technologies involve the development and use of surgical robot hands, trauma care technology, and remote teledecontamination of biologically polluted environments.

Appropriate control measures in combating bio- and chemical terrorism, and the production of bioweapons would involve:

- Enactment of national laws that criminalize the production, stockpiling, transfer and use of chemo - and bioweapons
- Enactment of national laws that monitor the use of precursor chemicals that lend themselves to the development of chemical and bio-weapons
- Establishment of national and international databanks that monitor the traffic of precursor chemicals, their use in industry outreach programmes, and their licensed availability in national, regional and international markets
- Establishment and use of confirmatory protocols in the destruction and dispersal of outdated stockpiles, and chemical precursor components.

Tissue-based biosensors provide reliable alerts and assessments of human health risks in counteracting bioterrorism and biowarfare. Comprised of multicellular assemblies, and wide-ranging antibody templates, such sensors detect and predict physiological consequences arising from biological agents that have not been fingerprinted nor identified at the molecular level. Alerts and assessments are made through the use of reporting molecules that express themselves through the phenomena of luminescence, fluorescence, etc. For example, the pigment bacteriorhodopsin obtained from the photosynthetic Halobacterium salinarum is used as a sensor for optical computing, artificial vision, and data storage. Defensive and deterrent technologies are being developed to afford maximum protection to civilian and military personnel; and to reduce to a minimum the fallout damage resulting from bioweapons that use unconventional pathogen countermeasures, controlled biological systems and biomimetics in the defence against biowarfare and bioterrorism (Table 7 a-c).

Table 7a : Example of devices for use in developing biodefence programmes

Bacillus Microchip	Detects *Bacillus anthracis* and identifies it from amongst other generic members such as *B. thuringiensis, B. subtilis* and. *B. cereus*
BIDS ·	Biological Integrated Detection System detects through a laser-based sensor large areas under biological attack. Also functions as a warning system. BIDS is also capable of speeding up treatment of biowarfare casualties by narrowing down the range of identities of specific biological agents used as bioweapons. Variations of the system allow for the detection of between to 4 and 8 biological warfare agents in lees than an hour. The system is transportable for use by vehicle and laboratory -designed aircraft
CRP ·	Critical Reagent Program designed to provide a ready available resources of antibodies, antigens, and gene probes for use in field detection and neutralization of biological warfare agents
IBAD ·	Interim Biological Detector designed as a manual hand-held assay for use on ships with links to aural and visual alarms, IBAD provides advance warning of the presence of biological warfare agents through immunochromatographic analysis

IOTA ·	Voltametric instrument comprised of miniaturized electrodes for optional use with antibodies, enzymes, organic dyes, and molecules for detection of heavy metals in body fluids, microorganisms, pesticide contaminants in foods and potable water, etc accompanied by graphic computation
JBPDS ·	Joint Biological Point Detection System is designed for use in protecting ports, naval ships, airfields, and as a portable warning system in conjunction with meteorological data. Automatic detection and identification of up to 10 biological warfare agents in less than 30hrs.is feasible. Enhanced versions of the systems focus on providing rapid facilities for the identification of 25 biological warfare agents thus speeding up choice of treatment of casualties
LRBSDS ·	Long Range Biological Standoff Detection System possesses a detection range of 50 kms, and through a laser eye distinguishes between artificial and natural aerosol clouds. The system has also been designed for complementary use with BIDS
LIBRA ·	Comprised of quartz crystal resonators coated with optional layers of antibodies, enzymes, etc for use in identification of microorganisms, pesticides, and other dangerous organic molecules and chemical gases with computer prints
MAGIChip ·	Micro-array of gel-immobilized compounds that identify simultaneously numerous biological agents through reliance on microbe-specific gene sequences, and microbespecific sequences of ribosomal ribonucleic acids (rRNAs)
PAB	Biosensor system with potentiometric alternating biosensing silicon chip which Interacts with a biological element such as cells, enzymes, etc with measured pH rates or redox potential variation. Used in determining metabolic variations in bacterial cells in response to presence of pollutants, drugs, hormones, pesticides, etc., with graphic computation
Portal Shield ·	Used in the Southeast Asian region for the protection of harbours and airfields, this biodefence system facilitates biological detection and identification, decontamination of biosensor equipment and reduction of casualties

Table 7b: Examples of biodefence programmes conducted by university, industry and Governmental agencies

Category	Characteristics
Antibacterials ·	Development of common signatures of infected eucaryotic cells; use of celldivision proteins as broad-spectrum antibacterial targets; development of drugs against bioengineered biological warfare bacteria; use of gene-based broadspectrum antimicrobial agents; and identification of novel targets that enhance pathogen vulnerability and neutralization
Animal Systems ·	Use of insect vectors as early warning systems e.g. detection of chemical signals by parasitic wasps, exploitation of arthropod interaction with biomolecular stimuli, and engineered bee-colonies for detection of harmful of biowarfare agents

Antitoxins ·	Determination of structural biology of Toxins, development of vaccines and potent toxoids, and rapid genetic identification of Gram-positive pathogens
Antivirals ·	Development of protein -based protective agents; invasive intracellular antibiotics, identification of common target in RNA viruses; disruption of cell transport with non-peptide antiviral agents; and rapid drug responses to biological warfare and bioterrorism without loss in potency and effectiveness during stockpiled storage

Table 7c: Biodefence programmes – R & D areas

Type	Features
Casualty Care ·	Programme depends on novel diagnostic non-invasive technologies coupled to rapid medical and surgical intervention in far forward battlefield areas thus reducing traumatic shock and speeding up containment of biowarfare agents through use of hand-held devices fitted up with ultrasonic imaging and remote telesurgical protocols
Tissue-based Biodefence ·	Uses functional biosensors providing assessments of dangers and risks to civilian and military personnel through detection of biowarfare agents in low concentrations through industrialresearch projects such as: - vascularized tissue sensors for detection of generic toxins and pathogens - rapid sensitive detection system for biological agents of mass destruction

DARPA's Unconventional Pathogen Countermeasures program focuses on the development of a powerful and effective deterrent force that limits, reduces and eliminates damage and spread out resulting from use of bioweapons. Such countermeasures focus on:

- Impeding and eliminating the invasive mechanisms of pathogens that facilitate their entry through inhalation, ingestion, and skin tissue
- Devising broad-spectrum medical protocols and treatments that are effective against a wide range of pathogenic organisms and their deleterious products
- Enhancement of external protection using polyvalent adhesion inhibitors in protective clothing, biomimetic pathogen neutralising materials, and personal environmental hygienic protection systems
  - A novel challenge for the biotechnological industry is the development of effective biological defence programmes based on novel fundamental research in biotechnology, genetics and information technology. Biosensor technology is the driving force in the development of biochips for the detection of pesticides, allergens, and micro-organisms;
- Gaseous pollutants e.g. ammonia, methane, hydrogen-sulphide, etc
- Heavy metals, phosphate and nitrates in potable water
- Biological and chemical pollutants in the dairy, food and beverage industries

Using the tenets of reliability, selectivity, range of detection, reproducibility of results, and, standard indices of taxonomy, contamination and pollution, biodefense programs are now being developed around the unique sensorimotor properties of biological entities. Bees, beetles, and other insects are being recruited as sentinel species in collecting real-time information about the presence of toxins or similar threats.

Biosensors, using fibre optic or electrochemical devices, have been developed for detecting micro-organisms in clinical, food technology, and military applications (Edgar J. DaSilva, 1999). An immunosensor is used for the detection of Candida albicans. *Bacillus anthracis*, and bacteria in culture are detected by optical sensors. In addition, several systems have been developed in the USA to detect biological weapons. Generic and polyvalent immunosensors have been devised to detect biological agents that cause metabolic damage and whose antigenic structure has been specifically genetically altered to avoid detection by antibody-based detection systems. Other biodetection systems functioning as early warning/ alert systems involve the detection of biological particle densities by laser eyes and electronic noses with incorporated alarms Emphasis in such systems is less on the identity of the biological agent, and more on the early warning aspect which constitutes an effective arm in counteracting the threat of bioterrorism in daily and routine peace time environments.

Such electronic noses result from a combination of neural informational networks with either chemical or biological sensor arrays or miniaturised spectral meters. Compact, automated and portable, electronic noses offer inexpensive on-the-spot real-time analysis of toxic fuel and gas mixtures, and identification of toxic wastes, household gas, air quality, and body odours. The goal of such programmes is to prevent unpleasant technological surprises arising from misuse of biological agents, chemicals, ethical pharmaceuticals, and obnoxious gases. The preparedness involves the intelligence. monitoring of the capabilities, intentions, and resource materials of potential opponents, and terrorists. In testimony to the U.S. Senate Public Health and Safety Committee, it was emphasised that:

- The strategy of developing and producing dual purpose diagnostics, therapeutics, and vaccines that protects public health and defends against biological weapons
- The control and elimination of infectious diseases through improved surveillance, early warning, communication and training networks, and
- The availability of front line preparedness and response in responding to bioterrotism and biological warfare are integral constitutive elements of a preparedness domestic capacity against bioterrorism.

Biological warfare can be used with impunity under the camouflage of natural outbreaks of disease to decimate human populations, and to destroy livestock and crops of economic significance. Attempts to regulate the conduction of warfare and the development of weaponry using harmful substances such as poisons and poisoned weapons are enshrined in conventions drawn up with respect to the laws and customs on land. These early instruments of war –prevention measures, and eventual confidencebuilding and peace-building measures, have evolved from normal practices and characteristic usages established amongst, civilised peoples; from the basic laws of humanity; the tenets of long established and widely accepted faiths, and the dictates of public conscience. In that context, the conventions outline steps and measures to safeguard buildings and historic monuments dedicated to art, religion and science, and to clinics and hospitals housing the sick and wounded, provided they are not engaged in combat. Use of such personnel in experiments designed to enhance the lethality of weaponry containing harmful substances such as poisons, disabling chemicals and ethical pharmaceuticals is implicitly and strictly prohibited. In the history of the interactions between science, culture and peace, the term Unit 731 is associated with the demeaning of science and humanity, their values and ethics. The activities carried out by Unit 731 in World War II were prohibited as far back as 1907.

In neutralizing the effects of biological agents and rendering them ineffectual for use as bioweapons, bioindustries are now concentrating on the development of a wide range of biotherapeutics – antibiotics and vaccines through development of biologically based defence science and technology programmes. Current bioweapons defence research is now focusing on developing biosensors containing specific antibodies to detect respiratory pathogens likely to be dispersed through sprays and air cooling systems (Edgar J. DaSilva, 1999). Also contract research centres around the use of biotechnologies to remediate environmental areas contaminated with heavy metals, herbicides, pesticides, radioactive materials, and other toxic wastes.

The genetic screening of human diseases and drug discovery have been facilitated by research advances in the field of bioinformatics . The automated and computerised study of shared information in the genomic DNA of biological resources in tandem with digital processing and graphic computation techniques, offers a base for the development of devices for monitoring environmental degradation and development of biodefense programmes (Table 4 a-c). The aim of such research in developing sensors for the timely detection and neutralization of biological weapons is reflected in “Sherlock Holmes’ dog that doesn’t bark”, i.e the silence of the sensor indicates the presence of a biological agent . Development of national preparedness and emerging responses to biological agents, either in bioterroristic or combat situations, is dependent upon the rapidity of intervention by trained antiterroristic personnel comprised of microbiologists, doctors, hospital staff, psychologists, military or law-enforcing forces, and public health personnel. In this regard, the economic impact of a bioterroristic attack has recently been assessed. Investing in public health surveillance helps enhance domestic preparedness in dealing with, bioterrorism, emerging diseases and food borne infections.

The likelihood of genetically engineered micro-organisms contributing to the emergence of new infections cannot be ignored. Public reaction to the introduction of genetically engineered crops into Europe, at this time, is accompanied by controversy and fears for environmental safety. The uncertainty accompanying the potential outbreaks of new scourges is another complicating factor. Increasing public awareness and understanding of safety issues and the release of genetically engineered organisms into the environment helps to overcome unsubstantiated fears and misconceptions, and to secure confidence through a state of preparedness. On such strategies, a ready and effective response exists to combat potential catastrophes and outbreaks of emerging diseases. The science and value of environmental safety evaluations constitute a right step in this direction.

New threats from weapons of mass destruction continue to emerge as a result of the availability of technology and capacity to produce, world-wide, such weapons for use in terrorism and organised crime. Novel and accessible technologies give rise to proliferation of such weapons that have implications for regional and global security and stability. In counteraction of such threats, and in securing the defence of peace, the need for leadership and example in devising preventive and protective responses has been emphasized through the need for training of civilian and non-civilian personnel, and their engagement in international cooperation. These responses emphasize the need for the reduction and elimination of bioterrorism threats through consultation, monitoring and verification procedures; and deterrence, through the constant availability and maintenance of a conventional law and order force that is well-versed in counter proliferation controls and preparedness protocols. Adherence to the Biological and Toxin Weapons Convention, reinforced by confidence-building measures is indeed, an important and necessary step in reducing and eliminating the threats of biological warfare and bioterrorism.

Primary prevention rests on creating a strong global norm that rejects development of such weapons. Secondary prevention implies early detection and prompt treatment of disease. The medical community plays an important role in secondary prevention by participating in disease surveillance and reporting and thus providing the first indication of biological weapons use. In addition, continued research to improve surveillance and the search for improved diagnostic capabilities, therapeutic agents, and effective response plans will further strengthen secondary prevention measures. Finally, the role of tertiary prevention, which limits the disability from disease, shall not be forgotten. Unfortunately, the tools of primary and secondary prevention are imperfect. While the BWC is prepared to assist those nations that have been targets of biological weapons, the medical community must be prepared to face the sequelae should the unthinkable happen.

3.8 Coronavirus : A Biological Warfare Weapon?

Coronaviruses (SARS, MERS, and COVID-19) are a family of RNA viruses that typically cause mild respiratory disease in humans. However, the 2003 emergence of the severe acute respiratory disease coronavirus (SARS-CoV) demonstrated that CoVs are also capable of causing outbreaks of severe infections in humans. A second severe CoV, Middle East respiratory syndrome coronavirus (MERS-CoV), emerged in 2012 in Saudi Arabia. More recently, a novel coronavirus was identified in Wuhan, China, in December 2019.

The issue of biological weapons has been on the back burner for decades. Biological weapons did not gain enough traction because of moral and ethical problems but most importantly because of the possibility of uncontrolled escalation and the disease engulfing friendly forces. As technology develops further, a resurgence of biological weapons to target specific genetic sequences is a possibility. Notwithstanding Biological and Toxin Weapons Convention, this area needs further deliberation and careful examination. With the global recorded deaths from COVID-19 surpassing one million, the biotechnological revolution has heightened the fear of future weaponized pathogens. The COVID-19 virus or its variant could be the most effective weapon for future biological warfare. The indiscriminate effect of such a weapon and its power to cripple economies and devastate the lives of people (Jha and Ratnabali , 2021).

The COVID-19 pandemic is wreaking havoc across the globe and so far, around 150 million people have been infected and about 3.2 million people have died from it with it globally. In India alone, the number of people who die from this pandemic is about to reach 2.5 lakhs. But in China, from which the virus spread all over the world, everything has become normal. In Wuhan city, the epicentre of the virus, people are now living without masks and social distancing. In such a situation, the question in everyone's mind is, what did China do that the coronavirus is now in control there? The answer to this question is revealed through a research paper (Abhishek Sharma, 2021).

This research paper has been published by a newspaper in Australia. It has been said that the discussion of using the coronavirus as a biological weapon started in China in 2015 itself. At that time, scientists of China's People's Liberation Army (PLA) and senior health officials in China had prepared a research paper, titled "The Unnatural Origin of SARS and New Species of Man-Made Viruses as Genetic Bio-weapons". This means that in the year 2019, when the first case of coronavirus came to light in the city of Wuhan, China, a research paper was already prepared 4 years before that and it was prepared by the Chinese army scientists and senior health officers.

In this research paper, then Chinese scientists had said that coronavirus could usher in a new era of biological weapons because they believed that if changes were made, then this virus could spread disease in humans on a large scale, as is happening now. In this research paper, Chinese scientists also wrote that if the virus was used as a weapon, then it would spread in the whole world so that it would be very difficult to control it. Just like the China scientists thought in 2015, it is happening in the world right now. When scientists from the PLA Army were researching this in 2015, they believed that the third world war would be fought on the basis of biological weapons. Because the world has used chemical and nuclear weapons in the first two world wars and many big countries have also developed modern technology to deal with them.

That is, in 2015, China understood that if it wants to compete with other countries, then it will have to develop the virus as a weapon because no country in the world has the technology to avoid the virus. The biggest thing in this research paper is that China had accepted that the coronavirus can be used in war even 6 years ago and it also has the ability to keep it alive for a long time. The newspaper of Australia, which published the report, states that the US intelligence agencies had obtained this research paper during their investigation, which was also shared with other countries and now this research paper has been revealed to the whole world.

These questions also arise on China because it never cooperated with the WHO to investigate it. If it is right and this virus is not spread among humans from its lab, then why is China afraid of investigation? This is a big question. Searching for an answer to this question, the US intelligence agencies reached this research paper and if we consider this research paper completely correct, then it would not be wrong to say that 3.2 million people who died by the coronavirus were murdered and China is responsible for this massacre (Abhishek Sharma, 2021).

Recently a number of reports have come to the fore indicating that the Coronavirus was manufactured in Wuhan Institute of Virology as a bioweapon. Brazil President Jair Bolsonaro, raising serious questions over the Chinese handling for Covid-19, suggested that China could have developed the pathogen in a laboratory and disseminated it as a “biological warfare” for economic gain (Pradhan, 2021). But he is not alone in this. US has recently come out with this theory. According to ‘The Sun’ newspaper in the UK, quoting reports first released by ‘The Australian’, the “bombshell” documents obtained by the US State Department reportedly show the Chinese People’s Liberation Army (PLA) commanders making the sinister prediction that the next World War would be fought with bioweapons. The US has reportedly obtained the papers, which were written by military scientists and senior Chinese public health officials in 2015 as part of the US investigation into the origins of COVID-19. The military document entitled “The Unnatural Origin of SARS and New Species of Man-Made Viruses as Genetic Bioweapons” clearly reveal the Chinese military’s plans with regards to bioweapons. The document is PLA’s bioweapon textbook by General Dezhong Xu, which points out two significant dimensions of the biological war. First, the ability to freeze-dry micro-organisms has made it possible to store biological agents and aerosolize them during attacks. Second, a bioweapon attack could cause the "enemy's medical system to collapse".

There are credible reports that those scientists from the Wuhan lab had initially brought out this fact were told to keep quiet. Among the whistle-blower doctors was Li Wenliang, an ophthalmologist. The eight were hauled up by police for "spreading rumours" and forced to sign statements withdrawing their claims. Dr Li later died in February from the Covid-19. Media reports also suggest that Chinese labs studying the novel coronavirus in late December 2019 and early January 2020 received orders to destroy their samples. China had been opposing independent enquiry. While the above statements and reports suggest that the virus was created in the Wuhan Institute of Virology, an article in the Global Times blamed that these are attempts to tarnish the image of China.

Since the pandemic emerged, Beijing has relied even more on disinformation and influence operations abroad (Mark Kortepeter, 2020). Beijing uses large numbers of fake social media accounts to push its messages. ProPublica had tracked more than 10,000 suspected fake Twitter accounts involved in a coordinated influence campaign with ties to the Chinese government/CCP. Among those are the hacked accounts of users from around the world that post the Chinese propaganda and disinformation about the Coronavirus outbreak. The Chinese influence operations are meant to capture mind and push individuals to project the Chinese propaganda. The big data is collected and programmes are carefully planned for this purpose. Individuals are so heavily bombarded with propaganda that their independent thinking gets impaired and they fail to logically think. It may be added that it is difficult to distinguish between the lab made and natural virus as in both the process is the same. In the lab the scientists can tweak for faster spread but that can happen in the non-lab virus as well. The use of artificial intelligence is possible.

India had been aware of the fact like some other countries that countries in our neighbourhood were working to have this capability. In the Nuclear Doctrine, the trigger for retaliation was broadened by including "a major attack against India, or Indian forces anywhere, by biological or chemical weapons". This was considered necessary as in India's neighbourhood training to soldiers in chemical and biological warfare was being imparted. The inclusion in the Nuclear Doctrine was aimed at deterring the adversaries. That appears to have not worked. Under the current conditions, it is imperative for India to take necessary steps to be prepared to face such a challenge. These demand creation of a robust infrastructure for health care and having BSL-4 labs to have necessary vaccines and medicines for dangerous viruses.

With the emergence of more infectious variants mounting a new challenge in the fight against the COVID-19 pandemic, excerpts from a 2015 Chinese research document published in The Weekend Australian have reignited the debate over the origin of the SARS-CoV-2 virus. The document reportedly titled 'The

Unnatural Origin of SARS and New Species of Man-Made Viruses as Genetic Bioweapons' suggests Chinese scientists were discussing the possibilities of weaponizing coronaviruses five years before the COVID-19 pandemic struck the worldAccording to The Weekend Australian, the document was authored by Chinese scientists, public health officials as well as weapons experts and was unearthed by the US State Department during its own investigation into SARS-CoV-2's origin (Anonymous, 2021). The publication added that it asked Robert Potter, a cybersecurity analyst specializing in leaked Chinese government documents, to verify the paper, who is reported to have said: "the document definitely isn't fake".

While the bioweapon hypothesis around the origin of the virus has sparked public imagination since the start of the pandemic, scathing excerpts from a purported Chinese government document in a mainstream Australian publication have legitimized the argument. As per the news report, the paper states that Chinese scientists "predicted a third world war would be fought with biological weapons" and were discussing how corona viruses could be "artificially manipulated into an emerging human-disease virus, then weaponized and unleashed in a way never seen before". It is important to note here that the Wuhan Institute of Virology is at the heart of China's biochemical research. The facility has Asia's first P4 lab built at a cost of USD 42 million and houses the largest virus bank in the continent with more than 1,500 strains. The 'P4' label is indicative of the highest possible biosafety rating, which is determined by the level of danger and resulting security measures posed by the pathogens studied there.

In June 2020, Shi Zhengli, Deputy Director of the P4 lab at Wuhan, raised eyebrows in an interview with a US magazine when she said she was initially anxious over whether the virus had leaked from her lab. However, subsequent checks revealed that its gene sequence of SARS-CoV-2 differed from the virus studied at the lab and Chinese state media reported Shi Zhengli claiming there was no leak but the damage was done. Some were quick to point out that the Wuhan lab is just a few miles away from the site of the initial outbreak. The Australian publication mentions two names that stand out in the list of authors - Lee Feng, former deputy director of China's Bureau of Epidemic Prevention, and Xu Dezhong, the former chief of China's SARS Epidemic Analysis Expert Group, which raises serious questions about the country's biochemical programs and its lack of transparency over the origin of the SARS-CoV-2 virus.

China was also reluctant to allow foreign experts into the country for an independent probe and stalled the entry of international experts for months (Anonymous, 2021). Therefore, after much back and forth, when a WHO panel reached Wuhan and ruled out the laboratory incident hypothesis as "extremely unlikely", the international community was quick to question if the team of experts conducting the probe had adequate access. China suffered another blow to its international reputation when an AFP report revealed that the team of experts spent only four hours at the virology institute, just an hour at the wet market, and several days inside their hotel without venturing out into the city. One of the leading advocates of the bioweapon theory, Chinese virologist Dr Li-Meng Yan, in an exclusive interview with Times Now, claimed that the coronavirus was purposely enhanced with a lot of caution to make it more harmful to humans. Dr Yan is currently living in exile after being allegedly harassed by the Chinese government.

The Weekend Australian's report comes ahead of the release of journalist Sharri Markson's investigative book What Really Happened in Wuhan, which is reported to have featured the aforementioned research document. On the other hand, the Chinese government mouthpiece Global Times slammed the Australian newspaper for "twisting" the contents of the document to support its own conspiracy theory against China.

3.8.1 The Corona Virus is not a Biological Weapon?

There are many reasons to be skeptical of conspiracy theories about the origins of the disease. Accusations that epidemics or pandemics are "biological warfare" are not new. Humans rightly have an innate fear of disease. "This plague is a deliberate attack" is a trope that is thousands of years old. Disease outbreaks have long been blamed on convenient scapegoats, from medieval plagues, which were often blamed on the Jews or heretics, to more recent conspiracy theories (Dan Kaszeta , 2020).

The current covid-19 crisis has included accusations of biological warfare. The presence of an advanced virology lab in Wuhan fed some theories and accusations that China had deliberately unleashed an attack. The answer is no, for a variety of reasons, both scientific and practical. It starts with the basic facts of virology and genetics: The very premise that a virus must be man-made simply because it is bad is the height of anthropocentric hubris. Nature has many millions of years making viruses and it is most capable of making viral horrors on its own without help. When it comes to man-made viruses, this is a field that is only decades old. Laboratories do not create viruses as neatly or as handily as mother nature does, and it's easy enough to demonstrate that a given virus isn't some Frankenstein creation stitched together out of RNA or DNA from other things. For that is what a man-made virus looks like. Genetic analysis shows this virus is not man-made. Beyond the science, however, there are practical reasons covid-19 biowarfare claims make no sense. We can look at history, military strategy and geopolitics as well as the life sciences. Biological warfare and biological weapons are an arcane subject little understood by the public. Indeed, public knowledge in this area seems to be far more based on science-fiction novels and films than on the actual history of the subject. During the Cold War, both East and West spent money and scientific effort on biological warfare in a clandestine arms race.

In science fiction, biological warfare agents are doomsday weapons designed to bring the globe to its knees. In the real world, those who developed biological weapons never aimed to produce pandemics. Instead, they expended time, money and effort toward concrete projects and understandable objects, not chaos for the sake of chaos (Dan Kaszeta , 2020). Even in the 1950s and 1960s, arguably the heyday of biowarfare research, the proponents of biological warfare understood and held to a basic principle: You don't want the bad disease you made to come back and make your own people sick. The world was less interconnected back then and the scenarios for warfare at the time did not foresee much, if any, travel between combatant countries. But the risk of accidental exposure to one's own troops was considered too high to risk messing around with things that could not be mitigated. Today, there is a massive amount of world travel and commerce, including between potential combatants, but those old lessons have not been forgotten. Economies and supply chains are complex and intertwined, and no country would willingly engineer a weapon that would risk disrupting them.

When we look at the agents that were developed in the old days, both in the East and West, we see things such as anthrax, botulism toxin, tularemia, Q-fever, Venezuelan equine encephalitis and a variety of pestilences that afflict agriculture. Indeed, a very high percentage of the U.S. program was anti-agricultural and not intended to make people ill. These agents could be delivered as aerosols mists of droplets or particles but were not easily contagious from person to person. This meant that they could be targeted. Even the best biological weapons were wildly inefficient and unpredictable, but not in ways that were likely to make them blaze wildly through whole populations. The vast majority of the microbes died in storage, in transit or upon dispersal, so there was a whole esoteric discipline of biological target analysis to make sure that one's expensive bioweapons were used properly. When one looks at the characteristics of an "ideal" biological warfare agent and compares this list to the features of this novel corona virus, there are glaring differences. Covid-19 isn't the sort of thing one would spend years developing. A biological weapon that disproportionately kills off or incapacitates the elderly and vulnerable, but leaves the economically productive fighting-age population mostly intact seems to be not well-thought through. An incubation period that is both long and variable would have been considered a poor characteristic for biological weapons.

A work in progress that leaked out before it was fine-tuned? No. It really doesn't work that way, not least of all because of its aforementioned genetic characteristics. This coronavirus is a fully fine-tuned end product, just one that is made by nature, not man. The only "leak" theory that is remotely plausible is that a Chinese lab was studying something that it found in nature, as one would logically do, which then made its way into the wider world through some horrible breach of safety protocols. This allegation has been made, as well, but there's little information to substantiate this claim at this point. Such an explanation would, in any case, be tragic, not sinister (Dan Kaszeta , 2020). Even if the science gave some hint of man-made origin, none of the practicalities make sense. This is a natural event. Not a man-made plague.

Questions remain over the origins of the deadly virus after a much derided World Health Organization (WHO) probe earlier this year, with the organization ordering a further investigation which factors in the possibly of a lab leak. Most scientists have said there is no evidence that COVID-19 is manmade but questions remain whether it may have escaped from a secretive biolab in Wuhan from where the pandemic originated (Anonymous, 2013).

3.8.2 Desirable Features

Overall, the SARS-CoV-2 virus has some "desirable" properties as a bioweapon, but probably not enough to make it a good choice for military purposes. Regardless, it has certainly reminded us of our vulnerabilities as a society to a new pathogen, and how crippling a pandemic can be, as we continue to watch the entire world grappling with how to contain it (Mark Kortepeter, 2020).

i. **Easy to access**: Except for smallpox, which has been eradicated and is now locked up in freezers at the CDC in Atlanta and in Russia, all of the category A threats are relatively easy to get a hold of. This is certainly the case for SARS-CoV-2, currently readily available across the world.

ii. **Easy to manufacture:** Most category A agents can be manufactured in large quantities so they can be sprayed over a battlefield or large population. Making biological weapons requires either fermentation technology (similar to what's used to make beer) or production in cell culture. Viruses like SARS-CoV-2 are harder to grow than bacteria (like anthrax spores), but it can be done.

iii. **Stable in the atmosphere.** This is a key property for a bioweapon in order for it to be used on a battlefield or against a large population (though not as important for smaller attacks or assassination attempts). SARS-CoV-2 fails when it comes to this criterion. Although it appears to spread very efficiently in indoor environments, it does not appear to survive well outdoors, especially in sunlight.

iv. **Only a small number is needed to spread widely.** If a small number of viruses, bacteria or fungi are needed to infect a single person and therefore cause a widespread infection, that's ideal for making a weapon that can cover a larger area. Right now, the jury is still out on how many organisms are needed to infect with SARS-CoV-2, so it's not clear how it lines up.

v. **A high percentage of infected people become Ill.** One key aspect of any weapon is predictability. If only a few people who are infected become ill, the effect of the pathogen is not reliable enough to base military response plans on it. SARS-CoV-2 virus doesn't fare well on this property. A high percentage, up to 40% or so, appear to have asymptomatic infection. In addition, individuals aged 18-24, which make up a large proportion of the military population, may only be mildly affected. That means, an army couldn't spray the opposing forces on a hilltop with the SARS-CoV-2 virus and expect them to get sick enough to allow for an easy attack.

vi. **Users of the bioweapon can be protected.** Use of biological weapons can be unpredictable. If one is released in the air and the wind blows in the wrong direction, one's own forces could be infected. Hence, there is a need to have a vaccine or treatment to protect forces. Currently we have no "magic bullet" for COVID-19 and no vaccine. Although Russia announced recently that it has a vaccine, whether it actually protects against infection is another story. Many other vaccine candidates are currently at various stages of human testing around the world.

vii. **The threat of use can cause panic**. This can be effective if the intent is to cause havoc from the threat alone. Certainly, we have seen the devastating social impact that has occurred as a result of SARS-CoV-2, as people fear getting infected in the workplace, schools, or in other community areas, so it gets a plus for this property.

viii. **It is contagious**. This property can be a double-edged sword. Once released, a bioweapon can be the "gift that keeps on giving" as it spreads efficiently through the opposing army or nation. It also means the pathogen can spread back to the country that released it if it doesn't have a countermeasure. We have witnessed this challenge firsthand with SARS-CoV-2, and how difficult it has been to contain once it gets into a population.

Investigation: An important thing this pandemic has demonstrated is that once the genie escapes the bottle, it is nearly impossible to put it back in. We lose control and the results are unpredictable. This is one reason the US got out of the bioweapon business when President Nixon shut down the weapons development program in 1969 and decided to focus solely on defensive measures. It is only a matter of time until we face this type of challenge again either from mother nature or an adversary. Now is the time to shore up the vulnerabilities in our preparedness and response that this pandemic has laid bare. Time is not on our side (Mark Kortepeter, 2020).

3.9 Biological Warfare Experiment in India -The Curious Case of Yellow Fever Mosquitoes

On July 17, 2020, the well-known civil rights activist and Supreme Court lawyer Nandita Haksar, in an article ("Stranger than fiction: Did the CIA conduct secret mosquito experiments in India in the 1970s?"), posed a question: Is scientific collaboration a "battle between politics for profits and politics for the people?" As the daughter of the late P.N. Haksar, a distinguished bureaucrat, Planning

Commission member and Principal Secretary to Prime Minister Indira Gandhi, she was aware of the controversial closing down of the Genetic Control of Mosquitoes Unit (GCMU) under the Indian Council of Medical Research (ICMR). She says: "It was in this room that I heard many stories of covert operations. That day, a young journalist came by and told my father of a strange experiment with mosquitoes being conducted right near Palam airport, as Delhi airport was then called. The man said it was an experiment on yellow fever. 'But we don't have yellow fever in India,' my father had exclaimed. The journalist said that this was exactly his point. He claimed it was a part of a biological warfare experiment. We all sat in shocked silence." The journalist, Chakravarthi Raghavan, went on to head the Press Trust of India (PTI). Dr K.S. Jayaraman, who did the investigations, was also no ordinary correspondent. With a PhD in nuclear physics from a university in the United States and journalism as an elective subject, he had resigned his government job as a scientist and joined PTI as its Chief Science Reporter (Rajagopalan, 2021).

The entire story of how the GCMU, established in 1970 by the World Health Organisation (WHO) to study the genetic control of mosquitoes, had to close down in 1975 is now forgotten. The guidelines issued by a committee of learned scientists such as Prof. M.S. Swaminathan and Prof. M.G.K. Menon in 1975 on international scientific collaboration have also been forgotten.

Mixed reactions: In India the Director General of Health Services (DGHS) admitted that the knowledge gained by the genetic control experiment could certainly be used for putting virus into mosquitoes and starting a focus on a disease like yellow fever (PAC, p. 135-137). In international scientific circles the biological warfare allegations against the WHO-ICMR project produced mixed reactions. New Scientist (October 9, 1975, p. 102) said the allegations were far less ridiculous. It quoted a biological warfare expert as saying that the GCMU data would be useful if one intended a yellow fever attack on India. It said the Indian data might have been useful in finding out why yellow fever had not occurred even though the vectors and monkeys were present. But even the critics admit that the WHO-ICMR project concerned an area where there was an overlap of public health and biological warfare interest. But the biological warfare implications were either ignored (in India) or were not pointed out at all when the project was mooted and many of the scientists became aware of it only after the investigative news report in 1974 and the subsequent PAC report (Rajagopalan, 2021).

There are, however, some pointers suggesting that the project was conceived with biological warfare as the main aim. The Sonepat site for the release of Aedes aegypti was selected by the WHO and the USPHS even before the GCMU formally took shape. Despite objections on scientific grounds from the local institute of health (NICD), the site was not changed (PAC, p.191). The testing of foreign strains of Aedes aegypti mosquitoes for their potential virulence of yellow fever vectors was considered unnecessary by the USPHS and by the WHO virologist (Dr Paul Bres), who (coincidentally?) happened to be a former colonel of the French Army. This is strange in a scientific project like this, particularly when experts had warned that it might be extremely serious if yellow fever were ever introduced in Asia or the Pacific Islands where the disease had never occurred ("Biological and Chemical Warfare policies of the U.S.", p.411). Also of concern was the unit's reluctance to change the priority from mosquitoes carrying malaria (problem number one not only in India but throughout the Indian subcontinent and neighbouring areas) to its obsession with studies on Aedes aegypti. It is well known that this species may be playing a beneficial role in the tropics by spreading the flu-like dengue fever which, in turn, protects the population against yellow fever.

Scientific espionage: Was the mosquito research in India part of a research programme in biological warfare which had been banned by world bodies but not from the minds of elite scientists and politicians? It is hard to say. But where the number of coincidences defies the law of averages they are not random occurrences, and probabilities must go the other way. Scientific espionage in developing countries is easy because scientists and scientific departments in these countries are starved of funds (Rajagopalan, 2021). Many of the key scientific figures have been trained in advanced nations which facilitates establishing contacts with would-be collaborators on a personal level. The inferiority complex and lack of suitable machinery to evaluate foreign-sponsored projects also expose the countries to evil designs of foreigners. Such an evaluation can be made by countries that are scientifically equal, but many developing countries are not in a position to make such an evaluation of foreign projects from security or economic angles, not even from the angle of utility to themselves.

The PAC, after considering the entire gamut of foreign financial or foreign collaborative research, recommended in its report-

"Government should identify a set of scientific or operation areas in which investigations by foreigners or by foreign assisted programmes should be subjected to the most careful and comprehensive scrutiny on a case by case basis before approval is given for the initiation of the project. The scientific areas selected at a particular point of time would need to be defined in the context of the prevalent international situating and advances in science and technology".

"To start with the committee would suggest the following areas: (a) Any and all aspects of oceanography and research related to ocean resources and our coastal areas; (b) any and all aspects relating to meteorology and weather, especially weather modification projects; (c) remote sensing by aircraft and satellites particularly for the assessment of natural resources; (d) areas in biology such as microbiology epidemiology, ecology and virology; (g) all aspects of toxicology of drugs, pesticides and other chemicals; (f) propagation of radio waves including studies aimed at collecting information about the ionosphere and other upper atmosphere layers over our country; (g) any and all scientific investigations in border areas such as 'Himalayan Geology'."

Did we learn the lessons?: The government should decide that all proposals for scientific investigations undertaken in these defined areas with the help of or in association with foreign organisations or with foreign monies from any source should be sent by the Ministry, agency laboratory or private institution concerned to a nodal point within the government for a comprehensive review and clearance. "The nodal point should be a high-power committee of scientists headed by the scientific adviser to the Ministry of Defence but include and perhaps ought to include other high security agencies of the government (Rajagopalan, 2021). The committee desires that once this mechanism has been set up it would also review all existing projects or of the type mentioned in preceding paragraph." But as far as I know, there is no such nodal point in existence, and our various institutions are having, even now, many foreign collaborative projects.

The kind of mechanism suggested by the PAC could at best deal with security and defence angles. But it cannot really deal with internal haemorrhage issue. In any event, unlike India, many developing countries do not even have the necessary scientific talent to assess these issues. Perhaps a solution for Third World countries is to set up their own organization secretariat or centre staffed by personnel selected for their integrity and ability and societal purpose and uses its resources for looking at and advising Third World countries in projects and proposals in the fields of S&T. The United Nations and its specialised agencies could have been the proper places to set up a watchdog agency in liaison with the plans for transfer of S&T to help developing nations consult and assess foreign projects in scientific research. However the U.N. and its agencies like the WHO, structured as of now, are really controlled by the Big Powers.

3.10 Bioterrorism

Bioterrorism is terrorism involving the intentional release or dissemination of biological agents. These agents are bacteria, viruses, insects, fungi or toxins, and may be in a naturally occurring or a human-modified form, in much the same way as in biological warfare. Further, modern agribusiness is vulnerable to anti-agricultural attacks by terrorists, and such attacks can seriously damage economy as well as consumer confidence. The latter destructive activity is called agrobioterrorism and is a subtype of agro-terrorism (Roberge Lawrence, 2019).

Definition: Bioterrorism is the deliberate release of viruses, bacteria, toxins or other harmful agents to cause illness or death in people, animals, or plants. These agents are typically found in nature, but could be mutated or altered to increase their ability to cause disease, make them resistant to current medicines, or to increase their ability to be spread into the environment. Biological agents can be spread through the air, water, or in food. Biological agents are attractive to terrorists because they are extremely difficult to detect and do not cause illness for several hours to several days. Some bioterrorism agents, like the smallpox virus, can be spread from person to person and some, like anthrax, cannot (Preston Richard, 2002).

3.10.1 Known Bioterror Agents and History

- A-Virus - A mutagenic virus that could tell from friend or foe developed by Glenn Arias and the remnants of Los Iluminados (Anonymous, 2021).
- C-Virus - A mutagen developed by merging "G" with the t-Veronica Virus which was known for it's arthropod origins. Created by Carla Radames.
- "G" - A mutagenic virus initially found in the body of Lisa Trevor. It caused numerous mutations in the body to guarantee survival from any injury. Created by William Birkin.
- "t" - A mutagen developed in the 1970s after a sample of Progenitor Virus mutated upon contact with leech DNA. Created by James Marcus.
- t-Abyss virus - A mutagenic weapon created by merging "t" with another virus known as "The Abyss". It was used in marine B.O.W.s in 2005.
- t-Phobos virus - A mutagen that only triggered when the subject experienced fear. Created by Alex Wesker.
- Plaga Type 2 - A genetically-modified parasite designed by TRICELL Africa, Inc. in 2008 and tested in Kijuju.
- Plaga Type 3 - Another genetically-modified parasite tested in Kijuju by TRICELL.

History: In 1916, the Russians arrested a German agent with similar intentions. Germany and its allies infected French cavalry horses and many of Russia's mules and horses on the Eastern Front. These actions hindered artillery and troop movements, as well as supply convoys. In 1972, police in Chicago arrested two college students, Allen Schwander and Stephen Pera, who had planned to poison the city's water supply with typhoid and other bacteria. Schwander had founded a terrorist group, "R.I.S.E.", while Pera collected and grew cultures from the hospital where he worked. The two men fled to Cuba after being released on bail. Schwander died of natural causes in 1974, while Pera returned to the U.S. in 1975 and was put on probation.

List of some bioterrorism incidents

Date	Incident	Organism	Details
1964-1966	Dr. Mitsuru Suzuki, physician with training, Japan	*Shigella dysenteriae* and *Salmonella typhi*	Objective: Revenge due to deep antagonism to what he perceived as a prevailing seniority system Dissemination: Sponge cake, other food sources Official investigation started after anonymous tip to Ministry of Health and Welfare. He was charged, but was not convicted of any deaths; later implicated in 200 – 400 illnesses and 4 deaths

Date	Incident	Organism	Details
1984	Rajneeshee religious cult attacks, The Dalles, Oregon	*Salmonella typhimurium*	Contaminated restaurant salad bars, hoping to incapacitate the population so their candidates would win the county elections 751 illnesses, Early investigation by CDC suggested the event was a naturally occurring outbreak. Cult member arrested on unrelated charge confessed involvement with the event
1987-1990	David J. Acer, Florida dentist	HIV	Infected 6 patients after he was diagnosed with HIV
1990s	Aum Shinrikyo attempts in Tokyo, Japan Tokyo subway sarin attack, Matsumoto incident	*Bacillus anthracis*, *Clostridium* botulinum	Dissemination: Aerosolization in Tokyo Shoko Asahara was convicted of criminal activity Aum Shinrikyo ordered C. botulinum from a pharmaceutical company and attempted to acquire from Zaire outbreak under guise of a "humanitarian mission" Resulted in around 20 deaths and more than 4000 injuries
1995	Larry Wayne Harris, a white supremacist, ordered 3 vials of *Yersinia pestis* from the ATCC	*Yersinia pestis*	
1996	Diane Thompson, clinical laboratory technician, Dallas, TX	*Shigella dysenteriae* Type 2	Removed Shigella dysenteriae Type 2 from hospital's collection and infected co-workers with contaminated pastries in the office breakroom Infected 12 of her coworkers, she was arrested, convicted, & sentenced to 20 years in prison
1998	Richard J. Schmidt, a gastroenterologist in Louisiana	HIV	Convicted of attempted second degree murder for infecting nurse Janice Allen with HIV by injecting her with blood from an AIDS patient
1999	Brian T. Stewart, a phlebotomist	HIV	Sentenced to life in prison for deliberately infecting his 11-month-old baby with HIV-infected blood to avoid child support payments
2001	"Amerithrax"	*Bacillus anthracis*	Letters containing anthrax spores were mailed to media offices and senators Suspected perpetrator was a US DOD scientist 22 infected, 5 deaths

Date	Incident	Organism	Details
2003	Thomas C. Butler, United States professor	*Yersinia pestis*	30 vials of Y. pestis missing from lab (never recovered); Butler served 19 months in jail

3.10.2 Major Global Bioterrorism Incidents

North America: At least five incidents took place by 2005 in the United States after the Raccoon City Incident - one in Alaska; two on the West Coast; one on the east and the Harvardville Airport outbreak. Another two incidents took place in Canada and Mexico. There was also a planned bioterrorist attack on the United States by the Plaga-worshipping cult Los Iluminados by kidnapping the President's daughter, Ashley Graham, injecting her with a Queen Plaga and releasing her to the United States. Then, in 2013, another bioterror event occurred in Tall Oaks leading to the deaths of over 70,000 people including President Adam Benford. Roughly three years later, Glenn Arias unleashed the A-Virus in New York City in retalation of the drone strike during his wedding that killed his wife (Anonymous, 2021).

South America : By 2002, there was an incident in southern Brazil done by Javier Hidalgo and in the north of the continent (Perhaps Ecuador).

Eurasia : This continent was notable for it's bioterror incidents. By 2005, there are incidents in the United Kingdom; Norway and the Mediterranean in Europe. Russia suffered from a further three. A bioterrorist attack by the organization, "Il Veltro" occurred on the island of Terragrigia in 2004. In the Middle East, there was an incident near the Sinai. In Southern and Eastern Asia, there was an incident near the Iranian/Pakistani border; one by Downing in India; two in China; one in Japan; one in Malaysia and a final incident in Indochina. In 2013, there was a third incident in China. In the year 2012 during the Edonian Civil War, the C-Virus was given to the members of the Edonian Liberation Army and resulted in an outbreak of the virus. Some time before and after Resident Evil 6, a bioterrorist attack occurred in a Middle East country which involved the Napads. Another bioterror attack occurred somewhere in Singapore in an isolated school named Marhawa School with a Lepotica being used to infect the entire academy.

Oceania : Two incidents occurred in Australia and a third took place in New Zealand.

Africa : South; Central and Southern Africa are involved in bioterrorist incidents in 2005. Also in 2009, the Kijuju Autonomous Zone Incident occurred and the B.S.A.A was sent there to suppress the outbreak.

Unknown : An incident lasting over a year happened in the nearly uncharted Sejm Island.

3.10.3 Bioterrorism Incidents in India

Experts in bioterrorism last week recommended that India expand its disease surveillance network and its ability to monitor bioterrorism. They want to ensure that bioterrorist attacks are not passed over as natural disease outbreaks or outbreaks of unknown origin, or classified as an emerging infectious disease (Sharma, 2001). There is no proof that biological warfare attacks have been carried out against India, but all the factors are very much in place, Colonel Anantasubramanian Nagendra, head of microbiology at the armed forces medical college in Pune, told a conference on pathogens of biological warfare at the National Institute of Communicable Diseases in New Delhi.

Terrorist incidents have occurred across India and insurgents are fighting in Kashmir, with the support of Pakistan and Osama bin Laden's terror network. "We have to wake up," Colonel Nagendra said. The threat of biological warfare has been engaging the attention of Indian defence and medical experts for a long time. During the Indo-Pakistan war of 1965, a scrub typhus outbreak in northeastern India came under suspicion. India's defence and intelligence outfits were alert to the outbreak of pneumonic plague well known in biological warfare in Surat and bubonic plague in Beed in 1994, which caused several deaths and sizeable economic loss.

Dr. Kamal Datta, director of the National Institute of Communicable Diseases, alluded at the conference to some "suspicious" outbreaks, such as a 1996 outbreak of dengue in Delhi (10252 cases; 423 deaths), and the outbreak of unidentified encephalitis in Siliguri, eastern India, in February this year (66 cases; 45 deaths). Both outbreaks have baffled Indian researchers. Molecular characterization studies on the dengue-2 virus samples were done by researchers from the institute and from the All India Institute of Medical Sciences. "We compared the 1996 samples with 1967 samples and found that there is roughly 10% divergence and nearly 30 mutations," Dr Syed Pasha of the institute said. "Whereas the earlier isolate was genotype 5, the 1996 isolate is genotype 4 and is more virulent." "We never had such a major episode like the 1996 dengue haemorrhagic fever outbreak from any part of the country with so many cases at one particular point in time," said Mr Datta. "Has it come from outside India? We are still not very sure." "We cannot even exclude that possibility," said Dr Mahendra Yadav, director of the Indian Veterinary Research Institute.

Indian health and defence experts are not even sure about the origins of the 1994 plague episodes. "The *Yersinia pestis* strains that are percolating in established plague foci in India are very much less virulent and definitely different from the samples we have seen from the plague outbreak region," said Dr Harsh Batra, joint director of the Defence Research and Development Establishment, who led the studies on outbreaks of plague. In the absence of more data and samples, he refused to attribute the 1994 plague conclusively to external biowarfare.

3.10.3.1 The History of Germ Warfare in India

The raging advance of Covid-19, the desperate search for a vaccine, clueless politicians, economic crisis and widespread panic could also teach us how to respond to bioterror attacks. Is India ready? (Ravi Shankar and Bipindra, 2020).

The dying months of 2001 were bad for America and the world. Less than a month after Islamic terrorists crashed airplanes into the World Trade Center, 62-year-old photojournalist Bob Stevens was admitted in a Florida hospital on October 2, 2001. The initial diagnosis was meningitis but it was soon found to be poisoning by anthrax, a weapon of bioterrorism. A few days later, in India, the Postal Department received 17 "suspicious" letters believed to be infected with anthrax spores. Though many individuals and institutions received the envelopes with white powder, none of them tested positive. It was dismissed as a copycat hoax.

"Biological attacks, both state-sponsored and otherwise, are a real threat despite the many treaties prohibiting them. Though the Indian Army is trained to prepare for chemical, biological, radiological and nuclear attacks, the programmes are on the back burner due to lack of resources," says Centre for Joint Warfare Studies Director Lieutenant General Vinod Bhatia (retired), who was previously Director General, Military Operations. India, with its vast disorganised population, dismal health facilities and poor connectivity, is sitting on a virus time bomb. Though the fatality, infection and recovery rate of Covid-19, as the novel coronavirus is called, is comparatively low, experts are not sure full data is available.

3.10.3.2 Is India Ready for Germ Warfare?

As far back as in December 1998, India began to train its medical personnel to deal with the eventualities of bioterror attacks. Since it had ratified the 1972 United Nation's Biological and Toxin Weapons Convention, India has not executed a bioweapon programme. However, the Army does maintain defensive biological warfare equipment at protected sites. With extensive help from the advanced dual-use pharmaceutical industry and defence labs, the military is researching ways to counter germ warfare. India has the scientific capability to carry out a bio-offensive in case of a first strike, using delivery systems ranging from crop dusters to ballistic missiles.

"India does not hold or believe in nuclear, biological and chemical weapons. However, the National Disaster Management Authority has resources and laboratories to counter bio-aggression by a hostile country. Selective attacks would catch the enemy by surprise, inflict a psychological blow and impose a drain on medical resources necessary to attend the victims," says Major General Kumar. Sources say that India has a sophisticated globally acknowledged biotechnology infrastructure, and sufficient well-trained and knowledgeable scientists, most of who are adequately experienced in handling epidemics. It has numerous pharma production facilities and biocontainment laboratories with Biosafety Levels 3 and 4, according to NTI, a Washington DC-based think-tank.

DRDO is India's biodefence industry's core, whose top laboratory is the Defence Research and Development Establishment (DRDE) located at Gwalior in Madhya Pradesh. It is India's go-to institution for studies in toxicology, biochemical pharmacology and the development of antibodies against bacterial and viral agents. The DRDO works and focuses on countering biothreats such as anthrax, brucellosis, cholera, plague, smallpox, viral hemorrhage fever and botulism. Additionally, the government has established nuclear, biological, and chemical (NBC) warfare directorates in the armed forces, as well as an inter-services coordination committee to monitor their training and preparation. The military has set up an NBC cell at Army Headquarters as well.

However, DRDO's massive failures of its indigenous weapons programmes do not paint an inspiring picture. Says former Indian Air Force Group Captain Sandeep Mehta, "India's preparedness to tackle a bioterror attack ranges from poor to pathetic, and its capability is limited to helping relief providers who are then expected to deliver." The Biosafety Level 2 laboratory at the Institute of Preventive Medicine in Hyderabad provides guidance in preparing the government for a biological attack. However, Indian Army's Medical Corps specialists have publicly expressed reservations that Indian hospitals are inadequately prepared. CISF has been enabled to deploy specially trained first responders. In January 2003, the government announced changes in India's nuclear use doctrine, which now retains "the option of retaliating with nuclear weapons", after the discovery that al-Qaeda manuals taught the production and use of toxins.

After the December 2002 Parliament attack, an Indian parliamentary committee considered plans to make underground bunkers to protect MPs from nuclear and biological attacks. Then defence minister George Fernandes indicated that "the government has initiated necessary steps to ensure protection from a nuclear and bio-attack." In an apparent follow-up in August 2004, the then Home Minister Shivraj Patil indicated that Indian scientists were formulating a response to potential biological, chemical, and other non-conventional forms of terrorism. India has stringent export control regulations outlined in the Special Chemicals, Organisms, Materials, Equipment, and Technologies (SCOMET) guidelines. Its national export product control list, which identifies goods, technologies and services are subject to dual-use licensing requirements.

However in 2003, the US sanctioned two Indian companies charged with violating government regulations by supplying dual-use plant equipment to the Saddam Hussein regime for its chemical and biological weapons programmes, the NTI website says. In June 2015, India and the US signed a 10-year defence framework agreement for cooperation in the development of defence capabilities, including "a lightweight protective suit effective in chemical and biological hazard environments." In September last year, Defence Minister Rajnath Singh warned that bioterrorism is among the new threats facing the country and asked the Armed Forces Medical Services to find effective ways to deal with new threats posed by advancing battlefield technologies. Whether Covid-19 is a bioterror weapon which went awry or a virus that got away, the real threat of a humanity ending, manufactured contagion unleashed by hostile countries for world domination haunts governments, military leaders, scientists and security experts worldwide. In spite of sophisticated electronic surveillance, countermeasures, scientific research and human intelligence, the corona virus proves that the bugs are never too far to arrive at a location near you soon.

3.10.3.3 Preparedness Strategies Against Bioterrorism

Recently, in the wake of 9/11 and subsequent frequent terrorist attacks, the medical fraternity is concerned about some biological agents being used as weapons in future. Globally over the years the weapons used have shifted from swords to bullets and bombs and are now heading towards biological weapons. The considerable ease of production along with the immense capacity to create panic has attracted the so-called terrorists towards the use of biological agents as future weapons (Kumath Manish and Kumar Adarsh, 2009). If such weapons are used in densely populated areas, the risk of massive destruction in the form of life is too high. To combat these bio-threats, conventional principals of outbreak investigation supported with efficient laboratory systems are required to fall in place.

Globally various agencies are working hard to curb such problems in future. In India recently, after the formation of a national disaster management authority and the establishment of the apex trauma centre, gradually the awareness, preparedness and mitigation of these threats are being discussed frequently for tackling them. We are presenting the problem of bioterrorism with suggested remedial measures for this emerging threat through this poster presentation highlighting modern bioterrorist incidences, types of biological agents, bio-surveillance strategies, public health infrastructures, risk communication, public private partnership and medico-legal issues.

India follows differing approaches to tackle biological threats emerging from both natural and human-made sources. However, there are some glaring gaps in its ability to manage these risks. Our response to naturally occurring outbreaks has exposed deep fault lines to tackle biological threats (Shruti sharma, 2021). There are many reasons for this: We have a poor disease-surveillance network, which makes timely detection of outbreaks difficult. Inadequate coordination among ministries to prevent zoonotic infections complicates the response. Dismal investment in scientific research disincentivises researchers involved in the public health sector, who could help by developing capacities to identify, treat and vaccinate against threatening organisms.

India may have developed comprehensive guidelines to ensure safety of biotechnological research, but implementation of biosafety guidelines falls under the ambit of the Union Ministry of Science and Technology and the Ministry of Environment, Forest and Climate Change (MoEFCC). Researchers though are often affiliated to laboratories supported by the Indian Council of Medical Research and the Indian Council of Agricultural Research , research bodies set up under the Ministry of Health and Family Welfare (MoHFW) and the Ministry of Agriculture and Farmers' Welfare. This multiplicity of organisations operating under different ministries makes it difficult to ensure the implementation of biosafety guidelines across the country. With regard to potential bioterrorism, the country has no dedicated policy that deals with risks of intentional release of dangerous organisms. Multiple ministries are empowered to ensure protection of plants, animals and humans from disease-causing organisms. A full-time office of biological threats preparedness and response under the National Disaster Management Authority can be one possible alternative in this regard. This office could become the nodal agency that brings together experts from the various ministries, representatives from the private sector, and professionals from the academic and the scientific communities.

Early detection and rapid response to bioterrorism hinges upon close cooperation between public health authorities and law enforcement. However, we seem to be lacking such collaboration at the moment.

To keep India battle ready to counter a bioterrorism attack, the National Disaster Management Authority (NDMA), Govt. of India (GoI) has proposed a model instrument where participation of both government and private sectors is a sine qua non to defeat any such attack. As epidemics have the potential to wreak havoc on a large scale like chemical and nuclear weapons, a multi-sector approach has been envisaged to be adopted (Jayanth Murali, 2019). In India, several nodal ministries have been earmarked for dealing with epidemics caused by bioterrorism.

The central departments which are involved are Ministry of Health and Family Welfare (MoH and FW) which is one of the main ministries tasked with providing directions and technical support for capacity building, surveillance and early detection of an outbreak. Health ministry also helps in the deployment of Rapid Response Team's, human resources and logistic support. The Ministry of Home Affairs (MHA) is another nodal ministry which works in conjunction with MoH and FW. MHA is responsible for the assessment of the threat, intelligence inputs and implementation of preventive mechanisms. National Disaster Response Force (NDRF) is a specialised force constituted under MHA to deal with chemical, biological, radiological and nuclear (CBRN) attacks. It consists of 12 battalions, three each from the BSF and CRPF and two each from CISF, ITBP and SSB.

Each battalion has 18 independent specialist search and rescue parties of 45 personnel each including engineers, technicians, electricians, dog squads and medical/paramedics. The total strength of each battalion is 1,149. All the 12 battalions have been equipped and trained to respond to natural as well as human-made disasters. Battalions are also trained and equipped for response during

CBRN emergencies. Besides, the Ministry of Defence (MOD) manages the matters and consequences of biowarfare. Clinical case management is backed by the Indian Army, as they have several hospitals nationwide. They use ambulances, aircraft and ships to handle casualties. The Defence R&D Organization (DRDO) is actively pitched into developing protective systems and equipment for troops to contend against nuclear, biological and chemical warfare.

The other ministries which are also responsible are the Ministry of Environment, Forests and Climate Change for evaluation of short and long-term consequences, the Ministry of Agriculture, the Department of Animal Husbandry, Dairying and Fisheries, Urban or Rural Development Ministry and Department of Drinking Water Supply, Indian Railways etc. NDMA has been made responsible for promulgating policies on management and approving plans of different ministries (Jayanth Murali, 2019). And the National Crisis Management Committee (NCMC) coordinates and monitors responses in crises especially in disasters. It provides strong coordination and implementation of relief measures during disasters.

In order to enhance the preparedness of India towards bio-terror attacks, we should emulate the West and formulate training exercises and implement technologies that are being adopted in countries like the USA and Europe. For instance, the state of New York in September 2016, held the fourth edition of its massive emergency response training exercise called the Excelsior Challenge, which is a training exercise designed for police and first responders to become familiar with techniques and practices should a real incident occur. "In 1999, the University of Pittsburgh's Center for Biomedical None Informatics deployed the first automated bioterrorism detection system, called Real-Time Outbreak Disease Surveillance (RODS), which collects data from many data sources and uses them to detect a possible bioterrorism event at the earliest possible moment.

The city of New York has developed unique software to combat bioterrorism. The tool called the New York City Syndromic Surveillance System tracks disease progression throughout the City of New York. We would do better if we conduct exercises like "Dark Winter" to know our preparedness and inadequacies in overcoming bio-terror attacks (Jayanth Murali, 2019).

Bioterrorism is a low-probability, high-impact event. Biological agents imperil human, livestock and crop health, and thereby hurt the Indian economy. It's therefore imperative that we enhance our understanding of them with the intention of dealing with them effectively. Political awareness and public participation are indispensable for threat alleviation. In a country like India, where the population is on the increase by the day and has exceeded a billion, preventive strategies and efforts against bioterrorism require to be strengthened, improved and made useful. Such preparedness against bioterrorism will also capacitate our populace against natural epidemics, thus transforming India into a resilient society.

3.11 Categories of Agents

3.11.1 Category A

These high-priority agents pose a risk to national security, can be easily transmitted and disseminated, result in high mortality, have potential major public health impact, may cause public panic, or require special action for public health preparedness. This category includes

Agro-terrorism, Alliance for Biosecurity, 2001 anthrax attacks, Anthrax weaponization and Atlantic Storm.

Severe Acute Respiratory Syndrome (SARS) virus: SARS, though not as lethal as other diseases, was concerning to scientists and policymakers for its social and economic disruption potential. After the global containment of the pandemic, the United States President George W. Bush stated "...A global influenza pandemic that infects millions and lasts from one to three years could be far worse" (Cook Alethia and David B Cohen, 2008).

Tularemia or "rabbit fever": Tularemia has a very low fatality rate if treated, but can severely incapacitate. The disease is caused by the Francisella tularensis bacterium, and can be contracted through contact with fur, inhalation, ingestion of contaminated water or insect bites. Francisella tularensis is very infectious. A small number of organisms (10–50 or so) can cause disease. If F. tularensis were used as a weapon, the bacteria would likely be made airborne for exposure by inhalation. People who inhale an infectious aerosol would generally experience severe respiratory illness, including life-threatening pneumonia and systemic infection, if they are not treated. The bacteria that cause tularemia occur widely in nature and could be isolated and grown in quantity in a laboratory, although manufacturing an effective aerosol weapon would require considerable sophistication.

Anthrax: Anthrax is a non-contagious disease caused by the spore-forming bacterium *Bacillus anthracis*. The ability of Anthrax to produce within small spores, or bacilli bacterium, makes it readily permeable to porous skin and can cause abrupt symptoms within 24 hours of exposure. The dispersal of this pathogen among densely populated areas is said to carry less than one percent mortality rate, for cutaneous exposure, to a ninety percent or higher mortality for untreated inhalational infections. An anthrax vaccine does exist but requires many injections for stable use. When discovered early, anthrax can be cured by administering antibiotics (such as ciprofloxacin). Its first modern incidence in biological warfare were when Scandinavian "freedom fighters" supplied by the German General Staff used anthrax with unknown results against the Imperial Russian Army in Finland in 1916. In 1993, the Aum Shinrikyo used anthrax in an unsuccessful attempt in Tokyo with zero fatalities.

Smallpox: Smallpox is a highly contagious virus. It is transmitted easily through the atmosphere and has a high mortality rate (20–40%). Smallpox was eradicated in the world in the 1970s, thanks to a worldwide vaccination program. However, some virus samples are still available in Russian and American laboratories. Some believe that after the collapse of the Soviet Union, cultures of smallpox have become available in other countries. Although people born pre-1970 will have been vaccinated for smallpox under the WHO program, the effectiveness of vaccination is limited since the vaccine provides high level of immunity for only 3 to 5 years. Revaccination's protection lasts longer.[33] As a biological weapon smallpox is dangerous because of the highly contagious nature of both the infected and their pox. Also, the infrequency with which vaccines are administered among the general population since the eradication of the disease would leave most people unprotected in the event of an outbreak. Smallpox occurs only in humans, and has no external hosts or vectors.

Botulinum toxin: The neurotoxin Botulinum is the deadliest toxin known to man, and is produced by the bacterium *Clostridium botulinum.* Botulism causes death by respiratory failure and paralysis. Furthermore, the toxin is readily available worldwide due to its cosmetic applications in injections.

Bubonic plague: Plague is a disease caused by the *Yersinia pestis* bacterium. Rodents are the normal host of plague, and the disease is transmitted to humans by flea bites and occasionally by aerosol in the form of pneumonic plague. The disease has a history of use in biological warfare dating back many centuries, and is considered a threat due to its ease of culture and ability to remain in circulation among local rodents for a long period of time. The weaponized threat comes mainly in the form of pneumonic plague (infection by inhalation) It was the disease that caused the Black Death in Medieval Europe.

Viral hemorrhagic fevers: This includes hemorrhagic fevers caused by members of the family Filoviridae (Marburg virus and Ebola virus), and by the family *Arenaviridae* (for example Lassa virus and Machupo virus). Ebola virus disease, in particular, has caused high fatality rates ranging from 25–90% with a 50% average. No cure currently exists, although vaccines are in development. The Soviet Union investigated the use of filoviruses for biological warfare, and the Aum Shinrikyo group unsuccessfully attempted to obtain cultures of Ebola virus. Death from Ebola virus disease is commonly due to multiple organ failure and hypovolemic shock. Marburg virus was first discovered in Marburg, Germany. No treatments currently exist aside from supportive care. The arenaviruses have a somewhat reduced case-fatality rate compared to disease caused by filoviruses, but are more widely distributed, chiefly in central Africa and South America.

3.11.2 Category B

Category B agents are moderately easy to disseminate and have low mortality rates.

- Bipartisan Commission on Biodefense
- Brucellosis (*Brucella species*)
- Epsilon toxin of *Clostridium perfringens*
- Eugenio Berríos
- Food safety threats (for example, *Salmonella species, E coli* O157:H7, Shigella, Staphylococcus aureus)
- Glanders (*Burkholderia mallei*)
- Melioidosis (*Burkholderia pseudomallei*)
- Psittacosis (*Chlamydia psittaci*)
- Q fever (*Coxiella burnetii*)
- Ricin toxin from *Ricinus communis* (castor beans)
- Abrin toxin from *Abrus precatorius* (Rosary peas)
- Staphylococcal enterotoxin B
- Typhus (*Rickettsia prowazekii*)
- Viral encephalitis (alphaviruses, for example,: Venezuelan equine encephalitis, eastern equine encephalitis, western equine encephalitis)
- Water supply threats (for example, *Vibrio cholerae*, *Cryptosporidium parvum*)

3.11.3 Category C to W

Category C agents are emerging pathogens that might be engineered for mass dissemination because of their availability, ease of production and dissemination, high mortality rate, or ability to cause a major health impact : Nipah virus, Hantavirus and 1989 California medfly attack; Category E- Ethnic bioweapon; Category O - Operation Dark Winter; Category P - Project Bacchus and Project Coast; Category R - 1984 Rajneeshee bioterror attack, 2003 ricin letters and April 2013 ricin letters; Category S - Allen Shofe; Category W-Wood Green ricin plot.

3.12 Biosurveillance

It is an aspect of biodefense relating to the detection of biological threats, including bioterrorist threats. Biosurveillance primarily focuses on developing effective surveillance, prevention and operational capabilities for detecting and countering biological threats (Anonymous, 2020). A comparison of syndromic surveillance with traditional clinical recognition is presented in Table 8.

Table 8: Characteristics of bioterrorism-related epidemics that affect detection through clinical recognition versus syndromic surveillance.

Characteristics[a]	Clinical recognition[b]	Syndromic surveillance[c]
Duration and variability of incubation period	Broader distribution of incubation period increases likelihood that patients with short incubation-period disease would be diagnosed before a statistical threshold of syndromic cases is exceeded.	More narrow distribution of incubation period which leads to a steeper epidemic curve in the initial phases increase likelihood that statistical threshold would be exceeded sooner.
Duration of nonspecific prodromal phase	Shorter prodrome increases likelihood of recognition or diagnosis at more severe or fulminant stage.	Longer prodrome increases likelihood that increase in syndromic manifestations would be detectable and that recognition of more severe stage (at which a diagnosis is more apt to be made) would be delayed.
Presence or absence of clinical sign that would heighten suspicion of diagnosis	Presence increases likelihood of earlier clinical recognition and diagnosis (e.g., mediastinal widening on chest X-ray in inhalational anthrax or multiple cases of rare disease presenting at similar time).	Absence decreases likelihood that diagnosis would be considered clinically, increasing opportunity for earlier detection by means of syndromic surveillance.
Likelihood of making diagnosis in the course of routine clinical evaluation	If diagnosis is apt to be made in the course of a routine diagnostic evaluation (not dependent on clinical suspicion of specific bioterrorism infection), early diagnosis through clinical care is likely.	If diagnosis is dependent on the use of a special test that is unlikely to be ordered in the absence of clinical suspicion of diagnosis, then diagnosis in clinical care may be delayed, increasing the opportunity for early detection through syndromic surveillance.

[a]Infection or disease attributes that may affect detection of an epidemic.
[b]Increases likelihood of initial detection through routine clinical care and reporting.
[c]Increases likelihood of initial detection through syndromic surveillance.

3.13 Detection Systems

i. **Remote Detection Systems:** One way that remote detection systems monitor for potential biothreats from a distance is by the observation of aerosolized masses or clouds. Finding and evaluating the contents of a cloud is referred to as "standoff" detection (Nicholas E. Kman and Daniel J. Bachmann, 2012). On its most basic level, these detectors aim to alert military or civilian public health personnel to the presence of an approaching cloud.

ii. **Point detection systems:** Those that sample an environmental source, attempting to detect and identify the agent. Specific identification of a biologic agent by rapid diagnostics at the site of the attack can be done using immunologic assays, genetic assays, and mass spectrometry. These systems can further be differentiated by the type and location of sample collected.

iii. **BioWatch:** BioWatch is a US government program to detect biological agents present in the environment through a system of filters..

iv. **The Enhanced Passive Surveillance Program:** It is working to further develop, demonstrate and deliver a proof-of-concept surveillance system for identifying endemic, trans-boundary and emerging disease outbreaks in livestock. This expanded system will integrate multiple surveillance data streams from varied sources in real-time, helping to determine baseline prevalence rates for observed animal health issues, track variations in animal health status and identify trigger points to alert (Anonymous, 2020).

v. **The Biothreat Awareness APEX Program :** It will develop affordable, effective and rapid detection systems and architectures to provide advance warning of a biological attack at indoor, outdoor and national security events. The project will also engage local jurisdictions in the testing of integrated information sharing systems that ensure relevant data is quickly available to key decision makers during an event.

vi. **The Biosurveillance Information and Knowledge Integration Program:** It seeks to develop a Community of Practice (COP) Platform prototype that integrates multiple data streams to support decision making during a biological event as well as inform training tools for state responders. This effort uses New York City as a focus, while working toward national implementation. By examining workflow processes and developing a COP platform to include their highest priority information streams, we will not only enable a near term prototype capability for a highly motivated set of stakeholders, but serve as a potential template for other metropolitan areas. By examining NYC biosurveillance, one can gain an understanding of how to provide a biodefense solution given finite resources as well as extreme levels of competing information and priorities.

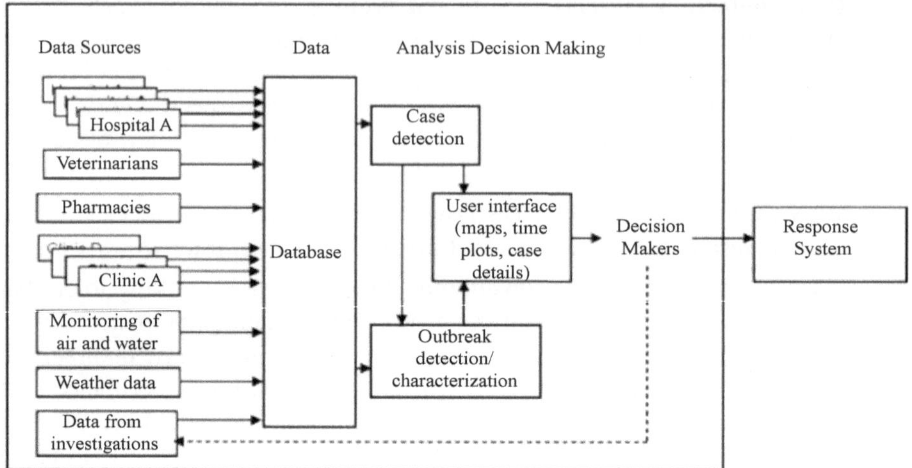

Fig. 5 : A generic biosurveillance system. The key elements of a biosurveillance system are data sources, a database, analysis, and decision making.

In 2000, Michael Wagner, the codirector of the RODS laboratory, and Ron Aryel, a subcontractor, conceived of the idea of obtaining live data feeds from "non-traditional" (non-health-care) data sources. The RODS laboratory's first efforts eventually led to the establishment of the National Retail Data Monitor, a system which collects data from 20,000 retail locations nation-wide (Wagner Michael and Espino Jeremy, 2004). Fig. 5 provides a generic biosurveillance system. The key elements of a biosurveillance system are data sources, a database, analysis, and decision making (Michael M. Wagner, 2006).

3.14 Protection Against Bioterrorism

Export controls on biological agents are not applied uniformly providing terrorists a route for acquisition (Kuntz Carol *et.al.*, 2013).

Aspects of protection against bioterrorism in the United States include:

i. **Detection and resilience strategies in combating bioterrorism:** This occurs primarily through the efforts of the Office of Health Affairs (OHA), a part of the Department of Homeland Security (DHS), whose role is to prepare for an emergency situation that impacts the health of the American populace.

ii. **Implementation of the Generation-3 automated detection system:** This advancement is significant simply because it enables action to be taken in four to six hours due to its automatic response system, whereas the previous system required aerosol detectors to be manually transported to laboratories.

iii. **Enhancing the technological capabilities:** Enhancing the technological capabilities of first responders is accomplished through numerous strategies.

The first of these strategies was developed by the Science and Technology Directorate (S&T) of DHS to ensure that the danger of suspicious powders could be effectively assessed, (as many dangerous biological agents such as anthrax exist as a white powder).

iv. **Enhanced equipment for first responders:** One recent advancement is the commercialization of a new form of Tyvex™ armor which protects first responders and patients from chemical and biological contaminants.

v. **Excelsior Challenge:** In the second week of September 2016, the state of New York held a large emergency response training exercise called the Excelsior Challenge, with over 100 emergency responders participating. According to WKTV, "This is the fourth year of the Excelsior Challenge, a training exercise designed for police and first responders to become familiar with techniques and practices should a real incident occur

3.15 Bioethics

Bioethics is the study of the ethical issues emerging from advances in biology and medicine. It is also moral discernment as it relates to medical policy and practice. Bioethics are concerned with the ethical questions that arise in the relationships among life sciences, biotechnology, medicine and medical ethics, politics, law, theology and philosophy. It includes the study of values relating to primary care and other branches of medicine ("the ethics of the ordinary"). Ethics also relates to many other sciences outside the realm of biological sciences (Wynar Bohdan,2007). Although bioethics began as a multi-disciplinary field of study, it is now a full-fledged discipline in its own right. As technology advances ever more quickly, and questions involving its implementation become more complex, bioethics will continue to grow and become increasingly important.

UNESCO as the lead UN agency in bioethics - has been promoting universal bioethical norms and principles, and assisted countries in the translation of those principles to concrete policy outcomes for their citizens. Stem cell research, genetic testing, cloning: progress in the life sciences is giving human beings new power to improve our health and control the development processes of all living species. Concerns about the social, cultural, legal and ethical implications of such progress have led to one of the most significant debates of the past century. The term coined to encompass these various concerns is bioethics.

3.15.1 Ethics Education Programme

By joining forces with UNESCO, partners can help to ensure support for the establishment and consolidation of the National Bioethics Committeesand to develop the ethics Education Programme which includes

i. **Core curriculum in bioethics:** Promoting the core bioethics curriculum developed by UNESCO to be introduced in Universities across a given region;

ii. **Ethics teachers training course:** Training a new generation of young scientists and professionals in the teaching of ethics at the University and School levels.

iii. Establish and link networks to promote the teaching of ethics.

The UNESCO Bioethics Programme ties together three key areas of work:

i. **Standard-setting**: the three declarations in this field, which have served as the blueprint for many regional and national legal instruments;

ii. **Global reflection**: the International Bioethics Committee that guides policymakers through a complex maze of ethical principles; and

iii. **Capacity-building**: using education and technical assistance for bioethics committees to build robust national bioethics infrastructures around the world.

Bioethics is concerned with a specific area of human conduct concerning the animate (for example, human beings and animals) and inanimate (for example, stones) natural world against the background of the life sciences and deals with the various problems that arise from this complex amalgam. Furthermore, bioethics is not only an inter-disciplinary field but also multidisciplinary since bioethicists come from various disciplines each with its own distinctive set of assumptions. While this facilitates new and valuable perspectives, it also causes problems for a more integrated approach to bioethics (Michael Boylan, 2020).

There are undesirable consequences, such as nuclear waste, water and air pollution, the clearing of tropical forests, and large-scale livestock farming, as well as particular innovations such as gene technology and cloning, which have caused qualms and even fears concerning the future of humankind. Lacunae in legal systems, for example, regarding abortion and euthanasia, additionally are a cause of grave concern for many people. Furthermore, moral problems which stem from a concrete situation, for example, gene-manipulated food, have given rise to heated public debates and serious public concerns with regard to safety issues in the past. There was and still is a need for ethical guidance which is not satisfied simply by applying traditional ethical theories to the complex and novel problems of the twenty-first century.

Policy making : The importance of the social and legal issues addressed in bioethics is reflected in the large number of national and international bodies established to advise governments on appropriate public policy. At the national level, several countries have set up bioethics councils or commissions, including the President's Council on Bioethics in the United States, the Det Etiske Råd (Danish Council of Ethics) in Denmark, and the Comité Consultatif National d'Ethique (National Consultative Bioethics Committee) in France. Elsewhere, as in the United Kingdom, there are a variety of different bodies that consider bioethical issues. The Nuffield Council on Bioethics has taken on the role of a national bioethics committee to a certain extent, but there also are national bodies that deal with specific fields, such as the Human Genetics Commission

Several international organizations also are involved in policy making on bioethical issues. The United Nations Educational, Scientific and Cultural Organization (UNESCO), for example, has an International Bioethics Committee; the Human Genome Organization has an Ethics Committee and the Council of Europe has issued the Convention on Human Rights and Biomedicine. The proliferation of such committees is evidence of the increasing political influence of the work performed by bioethicists. Indeed, acquaintance with developments in bioethics arguably is becoming an important aspect of national and global citizenship. At the same time, however, the role of bioethical experts on advisory or decision-making bodies has itself become a topic of study in bioethics.

3.15.2 International Bioethics Committee (IBC)

The International Bioethics Committee (IBC) is a body of 36 independent experts that follows progress in the life sciences and its applications in order to ensure respect for human dignity and freedom. It was created in 1993. The IBC provides the only global forum for reflection in bioethics. The Director-General of UNESCO convenes the IBC at least once a year. Through its sessions and working groups, the Committee produces advice and recommendations on specific issues that are adopted by consensus and are widely disseminated and submitted to the Director-General for transmission to the Member States, the Executive Board and the General Conference.

Who can participate in or attend IBC sessions?

- Member States, Associate Members of UNESCO may take part as official observers in the meetings of the IBC, while non-Member States that have set up a permanent observer mission may do so at the invitation of the Director-General.
- The United Nations and the other organizations of the United Nations system that have an agreement with UNESCO for reciprocal representation may take also part as observers in the meetings of the IBC.
- International governmental or non-governmental organizations with similar objectives to those of the IBC may be invited to take part as observers in the meetings of the IBC.
- Specialists or other relevant persons or groups may be consulted on matters within the competence of the IBC.
- Any individual or representative of an institution who wishes to attend a public session of the IBC should contact the Secretariat of the IBC to receive an invitation.

How does the IBC work?: Since 1998, the IBC has had Statutes defining its mandate, composition, etc.

The Director-General of UNESCO convenes the IBC at least once a year. Through its sessions and working groups, the Committee produces advice and recommendations on specific issues that are adopted by consensus and are widely disseminated and submitted to the Director-General for transmission to the Member States, the Executive Board and the General Conference.

Committee tasks

1. To promote reflection on the ethical and legal issues raised by research in the life sciences and their applications.
2. To encourage the exchange of ideas and information.
3. To encourage action to heighten awareness among the general public, specialized groups and public and private decision-makers involved in bioethics.
4. To co-operate with the international governmental and non-governmental organizations concerned by the issues raised in the field of bioethics as well as with the national and regional bioethics committees and similar bodies.
5. To contribute to the dissemination of the principles set out in the UNESCO Declarations in the field of bioethics, and to the further examination of issues raised by their applications and by the evolution of the technologies in question.

3.15.3 National Bioethics Committee

The Indian Council of Medical Research (ICMR) is the apex body in India for the formulation, coordination and promotion of biomedical research and has always been on the forefront promoting ethics in biomedical and health research in the country. ICMR Bioethics Unit located at National Centre for Disease Informatics and Research (NCDIR), Bangalore supports and foster initiatives towards ethical conduct of biomedical and health research in India. It is involved in development of ethical guidelines, policy and supplementary guidance related to various types of biomedical research conducted in the country. It is also involved in preparing tools and extends support for conducting programs for training and capacity building of ethics committees in the country. It also works losely with various government organizations, regulatory agencies and ministries in guiding as well as supporting initiatives for implementing ethical conduct of research in the country.

The Indian Council of Medical Research (ICMR) recently published the third revised guidelines "National Ethical Guidelines for Biomedical and Health-Related Research Involving Human Participants" in 2017 (Behera *et.al.*, 2019). The changes to the guidelines were needed to acculturate the rapid advances in the research environment and advances in science and technology. The revised guidelines propose substantial changes/ modifications compared to the previous version. These include the introduction of broad consent, ethical issues related to deception, review of multi-centric research by a single ethics committee and ethical issues involved in implementation research and other issues related to public health research. The revised guidelines also incorporate modifications and minor changes to the previous version. Although most of the changes in the revised guidelines are in parallel to most of the international guidelines, we have also highlighted the minor differences compared to other international guidelines.

3.16 Global Bioethics

Bioethics today is more and more a global enterprise. Not only challenges and problems are crossing borders (for example medical tourism, brain drain, international research, health disparities, marginalization of vulnerable populations, and climate change) but also approaches and solutions require international cooperation. The recent Ebola epidemic is a good example of the need for global governance (Henk A.M.J. ten Have and Bert Gordijn, 2013). This raises the question about the ethical framework and ethical methodologies for global bioethics. This long-term area of research is focused on examining the development of healthcare ethics into global bioethics, exploring not only the conceptual and methodological challenges of the globalization of bioethics, but also analyzing and clarifying the many new issues and problems that arise within a global perspective.

Globalization has meant that our ethical problems cross international borders, and consequently cultural, political and moral boundaries. Have leaves no doubt that global bioethics is of upmost relevance to 21st century living and conventional bioethical topics are discussed in a refreshing and relevant context. However there is no bioethical consensus in this new territory of global bioethics. For how can a comprehensive, binding consensus be reached on such diverse and complex topics as reproductive technology, commercial surrogacy and the rights of the dying? Whilst the laws in some countries may be different for cultural or religious reasons (for example, women travelling from Ireland to other countries in order to obtain an abortion), in many cases the fundamental driver for medical tourism is a clear discrepancy in economic power between the 'tourists' and their chosen country. This forms the very essence of medical tourism and, especially due to the implied risk of exploitation, should form a significant consideration for all concerned with global bioethics. Bioethics goes beyond the individual. The response to international disasters, regulation of tissue products and genetic research, all fall under the realm of global bioethics.

The field of bioethics has grown most rapidly in North America, Australia and New Zealand, and Europe. Cross-cultural discussion also has expanded and in 1992 led to the establishment of the International Association of Bioethics. A significant discussion under way at the start of the 21st century concerned the possibility of a "global" bioethics that would be capable of encompassing the values and cultural traditions of non-Western societies. Some bioethicists maintained that a global bioethics could be founded on the four-principle approach, in view of its apparent compatibility with widely differing ethical theories and worldviews. Others argued to the contrary that the four principles are not an appropriate basis for a global bioethics because at least some of them—in particular the principle of autonomy—reflect peculiarly Western values. Although the issue remains unresolved, the field as a whole continues to grow in sophistication. At the same time, the increasing pace of technological advances in medicine and the life sciences demands that bioethicists continually rethink the basic assumptions of their field and reflect carefully on their own methodologies.

3.16.1 Principles and Goals of Bioethics

One of the first areas addressed by modern bioethicists was that of human experimentation. The National Commission for the Protection of Human Subjects of Biomedical and Behavioral Research was initially established in 1974 to identify the basic ethical principles that should underlie the conduct of biomedical and behavioral research involving human subjects. However, the fundamental principles announced in the Belmont Report (1979), namely, respect for persons, beneficence and justice have influenced the thinking of bioethicists across a wide range of issues. Others have added non-maleficence, human dignity, and the sanctity of life to this list of cardinal values. Overall, the Belmont Report has guided research in a direction focused on protecting vulnerable subjects as well as pushing for transparency between the researcher and the subject. Research has flourished within the past 40 years and due to the advance in technology, it is thought that human subjects have outgrown the Belmont Report and the need for revision is desired.

Goals of bioethics: The field of bioethics has addressed a broad swath of human inquiry; ranging from debates over the boundaries of life (e.g. abortion, euthanasia), surrogacy, the allocation of scarce health care resources (e.g. organ donation, health care rationing), to the right to refuse medical care for religious or cultural reasons.

As a discipline of applied ethics and a particular way of ethical reasoning that substantially depends on the findings of the life sciences, the goals of bioethics are manifold and involve, at least, the following aspects:

1. **Discipline:** Bioethics provides a disciplinary framework for the whole array of moral questions and issues surrounding the life sciences concerning human beings, animals, and nature.
2. **Inter-disciplinary Approach:** Bioethics is a particular way of ethical reasoning and decision making that: (i) integrates empirical data from relevant natural sciences, most notably medicine in the case of medical ethics, and (ii) considers other disciplines of applied ethics such as research

ethics, information ethics, social ethics, feminist ethics, religious ethics, political ethics, and ethics of law in order to solve the case in question.

3. **Ethical Guidance:** Bioethics offers ethical guidance in a particular field of human conduct.
4. **Clarification:** Bioethics points to many novel complex cases, for example, gene technology, cloning, and human-animal chimeras and facilitates the awareness of the particular problem in public discourse.
5. **Structure:** Bioethics elaborates important arguments from a critical examination of judgements and considerations in discussions and debates.
6. **Internal Auditing:** The combination of bioethics and new data that stem from the natural sciences may influence–in some cases –the key concepts and approaches of basic ethics by providing convincing evidence for important specifications, for example, the generally accepted concept of personhood might be incomplete, too narrow, or ethically problematic in the context of people with disability and, hence, need to be modified accordingly.

3.16.2 Brief History and Origin of Bioethics

Historically speaking, there are three possible ways at least to address the history of bioethics. Each focuses on different aspects concerning the history of bioethics; however, one can only understand and appreciate the whole picture if one takes all three into account (Michael Boylan, 2020).

The origin of the notion of bioethics : It is commonly said that the origin of the notion of bioethics is twofold: (i) the publishing of two influential articles; Potter's "Bioethics, the Science of Survival" (1970), which suggests viewing bioethics as a global movement in order to foster concern for the environment and ethics, and Callahan's "Bioethics as a Discipline" (1973), in which he argues for the establishment of a new academic discipline, and (ii) discussions between Shriver and Hellegers about the need for an institute in which researchers should examine and analyze medical dilemmas by appealing to moral philosophy (1970). This institute was created in 1971 as the Joseph and Rose Kennedy Center for the Study of Human Reproduction and Bioethics, and is now known as the Kennedy Institute of Ethics. However, this oft-repeated story about the origin of the term bioethics is incorrect. The German theologian Fritz Jahr published three articles in 1927, 1928, and 1934 using the German term "Bio-Ethik" (which translates as "Bio-Ethics") and forcefully argued, both for the establishment of a new academic discipline, and for the practice of a new, more civilized, ethical approach to issues concerning human beings and the environment. Jahr famously proclaimed his bioethical imperative: "Respect every living being, in principle, as an end in itself and treat it accordingly wherever it is possible".

The origin of the academic discipline and institutionalization of bioethics : At the beginning, bioethics was seen as more or less identical with medical ethics– the latter notion is first mentioned by Thomas Percival (1803) and was mainly conducted by philosophers, theologians, and a few physicians. Animal ethics and environmental ethics are sub-disciplines which emerged at a later date. In the beginning, the great demand for medical ethics was grounded in reaction to some negative events, such as the research experiments on human subjects committed by the Nazis and the Tuskegee Syphilis Study (1932–1972) in the USA. At that time, bioethics was rather driven by urgent cases ("putting out fires") and did not consider systematic problems in healthcare such as the access to quality care. However, in reaction to these horrible events, the Nuremberg Code (1947) and the Declaration of Helsinki (1964) were created in order to provide researchers and physicians with ethical guidelines. In the case of the Tuskegee Syphilis Study (Belmont Report 1979) and other experiments in clinical research (Beecher 1966), one has to concede however that they were performed in the full knowledge of both sets of guidelines (and hence against the basic and most important idea of individual informed consent).

In particular, the idea of individual informed consent is due to the Prussian and German bureaucratic regulations of 1900/01 that appeal to the case of Dr. Albert Neisser in 1896 who publicly announced his concern about the possible dangers to the experimental subjects whom he vaccinated with an experimental immunizing serum (Zentralblatt der gesamten Unterrichtsverwaltung in Preussen 1901: 188). Additionally, the investigation of the death of 75 German children caused by the use of experimental tuberculosis vaccines in 1931 revealed that the mandatory informed consent was not obtained (Rundschreiben des Reichsministers des Inneren 28.2.1931, in: Sass 1989: 362-366). Baker rightly states that "the informed consent doctrine was thus originally a regulatory innovation created by Prussian bureaucrats; it was not an artifact of American legal or philosophical culture but of German bureaucratic culture. It was a German solution to problems created by the advances of German biomedical science" (Baker 1998).

Furthermore, influential books such as *Morals* and *Medicine*: The Moral Problems of the Patient's Right to Know the Truth, Contraception, Artificial Insemination, Sterilization, and Euthanasia and Ramsey's ground breaking book, The Patient as Person: Explorations in Medical Ethics (1970) argued that there was a serious and urgent need for thinking about complex moral issues in medicine and thereby facilitated the creation of the new academic discipline of medical ethics (also known as bioethics) (Michael Boylan, 2020).

3.17 Sub-disciplines in Bioethics

Bioethics is a discipline of applied ethics and comprises three main sub-disciplines: medical ethics, animal ethics, and environmental ethics. Even though they are "distinct" branches in focusing on different areas namely, human beings, animals, and nature, they have a significant overlap of particular issues, vital conceptions and theories as well as prominent lines of argumentation. Solving bioethical issues is a complex and demanding task (Michael Boylan, 2020).

Medical ethics : The oldest sub-discipline of bioethics is medical ethics which can be traced back to the introduction of the Hippocratic Oath (500 B.C.E.). Of course, medical ethics is not limited to the Hippocratic Oath; rather that marks the beginning of Western ethical reasoning and decision making in medicine. In medical ethics, one is concerned with the general ethical question of "what should one do" under the particular circumstances of medicine. In this respect, medical ethics is not different from basic ethics but it is limited to the area of medicine and deals with its particular state of affairs.

Animal ethics : The history of ethics is to some extent a history of who is and should be part of the moral community. Roughly speaking, in Antiquity only men of a particular social status were part of the moral community; several hundred years later, after a long and hard social struggle women achieved equal status with men, even though there is still a long way to go in many parts of most societies (for example, in the job market and equal pay for equal work).

Environmental ethics : Generally speaking, environmental ethics deals with the moral dimension of the relationship between human beings and non-human nature–animals and plants, local populations, natural resources and ecosystems, landscapes, as well as the biosphere and the cosmos. Strictly speaking, human beings are, of course, part of nature and it seems somewhat odd to claim that there is a contrast between human beings and non-human nature. At second glance, however, it seems reasonable to make this distinction because human beings are the only beings who are able to reason about the consequences of their actions which may influence the whole of nature or parts of nature in a positive or negative way.

Environmental ethics is commonly divided into two distinct areas: (i) anthropocentrism and (ii) non-anthropocentrism (or physiocentrism).

Anthropocentric approaches : Approaches such as virtue ethics and deontology stress the particular human perspective, and claim that values depend on human beings only. Anthropocentrism is faced with the objection of speciesism, the view that the mere affiliation to the species of Homo sapiens is sufficient to grant a higher moral status to human beings in comparison with animals. Singer has powerfully claimed, however, that the "mere difference of species in itself cannot determine moral status".

Non-anthropocentrism (or physiocentrism) approaches: These mainly consist of three main branches, viz., pathocentrism, biocentrism and ecocentrism, which can be further divided into an individualistic and holistic version. All non-anthropocentric approaches share the common claim that there are "objective" or more straightforward naturalistic values which are non-relational (intrinsic) and do not presuppose rational human beings. Nature (including animals) itself is valuable, independently of whether there are any human beings or not (non-instrumental view), even though one has to acknowledge the fact that many arguments about intrinsic value also have instrumental underpinnings.

3.17.1 The Four-principles

One of the most important approaches in bioethics or medical ethics is the four-principle approach developed by Tom Beauchamp and James Childress (1978, latest edition 2009). Since then they have continually refined their approach and integrated the points of criticism raised by their opponents. The four-principle approach, often simply called principlism, consists of four universal prima facie mid-level ethical principles, viz., autonomy, non-maleficence, beneficence and justice. Together with some general rules and ethical virtues, they can be seen as the starting point and constraining framework of ethical reasoning and decision making ("common morality"). Casuistry can in some sense be understood as a critical response to early versions of Beauchamp and Childress' four principles approach. The originators of these principles claim that none is more important than another, yet challenges have been laid against these principles on that basis as well as on other areas of disagreement. This paper looks at the nature of the most significant of those challenges. The four principles have withstood challenge now for nearly 30 years and still form the basis for most decision making in both the research setting and in clinical practice within the chiropractic profession (Dana J. Lawrence, 2007). However, professional understanding of the principles is not known and may provide a fertile area for further investigation.

The four principles currently operant in health care ethics had a long history in the common morality of our society even before becoming widely popular as moral action guides in medical ethics over the past forty-plus years through the work of ethicists such as Beauchamp and Childress (Thomas R. McCormick, 2020). In the face of morally ambiguous situations in health care the nuances of their usage have been refined through countless applications. Some bioethicists, argue that with the exception of nonmaleficence, the principles are flawed as moral action guides as they are so nonspecific, appearing to simply remind the decision maker of considerations that should be taken into account. Indeed, Beauchamp and Childress do not claim that principlism provides a general moral theory, but rather, they affirm the usefulness of these principles in reflecting on moral problems and in moving to an ethical resolution. Gert also charges that principlism fails to distinguish between moral rules and moral ideals and, as mentioned earlier, that there is no agreed upon method for resolving conflicts when two different principles conflict about what ought to be done. He asserts that his own approach, common morality, appealing to rational reflection and open to transparency and publicity is a more useful approach. In order to rigorously apply these principles in clinical situations their applicability must start with the context of a given case.

i. **Principle of respect for autonomy:** Any notion of moral decision-making assumes that rational agents are involved in making informed and voluntary decisions. In health care decisions, our respect for the autonomy of the patient would, in common parlance, imply that the patient has the capacity to act intentionally, with understanding, and without controlling influences that would mitigate against a free and voluntary act. This principle is the basis for the practice of "informed consent" in the physician/patient transaction regarding health care.

ii. **The Principle of Nonmaleficence:** The principle of nonmaleficence requires of us that we not intentionally create a harm or injury to the patient, either through acts of commission or omission. In common language, we consider it negligent if one imposes a careless or unreasonable risk of harm upon another. Providing a proper standard of care that avoids or minimizes the risk of harm is supported not only by our commonly held moral convictions, but by the laws of society as well (see Law and Medical Ethics). This principle affirms the need for medical competence. It is clear that medical mistakes may occur; however, this principle articulates a fundamental commitment on the part of health care professionals to protect their patients from harm.

iii. **The Principle of Beneficence:** The ordinary meaning of this principle is that health care providers have a duty to be of a benefit to the patient, as well as to take positive steps to prevent and to remove harm from the patient. These duties are viewed as rational and self-evident and are widely accepted as the proper goals of medicine. Â This principle is at the very heart of health care implying that a suffering supplicant (the patient) can enter into a relationship with one whom society has licensed as competent to provide medical care, trusting that the physician's chief objective is to help. The goal of providing benefit can be applied both to individual patients, and to the good of society as a whole. For example, the good health of a particular patient is an appropriate goal of medicine, and the prevention of disease through research and the employment of vaccines is the same goal expanded to the population at large.

    It is sometimes held that nonmaleficence is a constant duty, that is, one ought never to harm another individual, whereas beneficence is a limited duty. A physician has a duty to seek the benefit of any or all of her patients, however, a physician may also choose whom to admit into his or her practice, and does not have a strict duty to benefit patients not acknowledged in the panel. This duty becomes complex if two patients appeal for treatment at the same moment. Some criteria of urgency of need might be used, or some principle of first come first served, to decide who should be helped at the moment.

iv. **The Principle of Justice :** Justice in health care is usually defined as a form of fairness, or as Aristotle once said, "giving to each that which is his due." This implies the fair distribution of goods in society and requires that we look at the role of entitlement. The question of distributive justice also seems to hinge on the fact that some goods and services are in short supply, there is not enough to go around, thus some fair means of allocating scarce resources must be determined.

It is generally held that persons who are equals should qualify for equal treatment. This is borne out in the application of Medicare, which is available to all persons over the age of 65 years. This category of persons is equal with respect to this one factor, their age, but the criteria chosen says nothing about need or other noteworthy factors about the persons in this category. In fact, our society uses a variety of factors as criteria for distributive justice, including the following:

- To each person an equal share
- To each person according to need
- To each person according to effort
- To each person according to contribution
- To each person according to merit
- To each person according to free-market exchanges

3.17.2 Virtue Ethics

Virtue ethics is currently one of three major approaches in normative ethics. It may, initially, be identified as the one that emphasizes the virtues, or moral character, in contrast to the approach that emphasizes duties or rules (deontology) or that emphasizes the consequences of actions (consequentialism). Suppose it is obvious that someone in need should be helped. A utilitarian will point to the fact that the consequences of doing so will maximize well-being, a deontologist to the fact that, in doing so the agent will be acting in accordance with a moral rule such as "Do unto others as you would be done by" and a virtue ethicist to the fact that helping the person would be charitable or benevolent. While virtue ethics does not necessarily deny the importance of goodness of states of affairs or moral duties to ethics, it emphasizes moral virtue, and sometimes other concepts, like eudaimonia, to an extent that other theories do not.

In virtue ethics, a virtue is a morally good disposition to think, feel, and act well in some domain of life (Hursthouse Rosalind and Pettigrove Glen, 2018). Similarly, a vice is a morally bad disposition involving thinking, feeling, and acting badly. Virtues are not everyday habits; they are character traits, in the sense that they are central to someone's personality and what they are like as a person. A virtue is a trait that makes its possessor a good person, and a vice is one that makes its possessor a bad person. In ancient Greek and modern eudaimonistic virtue ethics, virtues and vices are complex dispositions that involve both affective and intellectual components. That is, they are dispositions that involve both being able to reason well about what the right thing to do is and also to engage our emotions and feelings correctly. For example, a generous person can reason well about when to help people, and also helps people with pleasure and without conflict.

In this, virtuous people are contrasted not only with vicious people (who reason poorly about what to do and are emotionally attached to the wrong things) and the incontinent (who are tempted by their feelings into doing the wrong thing even though they know what is right), but also the continent (whose emotions tempt them toward doing the wrong thing but whose strength of will lets them do what they know is right).

The revival of virtue ethics in moral philosophy in the last century was most notably spearheaded by Anscombe (1958), MacIntyre (1981), Williams (1985), Nussbaum (1988, 1990), and more recently Hursthouse (1987, 1999), Slote (2001), Swanton (2003), and Oakley (2009). This approach also deeply influenced the ethical reasoning and decision making in the field of bioethics, particularly in medical ethics (Michael Boylan, 2020).

Common objections to virtue ethics: Its theories provide a self-centered conception of ethics because human flourishing is seen as an end in itself and does not sufficiently consider the extent to which our actions affect other people. Virtue ethics also does not provide guidance on how we should act, as there are no clear principles for guiding action other than "act as a virtuous person would act given the situation." Lastly, the ability to cultivate the right virtues will be affected by a number of different factors beyond a person's control due to education, society, friends and family. If moral character is so reliant on luck, what role does this leave for appropriate praise and blame of the person?

3.18 Bioethics-Indian scenario and Feminist bioethics

The Indian Council of Medical Research formulates, coordinates and promotes biomedical research in India. In 1980, they formulated the first national ethical guidelines. They offer a number of different training programmes, from 1 day to 6 months. The council is developing a core curriculum for teaching bioethics, which would be applied uniformly in medical schools throughout the country. Drug development and ethics is also important in India, particularly now that the local pharmaceutical industry is expanding and so many drugs trials are outsourced to the country (Kumar Nandini, 2006). The council is also very active in encouraging the development of ethics review committees. Bioethics started emerging in the late 1960s and early 1970s as a way to push back, change the culture of medicine, and prevent future atrocities, but has now evolved into a full-fledged academic and professional discipline (Anonymous, 2021).

Indian Bioethics Project : It is an independent project under Gujarat National Law University, which aims at contributing to the Indian discourses on the intersections between Bioethics and Law. There has hardly been a time in Indian history when the Indian traditions have not been in transition, except perhaps hypothetically at the very beginning of Indian civilisation. India may appear to be a culture wrapped in its past, possessed by religion, and aglow in its mystical tradition. By dealing with questions of right and wrong, ethics operates quite closely with religion, and historically, there have indeed been many intersections between religion and ethics. Given the religious plurality of India, which also adds to the cultural diversity, Indian ethics and particularly Indian Bioethics may have lessons for the pluralistic world.

India speaks in many voices, and this polyphony makes it difficult to articulate a singular judgment. An evocative metaphor often employed to capture the spirit of the tradition is that of a palimpsest, an ancient parchment upon which generation after generation inscribe their messages layer upon layer, and yet no succeeding layer completely erases or hides what was previously written. Religion has historically influenced Indian society on a political, cultural and economic level. There is a sense of pride associated with the country's rich religious history as the traditions of Hinduism, Buddhism, Sikhism and Jainism all emerged out of India. The study shall serve as a guide for academics as an unprecedented teaching tool. It will help the policy makers frame effective policies by taking into account that the society is not homogenous and that religion plays a major role in ethical decision making, and even more so in the Indian Society.

Activities: The research will aim to enhance and advance conversations in bioethics in India by conducting the following activities:

- Conducting research in bioethics not only from a philosophical but a legal, social and theological and epistemological method as well
- Raise awareness and host discussions around bioethical issues among academics through conferences, round tables etc.

Feminist bioethics and criticism : Feminist bioethics can only be fully appreciated if one understands the context in which this increasingly important approach evolved during the late twentieth century. The social and political background of feminist bioethics is feminism and feminist theory with its major social and political goal to end the oppression of women and to empower them to become an equal gender (Michael Boylan, 2020).

Feminist bioethics developed from the early 1970s on and was initially focused on medical ethics; proponents later extended the areas of interest to issues in the fields of animal and environmental ethics . Important topics in feminist bioethics are concerned with the correct understanding of autonomy as relational autonomy, a strong focus on care, the claim for an equal and just treatment of women in order to fight against discrimination within healthcare professions and institutions on many different levels.

Without any doubt, feminist bioethics initiated discussion of important topics, provided valuable insights, and caused a return to a more meaningful way of ethical reasoning and decision making by, for example, not only adhering to universal moral norms. On the other hand, it can be doubted whether feminist bioethics, all things considered can be seen as a well-equipped and full moral theory. It may be that feminist bioethics complements the traditional ethical theories by adding an important and new perspective (that is, the feminist standpoint) to the debate. Feminist bioethics adds valuable insights to debates on various bioethical topics, but may not be a well-equipped full moral theory yet.

Bibliography

Abhishek Sharma, 2021. DNA Special: Did China plan use of coronavirus as biological weapon in 2015? May 11, 2021, 06:29 AM IST. SOURCE- DNA webdesk

Ainscough, M., 2012. Next Generation Bioweapons: Genetic Engineering and Biowarfare (April 2002). Available at http://www.au.af.mil/au/awc/awcgate/cpc-pubs/biostorm/ainscough.pdf (28 December 2012).

Anonymous, 2008. "Loner Likely Sent Anthrax, FBI Says". *Los Angeles Times*. Archived from the original on 7 April 2008. Retrieved 30 March 2008

Anonymous, 2012. Advances in Genetics Could Create Deadly Biological Weapons, Clinton Warns (07 July 2011). Available at http://www.breakingnews.ie/world/advances-in-genetics-could-create- deadly-biological- weapons-clinton -warns-531347.html (28 December 2012).

Anonymous, 2013. Anthrax Facts UPMC Center for Health Security". Upmc-biosecurity.org. Archived from the original on 2 March 2013. Retrieved 5 September 2013

Anonymous, 2013. Genetically Engineered Bioweapons: A New Breed of Weapons for Modern Warfare. Dartmouth Undergraduate. J.Sci., March 10, 2013- Applied Sciences, Winter 2013

Anonymous, 2020. "The World's Most Dangerous Weapon". *Washington Examiner*. 8 May 2017. Retrieved 15 April 2020.

Anonymous, 2020. Biosurveillance. CBD Focus Areas – Biosurveillance. The Science and Technology Directorate, US.

Anonymous, 2020a. "Biowarfare Against Agriculture". fas.org. Federation of American Scientists. Retrieved 15 February 2020.

Anonymous, 2021. Bioterrorism. *In: Miscellaneous Lore -Resident Evil WIKI*

Anonymous, 2021. COVID-19: What is the bioweapon theory all about?. Times Now Digital. Updated May 13, 2021 | 12:17 IST

Anonymous, 2021. GNLU Indian Bioethics Project. Gujarat National Law University, Gujarat, India

Anonymous, 2021. History of the Biological Weapons Convention". United Nations Office for Disarmament Affairs. Archived from the original on 16 February 2021. Retrieved 2 March 2021

Anonymous, 2021a Text of the 1925 Geneva Protocol". *United Nations Office for Disarmament Affairs*. Archived from the original on 9 February 2021. Retrieved 2 March 2021

Anonymous, 2021b. 1540 Committee". United Nations. Archived from the original on 2 March 2021. Retrieved 2 March 2021

Anonymous, 2021c. Biological Weapons Convention". United Nations Office for Disarmament Affairs. Archived from the original on 15 February 2021. Retrieved 2 March 2021.

Baker, R.,1998. A Theory of International Bioethics: The Negotiable and Non-Negotiable, Kennedy Instt. Ethics J., 8 : 233-273.

Barras, V. and Greub, G., 2014. "History of biological warfare and bioterrorism". *Clin. Microbiol. & Infection,* 20: 497–502

Baumann Peter, 2008. "Warfare gets the creepy-crawlies", *Laramie Boomerang,* October 18, 2008, accessed December 23, 2008.

Behera, S.K., Das, S., Xavier,A.S., Selvarajan,S. and Anandabaskar, N., 2019. Indian Council of Medical Research's national Ethical Guidelines for biomedical and health research involving human participants: The way forward from 2006 to 2017. *Perspect.Clin.Res.*, 10: 108-114

Carus, W.S., 2015. "The history of biological weapons use: what we know and what we don't". *Health Security*, 13: 219–55

Cook Alethia, H. and David B Cohen, 2008. "Pandemic Disease: A Past and Future Challenge to Governance in the United States." *The Rev. Policy Res.*, 25: 449-471

Covert, N.M.,2000. A History of Fort Detrick, Maryland (4th ed.). Archived from the original on 21 January 2012. Retrieved 20 December 2011.

Dan Kaszeta , 2020. No, the coronavirus is not a biological weapon. The Washington Post. April 27, 2020 at 6:30 p.m. GMT+5:30

Dana J. Lawrence, 2007. The Four Principles of Biomedical Ethics: A Foundation for Current Bioethical Debate. *J. Chiropractic Humanities*,14 : 34-40,

Edgar J. DaSilva, 1999 . Biological warfare, bioterrorism, biodefence and the biological and toxin weapons convention. EJB Electr. *J. Biotechnol.*, 2: 109-139

Franz, D., 2018. "The U.S. Biological Warfare and Biological Defense Programs" (PDF). Arizona University. Archived(PDF) from the original on 19 February 2018. Retrieved 14 June 2018

Frischknecht, F., 2003. The history of biological warfare. Human experimentation, modern nightmares and lone madmen in the twentieth century. *EMBO* Rep. 2003; 4 Spec No (Suppl 1):S47-S52. doi:10.1038/sj.embor. embor 849

Genuth. I. and Fresco-Cohen, L., 2006 . "BioPen Senses BioThreats". *The Future of Things.* Archived from the original on 30 April 2007

Gregory, B and Waag, D., 1997. Military Medicine: Medical aspects of biological warfare (PDF), Office of the Surgeon General, Department of the Army, Library of Congress 97-22242,

Hassani, M., Patel, M.C. and Pirofski, L.A., 2004. "Vaccines for the prevention of diseases caused by potential bioweapons". *Clin. Immunol.*, 111 : 1–15.

Henk A.M.J. ten Have and Bert Gordijn, 2013. *Handbook of Global Bioethics*. Springer Publishers, Dordrecht, 4 volumes, 1650 pages, ISBN 978-94-007-2511-9.

Hursthouse Rosalind and Pettigrove Glen, 2018. "Virtue Ethics". In Zalta, Edward N. (ed.). *Stanford Encyclopedia of Philosophy* (Winter 2018 ed.). Metaphysics Research Lab, Stanford University. Retrieved 2021-02-19

Hutchinson Robert C., 2017. "Homeland Security Today: EXCLUSIVE: Pandemic Crossroads". www.hstoday.us. Retrieved 2017-09-12

Hylton Wil, S., 2011. "How Ready Are We for Bioterrorism?" The New York Times. The New York Times Company, 26 Oct. 2011. Web.

Jayanth Murali, 2019. Bioterrorism in India. Deccan Chronicle. Mar 18, 2019

Jha, U.C. and Ratnabali, K., 2021. Biological Weapons: Coronavirus, Weapon of Mass Destruction? Kindle Edition, VIJ Books (India) Pty Ltd, 306 pp

Kelle, A., 2007. Synthetic Biology & Biosecurity Awareness *In: Europe. Bradford Science and Technology* Report No.9

King, J. C. H.,2016. Blood and Land: The Story of Native North America. Penguin UK. p. 73. ISBN 9781846148088.

Kirby Reid, 2008. "Using the flea as weapon", (Web version via findarticles.com), Army Chemical Review, July 2005, accessed December 23, 2008.

Kumar Nandini, 2006 . Bioethics activities in India. Eastern Mediterranean health journal = La revue de santé de la Méditerranée orientale = al-Majallah al-Sihhīyahyah li-sharq al-mutawassi. 12 Suppl 1. S56-65.

Kumath Manish and Kumar Adarsh, 2009. Preparedness strategies against Bioterrorism – An Indian Scenario . Pathology - *Journal of the RCPA*, 41: 83

Kuntz Carol, Salerno Reynolds and Jacobs Eli, 2013 . A Biological Threat Prevention Strategy: Complicating Adversary Acquisition and Misuse of Biological Agents. Center for Strategic & International Studies. ISBN 978-1-4422-2474-2

Kyle Brinson, 2019. The Bug Battalion. A descriptive of entomological warfare.

Lockwood, J.A.,2008. Six-legged Soldiers: Using Insects as Weapons of War. Oxford University Press. pp. 9–26. ISBN 978-0195333053

Mark Kortepeter, 2020. . A Defense Expert Explores Whether The Covid-19 Coronavirus Makes A Good Bioweapon. Forbes, Editors' Pick , Aug 21, 2020,10:08am EDT

Mayor, A.,2003.. Greek Fire, Poison Arrows & Scorpion Bombs: Biological and Chemical Warfare in the Ancient World. Woodstock, N.Y.: Overlook Duckworth.

Michael Boylan, 2020. Bioethics.*The Internet Encyclopedia of Philosophy* (IEP) (ISSN 2161-0002) https://www.iep.utm.edu/,

Michael M. Wagner, 2006. Biosurveillance. P. 3-12. *In: Handbook of Biosurveillance, Michael M. Wagner*, Andrew W. Moore, Ron M. Aryel (Edrs.) Academic Press, ISBN 9780123693785

Nicholas E. Kman and Daniel J. Bachmann, 2012. Biosurveillance: A Review and Update. Adv. Pre. Med., Vol. 2012, Article ID 301408, 9 pages

Novak Matt, 2016. "The Largest Bioterrorism Attack In US History Was An Attempt To Swing An Election". Gizmodo. Retrieved 2016-12-02

Pellerin Cheryl, 2011. "Global Nature of Terrorism Drives Biosurveillance." American Forces Press Service, 27 October 2011.

Petersen Carolyn, 1995. "Cryptosporidium and the food supply". *The Lancet,* 345: 1128–1129

Pradhan, S.D., 2021. Reports on the origin of Coronavirus: 'Smoking Gun' proof of the Chinese biological weapon. May 12, 2021, 5:25 PM IST in Chanakya Code, World, TOI

Preston Richard,2002. The Demon in the Freezer, Ballantine Books, New York. ISBN 9780345466631.

Rajagopalan, P.K., 2021. Biological Warfare Experiment- Biological warfare experiment in India and the curious case of yellow fever mosquitoes. Frontline –India's national magazine, January 29

Ravi Shankar and Bipindra, N. C., 2020. The history of germ warfare and how prepared India is. Indian Express, 22nd March 2020

Reeves, R. G., Voeneky, S., Caetano-Anollés, D., Beck, F. and C. Boëte, "Agricultural research, or a new bioweapon system?", Science, Retrieved 20th May 2019

Riedel, S., 2004. Biological warfare and bioterrorism: a historical review. Proc (*Bayl Univ Med Cent*). 17:400-406.

Roberge Lawrence, F., 2019. "Agrobioterrorism". Defense Against Biological Attacks: Volume II. Springer International Publishing: 359–383

Ryan C. Gott, 2018. The Sting of Defeat: A Brief History of Insects in Warfare. *Entomology Today*. June 13, 2018.

Sharma, R., 2001. India wakes up to threat of bioterrorism. *BMJ.*, 323 (7315):714.

Shruti sharma, 2021. Untold risks: India needs full-time body to tackle biological threats. DownToEarth, 30 March 2021

Skopec, R., 2020. Coronavirus is a Biological Warfare Weapon. *J. Clin. Stud. Med. Case. Rep.*, 7: 103.

Skopec, R.,2019. The Body Electric: Humans Have A 'Force Field' Around Their Bodies. Cancer Research Therapy.

Thavaselvam, D. and Swaran S. Flora, 2014. Chemical and biological warfare agents. *In: Biomarkers in Toxicology*, Ramesh C. Gupta (Ed.), Academic Press,485-519pp.

Thomas R. McCormick, 2020. Principles of Bioethics. UW Medicine, Dept. Bioethics and Humanities, School of Medicine, University of Washington

Treadwell, T.A., Koo, D., Kuker, K. and Khan, A.S., 2003. "Epidemiologic clues to bioterrorism". *Public Health Reports.* 118: 92–8.

Uzi Mahnaimi and Marie Colvin, 1998. "Israel planning 'ethnic' bomb as Saddam caves in". The Sunday Times

van Aken, J. and Hammond, E., 2003. Genetic engineering and biological weapons. New technologies, desires and threats from biological research. *EMBO Rep.* 2003;4 Spec No (Suppl 1):S57-S60. doi:10.1038/sj.embor.embor860

van Aken. J. and Hammond, E.. 2003. *EMBO Rep.*, 4: S57–S60.

Wagner Michael, M. and Espino Jeremy, 2004, "The role of clinical information systems in public health surveillance", *Healthcare Information Management Systems* (3 ed.), New York: Springer-Verlag, pp. 513–539

Wheelis Mark, Casagrande Rocco, and Madden Laurence, V., 2002. "Biological Attack on Agriculture: Low-Tech, High-Impact Bioterrorism Because bioterrorist attack requires relatively little specialized expertise and technology, it is a serious threat to US agriculture and can have very large economic repercussions". *BioScience.* 52 : 569–576.

William Cohen, 2006.. "Terrorism, Weapons of Mass Destruction, and U.S. Strategy". *Sam Nunn Policy Forum, University of Georgia.* Archived from the original on 2004-11-18. Retrieved 2006-07-12.

Wynar Bohdan, S.,2007. *American Reference Books Annual, Volume* 38. Libraries Unlimited.

Zanders Jean Pascal, 2009. "Research Policies, BW Development & Disarmament Archived 2008-10-06 at the Wayback Machine", Conference: "Ethical Implications of Scientific Research on Bioweapons and Prevention of Bioterrorism", European Commission, via BioWeapons Prevention Project.

4

Early Warning and Forecasting System

Forecasting and warning systems for pests and diseases play an important role in Integrated Pest Management (IPM) and in farm advisory work. Forecasting pest outbreaks during the growing season and applying control measures only when necessary is very important in conventional farming, but also in organic farming, to increase profitability while, at the same time, minimising negative effects on flora, fauna, and groundwater (Sigvald, 2014). Number of disease-warning systems have been developed and validated for dozens of crops. A few are in wide use by growers, and represent encouraging success stories in IPM. Most warning systems, however, have not made the transition from scientific validation to real-world application. his implementation shortfall is not limited to disease-warning systems, but also characterizes most other types of agricultural decision support systems (DSSs), whose rate of grower adoption is widely perceived as disappointing (Mark L. Gleason *et.al.*, 2008). One reason for this lack of implementation is failure of system developers to adequately involve growers and other end users in development and testing of the systems. The result of this disconnection between system developers and system users can be failure of the systems to meet growers' needs. One way to make DSSs more user-friendly is to streamline the logistics of using them. Several logistical barriers were described to grower adoption of disease-warning systems - inconvenience, added cost and labor, and difficulty in responding to advisories in a timely way - that are often encountered in the process of obtaining and handling weather data inputs. Grower adoption of warning systems is unlikely to increase unless these barriers come down.

Weather parameters that are commonly used as inputs to disease-warning systems include air temperature, rainfall, relative humidity (RH), and LWD. Additional variables, such as wind speed, wind direction, and solar radiation, are inputs to only a few warning systems. Measurements made at weather stations are the foundation for all types of weather inputs to warning systems. These data are sometimes measured on individual farms, either within or near a crop canopy of interest. Alternatively, data measured at regional networks of weather stations may be localized to a particular farm using Global Positioning Systems (GPS), Geographic Information Systems (GIS) methods, and meteorological models. Disease forecasting involves well organized team work and expenditure of time, energy and money. It is used as an aid to the timely application of chemicals. Among the first spray warning services to be established for growers, were the grapevine downy mildew forecasting schemes in France, Germany and Italy in the 1920s. Disease forecasting methods are available for several plant diseases including Grapevine downy mildew, Cucurbit downy mildew, Potato late blight, Tobacco blue mould, Apple and pear scab, Sugar beet root rot, Wheat brown (Leaf) rust, Corn bacterial wilt, Sugarbeet curly top etc.,(Table 9).

Table 9: Examples of forecast systems

Forecast system	Diseases
EPIDEM & TOMCAST / FAST	Early blight of potato and tomato caused by *Alternaria solani*
MYCOS	*Mycosphaerella* blight of Chrysanthemum
EPIVEN	Apple scab caused by *Venturia inaequalis*
PLASMO	Downy mildew of grapevine caused by *Plasmopara viticola*
EPICORN	Southern corn leaf blight caused by *Helminthosporium maydis*
BLIGHTCAST, USABlight, Indo-BlightCast & Phytoprog	Late blight of potato caused by *Phytophthora infestans*
NDAWN	Late blight and early blight
CERCOS	Cercospora blight of celery
EPIDEMIC	Designed for stripe rust of wheat, but could be modified for other diseases
MARYBLIGHT	Fire blight on apple caused by *Erwinia* amylovora
MELCAST	Watermelons (Anthracnose, Gummy stem blight), Muskmelons (*Alternaria*)

4.1 Information Needed for Disease Forecasting

Forecasting diseases is a part of applied epidemiology. Hence, knowledge of epidemiology (development of disease under the influence of factors associated with the host, pathogen) is necessary for accurate forecasting. The factors of epidemic and its components should be known in advance before forecasting is done. The information required for forecasting are as follows.

i. **Host Factors**

   a. Prevalence of susceptible varieties in the given locality

   b. Response of host at different stages of the growth to the activity of pathogen e.g. Some diseases are found during seedling stages while others attack grown up plants and

   c. Density and distribution of the host in a given locality. Dense populations of susceptible variety invite quick spread of an epidemic. Growing susceptible varieties in scattered locations and that too in a limited area are less prone to epiphytotic.

ii. Pathogen factors : a. Amount of primary (initial) inoculum in the air, soil or planting material b. Dispersal of inoculum c. Spore germination d. Infection e. Incubation period f. Sporulation on the infected host g. Re-dispersal / Dissemination of spores h. Perennating stages i. Inoculum potential and density in the seed, soil and air.

iii. Environmental factors a. Temperature b. Humidity c. Light intensity d. Wind velocity

Requirements or conditions for disease forecasting: There are five main requirements which must be satisfied before a useful and successful disease forecast is made.

i. The disease must cause economically significant damage in terms of yield loss or quality. Damage assessment is essential to develop strategy for controlling a disease. e.g., Annual estimation of yield loss caused by barley powdery mildew (*Erysiphe polygoni*) in England and Wales had ranged from 6 to 13 %. Potato late blight can cause a yield loss of 28% if the disease reaches the 75% stage by mid- August. Diseases like apple scab and potato common scab reduces the quality of the produce lower the value of the harvested crop and cause considerable financial loss to the growers.

ii. Control measures must be available at an economically acceptable cost.

iii. The disease must vary each season in the timing of the first infections and its subsequent rate of progress. If it does not, there is no need for forecasting.

iv. The criteria or model used in making a prediction must be based on sound investigational work carried out in the laboratory and in the field and tested over a number of years to establish its accuracy and applicability in all the locations where its use is envisaged.

v. Growers must have sufficient man power and equipments to apply control measures when disease warning is given. Long-term warnings or predictions are more useful than short-term warning or predictions

4.2 Forecasting Systems and Disease Examples

Disease forecasting requires field observations on the pathogen characters, collection of weather data, variety of the crop and certain investigations and their correlations. Usually the following methods are employed in disease forecasting

i. Forecasting based on primary inoculum: Presence of primary inoculum, its density and viability are determined in the air, soil or planting material. Occurrence of viable spores or propagules in the air can be assessed by using different air trapping devices (spore traps). In the case of soil-borne diseases the primary inoculum in the soil can be determined by monoculture method. Presence of loose smut of wheat, ergot of pearlmillet and viral diseases of potato can be detected in the seed lots at random by different seed testing methods. Seed testing methods can be used to determine potential disease incidence and enable decision to be made on the need for chemical seed treatment. The extent of many virus diseases is dependent on the severity of the preceding winter which affects the size of vector population in the growing season. e.g., Sugarbeet yellows virus.

ii. **Forecasting based on weather conditions:** Weather conditions viz., temperature, relative humidity, rainfall, light, wind velocity etc., during the crop season and during the intercrop season are measured. Weather conditions above the crop and at the soil surface are also recorded.

iii. **Forecasting based on correlative information:** Weather data of several years are collected and correlated with the intensity of the diseases. The data are compared and then the forecasting of the disease is done. Forecasting criteria developed from comparisons of disease observation with standard meteorological data have been provided for diseases like Septoria leaf blotch of wheat, fire blight of apple and barley powdery mildew.

iv. **Use of computer for disease forecasting :** In some advanced countries forecasting of disease is made by the use of computers. This system gives the results quickly. One such computer based programmes in the USA is known as 'Blitecast' for potato late blight.

Examples of well developed forecasting systems are given below.

a. **Early and late leaf spots of groundnut :** A technique has been developed for forecasting early and late leaf spots of groundnut in the U.S.A. When the groundnut foliage remains wet for a period greater than or equal to 10 h and the minimum temperature is 21°C or higher for two consecutive days or nights, the disease development is forecasted. A computer programme has been developed in the USA. This is accurate and is widely used in the USA. The data on hours for day with relative humidity (RH) of 95% and above and minimum temperature (T) during the RH observations for the period, for the previous 5 days are fed to the computer. Calculations are rounded to whole numbers. The T/RH index for each of the five days is calculated e.g., when hours of the RH 95% equal 10 and the minimum temperature during the period equals 21.1°C the T/RH index is 2.0 .The T/RH indices for days 4 and 5 are summed. If the total index exceeds 4 disease is forecasted. If the index is 3 or less no disease is forecasted.

b. **Late blight of potato:** In the USA a forecasting programme has been developed for late blight of potato (*Phytophthora infestans*). The initial appearance of late blight is forecasted 7 to 14 days after the occurrence of 10 consecutive blight favourable days. A day is considered to be blight favourable when the 5 day average temperature is 25.5°C and the total rainfall for the last 10 days is more than 3.0 cm. A computerized version (Blitecast) has also been developed in the U.S.A for forecasting potato late blight. Blitecast is written in Fortran IV. When a farmer desires blight cast (blitecast) he telephones the blight cast operator and reports the most recently recorded environmental data. The operator calls for the blight cast programmes in the computer viz., typewriter terminal and feeds the new data into the computer. Within a fraction of second the computer analyses the data and series of a forecast and spray recommendations to the operator who relays it to the farmer. The entire operation can be completed during standard three minutes telephone call. The system makes one of the four recommendations viz., no spray, late blight warning, 7 days spray schedule or 5 days spray schedule. The last 5 days spray schedule is issued only during severe blight weather. In West Germany, 'Phytoprog' is the programme used. It is based on measurements of temperature, relative humidity and rainfall. Phytoprog provides a negative prognose (an indication of when the usual routine spray application should be dispensed with).

c. **Blister blight of tea :** A system for predicting epidemics of blister blight of tea (Exobasidium vexans) has been developed based on the number of spores in the air in the tea plantation and the duration of surface wetness on the leaves. The duration of sunshine is negatively correlated with the duration of surface wetness.

   The following prediction equation has been developed.

   Y = 1.8324 + 0.8439 X1 + 0.9665 X2 – 0.1031 X3where, X1 = log % infection t2X2 = log % infection t2 – log infection t1Y = log of the number of spores in the air and t1 – t2 three weeksX3 = mean daily sunshine for a 7 days period preceding t2.

d. **Southern corn leaf blight:** 'Epimay' is a system for forecasting Southern corn leaf blight (Bipolaris maydis) based on conceptual model.

e. Rice blast: In India, forecasting rice blast (*Pyricularia ozyzae*) is done by correlative information method. It is predicted on the basis of minimum night temperature 20 to 26°C in association with high relative humidity of 90% or above. Computer based forecasting system has also been developed for rice blast in India.

f. **Wheat stem rust :** Forecasting wheat stem rust epidemic is done by analyzing therein samples which give precise data for inoculum present in the air. Moreover several wind trajectors are also prepared to survey the air-borne primary inoculums and its deposition. It has been observed that primary inoculum comes from South India, to the plains of Central and North India.

g. **Brown stripe downy mildew of corn:** The forecasting of brown stripe downy mildew of corn (*Sclerophthora raysise* var. *zeae*) which is restricted to India is done on the basis of average rainfall 100 to 200 cm or more accompanied by low temperature (25°C or less). Spore trapping techniques of acquisition of biological data for consecutive forecasting models are important. Spore traps have been widely used in to complete disease with weather conditions. Spore trapping is useful for understanding epidemiology of a disease and behaviour of the pathogens. This helps in developing models on dispersal of pathogens or on epidemiology of the disease and to formulate methods of management.

Methodology of spore trapping depends on the objectives including the biology of the pathogen, for infection forecasting, spore dispersal gradients and the management of the disease.

In epidemics of air-borne plant diseases the number of spores of the pathogen landing on the plant which depends on the number of spores in the atmosphere above the crop is an important factor for the quantitative sampling of the atmosphere (number of spores per unit volume of air).For trapping and estimating these studies different types of traps are used. The following spore traps are usually employed in trapping of fungal spores. Cylindrical rods or microscopic glass slides: It helps to gather data on the spore arrival in a locality. In this, the surface of microscopic slide is smeared with grease and made sticky. In the method, quantitative estimation is not possible as number of spores collected is very low.

i. **Hirst's volumetric spore trap :** In this instrument, air is sucked into at a controlled rate and impinged on to a glass slide moved by a clockwork mechanism past the orifice (Hirst 1952). It gives continuous count of spores in 24 h. The number of spores per unit volume of air at any given time can thus be calculated. 2. Rotorod sampler or rotorod spore trap (Sutton and Jones 1976) It comprises of a 'U' shaped rod attached at its mid point to the shaft of a small battery operated electric motor. In this equipment the surface of the rod is covered with a Vaseline strip of transparent cellophanes to catch spores which can be taken off and mounted on a glass slide. From the area of the strip and the speed of rotation, the volume of air samples can be calculated.

ii. **Anderson cascade spore sampler:** It is a device where Petri plates with nutrient agar are used to collect the spores.

iii. **Bourdillon slit sampler:** Air is sucked in a chamber by vacuum pump which strikes the rotating Petri dish containing agar medium. The agar medium retains the spores sucked in the air. Concentration of viable spores is calculated after counting germinated spores in the medium.

iv. **Burkard's 7 day volumetric spore trap :** This device records spores in the air drawn by a pump on 7 days basis on a cellophane strip wrapped on a drum rotating inside a chamber.

v. **Jet spore trap:** In the above sampling methods, the viability of the spores cannot be determined. To overcome this, living plants have been used as spore traps. A jet spore trap has spores impacted in an air jet into a column of still air through which they fall, to settle on leaf segments exposed at the base of the chamber. In this trap, suitable cultivars of host plants can be employed to determine number of viable spores.

An early warning system is an instrument for communicating information about impending risks to vulnerable people before a hazard event occurs, thereby enabling actions to be taken to mitigate potential harm, and sometimes, providing an opportunity to prevent the hazardous event from occurring. Early warning systems are routinely used for hazardous natural events such as hurricanes and volcano eruptions (Anonymous, 2021). In contrast, to date very little attention has been paid to the development of such systems for infectious disease epidemics. The goal of a disease early warning system would be to provide public health officials and the general public with as much advance notice as possible about the likelihood of a disease outbreak in a particular location, thus widening the range of feasible response options. The inherent dilemma of an early warning system, however, is that more lead time usually means less predictive certainty, as illustrated in figure 6.

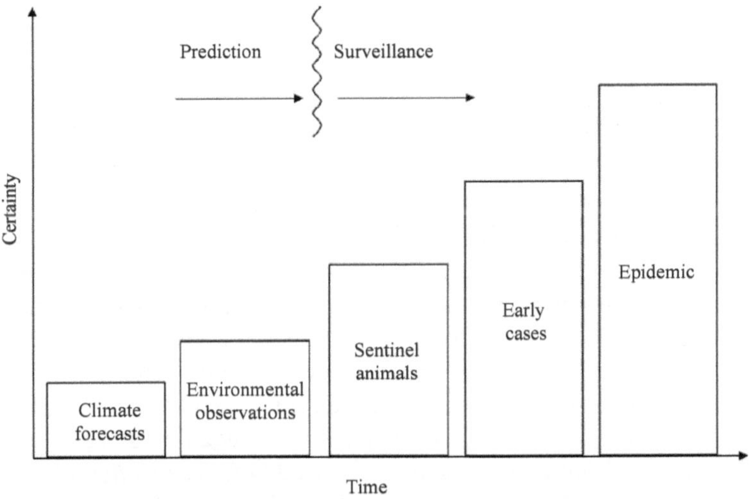

Fig. 6: Progression of the types of information that can used to indicate an impending disease outbreak

The most commonly used bell weather of an impending epidemic is the appearance of early cases of the disease in a population. In some instances, "sentinel" animals are placed in high-risk locations and monitored for evidence of infection, since infections among these animals will typically presage human cases. These "surveillance and response" approaches provide fairly high predictive certainty of an impending disease outbreak but often leave public health authorities with little advance notice for mobilizing actions to prevent further spread of the disease agent.

In contrast, ecological observations and climate forecasts can potentially be used in efforts to predict the appearance of a pathogen and thus allow opportunities to minimize its transmission. This approach is likely to have a much lower predictive value, however, given the uncertainties associated with most climate/disease relationships and the confounding influences of other factors. It is highly unlikely that precise predictions of an epidemic could be made solely on the basis of climate forecasts and environmental observations. Yet, this information can feasibly be used as the basis for issuing an alert (or a "watch") that environmental conditions are conducive to disease outbreak, which in turn can trigger intensive surveillance efforts for the area in question. If surveillance data then confirm the presence of the pathogen or an increase in its abundance subsequent warnings could be issued as needed. This watch/warning approach is analogous to the system used by the U.S. National Weather Service to alert communities to impending severe weather. A benefit of this multi-staged early warning approach is that response plans can be gradually ramped up as forecast certainty increases. This would give public health officials several opportunities to weigh the costs of response actions against the risk posed to the public.

4.3 Forecasting Methods for Specific Diseases

4.3.1. Late Blight on Potato

Interest in reducing fungicide use in potato cropping has led to increased efforts to develop methods for forecasting the risk of late blight (*Phytophthora infestans*) in potatoes. Late blight attacks in potatoes in Sweden differ greatly between regions and years. During the past decade, a few models for late blight in Sweden have been evaluated. The daily weather data needed are automatically transferred from the Swedish Meteorological and Hydrological Institute to the Swedish University of Agricultural Sciences. The forecasting method indicates that treatments recommended by the model are close to actual requirements and less than routine spraying, but further studies will have to be carried out to make the forecasts more reliable (Sigvald, 2014). About 20 years ago, the mating type A2 of *Phytophthora infestans* was discovered in Europe and in 1986 in Sweden. During the past decade there has been increased interest in investigating the occurrence of A2 in a number of countries in Europe, including Sweden. Since 1995, observations in Sweden and other European countries have indicated the presence of soil-borne inoculum of late blight. In potato fields in Sweden, both mating types A1 and A2 were isolated in 1996 and 1997 and oospores were found on leaves and stems, as well as in stolons. The importance of soil-borne inoculum of late blight for early infections is of great interest.

The main objective of the plant disease forecasting is to accurately predict the disease in the plant when the three factors - host, environment, and pathogen - all interact in such a fashion that disease can occur and cause economic losses. For the forecast, necessary environmental parameter which affects the growth of fungus was studied and those parameters are temperature, humidity and carbon dioxide concentration. Sensors were interfaced with the Arduino comprising of AT mega 328 microcontroller and the severity index has been calculated with the help of the standard equation of Beta Regression Model (Aradhana Rai *et.al.*, 2016). The main objective of the work predicted the risk of occurrence of the Late Blight disease in potatoes on the basis of the calculation of the severity/ infection index. Thus, for the purpose of the evaluation of the severity index value, the three of the sensors that is temperature, humidity and CO_2 concentration is studied and then interfaced with the Arduino and the corresponding values is observed. Based on the standard equation of the Beta Regression Model, the severity index is calculated which indicates the risk level to the crop. This method poses a greater importance as compared to the other methods used, because it has low cost, low power consumption and most importantly it is eco-friendly, and there is no need to wait for the results, as the sensor after interfacing with the Arduino produces immediate calibrated results.

4.3.2 Sclerotinia Stem Rot

Sclerotinia stem rot (*Sclerotinia sclerotiorum*) is one of the most important diseases on spring sown oilseed rape. During years with high humidity, the yield reduction can reach 60% in heavily infected fields, and in such years chemical treatment is profitable in 40-60% of fields. Sclerotinia stem rot can be effectively controlled by fungicide treatment during full flowering. Routine spraying is not profitable, since the cost of chemical treatment is high and disease incidence varies greatly between years and regions and also between fields within a region, thus justifying a forecasting system. A method for forecasting the incidence of Sclerotinia stem rot has been developed in Sweden (Sigvald, 2014). Besides field experiments and laboratory studies, data from more than 800 fields have been collected to improve the method, as well as for validation of the risk assessment.

The method is mainly based upon a number of risk factors, such as crop density, crop rotation, level of previous Sclerotinia infection (estimation of inoculum in soil), time of apothecia formation from sclerotia, rainfall during early summer and during flowering, and weather forecast. An initial risk assessment showed very good agreement between risk points and Sclerotinia stem rot incidence. To further improve the model, specific field data were analyzed by logistic regression. The results showed that the model could be simplified and still give very good or even better predictions. The method has been used by farmers during the past decade. It has been available in the form of risk points, but during recent years also interactively via the internet. Since the method was introduced about 20 years ago, the proportion of fields profitable to spray has increased, and today there is very good agreement between the need for spraying and actual treatments. The great success of the forecasting method for Sclerotinia stem rot is probably to a great extent due to the fact that advisory services and farmers have participated in validation and implementation.

4.3.3 Soybean Rust

Early-warning systems for plant diseases are valuable when the systems provide timely forecasts that farmers can use to inform their pest management decisions. To evaluate the value of the systems, this study examines, as a case study, USDA's coordinated framework for soybean rust (*Phakopsora pachyrhizi*) surveillance, reporting, prediction, and management, which was developed before the 2005 growing season (Michael *et.al.*, 2006). The framework's linchpin is a website that provides real-time, county-level information on the spread of the disease. The study assessed the value of the information tool to farmers and factors that influence that value. The information's value depends most heavily on farmers' perceptions of the forecast's accuracy. The study found that the framework's information is valuable to farmers even in a year with a low rust infection. It was estimated that the information provided by the framework increased U.S. soybean producers' profits by a total of $11-$299 million in 2005, or between 16 cents and $4.12 per acre, depending on the quality of information and other factors. The reported cost of the framework was between $2.6 million and almost $5 million in 2005.

4.3.4 Rust Diseases of Wheat

Wheat rust diseases pose one of the greatest threats to global food security. The fungal spores transmitting wheat rust are dispersed by wind and can remain infectious after dispersal over long distances. The emergence of new strains of wheat rust has exacerbated the risks of severe crop loss. Clare Allen-Sader *et.al.*, (2019) described the construction and deployment of a near real-time early warning system (EWS) for two major wind-dispersed diseases of wheat crops in Ethiopia that combines existing environmental research infrastructures, newly developed tools and scientific expertise across multiple organizations in Ethiopia and the UK. The EWS encompasses a sophisticated framework that integrates field and mobile phone surveillance data, spore dispersal and disease environmental suitability forecasting, as well as communication to policy-makers, advisors and smallholder farmers. The system involves daily automated data flow between two continents during the wheat season in Ethiopia. The framework utilizes expertise and environmental research infrastructures from within the cross-disciplinary spectrum of biology, agronomy, meteorology, computer science and telecommunications. The EWS successfully provided timely information to assist policy makers formulate decisions about allocation of limited stock of fungicide during the 2017 and 2018 wheat seasons. Wheat rust alerts and advisories were sent by short message service and reports to 10 000 development agents and approximately 275 000 smallholder farmers in Ethiopia who rely on wheat for subsistence and livelihood security. The framework represents one of the first advanced crop disease EWSs implemented in a developing country. It provides policy-makers, extension agents and farmers with timely, actionable information on priority diseases affecting a staple food crop. The framework together with the underpinning technologies are transferable to forecast wheat rusts in other regions and can be readily adapted for other wind-dispersed pests and disease of major agricultural crops.

4.3.5 Rice Diseases

Construction and database design of the rice disease early warning information system was presented based on WebGIS, and briefly analyzes the development software of WebGIS (Xuan Xie and Si-qing Yang, 2018). On the basis of establishing database of main rice diseases, with the Internet as the system platform, using WebGIS technology, ASP network programming technology, network database technology, artificial intelligence technology and other modern information system development method, it designs plant diseases and insect pests of rice early warning information system based on WebGIS. The system realizes the rice disease information collection, data transmission, disease diagnosis, effect of real-time release.

i. Blast: The impact of blast fungus (Magnaporthe oryzae syn. Pyricularia oryzae) is a global rice production issue. Liu *et.al.*,(2021) established a machine learning (ML)-based rice blast predicting model to decrease the appreciable losses based on short-term environment data. The average, highest and lowest air temperature, average relative humidity, soil temperature and solar energy were selected for model development. The developed multilayer perceptron (MLP), support vector machine (SVM), Elman recurrent neural network (Elman RNN) and probabilistic neural network (PNN) were evaluated by F-measures. Finally, a sensitivity analysis (SA) was conducted for the factor importance assessment. The study result showed that the PNN performed best with the F-measure ($\beta = 2$) of 96.8%. The SA was conducted in the PNN model resulting in the main effect period is 10 days before the rice blast happened. The key factors found are minimum air temperature, followed by solar energy and equaled sensitivity of average relative humidity, maximum air temperature and soil temperature. The temperature phase lag in air and soil may cause a lower dew point and suitable for rice blast pathogens growth. Through this study's results, rice blast warnings can be issued 10 days in advance, increasing the response time for farmers preparing related preventive measures, further reducing the losses caused by rice blast.

Many attempts have been made to develop models to forecast rice blast. A review of literature of the rice blast forecasting models revealed that 52 studies have been published, with the majority capable of predicting only leaf blast (Katsantonis Dimitrios *et.al.*, 2017). The most frequent input variable has been air temperature, followed by relative humidity and rainfall. Critical factors for the pathogenesis, such as leaf wetness, nitrogen fertilization and variety resistance have had limited integration in the development of these models. This review reveals low rates of model application due to inaccuracies and uncertainties in the predictions.

Five models are part of current operational forecasting systems in Japan, Korea and India. Development of in-field rice-specific weather stations, along with integration of leaf wetness and end-user interactive inputs should be considered.

ii. Bacterial blight : Bacterial leaf blight caused by *Xanthomonas* oryzae pv. oryzae is the most destructive and it has the potential to induce severe yield loss. From the last few years this disease is appearing regularly, and significantly reducing the crop quality and quantity (Muhammad Aslam Khan, 2020). The bacterium needs mechanical injury or enters through natural opening like stomata's and hydathodes for causing infection. Heavy doses of nitrogenous fertilizer, rains with strong winds and 30-35^{0}C temperature favour the infection process by the bacterium. Movement of contaminated water from infected to healthy plant and field helps in dissemination of disease. Rice stubbles, which survive in the winter, also harbour the organisms in the base of stems and roots. The bacterium is not seed-borne but it has been reported to survive in the rhizosphere of *Leersia* species and *Zizania latifolia*. Bacterial leaf blight has been affecting almost all varieties of rice, especially super Basmati which is a valuable source of foreign exchange for Pakistan.. A disease predictive model is required to forecast the bacterial leaf blight ahead of time so that rice growers are in a better position to take management decisions.

iii. Sheath blight: A computerized forecasting system (BLIGHTASIRRI) was constructed to estimate the development of rice sheath blight disease and the yield loss associates with the disease in the Philippines (Takashi Kobayashi *et al.*, 1995). This system which describes disease development as vertical and horizontal development was incorporated into all known data acquired from previous epidemiological studies. Vertical development expressed by the relative height of the uppermost lesions to the plant height was estimated on the basis of temperature, relative humidity and the susceptibility of rice sheath to the rice sheath blight fungus. On the other hand, horizontal development expressed by the percentage of diseased hills was determined from the combination of temperature, relative humidity and the number of sclerotia and tillers. Whole disease incidence was calculated from vertical and horizontal development by Hashiba's formula. To estimate the final yield loss, the relation between disease incidence and fully ripened kernels was analyzed. The model curve of vertical and horizontal development estimated by BLIGHTASIRRI was slightly different from actual development in paddy fields in the wet season of 1993 and 1994. Development of this system should enable to determine the need for fungicide application.

4.3.6 Bacterial Wilt of Tomato

Basic relationships for early diagnosis of tomato disease (bacterial wilt disease) by use of infrared thermal imagery was investigated (Kenji Chiwaki *et.al.*, 2005). Infrared thermal images of un-grafted (diseased) and grafted (non-diseased) tomato plants were obtained over the period extending before the appearance of external symptoms to until the diseased conditions. The temperature differentials between diseased and non-diseased tomato plants were analyzed by two methods. First, visual diagnosis of the pseudo-color thermal images; and second, histogram analysis of leaf temperatures from the images. Relative temperature increases of the diseased plants could be recognized visually on the thermal images as early as 5 days before the appearance of external symptoms..

4.3.7 Papaya Ring Spot Virus

Papaya ring spot virus-P (PRSV-P) disease is a most destructive and devastating disease of Papaya (*Carica papaya* Linn.). All the growth stages of papaya are vulnerable to PRSV infection. PRSV is style borne and transmitted by many species of aphid vectors in a non-persistent manner and spreads rapidly in the field (Pushpa *et.al.*, 2019). Monitoring of transitory aphids using yellow sticky traps in papaya orchard revealed, trapping of eight major aphid species viz., melon or cotton aphid (*Aphis gossypii*), cowpea aphid (*Aphis craccivora*), milkweed aphid (*Aphis nerii*), bamboo aphid (Astegopteryx bambusae), green peach aphid (*Myzus persicae)*, eupatorium aphid (*Hyperomyzus* carduellinus), cabbage aphid (*Brevicoryne brassicae*) and banana aphid (*Pentalonia nigronervosa*). The number of efficient vectors decides the epidemiology of PRSV incidence. Hence, a very high incidence of PRSV was observed between 14th and 23rd week after transplantation which was coincided with increased aphid population. It was concluded that the aphid population was peak with rainfall of below 8.00mm, maximum temperature of 28-35^{0}C, minimum temperature of 17- 23^{0}C, maximum and minimum relative humidity range of 60-90% and 30-50% respectively.

4.3.8 Potato Virus Y

Many aphid species are important vectors of non-persistently transmitted viruses. Most viruses transmitted by aphids in a non-persistent manner are believed to be acquired and transmitted within fields through probing by aphids flying from plant to plant. PVY is transmitted mainly by aphid species that do not feed on the crop that they inadvertently infect (Sigvald, 2014). After probing for 5–10 seconds, the aphids acquire the virus and are immediately able to transmit it to other plants. However, most aphids only remain infective for about 30 min. During the past decade there has been increasing interest in developing methods for PVY forecasting. The main variables used when forecasting the incidence of PVY include the number of winged aphids and their vector efficiency, the time of aphid migration in relation to plant age, and the availability of virus sources. However, the degree of spread of PVY varies greatly between regions and years. Nevertheless, few fields are infected so severely as to warrant rejection of the seed potatoes produced. There are often great differences in trap catches between years and regions. Results from suction trap catches in Sweden show that there is a rather weak relationship between the total number of winged aphids and proportion of PVY-infected progeny tubers, but the relationship is stronger when only main vectors of PVY are taken into account. The relationship is very good when the effect of mature plant resistance and proportion of virus sources are also taken into account.

4.3.9 Root-knot Nematode in Tomato and Potato

Root-knot nematodes (*Meloidogyne fallax* and *M. hapla*) cause significant reductions in potato yield by reducing tuber quality. Concentrations of *M. fallax* and *M. hapla* DNA in soil were determined by quantitative polymerase chain reaction following sampling at planting and harvest within 78 fields across 3 years in Australia (Hay Frank *et.al.*, 2015). *Meloidogyne* spp. were also detected using a tomato bioassay. *M. fallax* was more prevalent than *M. hapla* and DNA concentrations of *M. fallax* in soil were significantly higher in samples collected at harvest compared with those at planting. In contrast, *M. hapla* DNA in soil did not significantly change from planting to harvest. Using receiver operating characteristic curve analysis, *M. fallax* DNA in soil at planting and harvest was a highly accurate predictor of tuber damage at harvest and galling on tomato. Prediction accuracy for tuber damage was highest for *M. fallax* DNA compared with *M. hapla* or *M. fallax* + *M. hapla*. Both *Meloidogyne* spp. were detected in the peel of asymptomatic certified seed. For *M. fallax*, the addition of seaborne inoculum did not improve tuber damage predictions. This suggested that soil borne *M. fallax* populations contributed most substantially to tuber damage. These findings highlight the utility of this approach for predicting risk of crop damage from nematodes.

4.3.10 Root-knot Nematode in Soybean

Aerial imagery offers great potential as a predictive scouting method and could allow growers to better understand crop performance over time. The objective of this work was to predict where these might best be applied to other fields (Boney Barry, 2018). In a soybean field infested with M. incognita, apparent electrical conductivity was highly correlated with sand content, and treatments were applied the total length of the field, across two soil textural zones. Fluopyram and abamectin seed treatments were compared to seeds without a nematicide seed treatment (control) and seeds without a nematicide seed treatment but planted within 1,3-D treated areas.

Historical satellite images, normalized difference vegetation index (NDVI) and near infrared (NIR), from Sentinel-2 at 10- meter resolutions were compared to yield to determine if correlations with crop performance were evident over time. In 2016, treatment yields were not significant by zone, but yield was greater in the 1,3-D strips than all other treatments, while fluopyram and abamectin were not different than strips lacking nematicide ($P=0.001$). In 2017, 1,3-D strips had higher yield than all other treatments in both zones except for residual 1,3-D treatments that were applied in 2016 in Zone 2 (P=0.01). Fluopyram, abamectin, and the control treatments were not significantly different in Zone 1. Treatment effects for all treatments differed between the two textural zones (P=0.01). The distribution of *M. incognita* at harvest was uniformly distributed by treatments (P=0.08), suggesting that 1,3-D could be used as a two-year control and would be economically beneficial as a whole-field application when a susceptible soybean is planted. In 2017, NDVI and NIR observations were clustered, meaning that data are significantly positive for local spatial autocorrelation (P <0.05). Seven surrounding fields were further observed and 100% of analyses were clustered using NDVI and NIR images from multiple snapshots throughout the 2017 growing season. Initial analyses indicated correlations with yield, suggesting opportunities for prediction and that site-specific application of 1,3-D in these fields might be beneficial when susceptible soybean varieties are planted.

4.3.11 Cereal Cyst Nematode

Several simple agronomic practices such as crop rotation, timing of sowing, no-till management, and resistant cultivars can reduce damage by the cereal cyst nematode (CCN), (*Heterodera avenae*), but the damage levels are unpredictable. The objective of the present study was to develop a fast assay that would forecast the potential damage to wheat by CCN in the fields (Bonfil, *et.al.*, 2004). This forecast procedure was based on the "SIRONEM" bioassay with several modifications in the sampling and incubation of field soil samples and cultivation in them of wheat seedlings. For CCN disease assessment, a damage model that takes account of CCN infection and its damage to roots was developed. Field experiments were established to validate the resulting damage model, and showed a correlation between the forecast damage and field rating of CCN infection. This assay has advantages over other CCN estimation methods because it directly forecasts the potential crop damage in the field and it requires a relatively short period to obtain the data. According to the forecast damage and weather conditions, the most appropriate crop management regime, especially of sowing, can be chosen for each field assayed.

4.4 Insect Pests

4.4.1 Rice Brown Plant Hopper

Rice brown planthopper (BPH), *Nilaparvata lugens* is one of the most harmful insect pests in rice paddy fields, which causes considerable yield loss and consequent economic problems, particularly in the central plain of Thailand. Accurate and timely forecasting of pest population incidence would support farmers in planning effective mitigation. In this study, artificial neural network (ANN), random forest (RF) and classic linear multiple regression (MLR) analyses were applied and compared to forecast the BPH population using weather and host-plant phenology factors during the crop dry season from 2006 to 2016 in the central plain of Thailand (Skawsang *et.al.*, 2019). Data from satellite earth observation was used to monitor crop phenology factors affecting BPH population density. An ANN model with integrated ground-based meteorological variables and satellite-derived host plant variables was more accurate for short-term forecasting of the peak abundance of BPH when compared with RF and MLR, according to a reasonably validating dataset (RMSE of natural log-transformed (ln) BPH light trap catches = 1.686, 1.737, and 2.015, respectively). This finding indicated that the utilization of ground meteorological observations, satellite-derived NDVI time series, and ANN have the potential to predict BPH population density in support of integrated pest management programs. These results can be applied in conjunction with the satellite-based rice monitoring system developed by the Geo-Informatic and Space Technology Development Agency of Thailand (GISTDA) to support an effective pest early warning system.

4.4.2 Cotton Bollworms

The occurrence of cotton pests and diseases has always been an important factor affecting the total cotton production. Cotton has a great dependence on environmental factors during its growth, especially climate change. Xiao *et.al.*, (2019) used the common Aprioro algorithm to find the association rules between weather factors and the occurrence of cotton pests. The problem of predicting the occurrence of pests and diseases was formulated as time series prediction, and an LSTM-based method was developed to solve the problem. The association analysis revealed that moderate temperature, humid air, low wind spreed and rain fall in autumn and winter were more likely to occur cotton pests and diseases. The discovery was then used to predict the occurrence of pests and diseases. Experimental results showed that Long Short Term Memory (LSTM) network performs well on the prediction of occurrence of pests and diseases in cotton fields, and yields the Area Under the Curve (AUC) of 0.97.

Suitable temperature, humidity, low rainfall, low wind speed, suitable sunshine time and low evaporation are more likely to cause cotton pests and diseases. Based on these associations as well as historical weather and pest records, LSTM network is a good predictor for future pest and disease occurrences. Moreover, compared to the traditional machine learning models (i.e., SVM and Random Forest), the LSTM network performs the best.

Pink bollworm: There is a global concern about the effects of climate change driven shifts in species phenology on crop pests. Using geographically and temporally extensive data set of moth trap catches and temperatures across the cotton growing states of India, predicted the phenology of cotton pink bollworm *Pectinophora gossypiella* (Saunders). The approach was centered on growing degree days (GDD), a measure of thermal accumulation that provides a mechanistic link between climate change and species' phenology.

4.4.3 Diamondback Moth

In order to implement an integrated pest management (IPM) program for the diamondback moth (DBM), Kazunobu Okadome (2016) attempted to develop a simulation model to forecast DBM population fluctuations in cabbage fields. The egg density was estimated by a regression equation between the number of moths caught by a pheromone trap and the egg density in cabbage fields. The development of immature stages was started from each date of oviposition. Developmental velocity was assessed by a thermal constant using the daily mean temperature. Survival of immature stages was assumed to be affected by rainfall and insecticide spraying. The effects of the mortality factors were numerically evaluated using previously reported data. The number of individuals in each stage was calculated on a daily basis with a matrix model using a personal computer. From the results of a validity test, the model predicted well the actual population fluctuations in a cabbage field treated with pesticides in the cultivation period from spring to early summer, but it was not successful in predicting these fluctuations in autumn. Although the model needs to be improved in the future, it will be a useful tool for implementation of an IPM system for the DBM.

4.5 Impact of Climate Change on Pests and Diseases of Field Crops

Climate has a great influence on the occurrence of pests and diseases of field crops. Dry conditions are often favourable for insect pests, and wet conditions for fungal diseases, but there are several other factors which have a great influence on pest and disease development. Calculations of the damage to crops as a function of climate are complicated, and research within this area is working to develop methods for both understanding and predicting the effects of climate change on the dynamics of insects and diseases (fungal, viral, bacterial and nematodes) and on the damage they cause to crops. Insect attacks on crops will probably increase in the future (Sigvald, 2014). There are several reasons for this. Higher temperature will increase the number of generations during the growing season, and warmer climate during the winter will be favourable to insects and they will probably survive and therefore be more numerous in the spring. Aphids are very important in many countries of the world not only because of direct damage since they feed on the crop, but also because of indirect damage through the ability to transmit various viruses. In Sweden there are more than 50 aphid species of great importance for various crops. Some of these species will probably survive during mild winters, and most regions of Sweden will probably experience increased problems with damage caused by insects and virus diseases, but the increase will be greatest in southern Sweden and in dry areas.

Insects will be active considerably earlier in the spring than at present since the growing period will be extended. However, in some regions more rain is expected in late winter and early spring, and therefore sowing time can be delayed and earlier attack by insects can be expected in comparison with developmental stage of the crop. The greater numbers of aphids at spring sowing and the fact that spring crops will be exposed to virus diseases at an earlier stage of development will increase the need for pesticides unless there is an increase in other methods, such as use of resistant varieties.

Aphids and virus attacks in autumn are currently limited, but the future climate will bring about great changes. The number of aphids (vectors of virus diseases) is relatively low at present, but milder autumns and 3–4°C higher winter temperature will have a great influence. Some aphid species such as *Rhopalosiphum padi* will probably survive during the winter on grasses and winter cereals. Winter wheat and winter barley can be infected with BYDV by R. padi, which is an important vector of BYDV. Sitobion avenae will also contribute to transmission of BYDV on winter cereals. This will increase damage by aphids and viruses and will increase the need for insecticides in winter cereals. Aphids can also transmit viruses of winter oilseed rape, and the future warmer climate will probably increase numbers of such vectors and increase problems with viruses in this crop.

In a future warmer climate, there will probably be more aphids in the autumn, while newly introduced spring-sown crops such as maize that grow long into the autumn can act as a green bridge for viruses from spring-sown to winter-sown crops. Therefore virus attacks will increase on winter wheat and winter barley. New insect pests will probably become established in Sweden, depending on the crops grown and on winter conditions, but it is difficult to predict the insect species involved and a monitoring system is needed to follow developments in this area. Colorado beetle is one example of an insect pest that will probably be introduced in the coming 20 years. This will cause great problems for potato growers.

In most crops, moisture and higher temperature are favourable to fungal diseases. In some regions of Sweden with more rain during the growing season in future, we can expect greater attacks of late blight in potatoes. On the other hand, there may be drier conditions in south-east Sweden and thereby less problems than today. Winter cereals will be particularly vulnerable, since they will have a long infection period in the autumn and thus diseases such as brown rust will increase. For spring cereals, the effect can be less than at present in areas with a predicted relatively dry early summer period, such as southern areas of the country. In northern Sweden, fungal diseases of cereals will probably increase due to the generally wetter and warmer climate. In the future warmer climate with more aphid species also in northern parts of Sweden, seed potatoes will run a greater risk of virus attacks than at present. The need may then arise to establish special areas for seed potato production in which cultivation of ordinary commercial potato crops with a high proportion of virus-infected potato plants is restricted. Increased incidence of different insect pests on most crops will increase the use of pesticides, an undesirable development from a number of perspectives. Improved cropping systems, increased use of resistant varieties, and a good crop rotation to decrease the occurrence of pests and diseases will therefore be of increasing importance.

4.6 Warning Systems for Insect Pests and Diseases

There is a great need for effective warning systems for pests and diseases. Growers need information about the risk of attack by different pests and diseases long before spraying. Warnings are especially critical when high levels of attack are expected, which would lead to severe losses if farmers were unprepared (Sigvald, 2014). For more than 25 years a warning system has been under development in Sweden in a partnership between the Swedish University of Agricultural Sciences and the Swedish Board of Agriculture. In each of five regions, there is a Regional Plant Protection Centre to organise the work and handle the local information about the actual situation, which is then presented via the internet for different pests and diseases. Such information and availability of forecasting methods for different pests and diseases is very important in IPM. In southern and central Sweden, more than 1200 fields representing a variety of crops are inspected weekly. When the weekly analysis is ready, a summary is made available to the farm advisory service in the region via the internet. The day after the field evaluations are received, each Plant Protection Centre holds a telephone conference with advisors in the region concerned.

An invention discloses a disease forecasting and warning system which comprises a slave computer, a single chip control module, a storage module, a clock module, a liquid crystal display module, a solar power supply module, a GPRS (general packet radio service) module, a meteorological data acquisition module and a principal computer, wherein the principal computer is orderly connected with a GSM module, a mobile communication network and a mobile phone of a peasant, and the meteorological data acquisition module is connected with a temperature and humidity sensor, a leaf surface humidity sensor and a rain gauge. Through the scheme, the disease forecasting and warning system can acquire meteorological information of different geographic positions, automatically forecast the invasion of common diseases of plants, and send out early disease warming to corresponding stations from a long distance. The meteorological information and the information of the disease invasion and the like can be automatically sent to the principal computer through a wireless network, and can be also sent to the mobile phone of the peasant in the form of text messages at the same time. Moreover, the disease forecasting and warning system can enable dispersive stations to accurately collect data in time, and has better capability of adapting to the environment and anti-jamming capability.

4.6.1 Prediction Caution Systems of Diseases-Case Studies

Fundamental purpose of the present invention is to provide a kind of prediction caution system of disease, this system is when suffering the infecting of disease, system can carry out early warning automatically, it sends in the phone number of host computer database according to setting the form of relevant information with character and data, also can send in peasant household's mobile phone in the mode of short breath simultaneously, and the website that disperses is carried out accurately and timely Data Collection, preferably adaptive capacity to environment and anti jamming capability are arranged.

4.6.1.1 Single Chip Control Module

For addressing the above problem, the present invention adopts following technical scheme: a kind of prediction caution system of disease, comprise slave computer, and be arranged on single chip control module in the slave computer, and respectively with the memory module of single chip control module parallel join, clock module, LCD MODULE, solar powered module, GPRS module and weather data acquisition module, reach the host computer that is connected by mobile Internet with slave computer, described host computer is connected with gsm module in turn, mobile communication network and peasant household's mobile phone, described meteorological data collection module is connected with Temperature Humidity Sensor, blade face humidity sensor and digital rain gage. Further, include more than one disease forecast models in the system of described host computer. Further, described host computer includes following functions: website control, database information operate and empty database, website control, database information operate, empty database storage, the prediction of disease and the transmission of information of forecasting, and the expert intervenes the sending function of information immediately. Further, the information that described peasant household mobile phone receives can be medial temperature, average air humidity, whole day accumulative total rainfall, blade face humidity, disease infect dose, infects grade, and information is intervened in the suggestion that the expert provides immediately. Further, described slave computer quantity is more than one.

The beneficial effect of the prediction caution system of disease of the present invention is: pass through technique scheme, this system is when suffering the infecting of disease, system can carry out early warning automatically, it is according to carrying out access in the database of setting the form of relevant information with character and data and send to host computer, also can regularly send in peasant household's mobile phone in the short mode that ceases simultaneously, and the website that disperses carries out Data Collection accurately and timely, and preferably adaptive capacity to environment and anti jamming capability are arranged.

The beneficial effect of the prediction caution system of disease of the present invention is: pass through technique scheme, this system is when suffering the infecting of disease, system can carry out early warning automatically, it sends in the phone number of host computer database according to setting the form of relevant information with character and data, also can send in peasant household's mobile phone in the mode of short breath simultaneously, and the website that disperses carries out Data Collection accurately and timely, and preferably adaptive capacity to environment and anti jamming capability are arranged.

4.6.1.2 Yuktix GidaBits

Yuktix a provider of data intelligence tools for agriculture has launched its plant disease forecast platform, Yuktix GidaBits. The platform provides advance disease warning and management advice to farmers (Rajeev Jha, 2021). The GidaBits platform uses Yuktix IoT technology and ankiDB data analysis software to predict weather linked diseases in advance. Most of plant pathogens thrive in certain conditions. GidaBit used the accurate 24x7 weather and soil monitoring along with crop specific disease models to deliver advance warning about impending diseases to grower phones. Yuktix mobile app provides right chemical recommendations and best R&D package of practices for the specific crop. Farmers can also take pictures and share with a researcher via Yuktix platform for disease management recommendations. With a changing climate and shifting weather patterns, the problems are only going to be acerbated in the future. There is a significant body of research linking changing weather patterns to occurrence of diseases. There are many early warning systems hosted by research organizations using statistical models.

However, the current systems are unidirectional. Research organizations have websites dispensing the advice but there is no feedback coming from the farm. The advice is dispensed and forgotten. There are no feedback loops and hence there is no learning in the system. The cost of delivery is high and there is no out-of-the-box experience. Yuktix Gidabits is an interactive system. The forecast is half the story. A grower also needs the right recommendations from the system as well as a trained researcher. A system with feedback will also improve the forecast accuracy. Yuktix Gidabits is a hassle free experience for the growers. They just have to call a help line number (available in Kannada and English) to get the installation done. The Yuktix IoT device is a proven technology that is in use at major agri companies and NGO setups. The Yuktix IoT devices are solar powered sensor devices that relay accurate field data 24x7 to Yuktix ankiDB cloud. The IoT devices are like a black box that just needs to be put in the field and does not require any expertise to operate. This is the first system to provide disease forecast in snack sized subscription so anyone can join. The on boarding is extremely simple. Farmers just have to call a number. Further, all the interactions for grower is via a mobile app only. We are using sage ML pipeline on AWS for analyzing the images and data along with the statistical packages available in Yuktix ankiDB sdk. We are a firm believer in the need of domain expertise so the app allows the growers to interact directly with a research expert if required. The tools also allow an expert to quickly shift through the farms data along with recommendations and scribble his own. The system is going online with tomato, capsicum and guava.

4.6.1.3 SuperMap IS

A study was conducted to establish the warning and prediction system for crop diseases and pests based on SuperMap IS. NET geographic information system

(GIS), which was developed by Supermap company. In this system, the transform data information into a geographical information map to show the occurrence degree and distribution on various diseases and pests. Luo *et.al.*, (2009) described mainly warning flow, database design and the main functions of the system. Finally, the system realized successfully the warning of the wheat stripe rust in Xifeng region of Qingyang city in Gansu province in 2002, and the prediction result was satisfactory. It indicated that we could classify and predict diseases and pests, and select right time and technology to control the diseases and pests by this GIS system.

4.7 The Basic Goals and Design of the Warning Dystem

The warning on diseases and pests was to predict the occurrence condition and the trend of development on diseases and pests in certain range and period, and made some strategy to control diseases and pests according to the pre-warning results (Luo *et.al.*, 2009). The basic goals of the warning system include, by using the warning system, data information could be transformed into a geographical information map to show the occurrence and distribution on variety of diseases and pests by using visualization and spatial analysis with GIS; the warning system could obtain fast and real-time the information of diseases and insects by using RS, and finally it could accurately monitor the occurrence and predict the development of the diseases and pests in large area ; some standards and measures could be made for the investigation and controlling diseases and pests according to the pre-warning results of the warning system.

The Design and Realization the Warning System

Design route : At first, it was important to choose feasible GIS platform and the prediction model on diseases and pests after requirement analysis. Second, database was established according to parameter of model, and function modules of the system were also designed. Finally, the warning system was established and published after debugging the system .

Platform selection : SuperMap IS. NET was chosen as the system development platform, because components of design could be easily managed ; the multi-source data could be integrated and the massive image could be quickly accessed to ; server was clustering; it was with a high degree of flexibility; in the platform, Client and server belonged multi-level cache structure, which could support a variety of map engine work together.

Database design

i. **Attribute database :** The attribute database of the system was composed with monitoring data, basic data on diseases and pests, national meteorological observation data and the data table about latitude and longitude of meteorological observation site, and so on. Monitoring data on diseases and pests: field investigation data, monitoring data using remote sensing, field experiments data, and so on. Basic data on diseases and pests:

diseases and pests species data, damage characteristics data, control and preventive methods data, and so on. National meteorological observation data: temperature, moisture, rainfall and sunshine number, and so on.

ii. **Spatial databases:** The spatial database of the warning was composed of basic map, thematic map and warning information map. Basic map: national administrative division map, digital elevation map. Thematic map was got by basic map according to some goal, which included weather map, remote sensing imagery and crop division map. Warning information map was obtained by spatial analysis on the basic of few thematic maps and analysis result of prediction model, by which the system could show the occurrence and distribution on variety of diseases and pests. And ultimately, some effective guidance can be given to prevent and control the diseases and pests according the warning information map.

4.7.1 Basic Functions of the System

The functions of warning system was composed of special functions and general functions which were offered by GIS platform, including the translating, mitigating and amplificating grahics, and adding and deleting map layer, and so on.

While the special functions were as follows

i. **Retrieval of diseases and insect pests :** Users could search all kinds of knowledge about diseases and pests, for example, latin name and English name of disease and pest, harm symptom and epidemic law on diseases and pests and control methods, and so on.

ii. **Diagnosis of diseases and pests :** Users could obtain some papers on warning results and prediction methods after the operations of systems by various data and prediction models of diseases and pests.

iii. **Model base management :** Administrators could add, delete, or modify the models in the system by Model base management (Liu Shuhua et al, 2003).

iv. **Monitoring and evaluation using remote sensing :** The function was realized by the module of pests monitoring. Multiple and multi-temporal remote data was applicated in the module, and assessment model based on RS was call in the module. By the function module, users could develop monitoring and assessment the diseases and pests. And related remote sensing image s could also were call from spatial database and displayed for users.

v. **Visualization of warning result :** By making visualization and spatial analysis with GIS, the system could transform prediction result and emergence grade into a geographical information map to display visually the occurrence and distribution on variety of diseases and pests. And ultimately, some effective guidance could be given by the system to prevent and control the diseases and pests (Si Lili and Cao Keqiang, 2006).

Warning flow: The warning system was established on the basic on a database which was composed of attribute database and spatial database and a model base including diseases models and pests prediction models. At first, the result which was obtained by prediction model form model base and data from database was displayed in the GIS platform and was transformed into information diagram, meanwhile, some thematic maps from spatial database were overlaid on the information map. Finally, the system could offer a clearly electronic information map which could show the occurrence and distribution on variety of diseases and pests.

The application instance of the Warning System: The process how to realize the function of the warning system was displayed by taking the wheat stripe rust disease in Xifeng of Qingyang region in Gansu province as an example.

Summary of study area : The wheat stripe rust disease was one of the most serious diseases in China. Xifeng of Qingyang region in Gansu province was taken as an example to display the process how to realize the function of the warning system. It was reported that the occurrence of the wheat stripe rust disease was mildly severe and severe symptoms in 1986,1989,1990,1991,2002,2003, outbreak in 1985, moderate in 1993, and mild or none occurrence in other years. The wheat stripe rust disease was intermittent disease, but the occurrence was continuous and frequent, and becoming more and more serious in Xifeng of Qingyang region in Gansu province in recent ten years.

Data sources : The data of wheat stripe rust disease was from plant protecting station in Qingyang region in Gansu province; the weather data was offered by meteorological observing station in Qingyang region in Gansu province.

Model selection : At present, the models in the system were all from literature data. The model which was selected according to expert experience and model characteristics was $y = 0.3251+ 0.0414x_1 +0.0243\ x_2\ 0.0667\ x_3$. In this prediction model, y is emergence grade, 1 x is average temperature from January to March, 2 x is index of temperature and rainfall, 3 x is index of the disease.

4.7.2 The pre-warning Results on Study Area

System interface : The left page of the warning system concluded the field of user ' logging and operational area of prediction, while the data about the warning was display on the right of system interface with victor diagram.

The operating process of the warning on the wheat stripe rust disease : At first, the users could enter the main interface of the system by login; second, users could enter the main interface of diseases prediction by clicking the diseases prediction model. Finally, used could develop step-bystep operation in the operational area of prediction according.

The pre-warning results on study area: By the above operation, users could call,, in the model from the database of the system, and got the warning result by calculation. Finally, the prewarning results were displayed on the right of the main interface by different color by electrical information map, at the same time, the warning results were also display by character form on the left of main interface. It was displayed that the emergence grade of the wheat stripe rust was five and occurrence degree was outbreak in Xifeng region of Gansu province. The fact that the warning results coincided well with report results from plant protecting station of Gansu province showed that this warning system and prediction method were all feasible. A warning and information system for diseases and pests of crops was established base on Geography Information System (GIS). The occurrence and damage of major diseases and pests of crops could be monitored and forecasted in real time by using the system, based on the forecasting model. And the system could submit warning maps with 6 colors and words information to the users. By using the system, we could standardize the forecasting data collection, transmit information through network and forecasting results viewable. Users could use it to classify and predict diseases and pests, also could use it to select fitting time and technology to control.

Leaf wetness duration: Disease-warning systems are decision support tools designed to help growers determine when to apply control measures to suppress crop diseases. Weather data are nearly ubiquitous inputs to warning systems. A contribution reviewed ways in which weather data are gathered for use as inputs to disease-warning systems, and the associated logistical challenges (Mark L. Gleason *et.al.*, 2008). Grower-operated weather monitoring was contrasted with obtaining data from networks of weather stations, and the advantages and disadvantages of measuring vs. estimating weather data were discussed. Special emphasis was given to leaf wetness duration (LWD), not only because LWD data are inputs to many disease-warning systems but also because accurate data were uniquely challenging to obtain. It was concluded that there is no single " best" method to acquire weather data for use in disease-warning systems; instead, local, regional, and national circumstances are likely to influence which strategy is most successful.

Leaf wetness duration is the period of time during which free water – from dew, rainfall, fog, or irrigation is present on the aerial surfaces of crop plants. LWD is a non-standard meteorological parameter because it is a property of surfaces as well as the atmosphere. There is no widely accepted standard for calibrating LWD sensors, and the vast majority of automated weather stations worldwide do not include these sensors. Nevertheless, LWD is an important determinant of the development of many foliar and fruit diseases through its influence on such key processes as pathogen germination, infection, and sporulation. As a result, LWD is an input to numerous disease-warning systems. Leaf wetness duration is the most spatially heterogeneous weather input to warning systems. It varies not only with weather conditions but also with the type of crop, its developmental stage, and the position, angle, and geometry of individual leaves. During dew periods, different micro-sites on a single leaf can vary in LWD by several hours per day. This immense heterogeneity poses a formidable challenge for measuring or estimating LWD. If using LWD sensors, where should they be placed? If using estimates, what part of the canopy should be used as the reference point?

Measuring LWD: Many types of sensors have been employed to measure LWD. Among the first were mechanical sensors utilizing strings that were maintained under slight tension during exposure. As it became wetter or drier the string would contract or relax, resulting in movement of an attached ink-pen marker on a revolving paper chart. By the mid-1980s, mechanical LWD sensors and strip charts were superseded by electronic sensors and automated data-loggers. Most electronic sensors were based on a design developed by Davis & Hughes (1970), which sensed the presence of wetness as a drop in electrical resistance across two adjacent circuits etched onto a printed-circuit board. Later workers found that coating these sensors with latex paint enhanced their sensitivity to wetness, that height and angle of deployment influenced sensitivity, and that electronic sensors could also be fabricated as cylinders rather than flat plates. A recent innovation in flat-plate sensor design operates on the principle of electrical capacitance rather than resistance.

GovTech: GovTech can prove to be a great enabler for early detection, Monitoring & Early Warning System of Wheat Rust (Priyadarshi Nanu Pany, 2020). The use of intuitive, simple to use mobile applications by farmers, agriculture extension workers, NGOs and researchers has enabled the following.

- Digitization of important farm parameters such as geo-location co-ordinates, cropping pattern & type, time of plantation, and application of inputs such as fertilizers & irrigation
- Capture photographs depicting the type & severity of fungal disease that has affected the plant or the stage of pest larvae on breeding grounds. Currently, the analysis is mostly done manually by agricultural scientists. Leveraging machine learning algorithms on a broad set of photographs, this analysis can also be system-generated.
- Codification of samples collected from farms & sent to labs, which helps to trace the origin of any new strains of diseases found in samples to the farms they came from.

This digitized survey data also forms the backbone of global forecasting models for wheat rust managed by International Maize and Wheat Improvement Centre (CIMMYT). This international organization combines this survey input with meteorological models (on wind speed, temperature & humidity) developed by UK Metrology & University of Cambridge to analyze & predict the timing & areas of next outbreak. The Ministry of Agriculture in Ethiopia issues official advisories based on the insights provided by the model. Via the communication platform in the Early Warning System, Ethiopia has also been swift to reach out to local administrators & farmers and advise them on using pesticide usage in very early stages. Opportunity to integrate these advisories with drones for efficient pesticide application are also being explored.

Seasonal forecast-based disease intervention of the wheat blast outbreaks: Seasonal disease risk prediction using disease epidemiological models and seasonal forecasts has been actively sought over the last decades, as it has been believed to be a key component in the disease early warning system for the pre-season planning of local or national level disease control (Kwang-Hyung Kim and Eu Ddeum Choi, 2020). A retrospective study was conducted using the wheat blast outbreaks in Bangladesh, which occurred for the first time in Asia in 2016, to study a what-if scenario that if there was seasonal disease risk prediction at that time, the epidemics could be prevented or reduced through prediction-based interventions. Two factors govern the answer: the seasonal disease risk prediction is accurate enough to use, and there are effective and realistic control measures to be used upon the prediction. In this study, we focused on the former.

To simulate the wheat blast risk and wheat yield in the target region, a high-resolution climate reanalysis product and spatiotemporally downscaled seasonal climate forecasts from eight global climate models were used as inputs for both models. The calibrated wheat blast model successfully simulated the spatial pattern of disease epidemics during the 2014–2018 seasons and was subsequently used to generate seasonal wheat blast risk prediction before each winter season starts. The predictability of the resulting predictions was evaluated against observation-based model simulations. The potential value of utilizing the seasonal wheat blast risk prediction was examined by comparing actual yields resulting from the risk-averse (proactive) and risk-disregarding (conservative) decisions. Overall, our results from this retrospective study showed the feasibility of seasonal forecast-based early warning system for the pre-season strategic interventions of forecasted wheat blast in Bangladesh.

4.7.3 Early Warning System Framework and Models

Wheat rust diseases pose one of the greatest threats to global food security, including subsistence farmers in Ethiopia. The fungal spores transmitting wheat rust are dispersed by wind and can remain infectious after dispersal over long distances (Clare Allen-Sader *et.al.*, 2019). The emergence of new strains of wheat rust has exacerbated the risks of severe crop loss. The construction and deployment of a near real-time early warning system (EWS) was described for two major wind-dispersed diseases of wheat crops in Ethiopia that combines existing environmental research infrastructures, newly developed tools and scientific expertise across multiple organizations in Ethiopia and the UK. The EWS encompasses a sophisticated framework that integrates field and mobile phone surveillance data, spore dispersal and disease environmental suitability forecasting, as well as communication to policy-makers, advisors and smallholder farmers.

The system involves daily automated data flow between two continents during the wheat season in Ethiopia. The framework utilizes expertise and environmental research infrastructures from within the cross-disciplinary spectrum of biology, agronomy, meteorology, computer science and telecommunications. It involved Open data kit (ODK) field surveying, ATA mobile phone platform, Meteorological data from the UK Met Office's Unified Model (UM), the Numerical Atmospheric-dispersion Modeling Environment (NAME) mode, Algorithm to identify source locations for NAME from survey data, Environmental suitability models and EWS dissemination of information.

The EWS successfully provided timely information to assist policy makers formulate decisions about allocation of limited stock of fungicide during the 2017 and 2018 wheat seasons. Wheat rust alerts and advisories were sent by short message service and reports to 10 000 development agents and approximately 275 000 smallholder farmers in Ethiopia who rely on wheat for subsistence and livelihood security. The framework represents one of the first advanced crop disease EWSs implemented in a developing country. It provides policy-makers, extension agents and farmers with timely, actionable information on priority diseases affecting a staple food crop. The framework together with the underpinning technologies are transferable to forecast wheat rusts in other regions and can be readily adapted for other wind-dispersed pests and disease of major agricultural crops.

SISALERT- A generic web-based plant disease forecasting system: The revolution in web-based technologies has led to great strides in the development and employment of decision support systems for growers and pest management specialists. The present work illustrates an approach towards that direction by the use of novel programming languages and technology for the development of a web-based system for model implementation and delivery. SISALERT is a multi-model platform that unleashes the power of hourly weather station data and hour-by-hour weather forecast information using sophisticated disease risk assessment models (José Maurício Cunha Fernandes *et.al.*, 2011). These models interpret weather data giving information on past or recent disease behavior as well as predicted disease risk. The uniqueness of SISALERT format is the modularity that allows coupling crop and disease models depending on which model is under run.

A major obstacle in disease forecasting is the validity of the disease forecasting models. Although accuracy weather information could be improved by using on site weather data associated to weather forecast data, the model itself may cause inaccurate forecasting due to its limitation of validity. There is a need for continuous and systematic effort to develop better models. In the apple production area, the current advisory has enabled growers to manage fungicide use, time, and resources much more efficiently and profitably. Improvements to this system could only increase profit and efficiency, and reduce unnecessary dependency on fungicides. Besides extending risk information for a large geographical region the use of intuitive images representing epidemic risks may facilitate dissemination and understanding of risks to guide decision-making on plant disease management. In addition, maps may be useful for the fine tuning of crop zoning and for the identification of quality aspects in post-harvested areas.

webGIS technology: With the computer and the rapid development of information technology, The plant disease intelligence diagnosis and the plant disease early warning network is possible. Relying on the Internet, the realization of intelligent online plant disease diagnosis, early warning, prevention and other information query and forecasting capabilities were discussed (Shi Ming Wang, and Yan Zhou.2011). The database base on Internet-based wide area network environment and the help of webGIS technology, including diagnostic module, check module and the prediction module. The intelligent prognosis is tested to the extent and happening of plant diseases, for example, Wheat, cotton, tomatoes and other 12 kinds of plant diseases, and epidemic curve is given. To disease diagnosis and prediction of the professional, technical problems become easy, for the effective control of plant diseases, provide a reference.

Forecasting using fuzzy logic: Early detection of pest and its control is one of the aspects of IPM. Weather based forecasting is well accepted method for this (Tilva *et.al.*, 2013). Various meteorological data like-temperature, humidity, leaf wetness duration (LWD) plays the vital roles in the growth of microorganism responsible for disease. Effective forecasting of such diseases on the basis of climate data can help the farmers to take timely actions to restrain the diseases.

Low Cost Early Warning System: A low cost early warning system was proposed for the protection of wheat crop against the various diseases that are developed through pests and pathogens on the crop. a standard Logistic regression model has been developed to predict the risk. The infection can be predicted using four parameters.

A field monitoring device that will collect environmental parameters and the collected data will be processed to find out the severity of infection that has occurred on crop (Gaurav Sharma,2014). Through this system farmers can get the information about the severity rate of crop infection on their cell phones. Timely warning of the plant diseases can protect the crop from pesticides infection. Weather sensors like temperature, humidity, leaf wetness, and solar radiation are interfaced through ARM-32bit LPC2129 microcontroller. Whole system is used to collect the environment parameters. These collected environment parameters are processed to forecast the likeliness of disease to infect the crops. The disease forecasting can protect the crop from damaging at very early stage and this also saves the amount of pesticides used.

CARTs-Weather-data-based model: Classification and regression trees (CARTs) for data analysis, an hourly weather dataset, and a 3 year field incidence and severity dataset of winter wheat rust were integrated to forecast pathogens' presence/absence. The field dataset of incidence and severity was collected for three production cycles (Victor M. Rodríguez-Moreno *et.al.*, 2020). Measured records of 88 Automatic Meteorological Stations and the indirect weather dataset generated in the Weather Research and Forecasting environment interpolated to each Automatic Meteorological Station location were analyzed in the Python ecosystem. The focal point of the analysis was the severity of the disease.

Disease cycle approach: Plant disease cycles represent pathogen biology as a series of interconnected stages of development including dormancy, reproduction, dispersal, and pathogenesis. The progression through these stages is determined by a continuous sequence of interactions among host, pathogen, and environment (Erick D. De Wolf and Scott A. Isard, 2007). The stages of the disease cycle form the basis of many plant disease prediction models.

Need for replicating successful models: Quite similar to epidemics affecting humans (such as the novel Coronavirus), infections and pests affecting crops also spread like wildfire. Often termed as the "polio of agriculture", the airborne wheat rust has aggressively spread from Africa to Middle East and is now about to severely affect Europe & Asia (Priyadarshi Nanu Pany, 2020). Many mutations of the infection have also sprung up in different regions and developing resistant varieties is expected to take significant time. Hence, investing in Early Warning Systems is the most feasible & effective option to arrest epidemic spread. This system may have been pioneered by Ethiopia but its replication globally could save 1 billion people directly dependent on wheat for food & livelihood.

In fact, the framework of early warning systems, mobile based surveillance, forecasting models & communication platforms, is similar across all types of diseases & pests afflicting life on the planet globally including the locust outbreak. Amidst millions going hungry, a locust swarm spread over an area of one square kilometre nearly eats the same amount of food as 35,000 people every day. The problem is as real as it gets and needs to be nipped in the bud early. Governments worldwide need to actively scout for such best practices and invest significantly to replicate similar systems. Data from such systems feeding into global forecasting models shall further strengthen their robustness and accuracy of predictions. The time to prioritize early warning systems is now.

4.8 Mathematical Concepts and Methods of Disease Forecasting

Prediction of disease outbreaks enables the effective use of control measures, such as chemical or biological treatments, the prediction of crop yields and of the market potential for that crop. Disease forecasting involves the use of weather data and biological data to predict disease incidence. Usually, disease forecasting is only performed on economically important diseases, and as a method of cost reduction. If controlling a particular disease involves an expensive or time-consuming treatment, being able to predict outbreaks of the disease allows the treatment to be timed correctly, increasing its effectiveness, and reducing the cost compared to repeated treatments. Because environmental conditions vary from season to season, disease forecasting is necessary to predict the chance of disease in a certain set of conditions. Disease can be forecast using computer modeling and empirical correlations relating to weather conditions, levels of inoculum, test plots and site factors and the predictions can then be communicated to growers. Computer modeling of plant diseases uses systems analysis to accumulate all the factors that affect the development of a certain disease into a computer-based model, and make predictions of disease under different environmental conditions. A disease needs to be well understood in order to formulate an accurate model, and models based on diseases that we know little about are generally not very accurate. The more straightforward approach of developing empirical correlations between particular weather factors and disease has had considerable success. This does not attempt a complete modeling of all factors involved in a disease, but only those most important in affecting the disease. The accuracy of the model can then be measured statistically by comparing its predictions to what actually happens.

Monitoring the weather is the most important consideration in disease forecasting because of the overriding effect that weather has on disease development. While broad scale weather data has been used for disease forecasting, it is well known that microclimate within the crops has a more direct impact on disease. Devices have been developed to monitor microclimate factors like duration of leaf wetness and temperature and with time, they will be affordable and accurate enough for widespread use on individual farms. Synoptic weather forecasting charts can be used to predict 'critical periods' , the occurrence conditions favourable for disease development, so that farmers can spray their crop before it happens. There are now several self-calculating disease forecasting monitors available commercially that use environmental data and past season data to predict outbreaks of particular disease.

Some disease forecasting methods are based solely on monitoring inoculum levels, often as indicated by the amount of disease already present. This method can be successful when disease is developing steadily under relatively uniform or predictable weather conditions, but not for diseases that can spread explosively in favourable conditions. Monitoring the amount of disease present can indicate whether the amount of disease is likely to exceed a certain threshold, at which point control measures become economical. There are numerous methods of directly monitoring the concentration of spores in the air as an indication of the chance of disease. Trapping vectors of diseases can also be useful in predicting the occurrence of viral diseases. In addition, estimation of populations of soil-borne pathogens by examining soil samples is necessary for predicting the outbreak of the diseases they cause. Monitoring systems can be combined with data specific to the site and the crop, such as soil type, topography and irrigation levels, in order to increase the accuracy of predictions.

Test plots (or trap plots) of susceptible cultivars can be planted throughout a cropping area to give early warning of the arrival of inoculum or disease vectors.

Alternatively, inoculation of the test plot with the pathogen can give an indication of favourable environmental conditions for disease development. Test plots are also useful for monitoring the occurrence of minor diseases on new cultivars. The formation of a prediction is useful only if it can be communicated to the growers who will be affected by it. General warnings for areas are broadcast over the radio or internet. Predictions based on monitoring by individual farmers or groups of farmers in an area remove the need for widespread communication systems. Computerized decision support systems based on local monitoring can educate and empower farmers when making decisions about their crops.

4.8.1 Maximum Likelihood Based Methods

With the development of more rapid computing technologies, it is possible to compute a large number of simulations to support new modeling methods for plant disease forecasting. The development of disease forecasting models is based on determining the value of unknown parameters for an appropriate model.

One method for estimation of parameters is maximum likelihood estimation (MLE), first developed by R.A. Fisher (Myung 2003). The basic idea behind MLE is that one can calculate the probability of observing a particular set of data given the model of interest and this is called the likelihood of a sample. The likelihood of observing the actual data for each of the potential parameter values is calculated to determine which set of parameter values maximizes the likelihood. Schmidt, (1993) studied the modeling of the population dynamics of the sugar beet cyst Nematode, Heterodera schachtii (Fig. 7).

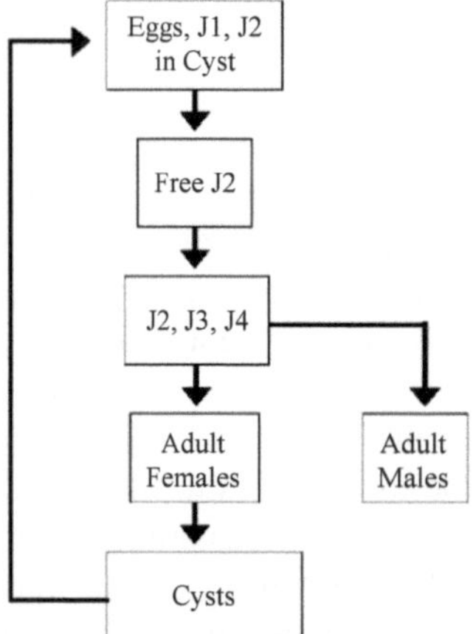

Fig. 7: Simple model of the sugar beet cyst nematode life cycle

Heterodera schachtii, the sugar beet cyst nematode, causes significant crop damage (Ravichadnra, 2014). Female nematodes form cysts that, when mature, are able to over-winter in the soil. These cysts consist of the protective ectoderm of the female and several eggs. As the season progresses, the eggs hatch in the female to form J1 stage juveniles which then molt once and form J2 juveniles. When environmental conditions are conducive, J2 juveniles leave the cyst and find a plant root to feed on. The juveniles develop through J3 and J4 stages until adult males and females are formed. Adult males leave the roots and search for females. Adult females, when fertilized, form cysts. The number of nematodes in a field is highly dependent on two things temperature and the current and previous crop species. Relatively hot years may result in several generations of nematodes whereas cool years may only allow one generation of reproduction. Though the sugar beet nematode is somewhat generalist and can survive to some degree on many different hosts, it is greatly favored by planting sugar beets. Crop rotation may be an effective way to reduce the population of the sugar beet cyst nematode. Knowledge of previous crop rotation and temperature may be used to forecast the amount of disease in a given sugar beet field over a season (Schmidt *et al.* 1993). In the model developed by Schmidt *et al.* (1993), probabilities are calculated for movement between each of the stages of a nematode life cycle..

Before the model is more thoroughly described, please note that the θ parameter is a crop-specific parameter that may take on a different value for each type of crop grown. In a more complete model set forth by the authors, θ is calculated separately for each component of the life cycle (the crop specific parameter is not necessarily equal for all components of the life cycle).

Factors affecting the ease or difficulty of forecasting and modeling cyst, root-ectoparasitic and soil surface nematodes were discussed (Jones and Kempton, 1980). To relate pre-planting to post-harvest populations of cyst nematodes, a version of the logistic curve was derived from first principles and modified to take account of the injury done to root systems as numbers increase, and of eggs carried over to the following year. Egg hatch, when potato or other crops are grown, appeared to be independent of population density. So also did root invasion over the population densities encountered in field plots. In the absence of enemies, population density was determined by competition for feeding sites induced by females in the root system, and by the size of the root system. Population oscillations about the equilibrium point were affected mainly by the proportion of the population that failed to hatch and the damage to the root system. The process of model building and the parameters required were outlined. The model included hypotheses relating to intraspecific (genetic) competition and interspecific competition. The limitations of models and the need to make them stochastic were discussed.

Multi-spectral remote sensing allows a wide range of application in precision farming. Remote sensing provides growers with yield assessments, shows yield variations across fields and give information about the growth rate at important development stages. This includes detection of stress due to drought and nutrient deficiency as well as a result of plant diseases or animal pests.

A study by Jacobi *et.al.*, (2011) was focused on the spatial distribution of fungal infestations in winter wheat and nematodes in sugar beet at different crop growth stages. Of particular interest was the detection of fungal diseases in early growth states by the use of very high resolution remote sensing (VHRRS) imagery. These images were taken from the Quick Bird, the satellite with the highest available resolution up to 0,6 m in the panchromatic mode and 2,4 m in the multispectral mode. Due to the rare availability of the QuickBird images, alternatives to the satellite images were tested, which were aerial hyper spectral images (HyMap-sensor with 126 spectral channels), images taken by a NIR digital camera, which was carried by a model helicopter for aerial view or fixed to a stand to take images in close-up range. With Image analysis, interpretation in relation to ground truth data and statistical analysis with GIS software the spatial spreading and infestation level of the fungal diseases in wheat and nematodes in sugar beet was mapped.

4.8.2 Feature Selection Techniques

Earlier, crop cultivation was undertaken on the basis of farmers' hands-on expertise. However, climate change has begun to affect crop yields badly. Consequently, farmers are unable to choose the right crop/s based on soil and environmental factors, and the process of manually predicting the choice of the right crop/s of land has, more often than not, resulted in failure (Suruliandi *et.al.*, 2021). Accurate crop prediction results in increased crop production. This is where machine learning playing a crucial role in the area of crop prediction.

4.8.3 Random Forest Method

With the availability of satellite sensor technology, thematic maps become a very attractive tool to the researchers, to understand the land use/cover characteristics of the earth's surface (Asli Ozdarici Ok *et.al.*, 2021). In agricultural perspective, the most interesting applications include classifying agricultural landscapes, determining land use areas, soil moisture estimation and revealing crop residues in agricultural fields which is an important consideration to reduce soil erosion, estimating the effect of crop classification error, land cover mapping in an agricultural setting using classification. In this light, various image classification algorithms are developed to satisfy the needs of different applications and classification problems. One of the most widely used algorithms is pixel-based Maximum Likelihood Classification (MLC) algorithm. It is also known as a statistical approach since it relies on a statistical model. In statistical approaches, an appropriate data model is used and parameters of the model can be approximated from the data.

To examine the performance of Random Forest (RF) and Maximum Likelihood Classification (MLC) method to crop classification through pixel-based and parcel-based approaches, Asli Ozdarici Ok *et.al.*, (2021) conducted a study. Analyses were performed on multispectral SPOT 5 image. First, the SPOT 5 image was classified using the classification methods in pixel-based manner. Next, the produced thematic maps were overlaid with the original agricultural parcels and the frequencies of the pixels within the parcels were computed. Then, the majority of the pixels were assigned as class label to the parcels. Results indicated that the overall accuracies of the parcel-based approach computed for the Random Forest method was 85.89%, which was about 8% better than the corresponding result of MLC.

4.9 The Future of Site-specific Weather Data Estimation

Additional technologies appear likely to enhance the accuracy of site-specific weather estimates in the near future. For example, ground-based radar, currently used primarily in avionics, also has the capability of locating areas of rainfall in near-real time (Mark L. Gleason *et.al.*, 2008). Local areas of rainfall, caused by small-scale convective systems, are among the most elusive phenomena for weather station networks, since these systems can drop substantial rainfall on swaths that are only a few kilometers wide, and thus may be invisible to stations situated on a broader-scale grid. By timing the return rate of radar reflections from precipitation, it is possible to track the passage of rainfall across a landscape. Although rainfall amount is more difficult to determine than its timing of occurrence due to radar attenuation errors, rainfall occurrence can help to pinpoint leaf wetness duration, which is a key input to disease-warning systems. Preliminary attempts have been made to use radar estimates of rainfall as inputs to disease-warning systems.

Satellite-based scans of the earth's surface may eventually enhance site-specific weather estimation. As the spatial resolution of these scans increases, they may be able to fill gaps in cloud-cover or albedo data that currently hamper estimation of net radiation, and thus the application of energy balance models to estimating LWD associated with dew periods.

Plant pathologists are well aware of the role weather plays in the epidemiology of many important pathogens. In particular, temperature, surface wetness, relative humidity, and sometimes rain are commonly used for disease management. The acquisition of on-farm or site-specific weather data can be problematic due to the costs of purchasing and maintaining farm based automated weather stations (AWSs). Where these AWSs have been lacking, weather data may only be available at spatial and temporal scales too crude for use in site-specific decision making. This applies especially to the use of forecast weather data, which has potential to convert disease warnings into true disease forecasts. Magarey *et.al.*, (2001) described the emerging technologies for obtaining site-specific weather data without on-site sensors. We will present a brief history of the development of these technologies, their present limitations, and their prospects for future improvement.

Global weather data and simulated crop yields: Crop simulation models can be used to estimate impact of current and future climates on crop yields and food security, but require long-term historical daily weather data to obtain robust simulations. In many regions where crops are grown, daily weather data are not available (van Wart *et.al.*, 2013). Alternatively, gridded weather databases (GWD) with complete terrestrial coverage are available, typically derived from global circulation computer models; interpolated weather station data; or remotely sensed surface data from satellites. van Wart *et.al.*, (2013) evaluated capacity of GWDs to simulate crop yield potential (Yp) or water-limited yield potential (Yw), which can serve as benchmarks to assess impact of climate change scenarios on crop productivity and land use change. Three GWDs (CRU, NCEP/DOE, and NASA POWER data) were evaluated for their ability to simulate Yp and Yw of rice in China, USA maize, and wheat in Germany. Simulations of Yp and Yw based on recorded daily data from well-maintained weather stations were taken as the control weather data (CWD). Agreement between simulations of Yp or Yw based on CWD and those based on GWD was poor with the latter having strong bias and large root mean square errors (RMSEs) that were 26–72% of absolute mean yield across locations and years. In contrast, simulated Yp or Yw using observed daily weather data from stations in the NOAA database combined with solar radiation from the NASA-POWER database were in much better agreement with Yp and Yw simulated with CWD (i.e. little bias and an RMSE of 12–19% of the absolute mean). We conclude that results from studies that rely on GWD to simulate agricultural productivity in current and future climates are highly uncertain. An alternative approach would impose a climate scenario on location-specific observed daily weather databases combined with an appropriate up scaling method.

Dataset of future daily weather data: Coupled atmosphere-ocean general circulation models (GCMs) simulate different realizations of possible future climates at global scale under contrasting scenarios of land-use and greenhouse gas emissions. Such data require several additional processing steps before it can be used to drive impact models (Duveiller *et.al.*, 2017). Spatial downscaling, typically by regional climate models (RCM), and bias-correction are two such steps that have already been addressed for Europe.

4.10 Ensemble-mean Climate Scenarios for Future Crop Yield Projections

Using climate scenarios from only 1 or a small number of global climate models (GCMs) in climate change impact studies may lead to biased assessment due to large uncertainty in climate projections. Ensemble means in impact projections derived from a multi-GCM ensemble are often used as best estimates to reduce bias (Di Ma *et.al.*, 2021). However, it is often time consuming to run process-based models (e.g. hydrological and crop models) in climate change impact studies using numerous climate scenarios. It would be interesting to investigate if using a reduced number of climate scenarios could lead to a reasonable estimate of the ensemble mean. In this study, we generated a single ensemble-mean climate scenario (En-WG scenario) using ensemble means of the change factors derived from 20 GCMs included in CMIP5 to perturb the parameters in a weather generator, LARS-WG, for selected locations across Canada. We used En-WG scenarios to drive crop growth models in DSSAT ver. 4.7 to simulate crop yields for canola and spring wheat under RCP4.5 and RCP8.5 emission scenarios. We evaluated the potential of using the En-WG scenarios to simulate crop yields by comparing them with crop yields simulated with the LARS-WG generated climate scenarios based on each of the 20 GCMs (WG scenarios). Our results showed that simulated crop yields using the En-WG scenarios were often close to the ensemble means of simulated crop yields using the 20 WG scenarios with a high probability of outperforming simulations based on a randomly selected GCM. Further studies are required, as the results of the proposed approach may be influenced by selected crop types, crop models, weather generators, and GCM ensembles.

A stochastic weather generator, LARSWG, was used to develop future climate scenarios (i.e. WG and En-WG scenarios) from multiple GCMs under 2 forcing scenarios, RCP4.5 and RCP8.5. Stochastic weather generators have attracted attention in the past decades as a convenient tool for producing daily climate scenarios in climate change impact studies. However, stochastic weather generators suffer from the problem known as over dispersion, i.e. they underestimate the interannual variability of climate. It has been reported that multi-model averaging can enhance the reliability of climate projections, largely re - sulting from the cancellation or compensation of errors in the individual models even on the regional scale. However, directly averaging daily climate outputs from multiple climate models to drive process-based models is not applicable because averaging smooths climate variability. It is even more difficult, if not impossible, to develop such a single scenario by averaging GCMs, as the relationship between climate variables and their impacts may not be linear. These results support the need for corrections or improvements in stochastic weather generation algorithms, such as LARS-WG used in this study, to remedy the underestimation of interannual climate variability, as several previous studies do. Efforts have been made to reduce overdispersion in stochastic weather generators.

However, climate scenarios generated by LARS-WG usually resulted in the underestimation of interannual variability in the simulated crop yields, which is common for many weather generators. The En-WG scenarios have the potential of efficiently estimating the ensemble means of future crop yield projections in a multi-GCM ensemble when a stochastic approach is applied to each of the individual GCMs, in terms of much less time (only 5% of a 20 GCM ensemble) for running simulations and a reasonable accuracy (approximately a 2% error on average).

4.11 The Challenge of Timeliness

The need for timeliness cuts across all methods of obtaining weather data inputs for warning systems. A key question is whether growers have sufficient time to respond effectively to spray advisories once they are issued. A fungicide-spray warning is of little use to a grower who cannot apply the needed spray in a timely manner due to rainfall, impassably muddy terrain, excessive wind, or re-entry interval restrictions on fungicides (Mark L. Gleason *et.al.*, 2008). On some farms, applying a single spray to the entire crop may require several days to a week due to the large size of the planting, speed and number of sprayers, labor availability, or other factors. Another constraint on timeliness is the mode of action of the pesticides used. Many of the most affordable fungicides and bactericides have a contact mode of action, meaning that they are effective against target microorganisms only on the outer surfaces of plants. Once a fungus or bacterium has invaded a plant, contact fungicides cannot stop the infection process. Pesticides with so-called penetrant (also known as systemic) modes of action can stop infections that have already started, but the time window for this eradicative activity is limited to a few hours or days.

These logistical constraints on timeliness mean that warning systems will be accepted only when growers have adequate time to respond to advisories. When growers operate their own weather stations, they may overshoot action thresholds if, as sometimes happens, they do not retrieve data sufficiently often. Although networks of stations may process data more frequently, a lag of 8 to 12 hours between weather measurements and issuance of advisories is not uncommon due to time requirements of data retrieval and processing from multiple stations. These time constraints have prompted increased efforts to use forecasted weather data, from one to several days into the future, as warning system inputs (Johnson *et al.*, 2004; Kim *et al.*, 2006; Thomas *et al.*, 2002; Magarey *et al.*, 2001, 2002). Although forecasting is gradually becoming more accurate as weather-monitoring technology improves, predicting the future entails an additional error factor beyond those inherent in making weather measurements and model-based estimates. The further into the future a prediction is made, the less accurate it can be expected to be. Consequently, using forecasts to operate disease-warning systems necessitates a trade-off between timeliness and accuracy.

To conclude, there is no single " best" method to acquire weather data for use in disease-warning systems. Instead, local, regional, and national circumstances may influence the sustainability of each strategy. A vision was presented for implementing agricultural decision-support systems that is relevant to the narrower focus (Mark L. Gleason *et.al.*, 2008). They point out that consortia of public-sector researchers and extension educators, private-sector service providers, and growers are likely to be the most sustainable arrangements for introducing and sustaining decision-support systems in North America, and the same generalization may hold for obtaining weather data inputs to warning systems. Researchers can undertake studies that lay the groundwork for obtaining weather data - by developing sensors and validating their performance in the environment, pioneering models that estimate weather data, and developing new ways to extend the spatial domain of weather station measurements. Extension educators can help to insure that growers' voices are heard as a warning system's design develops and matures. Private-sector companies can package the products of academic research and extension into formats that fit grower needs, while profiting from the services they provide. Growers determine which commercial services are sufficiently valuable and practical to succeed, and provide feedback for improvements to extension personnel, researchers, and service providers.

This is a compelling vision, but other modes of obtaining weather data may be more successful in some circumstances. The do-it-yourself model may flourish for growers who are committed to meeting quality-control and other management requirements. Alternatively, a predominantly public-sector model may be more practical where the private sector does not offer suitable weather-estimation services.

Reliability of weather data inputs is the backbone for sustainability of any of these schemes. Proper installation and timely maintenance of weather instrumentation, along with diligent error-checking of weather data, are unglamorous but inescapable requirements for implementing any disease-warning system successfully. This means that adequate training of weather-monitoring personnel is essential. In addition, convenience and simplicity of use will determine whether a warning system makes the transition from the toolbox to the real world of crop disease management.

Bibliography

Anonymous, 2021. Toward the Development of Disease Early Warning Systems. I*n: Under the Weather*: Climate, Ecosystems and Infectious Disease. The National Academies Press, 86-146.

Aradhana Rai, Sudhir Kumar Sharma and Shivling, V. D., 2016. Design and Development of an Early Warning System for Potato Disease- Late Blight. Intl.J. Innov. Res.in Engg & Manage.,(*IJIREM*) 3: 349-352

Asli Ozdarici Ok, Ozlem Akar and Oguz Gungor, 2012. Evaluation of random forest method for agricultural crop classification. *European J. Remote Sensing*, 45: 421-432

Boney Barry, 2018. "Understanding and Predicting Nematode Damage on Soybean using Spatially Weighted Analysis". Theses and Dissertations. 3034, University of Arkansas, Fayetteville

Bonfil, D.J., Dolgin, B., Mufradi, I., 2004. Bioassay to Forecast Cereal Cyst Nematode Damage to Wheat in Fields. *Precision Agri.*, 5: 329–344

Changzhen Zhang Id , Jiahao Cai, Deqin Xiao, Yaowen Ye and Mohammad Chehelamirani, 2018. Research on Vegetable Pest Warning System Based on Multidimensional Big Data. *Insects*, 9: 2-13

Clare Allen-Sader, William Thurston, Marcel Meyer, Elias Nure, Netsanet Bacha, Yoseph Alemayehu, Richard O J H Stutt, Daniel Safka, Andrew P Crai and, Eshetu Derso, 2019. An early warning system to predict and mitigate wheat rust diseases in Ethiopia. *Environ. Res. Lett,.* 14: 115004

Di Ma, Qi Jing , Yue-Ping Xu , Alex J. Cannon , Taifeng Dong , Mikhail A. Semenov , Budong Qian, 2021. Using ensemble-mean climate scenarios for future crop yield projections: a stochastic weather generator approach. *Clim. Res.,* 83: 161–171

Duveiller, G., Donatelli, M. and Fumagalli, D., 2017. A dataset of future daily weather data for crop modelling over Europe derived from climate change scenarios. *Theor. Appl. Climatol.*, 127: 573–585

Erick D. De Wolf and Scott A. Isard, 2007. Disease Cycle Approach to Plant Disease Prediction. *Ann.Rev.Phytopath.*, 45: 203-220

Gaurav Sharma, 2014. A Low Cost Early Warning System for the protection of Wheat against Brown Rust. Journal of Basic and Applied Engineering Research Print ISSN: 2350-0077; Online ISSN: 2350-0255; Volume 1, Number 10; October, 2014 pp. 79-94

Hay Frank, Herdina Herdina, Ophel-Keller, K. Hartley Diana and Pethybridge Sarah, 2015. Prediction of Potato Tuber Damage by Root-Knot Nematodes using Quantitative DNA Assay of Soil. Pl.Dis., 100. 150803104216002.

Jones, F.G.W. and Kempton, R.A., 1980, Forecasting Crop Damage by Nematodes: Nematode Population Dynamics. *EPPO Bull.*, 10: 169–180

José Maurício Cunha Fernandes , Willingthon Pavan and Rosa Maria Sanhueza, 2011 . SISALERT - A generic web-based plant disease forecasting system. In: Proceedings of the International Conference on Information and Communication Technologies for Sustainable Agri-production and Environment (HAICTA 2011), M. Salampasis, A. Matopoulos (eds.). Skiathos, 8-11 September, 2011.

Justin Van Wart, Patricio Grassini, Haishun Yang, Lieven Claessens, Andrew Jarvis and Kenneth G. Cassman, 2015. Creating long-term weather data from thin air for crop simulation modeling, Agricultural and Forest Meteorology, Volumes 209–210, Pages 49-58,

Katsantonis Dimitrios, Kadoglidou Kalliopi, Dramalis Christos and Puigdollers., 2017. Rice blast forecasting models and their practical value: A review. *Phytopathologia Mediterranea*, 56: 187-216.

Kazunobu Okadome, 2016. Simulation model for forecasting population fluctuations of the diamondback moth in cabbage fields Kyoto Prefectural Agricultural Research Institute, 9 Wakunari Amarube-cho Kameoka-shi Kyoto 621, Japan.

Kenji Chiwaki, Shigeyuki Nagamor and, Yoshio Inoue, 2005 .Predicting Bacterial Wilt Disease of Tomato Plants using Remotely Sensed Thermal Imagery. *J. Agri.metereo.*, 61: 153-164

Kim, K.H. and Jung, I., 2020. Development of a Daily Epidemiological Model of Rice Blast Tailored for Seasonal Disease Early Warning in South Korea. *Pl. Pathol. J.*, 36: 406-417.

Kwang-Hyung Kim and Eu Ddeum Choi, 2020. Retrospective Study on the Seasonal Forecast-Based Disease Intervention of the Wheat Blast Outbreaks in Bangladesh. *Front. Pl. Sci.,* 11: 1738

Liu Shuhu and Yang Xiaohong, 2003. Decision support system for crop diseases and insect pests prevention and control based on GIS. J. Agril. Engg. Res., 19:148-150

Liu, L.W., Hsieh, S.H., Lin, S.J., Wang, Y.M. and Lin, W.S., 2021. Rice Blast (Magnaporthe oryzae) Occurrence Prediction and the Key Factor Sensitivity Analysis by Machine Learning. Agronomy, 11: 771

Luo, J., Huang, W., Wang, J. and Wei, C., 2009, The Crop Disease and Pest Warning and PREDICTION SYSTEM. *In: IFIP International Federation for Information.* Volume 294, Computer and Computing Technologies in Agriculture II, Volume 2, eds. D. Li Z. Chunjiang, (Boston: Springer), pp. 937–945.

Magarey, R. D., Seem, R. C., Geneva, J. M., Russo, J. W., Zack, K. T., Waight, Travis J. W. and Oudemans, P. V., 2001. Site-specific weather information without on-site sensors. Pl. Dis., 85: 1216-1226

Mark L. Gleason, Katrina B. Duttweiler, Jean C. Batzer, S. Elwynn Taylo, Paulo Cesar Sentelhas, José Eduardo Boffino Almeida Monteiro and Terry J. Gillespie, 2008. Obtaining weather data for input to crop disease-warning systems: leaf wetness duration as a case study. Sci. Agric. (*Piracicaba, Braz.*),.Volume 65, No.spe Piracicaba Dec. 2008

Michael J. Roberts, David Schimmelpfennig, Elizabeth Ashley, Michael Livingston, Mark Ash and Utpal Vasavada,2006 . The Value of Plant Disease Early-Warning Systems-A Case Study of USDA's Soybean Rust Coordinated Framework. The Value of Plant Disease Early-Warning Systems: A Case Study of USDA's Soybean Rust Coordinated Framework/ERR-18 Economic Research Service/USDA

Muhammad Aslam Khan, 2020. A Bacterial Leaf Blight disease predictive model to issue advance warning forecasts to rice growers. July 2020, DOI:10.13140/RG.2.2.31199.69288. Project: Plant Disease Epidemiology

Myung, J., 2003. Tutorial on maximum likelihood estimation. *J. Math.Psychol.*, 47: 90-100

Nathaniel K. Newlands, 2018. Model-Based Forecasting of Agricultural Crop Disease Risk at the Regional Scale, Integrating Airborne Inoculum, Environmental, and Satellite-Based Monitoring Data. Front. *Environ. Sci.*, 27: 63 .

Parkes, B., Higginbottom, T.P., Hufkens, K., Ceballos, F., Kramer, B. and Foster, T., 2019. Weather dataset choice introduces uncertainty to estimates of crop yield responses to climate variability and change. Env. Res. Lett.Volume 14, Number 12, 124089

Priyadarshi Nanu Pany, 2020. Early warning systems for effective pest control. Founder & CEO of CSM Technologies. https://medium.com/@nanupany/early-warning-system-for-plant-protection-fc228a5eec18

Pushpa, R.N., Nagaraju, N., Sunil Joshi and Jagadish, K.S., 2019. Epidemiology of Papaya ring spot virus-P (PRSVP) infecting papaya (Carica papaya Linn.) and influence of weather parameters on population dynamics of predominant aphid species. *J. Entomol. & Zool. Stu.*, 7: 434-439

Rajeev Jha, 2021. Yuktix GidaBits, early warning system for plant diseases is launched for farmers in Karnataka. Yuktix Technologies Private Limited. Cision Communication Clouds.

Ravichandra, N.G., 2014. Horticultural Nematology, Springers Publications, Netherlands, 565 pp

Schmidt, K., Sikora, R. A. and Richter, O., 1993. Modeling the population dynamics of the sugar beet cyst nematode, Heterodera schachtii. *Crop Protn.*, **12:** 490-496.

Shi Ming Wang, and Yan Zhou.2011. "WebGIS Intelligent Diagnosis and Forecasting System of Plant Diseases." Key Engineering Materials, vol. 480–481, Trans Tech Publications, Ltd., June 2011, pp. 1603–1606.

Si Lili and Cao Keqiang, 2006.. Establishment of a real-time monitoring and forecasting system on main crop diseases and pests of China based on GIS. *J. Pl.Protn.*, 33: 282-286

Sigvald, R., 2014. Forecasting and warning systems for pests and diseases of field crops. Academic Agricultural Scince in Lativa-150. Proceedings.Internaitonal Scientific Conference, September 19-21, 2013.Jelgava, Latvia, Rivza, P. Skujans, J. 80-90

Skawsang, S., Nagai, M. K., Tripathi, N. and Soni, P., 2019. Predicting Rice Pest Population Occurrence with Satellite-Derived Crop Phenology, Ground Meteorological Observation, and Machine Learning: A Case Study for the Central Plain of Thailand. *Appl. Sci.*, 9: 4846.

Suruliandi, A.., Mariammal, G. and Raja, S.P.,2021. Crop prediction based on soil and environmental characteristics using feature selection techniques. *Mathematical and Computer Modelling of Dynamical Systems*, 27: 117-140,

Takashi Kobayashi, Tsutomu Ijiri, Twng Wah Mew, Gerard Maningas and Teruyoshi Hashiba, 1995.Computerized Forecasting System (BLIGHTASIRRI) for Rice Sheath Blight Disease in the Philippines. *Japanese J. Phytopathol.*, 61: 562-568

Tilva, V., Patel, J.and Bhatt, C., *2013.* Weather based plant diseases forecasting using fuzzy logic. *Nirma University International Conference on Engineering (NUiCONE)*, 2013, pp. 1-5, doi: 10.1109/NUiCONE.2013.6780173

van Wart J, Grassini P, Cassman KG. Impact of derived global weather data on simulated crop yields. *Glob Chang Biol.*, 19: 3822-3834.

Victor M. Rodríguez-Moreno, Alejandro Jiménez-Lagunes, Juan Estrada-Avalos, Jorge E. Mauricio-Ruvalcaba and José S. Padilla-Ramírez, 2020. Weather-data-based model: an approach for forecasting leaf and stripe rust on winter wheat. *RMetS*, Volume 27, Issue2, March/April 2020

Victor, M., Rodríguez-Moreno,Alejandro Jiménez-Lagunes, Juan Estrada-Avalos, Jorge E. Mauricio-Ruvalcaba andJosé S. Padilla-Ramírez, 2020. Weather-data-based model: an approach for forecasting leaf and stripe rust on winter wheat. *Meteorol. Appl.,* 27:e1896.

Xiao, Q., Li, W. and Kai, Y., 2019. Occurrence prediction of pests and diseases in cotton on the basis of weather factors by long short term memory network. *BMC Bioinformatics*, 20: 688,.

Xuan Xie and Si-qing Yang, 2018. An Warning Information System for Rice Disease Based on WebGIS. 8th International Conference on Applied Science, Engineering and Technology (ICASET 2018). Advances in Engineering Research, volume 159.

5

Emerging Resurgence of Pests and Diseases

Pest resurgence is defined as the rapid reappearance of a pest population in injurious numbers following pesticide application. Use of persistent and broad spectrum pesticides that kills the beneficial natural enemies is thought to be the leading cause of pest resurgence. However, resurgence is known to occur due to several reasons, for instance, increase in feeding and reproductive rates of insect pests, due to application of sub lethal doses of pesticides and sometimes elimination of a primary pest provides favorable conditions for the secondary pest to become primary pests (Harsh Garg, 2014). There are several pesticide induced pest outbreaks reported in walnut, hemlock, cotton and soybean. Among these, brown plant hopper in rice cultivation has gained a major importance in Asian countries. In general, natural BPH populations were kept under check by natural enemies including mired bugs, ladybird beetles, spiders and other pathogens. Pesticides, however, not only have destroyed the natural enemies, but have influenced the fecundity of BPH females further enhancing their resurgence. Additionally, the resurgence of bed bug and cotton bollworm has been reported due to insecticide resistance and indiscriminate use of pesticides.

In nature, pest populations are balanced by the relationship between predators and pests. If predators are removed from or migrate away from a particular environment, opportunistic pest species are able to breed unchecked. It is virtually impossible to annihilate an entire pest population from an area. Survivors may be safe in areas not covered by the pesticide application or survive sub-lethal doses as a result of uneven pesticide application. With predators no longer a factor, they are able to multiply very rapidly causing infestations that are worse than ever before. The problem can be addressed through Integrated Pest Management (IPM) strategies.

5.1 Pesticide Resistance

It describes the decreased susceptibility of a pest population to a pesticide that was previously effective at controlling the pest. Pest species evolve pesticide resistance via natural selection: the most resistant specimens survive and pass on their acquired heritable changes traits to their offspring. Cases of resistance have been reported in all classes of pests (i.e. crop diseases, weeds, rodents, etc.), with 'crises' in insect control occurring early-on after the introduction of pesticide use in the 20th century. The Insecticide Resistance Action Committee (IRAC) definition of insecticide resistance is 'a heritable change in the sensitivity of a pest population that is reflected in the repeated failure of a product to achieve the expected level of control when used according to the label recommendation for that pest species'.

Pesticide resistance is increasing. Farmers in the US lost 7% of their crops to pests in the 1940s; over the 1980s and 1990s, the loss was 13%, even though more pesticides were being used. Over 500 species of pests have evolved a resistance to a pesticide. Other sources estimate the number to be around 1,000 species since 1945 (Miller, 2004). Although the evolution of pesticide resistance is usually discussed as a result of pesticide use, it is important to keep in mind that pest populations can also adapt to non-chemical methods of control. For example, the northern corn rootworm (*Diabrotica barberi*) became adapted to a corn-soybean crop rotation by spending the year when the field is planted with soybeans in a diapause. As of 2014, few new weed killers are near commercialization, and none with a novel, resistance-free mode of action.Similarly, as of January 2019 discovery of new insecticides is more expensive and difficult than ever.

Causes: Pesticide resistance probably stems from multiple factors.

- Many pest species produce large numbers of offspring, for example insect pests produce large broods. This increases the probability of mutations and ensures the rapid expansion of resistant populations.
- Pest species had been exposed to natural toxins long before agriculture began. For example, many plants produce phytotoxins to protect them from herbivores. As a result, coevolution of herbivores and their host plants required development of the physiological capability to detoxify or tolerate poisons.
- Humans often rely almost exclusively on pesticides for pest control. This increases selection pressure towards resistance. Pesticides that fail to break down quickly contribute to selection for resistant strains even after they are no longer being applied.
- In response to resistance, managers may increase pesticide quantities/ frequency, which exacerbates the problem. In addition, some pesticides are toxic toward species that feed on or compete with pests. This can paradoxically allow the pest population to expand, requiring more pesticides. This is sometimes referred to as the pesticide trap or a pesticide treadmill, since farmers progressively pay more for less benefit.
- Insect predators and parasites generally have smaller populations and are less likely to evolve resistance than are pesticides' primary targets, such as mosquitoes and those that feed on plants. Weakening them allows the pests to flourish. Alternatively, resistant predators can be bred in laboratories.

- Pests with limited viable range (such as insects with a specific diet of a few related crop plants) are more likely to evolve resistance, because they are exposed to higher pesticide concentrations and has less opportunity to breed with unexposed populations.
- Pests with shorter generation times develop resistance more quickly than others.
- The social dynamics of farmers: Farmers following the common practices of their peers is sometimes problematic in this case. Overrelying on pesticides is a popular mistake and becomes increasingly popular as farmers conform to the practices around them (Jorgensen Peter Sogaard *et.al,* 2019).
- Unfamiliarity with variation in regulatory enforcement can hamper policy makers' ability to produce real change in the course of resistance evolution.

5.1.1 Major Examples

Resistance has evolved in multiple species: resistance to insecticides was first documented by A. L. Melander in 1914 when scale insects demonstrated resistance to an inorganic insecticide. Between 1914 and 1946, 11 additional cases were recorded. The development of organic insecticides, such as DDT, gave hope that insecticide resistance was a dead issue. However, by 1947 housefly resistance to DDT had evolved. With the introduction of every new insecticide class – cyclodienes, carbamates, formamidines, organophosphates, pyrethroids, even *Bacillus thuringiensis* – cases of resistance surfaced within two to 20 years.

- Studies in America have shown that fruit flies that infest orange groves were becoming resistant to malathion.
- In Hawaii, Japan and Tennessee, the diamondback moth evolved a resistance to *Bacillus thuringiensis* about three years after it began to be used heavily.
- In England, rats in certain areas have evolved resistance that allows them to consume up to five times as much rat poison as normal rats without dying.
- DDT is no longer effective in preventing malaria in some places. Resistance developed slowly in the 1960s due to agricultural use. This pattern was especially noted and synthesized by Mouchet 1988 (Roberts Donald *et.al.*, 2000).
- In the southern United States, *Amaranthus palmeri*, which interferes with cotton production, has evolved resistance to the herbicide glyphosate and overall has resistance to five sites of action in the southern US as of 2021.
- The Colorado potato beetle has evolved resistance to 52 different compounds belonging to all major insecticide classes. Resistance levels vary across populations and between beetle life stages, but in some cases can be very high (up to 2,000-fold).

- The cabbage looper is an agricultural pest that is becoming increasingly problematic due to its increasing resistance to *Bacillus thuringiensis*, as demonstrated in Canadian greenhouses. Further research found a genetic component to Bt resistance.
- The widespread introduction of *Rattus norvegicus* combined with the widespread use of anticoagulent rodenticides such as warfarin has produced almost equally widespread resistance to vitamin K antagonist rodenticides around the world.

5.1.2 Adaptation

- Pests becomes resistant by evolving physiological changes that protect them from the chemical.
- One protection mechanism is to increase the number of copies of a gene, allowing the organism to produce more of a protective enzyme that breaks the pesticide into less toxic chemicals. Such enzymes include esterases, glutathione transferases, and mixed microsomal oxidases.
- Alternatively, the number and/or sensitivity of biochemical receptors that bind to the pesticide may be reduced.
- Behavioral resistance has been described for some chemicals. For example, some *Anopheles* mosquitoes evolved a preference for resting outside that kept them away from pesticide sprayed on interior walls.
- Resistance may involve rapid excretion of toxins, secretion of them within the body away from vulnerable tissues and decreased penetration through the body wall (Dhole Sumit *et.al.*, 2020).
- Mutation in only a single gene can lead to the evolution of a resistant organism. In other cases, multiple genes are involved. Resistant genes are usually autosomal. This means that they are located on autosomes (as opposed to allosomes, also known as sex chromosomes). As a result, resistance is inherited similarly in males and females. Also, resistance is usually inherited as an incompletely dominant trait. When a resistant individual mates with a susceptible individual, their progeny generally has a level of resistance intermediate between the parents.
- Adaptation to pesticides comes with an evolutionary cost, usually decreasing relative fitness of organisms in the absence of pesticides. Resistant individuals often have reduced reproductive output, life expectancy, mobility, etc. Non-resistant individuals sometimes grow in frequency in the absence of pesticides - but not always, so this is one way that is being tried to combat resistance.

Blowfly maggots produce an enzyme that confers resistance to organochloride insecticides. Scientists have researched ways to use this enzyme to break down pesticides in the environment, which would detoxify them and prevent harmful environmental effects. A similar enzyme produced by soil bacteria that also breaks down organochlorides works faster and remains stable in a variety of conditions.

- Resistance to gene drive forms of population control is expected to occur and methods of slowing its development are being studied.
- The above adaptations to pesticides are unusually rapid and may not necessarily represent the norm in wild populations, under wild conditions. Natural adaptation processes take much longer and almost always happen in response to gentler pressures.

5.2 Emerging and Re-emerging Pests and Diseases-Current Scenario

It is of utmost importance to appraise the impact of new / reemerging pests and diseases burgeoned in the recent past and to develop novel technologies for their management. To devise an effective preventive and eradicative strategy for containing these biotic stresses, new research innovations need to be practiced such as deciphering basic/molecular mechanism of host-pathogen/insect interactions; endophytic mechanisms of plant protection; nanotechnology in pest management; host resistance strengthening by gene cloning, recombinant DNA technologies, RNA biology, utilizing gene editing technologies such as CRISPR/Cas9, etc (Kumar *et.al.*, 2021). This article presents a comprehensive account of new biotic stresses of agricultural crops built up in the country and also reviews the novel scientific inventions made worldwide which can be further employed to devise more efficient methods for alleviating impact of these biotic stresses of food crops in the country.

Environmental aberrations often may cause emergence of new pathogenic races/ biotypes. Pathogens and insect pests prevailing with minor status acquire major proportion due to enhanced adaptation capabilities owing to mutational changes for survival. The agricultural crops are also vulnerable to threats from exotic pests/ pathogens/ weeds which may get purposefully or accidentally introduced to the country or new areas in the country in addition to natural dispersal. Insect pests and other pathogens of important crops in India have built up in the recent times under the influence of climate change, mutational changes for survival, acquired resistance to pesticides, transboundary invasions, etc.

5.2.1 Emerging Insect Pests

A pest reported from an area on a particular crop showing considerable increase in its population and potential to cause economic damage over a period of time is termed as emerging insect pest (Kumar *et.al.*, 2021). Sucking insect pests, mirid bug, mealybug, whitefly, aphids and plant hoppers on major crops; *Helicoverpa armigera* on vegetables and pulses; *Spodoptera litura* on vegetables, cotton and oilseeds; Pieris brassicae on crucifers; *Liriomyza trifolii* on vegetables; *Atherigona* spp. on spring maize; aphid complex like Sitobion avenae, *Rhopalosiphum maidis* and *Schizaphis graminum* on wheat, barley and oat; green mirid bug,

Creontiades biseratense on cotton in states such as Karnataka, Tamil Nadu, Maharashtra, Andhra Pradesh; eriophyiid and tetranychid mites on bean, brinjal, cotton, cucurbits, okra, apple, ber, citrus, mango in North India, *Maruca vitrata* on pigeonpea and cowpea in Andhra Pradesh, sugarcane pyrilla on wheat and oat in Chhattisgarh, etc. have been reported to emerge in India3–6. The possible reasons for the emergence of insect pests are climate change, physiological and ecological impacts, change in feeding habit of herbivory, increased overwintering survival, increased number of generations of pest, breakdown of host resistance, change of genotypes/ impact of transgenics, injudicious use of pesticides, modification of cultural practices/tillage, etc. These genetic groups have differed in their ability to transmit begomoviruses and on many occasions this became a reason for the emergence of begomoviral disease epidemics.

Invasive insect pests: Invasive insect species cause huge economic losses and their threat is on continuous rise in the areas of Indian agricultural biodiversity, livelihoods, human and animal health, forestry and biodiversity. Since 1889, a total of 24 insect species have been reported to invade India. India witnessed its first ever invasive insect pest San Jose scale (*Quadaraspidiotus perniciosus* Comstock) in 1879 from China. As on 2019, the tally rose to a total of 23 insect pest species and since 1879 they are affecting the agro-ecological balance of the country and causing huge economic loss over the years. Tomato leaf miner (*Tuta absoluta* Meyrick) has become the most destructive major nocturnal pest of tomato production as well as other solanaceous crops due to its invasions in different parts of the world11. Since it originated in South America, it is known as South American tomato moth, tomato borer or American tomato pinworm. It is also reported to attack potato, eggplant, hot pepper, some weeds (*Datura stramonium* and *Nicotiana glauca*) and some other nonsolanaceous crop plants like green beans or *Malva* spp. Melon thrips (*Thrips palmi* Karny) is a polyphagous pest with wide host range including Solanaceae, Cucurbitaceae and Leguminosae plants in tropical and subtropical countries14. Another notorious invasive pest, the fall armyworm (*Spodoptera frugiperda*) was first noticed at Shivamogga, Karnataka during May 2018. This has been spreading fast into new territories in the states of Karnataka, Telangana, Andhra Pradesh, Maharashtra, Gujarat and Tamil Nadu.

5.2.2 Plant Pathogens

Burgeoning fungal pathogens of plants: Prevalence and distribution patterns of several fungal pathogens have changed over the years. Blast of rice remains the most important fungal disease in India, however, sheath blight's (*Rhizoctonia solani*) severity and drastic spread and increase in almost all rice growing regions during the past few years caused major losses to the yield as well as increasing fungicidal sprays. Similarly, increased occurrence of false smut of rice has been observed in recent years. Among other cereals, large areas under wheat are continuing to be vulnerable to yellow rust pathogen (*Puccinia striiformis*) which is evolving regularly rendering wheat varieties susceptible. Stalk rot, downy mildew and leaf spots are the major constraints in maize crop in India, and sugarcane red rot with continuous emergence of variability in the pathogen is posing a threat to sugarcane yield (Kumar *et.al.*, 2021).

The emergence of virulent Tropical race 4 (TR4) of Fusarium wilt has a devastating effect on banana cultivation in the world and has also been detected in India in the recent past with high incidence and crop damage. Soil-borne fungal pathogens are becoming increasingly devastating to the crops which are difficult to control. These pathogens cause heavy losses in economically valuable crops like pulses, vegetables, rice, etc. In recent years, dry root rot of chickpea (*Rhizoctonia bataticola*) and Phytophthora stem blight of pigeon pea (*Phytophthora drechsleri* f. sp. *cajani*) have emerged as potential threats. Soil-borne diseases caused by fungal pathogens *Phytophthora, Pythium, Rhizoctonia solani* and *Sclerotium rolfsii* are favoured by excess soil moisture, especially in pulses.

The inevitable climatic changes also affected the reproduction, spread and severity of fungal pathogens resulting in the increased frequency of occurrence and severity of different diseases. Climate change reduced the efficiency of many Sr genes in wheat which were governing resistance against Ug99 race of *Puccinia graminis* f. sp. *tritici*. Also, elevated temperature and CO_2 aggravated virulence of *Phytophthora infestans* causing late blight of potato, *Pyricularia oryzae* causing rice blast and *Rhizoctonia solani* causing sheath blight of rice.

Plant health is globally endangered through introduction of exotic pathogens as consequences of increasing globalization, travel and international trade. Certain virulent races of fungal pathogens, viz. Ug99 group of races of wheat stem rust pathogen, wheat blast pathogen, and soybean downy mildew, though not present in India, are a serious threat to our crop production programmes if they occur in India. A severe outbreak of wheat blast in the neighbouring country Bangladesh in early 2016 had roused a serious phytosanitary concern in India.

5.2.2.1 Emerging and Re-emerging Fungi and Oomycete Soil-borne Plant Diseases

Major driving factors of the emergence of plant diseases caused by soil borne fungi and oomycetes (here indicated as "fungi"), in Italy during recent years were discussed (Cacciola and Gullino, 2019). These factors include: accidental introduction of alien pathogens by human activities; effect of climate change; unusually severe weather events; favourable environmental and ecological conditions; pathogen genetic variation; host shifts and expansion of host ranges; introduction or expansion of the geographic range of a susceptible plant species or variety; limited availability of fungicides or development of fungicide-resistance pathogen strains; changes of cropping systems; and/or increased pathogen in soil as a consequence of intensive monoculture of crops. Although in most cases more than a single driving factor contributes to the emergence of an infectious disease, there are examples where a determinant may prevail over others. The case studies reviewed include pathogens belonging to major genera of soil-inhabiting fungi and oomycetes, including *Armillaria, Calonectria, Coniella, Fusarium sensu lato, Ilyonectria, Monosporoascus, Plectosphaerella, Rhizoctonia, Rosellinia, Sclerotinia, Sclerotium, Verticillium, Pythium* and *Phytophthora*. The examples encompass natural and forest ecosystems, economically important agricultural crops including citrus, fruit trees, olive, legumes, vegetables, and ornamentals, as well as exotic or expanding minor crops, such as avocado, goji berry, and pomegranate. Whatever the prevailing driving factor(s), these case studies all show that the large-scale emergence of soil-borne fungal diseases of plants is the consequence of human activities.

i. Phytophthora cinnamomi (Dieback / Root rot): Major determinants of the emergence of infectious plant diseases include anthropogenic introduction of alien pathogens, climate change, severe weather events, favourable environmental and ecological conditions, pathogen genetic recombination or mutation, host shifts or expansion of host range, introduction or expansion of the geographic ranges of susceptible plant species or varieties, limited availability of fungicides and development of resistance, changes of cropping systems, and increases in soil inoculum consequences of monoculture. These factors may be involved individually or together, instantaneously or continuously, simultaneously or in succession.

There is a body of direct or circumstantial evidence indicating that emergence of infectious plant diseases results from human activities and several driving factors usually concur to cause emergence of a disease . A recent study of the genetic structure of *Phytophthora cinnamomi* isolates of worldwide origin, using four microsatellite markers, showed that identical genotypes of this pathogen were associated with the same hosts on different continents (Cacciola and Gullino, 2019). This indicated long-distance transport by man, while the presence of identical genotypes in agricultural settings and neighbouring wildlands would suggested that specific commodities may have been the common sources of recent infestations caused by new invasive genotypes.

Climate change, whose most evident effect is the rise in temperatures, has often been presumed to be responsible for the emergence of soil-borne diseases, simply because they were caused by thermophilic or mesophilic pathogens. This over simplification causes direct and indirect effects of climate change to be underestimated and disease emergence drivers such as severe weather events or conducive environmental conditions, which have been considered distinct, may themselves be a consequence of climate change. It is generally assumed that biodiversity of natural ecosystems give them resilience against invasive exotic pathogens, but if the invader is a polyphagous pathogen, such as *P. cinnamomi* or *P. ramorum*, the result is a loss of biodiversity. It is, therefore, not easy to predict and evaluate the impacts of emerging diseases, as the emergence driving factors are numerous and complex, often interacting with each other.

Cucurbit Downy Mildew: In 2004, an outbreak of cucurbit downy mildew (CDM) caused by the oomycete *Pseudoperonospora cubensis* (Berk. & M. A. Curtis) Rostovzev resulted in an epidemic that stunned the cucumber (*Cucumis sativus* L.) industry in the eastern United States (Gerald J. Holmes *et.al.*, 2015). The disease affects all major cucurbit crops, including cucumber, muskmelon, squashes, and watermelon. The resurgence of CDM represents a significant threat to cucurbit production in the United States. Although host resistance provided protection until 2004, severe epidemics have subsequently occurred each year. Whether this resurgence of the disease is due to a change in the local pathogen population or entry of a new population into the United States, climate change that has allowed for expansion of the geographic area of the pathogen, or a combination of these factors is currently unknown. Nevertheless, significant progress has been made in understanding host-pathogen interactions, the genetic structure of pathogen populations, environmental factors that affect the spread and outbreak of CDM, and resistance management approaches to improve the efficacy of fungicides.

ii. *Bipolaris sorokiniana* (Black Point, Common Root Rot and Spot Blotch diseases of Wheat): Wheat is among the ten top and most widely grown crops in the world. Several diseases cause losses in wheat production in different parts of the world. Bipolaris sorokiniana (teleomorph, *Cochliobolus sativus*) is one of the wheat pathogens that can attack all wheat parts, including seeds, roots, shoots, and leaves. Black point, root rot, crown rot and spot blotch are the main diseases caused by *B. sorokiniana* in wheat (Abdullah M. Al-Sadi, 2021). Seed infection by *B. sorokiniana* can result in black point disease, reducing seed quality and seed germination and is considered a main source of inoculum for diseases such as common root rot and spot blotch. Root rot and crown rot diseases, which result from soil-borne or seed-borne inoculum, can result in yield losses in wheat. Spot blotch disease affects wheat in different parts of the world and cause significant losses in grain yield. This review paper summarizes the latest findings on B. sorokiniana, with a specific emphasis on management using genetic, chemical, cultural, and biological control measures. The search for new sources for resistance should consider finding less susceptible cultivars to all diseases caused by B. sorokiniana, instead of focusing on one disease.

iii. *Neopseudocercosporella capsellae* (White leaf spot in Brassicaceae) : White leaf spot can cause significant damage to many economically important *Brassicaceae* crops, including oilseed rape, vegetable, condiment, and fodder *Brassica* species, and recently has been identified as a re-emerging disease. The causal agent, *Neopseudocercosporella capsellae*, produces foliar, stem, and pod lesions under favorable weather conditions. *N. capsellae* secretes cercosporin, a non-host specific, photo-activated toxin, into the host tissue during the early infection process (Niroshini Gunasinghe *et.al.*, 2020).

5.2.2.2 Emergence of New Bacterial Diseases

Considerable yield losses occur in agricultural crops in India due to infection of phytopathogenic bacteria (Kumar *et.al.*, 2021). Some of them which have emerged with significant importance are bacterial blight of rice (*Xanthomonas oryzae* pv *oryzae*), bacterial blight of pomegranate (*X. axonopodis* pv *punicae*), bacterial wilt of vegetables (Ralstonia solanacearum), black spot of mango (*X. citri* pv mangiferae indicae), and bacterial blight of cotton (*X. citri* sub sp. *malvacearum*). Apart from these, panicle blight of rice (*Burkholderia glumae*) and stalk rot of maize (*Pectobacterium chrysenthemi* var. *zeae*) have also emerged in the recent years as potential threat to yield of respective food grains in India.

i. Bacterial blight of rice: Bacterial diseases cause substantial yield losses to crop plants worldwide. In India the major bacterial diseases that cause considerable damages to crop plants include bacterial blight of rice (*Xanthomonas oryzae* pv. *oryzae*), bacterial blight of pomegranate (*Xanthomonas axonopodis* pv. *punicae*), bacterial wilt of solanaceous vegetables including tomato, capsicum, potato and eggplants (*Ralstonia solanacearum*), black spot of mango (*Xanthomonas citri* pv. *mangiferae indicae*), and bacterial blight of cotton (*Xanthomonas citri* subsp. *malvacearum*) (Kalyan K Mondal, 2016). Besides, bacterial problems that have recently emerged as potential threat include panicle blight of rice (Burkholderia glumae), stalk rot of maize (*Pectobacterium chrysanthemi* var. *zeae*). A detailed account on these bacterial diseases with reference to the research status and future challenges was discussed based on the published literatures in the respective areas. The very purpose of the discussion is to bring an updated status of the research with particular reference to major bacterial diseases and to identify the areas of immediate attention for the future bacteriologists to effectively combat the emerging diseases.

ii.Non-fluorescent oxidase pseudomonads : Several bacteria, previously classified as non-fluorescent, oxidase positive pseudomonads, *Ralstonia*, *Acidovorax* and *Burkholderia* have emerged as serious problems worldwide (Schaad, 2008). Perhaps the most destructive is *R. solanacearum* (RS), a soil borne pathogen with a very wide host range. RS race 3, biovar 2 infects potato and geranium during cooler weather making it an additional threat. Acidovorax avenae subsp. avenae has emerged as a disease of upland rice in Southern Europe during periods of high temperatures and *B. gladioli* has emerged as a serious pathogen of orchids in Thailand. *B. andropogonis* has been identified for the first time on jojoba in eastern Australia; the plant grows under very high temperatures and the disease occurs in nursery stock grown under overhead watering.

Burkholderia glumae is emerging as a problem on rice in the southern United States and the pathogen has become much more prevalent in southern South Korea during periods of high temperatures.

Candidatus Phytoplasma dypsidis: In 2016, several ornamental palms within a conservatory in the Cairns Botanic Gardens, Queensland, died mysteriously. A sample was taken from one of the diseased plants and investigated from the Australian Government Department of Agriculture, Water and the Environment, and state and local government (Lynne M. Jones *et.al.*, 2021). They compared the characteristics and genome of the bacterium identified as the cause of the wilt disease and found the bacterium was similar to other species of Candidatus Phytoplasma, many of which are responsible for disease epidemics in palms elsewhere but was different enough to be an independent species. So far, infection with *Candidatus Phytoplasma* dypsidis has been found to cause disease in 12 different species of palms, including *Cocos nucifera*, which produces coconuts.

Outbreaks of exotic plant pathogens in Australia are rare due to the country's stringent biosecurity measures. "Australia, New Zealand and the Pacific island countries and territories have an enviable plant and animal health status compared to much of the rest of the world. This new disease could spread outside of Cairns and affect palm populations further north. It is important to continue to monitor the spread of *Candidatus Phytoplasma* dypsidis. A number of questions remain, including which insect vectors are spreading the disease, and whether the bacterium is capable of infecting other types of plant, including important crops such as bananas.

iii. *Erwinia tracheiphila: Erwinia tracheiphila* is a bacterial plant pathogen that causes a fatal wilt infection in some cucurbit crop plants. Wilt symptoms are thought to be caused by systemic bacterial colonization through xylem that impedes sap flow (Rocha *et.al.*, 2020). However, the genetic determinants of within-plant movement are unknown for this pathogen species. It was found that *E. tracheiphila* has horizontally acquired an operon with a microbial expansin (*exlx*) gene adjacent to a glycoside hydrolase family 5 (gh5) gene.

5.2.2.3 Emergence of New Viral Diseases

In recent decades, the issue of emerging and reemerging infectious diseases, especially those related to viruses, has become an increasingly important area of concern in plant health. Such diseases in a plant context are generally insect- or seed-transmitted and changes associated with global warming, and accidental introduction of vectors or infected materials in new areas facilitated by global trade, may affect their incidence, severity and diffusion. In particular, the current rapid expansion in human activity, combined with climate change, provides conditions permitting the (re)emergence of plant viruses at the interface between managed and natural vegetation, and outbreaks of emerging viruses therein. Both of these factors are involved in the increasing instability within virus-plant pathosystems.

The 'disease triangle' concept in plant pathology highlights the interaction of both pathogens and plants with the environment. This model can be used to predict epidemiological outcomes in plant health, both at local and global levels.

In particular, for disease to occur, a susceptible plant host, a virulent pathogen, and proper environmental conditions are required. The effect of environmental variables (such as fluctuations in mean temperatures) on pathogens and plants can have favorable outcomes on plant disease progression. Furthermore, climate change and the increase of international trade have given rise to the expansion of vector populations, which carry and spread viruses to new areas. This is the case of begomoviruses transmitted by various cryptic species of the insect pest *Bemisia tabaci* (whitefly). In recent years, several outbreaks caused by tomato yellow leaf curl New Delhi virus (ToLCNDV), in particular on cucurbits, have been reported in different countries in the Mediterranean basin. ToLCNDV had previously remained confined to Asia for more than 20 years and its recently accidental introduction into the Mediterranean coincided with the invasion of the Q2 mitochondrial variant of the *B. tabaci* MED genotype in the same geographical area. Furthermore, the Q2 variant of *B. tabaci* has shown to have higher fitness compared to other genotypes, enhanced by the increase in average temperatures. Therefore, it is of significance to anticipate future epidemics by gathering knowledge through relevant research, together with monitoring the emergence of both viruses and their potential vectors.

Climate change and human population pressure also favour emergence of new viral diseases in plants. Several viruses and their strains or isolates or their combinations caused diseases in crops especially vegetable crops due to introduction of hybrids. Recently, Tiranga disease was reported from Maharashtra and the diseased samples were found to be associated with five different viruses (Kumar *et.al.*, 2021). Rice tungro, ground nut bud necrosis, sunflower necrosis, yellow mosaic of legumes, pigeon pea sterility mosaic, soybean bud blight, cotton leaf curl, cassava mosaic, potato apical leaf curl, banana bunchy top, banana bract mosaic, papaya leaf curl, papaya ring spot, chlorotic leaf spot in peach, piper yellow mottle are some of the emerging viral diseases which have built up in the recent times. The incidence of papaya ring spot virus has also been recorded all over the country.

The number of recognized plant viruses has grown exponentially, in part by the discovery of new virus diseases, in part by the recognition of existing problems as virally caused diseases, in part by the origin of new virus entities and, in part by refining our techniques for separating and identifying the correct causal agents (Damsteegt, 1999). Viruses require avenues of entry into their plant hosts and arthropods of many types play the role of vectors.

Nature requires variability for selection to occur. Variability may be driven by the host component as new resistance genes are developed against a population of vectors and viruses, by the vector component as they mutate and adapt to new host systems, or by the virus as mutations and recombinations occur. Man plays an instrumental role in genetically modifying the host plants, by eliminating weeds and insects with selective pesticides and cultural practices, by rapid dissemination of germplasm to new areas, and by inadvertently or intentionally introducing vectors, viruses, and potential hosts into existing crop systems. These combinations may be detrimental to survival of the viral component or expand opportunities for increase. Examples include: 1) cacao swollen shoot in Africa after cacao was introduced into an environment where the virus had existed in equilibrium with native species; and 2) soybean dwarf in Northern Japan after soybeans replaced rice over large areas of Hokkaido bringing together a susceptible host with an aphid biotype which preferred it and a virus which had been in equilibrium with native legumes.

A list of potential new and emerging diseases is presented in Table 10.

Table 10: New, emerging, re-emerging, and threatening plant virus disease examples

Disease	Category	Major hosts	Vector(s)	Geographic distribution	Potential loss
Sharka (Plum pox potyvirus) D, M, C, Ea serogroups	Threatening	*Prunus* spp.	Aphids	Most of Europe, India, Egypt, Syria, and Chile	Devastating to Prunus production;
Lettuce infectious yellows crinivirus	Emerging	lettuce	*Bemisia tabaci*	SW USA	Serious economic losses in epidemics
Tomato infectious yellows crinivirus	New/emerging	tomato	*T. vaporariorum*	USA, Europe, Mid-East	
Tomato chlorosis crinivirus	New/emerging	tomato	*T. vaporariorum*	USA	
Lettuce chlorosis crinivirus	New/emerging	lettuce	*Bemisia argentifolii*	SW USA	
Cucurbit yellow stunting disorder crinivirus	Threatening	cucurbits	*Bemisia argentifolii*	Europe, Mid-East	High
Sweet potato chlorotic stunt crinivrus		sweet potato	*Bemisia tabaci*	Worldwide	
High Plains virus (sometimes complexed with wheat streak mosaic virus)	New/emerging	cereals	*Aceria tosichella*	Great Plains, USA	High

Disease	Category	Major hosts	Vector(s)	Geographic distribution	Potential loss
Citrus tristeza closterovirus	Emerging	citrus	several aphid species	Worldwide	High when Toxoptera citricida (BCA) are present
Citrus tatterleaf capillovirus	Threatening	citrus	unknown	Unknown	Moderate
Citrus chlorotic dwarf (unknown virus)	Threatening	Citrus	*Parabemisia myricae*	Turkey	Severe
Tomato spotted wilt tospovirus	Re-emerging	Ornamentals/ vegetables	thrips	Worldwide	Severe
Impatiens necrotic spot tospovirus	Emerging	Ornamentals	thrips	Worldwide	Moderate
Tomato yellow leaf curl geminivirus	Emerging	Tomato	*Bemisia* spp.	Widespread, FL, GA	Severe
Raspberry bushy dwarf idaeovirus	Re-emerging	Raspberries	pollen	Pacific Northwest	Moderate
Blueberry shock ilarvirus	New	Blueberries	pollen	Pacific Northwest	Severe

5.3 Factors Contributing to an Increase in New and Emerging Viruses

There are many causes for the apparent increase in new and emerging viruses and in the knowledge gained of viruses not found on the North American continent. In almost all the above scenarios, man plays an important role (Damsteegt, 1999). As misuse of antibiotics has led to the development of bacteria with multiple resistance, misplaced confidence in our ability to control vectors with pesticides has resulted in multiple chemical resistances. Germplasm collecting expeditions, international movement of plant material and, lowered trade barriers offers opportunities for unknown viruses to be introduced into new areas. As we increase the use of transgenic crops carrying viral genes into new vegetative environments, will we continue to increase the rate of new virus evolution, or will we be able to control the viruses now present and create a stable, balanced ecological system?

Three components in the virus disease triangle, plant host, arthropod vector and plant virus all play a role in the rise and fall in occurrence and spread of plant virus diseases. However, super-imposed over all these is the involvement of man as he manipulates the components and the environment in which they interact.

5.3.1 Emergent Plant Viruses

Emergent plant viruses can be placed into two broad categories: entire groups of viruses (e.g., genera or families) or individual viruses (Maria R. Rojas and Robert L. Gilbertson, 2008). Some examples of plant virus groups that are emergent, on a global level, include (1) the whitefly-transmitted begomoviruses (genus Begomovirus, family Geminiviridae), (2) the thrips-transmitted tospoviruses (genus Tospovirus, family Bunyaviridae) and (3) the criniviruses (genus Crinivirus, family Closteroviridae). Some individual viruses that have emerged relatively recently include the potexvirus, Pepino mosaic virus (PepMV), an emergent tomato virus; and the sobemovirus, Rice yellow mottle virus. In addition, there also can be outbreaks of "new" viruses, such as those causing necrosis-associated diseases of tomato in Mexico, Spain and Guatemala. Another example is the emergence of a novel whitefly-transmitted potyvirus infecting cucurbits in Florida.

This emergence can involve a range of mechanisms, depending on the virus(es) involved and the environment. Existing viruses can be moved long distances via human activities, allowing for subsequent establishment and spread in compatible agroecosystems. Irrespective of whether the virus has a DNA or an RNA genome, mechanisms of variability such as mutation, reassortment and recombination allow for the generation of new forms of existing viruses that have the potential to emerge as important pathogens. This is facilitated by new selection pressures placed on the viral population, such as those associated with modifications of existing agroecosystems. However, the rate at which such variants are generated and their economic impact is a function of multiple factors (efficiency of vector transmission, the existing pool of viral genetic variability in a region, nature of the host, aspects of the agroecosystem, etc.).

Novel viruses or viruslike entities can also emerge, and these often appear at interfaces of agricultural and undeveloped lands. In the case of plant viruses, emergence of variants of known viruses or novel agents is greatly facilitated by increases or emergence of insect vector populations. Entire groups of viruses (e.g., the begomoviruses, criniviruses and tospoviruses) have emerged following the worldwide emergence of their insect vectors. Moreover, given the continued global movement of plant materials and seeds, expansion of agriculture into new areas, and the propensity of large-scale commercial agriculture to favor development of large populations of insect pests, it is highly likely that new plant viruses will continue to emerge.

The challenge in dealing with emergence of new viruses is significant, but it has been lessened by improvements in technology and increased understanding of viral genetic diversity, ecology and genetics. Thus, with new detection technologies potentially new emergent viruses can be identified sooner. However, identification of novel viruses can still be very challenging, and often requires a combination of new technology and innovative approaches. Once an emerging plant virus has been identified, effective detection tools need to be developed and applied to answer questions about the biology of the virus. This information can then be used to develop effective management strategies. Ideally, multiple approaches will lead to development of an integrated pest management (IPM) approach. Such an approach has led to the effective management of the emergent TYLCV in the Dominican Republic, and tomato production has actually increased since the introduction of the virus.

5.3.2 Emerging Viral Diseases of Tomato

Viral diseases are an important limiting factor in many crop production systems. Because antiviral products are not available, control strategies rely on genetic resistance or hygienic measures to prevent viral diseases, or on eradication of diseased crops to control such diseases (Inge M Hanssen *et.al.*, 2020). Increasing international travel and trade of plant materials enhances the risk of introducing new viruses and their vectors into production systems. In addition, changing climate conditions can contribute to a successful spread of newly introduced viruses or their vectors and establishment of these organisms in areas that were previously unfavorable. Tomato is economically the most important vegetable crop worldwide and many viruses infecting tomato have been described, while new viral diseases keep emerging. Pepino mosaic virus is a rapidly emerging virus which has established itself as one of the most important viral diseases in tomato production worldwide over recent years. Begomovirus species and other whitefly-transmitted viruses are invading into new areas, and several recently described new viruses such as Tomato torrado virus and new Tospovirus species are rapidly spreading over large geographic areas. In this article, emerging viruses of tomato crops were discussed.

5.3.3 Emerging Virus Diseases Transmitted by Whiteflies

Virus diseases that have emerged in the past two decades limit the production of important vegetable crops in tropical, subtropical, and temperate regions worldwide, and many of the causal viruses are transmitted by whiteflies (order Hemiptera, family Aleyrodidae) (Jesús Navas-Castillo *et.al.*, 2011). Most of these whitefly-transmitted viruses are begomoviruses (family Geminiviridae), although whiteflies are also vectors of criniviruses, ipomoviruses, torradoviruses, and some carlaviruses. Factors driving the emergence and establishment of whitefly-transmitted diseases include genetic changes in the virus through mutation and recombination, changes in the vector populations coupled with polyphagy of the main vector, *Bemisia tabaci*, and long distance traffic of plant material or vector insects due to trade of vegetables and ornamental plants. The role of humans in increasing the emergence of virus diseases is obvious, and the effect that climate change may have in the future is unclear.

5.3.4 Other

Many plant viruses are threatening the global food security such as Rice yellow mottle virus in rice, Maize streak virus in maize, Cacao swollen shoot virus in cacao, Groundnut rosette virus in peanut, Banana bunchy top virus in banana, Rice tungro bacilliform virus in South and South-East Asia, Tomato yellow leaf curl virus of tomato, potyviruses of cucurbits, and soil-borne viruses of cereals (Virginia Rodrıguez Alfredo, et.a.l., 2020). *Potyviruses, Ipomoviruses* (Potyviridae), *Tospoviruses* (Bunyaviridae), *Begomoviruses* (Geminiviridae), Criniviruses (Closteroviridae), *Carlaviruses* (Betaflexiviridae), and *Torradoviruses* (Secoviridae) are seven large and economically important emerging group of viruses, although they differ in many aspects, including particle size, shape, types of vector, and infection modes.

5.4 Emergence of New Nematode Diseases

Our country is also facing a plethora of nematological problems in cereals, pulses, oilseeds, vegetables, fruits, fibre and plantation crops. The root-knot nematodes belonging to the genus of Meloidogyne have gained importance and are responsible for causing severe yield losses to vegetable crops under intensive cultivation (Kumar *et.al.*, 2021). They also became a major threat to protected cultivation of vegetables and ornamental crops. Rice root-knot nematode, *Meloidogyne graminicola* has established in low land rice cultivation. *Pratylenchus* spp. and *Rotylenchulus* spp. are noticed in high frequencies in important pulse and oilseed crops of the country. Guava and pomegranate crops are also facing severe threat from an invasive nematode pest, viz. *M. enterolobii* and *M. incognita*, respectively.

5.4.1 *Meloidogyne Enterolobii*

Meloidogyne enterolobii (synonym *M. mayaguyensis*) has been reappearing in a severe manner in crops like guava across the globe. It is recognized as a cryptic nematodes species widespread in southern Florida (Anonymous, 2013). It is representative of a category of emerging pest species that are initially recognized based on their ability to reproduce on a host or cultivar believed to be resistant to the species. *M. enterolobii*, a phenotypically variable species, was most likely misdiagnosed in Florida as *M. incognita* or *M. arenaria* due to strong morphological similarity to both species. In addition to Florida, it has been reported from south, west and eastern countries in Africa, China and Vietnam, Central and South America, Europe, and recently Mexico. *M. enterolobii* is a particularly aggressive species that can also infect *M. incognita* resistant soybeans and sweet potatoes, and peppers containing the N-resistance gene. This species is now recognized as a major pest of many plant species throughout tropical and subtropical regions of the world. More than 50 host species are known, but many more species are expected to be suitable hosts. The host range includes many of the vegetables grown in the U.S., as well as ornamentals commonly transported by the nursery industry. *M. enterolobii* is thought of as a tropical or subtropical nematode species and it has been frequently intercepted on plant species shipped from tropical countries. *M. enterolobii* has emerged as a major constraint in guava cultivation in India (Poornima *et.al.*, 2016; Ravichandra, 2019).

The resurgence of a pest, Guava root knot nematode (GRKN), known a few years ago to only occur in Florida and North Carolina (NC), has resulted in impacts to South Carolina (SC) nursery stock and sweet potato imports and exports. GRKN was detected in Darlington County fields during a routine survey by Clemson University's Department of Plant Industry (DPI) in September 2017 (Matthew Howle, 2021).

The following shows how each state has responded to this resurgence and how it impacts exports and imports in South Carolina:

- **North Carolina:** Internally quarantined the entire state of NC. Sweet potato planting stock is not allowed to be exported from NC without inspection, sampling, and certification for freedom from GRKN. Impacts to SC growers might include a reduction in available sweet potato planting stock and an increase in the cost of this planting stock.
- **Louisiana:** Externally quarantined the entire state of SC. Their quarantine includes prohibition of imports of fresh market sweet potatoes and sweet potato seeds and slips from SC. Further restrictions impact nursery stock, soil, and equipment movement from the entire state of SC. All SC nursery stock or equipment entering the state of Louisiana must have a DPI-issued certificate indicating the sample is free of GRKN.
- **Mississippi:** Externally quarantined Darlington County, SC. Their quarantine includes regulation of sweet potatoes and sweet potato seeds and slips from SC. Further restrictions impact nursery stock, soil, and equipment movement from the entire state of SC. All SC nursery stock

or equipment entering the state of Mississippi must have a DPI-issued certificate indicating the sample is free of GRKN.

- **Arkansas :** Externally quarantined Darlington County, SC. Their quarantine includes regulation of sweet potatoes and sweet potato seeds and slips from SC. Further restrictions impact nursery stock, soil, and equipment movement from the entire state of SC. All SC nursery stock or equipment entering the state of Arkansas must have a DPI-issued certificate indicating the sample is free of GRKN.

5.5 Invasive Insect Pests

One of the major challenges to humankind is threat to food security due to emerging and invasive pests. Increased global trade in agriculture has increased the chances of the introduction of exotic pests. Papaya mealybug (*Paracoccus marginatus*), cotton mealybug (Phenacoccus solenopsis), coconut mite (*Aceria guerreronis*), serpentine leaf miner (*Liriomyza trifolii* and tomato leaf miner (*Tuta absoluta*) are some examples (Mandeep Rathee and Pradeep Dalal, 2018). Insect pests on an average are estimated to cause 15–20% yield losses in principal major food and cash crops. Pest whose status has been changing from minor to major or secondary to primary pest is termed as an emerging pest. *Bemisia tabaci* on cotton, *Helicoverpa armigera* on vegetables and pulses, *Spodoptera litura* on vegetables, cotton and oilseeds, *Pieris brassicae* on crucifers, *L. trifolii* on vegetables and *Atherigonia* spp. on spring maize, have become increasingly severe during last decade. Increasing incidence of aphid complex, comprising of *Sitobion avenae, Rhopalosiphum maidis* and *Schizaphis graminum* is now observed on wheat, barley and oat. Mites of the Eriophyiidae and Tetranychidae family have emerged as major pests of bean, brinjal, cotton, cucurbits, okra, apple, ber, citrus and mango in Northern India. *Maruca vitrata* has emerged as a predominant pest in recent years in all pigeon pea and cowpea growing areas of India causing up to 42% damage in cowpea in Andhra Pradesh. The invasive pest, coconut eriopyhid mite, Aceria gurrreronis caused 64.16–89.42% nut infestation in at Thane, Maharashtra in 2014. During 2015–16 an epidemic of whitefly was noticed during August in the cotton growing areas of Haryana and Punjab. The situation of emerging insect pests of crops was discussed along with the probable reasons for their changing pest status.

5.5.1 Insect Pest Resurgence- Case Studies

5.5.1.1 Pesticide-induced Plant Hopper Population Resurgence in Rice

Plant hoppers are serious rice pests in Asia. Their population resurgence was first reported in the early 1960s, caused mainly by insecticides that indiscriminately killed beneficial arthropods and target pests. The subsequent resurgence involved two mechanisms, the loss of beneficial insects and insecticide-enhanced planthopper reproduction. Jincai Wu *et.al.*, (2020) identified two forms of resurgence, acute and chronic. Acute resurgence is caused by traditional insecticides with rapid resurgence in the F_1 generation. Chronic resurgence follows application of modern pesticides, including fungicides and herbicides, with low natural enemy toxicity, coupled with stimulated planthopper reproduction. The chemical-driven syndrome of changes leads to later resurgence in the F_2 or later generations. Chronic resurgence poses new threats to global rice production. Findings on the physiological and molecular mechanisms of chronic planthopper resurgence and suggest research directions that may help manage these new threats were reviewed.

Factors contributing to brown plant hopper resurgence: There have been significant advances in noninsecticidal pest management. Insecticides, however, are still indispensable to rice farmers because they are effective, easy to use, and provide immediate results. Synthetic organic insecticides provide effective insect control, but their widespread use has resulted in toxicity to natural pest enemies, toxic residues in plants and the environment, and insect resistance. Resurgence of some pests after insecticide application on rice is becoming common. Such an abnormal increase in the pest population after insecticide application often far exceeds the economic injury level. Insecticide-induced pest outbreaks have been reported in walnut, hemlock, soybeans and cotton. Among the pests infesting rice, the brown plant hopper (BPH), *Nilaparvata lugens* has gained major importance in several Asian countries (Chelliah and Heinrich, 1994). Introduction of high yielding, BPH-susceptible rice varieties, use of high levels of nitrogen fertilizers, continuous cropping, staggered planting, and use of some insecticides are the reported causes for increased BPH populations. Throughout Asia, insecticide is an important component of BPH control, especially in countries where commercial, resistant varieties are not available. Continuous use of insecticides has resulted in BPH resistance to insecticides in Taiwan, Japan and the Philippines. After application of insecticides, BPH resurgence was reported in Bangladesh, India, Indonesia, the Philippines and the Solomon lslands. Most of the hopper burned fields reported or observed in India, Indonesia, Philippines, and Sri Lanka received insecticides before the outbreak. Detailed investigations have been made in the past few years on the insecticide induced BPH resurgence in rice.

Factors contributing to resurgence : Research indicates that a variety of factors contribute to BPH resurgence. The degree of resurgence is dependent on the method, timing, and number of insecticide applications and the level of varietal resistance to BPH (Chelliah and Heinrich, 1994).

Suppression of natural enemies : Suppression of natural enemies following intensive broad-spectrum insecticide application was suggested as an important factor for BPH resurgence in rice. Reduction in the population of the mind predator *Cyrtorhinus lividipennis* after regular spraying with methyl parathion caused BPH resurgence. However, extensive field studies conducted later did not show adequate evidence that reduction in the *C. lividipennis* population caused the resurgence. When BPH resurgence occurred in the field, the population of the most important predators such as spiders *C. lividipennis* and *Microvelia atrolineata* could not increase to a sufficient level to suppress the increasing BPH population. Their investigations further indicated that when resurgence-inducing insecticides were applied in the field, they stimulated BPH population growth regardless of their relative toxicity to natural enemies. Natural enemy destruction was a minor factor.

Insecticide-induced plant growth : In rice, paddy water application of granular formulations of carbofuran, isazophos, ethoprop, and acephate significantly increases plant height. Spraying rice plants with methyl parathion induced tillering. Foliar application of deltamethrin and methyl parathion resulted in increased number of tillers and leaves and increased plant height. The phytotonic effect (healthy, green plants) of certain insecticides may attract more macropterous hoppers immigrating into rice fields. The alighting followed by increased feeding, reproduction, and longevity would increase BPH resurgence.

Feeding rate : The feeding rate of the BPH was observed to be significantly increased at sublethal doses of resurgence-inducing insecticides. In rice plants sprayed with three resurgence-inducing insecticides viz deltamethrin, methyl parathion, and diazinon, the BPH feeding rate was higher than that of the check by 61, 43, and 33% respectively. They further reported that, although the first two insecticides induced plant growth, the improved growth did not compensate for the higher rate of BPH feeding leading to hopperburn of these plants much earlier. Other resurgence inducing insecticides such as quinalphos, cypermethrin, fenthion, permethrin, and fenvalerate also increased the BPH feeding rate.

Influence on BPH biology : BPH population increase after insecticide application was due to stimulation of hopper reproduction. The BPH reproductive rate was enhanced when the hopper was feeding on rice plants sprayed with deltamethrin, methyl parathion, quinalphos, cypermethrin, permethrin, and fenvalerate. The increased reproduction might be due to the action of insecticide residues or their metabolites, chemical changes in the host plant receiving insecticides, or a combination of these factors. Reduction in the length of the nymphal stage resulting in a shortened life cycle and increased adult longevity resulting in a longer oviposition period were additional factors contributing to resurgence.

Changes in the nutrient content of the rice plant: A low concentration of calcium and high levels of nitrogen and phosphorus in insecticide-applied plants contributed to resurgence of the blue leafhopper Zygina maculifrons. However, in chemical analyses of major and minor nutrients in rice plants protected with insecticides, any marked BPH resurgence could not be observed. This area needs further investigation because other nutrients in the plants might provide nutritional evidence on BPH resurgence.

Effect of sublethal doses: To save money, farmers are using low insecticide doses. This practice, combined with the short residual toxicity of many commercial insecticides, will often cause the BPH to be exposed to sublethal insecticide doses. Low doses of resurgence-inducing insecticides increased the reproductive rate of the BPH and reduced the nymphal duration, eventually leading to resurgence.

Insecticide classes and resurgence: Insecticides causing resurgence include some synthetic pyrethroids, organophosphates, and carbamates. No single class of insecticide has been identified to be free from resurgence-inducement.

Insecticide rates: Insecticide rates had distinct influence on the degree of BPH resurgence. Deltamethrin at 30 g ai/ha induced significantly high BPH population compared with 20 and 10 g ai/ ha. With methyl parathion, resurgence rate was highest at 750 g ai/ ha compared with 500 and 250 g ai/ha.

Timing and number of application: The timing and number of insecticide applications ultimately govern BPH resurgence. Foliar sprays applied at 50 and 65 DT (days after transplanting) resulted in a high BPH egg and subsequent nymphal population, reaching a peak at 80 DT.

Method of insecticide application: Foliar spraying induced more BPH resurgence than root zone placement and broadcasting.

Influence of insecticide on nymphal and adult stages : The reproductive rates of BPH exposed to insecticide-sprayed plants during nymphal or adult stage or both varied significantly. The reproductive rate was significantly higher when the BPH was exposed to plants sprayed with resurgence-inducing insecticides at the fourth- and fifth-instar stage as well as at the adult stage.

Genetic resistance of rice varieties : In BPH management, resistant rice varieties, planted over large areas in Asia, play an important role. The extent of BPH resurgence after insecticide application decreased as varietal resistance increased. Populations in insecticide-treated plots compared with untreated plots were 74- 50- and 5-fold for susceptible, moderately resistant, and resistant varieties, respectively. Resurgence-causing insecticides should not be used indiscriminately on moderately resistant varieties because they contribute to a BPH population increase above economic injury levels and accelerate the biotype selection process.

Investigations have indicated that many factors are involved in inducing BPH resurgence. Some insecticides contribute to a favorable environment in the rice ecosystem for the BPH to alight, feed, and survive(Chelliah and Heinrich, 1994). This stimulates BPH reproduction leading to a high population buildup and severe damage. Rice production in Asia is being limited by insecticide-induced BPH outbreaks. To prevent outbreaks, a more natural pesticide management program must be adopted. National programs must thoroughly evaluate candidate insecticides to identify those causing BPH resurgence. Although identifying insecticides that induce resurgence is important in an insecticide evaluation program, their use for increasing field population of rice insects in varietal screening is also valuable.

Target pest resurgence: A pesticide gets a bad name if, as is usually the case, it kills more species than just the one at which it is aimed. However, in the context of the sustainability of agriculture, the bad name is especially justified if it kills the pests' natural enemies and so contributes to undoing what it was employed to do (Species Richness, 2020). Thus, the numbers of a pest sometimes increase rapidly some time after the application of a pesticide. This is known as 'target pest resurgence' and occurs when the treatment kills both large numbers of the pest and large numbers of its natural enemies. Pest individuals that survive the pesticide or that migrate into the area later find themselves with a plentiful food resource but few, if any, natural enemies. The pest population may then explode. Populations of natural enemies will probably eventually reestablish but the timing depends both on the relative toxicity of the pesticide to target and nontarget species and the persistence of the pesticide in the environment, something that varies dramatically from one pesticide to another. The pest bounces back because its enemies are killed.

5.5.1.2 Secondary Pests

The after-effects of a pesticide may involve even more subtle reactions. When a pesticide is applied, it may not be only the target pest that resurges. Alongside the target are likely to be a number of potential pest species that had been kept in check by their natural enemies ( ). If the pesticide destroys these, the potential pests become real ones - and are called secondary pests. A dramatic example concerns the insect pests of cotton in the southern part of the USA. In 1950, when mass dissemination of organic insecticides began, there were two primary pests: the Alabama leaf worm and the boll weevil (*Anthonomus grandis*), an invader from Mexico. Organochlorine and organophosphate insecticides were applied fewer than five times a year and initially had apparently miraculous results - cotton yields soared. By 1955, however, three secondary pests had emerged: the cotton bollworm, the cotton aphid and the false pink bollworm. The pesticide applications rose to 8-10 per year. This reduced the problem of the aphid and the false pink bollworm, but led to the emergence of five further secondary pests. By the 1960s, the original two pest species had become eight and there were, on average, an unsustainable 28 applications of insecticide per year. A study in the San Joaquin Valley, California, revealed target pest resurgence (in this case cotton bollworm was the target species) and secondary pest outbreaks in action (cabbage loopers and beet army worms increased after insecticide application against another target species, the lygus bug). Improved performance in pest management will depend on a thorough understanding of the interactions amongst pests and nonpests as well as detailed knowledge, through testing, of the action of potential pesticides against the various species.

Sometimes the unintended effects of pesticide application have been much less subtle than target pest or secondary pest resurgence. The potential for disaster is illustrated by the occasion when massive doses of the insecticide dieldrin were applied to large areas of Illinois farmland from 1954 to 1958 to 'eradicate' a grassland pest, the Japanese beetle. Cattle and sheep on the farms were poisoned,

90% of cats and a number of dogs were killed, and among the wildlife 12 species of mammals and 19 species of birds suffered losses (Luckman & Decker, 1960). Outcomes such as this argue for a precautionary approach in any pest management exercise. Coupled with much improved knowledge about the toxicity and persistence of pesticides, and the development of more specific and less persistent pesticides, such disasters should never occur again.

5.6 Modelling Pest Population Resurgence

Modelling pest population resurgence due to recolonization of fields following an insecticide application. To investigate the effect of insect recolonization on the insecticide!induced resurgence of crop pests a modified Lotka Volterra predator-prey model was partitioned into two areas sprayed and unsprayed. The unsprayed area provided a source of insects to recolonize the sprayed area, resulting in a change in pest dynamics in both areas following an insecticide application (Trumper and Holt, 1998). Model sensitivity to insecticide selectivity and rates of predation, insect dispersal and pest population increase were examined. Resurgence risk in the sprayed area increased with increasing pest dispersal rate, but decreased with increasing predator dispersal rate. Pest resurgence could also occur in the unsprayed area, especially when prey dispersal rates were low. The extent of resurgence in the unsprayed area could in some circumstances be worse than in the sprayed area itself. The more efficient and longer lived the predators, the greater the level of pest resurgence in both areas following insecticide use. More selective insecticides killing the pest but not the predator reduced resurgence provided that the average life span of the predators was reasonably long[Even highly selective insecticides could cause resurgence of the pest in the unsprayed area. When the prey carrying capacity of the unsprayed area was increased relative to that of the sprayed area resurgent effects in the unsprayed area were reduced but could still be significant under some circumstances.

Bt *cotton* area contraction drives regional pest resurgence, crop loss and pesticide use: Genetically modified crops expressing *Bt proteins* have been widely cultivated, permitting an effective non-chemical control of major agricultural pests. While their establishment can enable an area-wide suppression of polyphagous herbivores, no information is available on the impact of *Bt crop* abandonment in entire landscape matrices (Lu Yanhui *et.al.*, 2021). A resurgence of the cosmopolitan bollworm, Helicoverpa armigera was detailed, following a contraction of Bt *cotton area* in dynamic agro-landscapes of ambient H. armigera population levels, culminating in 1.5-2.1 fold higher yield loss and a 2.0 to 4.4 fold increase in pesticide use frequency in non *Bt* crops like maize, peanut and soybean. This work unveiled the fate of herbivorous insect populations following a progressive disuse of insecticidal crop cultivars and hints at how tactically deployed *Bt* crops could be paired with agroecological measures to mitigate the environmental footprint of crop production.

5.7 Emerging Insect Pests in Indian Agriculture

One of the major challenges to humankind is threat to food security due to emerging and invasive pests. Increased global trade in agriculture has increased the chances of the introduction of exotic pests (Mandeep Rathee and Pradeep Dalal, 2018). Papaya mealybug (*Paracoccus marginatus* Williams and *Granara de* Willink), cotton mealybug *Phenacoccus solenopsis* (Tinsley), coconut mite (*Aceria guerreronis* Keifer), serpentine leaf miner (*Liriomyza trifolii* Burgess) and tomato leaf miner [*Tuta absoluta* (Meyrick)] are some examples. Insect pests on an average are estimated to cause 15-20% yield losses in principal major food and cash crops. Pest whose status has been changing from minor to major or secondary to primary pest is termed as an emerging pest. *Bemisia tabaci* (Gennadius) on cotton, *Helicoverpa armigera* (Hubner) on vegetables and pulses, *Spodoptera litura* (F.) on vegetables, cotton and oilseeds, *Pieris brassicae* L. on crucifers, *L. trifolii* on vegetables and *Atherigonia* spp. on spring maize, have become increasingly severe during last decade. Increasing incidence of aphid complex, comprising of *Sitobion avenae* (F.), *Rhopalosiphum maidi*s (Fitch) and *Schizaphis graminum* (Rondani) is now observed on wheat, barley and oat. Mites of the Eriophyiidae and Tetranychidae family have emerged as major pests of bean, brinjal, cotton, cucurbits, okra, apple, ber, citrus and mango in Northern India. Maruca vitrata Geyer has emerged as a predominant pest in recent years in all pigeonpea and cowpea growing areas of India causing up to 42% damage in cowpea in Andhra Pradesh. The invasive pest, coconut eriopyhid mite, Aceria gurrreronis Keiffer caused 64.16-89.42% nut infestation in at Thane, Maharashtra in 2014. During 2015-16 an epidemic of whitefly was noticed during August in the cotton growing areas of Haryana and Punjab (Kranthi, 2015). The situation of emerging insect pests of crops was discussed along with the probable reasons for their changing pest status.

In the present scenario of rice pest management, host plant resistance is playing a pivotal role. It is the easiest way to reduce pesticide load in rice ecosystems as well as to mitigate pest problem under adverse environmental conditions (Mayabini Jena *et.al.*, 2018). With the development of nitrogen responsive high yielding rice varieties and hybrids, several insect pests and diseases assumed major status. Significant effort has been made in the past to develop resistant varieties against different pests, particularly, brown plant hopper, gall midge, blast and bacterial blight with identified resistant genes. But gradually, most of the developed varieties have lost their resistance resulting in more frequent and severe pest outbreaks. Therefore, again the search has began for new resistant donors of different pests from the existing vast germplasm collection of the country including those identified in the past. The mechanism of resistance in identified genotypes has to be studied by assessing its effect on pest biology and behavior or disease infection pattern. Molecular marker analysis of resistant genotypes (donors) will give an indication of the presence of a known or new gene effective against the particular pest. A set of donors should always be in place to be utilized in resistance breeding programme for the development of pest resistant varieties, which will benefit rice farmers of the country.

5.7.1 Insecticide Resistance Status in *Aphis gossypii* and *A.spiraecola*

Most commonly followed practice to manage sucking insect pests during flushing season in citrus by citrus growers is foliar application of insecticides. Leaf dip bioassays for insecticides, viz., acephate 75 SP, dimethoate 30 EC, quinalphos 25 EC, chlorpyriphos 20 EC, spinosad 45 SC, imidacloprid 17.8 SL and thiamethoxam 25 WG, were conducted to assess the susceptibility of field collected *Aphis gossypii* and *Aphis spiraecola* (Hemiptera–Aphididae) adults (George *et.al.*, 2019). In general, the resistance ratio (RR) values indicated that the current levels in aphid population from major citrus belts of Maharashtra are in very low resistance category. Among seven insecticides tested on adult aphids collected from Nagpur, Amravati and Wardha districts during 2013–2015, neonicotinoid group proved the most effective in causing mortality as indicated by the lower LC_{50} values (0.01–0.04 ppm for imidacloprid; 0.03–0.05 ppm for thiamethoxam). Among the three locations, Amravati population has registered comparatively higher RR values indicating less susceptibility to insecticides. RR values calculated based on the base population indicated that the current resistance levels are between no and very low levels ($1 < RR < 10$).

Resistance to insecticides mainly occurs due to changes in insect metabolic enzymes or due to the development of insecticide insensitive target sites in the insect nervous system (George *et.al.*, 2019). Increased metabolism is often caused by qualitative and/or quantitative change of esterases, GSTs and monooxygenases. Increase of carboxylesterase activity , rarely decrease of carboxylesterase activity, excessive production of different forms of carboxylesterase, overproduction of esterase and qualitative changes in enzyme structure are responsible for esterase-mediated metabolic resistance. Higher GST and lower AChE values in Amravati adult aphid population substantiate the chances of development of insecticide resistance. A GST value of 0.82 ± 0.59 µmol/min/mg and esterase activity of 0.71 ± 0.54 µmol/min/mg in A. gossypii through biochemical assays. AChE terminates nerve impulse by catalyzing the hydrolysis of the neurotransmitter acetylcholine. Organophosphates and carbamates are among the most commonly used (Aldridge, 1950); however, several insect strains escape poisoning because they possess an altered AChE which is less sensitive to the active metabolite of the organophosphate. Our susceptible culture indicated maximum AChE which means that is a highly susceptible population. Possibilities of cross-resistance should be further confirmed using molecular techniques for the same population. This study aids in providing a baseline data to be aware of the current insecticide resistance levels in aphids to further support the development of effective resistance management strategies for tackling such sucking insect pests which is also an efficient vector for CTV in the major citrus growing belts in India.

5.7.2 Insecticide Resistance Status in the Whitefly, *Bemisia tabaci*

The current level of the insecticide resistance to selected organophosphates, pyrethroids, and neonicotinoids in seven Indian field populations of *Bemisia tabaci* genetic groups Asia-I, Asia-II-1, and Asia-II-7 was summed up (Naveen *et.al.*, 2017). Susceptibility of these populations was varied with Asia-II-7 being the most susceptible, while Asia-I and Asia-II-1 populations were showing significant resistance to these insecticides. The variability of the LC_{50} values was 7x for imidacloprid and thiamethoxam, 5x for monocrotophos and 3x for cypermethrin among the Asia-I, while, they were 7x for cypermethrin, 6x for deltamethrin and 5x for imidacloprid within the Asia-II-1 populations.

5.7.3 Mite Pests in North India

The results generated by the Multi Locational Project on Agricultural Acarology, All India Coordinated Project on Agricultural Acarology, Network Project on Agricultural Acarology and the Network Project on Insect Biosystematics since 1983 are described, highlighting the most important mite pests of north India (Janardan Singh and Mahadevan Raghuraman, 2011). The following species are considered major pests in that region: Tetranychidae.

- *Eutetranychus orientalis* (Klein), *Oligonychus coffeae* (Nietner), *Tetranychus ludeni* Zacher, *Tetranychus neocaledonicus* André and *Tetranychus urticae* Koch; Eriophyidae - *Aceria litchii* (Keifer) and *Aceria mangiferae* (Sayed); Tarsonemidae - *Polyphagotarsonemus latus* (Banks). Other 16 species in those families as well as in the Tenuipalpidae are also considered important as plant pests in this area of India. Among the tetranychids, *T. ludeni* was identified as an alarming problem in 1987. Many outbreaks of this pest were recorded from 1988 to 1990 on cowpea, an important summer vegetable of eastern Uttar Pradesh. Okra and eggplant, particularly when grown in the summer, have serious problems with *T. urticae* and *Tetranychus macfarlanei* Baker & Pritchard. *Panonychus ulmi* (Koch) has emerged as a serious problem on the expanding cultivation of apple in Himachal Pradesh, whereas *Petrobia latens* (Muller) populations are increasing in dryland cultivation of Rajasthan, attaining serious pest status mainly on wheat and coriander.

Among the tarsonemids, a serious increase in *P. latus* on chilli has coincided with the growing cultivation of this crop, whereas increasing population levels of Steneotarsonemus spinki Smiley have caused severe damage to rice since its recent discovery in northern India. Serious problems have also been caused to tomato by the eriophyid *Aceria lycopersici* (Wolf) and to ber (*Ziziphus mauritiana* Lam.) by the tenuipalpid *Larvacarus transitans* (Ewing), an emerging serious pest of in Rajasthan area. The reason attributed to the increasing mite infestations is the widespread and continuous use of synthetic pyrethroid pesticides, which negatively affect the predatory mite fauna. The paper focused on problems of mite outbreak and suggests future thrust for use of predatory mites as bio-agents for integrated mite control.

5.8 Genome Editing for Resistance to Insect Pests

Plants are challenged incessantly by several biotic and abiotic stresses during their entire growth period. As with other biotic stress factors, insect pests have also posed serious concerns related to yield losses due to which agricultural productivity is at stake. In plants, trait modification for crop improvement was initiated with breeding approaches followed by genetic engineering. However, stringent regulatory policies for risk assessment and lack of social acceptance for genetically modified crops worldwide have incited researchers toward alternate strategies (Shaily Tyagi, *et.al.*, 2020). Genome engineering or genome editing has emerged as a new breeding technique with the ability to edit the genomes of plants, animals, microbes, and human beings.

Despite success in economically important crops to combat pathogens, their use in insect management has not been exploited to the fullest. Notwithstanding the intellectual exercise toward designing of strategies for insect pest resistance in both insects and plants, the success has been meager. Though editing insects is an intriguing option, it requires caution in the selection of traits that need to be environmentally friendly so that the food chain is not affected. The major setback has been the lack of availability of target genes in contrary to other stresses. Therefore, emphasis needs to be given by scientists globally to the identification of resistance sources which can form a platform for insect management. Various innovative targets are being envisaged on the basis of their suitability to reduce off-target effects on target insects. Extensive phenotyping and enrichment of the resistance gene pool is an important hint that needs to be addressed in a path breaking manner. This requires assessment of the available germplasm including wild relatives of specific crops for insect response and identification of stress responsive genes using multi omics approaches (Shaily Tyagi, *et.al.*, 2020). Multiplex editing of such identified resistance genes using high-throughput transformation techniques form the nitty gritty of insect management using genome editing. With such studies already in vogue, targeted mutations leading to the conversion of susceptible plants to those that are able to manage their respective pests is not distant in space or time. These aspects together with regulatory considerations on the gene edited crops could lead to the success of the technology not only toward scientific progress but also societal acceptance.

5.9 Insect Resistance Management

Growers have rapidly adopted biotech-derived crops that have been improved to express proteins for insect control, because these crops provide excellent protection from key damaging insect pests around the world. Insect protected crops also offer superior environmental and health benefits while increasing grower income. However, insect resistance development is an important concern for all stakeholders, including growers, technology providers, and the seed companies that develop biotech-derived crops (Anonymous, 2011). Given the benefits associated with insect protected seed, insect resistance management (IRM) must be a consideration when cultivating these crops. It is important that concerns about resistance should not be allowed to prevent the use of biotech crops; rather, these concerns should result in stewardship and management programmes that effectively delay resistance while enabling the benefits of the technology to the environment and to agriculture to be realised. The technical data and practical experience accumulated by developers, researchers and growers with insect protected crops in many global regions can inform different aspects of resistance management leading to robust, science-based IRM plans. A range of elements should be considered in assembling any IRM strategy, including:

- Key target pest biology/ecology;
- Efficacy of and target pest sensitivity to the insectprotection traits;
- Pyramided versus single insecticidal traits;
- Product deployment patterns;
- Local cropping systems; and
- Availability of alternative pest management options, including biotech, chemical, biological and cultural control options.

In addition, insect susceptibility monitoring measures any changes in pest susceptibility to the insect protected crop, while stakeholder and grower communication and education inform the end-users of any IRM stewardship guidelines and requirements. Infrastructure should be developed such that a remedial action plan can be designed and implemented should resistance develop. Each of these elements is described in more detail, with specific examples of how they can be combined and tailored to local/regional environments and grower practices. Insect resistance management plans need to be suitable for the given production situation. What works for large monoculture production systems in North and South America is unlikely to be appropriate for the small, more diverse agriculture of Southeast Asia or Africa. Although it is clear that insect protected crops impart considerable value to growers, it is also clear that it is in the best interest of all stakeholders to preserve insect protected crops for the long-term benefits they provide.

Successful resistance management should begin before introduction of insect protected crops through:

- The establishment of a local infrastructure of experts who can provide input on key target pest biology and pest/crop interactions that can help to tailor IRM recommendations;
- Evaluation by scientific experts of the available data to determine if additional research is needed to support implementation of an initial IRM plan and to refine the deployment of an IRM strategy as new research data and experience with the technology becomes available;

- Provision by grower groups and farm advisor organisations of specific crop production practices; evaluation of how IRM components can be practically implemented; and how information on IRM can be effectively disseminated;
- Provision of technical information from commodity groups and technology developers, to enable the development of educational materials for stakeholders and users;

Should resistance in insect protected crops develop, insect damage would undoubtedly increase, leading to a reduction in crop yield and in the value of the insect control technology (Anonymous, 2011). Confirmed resistance would have negative consequences for the environment and growers, which might include product withdrawal from the market and expanded use of insecticides for pest control. However, consequences differ by pest: for key target pests, resistance would lead to reduced yields, but for secondary pests, the major benefit of the technology remains and over-reaction (e.g., complete product withdrawal) could unnecessarily lead to a complete return to synthetic chemistries, or could put more selection pressure on remaining pest management practices.

In order to identify the best tactics for a robust IRM strategy, to extend the durability of insect protected crops for a given country or region, the following elements should be evaluated, viz., pest biology and ecology; trait efficacy and dose; expected product deployment patterns; local cropping systems; use of insect protected crops within IPM programmes; insect baseline susceptibility and monitoring options; stakeholder and grower communication and education and a remedial action plan in the event of resistance development.

Each of these elements must be assessed with a specific focus on the local conditions and agricultural practices that should be considered in the development of an IRM plan, because no matter how detailed the scientific information, these local conditions are critical for successful IRM deployment. Effective stakeholder participation, including experts from technology providers, grower and commodity groups, academia, and government, is another important factor during the development and implementation of an IRM strategy. As new information is collected from research efforts on pest biology and the genetics of resistance and as more practical experience accumulates with insect protected crops in various regions and countries, IRM plans and tactics should be reviewed and refined to extend the durability of all insect protected crops globally.

5.9.1 Management of Resurgence

The basic principle in dealing with resurgence is to integrate various pest management tactics to reduce the insecticide applications. The required applications can be lnodified in a way so as to avoid hol-moligosis (i.e., reproductive stimulation by sublethal doses of pesticides) and destruction of natural enemies after insecticide treatments is the other way of resurgence management. The following practices may be adopted for avoiding the pest resurgences.

a) **Avoid hormoligosis:** Eliminate or rninirnize application of sub-lethal doses of pesticides particularly in those areas where this problem has occurred or is expected. For reduction of application of sub-lethal doses, apply insecticide on target and avoid drift and also follow rules of good pesticide practices including proper chemical and formulation selection.

b) **Avoid natural enemy destruction:** Reduction in destruction of natural enemies is mostly based on pesticide selectivity which ranges between insects and warmblooded animals or between species of arthropods. Use of selective chemicals that affects the pest population hormoligosis more than the natural enemies is termed r as Plzysiologicul selectivity. Another way is to use otherwise non-selective chemicals in a selective manner and is termed as Ecological selecti~lity. The ecological selectivity approaches include:

   i) Monitoring pest population and follow recommended economic threshold,

   ii) Treat with the lowest pesticide rates possible,

   iii) Avoid treating broad areas,

   iv) Adopt timely treatments to avoid natural enemy destruction, and

   v) Use the most-selective insecticide formulations or change to another insecticide formulation.

C) **Inoculative release natural Enemies:** This approach is being used to prevent pest upsets and to repopulate the area with inoculative releases of natural enemies after insecticide application. When insecticide residues have declined and pest populations are recovering, releases of insectry-reared natural enemies are made, which affect natural control once more.

Avoid pest resurgence: Pest resurgence is the rapid reappearance of a pest population in injurious numbers, usually brought about after the application of a broad-spectrum pesticide has killed the natural enemies which normally keep a pest in check.A well-known example in rice cultivation is the resurgence of brown plant hopper (BPH). If no pesticides are used, BPH is kept under control by its natural enemies (mirid bugs, ladybird beetles, spiders and various pathogens). Pesticides kill the beneficials and create a situation where populations of BPH can multiply rapidly and thus become a man-made pest.Resurgence can be easily avoided by not spraying pesticides. But for many farmers it is difficult to recognize that resurgence has occurred in their field. They spray regularly because they see pests in their fields, without realizing that it is actually the spraying which is causing the pest problem.Another example is about spider mites. Spider mites are normally kept under control by populations of predatory mites. If pesticides are used, the predatory mites get wiped out and the populations of spider mites can increase and become a problem. The farmer responds by spraying more (to control the spider mites) while the proper response would be to stop spraying so that predatory mites can come back to control the pest.

5.9.2 Insect Resistance Management in *Bt* cotton by MGPS (multiple genes pyramiding and silencing)

The introduction of *Bacillus thuringiensis* (*Bt*) cotton has reduced the burden of pests without harming the environment and human health. However, the efficacy of Bt cotton has decreased due to field-evolved resistance in insect pests over time. In a review, Zafar *et.al.*, (2020) have discussed various factors that facilitate the evolution of resistance in cotton pests. Currently, different strategies like pyramided cotton expressing two or more distinct *Bt* toxin genes, refuge strategy, releasing of sterile insects, and gene silencing by RNAi are being used to control insect pests. Pyramided cotton has shown resistance against different cotton pests. The multiple genes pyramiding and silencing (MGPS) approach has been proposed for the management of cotton pests. The genome information of cotton pests is necessary for the development of MGPS-based cotton. The expression cassettes against various essential genes involved in defense, detoxification, digestion, and development of cotton pests will successfully obtain favorable agronomic characters for crop protection and production. The MGPS involves the construction of transformable artificial chromosomes, that can express multiple distinct *Bt* toxins and RNAi to knockdown various essential target genes to control pests. The evolution of resistance in cotton pests will be delayed or blocked by the synergistic action of high dose of *Bt* toxins and RNAi as well as compliance of refuge requirement.

Different studies suggested that the insecticidal efficacy of *Bt* cotton varied owing to variable expression of *Bt* protein during the cotton growing season. The factors affecting the Bt concentration in transgenic cotton are discussed in Table 11.

Table 11: Factors reducing the Bt concentration in cotton

Factors	Effects	References
Salinity	Significantly decrease the concentration of Bt protein in cotton leaves and insecticidal activity against cotton bollworm hampered with enhancement of soil salinity (11.46 dS•m^{-1}). As for as soil salinity increases 9.1 dS•m^{-1}, significant reduction of insecticidal protein in squares was observed.	Wang *et al.* 2018
Waterlogging	The *Bt* protein content greatly decreased in squares by waterlogging. It reduced the Bt protein content frome 38% to 50% in leaves from the 1st to the 3rd week under stress.	Luo *et al.* 2008
Drought	In water deficit stress the content of Bt protein in boll shells and associated insect resistance decreased. In moderate water-deficit conditions Bt concentrations decreased in leaves, flowers and bolls.	Zhang *et al.* 2017

Humidity	High humidity reduced the Bt toxins in leaf of cotton	Yuan *et al.* 2012
Temperature	31–35 °C is the best range for maximum expression Cry1Ac toxin in transgenic cotton. High temperature reduced the leaf (37 °C) and square (above 38 °C) protein content, and notable reduction of insecticidal protein in boll shell was observed at 38 °C after 24 h.	Rana *et al.* 2015, Zhang *et al.* 2018
Genotype	Expression of *Bt* insecticidal protein varies among different genotypes of cotton.	Adamczyk and Sumerford 2001; Bakhsh *et al.* 2012; Khan *et al.* 2018b
Plant age	In *Bt* cotton, content of insecticidal protein varies with the age of plant, and after 110 days of planting, level of toxins falls below the threshold level.	Dong and Li 2007, Kranthi *et al.* 2009
Plant parts	Cry1Ac expression was the highest in leaves, followed by squares, bolls and flowers.	Chen *et al.* 2017a, Chen *et al.* 2018
Agronomic practices	Insect resistance and Bt expression were enhanced by nitrogen fertilizer. 14% increase of Bt toxins in leaf was observed by high nitrogen fertilizer whereas plant growth regulators enhance the Bt toxins in squares.	Chen *et al.* 2018, Chen *et al.* 2017a

The adoption of *Bt* cotton increase the yield, profit and reduced the application of pesticides as well as load of insect pests without harming the human health and environment. The development of resistance in insects and pests has reduced effectiveness of single and pyramided *Bt* cotton (Zafar *et.al.*, 2020). The modification in midgut receptors, lack of high dose/refuge, cross resistance and fluctuation in expression of *Bt* protein during growing season are major factors that facilitate in resistance development. Besides, the resistance development in cotton pest and the drastic increment in population of secondary pest due to less application of insecticides have become a major concern for *Bt* cotton growers. Currently, different strategies like pyramided cotton expressing two or more distinct genes, refuge strategy, releasing of sterile insects, seed mixture, and genome editing by CRISPR/Cas9 and RNAi are being used to control insect pests. Recently, studies have suggested that insect pests (i.e., *P. gossypiella, H. zea, S. frugiperda*) have developed tolerance against dual gene pyramided cotton, and refuge also lost its efficacy in case of non-recessive resistance, i.e., cotton bollworm. The insects are remarkably adaptable and can develop resistance to any control tactics, including transgenic plants containing multiple *Bt* toxins and RNAi. The innovations like genetically modified *Bt* toxins and discovery of insecticidal proteins from bacteria other than Bt will continue to provide new tools for pest control. The MGPS-based cotton will be more durable with compliance of high refuges and other control tactics.

5.9.3 Slowing and Combating Pest Resistance to Pesticides

Pesticides can be used to control a variety of pests, such as insects, weeds, rodents, bacteria, fungi, etc. Over time many pesticides have gradually lost their effectiveness because pests have developed resistance, a significant decrease in sensitivity to a pesticide, which reduces the field performance of these pesticides. Environmental Protection agency (EPA) is concerned about resistance issues. We believe that managing the development of pesticide resistance, in conjunction with alternative pest-management strategies and integrated pest management programs, is an important part of sustainable pest management. To address the growing issue of resistance and preserve the useful life of pesticides, EPA are embarking on a more widespread effort and set of activities aimed at combating and slowing the development of pesticide resistance. Our effort includes releasing and requesting comment on two Pesticide Registration Notices (PRNs) that focus on strategies to combat or slow pesticide resistance (Anonymous, 2021).

Conventional pesticides and resistance management: PRN 2017-1, Guidance for Pesticide Registrants on Pesticide Resistance Management Labeling, revises and updates PRN 2001-5. PRN 2017-1 applies to all conventional, agricultural pesticides (i.e., herbicides, fungicides, bactericides, insecticides, and acaricides). The updates in PRN 2017-1 focus on pesticide product labels. These updates are aimed at improving information about how pesticide users can minimize and manage pest resistance. PRN 2017-1 updates PRN 2001-5 with the following three general categories of changes, viz., additional guidance to registrants, and a recommended format, for resistance management statements or information to place on labels; references to external technical resources for guidance on resistance management; and updated instructions on how to submit changes to existing labels in order to enhance resistance-management language.

PRN 2017-2, Guidance for Herbicide Resistance Management Labeling, Education, Training, and Stewardship, applies to herbicides only. PRN 2017-2 communicates the Agency's current thinking and approach to addressing herbicide-resistant weeds by providing guidance on labeling, education, training, and stewardship for herbicides undergoing registration review or registration (i.e., new herbicide active ingredients, new uses proposed for use on herbicide-resistant crops or other case-specific registration actions). This guidance is part of a holistic, proactive approach to slow the development and spread of herbicide-resistant weeds and to prolong the useful lifespan of herbicides and related technology. PRN 2017-2 also provides 11 elements focused on labeling, education, training and stewardship strategies. Herbicides posing the least risk of developing herbicide-resistant weeds will have the fewest resistance management elements, and herbicides that pose the greatest risk of resistance will have the most elements. More holistic guidance is being focused for herbicides first because they are the most widely used agricultural chemicals; no new herbicide mechanism of action has been developed in last 30 years; and herbicide-resistant weeds are rapidly increasing.

Biopesticides and resistance management: Biopesticides are compounds derived from biological sources, such as plants, bacteria, fungi, or viruses. Many of these products are used in organic crop production, and they can be included in resistance-management programs by both conventional and organic growers.

Regulation of plant incorporated protectants : Plant-incorporated protectants (PIPs) are pesticidal substances produced by plants and the genetic material necessary for the plant to produce the substance. PIP crops are plants whose genomes include foreign genes that code for compounds that have pesticidal activity. For example, PIP crops such as corn and cotton have genes for the production of *Bacillus thuringiensis* (*Bt*) endotoxin (derived from the bacterium *Bacillus thuringiensis*) (Anonymous, 2021). EPA registers and regulates PIPs (the expressed protein and its genetic material) as pesticides under FIFRA but does not regulate the plant itself. Regulatory oversight includes human health and environmental fate risk assessments as well as resistance-management risk evaluations. According to the Code of Federal Regulations (Title 40, Part 174—Procedures and Requirements for Plant-Incorporated Protectants) the characteristics of PIP crops, such as their production and use in plants, their biological properties, and their ability to spread and increase in quantity in the environment distinguish them from traditional chemical pesticides. Therefore, PIPs are subject to different regulatory requirements and procedures as compared to traditional chemical pesticides.

A major component of EPA's regulation of PIPs is the requirement for an in-depth insect resistance management (IRM) strategy. The primary component of IRM for *Bt* PIPs is the use of refuges to reduce selection pressure for resistance. A refuge typically consists of non-PIP crops that are planted to serve as source of susceptible insects to mate with any resistant insects arising from the PIP crop. Types of refuges include:

- Block refuge (a dedicated non-PIP portion of the PIP crop).
- Seed blends (non-PIP refuge seed is comingled with PIP seed in the same seed bag).
- Natural refuge (weeds, wild hosts, and other crops that produce susceptible insects, used for *Bt* cotton PIPs).

Other aspects of IRM for PIPs include IPM-based stewardship, resistance monitoring, grower education, compliance monitoring (for refuge deployment) and mitigation strategies in the event of confirmed resistance.

5.9.4 General Recommendations Related to Resistance Management in Agriculture

The goal of successful resistance management is to reduce populations of pests, whether they are resistant or susceptible to pesticides. The general recommendation when using pesticides is to alternate or tank-mix pesticides of different chemistry and modes of action. This recommendation is based on the theory that the likelihood of a population developing resistance to two or more pesticides, each from a different chemical group with different modes of action, is significantly less than the likelihood that a population could develop resistance to only one of the pesticides. Therefore, individuals in the pest population resistant to one of the pesticides would be killed upon exposure to the (different) partner pesticide. However, there are some cases where even this strategy has resulted in resistance to both pesticide groups.

The best way to avoid resistance is to use IPM strategies to prevent the increase in resistant pest types, such as:

- Rotating crops to reduce the use of the same pesticides season after season.
- Reducing nutrient sources such as plant stubble that can harbor pathogens and insects.
- Using grazing animals in some cases to reduce (some) weed populations.
- If available, using scientifically verified methods to sample pest populations and correlate them to economic estimates of crop damage before applying pesticides.
- Timing chemical applications when the pest is most susceptible (e.g., waiting for insect emergence or using preventive fungicide applications when weather conditions appear to be conducive to disease development).
- Scouting before and after pesticide application to correctly identify the pest and to determine if the application provided effective control.

5.10 Resurgence / Emergence of Plant Diseases

A disease is recognized as emerging if it is new, it occurs in a new host, there is an unexpected outbreak, its economic importance increases, if it attracts public opinion and the scientific community regardless of economic importance, or if it appears in an area for the first time (referred to as geographic emergence).

In many countries or regions, incidence of emerging diseases increases quickly over a few months with higher emergence rates in some countries than others. This feature may be due to the combination of favorable conditions (Graciela Dolores Avila-Quezada *et.al.*, 2018)..

5.10.1 Novel Plant Pathogens listed in EPPO Quarantine/ Alert List, and their Recent Worldwide Distribution (Graciela Dolores Avila-Quezada *et.al.*, 2018).

i. ***Thekopsora minima*** : *Thekopsora minima* is a rust that causes premature defoliation on blueberries (*Vaccinium corymbosum* L.) and also affects Azalea spp.. The pathogen was recorded in North America and Japan and has been introduced on blueberries in other countries (e.g., in South Africa in 2006, Australia in 2012 and Germany in 2015). In 2012 in Tasmania all infected plants were eradicated. In Mexico the disease is listed as economically damaging. The risk in Mexico is high because the market for berries is very important for the country's economy. Rust spores are spread over longer distances by wind and hurricanes besides trade.

ii. ***Raffaelea lauricola*** **:** It is a fungus that causes mortality of redbay (Persea borbonia). The fungus and its vector, the ambrosia beetle *Xyleborus glabratus*, are originated from Japan and Taiwan. The disease has been observed in the Southeastern USA since 2003; more recently, the disease was reported in Florida causing avocado wilt disease. The risk for the Mexican avocado is high because it is one of the most important exportation fruit, and the domestic consumption is high.

iii. ***Candidatus Phytoplasma phoenicium*****:** It is a bacterium that causes almond witches′ broom and was reported for the first time in almond in Lebanon in the 1990s. Later, symptoms of almond witches′ broom were observed in Iran, although some genetic variability was observed among the strains. In 2001, this disease was added to the EPPO Alert list, but was removed from the list in 2006. However, because the new phytoplasma species *Candidatus Phytoplasma phoenicium* was later found in several major stone fruit trees in Lebanon and Iran, it was added again to the EPPO alert list. Perhaps its genetic variability led the pathogen to be more aggressive to its host and even to extend its pathogenicity to a larger number of hosts. These latter hypotheses represent the current major risk around the world. *Asymmetrasca decedens* Paoli is reported as the vector of this pathogen.

iv. **Tomato leaf curl New Delhi virus (ToLCNDV):** It was first described on tomatoes in 1995 in India then other countries in Asia found the virus on a wide range of crops. ToLCNDV was observed on courgette (Cucurbita pepo var. giromontiina) in 2012 in Spain. After Spain, it was detected in Tunisia in January 2015, causing high severity on cucumber (*Cucumis sativus* L.), melon (*Cucumis melo* L.) and courgette (*C. pepo* var. *giromontiina*). Draws attention that in Italy it was reported in autumn the same year. The virus is transmitted in a persistent mode by the whitefly *Bemisia tabaci* (Gennadius). The insect vector migration could be the reason for the spread of the disease in Spain, then in Tunisia and in Italy. The alert of this disease is by its rapid dispersion and wide range of crops can affects.

v. ***Euwallacea*** **sp.,:** It is a pest of tea associated with *Fusarium ambrosium* was reported in Sri Lanka since the mid-2000s. Currently, the complex has been reported in Africa (i.e., Comoros, Madagascar, and Reunion), in America in Costa Rica, Guatemala, Panama, and in the USA (California, Florida, and Hawaii) and many countries in Asia. It affects woody plants, such as avocado. *Euwallacea* sp. and *Fusarium euwallaceae*, its obligate symbiotic fungi, is causing mortality on several trees in California (USA) and Israel. It is still unknown the way both were introduced into these countries. Although, transport of diseased plants likely contributed to the introduction and movement of infested plant material is likely to ensure broad dispersion.

vi. *Puccinia graminis* **f. sp. *tritici* Ug99:** It is present in Uganda, Kenya, Ethiopia, Sudan, Yemen, Iran, Tanzania, Eritrea, Rwanda, Egypt, South Africa, Zimbabwe and Mozambique. It affects wheat causing losses of 70% or more. A new virulent strain was identified in wheat fields in Uganda, in 1999, it was designated Ug99. This new race broke the resistance conferred by the gene Sr31. Transmission, as other rusts, has been carried by wind or the movement of people carrying the fungus (e.g., contaminated clothing).

The list of invasive or emerging pathogens is increasing. In addition to the above mentioned pathogens, new and emerging pathogens include *Neofusicoccum parvum* (Pennycook & Samuels) Crous, Slippers & A.J.L. Phillips 2006 , *Emaravirus* sp. (Rose rosette virus) and *Pantoea stewartii*.

5.10.2 Possible Causes of Emerging Pathogens

Some hypotheses for the emergence of new pathogenic organisms (i.e., bacteria and fungi) are the organism may be endemic in the crop regions but the new host discovered recently; after being endemic the organism became pathogenic, due to an increase in the organism's virulence, or due to a decrease in the host´s defenses; the organism may have been recently introduced into a new area and previously unexposed hosts and the organism is pathogenic to novel plants (e.g., chili pepper); or insect vectors exploit (i.e., feed upon) new plants harboring the pathogenic organism and transmit the organism to subsequent plants.

Many authors mention that adverse factors can interact and "help" to subsequently cause complex diseases. For example, climatic factors could change the nature of microorganisms turning them into opportunistic pathogens (Graciela Dolores Avila-Quezada *et.al.*, 2018).

Changes in gene functionality can also contribute to risks of a new pathogen. The occurrence of lateral transfer is well-documented between prokaryotic species and from prokaryotic to eukaryotic species. This process is exemplified when isolated incongruent genes are absent in closely-related species and are in species more distantly related. However, the lateral gene transfer argument has some difficulties. The orthologous gene could not been detected in organisms that are related. Although it is possible that the genes may not be avowed as similar but may still develop a high rate of homologous relationships. Changes to gene functionality may also be caused by mutations over translocations and gene creep which can be a result of a high number of genes copied. Thus, in a lineage, an incongruent gene can be developed by selective loss of genes. Differences lateral gain and lineage loss would be more discernible if more genome sequences were available. Additional evidence of gene functionality is that some strains of a species may lack a gene. This could happen when a recently acquired gene or a gene that had marginal evolutionary benefits. These genes are an example where the microorganism possesses said genes with functions as avirulence and necrosis-inducing.

Almost always, species have a subset of the genetic variation. However this is possible since lateral transfer of ToxA was successfully evidenced by genetic variation(Graciela Dolores Avila-Quezada *et.al.*, 2018). Analysis of transferred genes codon may show the bias in the donor organism rather than in the recipient one, when distantly related microorganisms are compared. Transposons are mechanisms that mediate chromosomal rearrangements. Movement of DNA from a donor organism is more probable when the gene is associated with a transposon, because the tendency of transposons is to insert new sequences or recombination events that lead to new genes. For example, *Magnaporthe oryzae* transposons have been shown to generate high recombination rates, resulting in high rates of gene sequence evolution and gene duplications. The core of lateral gene transfer must examine the function of supernumerary or b-chromosomes.

Are UV-B, O_3, and CO_2 involved / linked to emergence or pathogenicity? : Pathogens may be reportedly affected by climate change but strong evidence is lacking. Ultraviolet B (UV-B), ozone (O_3), and carbon dioxide (CO_2), are climatic elements that have been commonly addressed relative to climate change and the emergence of pathogens (Graciela Dolores Avila-Quezada *et.al.*, 2018). Some species present different sensitivity to UV-B radiation; while some may be negatively ingfluenced others may be tolerant. Increased UV-B after inoculation reduced disease in wheat, presumably because direct UV-B damaged the pathogen. *Septoria tritici* in wheat was very sensitive to UV-B rays. Although the potential effect on pathogens of UV irradiation is uncertain, these pathogens are rarely exposed directly to sunlight. Ozone (O_3) affects pathogens and host plants differently. It is unlikely that O_3 has direct adverse effects on pathogens, but they assure O_3 adverse effects are mediated by host plants. Similarly, the same authors have proclaimed that main effects of increased irradiation on diseases would be via alterations in the host plant, and that pathogens are less possibly to be negatively affected by UV-B, CO_2, and O_3 , than their corresponding host are. Climate change has been inconclusively linked neither to emerging pathogen nor to changes in plant pathogenicity gene mutations. Very little in known regarding the direct impacts of climatic change factors on emerging diseases and we currently have limited knowledge of the way the pathogens manage to adapt to climate change. Plant pathologists should be encouraged to consider the relationship(s) of CO_2, O_3 and UV-B factors relative to changes in genes affecting pathogenicity of microorganisms.

5.10.3 What Causes Diseases to "Emerge"?

There are many reasons why a plant disease may "emerge," or increase in importance.

i. The introduction of a new pathogen species or a new type within a species may be the most obvious source of emergent pathogens. An example of the latter is race Ug99 of the wheat stem rust pathogen, discovered in East Africa and moving north through important wheat- growing regions (Karen A. Garrett *et.al.*, 2010). Wheat stem rust has caused little yield loss in the United States in recent years because of effective re sis tance in common wheat varieties. But U.S. varieties do not generally have resistance to Ug99, so there is great urgency to develop effective resistance in U.S. cropping systems before Ug99 arrives through contaminated plant materials or even on the clothes of tourists.

ii. An increase in the availability of susceptible crop acreage may lead to pathogen emergence. A classic example of a surprising emergent disease is southern corn leaf blight (SCLB). While the disease was known in the United States, it had typically not been a serious problem. When corn breeders switched to a par tic u lar form of male sterile cytoplasm for varieties grown throughout the United States in the 1970s, SCLB suddenly became much more important, and scientists realized that this cytoplasm conferred susceptibility to SCLB. Homogeneity of resistant plant varieties can also support disease emergence through the selection pressure for pathogen genotypes that can overcome that resistance.

iii. New pathogens may arise through hybridization of existing pathogen species when they come into contact through changing geographic distributions; hybridization may have contributed to Dutch elm disease epidemics

iv. In some cases, it is the introduction of an arthropod vector that makes a pathogen more important. For example, between 1927 and 1930, Citrus tristeza virus (CTV) was introduced into South America, but only in 1950 with the introduction of the aphid vector Toxoptera citricidus was this pathogen considered economically important. A similar situation was seen in California for Xylella fastidiosa. This bacterium caused Pierce's disease for a century in this state, but with the introduction of new insect vector species, Pierce's disease spread rapidly, causing significant losses.

5.10.4 Actions to Mitigate Pathogen Emergence

- Countries are making efforts to minimize threats by establishing regulatory measures to prevent, control, or eradicate diseases caused by pathogens potentially dangerous to crops (Graciela Dolores Avila-Quezada *et.al.*, 2018).
- National and international organizations for free information about dissemination among scientists, governments, and the public would be crucial to minimize the threat of emergent pathogens.

- Phytosanitary regulations are performed worldwide to prevent, combat, and eradicate pests affecting plants. For example, quarantines are an example to prevent or delay the movement of pests from areas where there are not known to exist; conversely, quarantines also enable delayed introduction of pathogens into new areas until risks have been evaluated. Foreign quarantines, prevent the introduction and presence of exotic pests, and domestic quarantines slow the spread, control, or eradicate any pest that has been introduced to a certain country.
- Phytosanitary programs such as those in Mexico include surveillance at ports, airports, and borders and are aided by trained dogs to locate fruit, seeds, and plant products. These programs have generated successful results to stop the introduction of diseases from abroad. Despite similar programs and regulations, pathogens manage to arrive in new crop areas via vectors, humans, or environmental factors.

To sum up, researchers globally are encouraged to focus in the major food threats and cultivate collaborations to establish public policies with competent authorities, design phytosanitary regulations and establish diagnostic protocols which would thereby strengthen government decisions. The protocols are very important to enable accurate detection of pathogens. Morphological analyses, coupled with DNA sequences data currently available, facilitate the identity on new pathogens or variants thereof. These kinds of diagnostics contribute to a rapid and accurate detection of new pathogens and should be consulted for the development of proper diagnostic protocols.

The occurrence of new and emerging plant fungal infections is on the rise but has gone unnoticed because of inadequate detection methods. This alarming increase in novel and emerging pathogens can be attributed to a combination of geographic expansion, climate change, modified land use and the increased use of immunosuppressive and antifungal drugs in agricultural practices (Samantha C. Karunarathna *et.al.*, 2021). Emerging fungal pathogens are an increasing threat for the ecosystems, global health, food security and world economy. Emerging plant pathogenic fungi have not been widely recognized as posing major threats to health despite their significant devastating impact on economically important crops because of lack of identification methods. There is published evidence that common saprophytic fungi, such as Cryptococcus, Aspergillus and Penicillium species are emerging now as potential plant pathogens and the latter can represent major threats to staple crops such as rice, wheat, maize and potatoes either during cultural practices or during the post-harvest/storage stages. If these pathogens are not detected on time and targeted disease-management strategies are not put in place, global food security will be dramatically affected.

To be able to detect and identify plant pathogens from a scientific perspective, new molecular methods including DNA based phylogenetic approaches and fingerprinting approaches have been developed for the rapid detection and identification of fungal pathogens, especially emerging ones. Novel molecular methods will undoubtedly strengthen our ability to distinguish between previously identified and novel pathogenic species. But most importantly, they will assist us to develop more reliable and effective control measures to emerging uncharacterized pathogenic species. These new methods primarily based on PCR and amplification of DNA fragments of a wide array of genes coupled with multigene phylogeny have contributed significantly to more accurate detection method as well as enable plant pathologists to design specific primers and barcodes. In addition, the use of molecular dating based on DNA sequence data is also useful to better understand evolutionary trajectories of emerging plant pathogen.

lp in disease prevention. We discuss emerging diseases as a threat to crops, identify future research areas, and encourage the establishment of research networks focused on quarantine pathogens to address the problem and minimize risks.

A current and burgeoning challenge for the discipline of plant pathology is the introduction and spread of pathogens to new locations, and emergence or re-emergence of new pathogens against a background of a changing climate (Jeger *et.al.*, 2021).

Quarantine, using sentinel trap plants, molecular diagnostics or canines to detect an organism may all currently be used, but the challenge to develop novel tools that may be part of early detection, warning and management to increase the effectiveness of dealing with exotic diseases is a challenge. Also managing a disease once identified may require complex coordination of resources and the support of local communities, and the environment, that can otherwise derail eradication efforts as happened with citrus canker in Florida (Jeger *et.al.*, 2021).

5.11 Novel Tools for Devising Innovative Strategies of Emerging Pest and Disease Management

Novel mitigation measures need to be contrived on priority for curtailing the huge yield losses caused by various biotic stresses of agricultural crops. In this context, various key research areas for future work include deciphering basic/ molecular mechanism of host–pathogen/insect interactions, nanotechnology in pest management, fumigants, volatile and acoustic techniques for stored grains, post-harvest and vertebrate pest management, utilizing gene editing technologies such as CRISPR/Cas9, elaborated tritrophic pheromone/kairomone studies, deployment of remote sensing technology for pest management, disease/pest prediction and forecasting, determination of factors leading to evolution of pests and pathogens, development of technologies to minimize the adverse effects of climate change on biotic stresses of plants, allelopathy based weed management etc. Efforts have been made worldwide to search novel and more efficient tools for mitigating losses due to biotic stresses and a brief account of the same is presented below (Kumar *et.al.*, 2021).

i. Transgenic crops with insecticidal toxins: With the advancement of molecular biology in the recent times, a new innovation of genetic engineering lure the plant scientists worldwide and so is also true with agricultural scientists as well. Plant protection scientists are now capable of manipulating the genetic material of a cell in order to produce pest resistant varieties of several crops. Genetic engineering technology has been successfully employed in developing cotton genotypes resistant to lepidopteran larvae (caterpillars) and maize genotypes resistant to both lepidopteran and coleopteran larvae (root worms) leading to reduction in pesticide usage and that of production costs.

The transgenic plants of these crops had Bt genes derived from the soil bacterium *Bacillus thuringiensis* (*Bt*). Genotypes with *Bt* genes show insecticidal activity toward pests, and constitute a large reservoir of genes encoding insecticidal proteins. *Bt* genes are responsible for accumulation of crystalline inclusion bodies produced by the bacterium on sporulation (Cry proteins, Cyt proteins), or expressed during bacterial growth (Vip proteins). Additional *Bt* Cry genes can be transgenically incorporated into the crop for achieving protection against a wider range of pests. Transgenic cotton containing two Bt genes was first used in 1999, three years after the release of the original single Bt variety. Cotton plants containing two genes corresponding to Cry1Ac and Cry2Ab proteins respectively, showed more toxicity towards bollworms (Helicoverpa zea; target pest) and two species of armyworms (*Spodoptera frugiperda* and *Spodoptera exigua*; secondary pests), than the cotton with single gene related to Cry1Ac.

Site-directed mutagenesis modifies *Bt* toxins resulting in increased toxicity towards target pests. Such an alteration due to chemical domain exchange in *Bt* genes is termed as 'domain swap' approach. The key role of domain II in three domain Cry proteins in mediating interactions with insect receptors has been exploited by mutation of amino acid residues in the loop regions of this domain. Mutation of Cry1Ab increased its toxicity towards larvae of gypsy moth (*Lymantria dispar*) by up to 40 folds55. Hemipteran plant pests are not affected by known Bt toxins. But these are susceptible to lectin toxicity conferred by lectin genes which can be potentially exploited to develop insect resistant transgenic plants. About 50% reduction in survival of brown plant hopper (*Nilaparvata lugens*) and other hemipteran pests could be achieved through expression of the Man-specific snowdrop lectin (GNA) in transgenic rice plants using constitutive or phloem-specific promoters.

ii. Nematodal bacteria : Symbiotic enterobacteria present in nematodes of *Heterorhabditis* species offer a small-scale biological control of insect pests. These nematodes with bacterial cells in their gut release bacteria into the insect circulatory system. The bacterial toxins inside the insect circulatory system cause cell death in the insect host. A large number of insecticidal components have been investigated in the bacterium *Photorhabdus luminescens.* A compound, cholesterol oxidase released by such bacteria exhibit insecticidal activity similar to *Bt* toxins through membrane destabilization.

iii. Secondary metabolites : By-products of some metabolic pathways in transgenic plants have been found to exhibit toxic properties against pest species. For example, coding sequence fused to a chloroplast-targeting peptide or expression constructs containing part or all of the coding sequence of the protein, resulted in production of toxic active enzyme in transgenic tobacco. Alkaloid caffeine produced in tobacco by the introduction of three genes encoding Nmethyl transferases is another example of a by-product of secondary metabolite with toxicity to pests in transgenic tobacco.

iv. Volatiles : Plants are known to emit volatiles with properties of inhibiting pests. Such volatiles are known to be synthesized by RNA interference (RNAi)-mediated suppression of a cytP450 oxidase gene expressed in tobacco and Arabidopsis by constitutive overexpression of a plastid dual linalool/nerolidol synthase. These volatiles may be used as attractants for predators and parasitoids. For example, transgenic plants of Arabidopsis having maize terpene synthase gene TPS10 emitted the sesquiterpene volatiles which attracted parasitoid wasps that attack maize pests. Another example of volatile employed in pest control can be coated with sesquiterpene (E)-βfarnesene used by insects to communicate with each other and can be used as an alarm pheromone in aphids for attracting aphid predators and parasitoids. Arabidopsis containing (E)-β-farnesene synthase gene from mint (Mentha piperita) exhibited significant levels of aphid deterrence and also acted as the chosen attractant to the aphid parasitoid Diaeretiella rapae.

v. RNAi and its complementing techniques: RNA applications made revolution in the field of crop protection due to their environmentally friendly approach. RNA interference (RNAi) utilizes the plant's natural process for mediating management of pest control and is emerging as an alternative to chemical spray. The RNA-based active ingredients will be used for biocontrol in the near future. This depends upon the quality of double stranded RNA (dsRNA), its stability and large scale production. Also, modification of an organism's genome is possible by employing genome-editing tool. One such tool based upon CRISPR/Cas9 systems has been utilized for modification of susceptibility genes in agricultural crops to enhance resistance to crop pests. Furthermore, RNAi has emerged as a well-established technique of disrupting gene function in insect genome based on delivery by injection into insect cells or tissues. Partial resistance to insect pests can also be achieved by reducing the level of gene expression, measured by mRNA level, when fed to insects. This led to the development of transgenic plants producing dsRNAs in transgenic maize corn rootworm by suppressing mRNA in the insect, reducing feeding damage compared to controls. Similarly, in transgenic tobacco and Arabidopsis, dsRNA directed against a detoxification enzyme (Cyt P450 gene CYP6AE14) for gossypol in cotton bollworm caused the insect to become more sensitive to gossypol in the diet.

To foster a healthier environment in modern agriculture, it has been advocated that plant viral defense strategies with the new revolutionary tools such as RNAi silencing, ZFN, TALENS, SmART and CRISPR/Cas9 have to be used to manage viral diseases. Several new viruses belonging to Poty, Cucumo, Illar, Gemini, Tospo and other groups have been identified in agricultural crops. Polymerase chain reaction (PCR) based techniques and Real-time PCR have been successfully used for detection of virus strains and their variants in various crops. Understanding molecular interaction between viruses and the resistant and susceptible plants leading to resistance or pathogenesis is the need of the hour, which has to be pursued by employing genetic engineering techniques and functional genomics. Apart from this, factors contributing to upsurge of polyphagous vectors like whitefly need to be understood which involve mixed cropping pattern, overlapping host range, suitable weather which supports the survival and prevalence of whitefly. Also, recombination is one of the genetic events which foster the evolution of diverse strains of whitefly in India.

The side effects of chemical pesticides caused by their poisonous components like heavy metals and non-biodegradable constituents necessitate searching safer alternatives for environmental friendly plant protection. Nano-based antimicrobials offer the best potential in this context.

The nanopesticide is a formulation that is characterized with (a) Entities in the nanometer size range (up to 100 nm), (b) Designated nano prefix (e.g. nanohybrid, nanocomposite) and/or (c) Novel properties associated with small size. A wide variety of nanoproducts employed to protect from diseases/pests have been described in detail by European Commission Joint Research Centre. In the field of plant protection, nanopesticides involving organic ingredients, polymer-based inorganic silica, titanium dioxide, nanoemulsions and nanoclays in various forms (e.g. particles, micelles) have been evaluated and applied. Application of nano-pesticides is still in the incipient stages. However, nanosized silica silver (Si-Ag) particles have been successfully used for controlling fungal pathogens, viz. Pythium ultimum, Magnaporthe grisea, *Botrytis cinerea*, *Colletotrichum gleosporioides*, *Rhizoctonia solani* and also bacterial pathogens such as *Xanthomonas campestris*. Similarly, iron nanoparticles coated with carbon have been used for treating localized infections in plants. A nano-based product 'Nano-Gro' has been launched as plant growth regulator and immunity enhancer. Another product 'Nano-5' is also available in the market as natural mucilage organic solution for controlling several plant pathogens and pests.

vi. Endophytes in plant protection : Endophytic microbes, mostly bacteria and fungi live inside the plants and provide protection against pathogen attack and enhance tolerance to abiotic stresses also. Usage of endophytes in plant protection significantly reduces the usage of agricultural chemicals. Endophytes have been reported to induce systemic acquired resistance (SAR) against a variety of pathogens by triggering metabolic pathways in host leading to synthesis of phytohormones, viz. salicylic acid, jasmonic acid and ethylene, leading to the formation of infection barriers and antimicrobial substances in plants. These microbes are also used as biocontrol agents in plant protection, as they restrict the growth of pathogens by producing a range of bioactive metabolites, hydrolytic enzymes, antibiotics and lipopeptides. Endophytes promise an eco-friendly and viable option as biocontrol agents and may thus contribute potentially to sustainable agriculture.

vii. Strengthening quarantine services : Government quarantine system offers services which are beyond the capabilities of individual beneficiaries due to cost factor. But in real sense, plant quarantine is supposed to serve at national level by creating barriers to the introduction of exotic pests and pathogens and also prevent their further spread. Quarantine measures can be successfully implemented with the active support of administrators, general public, farmers, scientists, communication media, customs and other stakeholders involved in the farm sector. Subsequently, an effective quarantine system depends on several important pre-requisite data such as updated lists of endemic pests, authentic data on country-wide survey and surveillance, as well as current literature, which are also vital in implementing the PRA procedures. Further, to achieve an effective output of quarantine processing and biosecurity measures, it is essential to synchronize the plant quarantine system with the global plant quarantine system standards.

To conclude, plant health needs to be viewed as an important perspective of measures to ensure food security which involves a multidisciplinary approach, reforms in human resources, natural resource management, agricultural research, rural infrastructure, etc. In this context, contriving novel mitigation measures of biotic stresses in farming sector by revamping the inadequacies of the ongoing national programmes for assuring national food and nutrition security menaced by viciously vibrant pests/pathogens due to climate change, mutations, invasions, etc. is the need of the hour. Various priority research areas for future work include novel measures for deciphering basic/molecular mechanism of host–pathogen/pest interactions, nanotechnology in pest and disease management, fumigants, volatile and acoustic techniques for stored grains, postharvest and vertebrate pest management, utilizing gene editing technologies such as CRISPR/Cas9 for strengthening transgenic resistant crops. Research should also supplement aspects of tritrophic pheromone/kairomone studies, identification of broad spectrum resistance, utilizing wild species gene pool for durable resistance, deployment of remote sensing technology and other artificial intelligence based measures for pest management, development of disease/pest prediction models, determination of factors leading to evolution of pests and pathogens, minimizing the effects of climate change on aggravation of biotic stresses of plants, innovative weed management, vertebrate pest control, etc.

Bibliography

Abdullah M. Al-Sadi, 2021. Bipolaris sorokiniana-Induced Black Point, Common Root Rot, and Spot Blotch Diseases of Wheat: A review published on 11 March 2021., *Front. Cell. Infect. Microbiol.* doi: 10.3389/fcimb.2021.584899

Anonymous, 2011. Practical Approaches to Insect Resistance Management for Biotech-Derived Crops. CropLife International, Belgium. 40p

Anonymous, 2013. Recovery Plan for Root-Knot and Cyst Nematodes Parasites of Agronomic and Horticultural Plants Throughout North America, USDA.

Anonymous, 2021. U.S. Environmental Protection Agency, Office of Pesticide Programs, Mail Code 7506C 1200 Pennsylvania Ave. NW Washington, DC 20460

Cacciola, S.O. and Gullino, M.L.,2019. Emerging and re-emerging fungus and oomycete soil-borne plant diseases in Italy. *Phytopathologia Mediterranea*, 58: 451-472

Chelliah and Heinrich, E. A., 1994. Factors contributing to Brown plant hopper resurgence. Tamil Nadu Agricultural University, Coimbatore, Tamil Nadu, India; and International Rice Research Institute, P. O. Box 933, Manila, Philippines.

Damsteegt, V., 1999. New and Emerging Plant Viruses. 1999. APS*net* Features. Online. doi: 10.1094/APSnetFeature-1999-0999

Dhole Sumit, Lloyd, Alun, L., Gould Fred, 2020. "Gene Drive Dynamics in Natural Populations: The Importance of Density Dependence, Space, and Sex". *Ann. Rev. Ecol. Evol. & Syst. Ann. Rev.*, 51: 505–531

Dutcher, J.D.,2007. A Review of Resurgence and Replacement Causing Pest Outbreaks in IPM. In: Ciancio A., Mukerji K.G. (Eds.) General Concepts in Integrated Pest and Disease Management. Integrated Management of Plants Pests and Diseases, vol 1. Springer, Dordrecht. https://doi.org/10.1007/978-1-4020-6061-8_2

George, A., Rao C. N. and Rahangadale, S., 2019. Current status of insecticide resistance in Aphisgossypii and Aphisspiraecola (Hemiptera: Aphididae) under central Indian conditions in citrus, *Cogent Biology*, 5: 1, 1660494,

Gerald J. Holmes, Peter S. Ojiambo, Mary K. Hausbeck, Lina Quesada-Ocampo and Anthony P. Keinath, Resurgence of Cucurbit Downy Mildew in the United States: A Watershed Event for Research and Extension.APS Publications. Published Online:1 Apr 2015

Graciela Dolores Avila-Quezada, Jesus Fidencio Esquivel , Hilda Victoria Silva-Rojas , Santos Gerardo Leyva-Mir , Clemente de Jesús Garcia-Avila , Andrés Quezada-Salinas , Lorena Noriega-Orozco , Patricia Rivas-Valencia , Damaris Ojeda-Barrios , Alicia Melgoza-Castillo, 2018. Emerging plant diseases under a changing climate scenario: Threats to our global food supply . *Emirates J. Food and Agri.*, 30: 443-450

Harsh Garg, 2014. Pesticides: Environmental impacts and management strategies. IntechOpen Book series,DOI:10.5772/57399

Inge M Hanssen, Moshe Lapidot, Bart P. H. J. Thomma, 2010. Emerging viral diseases of tomato crops. *Mol Plant Microbe Interact.*, 23: 539-48.

Janardan Singh and Mahadevan Raghuraman, 2011. Emerging scenario of important mite pests in north India. In: Acarology XIII: Proceedings of the International Congress. Zoosymposia, 6, 1–304. Moraes, G.J. de & Proctor, H. (Eds)

Jeger, M., Beresford, R., Bock, C., 2021. Global challenges facing plant pathology: multidisciplinary approaches to meet the food security and environmental challenges in the mid-twenty-first century. *CABI Agric. Biosci.*, 2: 20 .

Jesús Navas-Castillo, Elvira Fiallo-Olivé, Sonia Sánchez-Campos, 2011. Emerging Virus Diseases Transmitted by Whiteflies. *Ann. Rev. Phytopathol.*, 49: 219-248

Jincai Wu, Linquan Ge, Fang Liu, Qisheng Song, David Stanley, 2020. Pesticide-Induced Planthopper Population Resurgence in Rice Cropping Systems. Ann. Rev. Entomol., 65: 409-429

Jorgensen Peter Søgaard; Folke, Carl; Carroll, Scott P.,2019. "Evolution in the Anthropocene: Informing Governance and Policy". *Ann.Rev. Ecol. Evol. and Syst. Ann. Rev*., 50: 527–546.

Kalyan K Mondal, 2016. Emerging Phytobacterial Diseases in India: Research Status and Challenges. In: Perspectives of Plant Pathology in genomic era. Dr. P. Chowdappa, Pratibha Sharma, Dinesh Singh & A.K. Misra (Eds.) Pages 167-168

Karen A. Garrett, Ari Jumpponen, and Lorena Gomez Montano, 2010. Emerging Plant Diseases: What Are Our Best Strategies for Management? International Research and Development at Virginia Tech, and by the Kansas Agricultural Experiment Station.

Ko, M.P., Schmitt, D.P., Fleisch, H., 1997. Re-establishment and resurgence of plant-parasitic nematodes in fumigated pineapple fields at different elevations and irrigation regimes in Hawaii. *Australasian Pl. Pathol*., 26: 60–68.

Kumar, J. , Murali-Baskaran, R. K., Jain, S. K., Sivalingam, P. N., Mallikarjuna, J., Vinay Kumar, Sharma, K. C., Sridhar, J., Mooventhan, P., Dixit, A.and Ghosh, P. K., 2021. Emerging and re-emerging biotic stresses of agricultural crops in India and novel tools for their better management. *Curr.Sci.*, 121: 26-36

Lu, Yanhui, Wyckhuys, Kris A. G., Yang, Long, Liu, Bing, Zeng, Juan, Jiang, Yuying, Desneux, Nicolas, Zhang, Wei and Wu, Kongming. 2021. Bt cotton area contraction drives regional pest resurgence, crop loss, and pesticide use. Pl. Biotechnol. J., Article in press. First published online on October 9, 2021. https://doi.org/10.1111/pbi.13721

lvira Fiallo-Olivé and Jesús Navas-Castillo,2020. Molecular and Biological Characterization of a New World Mono-/Bipartite Begomovirus/Deltasatellite Complex Infecting Corchorus siliquosus. *Front. Microbiol*., 11:1755

Lynne M. Jones, Bradley Pease, Sandy L. Perkins, Fiona E. Constable, Wycliff M. Kinoti, David Warmington, Ben Allgood, Sonya Powell, Pieter Taylor, Ceri Pearce, Richard I. Davis, 2021. 'Candidatus Phytoplasma dypsidis', a novel taxon associated with a lethal wilt disease of palms in Australia. *Intl. J. Systematic and Evolutionary Microbiol., 71* (5) DOI: 10.1099/ijsem.0.004818

Mandeep Rathee and Pradeep Dalal, 2018. Emerging insect pests in Indian agriculture. *Indian J. Entomol*., 80: 267-281

Maria R. Rojas and Robert L. Gilbertson, 2008. Emerging Plant Viruses: a Diversity of Mechanisms and Opportunities. In: M.J. Roossinck (ed.), Plant Virus Evolution. Springer-Verlag Berlin Heidelberg 2008

Mark R. Hardin, Betty Benrey, Moshe Coll, William O. Lamp, George K. Roderick, Pedro Barbosa, 1995. Arthropod pest resurgence: an overview of potential mechanisms, *Crop Protn.*,14: 3-18

Matthew Howle, 2021. Guava Root Knot Nematode. Regulatory Services,Clemson.edu.

Mayabini Jena, Rath, P.C., Mukherjee, A.K., Raghu, S., Guru, P., Pandi, G., Basana Gowda, G., Prasanthi, G., Yadav, M.K., Baite, M.S., Prabhukarthikeyan, S.R., Bag, M.K., Srikant Lenka, Arvindan, S., Naveen Kumar Patil, Mohapatra, S.D., Annamalai, M. and Adak, T., 2018. Exploring New Sources of Resistance for Insect Pest and Diseases of Rice. In : Rice research for enhancing productivity, profitability and climate resilience , Pathak (Edr.) 367-381.

Miller, G.T.,2004. Sustaining the Earth, 6th edition. Thompson Learning, Inc. Pacific Grove, California. Chapter 9, Pages 211-216.

Muralidharan Govindaraju, Yisha Li and Muqing Zhang, 2019. Emerging Bacterial Disease (Leaf Scald) of Sugarcane in China : Pathogenesis, Diagnosis and Management. IntechOpen Book series, DOI: 10.5772/intechopen.88333.

Naveen, N., Chaubey, R., Kumar, D., 2017. Insecticide resistance status in the whitefly, *Bemisia tabaci* genetic groups Asia-I, Asia-II-1 and Asia-II-7 on the Indian subcontinent. *Sci. Rep.*, 7: 40634

Niroshini Gunasinghe, Martin J. Barbetti, Ming Pei You, Daniel Burrell and Stephen Neate, 2020. White Leaf Spot Caused by Neopseudocercosporella capsellae: A Re-emerging Disease of Brassicaceae. Front. Cell. Infect. Microbiol., 28 October 2020

Poornima, K., Suresh, P., Kalaiarasan, P., Subramanian, S. and Ramaraju, K. 2016. Root Knot Nematode, Meloidogyne enterolobii in Guava (*Psidium guajava* L.) A new record from India. Madras Agri. J., 103:359

Ranjit Kumar, Sanjiv Kumar, Sivaramane, N., Dhandapani, A. and Meena, P.C., 2019. Adoption of Insect Resistance Management Practices in *Bt Cotton* Cultivation in India. ICAR-National Academy of Agricultural Research Management, Hyderabad, India.

Ravichandra, N.G., 2019. New Report of Root-Knot Nematode (*Meloidogyne enterolobii*) on Guava from Karnataka, *India. EC Agriculture*, 5 (9), 504-506

Roberts Donald, R., Manguin, S. and Mouchet, J., 2000. "DDT house spraying and re-emerging malaria". *The Lancet. Elsevier*, 356 : 330–332

Rocha, J., Shapiro, L.R. and Kolter, R., 2020. A horizontally acquired expansin gene increases virulence of the emerging plant pathogen *Erwinia* tracheiphila. *Sci. Rep.*, 10: 21743

Samantha C. Karunarathna , Sajeewa S. N. Maharachchikumbura, Hiran A. Ariyawansa , Belle Damodara Shenoy and Rajesh Jeewon, 2021. Emerging Fungal Plant Pathogens. Front. Cell. Infect. Microbiol., Published on 17 September 2021. doi: 10.3389/fcimb.2021.765549

Schaad, N.W., 2008. Emerging Plant Pathogenic Bacteria and Global Warming. In: *Pseudomonas* syringae Pathovars and Related Pathogens – Identification, Epidemiology and Genomics. Fatmi M. *et al.* (Eds), Springer, Dordrecht. https://doi.org/10.1007/978-1-4020-6901-7_38

Shaily Tyagi, Karthik Kesiraju, Manjesh Saakre, Maniraj Rathinam, Venkat Raman, Debasis Pattanayak and Rohini Sreevathsa, 2020. Genome Editing for Resistance to Insect Pests : An Emerging Tool for Crop Improvement. ACS Omega 2020, 5: 20674–20683

Species Richness, 2020. Chemical pesticides target pest resurgence and secondary pests. Ecology Centre, Dec 2020

Trumper ,E.V. and Modelling, J. Holt., 1998. Pest population resurgence due to recolonization of fields following an insecticide application. *J. Appl.Ecol.*, 35: 273-285

Virginia Rodrıguez Alfredo, Lagares Heiser Arteaga Salim and Mattar Luis Carlos Rui, 2020. Viral Emerging Pathogen Evolution. In : Emerging and Reemerging Viral Pathogens. Elsevier Inc. P 35-51.

Zafar, M.M., Razzaq, A. and Farooq, M.A., 2020. Insect resistance management in *Bacillus thuringiensis* cotton by MGPS (multiple genes pyramiding and silencing). *J Cotton Res.*, 3: 33

6

National Regulatory Mechanism and International Agreements Conventions

The concepts of biosafety and biosecurity are central to efforts to protect human health against the risks posed by exposure to hazardous biological agents. Biosafety, a term used to describe the collection of technologies, processes and practices a term used to describe the collection of technologies, processes and practices aimed at preventing the unintentional exposure to biological agents, has in particularaccrued increasing importance in recent decades as a result of the trend towards globalization and concomitant growth in international communication, transport and trade (WHO, 2020). In this international context, the outbreaks of highly-infectious diseases that have occurred in the last years serve to underscore the urgent need for effective prevention and detection and response to biological risks, in accordance with the International Health Regulations (IHR).

The nature and scope of the regulatory arrangements for assuring biosafety and biosecurity in biomedical laboratories among WHO Member States is currently extremely diverse. Some counties have highly-developed regulatory systems, with detailed legislation backed up by robust networks of regulatory bodies and stakeholders, each with well-defined responsibilities and processes. At the other end of the spectrum, however, there are some countries that almost completely lack regulatory guidance in the field of laboratory biosafety and biosecurity.

There is also a wide variation in the context and primary orientation of regulatory frameworks even among the countries with well-developed systems. Some countries have systems that are geared towards the protection of occupational health while others are focused on the threats to security. Furthermore, many established regulatory frameworks currently lack the necessary flexibility to proactively manage and to respond adequately to risks and situations derived from new technologies and newly-evolving and emerging diseases, especially zoonotic diseases. In addition, biosecurity risks related to the misuse of advanced technologies are often neglected or entirely omitted. Finally, because biosafety and biosecurity are closely linked to issues related to animal health, environmental protection and the misuse of biological agents, there is a now a greater need for regulatory frameworks to develop more integrative approaches to managing biological risk to ensure they cover a wider range of topics and the full spectrum of potential threats to global public health.

6.1 National Regulatory Mechanism

Biosafety regulations cover assessment of risks and the policies and procedures adopted to ensure environmentally safe applications of biotechnology. The regulatory framework for transgenic crops in India consists of the following rules and guidelines.

a. **Rules and policies:** Rules, 1989 under Environment Protection Act (1986); Seed Policy, 2002
b. **Guidelines:** Recombinant DNA guidelines, 1990 ; Guidelines for research in transgenic crops, 1998

The two main agencies identified for implementation of the rules are the Ministry of Environment, Forests and Climate Change and the Department of Biotechnology, Government of India. The rules have also defined competent authorities and the composition of such authorities for handling of various aspects of the rules. There are six competent authorities as per the rules.

1. Recombinant DNA Advisory Committee (RDAC)
2. Review Committee on Genetic Manipulation (RCGM)
3. Genetic Engineering Approval Committee (GEAC)
4. Institutional Biosafety Committees (IBSC)
5. State Biosafety Coordination Committees (SBCC)
6. District Level Committees (DLC).

Out of these, the three agencies that are involved in approval of new transgenic crops are:

1. **IBSC :** Set-up at each institution for monitoring institute level research in genetically modified organisms.
2. **RCGM :** Set-up at DBT to monitor ongoing research activities in GMOs and small scale field trials.
3. **GEAC :** Set-up in the Ministry of Environment, Forests and Climate Change to authorize large-scale trials and environmental release of genetically modified organisms.

The Recombinant DNA Advisory Committee (RDAC) constituted by DBT takes note of developments in biotechnology at national and international level and prepares suitable recommendations. The State Biotechnology Coordination Committees (SBCCs) set up in each state where research and application of GMOs are contemplated, coordinate the activities related to GMOs in the state with the central ministry. SBCCs have monitoring functions and therefore have got powers to inspect, investigate and to take punitive action in case of violations. Similarly, District Level Committees (DLCs) are constituted at district level to monitor the safety regulations in installations engaged in the use of GMOs in research and application.

6.1.1 Regulatory Frameworks: An Overview of Current Practice: (WHO, 2020)

Many countries base their national biosafety legislation on one or more international treaties and agreements, such as:

- The International Health Regulations (IHR)
- World Health Assembly Resolutions such as the WHA 58.29 Enhancement of laboratory biosafety
- The Cartagena Protocol on Biosafety to the Convention on Biological Diversity
- The OIE Terrestrial Animal Health Code
- The Manual of diagnostic tests and vaccines for terrestrial animals

In the terms of the matter of biosecurity, countries often defer to the following international agreements and conventions:

- The Biological Weapons Convention
- The United Nations Security Council Resolution 1540
- The OIE Biological Threat Reduction Strategy

While these above-mentioned agreements and instruments are designed to promote the concept of, and need for improvements in, biosafety and biosecurity, they are fairly broad in their scope and in many cases provide insufficient practical guidance for the policy-maker charged with the task of developing a comprehensive set of national polices and laws to regulate the activities of biomedical laboratories. In short, the texts of international conventions and agreements establish the general concepts and principles, but do not cover of the specifics of biosafety and security regulation.

In most countries, regulatory control of biological risk comprises a mix of primary and secondary legislation (i.e. laws, acts, regulations) and so-called "soft" law, that is, non-legally binding guidelines and standards. Several international expert bodies, including WHO and the International Organization for Standardization (ISO), have developed guidelines and standards which cover many different aspects of laboratory biosafety and biosecurity and which have been adopted (with or without adaptation) by individual countries. These international guidelines and standards generally advocate best practice approaches that provide conformity with state-of the-art technology and scientific knowledge and international agreements, such as the IHR. Examples of international guidelines and standards include the WHO Laboratory biosafety manual , ISO 15190 (ISO's Medical laboratories – Requirements for safety and the CWA 15793 (CEN Workshop Agreement on Laboratory Biorisk Management System). The latter has recently been converted so as to make it comparable (but not identical) in scope and content to the ISO standard, ISO 35001. In the context of the "One Health" approach, the *Codex Alimentarius* might be considered as another relevant standard example.

National outreach: National biosafety and bio-waste activities are governed by legislation through State Pollution Control Boards. Under the Integrated Disease Surveillance Programme, a network of laboratories with biosafety practices and infrastructure is being set up. A field manual has also been developed for biosafety. All major hospitals have biosafety committees which meet monthly. The Ministry of Environment & Forests and the Department of Biotechnology (DBT) are responsible for implementation of rules 1989 under the EPI act. DBT released Recombinant DNA Safety Guidelines (1990), Revised Guidelines for Research in Transgenic Plants and Guidelines for Toxicity and Allergenicity Evaluation of Transgenic Seeds, Plants and Plant Parts (1998) and guidelines for "Generating pre-clinical and clinical data for r-DNA based Vaccines, Diagnostics and other biological" (Gandhi, 2014).

A dedicated dynamic & interactive website on biosafety by DBT reflects national and international guidelines, national rules & procedures with dynamic interaction with Institutional Biosafety Committees (IBSC) and advices regulation of modern biotechnology with objectives of protecting environment including human and animal health from the unintended adverse effects of GMOs and products thereof. Another website on "Indian GMO Research Information System (IGMORIS)" provides information on research work going on in Indian laboratories and IBSCs at various public funded institutions, universities, private R&D institutions and industries. An institutional framework comprising competent authorities overlooks the issues of biosafety in the country. Mashelkar Committee Task Force on r-pharma, 2006 recommended Procedure for Regulation of Recombinant Pharma Products derived from Living Modified Organisms (LMOs). The biosecurity programmes involving responsible conduct and oversight of life science s research is covered under the Prevention of Terrorism Act, 2002 and the Weapons of Mass Destruction and their Delivery Systems (Prohibition of Unlawful Activities) Bill, 2005. Export of SCOMET items requires a license under the Foreign Trade (Development and Regulation) Act, 1992 (FTDR Act).

M.S. Swaminathan Task Force (2003) recommendations to set up an independent National Biotechnology Regulatory Authority (NBRA), to regulate GMOs and the Agricultural Biosecurity Bill, 2013 to establish an integrated national biosecurity system covering plant, animal and marine issues may further combat threats of bioterrorism from pests and weeds. The Biomedical Waste Management & Handling) Rules, 1998; ICMR Guidelines on Code of Conduct for Research Scientists engaged in biomedical research involving microbial or other biological agents; the Ethical Guidelines (2000) for the biomedical researchers and the Ethical Policies by DBT on the Human Genome, Genetic Research and Services (2002) sets the guidelines on principles and integrity, management and handling of biomedical wastes. However, the implementation arms are not very strong.

6.1.2 Magnitude of Problem in India

In India, the task to ensure biosafety and biosecurity in all public and private sector institutions involved in research and development and human healthcare and agriculture including animal and marine sciences is enormous. In human healthcare itself, there are more than 800 medical and healthcare R&D institutions; 575 medical colleges, 350 universities, more than 15,000 pharmaceutical and biopharmaceutical industries and unlimited number of healthcare entities and diagnostics laboratories handling biomedical wastes (Gandhi, 2014). Laboratories handling research on highly pathogenic organisms follow good laboratory practices and decontamination and disinfection procedures and a number of containment laboratories have been set up under human and animal health sectors. However, the concept of biosafety and biosecurity is not fully understood by each and every establishment handling infectious pathogens. To implement biosafety guidelines and to mitigate biological threats, a holistic approach is required on part of the administrative and technical authorities to adopt a systemic approach for successful bio-risk management. A scientific, rigorous, transparent, efficient, predictable and consistent regulatory mechanism and protocol for bio-safety and bio-security evaluation and related system need to be followed to meet these objectives.

Biosafety during lab work is an important concern in developing countries. A cross sectional study on the safety measures being adopted in clinical laboratories of India showed that a substantial number of laboratories do not follow the best practices as prescribed. Another study in Pakistan showed that nearly 50 percent of the laboratories were lax as far as implementations of best practices are concerned.

Government efforts: The Ministry of Environment and Forests (MoEF), Government of India, is implementing a GEF/World Bank funded project on Capacity Building on Biosafety in context of the Cartagena Protocol. The project covers the assessment, management and long term monitoring and documentation of the risks to the sustainable use of biodiversity and to human health potentially posed by the introduction of Living Modified Organisms (LMOs). The project aims to strengthen the legislative framework and operational mechanisms for biosafety management, enhance capacity for risk assessment and monitoring, establish the biosafety database system and biosafety clearing house mechanism, support centres of excellence and a network for research, risk assessment and monitoring and establish the Project Coordination and Monitoring Unit (PCMU).

Biotech Consortium India Ltd. (BCIL) supported by DBT, organize workshops on National Consultation on Biosafety aspects related to Genetically Modified Organisms; supports programmes on capacity building activities in biosafety including preparation of research documents, organizing conferences, workshops on key policy issues and publishing monthly newsletter "The South Asia Biosafety Program (SABP)".

6.1.3 Biorisk Management

The bio-risk management approach comprises biosafety, laboratory biosecurity and ethical responsibility by reducing the risk of unintentional exposure to pathogens and toxins or their accidental release; reducing the risk of unauthorized access, loss, theft, misuse, diversion or intentional release of microorganisms; and those suitable measures have been adopted and effectively implemented (Gandhi, 2014). It is important that a framework for continuous awareness raising for biosafety, laboratory biosecurity and ethical code of conduct, and training is provided. Laboratory bio-risk management as defined under CWA 15793:2011 on laboratory biosafety and biosecurity includes the analysis of ways, development of strategies and their implementation to minimize the likelihood of the occurrence of bio-risks; methodology used to organize and analyze scientific information in order to estimate the probability and severity of an adverse effect (assessment) and measures to minimize this effect (mitigation); and establishing policies and practices for risk management in the lab (day to day as well as emergencies).

An effective bio-risk management programme would involve bio-risk assessment including, investigation into the nature of the biological materials and the procedures used to store, handle, and transfer and dispose those materials. This may include review of research proposals and identify hazards that the biological materials can cause in a laboratory. Bio-risk management procedures may determine the control measures, or mitigation strategies, to be used to minimize or eliminate the defined risk. Risk assessment and the concept of continual improvement on bio-risk management can efficiently address issues relating to biosafety and biosecurity. Bio-risk mitigation would include actions and control measures that are put into place to reduce or eliminate the risks associated with biological agents and toxins. It determines what can be done to manage the identified bio-risks. These control measures are grouped under the categories of elimination or substitution, engineering controls, administrative controls, practices and procedures, and personal protective equipment.

A critical mass of dedicated human resource is required to implement provisions of bio-risk management, which requires:

- Developing training curriculum for stakeholders, including policy makers, management bio-risk advisors, scientists, laboratory management, laboratory and industrial workforce,
- Organising training programs, workshops to train-the-trainers,
- Organising training programmes on bio-threats and bio-risk mitigation strategies for law enforcement official,
- Developing training implementation strategies including regional training centres,
- National Coordination Centre to oversee implementation of bio-risk management programmes

6.1.4 Dedicated Training Facility

A panel discussion at the International Meeting on Host Parasite Interactions held at National Institute of Animal Biotechnology, Hyderabad on July 15, 2014 highlighted the issues involved with biosafety and biosecurity procedures faced by various stakeholders (Gandhi, 2014). Experts stressed on issues of information and awareness; implementation of regulations and guidelines, education and outreach, capacity building, risk assessment and risk management. In order to increase awareness and training on safety & security in research institutes, academia & industry, it was strongly recommended to the Government of India to have a fresh look at the implementation of the existing laws governing the biosafety and biosecurity to enforce effective bio-risk management in the country and filling in necessary gaps in capacity building. It was further suggested to set up a Bio-risk Management Training Academy (BRMTA), which would provide India an advantage and understanding on bio-risk management in the biotechnology sector.

The Objective of BRMTA would be to promote capacity building in bio-risk management and promotion of good biosafety and biosecurity practices in India through education and training, information dissemination and knowledge sharing in pursuit of the benefits of life science research. This will also ultimately foster culture of responsibility and code of ethics among the researchers. The mandate of BRMTA may include the following.

- Ensuring bio safety measures so that research is conducted in accordance with the highest standards to promote safe, secure and responsible use of dangerous biological agents and toxins that have high-consequence for its potential to be dual use research of concern,
- Developing of programmes for awareness of the prohibition and requirements for bioterrorism preparedness to inter-alia industry, scientific and technological communities, armed forces,
- Developing specialized courses in bio-risk management; related policies, applicable regulations, audits, inspections etc.;
- Special courses to manage challenges and opportunities associated with facility risk assessment; laboratory certification; reporting laboratory-associated infections,
- Organising education & training programmes; short-term courses; workshops and symposia; and conferences (national and international),
- Preparing guidelines for medical surveillance and evaluation system,
- Wide dissemination of information though updated technical guidance documents, safety manuals, biosecurity plan, safe practices and standard operating procedures and emergency response plans,
- Outreach to large number of beneficiaries including individuals, institutions and industries through electronic media and periodic newsletters,

- Advising the Government of India about issues related to biological disaster management, international collaboration on bio-risk management and developing risk mitigation strategies,
- Establishing and strengthening collaborations with international bodies and other similar institutions involved with risk assessment and management programmes or developing curriculum in the area to improve understanding and management of the risks associated with accidental and deliberate misuse of biological agents.

6.2 International Agreements Related to Biosafety

Biosafety issues are being addressed through a range of policy and procedural measures at both international and national levels. These measures aim at minimizing the potential risks that biotechnology processes and products pose to the environment and human health and are essentially based on the principles of risk assessment and management. Risk assessment is the determination of potential risk associated with a specific activity while risk management is the use or application of procedures and means to reduce the negative consequences of a risk to an acceptable level. It is assumed that risks can be limited by proper handling and use of various preventive measures (Gupta *et.al.*, 2008). Discussions on the hazards of recombinant DNA cloning began in the early 1970s. The set of guidelines produced by the Recombinant Advisory Committee (RAC) of the US National Institutes of Health in 1975, were among the first to ensure safety of the laboratory workers involved in biotechnology research, the public and the environment (Office of Biotechnology Activities, 2006). The biosafety guidelines were voluntary and had no legal standing. Over the years many countries have adopted them, or a derivation thereof for purposes of biosafety management. Besides, several countries, including those in the Asia-Pacific region, have entered into international agreements that address biosafety issues related to transboundary movement of GM crops and their products.

International Instruments on Biosafety : Six organizations are directly or indirectly involved in international biosafety regulation:

(a) **The Convention on Biological Diversity (CBD)-1992:** Concerns conservation, sustainable use and fair and equitable sharing of benefits arising from the use of biological resources. The Cartagena Protocol on Biosafety regulates transboundary movement of living modified organisms (LMOs).

(b) **The World Trade Organization (WTO)-1995:** Deals with trade in goods and services and dispute settlement. The WTO Agreement on Application of Sanitary and Phytosanitary (SPS) Measures is related to procedures of risk analysis of plant and animal pests and diseases, and food safety.

(c) **The International Plant Protection Convention (IPPC)-1952:** Develops International Standards on Phytosanitary Measures (ISPMs) against pests of plants and plant products including GMOs.

(d) **The** *Codex Alimentarius* **Commission (CAC)-1972:** Develops international standards including those for food safety and food labeling

(e) **The World Organization for Animal Health (OIE)-1924:** Harmonizes trade regulations for animals and animal products, and develops standards on animal health including for infectious animal diseases.

(f) **The Organization for Economic Cooperation and Development (OECD)-1961:** Undertakes harmonization of international regulations, standards and policies.

The term "living modified organisms" (LMOs) is defined by the Protocol as any living organism that possesses a novel combination of genetic material obtained through the use of modern biotechnology". For the present report, the term has been used interchangeably with GMOs.The Cartagena Protocol on Biosafety and the WTO are directly involved in regulation of trade in products of agricultural biotechnology (Gupta *et.al.*, 2008). WTO-SPS recognizes three standard setting bodies, IPPC, CAC and OIE that address different biosafety aspects including environment, health and food safety. Under SPS rules, member countries can use other standards but they have to develop their standards on the basis of the principle of risk assessment. OECD develops documents, guidelines and recommendations on harmonized rules, policies and standards for its members.

6.2.1 Cartagena Protocol on Biosafety

The Convention on Biological Diversity of the United Nations adopted the Cartagena Protocol on Biosafety (hereinafter, the Protocol) in the year 2000 which entered into force on 11 September 2003. The Protocol has been ratified by 143 Parties (as on January 2008) including most countries from the Asia-Pacific region. Since these countries have either developed or are in the process of developing their biosafety regulatory systems in conformity with the Protocol, detailed information about its provisions is furnished hereunder.

i. Salient Features : The Protocol comprising 40 articles is a legally binding agreement to ensure adequate levels of protection for safe transfer, handling and use of LMOs resulting from modern biotechnology that may have adverse effects on human health and conservation and sustainable use of biological diversity. It covers transboundary movement, transit, handling and use of LMOs. However, it does not cover non-food or non-feed products derived from LMOs (e.g. paper from GM trees) and LMOs that are pharmaceuticals for humans. It allows the members to take decisions on the import of LMOs intended for direct use as food or feed, or for processing, under their domestic regulatory framework. The Protocol also promotes cooperation to help developing countries acquire resources and capacity to use biotechnology safely and more efficiently and encourages training to promote safe transfer of technology. It also includes a clause clarifying that it does not alter the rights and obligations of parties under the WTO or other international agreements.

ii. Advanced Informed Agreement (AIA) (Article 7): The Protocol's main mechanism is its AIA requirement. Under this procedure, the exporting party must first as specified in Article 8 provide written notification (which includes a full set of information specified in Appendix II to the Protocol) to the importing government that it is interested in exporting a new LMO into the importing country. The importing country government must then acknowledge receipt of the notification as per Article 9. A competent body within the importing country must then make a decision according to Article 10, using risk assessment procedures described in Article 15. Article 10 contains explicit support for the precautionary approach of risk assessment stating "lack of scientific certainty due to insufficient relevant scientific information and knowledge regarding the extent of the potential adverse effects of the LMO on the conservation and sustainable use of biological diversity in the Party of import, taking also into account risks to human health, shall not prevent that Party from taking a decision". The exporter must provide a notification to the importing country containing detailed information about the LMO, its previous risk assessments and its regulatory status in the exporting country. The importing country must acknowledge receiving the information within 90 days and whether the notifier should proceed under a domestic regulatory system or under the Protocol procedure. In either case, the importing country must decide within 270 days whether to allow the import, with or without conditions or deny it.

The AIA is meant only for first time shipments and consecutive shipments are exempt from it. Also, LMOs not intended for release into the environment, those in transit and destined for contained use are exempt from the requirement of AIA. The Protocol also sets up a separate procedure for LMOs intended for direct use as food or feed, or for processing, in Article 11. Under its provision, any party making a final decision regarding domestic use of LMOs including placing on the market must within 15 days notify other parties of the Convention on the Biological Diversity of this fact through the Biosafety Clearing House (BCH).

iii. Risk Assessment (Article 15) : The Protocol requires that decisions on proposed imports be based on risk assessments, which are undertaken in a scientific manner based on recognized risk assessment procedures, taking into account advice and guidelines developed by relevant international organizations. Lack of scientific data or consensus must not be interpreted as indicating acceptance of particular level of risk. The risks associated with LMOs or their products should be considered in the context of risks posed by the non-modified recipients or parental organisms in the potential receiving environment. Risk assessment is carried out on a case-by-case basis.

iv. Handling, Transport, Packaging and Identification (Article 18): This article concerns the measures to be taken to avoid risks during transboundary movement of LMOs for intentional introduction into the environment. The objective of this article is to make sure that the LMOs are handled and moved safely to avoid adverse effects on biodiversity and human health.

v. Biosafety Clearing House (BCH) (Article 20) : The BCH is an information sharing mechanism operated through a website and administered by the Secretariat to the Convention. It has been established to facilitate the exchange of scientific, technical, environmental and legal information on LMOs and assist members to implement the Protocol. Examples of information contained in the BCH include any existing laws, regulations, or guidelines for implementation of the Protocol, summaries of risk assessments or environmental reviews of LMOs and final decisions regarding the importation and release of LMOs.

vi. Capacity Building (Article 22): This article calls for cooperation in the development and/or strengthening of human resources among developing countries, island developing states and Parties with economies in transition for sharing resources and institutional capacities on biosafety including biotechnology for effective implementation of the Protocol. The Protocol recognizes the inability of some countries to copewith the nature and scale of known and potential risks associated with LMOs. Hence, cooperation for capacity building is a priority.

vii. Public Awareness and Participation (Article 23): Parties are obliged to promote and facilitate public awareness, education and participation concerning the safe transfer, handling and use of LMOs by, inter alia, providing access to information on LMOs that may be imported.

viii. Socio-economic Considerations (Article 26) : In making import decisions, parties may take into account socio-economic considerations arising from the import of LMOs on the conservation and sustainable use of biodiversity, especially with regard to the value of biological diversity to indigenous and local communities.

ix. Liability and Redress (Article 27): This article addresses issues of liability and redress for damage resulting from the transboundary movement of LMOs. The liability procedure is under negotiation.

x. Compliance (Article 34) : The compliance regime for the Protocol which is not yet finalized will provide procedures and mechanisms to promote compliance and address non-compliance.

6.2.2 CAC – Codex Guidelines for GM Foods

Codex guidelines for GM foods including the analysis of unintended effects are of imprtance. The Codex's aim is to anticipate not only the direct risks, but also the indirect/unanticipated risks that the products of modern agriculture might pose for human health. It states that all the methods including protoplast fusion and/or recombinant DNA technology have the potential to generate unanticipated effects in plants (Gupta *et.al.*, 2008).

i. **Scientific Principle**: Case-by-case premarket assessment includes an evaluation of both direct and unintended effects. The safety assessment of GM foods covers direct health effects (toxicity), tendency to provoke allergic reactions (allergenicity), specific components thought to have nutritional or toxic properties, the stability of the inserted gene, nutritional effects associated with genetic modification and any unintended effects that could result from the gene insertion.
ii. **Risk Identification:** Risk identification to encompasses not only health-related effects of the food itself, but also the indirect effects of food on human health (for example, potential health risks derived from out crossing).
iii. **Basis of Assessment :** The potential long term effect of any foods is difficult to identify. In many cases, this is further confounded by wide genetic variability in the population, such that some individuals may have a greater predisposition to food-related effects. It concludes that application of the substantial equivalence concept contributes to a robust safety assessment framework.
iv. **Monitoring and Review :** Codex principles are not binding on member nations, but are referred to specifically in the SPS Agreement of the WTO and can be used as a reference in case of trade disputes.

World Organization for Animal Health (OIE): The OIE ensures transparency in the global animal disease situation to improve the legal framework and resources of national veterinary services (Gupta *et.al.*, 2008). It establishes standards, guidelines and recommendations relevant to animal diseases and zoonoses in accordance with its statutes and as defined in the WTO-SPS Agreement.

i. **Scientific Principle :** The standards are based on the principle of validation, control of exotic diseases and certification of diagnostic assays (test methods) for infectious animal diseases by the OIE.
ii. **Risk Identification:** It is aimed to provide importing countries with an objective and defensible method of assessing the disease risks associated with the import of animals, animal products, animal genetic material, feedstuffs, biological products and pathological material. The risk identification criteria are case specific for the listed diseases requiring appropriate measures during import and export of animals and animal products. The exporting country is provided with clear reasons for the imposition of import conditions or refusal to import.
iii. **Basis of Assessment:** Risk assessment is categorized into quantitative and qualitative assessments. Quantitative assessments require mathematical modeling while qualitative assessments do not require mathematical modeling and are for routine decision making. No single method of risk

assessment is applicable to all situations and different methods are used in different circumstances. The requirements are elaborated case-by-case under the various standards- "Terrestrial Animal Health Code", "Aquatic Animal Health Code, Manual of Diagnostic Tests and Vaccines for Terrestrial Animals", and "Manual of Diagnostic Tests for Aquatic Animals"

iv. **Monitoring and Review :** The OIE has an information system for dissemination of early warning messages whenever epidemiologically significant events are officially reported. This alert system helps decisionmakers to take necessary preventive measures as quickly as possible. In order to improve transparency and animal health information quality, the OIE has also set up an animal health information search and verification system for non-official information from various sources on the existence of outbreaks of diseases that have not yet been officially notified to the OIE.

6.2.3 OECD – Safety Considerations for Biotechnology, 1986 and 1992

The 1986 OECD report was the first attempt to set international safety guidelines for industrial, agricultural and environmental applications of biotechnology. It presents scientific principles that could underlie risk management for the release of GMOs into the environment (Gupta *et.al.*, 2008). The 1992 report follows from this and defines "Good Development Principles" for the design of safe, small-scale field trials of GM plants and microorganisms.

i. **Scientific Principle:** Proposals to release GMOs are considered on a case-by-case basis. The development and assessment of GMOs takes place in a step-wise fashion moving from the laboratory to the greenhouse, to small-scale field trials and then large-scale field trials. Each step in the process generates information to predict the safety of the next step. Safety concerns focus on whether GMOs pose an "incremental risk" above and beyond the background risks of conventional agriculture.

ii. **Risk Identification :** The 1986 report identifies fault trees and event trees as a means to quantify probability of risk.

iii. **Basis of Assessment:** Fault trees, event trees and simulation can be used to quantify the probability and the magnitude of consequences in the first two stages of the assessment framework. The last stage can be analyzed by adapting/adopting conventional epidemiological or toxicological methods, although ecological consequence assessment is less well developed than its human counterpart. It suggests that qualitative risk assessment can be used to compare the propensities for survival, establishment and genetic stability under different environmental conditions, as much data are not available in these fronts. The reports do not clearly discuss uncertainty or the significance of the risk estimates.

iv. **Monitoring and Review :** The 1992 report states that scientifically acceptable and environmentally safe field researchrequires formulation of a hypothesis and statement of objectives, development of specific methodologies to introduce and monitor the organisms and mitigate the risk, a precise description of the design of experiments, including planting density and treatment pattern, and a description of specific data to be collected, and methods for analysis to test for statistical significance.

6.3 Biosafety, Biosecurity and Internationally Mandated Regulation

By considering national biosafety and biosecurity implementation obligations in the context of the International Health Regulations, United Nations Security Council Resolution (UNSCR) 1540, and, the Biological Weapons Convention, it is possible for States Parties to find a framework in which biosafety and biosecurity obligations overlap. The process of combining responses to all three regimes by addressing biosafety and biosecurity together in a cross-regime response will be especially helpful in defining a States Party's needs for education and training both now and in the future (Sture *et.al.*, 2013). Such a cross-regime response will also enhance compliance with all three regimes, and, importantly, ease the management of reporting expectations under each of these instruments. Further, it will facilitate and support a necessary inter-ministerial collaboration; with each of these areas falling under the purview of different national authorities, such interministerial collaboration will maximize the use of limited resources in economically constrained countries and enhance collaboration in more economically sound countries. At the same time, complying States Parties will be engaging in global governance activities involving more than just governmental bodies, supporting the exchange of ideas and progress beyond the constraints of the political arena.

6.3.1 International Health Regulations (IHRs)- Core Capacities

The purpose of the WHO International Health Regulations is to 'prevent, protect against, control and provide a public health response to the international spread of disease'. New obligations of State Parties under these regulations apply both inside and beyond their borders and cover a wide variety of public-health emergencies that are not solely confined to infectious threats. The International Health Regulations 2005, as an international legal agreement, is binding on 194 States Parties across the globe, including all Member States of the WHO (Sture *et.al.*, 2013).

The 2005 revised version of the IHRs entered into force on 15 June 2007 and they call for:

- Strengthened national capacity for disease surveillance and control, including during travel and transport;
- In country prevention capacity, the provision of effective outbreak and epidemic alert-and-response systems to international public health emergencies;

- Global partnership and international collaboration;
- Rights, obligations and procedures and progress monitoring.

In this connection, States Parties accept an obligation to develop 'core capacities' relating to public health and to complete their obligations for plans and infrastructure as early as June 2012 (with the possibility of two, two-year extensions). To monitor compliance, the WHO published the 'IHR Monitoring Framework: Checklist and Indicators for Monitoring Progress in the Implementation of IHR Core Capacities in States Parties' which establishes a framework that provides:

A set of 20 global indicators for monitoring the development of IHR core capacities for reporting annually to the World Health Assembly (WHA) by all States Parties (mandatory for all)

[and]

An additional 10 indicators for monitoring the comprehensive development, strengthening, and maintenance of States Parties' IHR core capacities (optional).

The WHO questionnaire sent to States Parties in 2011 to assess the status of national implementation of the IHRs consists of 13 sections, covering the eight core capacities required of States Parties (coordination, legislation policy, surveillance, response, preparedness, laboratory, human resource capacity), four hazards (zoonotic events, chemical events, food safety, radiation emergencies) and finally, points of entry. It also includes the following questions:

- Are biosafety guidelines accessible to individual laboratories?
- Do regulations, policies or strategies exist for laboratory biosafety?
- Has a responsible entity been designated for laboratory biosafety and biosecurity?
- Have biosafety guidelines, manuals or SOPs been disseminated to laboratories?
- Are relevant staff trained on biosafety guidelines?
- Has national classification of microorganisms by risk group been completed?
- Is there an institution or person responsible for inspection (could include certification of biosafety equipment) of laboratories for compliance with biosafety requirements?
- Are biosafety procedures implemented, and regularly monitored?
- Has a biorisk assessment been conducted in laboratories to guide and update biosafety regulations, procedures and practice, including for decontamination and management of infectious waste?
- Are diagnostic laboratories designated and authorized or certified BSL 2 or above for relevant levels of the health care system?

- Have country experience and findings related to biosafety been evaluated and reports shared with the global community?

Among the eight Core Capacities required by the IHRs, the 'Laboratory' capacity, as defined under 'Core Capacity 8', relates to:

- Those laboratory quality services relying on communication, specimen collection and transport, financial resources, biosafety and biosecurity best practices, trained personnel, suitable infrastructure, appropriate equipment and reagents, and the delivery of reliable results
- Building laboratory capacity to support a public health system cannot be done effectively without a strong focus on biosafety .... Laboratory-based surveillance and outbreak detection are essential to the prevention and mitigation of biological threats, and quality laboratory services are dependent on the implementation of biosafety and biosecurity best practices supported by an appropriate legal framework ....
- They argue that biosafety and biosecurity are 'pillars' that underpin a concept which they refer to as 'global health security'. This concept addresses a number of distinct but related international mechanisms that have overlapping and shared international health and non-proliferation objectives. By combining these under a cross-regime concept such as global health security, they argue that unique national concerns can be transcended. We would argue further, in this paper, that biosafety and biosecurity can also act together to both underpin and facilitate responses to these various regimes.
- While the WHO offers significant training and educational resources for strengthening those agencies engaged in setting-up and managing systems for securing global public health under the IHR implementation framework
- A step forward in the area of shared responsibility is represented by the development of the *Laboratory Biorisk Management Standard CWA 15793:2011*, developed by the European Committee on Standardization (CEN) (European Committee for Standardization 2011), superseding the original version of 2008. Further, in 2012, CEN has produced a set of guidelines for the implementation of CWA 15793 (European Committee for Standardization 2012). Among other provisions, the 2011 document provides a framework that can be used as the basis for training, raising awareness of laboratory biosafety and laboratory biosecurity guidelines and best practices within the scientific community. It emphasizes that the document may be useful in the development of new programmes as well as courses integrated into existing certified trainings, and this approach is to be highly recommended. It includes some specific training recommendations, with sections on planning, risk recognition, management, personnel training and competencies, communications, operations, emergency responses, performance monitoring and oversight.

- In a significant step towards its practical implementation, the CEN *Laboratory Biorisk Management* Standard has been used by the WHO in the inspection of the maximum containment facilities at the SRC VB VECTOR and CDC repositories for Variola virus (WHO 2009).
- The WHO has also taken steps forward by its release of a guidance document entitled '*Responsible life science research for global health security*' (WHO 2010) which identifies biosafety and laboratory biosecurity as one of the pillars supporting public health. It states:
- By emphasizing the public health perspective this guidance can achieve a broad acceptance of the need to raise awareness in this area and thus be better able to implement the objectives of promoting responsible life sciences research in general on a global level.

and:

- Research is essential for public health. Communication, international collaboration and openness, which are central to a public health perspective, are indispensable for global health security, scientific discovery and evidence-based measures.
- The self-assessment questionnaire that is part of this guidance is meant to identify strengths, weaknesses, and gaps and corrective actions to be taken. Of note, the document also emphasizes that:
- Good laboratory biosafety practices will mitigate the risks posed by laboratory accidents while laboratory biosecurity procedures will strengthen the accountability and responsibility of laboratory workers and their managers and thereby enhance public confidence in the responsible conduct of scientific experiments (Sture *et.al.*, 2013).

6.3.2 Biosafety, Biosecurity and International Health Security

Biosafety and biosecurity are essential pillars of international health security and cross-cutting elements of biological nonproliferation. The critical aspects of biosafety, biosecurity, and biocontainment have been in the spotlight in recent years. There have also been increased international efforts to improve awareness of modern practices and concerns with regard to the safe pursuit of life sciences research, and to optimize current oversight frameworks, thereby resulting in decreased risk of terrorist/malevolent acquisition of deadly pathogens or accidental release of a biological agent, and increased safety of laboratory workers. Since there is no single technology or process that could be applied to prevent or deter the use of biological agents as weapons, the implementation of international instruments for nonproliferation (such as the Biological Weapons Convention and United Nations Security Council Resolution 1540) and public health (such as the International Health Relations) summarized in Figure 8, as well as the establishment of regional and international partnerships in countering biological threats (whether natural, accidental or deliberate in nature), are critical factors in achieving global health security (Bakanidze *et.al.*, 2010). The pillars supporting the global health security are biosafety and biosecurity as they transcend unique national concerns and stand at the nexus of public health and security.

	WHO International Health Regulations (2005)	**UN Security Council Resolution 1540 (2004)**	**Biological Weapons Convention (1972)**
Applicability:	All 192 UN Member States	All 192 UN Member States	163 States Parties
Purpose:	"To prevent, protect, protect against, control and provide a public health response to the international spread of disease..."	To prohibit non-State actors from developing, acquiring, manufacturing, possessing, transporting, transferring or using nuclear, chemical or biological weapons and their delivery systems.	To prohibit the development, production, acquisition, transfer, stockpiling and use of biological and toxin weapons
Requirements:	8 core capacities "to detect assess, notify, and report events" [Laboratory core capacity includes biosafety/ biosecurity]	Domestic controls to prevent the proliferation of nuclear, chemical and biological weapons, their means of delivery, and related materials	Any necessary measures to prohibit and prevent the development production, stockpiling, acquisition, retention, transfer or use of biological weapons
Entry into force:	15 June 2007	28 April 2004	26 March 1975
Mandated reporting/where/ when:	Status of implementation / WHO/ *"As soon as possible but no later than five years from entry into force ..."*	Status of implementation /1540 Committee/ *"without delay"*	None "CBM voluntary reporting/ BWC ISU/ annually by 04/15

← Biosafety / Biosecurity →

Fig. 8: Biosafety and biosecurity are essential pillars of international health security and cross-cutting elements of biological nonproliferation

6.3.2.1 United Nations Security Council Resolution (UNSCR) 1540: Strategies for Prevention and Deterrence

United Nations Security Council Resolution (UNSCR) 1540 was unanimously adopted on 28 April 2004 (Sture *et. al.*, 2013). Acting under Chapter VII of the

United Nations (UN) Charter, in accordance with UNSCR 1540, States Parties accept an obligation to refrain from:

1. ... providing any form of support to non-State actors that attempt to develop, acquire, manufacture, possess, transport, transfer or use nuclear, chemical or biological weapons and their means of delivery;

And that:

2. .... all States, in accordance with their national procedures, shall adopt and enforce appropriate effective laws which prohibit any non-State actor to manufacture, acquire, possess, develop, transport, transfer or use nuclear, chemical or biological weapons and their means of delivery, in particular for terrorist purposes, as well as attempts to engage in any of the foregoing activities, participate in them as an accomplice, assist or finance them;
3. .... all States shall take and enforce effective measures to establish domestic controls to prevent the proliferation of nuclear, chemical, or biological weapons and their means of delivery, including by establishing appropriate controls over related materials and to this end shall:
   (a) Develop and maintain appropriate effective measures to account for and secure such items in production, use, storage or transport; and
   (b) Develop and maintain appropriate effective physical protection measures;

Under UNSCR 1540, biosafety and biosecurity capacities are implicit in the requirement that the agreement places upon States Parties, since they oblige states to establish a system – and by implication, relevant education and training capacities – within a national legal framework 'to account for and secure items in production, use, storage or transport' and to create provision for 'effective physical protection measures' (Sture *et.al.*, 2013). Thus as argued by Bakanidze, Imnadze and Perkins, UNSCR 1540 is seen as a framework that facilitates a 'strategy of prevention based on each individual state accepting responsibility for implementing measures against the proliferation of materials and weapons'. Again, we would argue that compliance with UNSCR 1540 involves not only biosafety and biosecurity measures being implemented in relation to hazardous biological, chemical or nuclear materials per se, but also to all data, records, paper materials and other communications relating to the knowledge and understanding of processes involved in handling these. Any measures implemented nationally under Operative Paragraph 3 (above) will be incomplete without the inclusion of materials under this definition as well as those materials already recognised. Further, unless measures are implemented to educate, train and support the relevant individuals, communities and organizations involved in working with materials as defined by UNSCR 1540, and with materials as we define them here, we argue that States Parties' capacity to meet their obligations will be seriously compromised, if not fatally flawed.

6.3.2.2 The Biological Weapons Convention (BWC)- Legally Binding Obligations

Article I of the Biological Weapons Convention outlines a series of measures that are underpinned by a catch-all obligation to do no harm that is referred to as the 'General Purpose Criterion'(Sture et.al., 2013). Under Article I of the BWC, States Parties agree that:

Each State Party to this Convention undertakes never in any circumstances to develop, produce, stockpile or otherwise acquire or retain:

(1) Microbial or other biological agents, or toxins whatever their origin or method of production, of types and in quantities that have no justification for prophylactic, protective or other peaceful purposes;

(2) Weapons, equipment or means of delivery designed to use such agents or toxins for hostile purposes or in armed conflict..

The General Purpose Criterion is referred as a 'device' that gives the treaty adequate scope by enabling it to keep up with scientific and technological change. It protects the beneficent, peaceful application of science and technology for prophylactic, protective or other peaceful purposes while establishing broad prohibitions. Other Articles of the Convention relate to a range of measures that demonstrate State Parties' compliance with their legally binding obligations under the Convention and give effect to the General Purpose Criterion.

At the Seventh Review Conference of the BTWC in December 2011, States Parties reaffirmed the importance of Article I of the Convention confirming that:

1. … the Convention is comprehensive in its scope and that all naturally or artificially created or altered microbial and other biological agents and toxins, as well as their components, regardless of their origin and method of production and whether they affect humans, animals or plants, of types and in quantities that have no justification for prophylactic, protective or other peaceful purposes, are unequivocally covered by Article I.
2. The Conference reaffirms that Article I applies to all scientific and technological developments in the life sciences and in other fields of science relevant to the Convention….
3. The Conference reaffirms that the use by the States Parties, in any way and under any circumstances of microbial or other biological agents or toxins, that is not consistent with prophylactic, protective or other peaceful purposes, is effectively a violation of Article I. The Conference reaffirms the undertaking in Article I never in any circumstances to develop, produce, stockpile or otherwise acquire or retain weapons, equipment, or means of delivery designed to use such agents or toxins for hostile purposes or in armed conflict in order to exclude completely and forever the possibility of their use ….

Article IV of the Biological Weapons Convention places States Parties to the Convention under an obligation to put national implementation measures in place to enhance domestic compliance with the Convention. Article IV thus reads as follows:

Each State Party to this Convention shall, in accordance with its constitutional processes, take any necessary measures to prohibit and prevent the development, production, stockpiling, acquisition or retention of the agents, toxins, weapons, equipment and means of delivery specified in Article I of the Convention, within the territory of such State, under its jurisdiction or under its control anywhere.

With respect to biosafety and biosecurity, the Final Declaration of the Seventh Review Conference called upon States Parties to:

(c) ensure the safety and security of microbial or other biological agents or toxins in laboratories, facilities, and during transportation, to prevent unauthorized access to and removal of such agents or toxins.

Although 'legislation' is not specifically referred to in the wording of Article IV of the BWC, it has become common to associate the obligation placed upon State Parties to put in place 'necessary measures' with the enactment of national implementation legislation. The latter refers to as 'domestic legal effect' to the scope of the prohibition as outlined in the Convention. In terms of the function of this Article of the Convention, Sims points out that national implementation legislation:

Ties the Convention into national legal systems in the clearest possible way. It contributes to the strengthening of compliance by expanding the constituency with an institutional interest in the success of the Convention. It also builds the treaty regime flowing from the Convention into normative structures at the national level, in the form of rules and expectations and procedures for upholding them. These rules, expectations and procedures in turn uphold their counterparts at the international level (Sture *et.al.*, 2013).

6.4 Biosafety and Biosecurity in Containment-A Regulatory Overview

When biosafety for contained use is addressed in international fora and discussions, often the topic is limited to working with genetically modified organisms (GMOs) in facilities such as laboratories, animal facilities, and greenhouses. However, the scope of biosafety in containment encompasses many other types of biological materials, such as human, animal and plant pathogens, nucleic acids, proteins, human samples, animals or plants, or by-products thereof, and overlaps often with the topic of biosecurity.

Many of these internationally accepted reference documents share the same basic principles: (1) a classification system for the biological agents or biological materials in so-called risk groups, often divided into four classes going from 1 (low) to 4 (high); (2) the understanding that increasing occupational and environmental risks require more stringent containment measures to work with that material, which is translated in a requirement for both risk assessment and risk management that is tailored to the activities performed with the biological materials, and (3) the description of containment measures, either result-oriented or more prescriptive as true containment or biosafety levels (Table 13) (Beeckman and Rüdelsheim, 2020).

Table 13: Overview of principles shared among internationally accepted reference documents for biosafety in containment.

Topic	WHO LBM	35001	BMBL	NIH G	CDC G	CBS + CBH
Risk groups	X	-	X	X	(X)	X
Activity based risk management	X	X	X	X	X	X
Containment measures – prescriptive	X	-	X	X	X	X
Containment measures – result oriented	-	X	-	-	X	X

Legend: WHO LBM, WHO Laboratory Biosafety Manual; 35001, ISO 35001:2019 Biorisk management for laboratories and other related organizations; BMBL, Biosafety in Microbiological and Biomedical Laboratories 5th ed.; NIH G, NIH Guidelines for Research Involving Recombinant or Synthetic Nucleic Acid Molecules; CDC G, CDC Guidelines for Safe Work Practices in Human and Animal Medical Diagnostic Laboratories; CBS, Canadian Biosafety Standard (CBS), 2nd Ed.; CBH, Canadian Biosafety Handbook (CBH), 2nd Ed.

Some of these reference documents have also served as the foundation for the development of national biosafety and biosecurity legislation, regulations and policies, either by including and refining the concepts mentioned in these documents or including the compliance with these documents as a requirement in the legislation.

Examples of Country- or Region-Specific Legislation: Due to the multiple objectives envisaged by biosafety and biosecurity (vide infra), regulatory requirements are most often part of legislation that is focusing on topics such as Worker Protection, Activities with Genetically Modified Organisms (GMO), Activities with Pathogens (human, animal, plant, quarantine), Waste or Biosecurity. The specific references to biosafety and biosecurity aspects in these themes for key countries were compiled. These overviews were prepared for Australia, Brazil, Canada, the European Union, Singapore and the United States of America, and they reflect the regulatory status at the time of compilation (period July – Dec 2019) as examples of different approaches (Beeckman and Rüdelsheim, 2020).

6.4.1 Overlaps with other Regulatory Frameworks

Overlaps with other regulatory frameworks that have provisions on handling biological material. Both on the international and the regional or local level, additional provisions for handling of biological materials are imbedded in diverse regulatory texts, several of which on first sight would not be immediately recognized as being relevant for biosafety and biosecurity in containment. Many of them are related to the topic of transboundary movement, traceability, transport and occupational hygiene, and their link to biosafety and biosecurity for contained use is explained here further for some concrete examples (Beeckman and Rüdelsheim, 2020).

Cartagena Protocol : The "Cartagena Protocol on Biosafety to the Convention on Biological Diversity"describes in its Article 18 that "LMOs [Living Modified Organisms] that are subject to intentional transboundary movement within the scope of the Protocol are [to be] handled, packaged, and transported under conditions of safety, taking into consideration relevant international rules and standards", thus clearly referring to existing rules and requirements for maintaining containment during transport. Specifically for LMOs that are destined for contained use it is stipulated that they should be clearly identified as LMOs, a requirement which is common to many GMO specific regulations in different countries, and shipment documentation should provide instructions for the safe handling, storage, transport and use, thereby ensuring containment. In addition, by means of Article 15 "Risk Assessment" and Article 16 "Risk Management", the Cartagena Protocol is aligned with the concepts described in different internationally accepted reference documents for biosafety in containment (see International Framework and Guidance Documents).

Nagoya Protocol: The "Nagoya Protocol on Access to Genetic Resources and the Fair and Equitable Sharing of Benefits Arising from their Utilization to the Convention on Biological Diversity" states that when benefits (either monetary or non-monetary) are arising from the utilization of genetic resources (e.g., in research) as well as during subsequent commercialization, that these benefits "shall be shared in a fair and equitable way with the Party providing such resources that is the country of origin of such resources or a Party that has acquired the genetic resources in accordance with the Convention" (Beeckman and Rüdelsheim, 2020).

Plant and Animal Health : In the section on "Protecting Animal and Plant Health" the efforts from the International Plant Protection Convention (IPPC) and the World Organization for Animal Health (OIE) in safeguarding containment when handling plant and animal pathogens, respectively, were highlighted (Beeckman and Rüdelsheim, 2020). However, many of the standards developed by these two organizations deal with topics such as import and export as well as traceability. This is especially important in case of newly emerging infections with the potential of world-wide epidemics. Checking the sanitary status of plant materials and animals prior to import or export reduces the risk of spreading diseases, while the recording of movements is imperative to allow for a quick and targeted response in case it does go wrong. Occupational Hygiene : Occupational Hygiene, as defined by the International Occupational Hygiene Association is "the discipline of anticipating, recognizing, evaluating, and controlling health hazards in the working environment with the objective of protecting worker health and well-being and safeguarding the community at large" (Beeckman and Rüdelsheim, 2020). Also known under the term of Industrial Hygiene, it is typically part of an Occupational Safety and Health program, where it focuses on chemical, physical and biological agents in the workplace possibly causing illness or discomfort, and aims to avoid health effects through risk assessment and management. Although occupational hygiene and biosafety go hand in hand in terms of both intended and unintended exposure to biological agents, there is a clear difference in scope, being the general workplace as a whole vs. specific activities with biological materials, respectively. A clear example in this respect in the prevention against Legionnaires disease (Legionella), which is a typical workplace biological exposure monitored and managed by occupational hygiene, and generally not in scope of biosafety.

Transport Regulations : The UN Model Regulations from the UN Economic and Social Council's Committee of Experts (UNECE) on the Transport of Dangerous Goods describe the recommendations for transport of dangerous goods to safeguard workers' health and safety, property, or environment protection during all modes of transport..

6.4.2 Biosafety Regulations of GMOs

National and International aspects and Regional Cooperation : Genetically modified organisms (GMOs) are organisms, produced by transferring, very often of one gene encoding a desirable trait, in one organism to another, through precise molecular biology technique, called genetic modification (in comparison with the classical breeding, in which an uncontrolled transfer of more than one genes occurs and many of the genes have unidentified effect on the resulting organism).. After the entering into force of the Cartagena Protocol on Biosafety on 11th of September 2003 there is an urgent need for acquiring specific information and capacity-building in this area for further implementation of biosafety policy and taking informed decisions by governments at national, regional and international level.

Epochal changes have been brought about by globalization, the increasing role of the private sector in third world agricultural economies, and the development and introduction of genetically modified (GM) organisms coupled with the possibility of introducing alien species, genotypes and plant pests into the environment. These changes justify the need for developing biosafety frameworks at national level as well as harmonized international instruments and guidelines..

With this context, Agenda 21 provides a blue print for international collaboration for the further development and application of biotechnology and biosafety. Another key agreement adopted at the Earth Summit in Rio was the Convention on Biological Diversity. The objectives of the Convention are the conservation of biological diversity, the sustainable use of its components and the fair and equitable sharing of the benefits arising out of the utilization of genetic resources. In 2000, the Cartagena Protocol on biosafety, established under The Convention of Biological Diversity, has been adopted. It aims at regulating the transboundary movement of living modified organisms (LMO) resulting from modern biotechnology in light of protecting biological diversity from potential risks that may be posed by the LMOs (Alexandrova, 2005).

i. **International agreements Convention on Biological Diversity (CBD):** One of the major international instruments relevant to biosecurity is the Convention on Biological Diversity (1992) and its Cartagena Protocol on Biosafety (2000). The scope of the Convention and its objectives are the conservation of biological diversity, the sustainable use of its components and the fair and equitable sharing of the benefits arising out of the utilization of geneticresources.
ii. **Cartagena Protocol on Biosafety to the Convention on Biological Diversity:** The Protocol is was adopted in 2000 and entered into force by September, 2003. It is the first global legally binding instrument focusing on LMOs. The purpose of the Protocol is to ensure adequate levels of protection in the field of safe transfer, handling and use of LMOs resulting from modern biotechnology that may have adverse effects on the conservation and sustainable use of biodiversity or pose a risk to human health (art. 1; art. 4). The Protocol is applicable to all LMOs apart from those that are pharmaceuticals for humans that are addressed by other international agreements (art. 5).
iii. **WTO-SPS Agreement :** The WTO oversees the implementation of the Agreement on the Application of Sanitary and Phytosanitary Measures (SPS Agreement). The SPS Agreement provides for a common approach to different sectors in the context of biosafety by applying to all sanitary and phytosanitary measures (SPMs), which may directly or indirectly affect international trade (art. 1). The SPS Agreement, while permitting governments to maintain appropriate sanitary and phytosanitary protection, reduces possible arbitrariness of decisions and encourages consistent decision-making
iv. **WTO-TBT Agreement :** The WTO also oversees the implementation of the Agreement on Technical Barriers to Trade (TBT Agreement). The Agreement tries to ensure that regulations, standards, testing and certification procedures do not create unnecessary obstacles. The WTO's version is a modification of the code negotiated in the 1973-79 Tokyo Round. The TBT Agreement may be of relevance to biosafety also for its relevance to biotechnology products because it generally applies to technical regulations and standards, including packaging, marking and labelling requirements.

v. ***Codex Alimetarius:*** In the food safety area, the *Codex Alimentarius* is the primary collection of internationally adopted food standards and as such is of great relevance to biosafety. The Codex has become the seminal global reference point for consumers, food producers and processors, national food control agencies and the international food trade. Both Codex subsidiary bodies and the Commission, an intergovernmental body developing and keeping under review the Codex, give the highest priority to consumer interests in the formulation of commodity and general standards. Instrument still being developed is the Proposed Draft Guidelines for the Labelling of Foods Obtained through Certain Techniques of Genetic Modification/ Genetic Engineering. In addition, in the animal feed area, the Codex Commission is developing a proposed Code of Practice on Animal Feeding aimed at establishing a feed safety system that covers the whole "'feed chain' from farm to table". This will eliminatepotential risks to human health, animal health and the environment. It will apply to the production and use of all materials of animal, plant and marine origin used in animal feed at all levels, whether produced industrially or on the farm.

vi. **International Plant Protection Convention (IPPC) :** The IPPC came into force in 1952 and has been amended once in 1979 and again in 1997. The IPPC regulates plant pests. It also regulates "any organism, object or material capable of harbouring pests or spreading pests that affect plants or plant products" (art. I(4)). The purpose of regulation is to prevent "the spread and introduction of these pests and promoting measures for their control" (ICPM, 2001a). IPPC application to plants is not limited only to the protection of cultivated plants or direct damage from pests..

vii. **Convention on access to information, public participation in decision-making and access to justice in environmental matters :** The Convention on access to information, public participation in decision-making and access to justice in environmental has been adopted in 1998 in Aarhus, Denmark and it is also known as Aarhus Convention. Theobjective of the convention is to contribute to the protection of the right of every person of present and future generations to live in an environment adequate to his or her health and well-being, each Party shall guarantee the rights of access to information, public participation in decision-making, and access to justice in environmental matters in accordance with the provisions of this Convention.

National regulations and legislation : According to the increasing economical importance of biotechnology products, developed and developing countries have instituted biosafety policies and procedures to ensure their safe use (Alexandrova, 2005). These protective measures are implemented though establishing biosafety systems that provide a mechanism for making informed decisions. For these regulatory systems to function successfully and effectively, they should be flexible to adapt amenably to the evolution of scientific knowledge, to ensure feedback mechanisms and to reflect the conditions of a given country as much as possible.

6.4.3 Biosafety Regulations- Indian Scenario

In India, the task to ensure biosafety and biosecurity in all public and private sector institutions involved in research and development and human healthcare and agriculture including animal and marine sciences is enormous. In human healthcare itself, there are more than 800 medical and healthcare R&D institutions; 575 medical colleges, 350 universities, more than 15,000 pharmaceutical and biopharmaceutical industries and unlimited number of healthcare entities and diagnostics laboratories handling biomedical wastes. Laboratories handling research on highly pathogenic organisms follow good laboratory practices and decontamination and disinfection procedures and a number of containment laboratories have been set up under human and animal health sectors (Gandhi, 2014). However, the concept of biosafety and biosecurity is not fully understood by each and every establishment handling infectious pathogens. To implement biosafety guidelines and to mitigate biological threats, a holistic approach is required on part of the administrative and technical authorities to adopt a systemic approach for successful bio-risk management. A scientific, rigorous, transparent, efficient, predictable and consistent regulatory mechanism and protocol for bio-safety and bio-security evaluation and related system need to be followed to meet these objectives.

Biosafety during lab work is an important concern in developing countries. A cross sectional study on the safety measures being adopted in clinical laboratories of India showed that a substantial number of laboratories do not follow the best practices as prescribed. Another study in Pakistan showed that nearly 50 percent of the laboratories were lax as far as implementation of best practices is concerned.

Containment Facilities: An important element of control for laboratory containment and research product protection is strict adherence to laboratory biosafety containment practices and good microbiological practice (GMP) based on widely accepted aseptic practices. In India a number of containment laboratories for human and animal health safety are being planned and about 30 such biosafety laboratories of the level of BSL3 or BSL2+ currently under operation, mostly at the laboratories of CSIR, ICMR and DRDO, vaccine industries and other research institutes. Two BSL4 facilities in India are already functional. While each laboratory requires a different level of management to bio-safety and bio-security, the premises, materials and workers in these labs, however, remain poorly supervised and managed by less trained personnel.

6.4.3.1 Risk Mitigation Strategies

Increased number of global incidence of mismanagement of dangerous pathogens at workplaces has necessitated revisiting the existing regulations on biosafety and biosecurity in the country and strict implementation of these laws. All out efforts should ensure that such research is conducted in accordance with the highest standards to focus on prevention of accidental or unintentional exposures to or releases of pathogens and toxins so as to protect workers including researchers, public, animals and the environment from accidents. Identification of technologies with potential to be misused and putting in place measures to protect against intentional theft, misuse, or release of biological materials must be prioritized.

Scientists involved in bio-medical sciences shoulder larger responsibilities and need to abide by a voluntary code of conduct of research, based on the recognised ethical principles and values and comply with the requirements of international conventions and treaties relevant to their research work.

In India, laboratory safety has to be an integral part of overall safety programme in hospitals and all this can be achieved by having a quality control program in hospitals in general and laboratories in particular. Accreditation has to be made necessary and all laboratories can be graded as per their performance against a set of predetermined standards.

Awareness to large number of people involved in research and development of practices and procedures for both laboratory biosafety and laboratory biosecurity and biological waste management would require a systematic and holistic approach that may include:

- Improved understanding and management of the risks associated with accidental and deliberate misuse of biological agents,
- Develop strategies for biorisk management for R&D in bio-sciences and biotechnology in public and private sectors,
- Understand need of large number of academic and research institutions in basic sciences, medicine, healthcare, agriculture, environment etc.,
- Outreach to large number of hospitals, laboratories and healthcare providers,
- Outreach to growing biotech and pharmaceutical industries and their scientists in R&D

A system of checks and balances could provide the assurance that the advances in life sciences are only used to protect life and not to destroy it. Issues requiring consideration may include:

- Increased awareness of risks of bio-terrorism, bio-safety, biothreat identification, prevention, and response among scientists and managers,
- Develop training programmes and materials for educating scientists on laboratory biorisk management for bio-safety and bio-security,

- Establish in universities and scientific institutions procedures to monitor research activities and mechanisms to prevent dissemination of information likely to be utilized for bio-terrorism,
- A bottom-up approach in formulation and implementation of bio-safety and bio-security policies through direct involvement of scientists,
- Adoption of policy of outreach to industry to inform and involve it in the process of evolution of bio-safety and bio-security policies,
- Training curriculum in bio-waste management and environment risk management of pharmaceuticals,
- Ensure good microbiological practices for responsible conduct and oversight of life sciences research that can threaten public health or national security,
- Awareness of and compliance with the requirements of international conventions and treaties relevant to their research work,
- Establish legal and institutional procedures and mechanisms for monitoring and regulation.

Government level oversight can be aimed at policy for bio-risk management; the international level structured Global Bio-risk Management Curriculum has been developed and is a good reference for this task.

6.4.3.2 Government Efforts and Biorisk Management

The Ministry of Environment and Forests (MoEF), Government of India, is implementing a GEF/World Bank funded project on Capacity Building on Biosafety in context of the Cartagena Protocol (Gandhi, 2014). The project covers the assessment, management and long term monitoring and documentation of the risks to the sustainable use of biodiversity and to human health potentially posed by the introduction of Living Modified Organisms (LMOs). The project aims to strengthen the legislative framework and operational mechanisms for biosafety management, enhance capacity for risk assessment and monitoring, establish the biosafety database system and biosafety clearing house mechanism, support centres of excellence and a network for research, risk assessment and monitoring and establish the Project Coordination and Monitoring Unit (PCMU).

Biotech Consortium India Ltd. (BCIL) supported by DBT, organizes workshops on National Consultation on Biosafety aspects related to Genetically Modified Organisms; supports programmes on capacity building activities in biosafety including preparation of research documents, organizing conferences, workshops on key policy issues and publishing monthly newsletter "The South Asia Biosafety Program (SABP)".

The bio-risk management approach comprises biosafety, laboratory biosecurity and ethical responsibility by reducing the risk of unintentional exposure to pathogens and toxins or their accidental release; reducing the risk of unauthorized access, loss, theft, misuse, diversion or intentional release of microorganisms; and those suitable measures have been adopted and effectively implemented. It is important that a framework for continuous awareness rising for biosafety, laboratory biosecurity and ethical code of conduct, and training is provided.

Laboratory bio-risk management as defined under CWA 15793:2011 on laboratory biosafety and biosecurity includes the analysis of ways, development of strategies and their implementation to minimize the likelihood of the occurrence of bio-risks; methodology used to organize and analyze scientific information in order to estimate the probability and severity of an adverse effect (assessment) and measures to minimize this effect (mitigation); and establishing policies and practices for risk management in the lab (day to day as well as emergencies). An effective bio-risk management programme would involve bio-risk assessment including, investigation into the nature of the biological materials and the procedures used to store, handle, and transfer and dispose those materials. This may include review of research proposals and identify hazards that the biological materials can cause in a laboratory. Bio-risk management procedures may determine the control measures, or mitigation strategies, to be used to minimize or eliminate the defined risk. Risk assessment and the concept of continual improvement on bio-risk management can efficiently address issues relating to biosafety and biosecurity (Gandhi, 2014).

Bio-risk mitigation would include actions and control measures that are put into place to reduce or eliminate the risks associated with biological agents and toxins. It determines what can be done to manage the identified bio-risks. These control measures are grouped under the categories of elimination or substitution, engineering controls, administrative controls, practices and procedures, and personal protective equipment.

A critical mass of dedicated human resource is required to implement provisions of bio-risk management, which requires:

- Developing training curriculum for stakeholders, including policy makers, management bio-risk advisors, scientists, laboratory management, laboratory and industrial workforce,
- Organising training programs, workshops to train-the-trainers,
- Organising training programmes on bio-threats and bio-risk mitigation strategies for law enforcement official,
- Developing training implementation strategies including regional training centres,
- National Coordination Centre to oversee implementation of bio-risk management programmes.

6.4.3.3 Dedicated Training Facility

A panel discussion at the International Meeting on Host Parasite Interactions held at National Institute of Animal Biotechnology, Hyderabad on July 15, 2014 highlighted the issues involved with biosafety and biosecurity procedures faced by various stakeholders. Experts stressed on issues of information and awareness; implementation of regulations and guidelines, education and outreach, capacity building, risk assessment and risk management. In order to increase awareness and training on safety & security in research institutes, academia & industry, it was strongly recommended to the Government of India to have a fresh look at the implementation of the existing laws governing the biosafety and biosecurity to enforce effective bio-risk management in the country and filling in necessary gaps in capacity building. It was further suggested to set up a Bio-risk Management Training Academy (BRMTA), which would provide India an advantage and understanding on bio-risk management in the biotechnology sector.

The Objective of BRMTA would be to promote capacity building in bio-risk management and promotion of good biosafety and biosecurity practices in India through education and training, information dissemination and knowledge sharing in pursuit of the benefits of life science research (Gandhi, 2014). This will also ultimately foster culture of responsibility and code of ethics among the researchers. The mandate of BRMTA may include:

- Ensuring bio safety measures so that research is conducted in accordance with the highest standards to promote safe, secure and responsible use of dangerous biological agents and toxins that have high-consequence for its potential to be dual use research of concern,
- Developing of programmes for awareness of the prohibition and requirements for bioterrorism preparedness to inter-alia industry, scientific and technological communities, armed forces,
- Developing specialized courses in bio-risk management; related policies, applicable regulations, audits, inspections etc.;
- Special courses to manage challenges and opportunities associated with facility risk assessment; laboratory certification; reporting laboratory-associated infections,
- Organising education & training programmes; short-term courses; workshops and symposia; and conferences (national and international),
- Preparing guidelines for medical surveillance and evaluation system,
- Wide dissemination of information though updated technical guidance documents, safety manuals, biosecurity plan, safe practices and standard operating procedures and emergency response plans,
- Outreach to large number of beneficiaries including individuals, institutions and industries through electronic media and periodic newsletters,

- Advising the Government of India about issues related to biological disaster management, international collaboration on bio-risk management and developing risk mitigation strategies,
- Establishing and strengthening collaborations with international bodies and other similar institutions involved with risk assessment and management programmes or developing curriculum in the area to improve understanding and management of the risks associated with accidental and deliberate misuse of biological agents.

Biosafety in India is primarily focused on genetically modified (GM) agricultural research and ensuring environmental safety. This is evidenced by the Indian definition of biosafety as "the need to protect the environment including human and animal health from the possible adverse effects of the Genetically Modified Organisms (GMOs) and products thereof derived from the use of modern biotechnology." The Environment Protection Act (EPA) of 1986 created room for the development of India's first biosafety regulations, the 1989 Rules for the Manufacture/Use/Import/Export and Storage of Hazardous Microorganisms, Genetically Engineered Organisms or Cells (Gigi Kwik Gronvall*et.al.*, 2016). Since the publishing of the "Rules 1989," several other key guidelines have been released to provide revised legislation and guidance for studies involving recombinant DNA, and transgenic plants.Each research institution working with rDNA or GMOs is required to have an Institutional Biosafety Committee (IBSC)that reports to the Ministry of Environment, Forests, and Climate Control (MoEFCC) and Department of Biotechnology (DBT), who together provide oversight for biological research institutions.

Pathogen characterization : According to the Recombinant DNA Safety Guidelines, 1990, microorganisms are categorized in to four risk groups based upon pathogenicity, transmissibility, host range, the availability of preventative or curative treatment, prevalence in the country, and its ability to cause disease in humans, animals and plants(Gigi Kwik Gronvall*et.al.*, 2016). Risk Group IV contains agents of the greatest risk to human safety, and Risk Group I contains those who pose the smallest risk. India uses the same BSL 1-4 containment facility categorization as the WHO and the United States.

6.4.3.4 Relevant Regulations and Legislation

Listed below are important biosafety guidelines and legislation in chronological order:

i. **Environment Protection Act (1986) and Environment (Protection) Rules (1986) :** The Act relates to the protection and improvement of environment and the prevention of hazards to human beings, other living creatures, plants and property (Gupta *et.al.*, 2008). The Act mainly covers the rules to regulate environmental pollution and the prevention, control, and abatement of environmental pollution. The Environment (Protection)

Rules cover management and handling of hazardous wastes, manufacture, storage and import of hazardous chemicals and rules for the manufacture, use, import, export and storage of hazardous microorganisms, genetically engineered organisms or cells.

ii. **Drugs and Cosmetics Rules, 1988 (8th Amendment):** It designates the Ministry of Health and Family Welfare (DoH) as the regulatory body overseeing the import or manufacturing of biological and biotechnological products.

iii. **Rules for the Manufacture, Use/Import/Export and Storage of Hazardous Microorganisms/Genetically Engineered Organisms or Cells. (Notified under the EP Act, 1986) (1989):** These Rules include the rules for pharmaceuticals, transit and contained use of genetically engineered organisms, microorganisms and cells and substances/products and food stuffs of which such cells, organisms or tissues form a part, LMOs for intentional introduction into the environment, handling, transport, packaging and identification. These rules are applicable to the manufacture, import and storage of microorganisms and gene technology products. The rules are specifically applicable to Sale, storage and handling; Exportation and importation of genetically engineered cells or organisms; Production, manufacturing, processing, storage, import, drawing off, packaging and repackaging of genetically engineered products that make use of genetically engineered microorganisms in any way.

iv. **Recombinant DNA Safety Guidelines (1990):** The Guidelines prescribe safety measures for research, field cultivation and also the environmental impact during field applications of genetically altered material products. They are applicable to research involving genetically engineered organisms originating from genetic transformation of green plants, rDNA technology in vaccine development, and also large scale production and deliberate/ accidental release of organisms, plants, animals and products derived by rDNA technology into the environment. The Guidelines also prescribe the criteria for ecological assessment on a case-by-case basis for planned introduction of rDNA organism into the environment.

v. **The Revised Guidelines for Research in Transgenic Plants & Guidelines for Toxicity and Allergenicity Evaluation of Transgenic Seeds, Plants and Plant Parts (1998):** It provides rules for recombinant DNA research on plants including molecular analysis and field evaluation. These guidelines also address importing and exporting of GM plants for research use.,

vi. **Foreign Trade (Development & Regulation) Act, 1992 (2006) (draft amendment):** The Act provides for the development and regulation of foreign trade by facilitating imports into and augmenting exports from India and for matters connected with it. In 2006, the government made draft amendment in the foreign trade policy, making labeling of imported GM products mandatory.

vi. **Revised Guidelines for Research in Transgenic Plants & Guidelines for Toxicity and Allergenicity Evaluation of Transgenic Seeds, Plants and Plant Parts (1998):** The Guidelines cover rDNA research on plants including the development of transgenic plants and their growth in soil for molecular and field evaluation. It also includes LMOs for contained use and intentional introduction into the environment, and LMOs for use as food or feed or for processing, pharmaceuticals and transboundary movement. The Guidelines also specify requirements for import and shipment of GM plants for research use only.

vii. **Guidelines for Generating Preclinical and Clinical Data for rDNA Vaccines, Diagnostics and other Biologicals (1999):** The Guidelines cover preclinical and clinical evaluations of rDNA vaccines, diagnostics and other biologicals/pharmaceuticals. The objectives of the preclinical studies are to define physiological, toxicological and efficacious potential of r-DNA products prior to initiation of human studies. Both in vitro and in vivo studies can contribute to evaluating the effects of r-DNA products. The Guidelines also cover safety, purity, potency and effectiveness of the rDNA products, in vitro diagnostic recombinant reagents and monoclonal antibodies, and describe in detail procedures for generating monoclonal antibodies. Sensitivity and specificity required for diagnostics of infections of widespread diseases like HIV-I/II are also prescribed.

viii. **Plant Quarantine (Regulation of Import into India) Order (2003):** The Order allows import of transgenics/GMOs into India for the purpose of agricultural research or experimentation purpose only. No commercial imports are allowed under this order.

ix. **The Seed Bill (2004) (draft):** The Bill provides for regulating the quality of seeds for sale, import and export and to facilitate production and supply of seeds of quality and other related matters. Apart from other provisions related to seed, the Bill has special provisions for registration of transgenic varieties. Clause 15 of the draft bill covers specific provisions for transgenic varieties requiring clearance under the provisions of the Environment (Protection) Act, 1986.

x. **The Food Safety and Standards Act (2006):** The objective of the Act is to bring out a single statute relating to food and to provide for a systematic and scientific development of food processing industry. The Act incorporates the salient provisions of the Prevention of Food Adulteration Act, 1954 (37 of 1954) and is based on international legislations, instrumentalities and *Codex Alimentarius* Commission Guidelines (Gupta *et.al.*, 2008). The Act is in tune with the international trend towards modernization and convergence of regulations of food standards with the elimination of multi-level and multi-departmental control. The emphasis is on (a) responsibility with manufacturers, (b) recall, (c) GM and functional foods, (d) emergency control, (e) risk analysis and communication and (f) food safety and good manufacturing practices and process control viz., hazard analysis and critical control point.

xi. **The Guidelines and Handbook for Institutional Biosafety Committees (2011):** It provides details on the necessary composition of the biosafety committees and each member's role, procedures for registration, the role of IBSCs in project approval, training requirements and committee meeting specifics, among other details..

xii. **The Biotechnology Regulatory Authority of India Act (BRAI Act) (2013):** It mandates the establishment of the Biotechnology Regulatory Authority of India to "regulate the research, transport, import, manufacture and use of organisms and products of modern biotechnology and for matters connected therewith or incidental thereto." It was initially prepared by the Department of Biotechnology (DBT) in 2008, the bill went through several revisions before it was passed in 2013.

Regulatory and Oversight Agencies: The Ministry of Environment, Forests, and Climate Change (MoEFCC) along with the Ministry of Science & Technology's Department of Biotechnology are responsible for enforcing the policies of the 1986 Environmental Protection Act. The Genetic Engineering Approval Committee (GEAC), the Review Committee on Genetic Manipulation, and Institutional Biosafety Committees (IBSCs) review and approve research projects. Both the State Biotechnology Coordination Committee (SBCC) and the District Level committee (DLC) monitor ongoing studies(Gigi Kwik Gronvall*et.al.*, 2016). The recombinants DNA Advisory Committee (RCAC) advises committees and government agencies when necessary.

6.5 Biosafety Officers and Institutional Biosafety Committees

The Guidelines and Handbook for Institutional Biosafety Committees identifies the need for a Biosafety Officer (BSO) "...if research is conducted on organisms that require special containment conditions (Biosafety Level 3 or 4)...or if large-scale rDNA research is conducted." According to these guidelines, the BSO must "also a member of the IBSC and act as a technical liaison between researchers and the IBSC. The Biosafety Officer should be adequately trained and be able to offer advice on specialized containment requirements." Specifics about the level of training are not provided, but the BSO must know enough to ensure that rDNA safety guidelines and good laboratory practices are followed. Institutional Biosafety Committees (IBSC) are required for every facility working with rDNA and GMOs.These committees are authorized to approve laboratory studies, excluding field trials and hazardous genetic experiments or methods.65 Each committee is comprised of academics and researchers from the institute, including the head of the institution and a medical expert, plus a member of the DBT.There are currently about 500 IBSCs within India.

Accident and Incident reporting: The 1990 Recombinant DNA Safety Guidelines specify that an emergency plan should be in place in every facility for incidents such as breakage and spillage; accidental exposure to a chemical, toxin or agent; and natural disasters. Specific instructions are given for cleanup, sharps containers and disposal of contaminated materials. The Guidelines and Handbook for Institutional Biosafety Committees states that all accidents, illnesses, or breaches in protocol must be reported to the Principle Investigator (PI) who then must report the incident to the IBSC Chairperson within 24 hours and to the RCGM within 48 hours of the incident. In anticipation of an accidental GMO release, the PI of a study is required to haveemergency protocol in place and distributed to appropriate regulatory committees, as is outlined in the "Rules, 1989."

Synthetic/Advanced Biology: To date, there are a limited number of institutions and research centers, and a handful of industrial or commercial companies engaged in research or work with synthetic biology (Gigi Kwik Gronvall*et.al.*, 2016). India's DBT is actively pushing to increase national interest and engagement in synthetic biology. The nation's 12th Five-Year Plan (2012-2017), with a budget of approximately EUR 277 million, has taken recommendations from the Task Force on Synthetic and Systems Biology Resource Network to boost interest in synthetic biology research. As the DPT pushes for greater investment in synthetic biology, the need for guidelines to accommodate these new endeavors has been expressed by many academics.

Training : The Guidelines and Handbook for Institutional Biosafety Committees state that biosafety training is required "for all individuals conducting research with GMOs/LMOs/rDNA materials." The training can be conducted by the research institution itself as long as the IBSC has documentation that all staff are trained or have appropriate experience in biosafety protocol and standard operating procedures. All staff must have "knowledge in handling and management of incidents/accidents in the facility and information on when and how to report laboratory incidents." Additionally, lab staff working in BSL 3 and BSL 4 facilities must have more detailed training, but the specifics of the training or who is to provide the training is not provided.

Laboratory numbers: In 2008, it was reported that India had 14 BSL-3 facilities with 6 planned for construction and 1 BSL-4 facility with 2 planned for construction.78 In conflict with this report are several 2013 sources which reported that in March of 2013, India's first BSL-4 facility was completed at the National Institute of Virology (NIV) in Pune (Gigi Kwik Gronvall *et.al.*, 2016). At this time, the NIV had already constructed a BSL-3 facility at its Microbial

Containment Complex campus. The new BSL-4 facility allows for extended research capabilities with agents like pandemic influenza, SARS, Nipah virus and Crimean-Congo hemorrhagic fever virus.

Research Status: India was listed 51st on the 2015 Scientific American Worldview Overall Scores for biotechnology innovation. Its highest scores were in productivity and intensity, and it lowest scores were in IP protection and enterprise support. India's Department of Biotechnology has dedicated significant effort increasing the nation's global contribution to biotechnology and scientific research. In the past few years, the DBT has worked to create an "Open Access Policy," developing a database where funded researchers can share and learn from each other's work. Much of Indian investment and effort in biotechnology is directed towards genetically modified agriculture. India's first transgenic crop, Bt cotton, was introduced in 2002 (Gigi Kwik Gronvall*et.al.*, 2016). Now, India produces 11.6 million hectares of Bt cotton and is the 4th highest producer of biotech crops in the world. Currently, India has developed 23 biotech crops with 67 biotech traits in different stages of development. These crops are developed by both the public and private sectors (39 public sector, 20 private, 8 autonomous institutes). Biopharmaceuticals, bioinformatics, and bioservices are also major parts of India's biotechnology industry. India's biopharmaceutical industry accounts for 62% of the total biotechnology sector, with bioservices at 18%. The Bioservices sector continues to grow in capacity for contract research, clinical trials, and manufacturing engagements.

6.6 Biosafety Guidelines

Biosafety regulations in India comprise biosafety rules and guidelines. The existing legislative framework in India for biosafety regulations has followed a disaggregated approach with regulatory powers imposed in a top-down fashion. The framework legislation for biosafety regulations in India is the EPA. Three provisions of the EPA form the basis of the biosafety regulations. These are sections 6, 8, and 25. While Section 6 of the Act empowers the Central Government to make rules on procedures, safeguards, prohibition and restrictions for handling of hazardous substances, Section 8 of the Act prohibits a per-son from handling hazardous substances, except in accordance with procedures and after complying with safeguards. Section 25 of the EPA empowers the Central Government to lay down rules regarding procedures and safeguards for handling hazardous substances. Thus, the biosafety rules in India are statutory in nature as they originate from the EPA. These provisions of the EPA led to the adoption of the 1989 Rules for the Manufacture, Use/Import/Export and Storage of Hazardous Micro organisms/ Genetically Engineered Organisms or Cells.

Guidelines for safety have been issued by the Department of Biotechnology (DBT) in 1990 covering research in biotechnology, field trials and commercial applications. DBT had also brought out separate guideline for research in transgenic plats in 1998 and for clinical products in 1999. Activities involving GMOs are also covered under other policies such as the Drugs and Cosmetics Act (8th Amendment), the Drug Policy, 2002, and the National Seed Policy, 2002 (Kannaiyan, 2006).

There are six competent authorities for implementation of regulations and guidelines in the country, *viz*., Recombinant DNA Advisory Committee (RDAC); Review Committee on Genetic Manipulation (RCGM) ;Genetic Engineering Approval Committee (GEAC) ;Institutional Biosafety Committee (IBSC) attached to every organization engaged in rDNA research; State Biosafety Coordination Committees (SBCC) and District Level Committee (DLC).

Of the above committees, the IBSC is constituted by organizations involved in research with GMOs with the approval of DBT. The IBSC is the nodal point for interaction within the institution for implementation of the guidelines. (Every research project using GMOs has an identified investigator who is required to get the research project approved from safety angle and inform the IBSC about the status and results of the experiments being conducted). The functions of IBSC include to:

- Reviewing and giving clearance to project proposals falling under restricted category as per DBT guidelines
- Recommending Category III risk or above experiments to RCGM approval
- Tailoring biosafety programme to the level of risk assessment
- Training of personnel on biosafety
- Adopting emergency plans

The role of IBSC assumes major importance since it is the only Statutory committee, which operate from the premises of institution and hence is in a position to conduct onsite evaluation, assessment and monitoring of adherence to the biosafety guidelines. The decisions taken by the next higher committee i.e Review Committee on Genetic Manipulation (RCGM), which operates from DBT are based on the applications submitted by the investigators with the approval of IBSC on the status of the project and its conformity with the regulatory guidelines.

The IBSC comprises head of the institution / nominee, three or more scientists engaged in DNA work or molecular biology, an outside expert in the relevant discipline, a member with medical qualifications and nominee of DBT. The DBT nominee oversees the adherence to biosafety guideline and server as the link between the Department of Biotechnology and the respective IBSC.

There are approximately 225 IBSCs in various institutions and laboratories in industries in India overseeing a range of research projects including development of genetically modified microorganisms, transgenic plants, transgenic animal and recombinant products. DBT has substantially funded research and development in biotechnology in the country, s also agencies like CSIR, ICMR and ICSR,

Several private companies have also initiated research either independently or in collaboration with Indian and foreign institutions/companies. As a result of these efforts, a few products are in advanced stages of development and are likely to e commercialized shortly. 18 recombinant therapeutic products are already in the market, out of which five therapeutic are being manufactured indigenously. In the area of agricultures, four varieties of Bt cotton have been approved for commercial cultivation in the country and several crops are in the approval pipeline.

Biosafety regulations in India comprise biosafety rules and guidelines. The existing legislative framework in India for biosafety regulations has followed a disaggregated approach with regulatory powers imposed in a top-down fashion. The framework legislation for biosafety regulations in India is the EPA. Three provisions of the EPA form the basis of the biosafety regulations. These are sections 6, 8, and 25. While Section 6 of the Act empowers the Central Government to make rules on procedures, safeguards, prohibition and restrictions for handling of hazardous substances, Section 8 of the Act prohibits a person from handling hazardous substances, except in accordance with procedures and after complying with safeguards. Section 25 of the EPA empowers the Central Government to lay down rules regarding procedures and safeguards for handling hazardous substances. Thus, the biosafety rules in India are statutory in nature as they originate from the EPA. These provisions of the EPA led to the adoption of the 1989 Rules for the Manufacture, Use/Import/Export and Storage of Hazardous Micro organisms/ Genetically Engineered Organisms or Cells.

Biosafety guideline refers to the policy proposed or adopted by the Government to avoid the risks of GEOs on environment and public health.

Aim of biosafety guidelines

- Regulating rDNA research with organisms that have least or no adverse effect.
- Minimizing the possibilities of occasional release of GEOs from the laboratory
- Banning the release of GEOs if they are supposed to be causing potential risks in the environment

6.6.1 The rDNA Biosafety Guidelines of India

In India, DBT has proposed "The recombinant DNA safety guidelines" in 1983 and amended in 1990. These guidelines deal with a set of rules for production, use, import, export and storage of hazardous organisms (Anonymous, 2017).

In India, the rDNA guidelines have been implanted through three committees.

- **Institutional biosafety committee (IBSC):** controls research activities at institutional level.

- **Review committee on genetic manipulation (RCGM):** reviews special situations where research with hazardous organisms in laboratory and grants permission to do that research.
- **Genetic engineering approval committee (GEAC):** approves the GEOs for large scale production and use in India.

To perform certain gene manipulation experiments, the workers need to acquire permission of review committee and approval committee before commencement.

- Toxin gene cloning
- Cloning of genes for vaccine production
- Cloning of mosquito and tickDNA
- Cloning of antibiotics resistance genes
- Cloning of oncogenes
- Experiments with infectious animal and plant viruses
- Transgenesis experiment in animal cell cultures
- Transfer of toxicity genes into plants
- Gene therapy for hereditary diseases

6.6.2 Containment

The term "Containment" is used in describing the safe methods for managing infectious agents in the laboratory environment where they are being handled or maintained.

Purpose of containments: To reduce exposure of laboratory workers, other persons, and outside environment to potentially.

Elements of containments: Laboratory Practice and Technique , Safety Equipment(primary barriers) and Design Facility(Secondary Barrier).

Types of containments

i. **Physical containment :** The physical methods being adopted inside the laboratories to prevent escaping the GEOs to the environment It works on the principal of physical barriers It helps tokeep the dirt in the laboratory itself. It includes Air filtration, Sterilization lights, Waste disposal and Protective handling.

ii. **Biological containment:** The biological principles used in the laboratories to prevent the escape of GEOs or microbes Biological containment makes the organisms unable to survivein the outside environment.

Implementation of biosafety guidelines: The rDNA biosafety guidelines are implemented for the government of India by the following four committees.

i. Recombinant DNA Advisory Committee (RDAC): Organized by the Department of Biotechnology (DBT) under the Ministry of science and technology.

It provides regulatory control to the implementation committees. The RDAC has been arranging meeting once in six months or sooner to discuss about the standards of safety regulations. These meetings help to

- Evolve long term policy for R/D in rDNAresearch
- Formulate suitable safety guidelines
- Train the research and technicians about the hazards and risks of rDNA research techniques

ii. Institutional Biosafety Committee (IBSC): This is a small committee established by every institution engaged in rDNA research and the related production activities. It monitors rDNA research activities at the institutional level. This committee is formed of head of institution, 3or more scientists, a medical officer and one DBT nominated person.

Role of IBSC

- Sends report to RCGM regarding observance of safety guidelines on accidents risks and on deviations if any
- Reviews the requirements of guidelines for safety new projects
- Allows some person to take training on biosafety inresearch activities
- Takes emergency plans in urgent situations Attempts to provide medical care to persons working in the laboratory

iii. Review Committee on Genetic Manipulation (RCGM): The RCGM is functioning under DBT It is formed of

- Department of Biotechnology (DBT)
- Indian Council of Medical Research (ICMR)
- Indian Council of Agricultural Research (ICAR)
- Council of Scientific and Industrial Research (CSIR)
- Department of Science and Technology (DST)

RCGM functions

- Establishes the procedural guidance manual for regulatory process with GEOs
- Reviews the risk potentials of GEOs in the laboratory and field experiments
- Decides which containment have to be followed forexperiments with risky hazardous microbes
- Advices custom authorities on import of GEOs and other biological materials from other nations
- Provides advice on IPR and patents
- Assists the Bureau of India standards (BIS) to evolve standards of products coming from rDNA technology

- The monitoring group of RCGM visit frequently to laboratories where rDNA works are going on and inspects safety conditions in those lab

iv. Genetic Engineering Approval Committee (GEAC): This is a higher level committee working under the Department of Environment and Forests. It has full power to permit

- Large scale use of genetically engineered organisms
- rDNA products
- R/D of rDNA technology
- Industrial production of rDNA products
- Release of GEOS in environment and field use

The GEAC gives approval for

- Import, export, transport, production and sale of GEOs and other organisms
- Release of GEOs from the laboratories to environment
- Large scale culture and use of GEOs and microbes inindustries
- Use of GEOs in field application and experimental trials
- Monitoring the risks and accidents due to GEOs

Due to the growing concerns arising from Genetically Modified Organisms (GMOs) throughout the globe the WHO has built an informal working group on biosafety in 1991. This group prepared the 'voluntary code for the release of organisms into the environment'. ICGEB (International Centre for Genetic Engineering and Biotechnology) has played a significant role in issues related to biosafety and the environmentally sustainable use of biotechnology. The main 'topic of concern' related to the release of GMO's are risks for human health, environment, and agriculture which is found on the website of ICGEB.

In India, DBT has evolved 'rDNA safety guidelines' to exercise powers conferred through the Environmental Protection Act 1986 for the manufacture, use, import, export and storage of hazardous micro organisms and genetically engineered organisms, cells etc., These guidelines are implemented and monitored by the Institutional Biosafety Committees (IBSCs), the Review Committee on Genetic Manipulation (RCGM) and the Genetic Engineering Approval Committee (GEAC) of the Ministry of Environment and Forest.

6.7 Intellectual Property Rights (IPR) and Protection (IPP)

The physical objects like household goods or land or properties of a person and the ownership and rights on these properties is protected by certain laws operating in the country. This type of physical property is tangible; but the transformed microorganisms, plants, animals and technologies for the production of commercial products are exclusively the property of the intellectuals. The discoverer or inventor has complete rights on his property or invention. The rights of intellectuals are protected by laws framed by a country. The intellectual property is an intangible asset. Legal rights or patents provide an inventor only a temporary monopoly on the use of an invention, in return for disclosing the knowledge to the others who may use the knowledge to develop further inventions and innovations.

The laws are formulated from time to time at national and international levels. Development of new crop varieties is also an intellectual property right. It is protected by 'plant breeders rights' (PBRs). PBRs recognize the fact that farmers and rural communities have contributed to the creation, conservation, exchange and knowledge of genetic and species utilization of genetic diversity. IPR and IPP are granted by the Government to plant breeders for producing a specific plant variety that is new and never existed before.

IPR is protected by different ways like patents, copyrights and trade marks.

Patents : The science of biotechnology involves the production of enormous number of commercial products of economic importance. The inventions include biotechnology products and processes. The products include living entities like micro organisms, animals, plants, cell lines, cell organelles, plasmids and genes and naturally occurring products like primary and secondary metabolites produced by living systems e.g. alcohol, antibiotics. The biotechnological processes involve isolation, purification, cultivation, bioconversion of novel, innovative, simple and cost effective processes, and creation of biotechnological products.

A patent is a Government issued document that allows the person for an exclusive right to manufacture, use or sell an invention for a defined period (usually 20 years). It is a legal document safeguarding the rights and privileges of an inventor / invention. The purpose of patenting in biotechnology ensures fair financial returns for those who have invested finances, ideas, time and hard work for an invention.

- The invention must be novel and useful;
- The product must be inventive and reproducible;
- The patent application should provide the full description of the invention and the invention must be patentable.

General agreement of tariffs and trade (GATT) and trade related IPRs (TRIPs): GATT was framed in 1948 by developed countries to settle dispute, among the countries regarding share of world trade. The benefits of GATT were enjoyed only by developed countries. In 1988 US congress enacted a law 'the omnibus trade and competitiveness act' (OTCA) which gave powers to US to investigate the laws related to trade.

Geographical indication (GI): A geographical indication is a name or sign used on products which correspond to a specific geographical origin and possess qualities or a reputation that are due to that origin. Geographical indications are typically used for agricultural products, food products, handicrafts and industrial products. Darjeeling tea was the first GI tagged product in India in 2004-05. In

Tamil Nadu, Kancheepuram silk, Coimbatore wet grinder, Thanjavur paintings, Madurai Malli and Temple jewellery of Nagercoil are GI tagged.

Copyright: The protection of authorships of published work comes under copyrights of IPRs. Copyright protection is given for form of expressions of ideas. For example the authors, editors, publishers or both the publisher/ editor of a book have copyrights. The content of the book cannot be reproduced or reprinted without written permission from copyright holders. Patents and trade secrets provide protection for the basic knowhow but copyright protects the expressed materials in printed, video recorded or taped forms. In the field of biotechnology the data base of DNA sequences or any published forms, photomicrographs, etc., are subject to copyright.

Trademarks: Any specific symbol or words to identify a particular product or process of a company constitute trademark. This enables the public to distinguish between a trader's goods from similar goods of other traders. Biotechnology as an independent discipline has drawn world wide attention from the Governments and the corporate world because of its limitless applications. It is looked upon as a panacea for treating diseases and genetic disorders. The global demand of the biotechnological products is on the increase. It is the science for the future with solutions to many of the problems related to health, agriculture, environment and industries.

6.8 Regulations and Guidelines on Biosafety of Recombinant DNA Research & Biocontainment

As mandated in the Rules, 1989 of Environment (Protection) Act, 1986, Review Committee on Genetic Manipulation (RCGM) administered by the Department of Biotechnology, Ministry of Science and Technology has updated "Recombinant DNA safety guidelines, 1990"; "Revised Guidelines for Safety in Biotechnology, 1994" and "Revised guidelines for research in transgenic plants, 1998" and prepared "Regulations and Guidelines on Biosafety of Recombinant DNA Research and Biocontainment, 2017". These guidelines are based on current scientific information, best practices and from the experience gained while implementing the biosafety frameworks within the country (Anonymous, 2017). A series of consultation with researchers, experts, academicians, concerned Ministries/departments and other stakeholders was carried out during preparing this guideline. This document specifies the practices for handling (Manufacture, Use, Import, Export, Exchange and Storage) of hazardous biological material, recombinant nucleic acid molecules and cells, organisms and viruses containing such molecules to ensure an optimal protection of public health and of the environment. The document provides clarity on biosafety requirements and recommendations for laboratory facilities such as facility design, biosafety equipment, personal protective equipment, good laboratory practices and techniques, waste management, etc.

Objectives

i. Outline the general principles of containment and establish a minimum standard for laboratories that must be adopted pan India for all handling of genetically engineered (GE) organisms (organism includes microorganisms, animals, plants, arthropods, aquatic animals, etc.) and non-genetically engineered (non-GE) hazardous microorganisms (microorganism includes parasites, protozoa, algae, fungi, bacteria, virus, prions, etc.).

ii. Identify the levels of risk(s) associated with GE organisms and non-GE hazardous microorganisms and classification of those organisms into their respective risk groups to select appropriate containment facilities. It also covers certification of containment facilities.

iii. Prescribe criteria for Manufacture, Use, Import, Export, Exchange and Storage of any hazardous microorganisms, GE organisms or cells and product(s) produce through exploration of such organisms.

iv. Ensure that national authorities, institutions and all other stakeholders involved in research & development are well informed or have access to information on safety thereby facilitating the safe use and handling of hazardous microorganisms, GE organisms or cells and product(s) produce through exploration of such organisms.

v. Emphasize the need and responsibility of all national authorities institutions and all other stakeholders involved in research to ensure that the public is well informed about the containment strategies followed in India.

Adoption of these guidelines shall be binding pan India for all public and private organizations involved in research, development and handling of GE organisms (organism includes microorganisms, animals, plants, arthropods, aquatic animals, etc.) and non-GE hazardous microorganisms (microorganism includes parasites, protozoa, algae, fungi, bacteria, virus, prions, etc.) and products produced through exploration of such organisms.

Scope of regulations: These regulations are to implement the provisions of Rules 1989 of Environment (Protection) Act, 1986 for the manufacture, use, import, export and storage of hazardous microorganisms, GE organisms or cells and products thereof which applies to the whole of India in the following specific cases:

i. Sale, offers for sale, storage for the purpose of sale, offers and any kind of handling over with or without a consideration

ii. Exportation and importation

iii. Production, manufacturing, processing, storage, import, drawing off, packaging and repacking of the GE Products

iv. Production, manufacture etc. of drugs and pharmaceuticals, food and food components, distilleries and tanneries, etc. which make use of hazardous micro-organisms or GE organisms one way or the other.

Definition applicable as per Rules, 1989: Definitions applicable to this guideline as per Rules, 1989 unless the context requires:

i. "Biotechnology" means the application of scientific and engineering principles to the processing of materials by biological agents to produce goods and services;

ii. "Cell hybridisation" means the formation of live cells with new combinations of genetic material through the fusion of two or more cells by means of methods which do not occur naturally;

iii. "Gene Technology" means the application of the gene technique called genetic engineering, include self-cloning and deletion as well as cell hybridisation;

iv. "Genetic engineering" means the technique by which heritable material, which does not usually occur or will not occur naturally in the organism or cell concerned, generated outside the organism or the cell is inserted into said cell or organism. It shall also mean the formation of new combinations of genetic material by incorporation of a cell into a host cell, where they occur naturally (self-cloning) as well as modification of an organism or in a cell by deletion and removal of parts of the heritable material;

v. "Microorganisms" shall include all the bacteria, viruses, fungi, mycoplasma, cells lines, algae, protozoans and nematodes indicated in the schedule and those that have not been presently known to exist in the country or not have been discovered so far.

Competent Authorities: The Rules 1989 are broad in scope and covers area of research as well as large scale handling of hazardous microorganisms, GE organisms or cells and products thereof. In order to implement the Rules in the entire country, six competent authorities and their roles have been notified (Table 14) for:

i. Regulation and control of contained research activities with hazardous microorganisms, and GE organisms.

ii. Regulation and control of large scale use of GE organisms in production activity.

iii. Import, export and transfer of hazardous microorganisms, GE organisms and products thereof.

iv. Release of GE organisms and products thereof in environmental applications under statutory provisions.

Table 14: Competent Authorities under Rules, 1989

Competent Authorities	Role
Recombinant DNA Advisory Committee (RDAC)	Advisory
Institutional Biosafety Committee (IBSC)	Regulatory/ Approval
Review Committee on Genetic Manipulation (RCGM)	
Genetic Engineering Appraisal Committee (GEAC)	
State Biotechnology Coordination Committee (SBCC)	Monitoring
District Level Committee (DLC)	

State Biotechnology Co-ordination Committee (SBCC): The State Biotechnology Co-ordination Committee (SBCC) is a monitoring committee at State level (Anonymous, 2017). It shall have powers:

i. To inspect, investigate and to take punitive action in case of violations of statutory provisions through the State Pollution Control Board (SPCB) or the Directorate of Health etc.

ii. To review periodically the safety and control measures established at various institutions handling GE organisms.

iii. To act as nodal agency at the State level to assess the damage, if any, due to release of GE organisms and to take on site control measures.

District Level Committee (DLC): There shall be a District Level Biotechnology Committee (DLC) in the districts wherever necessary under the District Collectors to monitor the safety regulations in installations engaged in the use of genetically modified organisms/ hazardous microorganisms and its applications in the environment. The District Level Committee/or any other person/s authorized in this behalf shall visit the installation engaged in activity involving hazardous microorganisms, GE organisms or cells, and , formulate information chart, find out hazards and risk(s) associated with each of these installations and coordinate activities with a view to meeting any emergency. They shall also prepare an off-site emergency plan. The District Level Committee shall regularly submit its report to the SBCC/ GEAC.

6.9 Types of Microbiological Biosafety Level Facilities

6.9.1 Biosafety Level 1 (BSL-1): BSL-1 will be applicable for

i. Isolation, cultivation and storage of Risk Group (RG) 1 microorganisms those are abundant in natural environment.

ii. Experiments on RG 1 microorganisms provided that the experiments will not increase environmental fitness and virulence of the microorganisms.

iii. Category I genetic engineering experiments on microorganism. This category includes experiments which generally do not pose significant risk(s) to laboratory workers, community or the environment and the modifications have no effect on safety concerns. Examples are:

a. Insertions of gene into RG 1 microorganism from any source, deletions, or rearrangements that have no adverse health, phenotypic or genotypic consequence. Modification should be well characterized and that the gene functions and effects are adequately understood to predict safety.

b. Experiments involving approved host-vector systems (refer to Annexure 3) provided that the donor DNA is originated from RG 1 microorganism, not derived from pathogens. The DNA to be introduced should be characterized fully and will not increase host or vector virulence.

c. Experiments involving the fusion of mammalian cells which generate a non-viable organism, for example, the construction of hybridomas to generate monoclonal antibodies.

d. Any experiments involving microorganism belonging to RG 1. For example, self-cloning, fusion of protoplasts between non-pathogenic RG 1 organism.

Before commencement of Category I GE experiments, the investigator should intimate the IBSC about the objective and experimental design of the study along with organisms involved. IBSC should review the same as and when convened for record or action if any. It is desirable to designate a separate area in the facility with proper labelling for Category I GE experiments to avoid any chances of contamination.

6.9.2 Biosafety Level 2 (BSL-2): BSL-2 will be applicable for:

i. Isolation, cultivation and storage of RG 2 microorganisms.

ii. Handling of environmental samples collected from environment that is unlikely to contain pathogens. Isolation of microorganisms from those samples and subsequent experiments.

iii. Experiments on RG 2 microorganisms or isolates from environment mentioned above, provided that the experiments will not increase environmental fitness and virulence of the microorganisms.

iv. Category II genetic engineering experiments on microorganism: These experiments may pose low-level risk(s) to laboratory workers, community or the environment. Examples are:

  a. Experiments involving the use of infectious or defective RG 2 viruses in the presence of helper virus.

  b. Work with non-approved host/vector systems where the host or vector either:

  - does not cause disease in plants, humans or animals; and/ or
  - is able to cause disease in plants, humans or animals but the introduced DNA is completely characterized and will not cause an increase in the virulence of the host or vector.

- experiments using replication defective viruses as host or vector

c. Experiments with approved host/vector systems, in which the gene inserted is:

- a pathogenic determinant;
- not fully characterized from microorganisms which are able to cause disease in humans, animals or plants; or an oncogene.

d. Modification leading to persistent transient disruption of expression of gene(s) that are involved directly or indirectly in inducing pathogenicity, toxicity, survival, or fitness. Modification should be well characterized and the gene functions and effects are adequately understood to predict safety.

e. Work involving fragments of Transmissible Spongiform Encephalopathy (TSEs) proteins or modified TSEs proteins that are not pathogenic and is not producing any harmful biological activity.

f. Experiments in which DNA from RG 2 or 3 organisms are transferred into non-pathogenic prokaryotes or lower eukaryotes. However, handling of live cultures of RG 3 organism should be performed in BSL-3 laboratory.

All category II GE experiments require prior authorization from IBSC before the commencement of the experiments through submission of information in the prescribed proforma. It is desirable to designate a separate area in the facility with proper labelling for Category II GE experiments to avoid any chances of contamination.

6.9.3 Biosafety Level 3 (BSL-3): BSL-3 will be applicable for

i. Isolation, cultivation and storage of RG 3 microorganisms.
ii. Handling of environmental samples collected from environment that is likely to contain pathogens of potential disease consequences. Isolation of microorganisms from those samples and subsequent experiments.
iii. Experiments on RG 3 microorganisms or isolates from environment mentioned above provided that the experiments will not increase environmental fitness and virulence of the microorganisms
iv. Category III and above genetic engineering experiments on microorganism: These kinds of experiments pose moderate to high risk(s) to laboratory workers, community or the environment. Examples are:

    a. Experiments on RG 2 and RG 3 microorganisms where insertion of gene directly involved in production of toxin or allergen or antimicrobial compounds.
    b. Insertions of gene into RG 3 microorganisms from any source, deletions, or rearrangements that affect the expression of genes, whose functions or effects are not sufficiently understood to determine with reasonable

certainty if the engineered organism poses greater risk(s) than the parental organism.

c. Insertions of nucleic acid from any source, deletions, or rearrangements that have known or predictable phenotypic or genotypic consequence in the accessible environment that are likely to result in additional adverse effects on human and/or animal health or on managed or natural ecosystems, e.g., those which result in the production of certain toxins.
d. Research involving the introduction of nucleic acids (recombinant or synthetic) into RG 3 organisms or organisms listed in SCOMET items.
e. Genetic engineering of organisms isolated from environment where there are reported cases of disease prevalence and possibility of presence of infectious microorganisms.

All category III and above GE experiments require prior authorization from IBSC and subsequent approval from RCGM before commencement of the experiments through submission of information in the prescribed proforma.

6.9.4 Biosafety Level 4 (BSL-4)

BSL-4 laboratory is the maximum containment laboratory. Strict training, strictly restricted access and supervision are required and the work must be done under stringent safety conditions and positive pressure personnel suits. BSL-4 will be suitable for:

i. Isolation, cultivation and storage of RG 4 microorganisms.
ii. Handling of samples collected from environment/patients that are likely infected with RG 4 organisms with serious/fatal health effects.
iii. Experiments on RG 4 microorganisms or isolates from environment/ patients mentioned above to find remedial measures.
iv. Category III and above genetic engineering experiments on microorganisms involving introduction of nucleic acids (recombinant or synthetic) into RG 4 microorganisms or exotic agents.

Note

i. BSL facilities are not meant for Permanently housing/keeping/rearing of any animals, arthropods or aquatic organisms for longer than the time required to complete laboratory procedures on them ; The growing of any plants, except those in tissue culture bottles or fully contained in a plant growth chamber.
ii. Genetic engineering experiments not covered under any of the above four categories will require case by case evaluation for selection of appropriate containment strategies. Prior approval and/or permission from the IBSC and/ or the RCGM shall be required to initiate such experiment. Few examples are:

a. Clubbing of experiments pertaining to different categories.

b. Any experiments involving primates, dogs, large animals, and human participants within the laboratory.

c. Experiments involving the use of infectious or defective RG 3 and above viruses in the presence of helper virus.

d. Experiments using DNA which encodes a vertebrate toxin having an LD50 of less than 100 μg/kg.

e. Experiments with genes that alter the growth status of cells such as oncogenes, cytokines and growth factors.

f. Experiments aimed at controlling natural populations.

iii. All existing BSL-3 and 4 facilities must be certified by RCGM. A format for certification is available in this guideline.

iv. The new BSL-3 and 4 facilities shall require certification at the time of commissioning operations as per the format.

Types of Plant Biosafety Level facilities : For experiments on plants, four biosafety levels laboratories are specified depending on nature of works. The levels, PBSL-1-4, include structures comprising greenhouses, screen houses and flexible film plastic structures.

Before commencement of Category I GE experiments, the investigator should intimate the IBSC of the objective and experimental design of the study along with organisms involved. IBSC should review the same as and when convened for record or action if any. It is desirable to designate a separate area in the facility with proper labelling for Category I GE experiments to avoid any chances of contamination. All category II GE experiments require prior authorization from IBSC before the commencement of the experiments through submission of information in the prescribed proforma. It is desirable to designate a separate area in the facility with proper labelling for Category II GE experiments to avoid any chances of contamination. All category III and above GE experiments require prior authorization from IBSC and subsequent approval from RCGM before commencement of the experiments through submission of information in the prescribed proforma.

Other Applicable Policies

i. Bio-medical Waste Management Rules, 2016.

ii. Biological Diversity Act, 2002.

iii. Disaster Management Act, 2005.

iv. Food Safety and Standards Act, 2006.

v. Hazardous Waste Rules, 2016.

vi. Industries (Development & Regulation) Act, 1951 - New Industrial Policy & Procedures, 1991.

vii. Plant Quarantine (Regulation of Import into India) Order, 2003.

viii. Protection of Plant Varieties and Farmers' Rights Act 2001, PPV & FR Regulations 2006.

ix. Seeds Act, 1966; Seeds Rules, 1968; Seeds (Control) Order, 1983; Seeds Policy 1988 & 2002.

6.10 List of Infective Microorganisms Corresponding to Different Risk Groups

The list is indicative but not exhaustive and will be updated periodically (Anonymous, 2017). For working with organisms not listed here, investigators should determine the appropriate containment level with consultation with IBSC. Depending upon the work envisaged, organisms listed in higher risk groups (RG 2/3) may be handled in lower containment laboratory provided, if the work does not involve genes known to be involved in pathogenesis/carcinogenesis/ production of toxins, etc. Accordingly, lower containment facility may be used with prior approval from IBSC and information to RCGM.

6.10.1 A. List of Risk Group 1 Microorganisms

Bacteria: *Acetobacter* spp. • *Actinoplanes* spp. • *Agrobacterium* spp. • *Alcaligenes aquamarinus* • *Aquaspirillum* spp. • *Arthrobacter* spp. • *Azotobacter* spp. • *Bacillus* spp., except cereus and anthracis • *Bifidobacterium*.spp., except dentium • *Bradyrhizobium* spp. • *Brevibacterium* spp. • *Caryophanon* spp. • *Clostridium* spp. i.e. *C. aceticum, C. acetobutylicum, C. acidiuric, C. cellobiparum, C. kluyveri, C. thermoaceticum, C. thermocellum, C. thermosulfurogenes* • *Corynebacterium* spp. i.e. *C. glutomicum, C. lilium* • *Enterococcus facium* • *Erwinia* spp. except *E. chrysanthemi, E. amylovora* and *E. herbicola* • *Gluconobacter* spp. • *Klebsiella planticola* • *Lactobacillus* spp. i.e. *L. acidophilus, L. bauaricus, L. breuis, L. bucneri, L. casei, L. cellobiosis, L. fermentum, L. helveticus, L. sake* • *Lactococcus lactis* • *Leuconostoc* spp. • *Lysobacter* spp • *Methanobacter* spp. • *Methylomonas* spp. • *Micrococcus* spp. • *Nonpathogenic Escherichia coli* e.g. ATCC 9637, NCIB 8743, K12 and derivatives • *Pediococcus* spp. • *Pseudomonas* spp. i.e. *P. fluorescens, P. gladioli, P. syringae* •*Ralstonia* spp. • *Rhizobium* spp. • *Rhodobacter* spp. •*RhodoPseudomonas* spp. • *Rickettsiella* spp. • *Staphylococcus carnosus* • *Streptococcus salivarius* • *Streptomyces* spp • *Thermobacteroides* spp. • *Thermus* spp. • *Thiobacillus* spp. • *Vibrio* spp. i.e *V. fischeri, V. diazotrophicus.*

Fungi: • *Agaricus bisporus* • *Acremonium* spp. i.e. *A. chrysogenum, A. elegans,* • *Actinomucor elegans* • *Ashbya gossypii* • *Aspergillus oryzae* • *Aureobasidum pullulans* • *Blakeslea trispora* • *Brettanomyces bruxellensis* • *Candida* spp. i.e. *C. boindinii, C. utilis* • *Chaetomium globosum* • *Cladosporium cladosporioides* • *Claviceps* spp. i.e. *C. purpurea,C. paspali* • *Coprinus cinereus* • *Cunninghamella* spp. i.e. *C. blakesleana, C. elegans* • *Cyathus stercoreus* • *Dacrymyces deliquescens* • *Debaryomyces hansenii* • *Engyodontium album*• *Hansenula* spp. i.e. *H. anomala, H. polymorpha* • *Hypholama* spp. i.e. *H. fasciculare, H. roseonigra* • *Lentinus edodes* • *Lipomyces lipofer* • *Metarhizium anisopliae* • *Monascus pupureus* • *Moniliella suaveolens* • *Mortierella vinacea* • *Mucor* spp. i.e, *M. mucedo, M. plumbeus, M. rouxii* • *Neurospora* spp. i.e. *N. crassa, N. sitophilla* • *Nigrospora sphaerica* • *Oxyporus populinus* • *Pachysolen tannophilus* • *Paecilomyces varioti* • *Penicillium* spp. i.e. *P. funiculosum, P. camemberti, P. chrysogenum* • *Phycomyces blakesleanus* • *Pichia* spp. i.e. *P. membranae faciens, P. farinosa, P. guilliermondii, P. stipitis*• *Pleurotus ostreatus* • *Rhizoctonia solani* • *Rhodosporidum toruloides* • *Rhodotorula glutinis* • *Saccharomyces cerevisiae* • *Schizosaccharomyces pombe* • *Schwanniomyces occidentalis* • *Sordaria macrosopra* • *Thanatephorus cucumeris*• *Trametes versicolor* • *Trichoderma* spp. i.e. *T. harzianum, T. viride* • *Trigonopsis variabilis* • *Verticillium lecanii* • *Volvariella volvacea* • *Wallernia sebi* • *Xeromyces bisporus* • *Zygorhynus moelleri* • *Zygosaccharomyces* spp. i.e. *Z. bailii, Z. rouxii*

Viruses: • Apathogenic, endogeneous, animal retroviruses • Attenuated viral strains which are accepted vaccines. Only a limited number of passages in defined cell culture or host-systems are allowed • Baculoviruses of insects • Newcastle disease virus - strains licensed for vaccine use • Influenza virus A/PR/8/34 • Poikilothermal vertebrate retrovirus • Rinderpest - attenuated virus strain (e.g. Kabatte-O) licensed for vaccine use. • Viral strains from fungal or bacterial systems, provided they do not contain virulence-factors and are described as apathogenic for higher animals and human beings.

6.10.2 B. List of Risk Group 2 Microorganisms

Bacteria: • *Acinetobacter* spp. i.e., *A. calcoaceticus, A. lwoffii* • *Actinobacillus* spp. • *Actinomadura* spp. i.e., *A. madurae, A. pelletieri* • *Actinomyces* spp. i.e., *A. israelii, A. bovis* • *Aeromonas hydrophila* • *Afipia* spp • *Aggregatibacter actinomycetemcomitans* • *Agrobacterium radiobacter* • *Alcaligenes* spp. • *Amycolata autotrophica* • *Anaplasma* spp. • *Arachnia propionica*• *Archanobacterium haemolyticum* • *Arizona hinshawii - all serotypes* • *Bacillus* spp. i.e., *B. anthracis, B. cereus* • *Bacteroides* spp. • *Bartonella* spp. i.e., *B. bacilliformis, B. henselae* • *Bifidobacterium dentium* • *Bordetella* spp. i.e., *B. avium, B. bronchiseptica, B. parapertussis, B. pertussis, B.quintana, B.vinsonii* • *Borrelia* spp. i.e., *B. burgdorferi, B.recurrentis, B.duttonii, B.vicenti* • *Brucella ovis* • *Burkholderia* spp. i.e., *B. cepacia B.mallei (Pseudomonas mallei)* • *Campylobacter* spp. i.e., *C. coli, C. fetus subsp. fetus, C. jejuni* • *Cardiobacterium hominis* • *Chlamydia* spp. i.e., *C. pneumonia, C. trachomatis* • *Chlamydophila pneumonia* • *Chlamydia trachomatis* • *Citrobacter* spp. • *Cladosporium (Xylohypha) trichoides* • *Clostridium* spp. i.e., *C. chauvoei, C. difficle, C. fallax, C.haemolyticum, C. histolyticum, C.novyi, C. perfringens, C. septicum* • *Corynebacterium* spp. i.e., *C. diphtheriae , C. minutissimum, C. pseudotuberculosis, C. renale, C. ulcerans* • *Coxiella burnetii - specifically the Phase II, Nine Mile strain, plaque purified, clone 4* • *Cytophaga* spp. *pathogenic to animals* • *Dermatophilus congolensis* • *Edwardsiella tarda* • *Eikenella coreodens* • *Enterobacter* spp. • *Enterococcus faecalis* • *Eperythrozoon* spp. • *Erysipelothrix* spp. i.e., *E. rhusiopathiae, E. tonsillarum*• *Escherichia coli (excluding non-pathogenic strains* • *Flavobacterium meningosepticum* • *Fluoribacter bozemanae* • *Francisecla tularensis*• *Fusobacterium* spp. *including F. necrophorum* • *Gardnerella vaginalis* • *Haemophilus* spp. i.e.,*H. ducreyi, H. influenzae* • *Helicobacter* spp. i.e., *H. pylori, H. hepaticus* • *Klebsiella* spp. i.e., *K. pneumonia, K. mobilis, K. oxytoca* • *Legionella* spp. i.e., *L. pneumophila* • *Leptospira interrogans - all serotypes* • *Listeria* spp. • *Moraxella* spp. • *Morganella morganii* • *Mycobacterium BCG vaccine strain* • *Mycoplasma* spp. i.e., *M. agalactiae, M. mycoides, M. caviae , M. hominis, M. pneumoniae* • *Neisseria* spp. i.e., *N. meningitidis, N. gonorrhoeae* • *Nocardia* spp. i.e., *N. otitidiscaviarum, N. brasiliensis, N. farcinica, N. nova, N. otitidiscaviarum, N. asteroids, N. transvalensis* • *Pantoea agglomerans* • *Pasteurella - all species except those listed in RG 3*• *Peptococcus* spp. • *Peptostreptococcus* spp. • *Porphyromonas* spp. • *Prevotella* spp. • *Proteus* spp. i.e., *P. mirabilis, P. penneri, P. vulgaris*• *Providencia alcalifaciens* • *Pseudomonas aeruginosa* • *Rhodococcus equi* • *Salmonella* spp. i.e., *S. abortusequi, S.abortusovis, S. arizonae, , S. choleraesuis, S.dublin, S. enteritidis, S. gallinarum, S. meleagridis, S. paratyphi, A, B, C, S. typhi, S.pullorum, S. typhimurium* • *Serpulina* spp. • *Serratia* spp. i.e., *S.liquefaciens, S. marcescens* • *Shigella* spp. i.e., *S.boydii, S. dysenteriae, S. flexneri, S. sonnei* • *Sphaerophorus necrophorus* • *Staphylococcus* spp. i.e., *S. aureus, S. epidermidis* • *Streptobacillus moniliformis* • *Streptococcus* spp. i.e., *S.agalactiae, S. dysgalactiae, S. pneumoniae, S. pyogenes, S. suis, S. uberis, S. equi*• *Streptomyces somaliensis* • *Treponema* spp. i.e., *T.pertenue, T. carateum,T. pallidum* • *Ureaplasma urealyticum* • *Veillonella* spp. • *Vibrio* spp. i.e., *V. parahaemolyticus, V. vulnificus, V. cholerae, V. fluvialis, V. metschnikovii, V. mimicus*

Fungi: • *Acremonium* spp. i.e., *A. kiliense, A. recifei, A. falsiforme, A. strictum* • *Arthroderma benhamiae/simiti*• *Aspergillus* spp. i.e., *A. fumigatus, A. flavus, A. parasiticus* • *Basidiobolus haptosporus* • *Blastomyces dermatitidis* • *Candida* spp. i.e., *C. albicans, C. tropicalis* • *Cryptococcus neoformans var gattii (Filobasidiella bacillispora)* • *Curvularia lunata* • *Dactylaria galopava* • *Emmonsia parva var crescens, var parva* • *Epidermophyton* spp. *including:E. floccosum* • *Exophilia* spp. i.e., *E. castelanii, E. dermatitidis, E. mansonii* • *Filobasidiella neoformans* • *Fonsecaea* spp. i.e., *F. compacta, F. pedrosoi* • *Fusarium coccophilum* • *Geotrichum candidum* • *Histoplasma capsulatum* • *Hortaea werneckii* • *Leptosphaeria* spp. i.e., *L. senegalensis, L. thompkinsii* • *Loboa loboi* • *Madurella* spp. i.e., *M. grisea, M. mycetomi*• *Microsporum* spp. i.e., *M. audouinii, M. canis, M. distortum, M. duboisii, M. equinum, M. ferrugineum, M. gallinae, M. gypseum, M. praecox, M. nanum, M. persicolor* • *Monosporium apiospermum* • *Myrothecium verrucaria* • *Nannizzia* spp. i.e., *N. gypsea, N. obtussa, N. otaeNeotestudina rosatii* • *Paceilomyces lilacinus* • *Paracoccidioides brasiliensis* • *Penicillium marneffei*

• *Phialophora verrucosa* • *Pseudollescheria boydii* • *Rhinocladiella* spp. i.e., *R. compacta, R. pedrosoi, R. spinifera* • *Rhinosporidium seeberi* • *Rhizomucor pusillus* • *Rhizopus* spp. i.e., *R. cohnii, R. microspous* • *Scedosporium* spp. i.e., *S. apiospermum, S. prolificans* • *Sporothrix schenckii* • *Trichophyton* spp. i.e., *T. cocentricum, T. equinum, T. erinacei, T. gourvilli, T. megninii, T. mentagrophytes, T. rubrum, T.schoenleinii, T. smii, T. soudanense, T. tonsurans, T. verrucosum, T. violaceum, T. yaoundei* • *Xylophora carrionii*

Viruses: • *Adeno-associated viruses (AAV)* • *Aino virus* • *All isolates of Orthoreovirus and Orbivirus* • *Alphaviruses* • *Animal adenoviruses* • *Animal papillomaviruses* • *Animal papillomaviruses* • *Astroviruses* • *Aura-virus* • *Avian viruses. i.e. adenovirus, encephalomyelitis, enterovirus, influenza, poxvirus, retrovirus, encephalomyelitis virus, reticuloendotheliosis virus, smallpox virus, sarcoma virus* • *Barmah forest virus* • *Batai virus* • *Bebaru virus* • *Bern-virus* • *BK-virus* • *Border disease virus* • *Borna virus* • *Bovine ephemeral-fever virus* • *Bovine foamy virus* • *Bovine herpesvirus 2, 3, and 4* • *Bovine mucosal disease virus* • *Bovine papilloma virus* • *Bovine polyomavirus (BPoV)* • *Bovine rhinoviruses (types 1-3)* • *Breda virus* • *Bunyamwera virus* • *Cache Valley virus*• *Caliciviruses* • *California encephalitis virus* • *Canine distemper virus* • *Canine parvovirus (CPV)* • *Chikungunya vaccine strain 181/25* • *Chikungunya virus (for studies except vector inoculation, transmission)* • *Chimpanzee herpesvirus* • *Chuzan virus* • *Coltivirus all types including Colorado tick fever virus*• *Coxsackie A and B viruses* • *Cytomegalovirus* • *Dengue virus (for studies except vector inoculation, transmission)* • *Drosophila X virus* • *Eastern equine encephalomyelitis virus* • *Echoviruses - all types* • *Ectromelia virus* • *Emsliki Forest virus* • *Encephalomyocarditis virus (EMC)* • *Enterovirus* • *Entomopoxviruses* • *Equine infectious anemia virus* • *Equine influenza virus 1 (H7N7) and 2 (H3N8)* • *Equine rhinopneumonitis virus* • *Exogenous retroviruses (i.e. murine mammary-tumor virus, feline immunodeficiency virus)* • *Feline calcivirus* • *Flanders virus* • *Fort morgan virus* • *Hart Park virus* • *Hepatitis A & D* • *Herpes simplex types 1* • *Herpes zoster* • *Human adenoviruses* • *Human herpesvirus types 6 and 7* • *Human papillomaviruses (HPV)* • *Human parvovirus (B 19)* • *Human rhinoviruses*• *HVJ virus (Sendai virus)* • *Infectious Bovine Rhinotraecheitis virus (IBR)* • *Infectious Bursal diseases of poultry* • *Infectious Laryngotraechitis (ILT)* • *Influenza viruses - all types except A/ PR/8/34, which is in RG 1* • *JC virus* • *Japanese encephalitis virus (for studies except vector inoculation, transmission)* • *Rhinoviruses - all types* • *Ross river virus* • *Rota virus* • *Rubella virus* • *Sandfly fever virus* • *Shope fibroma virus* • *Simian foamy virus* • *Simian virus 40*• *Simian viruses - all types except Herpervirus simiae (Monkey B virus) and Marburg virus which are in RG 4* • *Junin virus* • *Langat virus* • *Lassa virus* • *Lumpy Skin Disease (LSD) Virus* • *Lumpy skin disease virus* • *Lymphocytic choriomeningitis virus (nonneurotropic strains)* • *Lymphocytic choriomeningitis virus (nonneurotropic strains)* • *Mammalian retrovirus (except HIVm HTLV-1 (ATLV) and HTLV-II)* • *Marek's Disease virus*

• *Measles virus* • *Minute virus in mice* • *Molluscan contagiosum virus* • *Monkey (SV40, SA-12, STMV, LPV)* • *Mouse hepatitis virus* • *Mouse rotaviruses (EDIM, epizootic diarrhoea of infant mice)* • *Mumps virus* • *Murine penumoniae virus* • *Myxoma virus* • *NDV* • *O'nyong-nyong virus* • *Orbi virus* • *Other avipoxviruses* • *Parainfluenza viruses* • *Paramyxoviruses* • *Parvovirus* • *Pichinde virus* • *Pixuna virus* • *Polio viruses-all types, wild and attenuated* • *Polyoma virus* • *Porcine adenovirus* • *Poxviruses- - All types except Alastrim, Monkey pox, Sheep pox and White pox* • *Pseudorabies virus* • *Rabies (fixed, attenuated) virus* • *Rat rotavirus* • *Reovirus* • *Respiratory syncytial virus*• *Reticuloendotheliosis viruses (REV)* • *Rhabdoviruses* • *Shope fibroma virus* • *Simbu virus* • *Sindbis virus* • *Stomatitis papulosa virus* • *Swine vesicular disease virus* • *Tensaw virus* • *Turlock virus* • *Una virus* • *Uukuniemi virus* • *Vaccinia virus* • *Varicella virus* • *Venezuelan equine encephalomyelitis vaccine strains TC-83 and V3526* • *Vesicular stomatitis virus*• *Vesiculovirus* • *Yellow fever virus vaccine strain 17D* • *Sindibis virus*

Parasites: *Ancylostoma human hookworms* • *Acanthamoeba* spp. • *Ancylostoma* spp. i.e., *A. ceylanicum, A. duodenale* • *Angiostrongylus* spp. • *Anisakis simplex* • *Ascaris lumbricoides* • *B.microti* • *Babesia* spp. i.e., *B. divergens* • *Balantidium coli*• *Blastocystis hominis* • *Brugia filaria worms: B. malayi, and B. timori* • *Capillaria* spp. • *Coccidia* spp. • *Cryptosporidium* spp. • *Cysticercus cellulosae (hydatid cyst, larva of T. solium)* • *Dicrocoelium dendriticum* • *Dientamoeba fragilis* • *Dracunculus medinensis* • *Echinococcus* spp. i.e., *E. granulosus, E. multilocularis, E. vogeli* • *Entamoeba histolytica* • *Enterobius* spp. • *Enterobius vermicularis* • *Fasciola* spp. i.e., *F. hepatica, F. gigantica* • *Giardia lamblia* • *Heterophyes* spp. • *Hymenolepis* spp. i.e., *H. diminuta, H. nana* • *Isospora spp*• *Leishmania* spp. i.e., *L. braziliensis, L. donovani, L.ethiopia, L. peruvania, L. tropica, L. major, L. mexicana* • *Loa loa filaria worms* • *Mansonella* spp. i.e., *M. ozzardi, M. perstans, M. streptocerca* • *Metagonimus* spp. • *Microsporidium* • *Naegleria* spp. i.e., *N. fowleri, N. gruberi* • *Necator americanus, Necator human hookworms* • *Nosema bombycis* • *Onchocerca filaria worms* • *Onchocerca volvulus* • *Opisthorchis* spp. i.e., *O. felineus, O. sinensis (Clonorchis sinensis), O. viverrini (Clonorchis viverrini), O. filaria, O. volvulus* • *Plasmodium* spp. i.e., *P. cynomolgi, P. falciparum, P. malariae, P. ovale, P. vivax, P. westermani* • *Sarcocystis sui hominis* • *Schistosoma* spp. i.e., *S. mansoni, S. haematobium, S. intercalatum, S. japonicum, S. mekongi* • *Strongyloides stercoralis* • *Taenia solium* • *Toxocara canis* • *Toxoplasma gondii* • *Trichinella* spp. i.e., *T. nativa, T. nelso, T. pseudospiralis, T. spiralis* • *Trichomonas vaginalis* • *Trichostrongylus orientalis*• *Trichuris trichiura* • *Trypanosoma* spp. i.e., *T. brucei brucei, T. brucei gambiense, T. brucei rhodesiense, T. cruzi* • *Wuchereria bancrofti*

Bacteria – plant pathogens: • *Agrobacterium* spp. i.e., *A. rhizogenes, A. rubi, A. tumefaciens* • *Clavibacter michiganensis*• *Erwinia* spp. i.e., *E. carotovora subsp. betavasculorum, E. chrysanthemi pv. Chrysanthemi* • *Pseudomonas* spp. i.e., *P.*

cichorii, P. fluorescens, P. syringae subsp. syringae • *Rhodococcus fascians* • *Xanthomonas campestris pv. alfalfae*

Fungi–plant pathogens: • *Alternaria dauci* • *Botrytis* spp. i.e., *B. allii, B. elliptica* • *B. hyacynthi, B. tulipae* • *Cladosporium* spp. i.e., *C. phlei, C. variabile* • *Claviceps purpurea* • *Fusarium* spp. i.e., *F. arthrosporioides, F. culmorum, F. graminum, F. oxysporum f. sp. betae, F. oxysporum* f. sp. *pisi* • *Glomerella* spp. i.e., *G. cingulata (anamorph Colletotrichum gloeosporioides), G. graminicola, G.tucamanensis (anamorph Colletotrichum falcatum)* • *Mucor circnelloides* • *Penicillium* spp. i.e., *P. corymbiferum, P. cyclopium, P. digitatum, P. expansum, P. italicum* • *Phytophthora* spp. i.e., *P. infestans, P. megasperma* • *Rhizoctonia* spp. i.e., *R. carotae, R. fragariae, R. tuliparum* • *Rhizopus* spp. i.e., *R. arrhizus, R. stolonifer, R. oryzae* • *Sclerophthora* spp. i.e., *S. macrospora, S. graminicola*• *Sclerotinia minor* • *Sclerotinia trifoliorum* • *Septoria* spp. i.e., *S. azalea, S. lactucae* • *Trichoderma longibrachaiatum*

Viruses –plant pathogens : • Alfalfa mosaic virus • Apple chlorotic leaf spot virus • Apple mosaic virus • Apple stem grooving virus • Barley yellow mosaic virus • Beet western yellows virus • Carnation ringspot virus • Cucumber mosaic virus • Hop mosaic virus • Maize dwarf mosaic virus • Melon necrotic spot virus • Papaya ringspot virus • Pea early-browning virus • Potato leafroll virus • Potato virus A, M, S, X, and Y • Tobacco mosaic virus • Tobacco necrosis virus • Tobacco rattle virus • Tobacco stunt virus • Tomato mosaic virus • Tulip breaking virus • Turnip crinkle virus • Turnip mosaic virus.

6.10.3 C. List of Risk Group 3 Microorganisms

Bacteria : • *Actinobacillus mallei* • *Bacillus anthracis* • *Bartonella bacilliformis* • *Bordetella bronchiseptica* • *Brucella* spp. *(except B. ovis, listed in Risk Group 2) i.e. B. abortus, B.canis, B.melitensis, B.suis* • *Burkholderia (Pseudomonas) mallei* • *Campylobacter fetus subsp. venerealis* • *Chlamydia psittaci (avian strains)* • *Cladosporium* spp. i.e., *C. bantianum,C. (Xylohypha) trichoides* • *Coccidiodes immitis* • *Clostridium* spp. i.e. *C. botulinum, C. tetani* • *Coxiella burnetii* • *Ehrlichia* spp. *including E. sennetsu* • *Francisella tularensis* • *Mycobacterium* spp. i.e. *M. marinum, M. scrofuiaceum, M. ulcerans, M. africanum, M. avium, M. bovis (except the BCG strain), M. chelonae, M. leprae, M. tuberculosis, M. fortuitum* • *Mycoplasma* spp • *Pasteurella multocida* • *Pseudomanas mallei, P.* • *R. typhi pseudomallei* • *Yersinia pestis*

Viruses: • Alastrim, Monkeypox, Whitepox viruses (when used in vitro) • African Horse Sickness Virus (Attenuated strain except animal passage) • Akabane virus • Avian herpesvirus 1 (ILT) • Avian influenza virus A-Fowl plague • Avian leucosis viruses (ALV) • Blood-borne hepatitis viruses • Blue Tongue virus (only serotypes reported in India) • Border disease virus • Borna diseases virus (Gr. Bornaviridae) • Bovine diarrhoea virus • Bovine herpesvirus 1 • Bovine immunodeficiency virus (BIV) • Cabassou virus • California encephalitis virus • Chikungunya virus • Chimeric Viruses • Classical Swine Fever Virus • Contagious Bovine Pleuropneumonia Agent • Contagious Caprine Pleuropneumonia Agent • Dengue virus type 1-4• Eastern equine encephalitis virus • Enzephalitis virus • Epstein - Barr virus • Equine arteritis • Everglades virus • Feline immunodeficiency virus (FIV) • Feline Leukemia • Feline sarcoma virus (FeSV) • Flexal virus • Foot and Mouth Disease virus • Fowl plague virus • Gibbon Ape Lymphosarcoma • Hantaan virus (Korean haemorrhagic fever) • Hazara virus • Hepatitis virus (type B, C and E) • Herpes simplex 2 • Herpes simplex saimiri • Herpes virus ateles • Herpes virus B • Hog cholera virus • Human immunodeficiency viruses (HIV) types 1 & 2 • Human T-cell lymphotropic viruses (HTLV) types 1 & 2 • Infectious Equine Anaemia virus • Infectious pancreatic necrosis virus (Gr. Birnaviridae) • Japanese encephalitis virus• Las Crosse virus • Leukomogenic murine oncovirus (Murine lymphosarcoma virus: MuLV) • Lymphocytic choriomeningitis virus (LCM) (neurotropic strains) (Gr. Arenaviridae) • Lymphosarcoma viruses of nonhuman primates • Mayaro virus • Meningitis virus• Middle East respiratory syndrome coronavirus (MERS-CoV) • Middelburg virus • Mobala virus (Gr. Arenaviridae) • Monkey mammary tumor viruses (MPTV) • Mucambo virus • Murine mammary tumor viruses (MMTV) • Murine sarcoma viruses (MuSV) • Murray Valley encephalitis virus • Nairobi sheep disease virus • Newcastle disease virus (Asiatic strains) • Israel turkey meningocephalomyelits virus • Louping ill virus • Lumpy skin disease virus • Oropouche virus • Pappataci-fever virus • Porcine sarcoma virus • Pseudorabies virus • variant of vaccinia • Rabies street virus, when used inoculations of carnivores • Rabies virus• Rat lymphosarcoma virus (Murine lymphosarcoma virus: MuLV) • Rickettsia spp. i.e. R. bellii, R. canada, R. conori, R.akari, R. australis, R. montana, R. parkeri, R. sibirica, R. tsutsugamushi • Rift Valley fever virus • SARS-associated coronavirus (SARS-CoV) • Semliki Forest virus • Sheep pox (field strain) • Simian immunodeficiency viruses (SIV) • Simian sarcoma viruses (SSV) • St. Louis encephalitis virus • Swine Fever virus • Swine vesicular disease virus • Tacaribe virus • Vesicular Somatitis virus • West Nile virus (WNV) • Wooly Monkey fibrosarcoma • Wesselsborn disease virus • Yaba pox virus

Fungi: • *Ajellomyces capsulatus/dermatitides* • *Coccidioides immitis* • *Cladosporium trichoides* • *Histoplasma duboisii* • *Histoplasma farciminosum* • *Paracoccidioides brasiliensis*

Parasites: • *Eimeria* spp. i.e. *E. acervulina, E. burnetti, E. maxima, E. necratix* • *Theileria* spp. i.e. *T. annulata, T. hirei, T. parva* • *Trichomonas* spp. i.e. *T. foetus, T. brucei*

Prions including Bovine spongiform encephalopathy (BSE), transmissible spongiform encephalopathy (TSE), Gerstmann-SträusslerScheinker encephalopathy (BSE)

Bacteria– plant pathogens: • *Erwinia* spp. i.e. *E.salicis, E. tracheiphila* • *Pseudomonas syringae pv. phaseolicola and pv. Pisi* • *Xanthomonas* spp. i.e. *X.campestris pv. aberrans, X. populi*

Virus- plant pathogens: • Lettuce mosaic virus • Tobacco streak virus • Tomato bushy stunt virus • Tomato yellow leaf curl virus • Wheat dwarf virus • Wheat spindle steak mosaic virus

Fungi- plant pathogens: • *Alternaria solani* • *Botrytis fabae* • *Claviceps gigantea* • *Fusarium* spp. i.e. *F. coeruleum, F. oxysporum* f. sp. *lycopersici, F.oxysporum* f. sp. *trifolii, F.solani* f. sp. *cucurbitae, F.solani f. sp. phaseoli, F.solani* f. sp. *pisi* • *Mucor* spp. i.e. *M. circinelloides, M. piriformis, M. racemosus, M. strictus* • *Septoria* spp. i.e. *S.apiicola, S.chrysanthemella, S.lycopersici var. lycopersici*

Parasites/Nematodes- Plant Pathogen: Heterodera glycines

6.10.4 D. List of Risk Group 4 Microorganisms

• Alastrim, Monkeypox, Whitepox viruses (when used for transmission and animal inoculation experiments) • African horse sickness virus (serotype not reported in India and challenge strains) • African swine fever virus (Gr. Adenoviridae) • All Hemorrhagic fever agents • Besnoitia besnoiti • Crimean-Congo hemorrhagic fever virus (Gr. Bunyaviridae) • Ebola fever virus • Ephemeral fever virus • Equine morbillivirus (Hendra virus) • FMD virus (Exotic types) • Guanarito virus • Hanzalova virus • Herpesvirus simiae (Herpes B or Monkey B virus) • Junin virus (Gr. Arenaviridae) • Kyasanur forest virus • Lassa virus (Gr. Arenaviridae) • Machupo virus (Gr. Arenaviridae) • Marburg virus • Mopeia viruses (Gr. Arenaviridae) • Rift Valley fever virus • Sabia (Gr. Arenaviridae) • Tick-borne encephalitis virus complex including Russian spring-summer encephalitis viruses • Variola (major & minor) virus • Western equine encephalomyelitis virus • Yellow fever virus • Zika virus (Gr. Flavivirida).

6.11 Transgenic Crops-Biosafety Concerns and Regulations in India

Plant genetic engineering methods were developed over 30 years ago, and since then, genetically modified (GM) crops or transgenic crops have become commercially available and widely adopted in many countries. In these plants, one or more genes coding for desirable traits have been inserted. The genes may come from the same or another plant species, or from totally unrelated organisms. The traits targeted through genetic engineering are often the same as those pursued by conventional breeding. However, because genetic engineering allows for direct gene transfer across species boundaries, some traits that were previously difficult or impossible to breed can now be developed with relative ease. The first-generation GM crops have improved traits like Herbicide-resistant crops (soybeans and maize, Pest resistance (Cotton and corn). Second-generation GM crops involve enhanced quality traits, such as higher nutrient content. "Golden Rice," one of the very first GM crops, is biofortified to address vitamin

A deficiency. Other biofortification projects include corn, sorghum, cassava, and banana plants, with enhanced minerals and vitamins. Crops can also be modified to ward off plant viruses or fungi. Even though the seed is more expensive, these GM crops lower the costs of production by reducing inputs of machinery, fuel, and chemical pesticides. Important environmental benefits, such as controlling farm runoff that otherwise pollutes water systems, are associated with reduced spraying of chemical insecticides and highly toxic herbicides.

With the rapid advance research and development in agricultural biotechnology, countries are approving many genetically modified crops for commercial release and agricultural production. ISAA reported in 2017, the 21st year of commercialization of biotech crops, 189.8 million hectares of biotech crops were planted by up to 17 million farmers in 24 countries. From the initial planting of 1.7 million hectares in 1996 when the first biotech crop was commercialized, the 189.8 million hectares planted in 2017 indicates ~112-fold increase. In India, Bt cotton was approved by Government of India in March 2002 as the first transgenic crop for commercial cultivation for a period of three years. Apart from cotton, there are more than 20 crops under research and development in about 50 public and private sector organizations in India. Out of these, 13 crops have been approved for contained limited field trials in India. Though, it is widely claimed that transgenic crops offers dramatic promise for meeting some of greatest challenges but like all new technologies, it also poses certain risks, because of the fact that transgenic crops can bring together new gene combinations which are not found in nature having possible harmful effects on health, environmental and non-target species. However, as more and more transgenic crops are being released for field-testing and commercialization, concerns have been expressed about the potential risks associated with their impact to human health, environment and biological diversity.

6.11.1 Biosafety Concerns

As more and more transgenic crops are released for field-testing and commercialization, concerns have been expressed regarding potential risks to both human health and environment. Biosafety describes the principles, procedures and policies to be adopted to ensure the environmental and personal safety. Recognizing the need of biosafety in GE research and development activities, an international multilateral agreement on biosafety "the Cartagena Protocol on Biosafety (CPB)" has been adopted by 167 parties, including 165 United Nations countries, Niue, and the European Union (Usha Kiran Betha, 2019). The Protocol entered into force on 11 September 2003, and its main objectives are:

1. To set up the procedures for safe trans-boundary movement of living modified organisms
2. Harmonize principles and methodology for risk assessment and establish a mechanism for information sharing through the Biosafety Clearing House (BCH).

Research work in the area of GE and GMOs requires prior approval from the appropriate regulatory authorities of the country. Following guidelines provided for minimizing biosafety issues is mandatory. The primary regulatory body at research institute level is the Institutional Biosafety Committee (IBSC) or its equivalent body consisting of experts from different relevant disciplines. The IBSC ensures existence of the basic biosafety equipment required as per the safety level of the experiments to be conducted. There has been increasing awareness among the researchers, producers and users of GMOs, administrators, policy makers, environmentalists and general public about biosafety all over the world. Transgenic crops are not toxic nor are likely to proliferate in the environment. However, specific crops may be harmful by virtue of novel combinations of traits they possess. This means that the concerns associated with use of GMOs can differ greatly depending on the particular gene organism combination and therefore a case-by-case approach is required for risk assessment and management.

The major biosafety concerns falls into these categories:

Bio-safety of human and animal health:

1. Risk of toxicity, due to the nature of the product or the changes in the metabolism and the composition of the organisms resulting from gene transfer.
2. Newer proteins in transgenic crops from the organisms, which have not been consumed as foods, sometimes has the risk of these proteins becoming allergens.
3. Genes used for antibiotic resistance as selectable markers have also raised concerns regarding the transfer of such genes to microorganisms and thereby aggravate the health problems due to antibiotic resistance in the disease causing organisms.

Ecological concerns

1. Gene flow due to cross pollination for the traits involving resistance can result in development of tolerant or resistant weeds that are difficult to eradicate.
2. GM crops could lead to erosion of biodiversity and pollute gene pools of endangered plant species.
3. Genetic erosion has occurred as the farmers have replaces the use of traditional varieties with monocultures.

Environmental concerns

1. Effect of transgenic plants on population dynamics of target and non-target pests, secondary pest problems, insect sensitivity, evolution of new insect biotypes, environmental influence on gene expression, development of resistance in insect population, development of resistance to herbicide

2. Gene escape into the environment- accidental cross breeding GMP plants and traditional varieties through pollen transfer can contaminate the traditional local varities with GMO genes resulting in the loss of traditional varieties of the farmers.

Public attitude

1. Consumer response depends on perceptions about risks and benefits of genetically modified foods. The media, individuals, scientists and administrator, politicians and NGO have the responsibility to educate the people about the benefits of GM foods.

Socio economic and ethical consideration

1. Potential benefits to the consumers and farmers. Due to increasing seed market, the developing countries may get dependent on few suppliers.
2. GURT: Gene Use Restriction Technologies. India has totally banned the use of GURT in plant variety for registration under PPV & FRA, 2001

6.11.2 Regulatory Mechanisms in India

Biosafety regulations cover assessment of risks and the policies and procedures adopted to ensure environmentally safe applications of biotechnology. The regulatory framework for transgenic crops in India consists of the following rules and guidelines.

a) Rules and policies

1. Rules, 1989 under Environment Protection Act (1986)
2. Seed Policy, 2002

b) Guidelines

1. Recombinant DNA guidelines, 1990
2. Guidelines for research in transgenic crops, 1998

The two main agencies identified for implementation of the rules are the Ministry of Environment, Forests and Climate Change and the Department of Biotechnology, Government of India. The rules have also defined competent authorities and the composition of such authorities for handling of various aspects of the rules. There are six competent authorities as per the rules.

1. Recombinant DNA Advisory Committee (RDAC)
2. Review Committee on Genetic Manipulation (RCGM)
3. Genetic Engineering Approval Committee (GEAC)
4. Institutional Biosafety Committees (IBSC)
5. State Biosafety Coordination Committees (SBCC)
6. District Level Committees (DLC).

Out of these, the three agencies that are involved in approval of new transgenic crops are:

1. IBSC - set-up at each institution for monitoring institute level research in genetically modified organisms.
2. RCGM - set-up at DBT to monitor ongoing research activities in GMOs and small scale field trials.
3. GEAC - set-up in the Ministry of Environment, Forests and Climate Change to authorize large-scale trials and environmental release of genetically modified organisms.

The Recombinant DNA Advisory Committee (RDAC) constituted by DBT takes note of developments in biotechnology at national and international level and prepares suitable recommendations. The State Biotechnology Coordination Committees (SBCCs) set up in each state where research and application of GMOs are contemplated, coordinate the activities related to GMOs in the state with the central ministry. SBCCs have monitoring functions and therefore have got powers to inspect, investigate and to take punitive action in case of violations. Similarly, District Level Committees (DLCs) are constituted at district level to monitor the safety regulations in installations engaged in the use of GMOs in research and application.

6.11.3 Recombinant DNA Safety Guidelines, 1990

The guidelines cover areas of research involving genetically engineered organism. It also deals with genetic transformationof green plants, rDNA technology in vaccine development and on large scale production and dekliberate/ accidental release of organisms, plants, animals and products derived by rDNA technology into the environment. The issues relating to Genetic Engineering of human embryos, use of embryos and foetuses in research and human germ line gene therapy are excluded from the scope of the guidelines.The guidelines are in respect of safety measures for the research activities, large scale use and also the environmental impact during field applications of genetically altered material products.

Scope of the Revised Guidelines

1. Research: The levels of the risk and the classification of the organisms within these levels based on pathogenicity and local prevalence of diseases and on epidemic causing strains in India are defined in the guidelines. Some of the microorganisms not native to the country have been assigned to a special category requiring highest degree of safety. These include Lassa virus, Yellow fever virus etc. Appropriate practices, equipment and facilities are recommended for necessary safeguards in handling organisms, plants and animals in various risk groups. The guidelines employ the concept of physical and biological containment and also based upon the

principle of good laboratory practice (GLP). In this context, biosafety practices as recommended in the WHO laboratory safety Manual on genetic engineering techniques involving microorganisms of different risk groups have incorporated in the guidelines (Chapter IV).

2. Large scale operations: The concern does not diminish when it comes to the use of recombinant organisms scale fermentation operations on large scale fermentation operations or applications of it in the environment. As such, the guidelines prescribe criteria for good large scale practices (GLSP) for using recombinant organisms. These include measures such as proper engineering for containment, quality control, personnel protection, medical surveillance, etc.
3. Environmental risks: Application and release of engineered organisms into the environment could lead to ecological consequences and potential risks unless necessary safeguards are taken into account. The guidelines prescribe the criteria for assessment of the ecological aspects on a case by case basis for planned introduction of rDNA organism into the environment. It also suggests regulatory measures to ensure safety for import of genetically engineered materials, plants and animals. The recommendations also cover the various quality control methods needed to establish the safety, purity and efficacy of rDNA products.

6.11.4 Classification of Pathogenic Microorganisms

The classification of infective microorganisms are drawn up under 4 risk groups in increasing order of risk based on the following parameters:

- pathogenicity of the agent
- modes of transmission and host range of the agent
- availability of effective preventive treatments or curative medicines
- capability to cause diseases to humans/animals/plants
- epidemic causing strains in India

The above mentioned parameters may be influenced by levels of immunity, density and movement of host population, presence of vectors for transmission and standards of environmental hygiene.

An inventory of pathogenic organisms classified in different groups is provided in Chapter V: A1. The scientific considerations for assessment of potential risks in handling of pathogenic organisms include the following:

Characterisation of donor and recipient organisms

i) Characterisation of the modified organism
ii) Expression and properties of the gene product

Based on the risk assessment information, the probability of risk could be further assigned certain quantitative values (Chapter V: A7) for categorisation of experiments in terms of the following:

i) access factor of the organism
ii) expression factor of DNA
iii) damage factor of the Biologically active substance

6.11.5 Containment Facilities

Containment facilities are for different Risk Groups as per the recommendations of World Health Organization (WHO). The term "Containment" is used in describing the safe methods for managing infectious agents in the laboratory environment where they are being handled or maintained.

Purpose of containment: To reduce exposure of laboratory workers, other persons, and outside environment to potentially hazardous agents.

Types and elements of containment

i. **Biological containment (BC):** In consideration of biological containment, the vector (plasmid, organelle, or virus) for the recombinant DNA and the host (bacterial, plant, or animal cell) in which the vector is propagated in the laboratory will be considered together. Any combination of vector and host which is to provide biological containment must be chosen or constructed to limit the infectivity of vector to specific hosts and control the host-vector survival in the environment. These have been categorized into two levels - one permitting standard biological containment and the other even higher that relates to normal and disabled host-vector systems respectively (Chapter V: A3).

ii. **Physical Containment (PC):** The objective of physical containment is to confine recombinant organisms thereby preventing the exposure of the researcher and the environment to the harmful agents. Physical containment is achieved through the use of i) Laboratory Practice, ii) Containment Equipment, and iii) Special Laboratory Design. The protection of personnel and the immediate laboratory environment from exposure to infectious agents, is provided by good microbiological techniques and the use of appropriate safety equipment, (Primary Containment).

The protection of the environment external to the laboratory from exposure to infectious materials, is provided by a combination of facility design and operational practices, (Secondary Containment).

Elements of Containment: The three elements of containment include laboratory practice and technique, safety equipment and facility design.

i) Laboratory practice and technique:
    - Strict adherence to standard microbiological practices and techniques

- Awareness of potential hazards
- Providing/arranging for appropriate training of personnel
- Selection of safety practices in addition to standard laboratory practices if required
- Developing of adopting a biosafety or operations manual which identifies the hazards

ii) **Safety equipment (primary barriers):** Safety equipment includes biological safety cabinets and a variety of enclosed containers (e.g. safety centrifuge cup). The biological safety cabinet (BSC) is the principal device used to provide containment of infectious aerosols generated by many microbiological procedures. Three types of BSCs (Class I, II, III) are used in microbiological laboratories. Safety equipment also includes items for personal protection such as gloves, coats, gowns, shoe covers, boots, respirators, face shields and safety glasses, etc.

iii) **Facility Design (Secondary barriers):** The design of the facility is important in providing a barrier to protect persons working in the facility but outside of the laboratory and those in the community from infectious agents which may be accidentally released from the laboratory. There are three types of facility designs: viz, the Basic Laboratory (for Risk Group I and II), the Containment Laboratory (for Risk Group III) and the Maximum Containment Laboratory (for Risk Group IV).

Biosafety levels: It consists of a combination of laboratory practices and techniques, safety equipment and laboratory facilities appropriate for the operations performed and the hazard posed by the infectious agents. The guidelines for Microbiological and Biomedical Laboratories suggest four Biosafety levels in incremental order depending on the nature of work. Additional flexibility in containment levels can be obtained by combination of the physical with the biological barriers. The proposed safety levels for work with recombinant DNA technique take into consideration the source of the donor DNA and its disease-producing potential. These four levels corresponds to (P1<P2<P3<P4) facilities approximate to 4 risk groups assigned for etiologic agents.

These levels and the appropriate conditions are enumerated as follows

Biosafety Level 1: These practices, safety equipment and facilities are appropriate for undergraduate and secondary educational training and teaching laboratories and for other facilities in which work is done with defined and characterized strains of viable microorganisms not known to cause disease in healthy adult human. No special accommodation or equipment is required but the laboratory personnel are required to have specific training and to be supervised by a scientist with general training in microbiology or a related science.

Biosafety Level 2: These practices, safety equipment and facilities are applicable in clinical, diagnostic, teaching and other facilities in which work is done with the broad spectrum of indigenous moderate-risk agents present in the community and associated with human disease of varying severity. Laboratory workers are required to have specific training in handling pathogenic agents and to be supervised by competent scientists. Accommodation and facilities including safety cabinets are prescribed, especially for handling large volume are high concentrations of agents when aerosols are likely to be created. Access to the laboratory is controlled.

Biosafety level 3: These practices, safety equipment and facilities are applicable to clinical, diagnostic, teaching research or production facilities in which work is done with indigenous or exotic agents where the potential for infection by aerosols is real and the disease may have serious or lethal consequences. Personnel are required to have specific training in work with these agents and to be supervised by scientists experienced in this kind of microbiology. Specially designed laboratories and precautions including the use of safety cabinets are prescribed and the access is strictly controlled.

Biosafety level 4: These practices, safety equipment and facilities are applicable to work with dangerous and exotic agents which pose a high individual risk of life-threatening disease. Strict training and supervision are required and the work is done in specially designed laboratories under stringent safety conditions, including the use of safety cabinets and positive pressure personnel suits. Access is strictly limited.

A specially designed suit area may be provided in the facility. Personnel who enter this area wear a one-piece positive pressure suit that is ventilated by a life support system.

6.12 Mechanism of Implementation of Biosafety Guidelines

For implementation of the guidelines it is necessary to have an institutional mechanism to ensure the compliance of requisite safeguards at various levels. The guidelines prescribe specific actions that include establishing safety procedures for rDNA research, production and release to the environment and setting up containment conditions for certain experiments. The guidelines suggest compliance of the safeguards through voluntary as well as regulatory approach (Anonymous, 2016). In this connection, it is proposed to have a mechanism of advisory and regulatory bodies to deal with the specific and discretionary actions on the following:

a. Self regulation and control in the form of guidelines on recombinant research activities; and
b. Regulation of large scale use of engineered organisms in production activity and release of organisms in environmental applications under statutory provisions.

The institutional mechanism as proposed for implementation of guidelines is shown in organogram in Figure 2. Mainly it consists of the following:-

i) Recombinant DNA Advisory Committee (RDAC)

ii) Institutional Biosafety Committee (IBSC)

iii) Review Committee on Genetic Manipulation (RCGM)

iv) Genetic Engineering Approval Committee (GEAC)

Scope and functions of advisory committee and statutory body

i. Recombinant DNA Advisory Committee (RDAC): The Committee should take note of developments at national and international levels in Biotechnology towards the currentness of the safety regulation for India on recombinant research use and applications. It would meet once in 6 months or sooner for this purpose.

The specific terms of reference for Recombinant Advisory Committee include the following:

i) To evolve long term policy for research and development in Recombinant DNA research.

ii) To formulate the safety guidelines for Recombinant DNA Research to be followed in India.

iii) To recommended type of training programme for technicians and research fellows for making them adequately aware of hazards and risks involved in recombinant DNA research and methods of avoiding it.

ii. Implementation Committees

Institutional Biosafety Committee (IBSC): Institutional Biosafety Committee (IBSC) are to be constituted in all centres engaged in genetic engineering research and production activities. The Committee will constitute the following:

Head of the Institution or nominee

(i) 3 or more scientists engaged in DNA work or molecular biology with an outside expert in the relevant discipline.

(ii) A member with medical qualifications - Biosafety Officer (in case of work with pathogenic agents/large scale use).

(iii) One member nominated by DBT.

The Institutional Biosafety Committee shall be the nodal point for interaction within institution for implementation of the guidelines. Any research project which is likely to have biohazard potential (as envisaged by the guidelines) during the execution stage or which involve the production of either microorganisms or biologically active molecules that might cause bio-hazard should be notified to IBSC. IBSC will allow genetic engineering activity on classified organisms only at places where such work should be performed as per guidelines. Provision of suitable safe storage facility of donor, vectors, recipients and other materials involved in experimental work should be made and may be subjected to inspection on accountability.

The biosafety functions and activity include the following:

Registration of Bio-safety Committee membership composition with RCGM and submission of reports.

IBSC will provide half yearly report on the ongoing projects to RCGM regarding the observance of the safety guidelines on accidents, risks and on deviations if any. A computerised Central Registry for collation of periodic report on approved projects will be set up with RCGM to monitor compliance on safeguards as stipulated in the guidelines.

i) Review and clearance of project proposals falling under restricted category that meets the requirements under the guidelines.

   IBSC would make efforts to issue clearance quickly on receiving the research proposals from investigators.

ii) Tailoring biosafety programme to the level of risk assessment.

iii) Training of personnel on biosafety.

iv) Instituting health monitoring programme for laboratory personnel.

   Complete medical check-up of personnel working in projects involving work with potentially dangerous microorganisms should be done prior to starting such projects. Follow up medical checkups including pathological tests should be done periodically, at least annually for scientific workers involved in such projects. Their medical records should be accessible to the RCGM. It will provide half yearly reports on the ongoing projects to RCGM regarding the observance of the safety guidelines on accidents, risks and on deviations if any.

v) Adopting emergency plans.

iii. Review Committee on Genetic Manipulation (RCGM): The RCGM will have the following composition:

i) Department of Biotechnology

ii) Indian Council of Medical Research

iii) Indian Council of Agricultural Research

iv) Council of Scientific & Industrial Research

v) Three Experts in Individual capacity

vi) Department of Science & Technology

The RCGM will have the functions: To establish procedural guidance manual - procedure for regulatory process with respect to activity involving genetically engineered organisms in research, production and applications related to environmental safety.

i) To review the reports in all approved ongoing research projects involving high risk category and controlled field experiments, to ensure that safeguards are maintained as per guidelines.

ii) To recommend the type of containment facility and the special containment conditions to be followed for experimental trials and for certain experiments.

iii) To advise customs authorities on import of biologically active material, genetically engineered substances or products and on excisable items to Central Revenue and Excise.

iv) To assist Department of Industrial Development, Banks towards clearance of applications in setting up industries based on genetically engineered organisms.

v) To assist the Bureau of Indian Standards to evolve standards for biologics produced by rDNA technology.

vi) To advise on intellectual property rights with respect to rDNA technology on patents.

The RCGM would have a Research Monitoring function by a group consisting of a smaller number of individuals (3 or 4). The monitoring group would be empowered to visit experimental facilities in any laboratory in India where experiments with biohazard potential are being pursued in order to determine the Good Laboratory practice and conditions of safety are observed.

In addition, if the RCGM has reasons to believe that there is either actual or potential danger involved in the work carried out by any laboratory (which might or might not have obtained prior clearance for the project), the monitoring group would be empowered to inspect the facility and assess the cause of any real or potential hazard to make appropriate recommendation to the RCGM. RCGM would be empowered to recommend alteration of the course of experiments based on hazard considerations or take steps to cancel the project grant, in case of deliberate negligence and to recommend appropriate actions under the provisions of Environmental Protection Act (EPA) where necessary.

iv. Genetic Engineering Approval Committee (GEAC): Genetic Engineering Approval Committee (GEAC) will function under the Department of Environment (DOEn) as statutory body for review and approval of activities involving large scale use of genetically engineered organisms and their products in research and development, industrial production, environmental release and field applications.

The functions include giving approval from environmental angle on:

Import, export, transport, manufacture, process, selling of any microorganisms or genetically engineered substances or cells including food stuffs and additives that contains products derived by Gene Therapy.

i) Discharge of Genetically engineered/classified organisms/cells from Laboratory, hospitals and related areas into environment.

ii) Large scale use of genetically engineered organisms/classified microorganisms in industrial production and applications. (Production shall not be commenced without approval).

iii) Deliberate release of genetically engineered organisms. The approval will be for a period of 4 years.

Composition of the Committee : The composition of the Committee would be as follows.

Chairman - Additional Secretary, Department of Environment

Co-Chairman - Expert Nominee of Secretary, DBT.

Representatives of concerned Agencies and Departments

- Ministry of Industrial Development
- Department of Science & Technology
- Department of Ocean Development
- Department of Biotechnology

Expert Members

- Director-General, Indian Council of Agricultural Research
- Director General, Indian Council of Medical Research
- Director-General, Council of Scientific & Industrial Research
- Director-General, Health Services (Ministry of Health & Family Welfare)
- Plant Protection Adviser (Ministry of Agriculture)
- Chairman, Central Pollution Control Board
- 3 Outside experts in individual capacity.
- Member Secretary - Official of, DOEn:

GEAC will have the Biotechnology Coordination Committees under it which will functions as legal and statutory body with judicial powers to inspect, investigate and take punitive action in case of violations of statutory provisions under EPA. Review and control of safety measures adopted while handling large scale use of genetically engineered organisms/classified organisms in research, developmental and industrial production activities.

i) Monitoring of large scale release of engineered organisms/products into environment, oversee field applications and experimental field trials.

ii) To provide information/data inputs to RCGM upon surveillance of approved projects under industrial production, and in case of environmental releases with respect to safety, risks and accidents.

Statutory rules and regulations to be operated by the GEAC would be laid down under the Environment Protection Act, 1986.

Funding agency: The funding agency will be responsible for approval and clearing of research proposals for grants in aid in respect of rDNA research activities. The funding agency at the centre and state level will be advised to ensure that the guidelines are taken into account for compliance while supporting grants on research projects. Investigators will be required to submit as part of the project application an evaluation of biohazards that may arise and also the requirement on the type of containment facility, certified by IBSC. The funding agency should state clearly that support on approved projects will be withdrawn in case of deliberate violation or avoidable negligence of the rDNA guidelines. The investigators will also be asked to make a declaration in their publications that the work was carried out following the national guidelines. The funding agency will annually submit to RCGM the list of approved projects that come under high risk categories.

The concerned institutions will be instructed to the effect that initiation and execution of any research project, production activity and field trials should be preceded by necessary procedures of notification and approval of the competent authority including IBSC, GEAC depending on the nature of projects and activities. Initially, to familiarize the R&D groups in industry and other institutions the guidelines will be widely publicised through scientific journals and popular science magazines (Fig.9). Workshops and group discussions will be organised in R&D institutes, and other places to fulfill the need for public information on safety aspects of rDNA technology. Steps will be taken to introduce courses in biohazards and safety procedures for personnel working in areas which are likely to involve biohazards as part of the training programme.

6.13 Plants, Agriculture and Plant Pathopgens

The application of genetic engineering to agriculture is directed to deliver products whose research, evaluation and commercial use would require studies on introduction into the field. These products include genetically engineered plants, microbes, animal vaccines and animals.

Many of the scientific considerations described in earlier chapter are relevant to plants and animals derived by rDNA techniques. Additionally, the general considerations (Chapter V) describing the significance of the donor, recipient and modified organisms are also essential to safety assessment evaluation.The proposed regulation requires a permit for the introduction of any "regulated article" which is defined as "any organism or product which has been altered or produced through genetic engineering, if the donor organism, recipient organism, or vector or vector agent" is specifically listed in the regulation or which is determined by the competent agency as a plant pest/pathogen that cause disease to plants. The proposed regulated articles are grouped by class, order, genus, family and other groupings.

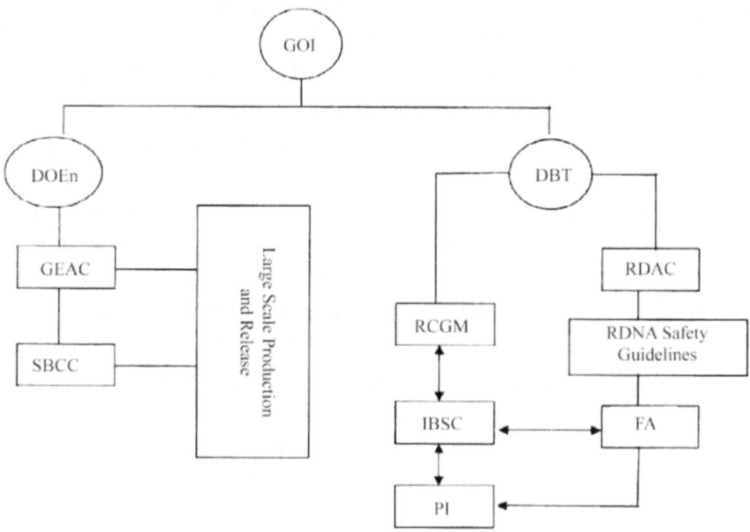

GOI - Government of India
DBT - Department of Biotechnology
RDAC - Recombinant DNA Advisory Committee
IBSC - Institutional Biosafety Committee
RCGM - Review Committee on Genetic Manipulation
DOEn - Department of Environment
GEAC - Genetic Engineering Approval Committee
SBCC - State Biotechnology Coordination Committee
PI - Principal Invstigator (R&D/Industry/Others)
FA - Funding Agency (Govt./Private & Public

Fig. 9: Institutional mechanism for implementation of guidelines frame work

6.13.1 Plant Pathogens

The taxa or group of organisms which are or contains plant pest are listed. Organisms belonging to all lower Taxa contained within the group listed are also included.

i. Viruses: All members of groups containing plant viruses and all other plant and insect viruses.

The viruses subjected to quarantine also include Bean Yellow Mosaic (Pea strain), Pea Early Browning, Pea Enation, Cowpea Mottle, Cowpea Mild Mottle, Cowpea Severe Mosaic, Cowpea Yellow Mosaic, Cowpea Ring spot, Soybean stunt, Cucumber Mosaic (Soybean strain & other), Tobacco Ring spot (Soybean strain), Tobacco Streak (Soybean strain), Tomato Ring spot, Bean Pod Mottle, Soybean Mild Mottle, Soybean stunt, Peanut stripe, Cowpea Mild Mottle, Peanut stunt, Cacao Necrosis Virus,Marginal Chlorosis (Soybean strain), Cowpea Mottle, Pea Seed-borne Mosaic (Bambara groundnut strain), Cucumber Mosaic (Green gram strain), Nuclear Polyhedrosis Virus, Black gram mottle, Cytoplasmic Polyhedrosis Virus, Bean Yellow Mosaic, Granular Virus-Baculo (Green gram strain), Gemini viruses, Cucumber Mosaic, Caulimoviruses (Groundnut strain)

ii. Bacteria: *Bacillus thuringiensis, Bacilus sphericus, Genus Pseudomonas, Genus Xanthomonas, Genus Azotobacter, Genus Rhizobium/Azorhizobium, Genus Bradyrhizobium, Genus Agrobacterium, Genus Phyllobacterium, Genus Erwinia, Genus Enterobacter, Genus Klebzieller, Genus Azospirillum, Genus Acquspirillum, Genus Oceonospirillum, Genus Streptomyces, Genus Nocardia, Genus Actinomyces, Coryneform group, Genus Clavibacter, Genus Arthrobacter, Genus Curtobacterium, Genus Bdellovibro,* Rickettsial - like organisms associated with insect diseases, Gram-negative phloem-limited bacteria associated with plant diseases, Gram-negative xylem-limited bacteria associated with plant diseases, Genus Spiroplasma, Mycoplasma - like organisms associated with plant diseases and Mycoplasma - like organisms associated with insect diseases.

iii. Algae: Families Chlorophyceae, Euglenophyceae, Pyrophyceae, Chrysophyceae, Phaephyceae and Rhodophyceae

iv. Fungi : Families Plasmodiophoraceae, Eurotiaceae, Chytridiaceae, Ophiostomataceae, Hypochytridiaceae, Ascophaeraceae, Olpidiopsidaceae, Onygeneaceae, Synchytriaceae, Microascaceae, Catenariaceae, Erysiphaceae, Coelomomycetaceae, Meliolaceae, Saprolegniaceae, Xylariaceae, Zoopagaceae, Diaporthaceae, Albuginaceae, Hypocreaceae, Peronosporaceae, Clavicipitaceae, Pythiaceae, Phacidiaceae, Leptolegniellaceae, Ascocorticiaceae, Mucoraceae, Hemiphacidiaceae, Choanephoraceae, Dermataceae, Mortierellaceae, Sclerotiniaceae, Endogonaceae, Cytarriaceae, Syncephalastracae, Helotiaceae, Dimargaritaceae, Kickxellaceae,Saksenaeaceae,Entomophthoraceae,Ecerinaceae,Protomycetaceae, Taphrinaceae, Endomycetaceae, Saccharomycetaceae, Elsinoeaceae, Myriangiaceae, Dothideaceae, Chaetothyriaeae, Parmulariaceae, Phillipsiellaceae, Hysteriaceae, Pleosporaceae, Melanomotaceae, Sacrosomataceae, Ganodeniatiaceae, Sarcoscyphaceae, Labonlbeniaceae, Auriculariaceae, Sphaeropsidaceae, Ceratobasidiaceae, Melanconiaceae, Corticiaceae, Tuberculariaceae, Hymenochaetaceae, Dematiaceae, Echinodontiaceae, Moniliaceae, Fistulinaceae, Aganomycetaceae, Clavariaceae, Polyporaceae, Tricholomataceae, y Ustilaginaceae, Sporobolomycetaceae, Uredinaceae, Agaricaceae, Graphiolaceae, Pucciniaceae and Melampsoraceae

v. Protozoa: Genus *Phytomonas* and all Protozoa associated with insected diseases

vi. Nematodes: Families Anguinidae, Belonolaimidae, Caloosiidae, Criconematidae, Dolichodoridae, Fergusobiidae, Hemicycliophoridae, Heteroderidae, Hoplolaimidae, Meloidogynidae, Neotylenchidae, Nothotylenchidae, Paratylenchidae, Tylenchidae, Tylenchulidae, Adhelenchoididae, Longidoridae and Trichodoridae.

vii. Mollusca : Superfamilies Planorbacea, Achatinacae, Arionaceae, Limacacea, Helicacea and Veronicellacea

viii. Arthropoda: Superfamilies Ascoidea, Dermanyssoiedea, Eriohyoidea, Tetranychoidea, Eupodoidea, Erythraenoidea, Trombidioidea, Hydryphantoidea, Tarsonemoidea, Hydryphantoidea, Tarsonemoidea, Pyemotoidea, Hemisarcoptoidea and Acaroidea

Order Polydesmida- Families Sminthoridae and Forticulidae.

Order Isoptera: Order Thysanoptera –Families Acrididae, Gryllidae, Cryllacrididae, Cryllotalpidae, Phasmatidae, Ronaleidae, Tettigoniidae, Tetrigidae and Thaumastocoridae.

Superfamilies Piesmatoidea, Lygaeoidea, Idiostoloidea, Coreoidea, Pentatomoidea, Pyrrhocoroidea, Tingoidea and Miroidea

Order Homoptera : Families Anobiidae, Torymidae, Apionidae, Xylocopidae, Anthribidae, Bostrichidae, Brentidae, Bruchidae, Buprestidae, Byturidae, Cantharidae, Carabidae and Cerambycidae.

Family Chrysomelidae: Subfamily Epilachninae

Families Curculionidae, Dermestidae and Elateridae: Genus Helophorous

Families Lyctidae, Family Melodiae, Family Mordellidae

Subfamilies Melolonthinae, Rutelinae, Cetoniinae and Dynastinae

Families Scolytidae and Seblytidae

Order Lepidoptera: Families Agromyzidae, Anthomyiidae, Cecidomyiidae, Chloropidae, Ephydridae, Lonchaeidae, Muscidae, Otitidae, Syrphidae, Tephritidae, Apidae, Caphidae, Chalcidae, Cynipidae, Eurytomidae, Formicidae, Psilidae, Siricidae and Tenthredinidae.

6.13.2 Genetic Manipulation of Plants and Plant Pathogens

The experiments that include the introduction of foreign nucleic acid into plants.

a) The introduction of foreign nucleic acid into any plant pathogen where pathogen is defined as "any living organism, other than a vertebrate animal which is injurious to any plant, and includes any culture of such organism."

Notification: Plant experiments that do not involve plant pathogen may, where appropriate be initiated once notification has been given to IBSC.

All experiments involving the genetic manipulation of plant pathogens and the use of such genetically manipulated plant pathogens will require approval of IBSC. Use of pathogenic vectors is mainly two, *Agrobacterium* tumefaciens and Cauliflower Mosaic Virus.

Agrobacterium tumefaciens is used to routinely that it must be considered analogous to *E.coli* K12. Apart from transfer of *B. thuringiensis* toxin gene to plants, a new class of experiments, involves transfer of sequences from plant viruses which impart their resistance to plants to infection of these viruses (e.g. Tobacco Mosaic Virus, Alfalfa Mosaic Virus etc.) Testing to this should be in a glasshouse. As much of information as possible should be provided about the pathogen including its host range, mode of dispersal and pathogenicity. Isolated plasmids from plant pathogens are not normally considered as pathogen per se, so that transformation of plant cells by isolated plasmids of plant pathogens would not normally require approval. The genetic manipulation of microbes (including plant pathogens) is adequately covered by the existing rDNA guidelines.

Plant experiments with no plant pathogens: The growth of whole plants will, however require special environmental conditions which may be achieved by using glasshouse containment.

Glasshouse Containment A is appropriate to plant experiments involving no plant pathogens and would be suitable for experiments involving non-pathogen DNA vector systems and regeneration from single cells. The minimal requirements for Glasshouses Containment A are :

i) Plants should be grown in a designated glasshouse or compartment, clearly marked with a bio-hazard sign indicating "glasshouse containment A".

ii) Any other plants grown in the designated glasshouse or compartment must be handled under conditions appropriate for the experimental plans.

iii) Plants should be managed by suitably trained personnel with the principles of good glasshouse hygiene.

iv) The IBSC should consider whether any additional factors such as pest control, screening to prevent ingress by vermin, birds and insects and destruction of surplus plants and seed are relevant to the particular experiment.

Plant experimentation involving plant pathogens: Prior approval for laboratory experiments involving genetically manipulated plant pathogens, such as the production of manipulated DNA vector systems for the transformation of cultured plant cells will normally be needed on the basis of containment categorization.

Glasshouse Containment B is appropriate for glasshouse experiments involving (i) genetically manipulated plant pathogens including plant viruses such as the propagation of genetically manipulated organism in plants and (ii) the growth of plants regenerated from cell transformed by genetically manipulated pathogen vector systems which will still contain the pathogen.

Glasshouse containment B conditions will be specified by the Committee (RCGM) and will vary with the pathogen, being particularly dependant on its mode of dispersal, host range and pathogenicity and they are to be worked out on case by case basis. Special conditions may be needed in addition to those given under `A' to prevent dissemination of the genetically manipulated plant pathogen especially during transfer between glasshouse and laboratory, during disposal of plants and equipment and through survival of pollen, seeds or other biological vectors.

a) Need for negative pressure and air filtration double doors etc. in cases where airborne dispersal is a potential hazard.
b) Need for effluent treatment plant where water borne dispersal is a hazard.
c) Need for suitable construction of glasshouse (floors, dwarf wall, threshold at door etc.) in cases where waterborne or soil borne dispersal are potential hazards.
d) Need to prevent pollination and seeding, or to contain pollen and seed in cases where pollen and seed-borne dispersal is a potential hazard.
e) Need for measures either to prevent contamination of, or to decontaminate the clothing of personnel or tools, pots, equipment etc., where mechanical transmissions is an above average hazard.
f) Need to limit the growing of host plants in the vicinity of the containment facility and to provide monitoring for escape.

Inspection of a 'Glasshouse Containment B' facility by IBSC will be required before approval.

6.14 Pre-release tests of Genetically Engineered Organisms on Agricultural Applications

Safety concerns focus on whether environmental and agricultural applications of organisms modified by rDNA technique pose an incremental risk. While at this time, the assessment of risk rests primarily on extrapolations from experiences with the introduction of naturally occurring organisms to eco-systems to which they are not native; evolution of noval traits in existing populations and manipulations of agricultural crops and plant-associated microbes.

No adverse consequences were noted on introduction of naturally occuring species, or the selected species evolved for agricultural applications. In analogy, it is expected that the impact on application of rDNA organism may be low as modified organism have greater predictability compared to species evolved by traditional techniques. The assessment may be conducted in small field trials upon clearance of GEAC as to those done with the introduction of selective species into the eco-system.

a. Strains improved by transfer of genes between rhizobia: Conventional tests should be sufficient. These may include the following.

Elucidation of genetic markers and host range and requirements for vegetative growth.

Effectivity tests using corresponding legume varieties under the variety of conditions to which host legume gets exposed in growth chamber and pot culture.

Tests on persistence and stability using isolated small plots.

Same as in (ii) in experimental field plots, for two years.

Trials in farmer fields.

The strain to be released should be highly specific, competitive and stable unless it has been produced for a special need.

b. Strains improved by transfer of genes from heterologous nonpathogenic bacteria

i. Use of foreign genes in *Agrobacterium*.

Tests to ensure that the strain is not tumour inducing and does not transform in host-cells.

Other test like in (a) above.

ii. When source of foreign genes is a non-Rhizobian prokaryote such as Escherichia coli

Tests for enteropathogenicity on selected animals.

Other tests like in (a) above.

c. Bacteria manipulated by any method should satisfy the following: The manipulated microorganisms should be tested for pathogenicity against its intended associative partner and also other crop plants.They should not eliminate useful microorganisms like VA-Mycorrhizae. Suitable testing should be done in this regard before releasing a manipulated organism.The microorganism should not stimulate unwanted plants like weeds.

d. Blue Green Algae: All the tests as given in A and B will be applicable except that rice will be the plant against which beneficial effects of concerned algal strain will be tested.

e. Crop Plants: When the improved plant has been derived by transfer of genes by DNA technique from wild species or a different organism, tests on the food product and residual presence of agents toxic to man, cattle and other animals must be done.

Bio-hazard evaluation of viral, bacterial, insecticidal agents for large scale application: World Health Organisation (WHO) has developed programmes for evaluation, testing and safe use of insecticides to control vector borne diseases of public health, veterinary and agriculture importance (WHO TRS 634). The criteria (Bull. WHO (1971), 44, 11-22) for assessment of the ecological impact involve controlled testing and evaluation under field conditions. There are growing number of tested and accepted biological insect control agents belonging to diverse group such as bacteria, fungi and certain viruses etc. Some of them are registered in global market. Any attempt at genetically altering, improving changing the host range, target specificity, differential pathogen toxicity, toxic agent productivity, factors affecting safety and efficacy, new formulations leading to newer uses of these biological control agents and related organisms and their products derived through genetic alteration would require the application of rDNA safety guidelines and regulations as per categorization scheme worked out based on risk assessment levels.

Whereas the testing and large scale use of biological control agents would itself require the normal course of approval from Directorate of Plant Protection and quarantine under Ministry of Agriculture, the production testing and use of these genetically altered agents would be strictly governed by the rDNA guidelines and regulations of the Government of India. Violations and non-compliance including non-reporting of the R&D work in this area would attract the punitive actions provided under the Environmental Protection Act.

Bacterial agents: Three main groups of bacteria, viz. *Bacillus popilliae*, *B.thuringiensis* and *B. sphericus* have been subject of extensive studies. Of these, B.thuringiensis particularly H-14 strain has been found to be most promising for the control of larvae of lepidoptera, mosquitoes and black flies. The protein crystal toxin (δendotoxin) of these bacilli acts as potent gut poison on ingestion by the larvae.The development of new mutants of *B. thuringiensis* (including the asporogenic strain) producing the same toxin would not require additional safety evaluation. However, a strain producing modified toxin, possibly with altered biological activity, would require some additional evaluation of the live microorganisms and fully safety evaluation of the toxin. When the toxin-producing gene is transferred to another microorganism complete safety evaluation would be required.

Insect viruses as control agents: Virus diseases have been reported from more than 800 species of insects and mites. Major groups of virus pathogens of insects are being currently studied as control agents for pests. Insect viruses for pest are produced on an industrial scale, viz., *Heliothis zea, H. virescens, Lymantira* dispar, *Neodiprion sertifier, Orgyia pseudot-sugata* and *Dendrolimus*.

The bio-hazard evaluation of promising viral agents which are naturally occurring entempopathogens should include the following:

- Elaborate tests conducted in the laboratory as well as under green-house conditions to understand potential physiological and/or genetical hazards for non-target organisms. The overall response of a species is likely to be polygenic.
- In stable ecosystems, where the potential of the viruses can be utilised most effectively, more information is required on the relationships between host density and susceptibility, virus production, persistence and transmission. Analytical approaches provide powerful ways of highlighting the importance of such factors in virus epizootiology.
- The safety measures for large-scale application of such products would require very careful evaluation since the combination of two or more types of biocides may affect the non-target organisms particularly those which

are beneficial. For example genus Apis which plays an important role in pollination of oil seeds, legumes, vegetables, forage and food crops.

- The characteristics of baculovirus that is more useful for identification is the profile produced by cleavage of the rival DNA with bacterial restriction endonucleases. Such techniques should be used to screen all production batches which should be preferably purified before release. These batches should also be examined for the presence of other 'occluded' or 'non-occluded' viruses.
- Bacteriological check and other safety tests as mentioned in the WHO guidelines are also needed.
- The purified virus should then be formulated in such a way that its stability both on the shelf and field is satisfactory.
- The biological activity of the propagation should be measured by reproducible and effective bio-assays to measure the responses in standard activity units which can be related to the activity of the other batch.
- The application of viruses can be most effective in those areas, where there is a good understanding of the ecology of the host-virus system. The most appropriate method of 'oryctes' virus introduction, was appreciated when the effects of virus replication on larvae and adults had been extensively studied. It is probable that alternative methods of virus introduction, such as the release of infected host, would become advantageous over other methods.

Recombinant insect viruses: *Autographa californica* nuclear polyhedrosis virus (AcNPV) is a registered insecticide in USA and is also now gaining importance for being employed as recombinant vector. The recombinant technology could be extended to the construction of noval AcNPVs with genes of *B. thuringiensis d*-endotoxin and insect neuropeptides for greater effectiveness. Thus, baculovirus recombinant vector containing s-toxin of *B. thuringiensis* and insect neuropeptides could be of immense use in planning overall strategy for insect control. Such products having multiple insect control features needs to be carefully assessed for the risk to the health and environment before it is licensed.

6.15 India at International Biosafety Fora

Despite backlogs in domestic biosafety policy and enforcement, India also has to deal with these issues in the international arena. The United Nations Environment Programme (UNEP) and World Health Organisation (WHO) have issued international guidelines for the safe use of GMOs. At UNEP, the 'Protocol on Biosafety' is being formulated under the Convention on Biological Diversity (Chaturvedi, 1997). This protocol includes public participation by making available the results of any testing and monitoring involving GMOs. However, this has not been adopted in the Indian policy. Parts of OECD guidelines on risk evaluation have also been included in the Indian guidelines. Some companies in India think that what has been adopted is very precautious and does not follow the familiarity principle in biosafety assessment. The familiarity principle involves the notion that the country has knowledge and experience in testing the crops involved; and thereby has the basis to predict how the crop will behave once it is released into the environment. In the precautionary principle approach each case is isolated from other experiences, and is separately reviewed. Given the commitments that India has made in international fora, the country will have to take a long and hard look at the existing mechanisms aimed at ensuring that biotechnology does not have an adverse impact on the environment. Aside from lack of participation and slow clearances, Indian regulatory bodies need political support and technical expertise to effectively enforce and monitor trials involving GMOs.

6.15.1 Criticism of the Policy

Biosafety regulations come under severe criticism from environmentalists and industries alike. Industries claim that the absence of a transparent system has led to unnecessary doubts in the minds of the public about the implications of GMOs. Moreover, the DBT deals with companies on a casebycase basis. This prevents the industries from pleading their cases collectively. However, this also presents some advantage to the industries. For example, they do not have to take a public stand on any controversies regarding transgenics. Additionally, the lack of clear guidelines enables industries to have more influence on the guideline formulation and implementation. Some industries claim that the regulatory authorities often come out with whimsical guidelines. When the guidelines need amendments, arbitrary arguments arise. One point of argument was the isolation distance of the trial field from related conventional crops. DBT earlier came out with a regulation distance of 100 metres. After disagreements with an industry, after eight, long months the DBT amended its stand to 70 meters. The usual debate between industrial firms and DBT stems from the lack of data to serve as models for local experiments. DBT's knowledge base and scientific base for determining protocol for field testing remains weak. Industrial firms prefer to use data from other countries and suggest that there is no need to "reinvent the wheel" in experiments involving GMOs.

Environmental organizations such as the Gene Campaign and the Research Foundation for Science Technology and Ecology (RESTE) criticize biosafety policy for ignoring specific Indian needs and environmental conditions. For instance, copying the USA legislation, as some critics accuse India of, neglects the dangers of cross fertilization. The danger is much higher in a biodiversityrich country like India compared to a biodiversity poor country like USA. Environmentalists criticise the RCGM for not involving any participation from either research level (universities, research institutes) or user level (industry, consumers, farmers). As a consequence, both the governing bodies and the system of policy making lack transparency. The absence of NGO representation in any of the regulatory committees has been particularly noted. According to Vandana Shiva of RESTE, they have the technical experts in biosafety assessment to merit participation in these committees.

In India the dangers of GMOs receive a lot of public and media attention. The apprehensions about GMOs are linked, amongst others, with concerns for biodiversity preservation. Cross fertilization may result in the loss of indigenous species because of competition in the ecological system. The potential dangers include the displacement or destruction of indigenous or endemic species; and exposure of species to new pathogens. In response to such criticisms, DBT plans to conduct activities to enhance awareness and technical expertise about biosafety. This includes training of scientists, and visits to countries with biosafety policies such as the USA. DBT hopes to improve their data base on biosafety regulations. On the other hand, NGOs such as RESTE are contributing to the information campaign. However, the overall institutional capacity building to effectively enforce and monitor biosafety policies remains inadequate.

6.15.2 Governing Biosafety in India

A paper was written as part of the Global Environmental Assessment Project, a collaborative, interdisciplinary effort to explore how assessment activities can better link scientific understanding with effective action on issues arising in the context of global environmental change (Gupta, Aarti, 2000). The Project seeks to understand the special problems, challenges and opportunities that arise in efforts to develop common scientific assessments that are relevant and credible across multiple national circumstances and political cultures. It takes a long-term perspective focused on the interactions of science, assessment and management over periods of a decade or more, rather than concentrating on specific studies or negotiating sessions. Global environmental change is viewed broadly to include not only climate and other atmospheric issues, but also transboundary movements of organisms and chemical toxins.

The paper examines emerging global and national governance regimes for biosafety or the safe use of biotechnology in agriculture. The central concern is with examining the nature of the transnational national interface in biosafety governance, i.e. the relationship between multilaterally negotiated rules and national-level biosafety decision-making. The paper examines the relevance of the recently concluded Cartagena Protocol on Biosafety, dealing with the transboundary movement of genetically modified organisms (GMOs), for biosafety governance in India. In its call for “informed consent” prior to transfer of certain GMOs, the Cartagena Protocol validates the need for national-level choice in biosafety decision-making. However, a competing imperative is standardization of rules governing such choice, in order to enhance predictability and reduce national differences in biosafety decision-making. I argue here that these potentially contradictory goals are reconciled in the Cartagena Protocol through reliance upon a minimalist scope and ambiguous decision-criteria for informed consent. In light of this, it is argued that the Cartagena Protocol can be relevant to national biosafety governance in three ways: first, it legitimizes the existence of domestic biosafety regulations; second, notwithstanding a minimalist scope, it shifts the burden for information sharing to producers of GMOs; and third, given ambiguous decision-criteria, it leaves unchanged national discretion in GMO decision-making. The paper then evaluates the likely relevance of such impacts for national-level biosafety governance in India, through examining the processes of biosafety decision-making and information sharing currently in place. I conclude with observations about the transnational-national interface, and the role for multilateral rule-making, in facilitating governance of contested decision-areas such as biosafety.

To conclude, the International Service for National Agricultural Research (ISNAR) and FAO have identified four elements that are required to be considered when developing a regulatory framework for biotechnology. The answer to the problems of biosafety regulations in India is neither simple nor straightforward. India's environmental regulatory mechanisms in the field of biotechnology were laid from above. The regulatory space is governmental (Damodaran, 2005). Uncertainties and gaps in the knowledge base in relation to high-tech disciplines formed the original rationale for a governmentalised biosafety regime..

The first presumption is that there can be no meeting ground between those who believe in the norm of case-to-case approaches to biosafety project clearance and those who are against it. The second assumption is that biosafety regulations in India are stand-alone in nature and do not face prospects of interference from the cognate legislations. The third assumption is that the single window approach to decision-making could lead to transparency and quick and effective results. In practical terms this means that single window initiatives in the field of biosafety regulations should not only be vertically stacked but also horizontally broad-based with both civil society and industry associations accorded their due role in decision-making processes.

Bibliography

Alexandrova, N., Georgieva, K. and Atanassov, A., 2005. Biosafety Regulations of GMOS: National and International Aspects and Regional Cooperation, Biotechnology & Biotechnological Equipment, 19:sup3, 153-172,

Anonymous, 2011. Guidelines and Handbook for Institutional Biosafety Committees (IBSCs), Department of Biotechnology, Ministry of Science & Technology & Biotech Consortium India Limited, New Delhi

Anonymous, 2016. Indian council of Agricultural Research Biosafety Portal, New Delhi

Anonymous, 2017. Regulations and guidelines for recombinant DNA research and biocontainment. Department of Biotechnology Ministry of Science and Technology Government of India

Bakanidze, L., Imnadze, P. and Perkins, D., 2010. Biosafety and biosecurity as essential pillars of international health security and cross-cutting elements of biological nonproliferation. BMC Public Health, 10, S12

Beeckman, D.S.A. and Rüdelsheim, P., 2020. Biosafety and Biosecurity in Containment: A Regulatory Overview. *Fr. Bioeng. Biotechnol.*, 8: 650.

Chaturvedi, S., 1997, "Biosafety Policy and Implications in India." Biotechnology and Development Monitor, No. 30, p. 1013.

Damodaran, A., 2005. Re-Engineering Biosafety Regulations In India: Towards a Critique of Policy, Law and Prescriptions. *Law Env. & Dev. J.*, 1:1

Gandhi,B.M., 2014. Recent Incidences of Global Biosafety and Biosecurity Lapses in Laboratories Need Relook at Implementation of National Policies. CBW Magazine. The Manohar Parrikar Institute for Defence Studies and Analyses (MP-IDSA),

Gigi Kwik Gronvall, Matthew P Shearer and Hannah Collins,2016. National Biosafety Systems. UPMC Center for Health Security Project Team. Case studies to analyze current biosafety approaches and regulations for Brazil, China, India, Israel, Pakistan, Kenya, Russia, Singapore, the United Kingdom, and the United States. Project was supported by the Naval Postgraduate School Project on Advanced Systems and Concepts for Countering WMD (PASCC) Grant No. N00244-15-1-0028 to provide funding support for research entitled "Improving Security Through International Biosafety Norms."

Gupta Aarti, 2000. "Governing Biosafety in India: The Relevance of the Cartagena Protocol." Belfer Center for Science and International Affairs (BCSIA) Discussion Paper 2000-24, Environment and Natural Resources Program, Kennedy School of Government, Harvard University, 2000

Gupta, K., Karihaloo, J.L. and Khetarpal, R.K., 2008. Biosafety Regulations of Asia Pacific Countries. Asia-Pacific Association of Agricultural Research Institutions, Bangkok; Asia Pacific Consortium on Agricultural Biotechnology, New Delhi and Food and Agricultural Organization of the United Nations, Rome, P 108 + i-x.

Kannaiyan, S., 2006. Biotechnology and Biosafety. Workshop on Biosafety for Science Journalists-Inaugural Address- Dec-22,23-2006.

Sture, J., Whitby, S. and Perkins, D., 2013. Biosafety, biosecurity and internationally mandated regulatory regimes: compliance mechanisms for education and global health security. *Med Confl Surviv.*, 29: 289-321.

Usha Kiran Betha, 2019. ICAR-Indian Institute of Oilseeds Research, Rajendranagar, Hyderabad, India

WHO, 2020. Who Guidance on implementing regulatory requirements for biosafety and biosecurity in biomedical laboratories - a stepwise approach World Health Organization Geneva, 2020

7

International Standards for Phytosanitary Measures

International Standards for Phytosanitary Measures (ISPMs) are standards adopted by the Commission on Phytosanitary Measures (CPM), which is the governing body of the International Plant Protection Convention (IPPC). The first International Standard for Phytosanitary Measures (ISPM) was adopted in 1993.

As of March 2021, there are 44 adopted ISPMs (ISPM 30 being revoked), 29 Diagnostic Protocols and 39 Phytosanitary Treatments. These international standards protect sustainable agriculture and enhance global food security, protect the environment, forests and biodiversity and facilitate economic and trade development. International Standards for Phytosanitary Measures are prepared by the Secretariat of the International Plant Protection Convention as part of the United Nations Food and Agriculture Organization's global programme of policy and technical assistance in plant quarantine. This programme makes available to FAO Members and other interested parties these standards, guidelines and recommendations to achieve international harmonization of phytosanitary measures, with the aim to facilitate trade and avoid the use of unjustifiable measures as barriers to trade (Anonymous, 2005).

7.1 International Standards for Phytosanitary Measures (ISPMs)

ISPM No. 1 (1993) : *Principles of plant quarantine as related to international trade*

ISPM No. 2 (1995) : *Guidelines for pest risk analysis*

ISPM No. 3 (2005) : *Guidelines for the export, shipment, import and release of biological control agents and other beneficial organisms*

ISPM No. 4 (1995) : *Requirements for the establishment of pest free areas*

ISPM No. 5 (2005) : *Glossary of phytosanitary terms*

ISPM No. 6 (1997) : *Guidelines for surveillance*

ISPM No. 7 (1997) : *Export certification system*

ISPM No. 8 (1998) : *Determination of pest status in an area*

ISPM No. 9 (1998) : *Guidelines for pest eradication programmes*

ISPM No. 10 (1999) : *Requirements for the establishment of pest free places of production and pest free production sites*

ISPM No. 11 (2004) : *Pest risk analysis for quarantine pests, including analysis of environmental risks and living modified organisms*

SPM No. 12 (2001) : *Guidelines for phytosanitary certificates*

ISPM No. 13 (2001) : *Guidelines for the notification of non-compliance and emergency action*

ISPM No. 14 (2002): *The use of integrated measures in a systems approach for pest risk management*

ISPM No. 15 (2002): *Guidelines for regulating wood packaging material in international trade*

ISPM No. 16 (2002) : *Regulated non-quarantine pests: concept and application*

ISPM No. 17 (2002) : *Pest reporting*

ISPM No. 18 (2003): *Guidelines for the use of irradiation as a phytosanitary measure*

ISPM No. 19 (2003) : *Guidelines on lists of regulated pests*

ISPM No. 20 (2004) : *Guidelines for a phytosanitary import regulatory system*

ISPM No. 21 (2004): *Pest risk analysis for regulated non-quarantine pests*

ISPM No. 22 (2005) : *Requirements for the establishment of areas of low pest prevalence*

ISPM No. 23 (2005) : *Guidelines for inspection*

ISPM No. 24 (2005): *Guidelines for the determination and recognition of equivalence of phytosanitary measures*

One of the key activities of the IPPC is to establish international standards for phytosanitary measures (ISPMs). The Commission on Phytosanitary Measures (CPM) is the governing body of the IPPC and it has adopted a number of ISPMs that provide guidance to contracting parties in meeting the aims and obligations of the Convention. The intention of ISPMs is to harmonize phytosanitary measures for the purpose of facilitating international trade. ISPMs can cover a wide range of issues including; surveillance, pest risk analysis, establishment of pest free areas, export certification, phytosanitary certificates and pest reporting. The IPPC encourages adoption of these standards, but they only come into force once contracting (members) and non-contracting parties to establish requirements in national legislative instruments. Most standards are developed initially by regional or national plant protection organizations (NPPO and RPPO, respectively), but are often adopted by the IPPC if they have an international relevance. IPPC standards generally fall into three categories:

1) Reference standards, such as the Glossary of phytosanitary terms.
2) Conceptual standards, such as the Guidelines of pest risk analysis.

3) Specific standards, which typically directed at a specific pest or pathogen (e.g. surveillance for citrus canker).

7.1.1 National Plant Protection Organization (NPPO)

Article IV of the Convention requires that contracting parties, to the best of their ability, establish a National Plant Protection Organization (NPPO) which implements many of the services and functions specific to obligations under the IPPC. The NPPO usually administers a country's phytosanitary laws and regulations, and is responsible for delivering the following activities:

- Issuance of phytosanitary certificates,
- Conducting surveillance and inspection of pests and pathogens,
- Implementing control measures to prevent the spread of pests and pathogens (e.g. treatments, disinfection, or disinfestation),
- Protecting habitat and endangered areas,
- Conducting pest risk assessment and analysis,
- Ensuring the phytosanitary security of consignments to be exported,
- Designation and maintenance of pest free areas and areas of low pest presences

In order to facilitate the exchange of information between the IPPC and the NPPO's of contracting members, each country must designate an official point of contact. Information about the various NPPO's can be found at the IPPC website.

7.1.2 Regional Plant Protection Organization (RPPO)

A Regional Plant Protection Organization (RPPO) is an inter-governmental organization functioning as a coordinating body for National Plant Protection Organizations (NPPO) on a regional level. Not all contracting parties to the IPPC are members of RPPOs, nor are all members of RPPOs contracting parties to the IPPC. Moreover, certain contracting parties to the IPPC belong to more than one RPPO. Article IX of the Convention encourages contracting parties cooperate on a regional basis and form Regional Plant Protection Organizations (RPPOs). The RPPO's usually perform a coordinating function to gather and disseminate information as well as address technical issues of a regional nature. This can include promoting harmonized phytosanitary measures for controlling pests and preventing their introduction and/or spread.

There are currently 10 RPPOs (Anonymous, 2020.):

- Asia and Pacific Plant Protection Commission (APPPC)
- Caribbean Agricultural Health and Food Safety Agency (CAHFSA)
- Comunidad Andina (CAN)
- Comite de Sanidad Vegetal del Cono Sur (COSAVE)

- European and Mediterranean Plant Protection Organization (EPPO)
- Inter-African Phytosanitary Council (IAPSC)
- Near East Plant Protection Organization (NEPPO)
- North American Plant Protection Organization (NAPPO)
- Organismo Internacional Regional de Sanidad Agropecuaria (OIRSA)
- Pacific Plant Protection Organization (PPPO)

Article IX of the IPPC provides for RPPO contributions to various activities that achieve the objectives of the IPPC. It extends the responsibilities of RPPOs to specify cooperation with the IPPC Secretariat and the Commission for Phytosanitary Measures in developing international standards. The RPPOs therefore play an important role in the cooperative endeavour to implement the IPPC.

Functions of RPPOs

The functions of RPPOs are mostly laid down in the Article IX of the IPPC and include:

- Coordination and participation in activities among their NPPOs in order to promote and achieve the objectives of the IPPC
- Cooperation among regions for promoting harmonized phytosanitary measures
- Gathering and dissemination of information, in particular in relation with the IPPC
- Cooperation with the CPM and the IPPC Secretariat in developing and implementing international standards for phytosanitary measures.

Each RPPO has its own activities and programme.

How a new RPPO can request recognition from the CPM

The procedure for the recognition of new RPPOs was adopted by the ICPM-04 (2002).

Technical Consultations among RPPOs : Technical Consultations among Regional Plant Protection Organizations (TC-RPPOs) are regularly convened (annual basis) by the Secretary of the Commission. The purpose of the TC-RPPOs is to inter alia, encourage inter-regional cooperation in promoting harmonized phytosanitary measures, and the development and use of relevant international standards for phytosanitary measures. In addition to being a forum for RPPOs to consult as a group, the TC can support regional programmes under the IPPC and contribute to the work programme of the CPM.

7.1.3 Asia & Pacific Plant Protection Commission (APPPC)

History: The Plant Protection Agreement for the Asia and Pacific Region (formerly the Plant Protection Agreement for South-East Asia and Pacific Region) was approved by the 23rd Session of the FAO Council in November 1955 and entered into force on 2 July 1956. Amendments to this Agreement are:

i) Article I(a) approved by the 49th Session of the FAO Council in November 1967;

ii) Title of the Agreement and the name of the Committee approved by the 75th Session of the FAO Council in 1979;

iii) Article I(a) of the Agreement approved by the 84th Session of the FAO Council in November 1983;

iv) Articles II, III, IV and XIV concerning financial obligations approved by the 84th Session of the FAO Council in November 1983;

v) At the 117th Session of the FAO Council approved, in November 1999, two sets of amendments to the Agreement designed to bring the Agreement into line with the New Revised Text of the International Plant Protection Convention (IPPC), and the Agreement on the Application of Sanitary and Phytosanitary Measures (SPS Agreement), as well as with modern requirements for plant protection and to strengthen the Asia and Pacific Plant Protection Commission. The Council agreed that such amendments did not involve new obligations for the Contracting Governments and, therefore, pursuant to Article IX.4 of the Agreement, will come into force with respect to all Contracting Governments as from the thirtieth day after acceptance by two-thirds of Members.

Agreement: Plant Protection Agreement for the Asia and Pacific Region (currently enforced). Revision of the Agreement for the Asia and Pacific Region. There are two sets of amendment of the revised agreement. The second set of amendment providing for the deletion of the detailed measure of the South American Leaf Blight is not circulated until the Director General is notified by the APPPC Secretary that a satisfactory regional standard on the disease has been adopted by the Commission. The first set of amendments has been circulated to APPPC members for acceptance.

Membership: Twenty five countries are currently members of the Commission:

Australia, Bangladesh, Cambodia, China, Democratic People's Republic of Korea, Fiji, France, India, Indonesia, Laos, Malaysia, Myanmar, Nepal, New Zealand, Pakistan, Papua New Guinea, Philippines, Republic of Korea, Samoa (Western), Solomon Islands, Sri Lanka, Thailand, Timor-Leste, Tonga, Viet Nam

Structure and Organization: The Plant Protection Agreement for Asia and Pacific Region is an intergovernmental treaty and administered by the Asia and Pacific Plant Protection Commission. The Commission consists of representatives of all member countries and elects amongst them a Chairperson who serves for a period of two years. The Director-General of Food and Agriculture Organization appoints and provides the secretariat that coordinates, organizes and follows up the work of the Commission. The Commission, according to its provisions convenes at least once for every two years and opens for participation to all member countries. For implementation of the Agreement, the Commission has established three standing committees, namely:

- Standing Committee on plant quarantine
- Standing Committee on integrated pest management
- Standing Committee on pesticides.

Members: Australia, Bangladesh, Cambodia, China, Democratic People's Republic of Korea, Fiji, France, India, Indonesia, Laos, Malaysia, Myanmar, Nepal, New Zealand, Pakistan, Papua New Guinea, Philippines, Republic of Korea, Samoa (Western), Solomon Islands, Sri Lanka, Thailand, Timor-Leste, Tonga, Viet Nam.

7.1.4 Caribbean Agricultural Health and Food Safety Agency (CAHFSA)

History: The Caribbean Agricultural Health and Food Safety Agency (CAHFSA), was created as an Inter-Governmental Organization by the signing of an Agreement among Member States of the Caribbean Community (CARICOM). Member States of CARICOM originally signed this Agreement establishing CAHFSA in Roseau, Dominica, on March 12, 2010. The Revised CAHFSA Agreement adopted on February 25, 2011 in St. George's, Grenada replaced the originally signed Agreement. In October 2014, CAHFSA became operational with the appointment of a Chief Executive Officer. Located in Paramaribo, Suriname, CAHFSA is governed by a Board of Directors comprising one representative of each Member State representing the specialized areas of focus of CAHFSA as well as a representative from the CARICOM Secretariat who is represented on the Board as an observer. A Chief Executive Officer and three (3) Department Heads manage the day-to-day activities of the Agency. The services offered by CAHFSA falls under three (3) main areas: Food Safety, Plant Health, and Animal Health.

Mandate: CAHFSA is mandated to

- perform a coordinating and organizing role in the establishment of an effective and efficient regional sanitary and phytosanitary (SPS) regime;
- execute on behalf of Member States such SPS actions and activities that can be more effectively and efficiently executed through a regional mechanism.

Vision: "To be a leading regional institution of international repute in the application of Agricultural Health and Food Safety Systems."

Mission: "To enhance regional development in agricultural health and food safety through the application of SPS Measures that meets the expectation of all stakeholders and contribute to the welfare of our citizens."

Primary Objectives

- To provide regional and national support to the Community in the establishment, management and operations of its national agricultural health and food safety systems as they relate to the SPS measures of the SPS Agreement;
- To execute on behalf of those countries such actions and activities that can be more effectively and efficiently executed through a regional mechanism.

Specific Objectives

- To provide a mechanism for the coordination and integration of technical support to stakeholders by relevant regional and international organizations;
- To provide a framework to continuously monitor and evaluate national agricultural health and food safety programmes and provide technical support
- To facilitate the development and use of regional, as well as international SPS standards, measures and guidelines;
- to strengthen the policy and legal framework for SPS related issues;
- to facilitate the harmonization of technical procedures in relation to SPS issues
- to provide a framework for the conduct of tests and the strengthening of laboratory services
- to provide a mechanism for achieving regional consensus on SPS issues that can be represented in international fora;
- to provide an effective regional mechanism to respond rapidly to emergencies and emerging issues;
- to establish a mechanism to facilitate the provision of technical assistance to stakeholders in the private sector by regional and international organizations to improve productivity and competitiveness in the market place.

Membership

- Full Membership of CAHFSA is opened to all members of the Caribbean Community. This includes Antigua and Barbuda. Bahamas. Barbados. Belize. Dominica. Grenada. Guyana. Haiti. Jamaica. Montserrat. Saint Lucia. St Vincent and the Grenadines. Suriname. Trinidad and Tobago.
- Associate Membership of CAHFSA is opened to Associate Members of the Community. These include Anguilla, Bermuda, British Virgin Islands, Cayman Islands, Turks and Caicos Islands.

Comunidad Andina : (Established 1969)

Members : Bolivia, Colombia, Ecuador, Peru

Comite Regional de Sanidad Vegetal del Cono Sur (COSAVE): (Established 1980)

Members: Argentina, Bolivia, Brazil, Chile, Paraguay, Perú, Uruguay

7.1.5 European & Mediterranean Plant Protection Organization (EPPO)

EPPO is an intergovernmental organization responsible for cooperation and harmonization in plant protection within the European and Mediterranean region. Under the International Plant Protection Convention (IPPC), EPPO is the regional plant protection organization (RPPO) for Europe. Founded in 1951 by 15 European countries, EPPO now has 51 members, covering almost all countries of the European and Mediterranean region. Its objectives are to develop an international strategy against the introduction and spread of pests that damage cultivated and wild plants, in natural and agricultural ecosystems (including invasive alien plants); Encourage harmonization of phytosanitary regulations and all other areas of official plant protection action; Promote the use of modern, safe, and effective pest control methods; and provide a documentation service on plant protection. As a RPPO, EPPO also participates in global discussions on plant health organized by FAO and the IPPC Secretariat. EPPO has produced a large number of regional standards and publications on plant pests, phytosanitary regulations, and plant protection products.

The EPPO Website https://www.eppo.int includes further information on the EPPO structure, planned meetings, the list of EPPO Standards, the links to Databases (including EPPO Global Database and PQR) and to publications. Since 2014, EPPO is hosting Euphresco, a network of organizations funding and coordinating research in plant health.

Members: Albania, Algeria, Austria, Azerbaijan, Belarus, Belgium, Bosnia and Herzegovina, Bulgaria, Croatia, Cyprus, Czech Republic, Denmark, Estonia, Finland, France, Georgia, Germany, Greece, Guernsey, Hungary, Ireland, Israel, Italy, Jersey, Jordan, Kazakhstan, Kyrgyzstan, Latvia, Lithuania, Luxemburg, Macedonia, Malta, Moldova, Morocco, Netherlands, Norway, Poland, Portugal, Romania, Russia, Serbia, Slovakia, Slovenia, Spain, Sweden, Switzerland, Tunisia, Turkey, Ukraine, United Kingdom, Uzbekistan.

Au Inter-African Phytosanitary Council (IAPSC) (Established 1954): Through the Maputo declaration, the profile of the Inter-African Phytosanitary Council (IAPSC) was designed as follows.

Vision: The Maputo declaration recognises the need to secure a common and effective action to prevent the spread and introduction of pests of plants and plant products as well as the need to promote appropriate measures for their control and therefore re echoes the IPPC in its mission to the IAPSC as follows:

- To prevent the introduction and spread of pests which attack and damage crops and forests in Africa.
- To develop a common strategy against the introduction and spread of pests particularly through the harmonisation of phytosanitary legislation
- To ensure co-operation and a harmonised approach in all areas of plant protection where governments take official measures (registration of pesticides, certification of plant materials, accreditation of people who apply pesticides etc)
- To provide a documentation service for provision and exchange of information in all areas of its activities. Activities as outlined in the Maputo Declaration:
- Plant protection information management
- Development of strategies against the introduction and spread of plant pests
- Promotion of safe and sustainable plant protection techniques
- Enlighten member states on the implications of the WTO-SPS Agreement on international agricultural trade
- Capacity building among Member states in phytosanitary and plant protection activities. Core Functions:
- Development and management of information to serve Africa and international Plant protection Organisations (IPPOs)
- Harmonization of Phytosanitary regulations in Africa
- Development of regional strategies against the introduction and spread of plant pests (insects, plant pathogens, weeds etc)
- Training of various cadres of NPPOs in Pest Risk Analysis(PRA), Phytosanitary inspections and treatment, field inspection and certification , laboratory diagnoses, pest surveillance and monitoring etc.

In line with the above activities, the following is the status of IAPSC as the African RPPO for the period ending June 2004. Members All African Union (AU) members i.e. all African countries, except Morocco.

7.1.6 Near East Plant Protection Organization (NEPPO)

The lack of a Regional Plant Protection Organization in the near East Region has negative impact on the regional collaboration in the area of plant protection, in particular in the development of a regional strategy to monitor and control the trans-boundry plant pests (including diseases and weeds). Pest outbreaks are an increasing trend and threat to the region in the last twenty years (particularly in the light of increasing international trade in agricultural plants and plant products), which require close attention from the region at minimum in the form of information exchange and the initiation of an effective regional monitoring and control strategy.

In response to a request made by the Near East Regional Commission on Agriculture at its Third Session held in Nicosia (Cyprus) from 11 – 15 September 1989, and following a recommendation made by a technical consultation held in Rome from 14- 16 April 1992, A conference of Plenipotentiaries on the establishment of the Near East Plant Protection Organization (NEPPO), organized by FAO, was held in Rabat Morocco 16-18 Feb.1993. The conference was attended by 17 countries from the FAO Near East region signed.

The NEPPO entered into force on January 9, 2009. The first Governing Council was held in Rabat (Kingdom of Morocco) in October, 2010, during which NEPPO established its structure, adopt rules and procedures, elected its President and Vice-president and nominated its Executive Director.

NEPPO was formally recognized as a Regional Plant Protection Organization (RPPO) under the IPPC on 23 March 2012. There are now ten RPPOs under the IPPC and Article IX of the New Revised Text of the IPPC which states RPPOs:

1. Shall function as the coordinating bodies in the areas covered, shall participate in various activities to achieve the objectives of this Convention and, where appropriate, shall gather and disseminate information, and
2. Shall cooperate with the Secretary in achieving the objectives of the Convention and, where appropriate, cooperate with the Secretary and the Commission in developing international standards.

Member States: Algeria, Egypt, Iraq, Jordan, Libya, Malta, Morocco, Pakistan, Sudan, Syria and Tunisia.

Signed but not ratified: Mauritania and Yemen.

7.1.7 North American Plant Protection Organization (NAPPO)

(Established 1976)

Members : Canada, Mexico, United States of America

7.1.8 Organismo Internacional Regional de Sanidad Agropecuaria (OIRSA)

Members: Belize, Costa Rica, República Dominicana, El Salvador, Guatemala, Honduras, México, Nicaragua, Panamá.

7.1.9 Pacific Plant Protection Organization (PPPO)

The Pacific Plant Protection Organisation (PPPO) was founded in October 1994 by the South Pacific Conference (now Pacific Community Conference) at its 34th Session in Port Vila, Vanuatu. The Land Resources Division of the Secretariat of the Pacific Community is the Secretariat of the PPPO and runs the day-to-day affairs of the organisation. The PPPO has the responsibility of coordinating harmonization of phytosanitary measures, foster co-operation in plant protection and other phytosanitary matters among and between Members and countries and organisations outside the Pacific region. The PPPO also act for the members in developing contacts with, and where appropriate providing input into, other global and regional organisations that have authority in such matters.

PPPO is one of the Regional Plant Protection Organisations recognised by the International Plant Protection Convention and exists to provide advice on phytosanitary measures in order to facilitate trade without jeopardizing the plant health status of the importing Members and countries and in particular: Its main objectives are,

- to ensure that the views and concerns of Pacific members are adquately taken into account in the development and implementation of global phytosanitary measure
- to assist in the development and implementation of effective and justified phytosanitary measure
- to provide a framework for regional and global co-operation in phytosanitary matters consistent with international principles for trade in plants and plant products
- to facilitate the flow of information among Members and with other regional plant protection organizations and
- to collaborate with the SPC Plant Protection Service (now as part of the Land Resources Division) on specific issues including pesticides and integrated pest management.

Members : All Members of the Pacific Community are Members of the Pacific Plant Protection Organisation. The Pacific Community consists of twenty seven (27) members including twenty two (22) Pacific Island Countries and Territories (PICTS) and 5 founding members. Pacific Island Countries and Territories Members are: American Samoa, Cook Islands, Federated States of Micronesia (FSM), Fiji islands, French Polynesia, Guam, Kiribati, Marshall Islands, Nauru, New Caledonia, Niue, Northern Mariana Islands (CNMI), Palau, Papua New Guinea (PNG), Pitcairn Islands, Samoa, Solomon Islands, Tokelau, Tonga, Tuvalu, Vanuatu, and Wallis and Futuna.

The four remaining founding countries are: Australia, France, New Zealand, and the United States of America. The United Kingdom withdrew at the beginning of 1996 from SPC (at the time of The South Pacific Commission), rejoined in 1998 and withdrew again in January 2005.

Application: International Standards for Phytosanitary Measures (ISPMs) are adopted by contracting parties to the IPPC through the Commission on Phytosanitary Measures. ISPMs are the standards, guidelines and recommendations recognized as the basis for phytosanitary measures applied by Members of the World Trade Organization under the Agreement on the Application of Sanitary and Phytosanitary Measures. Non-contracting parties to the IPPC are encouraged to observe these standards.

Review and Amendment: International Standards for Phytosanitary Measures are subject to periodic review and amendment. The next review date for each standard is five years from their endorsement, or such other date as may be agreed upon by the Commission on Phytosanitary Measures. Standards will be updated and republished as necessary. Standard holders should ensure that the current edition of standards is being used.

Distribution: International Standards for Phytosanitary Measures are distributed by the Secretariat of the International Plant Protection Convention to IPPC contracting parties, plus the Executive/Technical Secretariats of the Regional Plant Protection Organizations, viz., Asia and Pacific Plant Protection Commission, Caribbean Plant Protection Commission, Comité Regional de Sanidad Vegetal para el Cono Sur, Comunidad Andina, European and Mediterranean Plant Protection Organization, Inter-African Phytosanitary Council, North American Plant Protection Organization, Organismo Internacional Regional de Sanidad Agropecuaria and Pacific Plant Protection Organization

Notes on the Publication: International Standards for Phytosanitary Measures (ISPMs) were originally produced as separate booklets. The current book was produced by the IPPC Secretariat according to the decision made by the Interim Commission for Phytosanitary Measures at its Seventh session in 2005 (ICPM-7). It compiles all ISPMs without modification to their content, except in relation to the section Definitions, as decided by ICPM-7. The book is also available on line on the IPPC website at https://www.ippc.int. In addition, individual standards are available on the IPPC website as extracts from the book. Official control should be established or recognized by the national government or the NPPO under appropriate legislative authority; be performed, managed, supervised or, at minimum, audited/reviewed by the NPPO ; have enforcement assured by the national government or the NPPO; be modified, terminated or lose official recognition by the national government or the NPPO. Responsibility and accountability for official control programmes rests with the national government. Agencies other than the NPPO may be responsible for aspects of official control programmes, and certain aspects of official control programmes may be the responsibility of sub-national authorities or the private sector. The NPPO should be fully aware of all aspects of official control programmes in their country.

7.2 Principles of Plant Quarantine as Related to International Trade General Principles

International standards for phytosanitary measures are prepared by the Secretariat of the International Plant Protection Convention as part of the United Nations Food and Agriculture Organization's global programme of policy and technical assistance in plant quarantine (Anonymous, 1995). This programme makes available to FAO Members and other interested parties these standards, guidelines and recommendations to achieve international harmonization of phytosanitary measures, with the aim to facilitate trade and avoid the use of unjustifiable measures as barriers to trade.

Review and amendment: International standards for phytosanitary measures are subject to periodic review and amendment. The next review date for this standard is December 1996, or such other date as may be agreed upon by the Commission on Phytosanitary Measures. Standards will be updated and republished as necessary. Standard holders should ensure that the current edition of this standard is being used.

Distribution: International standards for phytosanitary measures are distributed by the Secretariat of the International Plant Protection Convention to all FAO Members, plus the Executive/Technical Secretariats of the Regional Plant Protection Organizations.

Scope: This reference standard describes the general and specific principles of plant quarantine as related to international trade.

Aim: The primary aim in formulating the following principles is to facilitate the process of developing international standards for plant quarantine. It is envisaged that implementation of these principles by the relevant phytosanitary authorities will result in the reduction or elimination of the use of unjustifiable phytosanitary measures as barriers to trade. Furthermore, in addition to general principles there are others specific to particular areas of quarantine activity. The general principles indicate the process of development of phytosanitary measures as applicable to international commerce. These general principles should be read as a single entity and not interpreted individually. The specific principles either directly support the International Plant Protection Convention (IPPC) or are related to particular procedures within the plant quarantine system. This relationship is indicated in the tabulation. It is expected that the principles will be subject to continuing review and should reflect changing quarantine concepts and technologies.

General Principles

1. **Sovereignty :** With the aim of preventing the introduction of quarantine pests into their territories, it is recognized that countries may exercise the sovereign right to utilize phytosanitary measures to regulate the entry of plants and plant products and other materials capable of harbouring plant pests.
2. **Necessity:** Countries shall institute restrictive measures only where such measures are made necessary by phytosanitary considerations, to prevent the introduction of quarantine pests.
3. **Minimal impact :** Phytosanitary measures shall be consistent with the pest risk involved, and shall represent the least restrictive measures available which result in the minimum impediment to the international movement of people, commodities and conveyances.

4. **Modification :** As conditions change, and as new facts become available, phytosanitary measures shall be modified promptly, either by inclusion of prohibitions, restrictions or requirements necessary for their success, or by removal of those found to be unnecessary.
5. **Transparency :** Countries shall publish and disseminate phytosanitary prohibitions, restrictions and requirements and, on request, make available the rationale for such measures.
6. **Harmonization :** Phytosanitary measures shall be based, whenever possible, on international standards, guidelines and recommendations, developed within the framework of the IPPC.
7. **Equivalence:** Countries shall recognize as being equivalent those phytosanitary measures that are not identical but which have the same effect.
8. **Dispute settlement:** It is preferable that any dispute between two countries regarding phytosanitary measures be resolved at a technical bilateral level. If such a solution cannot be achieved within a reasonable period of time, further action may be undertaken by means of a multilateral settlement system.

Specific Principles

9. **Cooperation:** Countries shall cooperate to prevent the spread and introduction of quarantine pests, and to promote measures for their official control.
10. **Technical authority:** Countries shall provide an official Plant Protection Organization.
11. **Risk analysis :** To determine which pests are quarantine pests and the strength of the measures to be taken against them, countries shall use pest risk analysis methods based on biological and economic evidence and, wherever possible, follow procedures developed within the framework of the IPPC.
12. **Managed risk:** Because some risk of the introduction of a quarantine pest always exists, countries shall agree to a policy of risk management when formulating phytosanitary measures.
13. **Pest free areas :** Countries shall recognize the status of areas in which a specific pest does not occur. On request, the countries in whose territories the pest free areas lie shall demonstrate this status based, where available, on procedures developed within the framework of the IPPC.
14. **Emergency action:** Countries may, in the face of a new and/or unexpected phytosanitary situation, take immediate emergency measures on the basis of a preliminary pest risk analysis. Such emergency measures shall be

temporary in their application, and their validity will be subjected to a detailed pest risk analysis as soon as possible.

15. **Notification of non-compliance:** Importing countries shall promptly inform exporting countries of any non-compliance with phytosanitary prohibitions, restrictions or requirements.
16. **Non-discrimination:** Phytosanitary measures shall be applied without discrimination between countries of the same phytosanitary status, if such countries can demonstrate that they apply identical or equivalent phytosanitary measures in pest management. In the case of a quarantine pest within a country, measures shall be applied without discrimination between domestic and imported consignments.

The Agreement on the Application of Sanitary and Phytosanitary Measures, also known as the SPS Agreement or just SPS, is an international treaty of the World Trade Organization (WTO). It was negotiated during the Uruguay Round of the General Agreement on Tariffs and Trade (GATT), and entered into force with the establishment of the WTO at the beginning of 1995 (Timothy J. Miano, 2006). Broadly, the sanitary and phytosanitary ("SPS") measures covered by the agreement are those aimed at the protection of human, animal or plant life or health from certain risks.

Under the SPS agreement, the WTO sets constraints on member-states' policies relating to food safety (bacterial contaminants, pesticides, inspection and labelling) as well as animal and plant health (phytosanitation) with respect to imported pests and diseases. There are 3 standards organizations who set standards that WTO members should base their SPS methodologies on. As provided for in Article 3, they are the *Codex Alimentarius* Commission (Codex), World Organization for Animal Health (OIE) and the Secretariat of the International Plant Protection Convention (IPPC).

The SPS agreement is closely linked to the Agreement on Technical Barriers to Trade, which was signed in the same year and has similar goals. The TBT Emerged from the Tokyo Round of WTO negotiations and was negotiated with the aim of ensuring non-discrimination in the adoption and implementation of technical regulations and standards.

History and framework: As GATT's preliminary focus had been lowering tariffs, the framework that preceded the SPS Agreement was not adequately equipped to deal with the problems of non-tariff barriers (NTBs) to trade and the need for an independent agreement addressing this became critical.[4] The SPS Agreement is an ambitious attempt to deal with NTBs arising from cross-national differences in technical standards without diminishing governments prerogative to implement measures to guard against diseases and pests (Tim Buthe, 2008).

Main Provisions

- Article 1 – General Provisions - Outlines the application of the Agreement.
- Annex A.1 – Definition of SPS measures.
- Article 2 – Basic Rights and Obligations. Article 2.2 - requires measures to be based on sufficient scientific analysis. Article 2.3 - states that Members shall ensure that their sanitary and phytosanitary measures do not arbitrarily or unjustifiably discriminate between Members where identical or similar conditions prevail, including between their own territory and that of other Members. Sanitary and phytosanitary measures shall not be applied in a manner which would constitute a disguised restriction on international trade.
- Article 3 – Harmonization. Article 3.1- To harmonize sanitary and phytosanitary measures on as wide a basis as possible, Members shall base their sanitary or phytosanitary measures on international standards, guidelines or recommendations, where they exist, except as otherwise provided for in this Agreement, and in particular in paragraph 3. Article 3.3 – allows Members to implement SPS measures higher than if they were basing them on international standards where there is a scientific justification or the Member determines the measure to be appropriate in accordance with 5.1-5.8.
- Annex A.3 – outlines the standard-setting bodies.
- Article 5 – Risk Assessment and Determination of the Appropriate Level of SPS Protection. Article 5.1 - Members shall ensure that their sanitary or phytosanitary measures are based on an assessment, as appropriate to the circumstances, of the risks to human, animal or plant life or health, taking into account risk assessment techniques developed by the relevant international organizations.
- Annex A.4 – outlines risk assessment process.
- Article 5.5 - each Member shall avoid arbitrary or unjustifiable distinctions in the levels it considers to be appropriate in different situations, if such distinctions result in discrimination or a disguised restriction on international trade. Article 5.7– echoes the 'Precautionary Principle' where there is no science available with which to justify a measure.

Cases: Some of the most important WTO 'cases' regarding the implementation of SPS measures include:

- EC – Hormones (Beef Hormone Dispute) (1998)
- Japan – Agricultural Products (1999)
- Australia – Salmon (1999)
- Japan – Apples (2003)

Genetically modified organisms: In 2003, the United States challenged a number of EU laws restricting the importation of Genetically Modified Organisms (GMOs) in a dispute known as EC-Biotech, arguing they are "unjustifiable" and illegal under SPS agreement. In May 2006, the WTO's dispute resolution panel issued a complex ruling which took issue with some aspects of the EU's regulation of GMOs, but dismissed many of the claims made by the USA (Anonymous, 2006).

Interaction with other World Trade Organization instruments : While Article 1.5 of the TBT precludes the inclusion of SPS measures from its ambit, in EC-Biotech, the panel recognised that situations could arise where a measure is only partly an SPS measure, and in those cases, the SPS part of the measure will be considered under the SPS Agreement. If a measure conforms with SPS, under Article 2.4 of the SPS Agreement, it is assumed that the measure falls within the scope of GATT, Article XX(b).

7.3 Sanitary and Phytosanitary Measures (SPS)

Sanitary and phytosanitary (SPS) measures are quarantine and biosecurity measures which applied to protect human, animal or plant life or health from risks arising from the introduction, establishment and spread of pests and diseases and from risks arising from additives, toxins and contaminants in food and feed.

These measures are governed by the World Trade Organization's (WTO) Agreement on the Application of Sanitary and Phytosanitary Measures (the SPS agreement), and its Committee of Sanitary and Phytosanitary Measures (the SPS committee).

The SPS agreement: The SPS agreement provides a framework of rules to guide WTO members in the development, adoption and enforcement of sanitary (human or animal life or health) and phytosanitary (plant life or health) measures which may affect trade. All WTO members are required to meet and uphold the principles and obligations of the SPS agreement.

The SPS agreement provides WTO members with the right to use SPS measures to protect human, animal or plant life or health. Each WTO member is entitled to maintain a level of protection it considers appropriate to protect human, animal or plant life or health within its territory. This is called the appropriate level of protection (ALOP).

The right to adopt SPS measures is accompanied by obligations aimed at minimising negative impacts of SPS measures on international trade. The basic obligations are that SPS measures must:

- be applied only to the extent necessary to protect human, animal or plant life or health and not be more trade restrictive than necessary
- be based on scientific principles and not maintained without sufficient scientific evidence, and

- not constitute arbitrary or unjustifiable treatment or a disguised restriction on international trade.

The SPS committee : The SPS committee oversees the implementation of the SPS agreement and provides a forum for discussion on animal and plant health and food safety measures affecting trade. The SPS Committee meets three times a year at the WTO headquarters in Geneva. It provides a forum for all WTO members to discuss the implementation of the SPS agreement, including sharing their experiences, raising concerns about other members' activities and developing further guidance on implementing the SPS agreement.

The use of international standard setting bodies: The SPS Agreement encourages WTO members to harmonise their measures by basing SPS measures on agreed international standards. These international standards are developed by organisations referred to as the 'three sisters'. The 'three sisters' develop international standards, recommendations and guidelines for plant and animal health and food safety. They are the:

- International Plant Protection Convention (IPPC)
- World Organisation for Animal Health (Office International des Epizooties, OIE)
- *Codex Alimentarius* Commission (Codex)

The department and DFAT work together to coordinate Australia's input to international SPS policy and influence its development through active participation in the three sisters (IPPC, OIE and Codex). For further information on the department's involvement in the three sisters, see the departments webpages on the IPPC, OIE and Codex.

Understanding the WTO Agreement on SPS measures: The Agreement on the application of Sanitary and Phytosanitary Measures (the "SPS Agreement") entered into force with the establishment of the World Trade Organization on 1 January 1995. It concerns the application of food safety and animal and plant health regulations. This introduction discusses the text of the SPS Agreement as it appears in the Final Act of the Uruguay Round of Multilateral Trade Negotiations, signed in Marrakesh on 15 April 1994. This agreement and others contained in the Final Act, along with the General Agreement on Tariffs and Trade as amended (GATT 1994), are part of the treaty which established the World Trade Organization (WTO). The WTO superseded the GATT as the umbrella organization for international trade. The WTO Secretariat has prepared this text to assist public understanding of the SPS Agreement. It is not intended to provide legal interpretation of the agreement.

Bibliography

Anonymous, 1995. International Standards for Phytosanitary Measures. Secretariat of the International Plant Protection Convention Food and Agriculture Organization of the United Nations Rome,

Anonymous, 2005. International Standards for Phytosanitary Measures. Produced by the Secretariat of the International Plant Protection Convention Food and Agriculture Organization of the United Nations Rome, 2006

Anonymous, 2006. Panel Report, European Communities. Measures Affecting the Approval and Marketing of Biotech Products, WTO Doc WT/DS291, WT/DS292/R, WT/DS293/R

Anonymous, 2020. International Plant Protection Convention Secretariat, FAO of UN, Italy.

Tim Buthe, 2008. 'The globalization of health and safety standards: delegation of regulatory authority in the SPS Agreement of the 1994 agreement establishing the World Trade Organization' 71(1) Law and Contemporary Problems, 219-255

Timothy J. Miano, 2006. "Understanding and Applying International Infectious Disease Law: U.N. Regulations during an H5N1 Avian Flu Epidemic" *Chi-Kent J. Int'l & Comp.*, 26: 42-48

8

Pest Risk Analysis, Risk Management Models and Pest Information System

8.1 Pest Risk Analysis

Pest Risk Analysis (PRA) is a form of risk analysis conducted by regulatory plant health authorities to identify the appropriate phytosanitary measures required to protect plant resources against new or emerging pests and regulated pests of plants or plant products. Specifically pest risk analysis is a term used within the International Plant Protection Conventions (IPPC) (Article 2.1) and is defined within the glossary of phytosanitary terms as "the process of evaluating biological or other scientific and economic evidence to determine whether an organism is a pest, whether it should be regulated, and the strength of any phytosanitary measures to be taken against it. In a phytosanitary context, the term plant pest, or simply pest, refers to any species, strain or biotype of plant, animal or pathogenic agent injurious to plants or plant products and includes plant pathogenic fungi, bacteria, fungus-like organisms, viruses, nematodes and virus like organisms as well as insects, mites and weeds.

Pest Risk Analysis is mandatory before any plant or plant materials being permitted to be imported into the country. The Import Permit issuing authorities shall issue import permits for commodities specified in Schedule-V, Schedule-VI and Schedule-VII of PQ Order, 2003 for which Pest Risk Analysis has already been done. Any new commodities which are not covered under the above mentioned Schedules, the import permit can not issued. Also for the import of commodity from a country other than listed in the above schedules the import permit cannot be issued. An Importer who intend to import a new commodity or from a country not covered under the list shall send a specific Pest Risk Analysis Request Form to the Plant Protection Adviser to the Government of India, Directorate of Plant Protection, Quarantine & Storage, N.H.-IV., Faridabad-121 001, Haryana.

Objectives of PRA assessment

- To identify and assess risks to plants from harmful exotic organisms that can spread internationally or be introduced to new areas.
- To identify management measures & support risk management decision making

- To comply with WTO rules (SPS Agreement)
- Is a public good (society benefits)

PRA complexity, application and scientific input is given hereunder.

Type of PRA	Main Applications	Inputs to assessment
Rapid PRAs (days)	Following new interceptions & new outbreaks	Rapid qualitative evaluation of the literature, online datasets and other evidence
PRAs (weeks / months)	To modify EU legislation (with EU partners?)	Detailed qualitative evaluation of the literature, online datasets and other evidence
Detailed analysis of PRA components (weeks / months) research projects	To resolve major uncertainties To support emergency action To guard against challenges to the PRA	Detailed quantitative assessment with modelling, & mapping of, e.g. interception data, climatic suitability, spread and potential impacts

The role of Pest Risk Analysis: The practice of risk analysis in plant quarantine is known as Pest Risk Analysis. The process is closely linked to the international regulatory framework formed by the World Trade Organization Agreement on the Application of Sanitary and Phytosanitary Measures (the WTO-SPS Agreement) and the International Plant Protection Convention (IPPC) with associated international standards for phytosanitary measures (ISPMs) (Burgman *et.al.*, 2014). Concepts, terms and definitions in this framework provide the foundation for international harmonization of the processes used by National Plant Protection Organizations (NPPOs) to support regulatory policies based on scientific principles and evidence. The methodologies described in standards distinguish risk assessment from risk management and focus analyses on the likelihood of pest entry, establishment, and spread, the consequences of pest introduction, the efficacy and feasibility of possible measures for avoiding, reducing, or eliminating the risk, and the uncertainty associated with all aspects of the analysis. Various qualitative and quantitative methods may be used in these analyses depending on the nature of the issue, data, and available resources. They include structured methods for combining language-based risk estimates, subjective assessments that use ratings scales, point scoring systems, logical rules, Bayesian networks, and Monte Carlo simulation. Examples from several countries serve to demonstrate a wide range of approaches and illustrate key aspects of the pest risk analysis process. Consequence assessments should include economic, social and environmental criteria. All methods should deal explicitly and thoroughly with uncertainty, be as precise as possible in defining terms and operations, and be consistent with the rules of probability.

8.1.2 Not Authorized Pending Pest Risk Analysis (NAPPRA)

Plants for planting can carry a wide variety of pests that are more likely to become established in the United States because they are already on a suitable host. In some cases, the plants themselves are the pest. To ensure U.S. import regulations provide adequate protection against the risk posed by plants for planting, the U.S. Department of Agriculture's Animal and Plant Health Inspection Service (APHIS) established a new regulated category called not authorized pending pest risk analysis (NAPPRA). NAPPRA allows APHIS to more fully protect U.S. agriculture from foreign pests while minimizing adverse economic and trade impacts. Before NAPPRA, APHIS' plants for planting regulations (also known as Q37) categorized imported plants as either prohibited (not allowed) or restricted (allowed under certain conditions). The regulations did not require a pest risk analysis prior to the importation of a new taxonomic group of plants. This differed from APHIS' fruits and vegetables regulations (Q56) where the importation of regulated articles is prohibited until APHIS completes a pest risk analysis.

The NAPPRA Process : If scientific evidence indicates that a taxon of plants for planting is a quarantine pest or a host of a quarantine pest and has little or no recent import history, APHIS will publish a proposed notice in the Federal Register proposing the taxon as NAPPRA. The notice will cite the scientific evidence we considered in making this determination, and give stakeholders the opportunity to comment on our determination. If the comments received do not lead us to revise the determination, the taxon will be added to the NAPPRA list. A plant taxon is added to NAPPRA for the listed pest and for all other quarantine pests for which the taxon is a host. This process allows us to take prompt action in response to evidence that the importation of a taxon of plants for planting may pose a risk while allowing public participation. Limited quantities of material may be imported under a controlled import permit for experimental, therapeutic or developmental purposes.

Requesting a Pest Risk Analysis : Importers who wish to import plants or plant materials on the NAPPRA list must submit a request per the instructions included in the U.S. Code of Federal Regulations: 7 CFR 319.5 (d). Upon receipt of a completed request, APHIS will develop the Pest Risk Analysis (PRA). Based on the PRA results, we will either remove the taxon from the NAPPRA list from the country or countries for which we had conducted the PRA, and then allow its importation subject to general requirements, allow its importation subject to specific restrictions, or continue to prohibit its importation.

Importing small quantities of plant taxa on the NAPPRA list : Importers who wish to import small quantities of plants or plant material on the NAPPRA list for experimental, therapeutic, or developmental purposes may apply for a controlled import permit (CIP, also referred to as PPQ Form 588).

Federal register publications

- Notice of New NAPPRA Candidates, Round 3 - *For Public Comment* - Plants for Planting Whose Importation is Not Authorized Pending Pest Risk Analysis; Data Sheets for Taxa of Plants for Planting that are Quarantine

Pests or Hosts of Quarantine Pests [Comment period closed January 24, 2020]

- Final Notice, Round 2 - Plants for Planting Whose Importation Is Not Authorized Pending Pest Risk Analysis; Notice of Addition of Taxa of Plants for Planting To List of Taxa Whose Importation Is Not Authorized Pending Pest Risk Analysis [Effective June 19, 2017]
- Extension of Approval of Plants for Planting Information Collection. Docket APHIS-2014-0038 [Published June 5, 2014]
- Notice of New NAPPRA Candidates, Round 2 - *For Public Comment* - Plants for Planting Whose Importation is Not Authorized Pending Pest Risk Analysis; Data Sheets for Taxa of Plants for Planting that are Quarantine Pests or Hosts of Quarantine Pests [Comment period closed July 5, 2013]
- Final Notice, Round 1 - Plants for Planting Whose Importation Is Not Authorized Pending Pest Risk Analysis; Notice of Addition of Taxa of Plants for Planting To List of Taxa Whose Importation Is Not Authorized Pending Pest Risk Analysis [Effective May 20, 2013]
- Final Rule: Importation of Plants for Planting; Establishing a Category of Plants for Planting Not Authorized for Importation Pending Pest Risk Analysis [Effective June 27, 2011]

NAPPRA lists

- Quarantine Pest Plants
  - Round 1 – Effective May 20, 2013
  - Round 2 – Effective June 19, 2017
  - Round 3 – Proposed on November 25, 2019
- Hosts of Quarantine Pests
  - Round 1 – Effective May 20, 2013
  - Round 2 – Effective June 19, 2017
  - Round 3 – Proposed on November 25, 2019

Pest datasheets

- Quarantine Pest Plant Datasheets
  - Round 1 – Final notice published April 18, 2013
  - Round 2 – Final notice published June 19, 2017
  - Round 3 – Noticed published November 25, 2019
- Hosts of Quarantine Pests Datasheets
  - Round 1 – Final notice published April 18, 2013
  - Round 2 – Final notice published June 19, 2017
  - Round 3 – Noticed published November 25, 2019

Supporting documents : Regulatory Impact Analysis; Importation of Plants for Planting; Establishing a Category of Plants for Planting Not Authorized for Importation Pending Pest Risk Analysis.

8.1.3 Pest Risk Analysis and International Plant Protection Convention

Introduced plant pests can lower crop yields and have environmental impacts. The spread of plant pests from one geographical area to another is an issue of international concern. The principal international agreement aimed at addressing the spread of plant pests through international trade is the International Plant Protection Convention, a multilateral treaty for international cooperation in plant protection aimed at preventing the spread of pests of plants and plant products, and promoting appropriate measures for their control (IPPC, Article I.1) (Pimentel, *et.al.* 2005) In accordance with the WTO Sanitary and Phytosanitary Agreement the IPPC aims to protect plants while limiting interference with international trade A key principle of the IPPC is that contracting parties (signatories) provide 'technical justification' to support phytosanitary decision making affecting trade. The IPPC recognises pest risk analysis as the appropriate format for such technical justification. The responsibility for conducting pest risk analysis sits within government, specifically within a country's National Plant Protection Organization (NPPO) and comes as an obligation when countries become contracting parties to the IPPC (IPPC Article IV, 2a). Devorshak (2012) describes the principles of pest risk analysis, how analyses can be performed and the use of pest risk analysis in regulatory plant protection. A general guide to the principles of pest risk analysis for plant pests and a description of some of the problems and difficulties that may be encountered when undertaking such analyses are included in text by Ebbels (2003) which also covers wider plant health issues.

PRA documentation requirements : The IPPC and the principle of "transparency" (ISPM No. 1: Principles of plant quarantine as related to international trade) require that countries should, on request, make available the rationale for phytosanitary requirements. The whole process from initiation to pest risk management should be sufficiently documented so that when a review or a dispute arises, the sources of information and rationale used in reaching the management decision can be clearly demonstrated. The main elements of documentation are purpose for the PRA, pest, pest list, pathways, PRA area, endangered area, sources of information, categorized pest list, conclusions of risk assessment, probability, consequences, risk management, options identified and options selected.

DEFRA's (Department for Environment, Food and Rural Affairs) approach to Pest Risk Analysis:

i. Plant pests include insects, other invertebrates, bacteria, fungi, viruses and other pathogens which affect the health of plants by feeding on them or causing disease.

ii. Pest Risk Analysis (PRA) is defined under the International Plant Protection Convention as "The process of evaluating biological or other scientific and economic evidence to determine whether a pest should be regulated and the strength of any phytosanitary [plant health] measures to be taken against it". PRAs may also be conducted on commodity imports to determine whether they provide a pathway for pests to enter the importing country or area.
iii. There are international standards that set out what to consider when conducting a PRA . PRAs vary in length from simple expert judgments ("does not feed on plants" or "tropical, no host plants outdoors in Europe") to long documents with assessment of biological factors, climate maps, and estimated costs and benefits of the measures needed to keep a pest out of an area, or to eradicate an outbreak. Different schemes are being developed for PRA. The UK uses both its own short scheme and a more detailed scheme from the European and Mediterranean Plant Protection Organization (EPPO). In most cases, the PRAs published on the Defra web site provide summaries of more detailed analyses.
iv. PRAs are written by plant health scientists to inform decisions by plant health policy makers. They are more technical than the information we prepare for the general public, but more accessible than most scientific papers. They only give a snapshot of the analysis at a particular time, and although we do revise them, they may go out of date as new information comes to light.
v. Specific treatments may be mentioned in a PRA as examples of control measures which are applied in areas of the world where the pest is established. These should not be regarded as recommendations: indeed the treatments may not be approved for use in the UK. All suspect findings of new pests in England or Wales should be reported to the Plant Health and Seeds Inspectorate (HQ tel 01904 455174).
vi. For some organisms the conclusion of the PRA is that the risk is low and statutory action by the plant health service to exclude or eradicate the organism is not warranted. However it is an offence under section 14 of the Wildlife and Countryside Act 1981 for any person to release or allow to escape into the wild any animal which is of a kind which is not ordinarily resident in and is not a regular visitor to Great Britain in a wild state, or which is listed on Part I of Schedule 9 to the Act. The release does not have to be deliberate or intentional, as this offence is one of strict liability.
vii. All risk analyses have to cope with areas of uncertainty where evidence is lacking, inconclusive or contradictory. For organisms that do not occur here our experts have to use their judgement in extrapolating from available data to assess how a pest might behave under our conditions on our crops and wild plants. If there is a wide range of uncertainty about the risk we may

commission research to fill gaps in the risk analysis. Until more is known we take a precautionary approach.

viii. Defra's PRAs focus on the risk to the UK and the European Community. Rules on imports and movement of plants are co-ordinated in the EC's Standing Committee on Plant Health and set out in the Plant Health Directive. The committee takes account of the risks from the pest, and the cost of risk management measures, across the whole of the EC. When the risk is confined to one area there is provision in the Directive for action to be restricted to that area by recognising it as a "protected zone".

ix. This is a very brief summary of Defra's approach to Pest Risk Analysis.

8.1.4 Action plan for Pest Risk Analysis

Pest risk analysis plays a key role in the new Order for plant quarantine in India. An action plan for pest risk analysis was drawn up, to take effect from December 2003. A major feature of the plan is the establishment of a national pest risk analysis unit. The action plan includes organizing PRA training, establishing working groups and holding a workshop attended by national and international experts to prioritize crops and commodities for pest risk analysis. Some 36 commodities (Table 15) were selected for which a pest database is under development. Detailed pest risk analyses for these commodities have begun, with the aim of completing the pest risk analysis for 13 commodities each year.

Table 15: India's priority crops for pest risk analysis

Strawberry	Banana	Kiwi
Musk melon	Watermelon	Pears
Mandarin	Cashew nut	Apple
Grape	Citrus fruits	Lentil
Red beans	Chickpea	Jute
Black gram	Green gram	Cotton
Wheat	Rice	Barley
Maize	Baby corn	Pearl millet
Sorghum	Lettuce	Garlic
Broccoli	Potato	Chinese cabbage
Mustard	Sunflower	Safflower
Linseed	Castor	Rape seed

The EPPO pest risk assessment scheme: The IPPC standards on pest risk analysis are useful general bases to the assessment of pest risks, containing all relevant and necessary steps and being recognized by the WTO via the SPS Agreement (WTO, 1994). Nevertheless, their applicability can be enhanced when transformed into a questionnaire. EPPO supplies a user-friendly questionnaire (EPPO, 1997) to assess the risks of organisms that are harmful to plants and to apply pest risk analysis in the framework of the IPPC. This scheme provides technical justification for the regulation of certain pests for the EPPO region (or parts of it) as the PRA area. It is based on ISPM 2, addressing environmental risks of plant pests only generally. The structure as well as the use of the EPPO scheme is described below and its adaptation to the revised ISPM 11, in particular to its applicability to plants as pests.

Structure of the EPPO PRA scheme : The (more or less) simple, clearly arranged scheme includes a sequence of questions for deciding whether an organism could present a pest risk and gives detailed instructions for the several stages of pest risk analysis for quarantine pests, viz., initiation, pest categorization, probability of introduction (entry and establishment) , economic impact assessment and final evaluation. In the initiation stage, the reasons for performing a pest risk analysis have to be given, the identity of the organism has to be indicated and the PRA area (i.e. the region for which the risk posed by the organism is to be assessed) has to be defined. Then follows a qualitative assessment to determine, if the organism could be a pest and present a risk in the PRA area (section A). Geographical and regulatory criteria have to be considered. An estimation is made of whether the organism would be able to establish in the PRA area and whether there is potential for economic, environmental and social damage and if the organism does not fulfill certain criteria, the assessment could be terminated (Anonymous, 2003).

8.1.5 PRA-tools, Resources and Key Challenges

The international standard for phytosanitary measures ISPM 11 [2001]: Pest risk analysis for quarantine pests (and its revisions) provides a clear description of the procedures to be followed in conducting pest risk analyses but, apart from statements such as "climatic modeling systems may be used", does not give further guidance on the tools and resources that can support the pest risk analyst. This paper summarizes the principal tools and resources available for the pest risk assessment component of pest risk analysis, giving examples of how they can be used. It also highlights the principal challenges for the future.

Of the international standards for phytosanitary measures, ISPM 11 [2001]: Pest risk analysis for quarantine pests (and its subsequent revisions to its current form, ISPM 11 [2004]: Pest risk analysis for quarantine pests, including analysis of environmental risks and living modified organisms) describes the procedures to be followed when conducting pest risk analyses, highlighting the key factors that need to be considered. Regional and national PRA schemes based on ISPM 11 have been developed to provide a logical structure for the analyst to follow. Examples are those by the European and Mediterranean Plant Protection Organization (Anonymous, 1997.) and the United States Department of Agriculture (Anonymous, 2000). However, there is little guidance for risk analysts when searching for the data required and the tools and other resources that can ease their task. As a result, the production of pest risk analyses may appear to be daunting.

Assessing entry potential : Although a variety of information is needed to assess the extent to which a pest will be able to pass through all the stages of the pathway from the origin to the PRA area, data on trade pathways and interceptions (detections in consignments) are most important. Unfortunately, no readily available compilations of these data sets exist and, for most purposes, specific enquiries for unpublished data have to be made. MacLeod and Baker (1998) analysed the European interception data for Thrips palmi, highlighted the role played by orchids from Thailand and showed how detections decreased once action had been taken to prevent pests travelling along this pathway.

Assessing establishment potential : Baker (2002) has reviewed the data required and the techniques that can be used both for assessing establishment potential and for predicting the limits to the distribution of quarantine pests once established in a country. Essentially, ecological factors (such as the suitability of the ecological environment, presence of hosts and natural enemies) and factors intrinsic to the pest itself (such as its reproductive strategy and genetic adaptability) should be considered. Even when a pest's responses to the abiotic and biotic environment are poorly understood and when little can be obtained from the literature concerning the intrinsic factors, some judgements can still be made.

If the area of origin and the host plants are known, climates in the area of origin can be compared with climates in the area under threat and the distribution of host plants determined. The computer program CLIMEX (Sutherst *et.al.*,1999) is particularly useful in predicting potential distribution based on current distribution even when an organism's responses to climate are unknown. Baker *et.al.*, (2003) showed how CLIMEX models predicting development based on temperature, maps of maize distribution, current climate data and climate change data can be used to predict the potential distribution of Diabrotica virgifera virgifera in the United Kingdom. Computer mapping software, known as a geographic information system (GIS), provides an extremely powerful method for analysing the different data sets and displaying the results.

Assessing economic, environmental and social impacts : Assessing the magnitude of the consequences for plants in the PRA area after establishment of a pest ideally requires knowledge of the pest's impacts in its current range, sufficient biological data to predict its spread and population dynamics in the PRA area coupled with financial, economic, environmental and social data for the enterprises, ecosystems and people likely to be affected. As for establishment potential, assessments can still be made even if data are lacking, for example by using expert opinion. Morgan and MacLeod (1996) showed how a population model for *Bemisia tabaci*, a transmission model for Tomato yellow leaf curl virus and a gross margin budget for glasshouse tomatoes in the United Kingdom could be combined to estimate the financial consequence of the introduction of the virus and its vector.

Estimating the environmental consequences of pest introductions is more challenging. Assessing the potential environmental impacts in Europe of *Phytophthora ramorum*, the pathogen responsible for sudden oak death in

California and Oregon, is particularly difficult. Key factors, such as those influencing infection and host damage, are poorly known (Jones *et.al.*, 2003). While research is being carried out, key areas for conservation where susceptible hosts grow in climatic conditions most closely matching those in affected areas of the United States can be identified, and maps made and used to help target surveys and generate contingency plans.

8.2 Pest Risk Management

The conclusions from pest risk assessment are used to decide whether risk management is required and the strength of measures to be used. Since zero-risk is not a reasonable option, the guiding principle for risk management should be to manage risk to achieve the required degree of safety that can be justified and is feasible within the limits of available options and resources. Pest risk management (in the analytical sense) is the process of identifying ways to react to a perceived risk, evaluating the efficacy of these actions, and identifying the most appropriate options. The uncertainty noted in the assessments of economic consequences and probability of introduction should also be considered and included in the selection of a pest management option. For a quarantine pest, pest risk management is the process of evaluation and selection of options to reduce the risk of introduction and spread of the pest. Conclusions from the pest risk assessment (Stage 2) are used to support decisions regarding the level of risk presented by the pest. If a pest is judged to present an unacceptable risk then phytosanitary measures should be identified that will reduce the risk to an acceptable level. Phytosanitary measures should accord with IPPC principles of necessity, managed risk, minimal impact, transparency, harmonization, non-discrimination and technical justification (Anonymous, 2006). ISPM 11 provides more information about each stage of pest risk analysis for quarantine pests.

i. **Level of risk :** The principle of "managed risk" (ISPM Pub. No. 1: Principles of plant quarantine as related to international trade) states that: "Because some risk of introduction of a quarantine pest always exists, countries shall agree to a policy of risk management when formulating phytosanitary measures." In implementing this principle, countries should decide what level of risk is acceptable to them. The acceptable level of risk may be expressed in a number of ways, such as reference to existing phytosanitary requirements, indexed to estimated economic losses, expressed on a scale of risk tolerance and compared with the level of risk accepted by other countries.

Level of detail required: The level of detail in a pest risk analysis will be limited by the amount and quality of information available, the tools, and time available before a decision is required. Quantitative and qualitative techniques are used in pest risk analysis but pest risk analysis need only be as complex as is required by the circumstances to support a phytosanitary decision and provide the necessary technical justification to defend decisions regarding phytosanitary measures.

Nevertheless, a pest risk analysis should be based on sound science, be transparent and consistent with other pest risk analyses conducted by the NPPO.

8.3 Risk Assessment Models and Approaches

Pest Risk Analyses (PRAs) are conducted worldwide to decide whether and how exotic plant pests should be regulated to prevent invasion. There is an increasing demand for science-based risk mapping in PRA. Spread plays a key role in determining the potential distribution of pests, but there is no suitable spread modelling tool available for pest risk analysts. Existing models are species specific, biologically and technically complex, and data hungry (Robinet *et al.*, 2012). A set of four simple and generic spread models that can be parameterized with limited data were presented. Simulations with these models generate maps of the potential expansion of an invasive species at continental scale. The models have one to three biological parameters. They differ in whether they treat spatial processes implicitly or explicitly, and in whether they consider pest density or pest presence/absence only. The four models represent four complementary perspectives on the process of invasion and, because they have different initial conditions, they can be considered as alternative scenarios. All models take into account habitat distribution and climate. An application of each of the four models was presented to the western corn rootworm, Diabrotica virgifera virgifera, using historic data on its spread in Europe. Further tests as proof of concept were conducted with a broad range of taxa (insects, nematodes, plants, and plant pathogens). Pest risk analysts, the intended model users, found the model outputs to be generally credible and useful. The estimation of parameters from data requires insights into population dynamics theory, and this requires guidance. If used appropriately, these generic spread models provide a transparent and objective tool for evaluating the potential spread of pests in PRAs. Further work is needed to validate models, build familiarity in the user community and create a database of species parameters to help realize their potential in PRA practice.

Fig.10 illustrates the classification of the models used for calculating scenarios of pest spread. heoretical models have been developed to quantify spread based on reaction-diffusion models. When these models do not fit the observed spread pattern because of long distance dispersal, stratified dispersal models that combine long distance jumps with local spread can be used. Some specific spread models have been developed to simulate the potential spread of a species taking into account human-assisted dispersal. These models address details of the life cycle and dispersal mechanisms, and they take considerable time and effort to develop, parameterise and test. It is not realistic to request the development of species-specific complex models in real world PRAs because risk assessors are generally not modellers, and they lack the time, resources and training to do it. Instead, there is a need for generic modelling tools in PRAs that can be used by risk assessors to capture the main processes driving the invasion process of alien species (Robinet *et al.*, 2012). While developments towards generalization and more unified application of complex modelling platforms in spread modelling for PRA are underway, there is as yet no modelling toolbox that risk assessors may use to conduct rapid appraisals of pest spread in the context of a PRA.

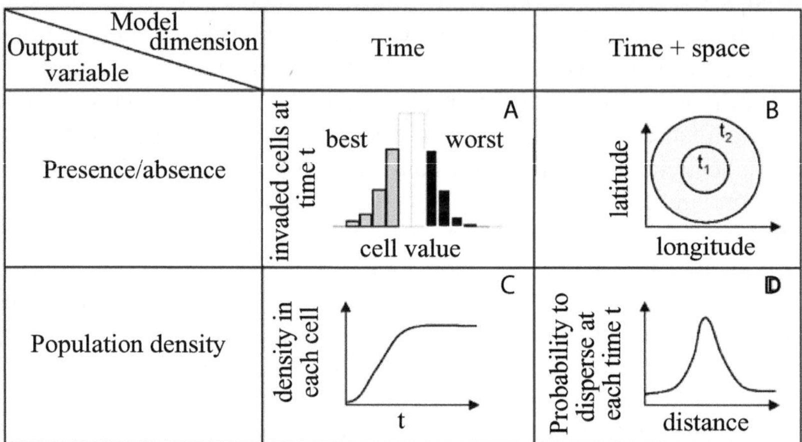

Fig. 10: Classification of the models used for calculating scenarios of pest spread

Models A and B are models for occupancy of cells (presence/absence) on the PRA area. Models C and D are models for pest density. They calculate pest abundance within cells at given times t. Within each class, one model considers the process of spread only in the time dimension (A, C) while the other model considers processes in both time and space (B, D).

Assessing the likelihood and magnitude of spread is one of the cornerstones of pest risk analysis (PRA), and is usually based on qualitative expert judgment. This paper proposes a suite of simple ecological models to support risk assessors who also wish to estimate the rate and extent of spread, e.g. when modelling the dynamics of invasion and the economic impacts that may result. Models are based on simple ecological principles, such as logistic growth, radial range expansion and population growth in combination with dispersal. Different models capture different perspectives of the spread process, being based on pest density or simply presence /absence, and they compare spatially explicit and spatially implicit approaches. A case study on Diabrotica virgifera virgifera is provided for illustration. The suite of models requires further development and testing with the risk assessment com-munity building familiarity before their more general application in PRA.

Guidance on quantitative pest risk assessment: This Guidance describes a two-phase approach for a fit-for-purpose method for the assessment of plant pest risk in the territory of the EU. Phase one consists of pest categorization to determine whether the pest has the characteristics of a quarantine pest or those of a regulated non-quarantine pest for the area of the EU (Michael Jeger *et.al.*, 2018). Phase two consists of pest risk assessment, which may be requested by the risk managers following the pest categorization results. This Guidance provides a template for pest categorization and describes in detail the use of modelling and expert knowledge elicitation to conduct a pest risk assessment. The Guidance provides support and a framework for assessors to provide quantitative estimates, together with associated uncertainties, regarding the entry, establishment, spread and impact of plant pests in the EU. The Guidance allows the effectiveness of risk reducing options (RROs) to be quantitatively assessed as an integral part of the assessment framework. A list of RROs is provided. A two-tiered approach is proposed for the use of expert knowledge elicitation and modelling. Depending on data and resources available and the needs of risk managers, pest entry, establishment, spread and impact steps may be assessed directly, using weight of evidence and quantitative expert judgement (first tier), or they may be elaborated in substeps using quantitative models (second tier). An example of an application of the first tier approach is provided. Guidance is provided on how to derive models of appropriate complexity to conduct a second tier assessment. Each assessment is operationalized using Monte Carlo simulations that can compare scenarios for relevant factors, e.g. with or without RROs. This document provides guidance on how to compare scenarios to draw conclusions on the magnitude of pest risks and the effectiveness of RROs and on how to communicate assessment results.

8.3.1 Formal Model

Model scope: The quantitative framework described in this section aims to provide a flexible framework for assessing quantitatively the risk of entry, establishment, spread and impact. The risk assessment area may comprise the whole EU, or in the case of a protected zone organism, the protected zone from which the organism is absent or under official control (Michael Jeger *et.al.*, 2018). Specific choices are made to simplify the assessment process.The spatial extent of the assessment is the whole EU if the organism does not occur in the EU, but could be limited to a protected zone within the EU if the organism already occurs in the EU. The temporal extent depends on the organism and the ToR and can be decided accordingly by the risk assessor. It could span time periods varying from ~ 5 to ~ 50 years.

Establishment: Establishment starts with the arrival of the pest in the territory and the transfer of inoculum or individuals to a host. The end-point is a pest population that will persist for the foreseeable future (Fig. 12). For the risk assessment, establishment is quantified in terms of the number of founder populations that are established. Founder populations are local populations of the pest, e.g. one or a few infected or infested trees in an orchard, a patch of nematodes in a field, a cluster of infested trees in a forest. They are localised in the sense that outbreak control would still be feasible (Michael Jeger *et.al.*, 2018). A delay is possible (from a few to many cycles of multiplication) between the initial introduction and transfer of a pest and the establishment of a founder population that will persist indefinitely and produce offspring populations (spread). This Guidance does not prescribe specific methods for assessing the establishment potential for a pest. There are many spatially explicit mapping approaches that may be used to show and estimate the area in which establishment may occur ('the area of potential establishment') and illustrate gradations in the suitability of areas according to their climate, presence of hosts and other relevant factors.Process-based (i.e. mechanistic) demographic models can provide meaningful information for assessing the establishment. They can produce a spatially explicit representation of an index that is a direct measure of the population abundance. This allows the description of the area of potential establishment as well as a point-based analysis of the habitat suitability. Demographic model are suited not only for assessing the establishment but also for the impact as the population abundance represents the main driver determining the pest impact on the cultivated plants and on the environment.

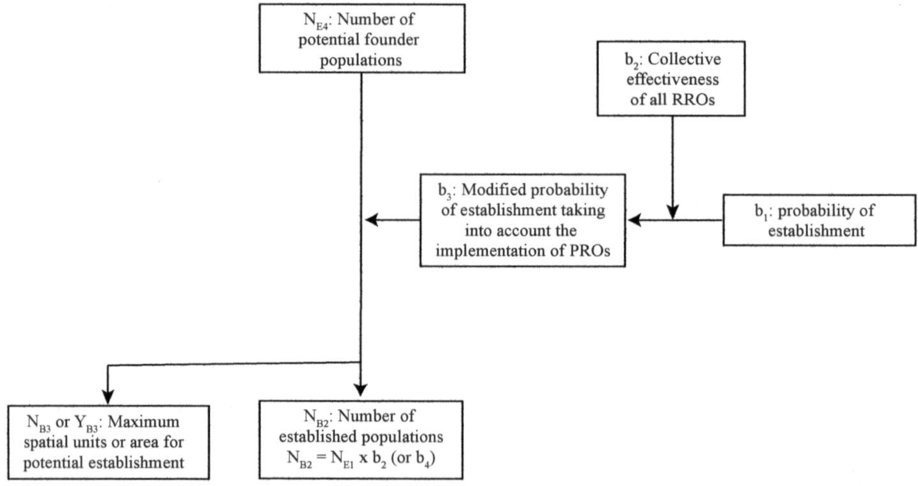

Fig. 12: Information flow in a conceptual model of the establishment step

Modelling establishment and parts of establishment in a spatially explicit manner is very informative for risk managers because it clarifies in which areas establishment and impact may occur. Such maps may be interpreted in a conditional way, 'if entry in this region happens, then the probability of establishment will be very high' (accompanied by quantification). Coupling of entry and establishment in a spatially explicit manner is not required to allow decision-making by risk managers that is spatially informed.

Spread: Spread is movement of a pest into a new area where it can persist. Essentially, the spread process is therefore the same process as entry +

establishment, with the difference that the term entry is normally defined as movement crossing a border of risk assessment area, whereas spread occurs within this area, without crossing an external border (Fig. 13) (Michael Jeger *et.al.*, 2018). An inventory of spread models was produced. These authors provided an overview of 468 models for plant pest spread and dispersal from the literature and assessed strengths and weaknesses of these models for risk assessment. A decision support scheme was provided to help the assessors find the most suitable model. A set of simple models was proposed and noted the epidemiological network modeling, which is potentially a powerful and mechanistically sound way to calculate spread processes. However, network modelling requires detailed information on trade pathways within the EU and this information is not officially collected, although it may be (partly) available in specific industries.

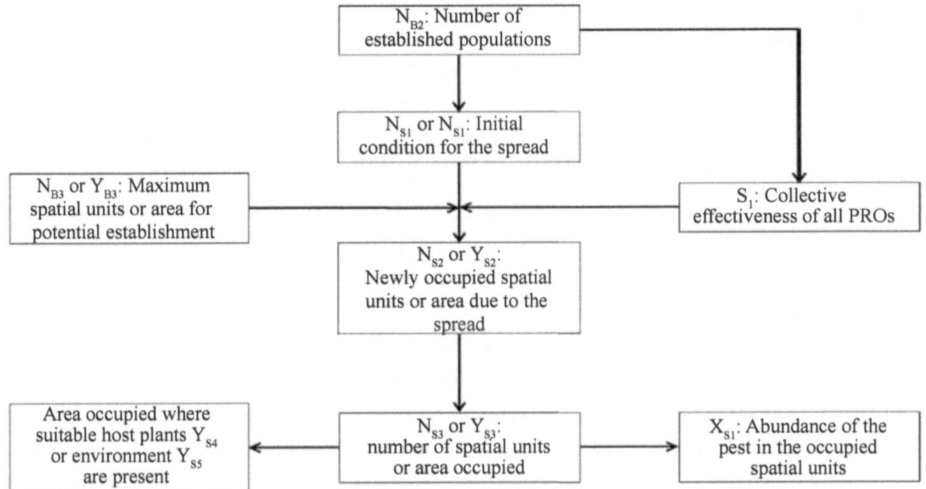

Fig. 13 : Information flow in a conceptual model of the spread step based on occupancy of spatial units, such as NUTS regions

8.3.2 Climatic and Ecological Niche Models

The most common and readily applied approaches to predicting the risk of invasive species occupying sites across a large region rely on biogeographical distribution models. These models are based on information about the biophysical factors that limit where a species can survive. Such models are known as bioclimatic envelope models, biogeographical distribution models, and (ecological) niche models (Michael Jeger *et.al.*, 2018). Such models are generally correlative and may be either statistically based or rule based. As applied to invasive species, such approaches typically attempt to map which parts of a region are suitable for the invading species, and suitability is typically based on habitat requirements. For pests and pathogens, the simplest approach is to map the presence or absence of suitable hosts. Such maps are typically developed from available regional data sets, which often provide relevant but not necessarily ideal data for a particular invasive species. Such maps may be useful for strategic planning at the regional scale, but may be of limited use for managing specific areas presuming that the managers of those areas already know where different species occur.

Niche models typically identify habitats for invasive species based on records of their presence at known locations. Such records can be obtained from museum collections such as herbaria, but currently, only 5 to 10 percentof such data are available in electronic form worldwide. To define the niche or bioclimatic envelope, biophysical data for each such location are often extracted from regional databases, usually in a GIS. The most important distinction among such approaches is whether they use absence data in addition to presence data. In other words, whether locations where the invasive species does not occur (absence) are used to define biophysical conditions that are outside of the niche. Either approach is problematic for invasive species because, typically, they have not yet occupied all possible sites. Thus, sites where the species doesn't occur may not necessarily provide information about the species niche or requirements; instead, those may be sites that the invasive species haven't yet reached.

Presence and absence data can be obtained from the native region of the invasive species, but the species may have a different niche in the part of the world it is invading, as compared with its native region. However, use of data from the native region may be the only reasonable choice for species that have not become widely established in the United States. Even so, there may be substantial uncertainty in such predictions until a species becomes widely established. For example, an analysis of purple loosestrife (a common invasive species in Eastern United States wetland areas) determined that a reliable prediction of the current nonnative distribution in North America was only possible 150 years after initial establishment.

Many variations of the niche approach are used to predict the niche of the invasive species including

1. Simple ranges for factors based on mean climatic variables such as the widely used BIOCLIM and DOMAIN models.
2. Fuzzy rather than crisp calculations of the niche (Robertson and others 2004).
3. The use of spatial statistical techniques and newer computational approaches, such as genetic algorithms and support vector machines.

8.3.3 GARP Niche Modeling Approach

In this family of approaches implemented in a software tool, the potential range of invasive species is predicted based on point data from the species native home range and spatial data including mean annual temperature, rainfall and elevation

(Michael Jeger *et.al.*, 2018). This approach shares many features with other approaches to predict ecological niches based on bioclimatic data, including climate envelope modeling and other methods for niche modeling. All of these approaches assume that bioclimatic predictor variables (for example, mean annual temperature and precipitation) control the native distribution of an invasive species, and these factors will also control the potential distribution in the United States This technique differs from others because it uses a machine learning method (also known as artificial intelligence) named Genetic Algorithm for Rule-Set Prediction (GARP). Based on only 15 to 20 records of locations of a species from its native home range (species input data), the method can predict the potential distribution (home range, or niche) of a species. This approach has been used by its developers to model the niche of both invasive species and noninvasive species. The user needs to provide species input data of known points where the species has been found in its native region. These data should be well distributed throughout the species native range and need to be georeferenced. The user also needs to provide environmental data covering the entire area for which predictions are desired, including mean annual temperature and precipitation (modeled surfaces). Potentially, many other input data could be used such as remote sensing images, but they might need to be available for both native region and the analysis region.

The software used is desktopGARP, which can be downloaded from the following Web site: http://nhm. ku.edu/desktopgarp/. The user selects a type of inferential tool, such as logistic regression, or bioclimatic rules. The input data are then divided into training data and validation data. The software generates pseudodata via resampling, and then iteratively tries a large number of rule sets, continuing either until there is no further improvement in the predictions, or 1,000 iterations. The output from the model is a map of species niche as presence/absence, with some confidence values. Modeling may be done for either counties or for grid cells (pixels). The primary prediction is whether a county or a pixel is contained in the species potential (fundamental) niche. A measure of likelihood is generated by using multiple models, and assigning higher likelihoods to counties or pixels predicted to be included in the niche by multiple models.

In the following citations, only one predicted value is made per county, although the approach could be extended for finer grain analyses if input data are available at finer scales. The methodology and its use to predict the distribution of four alien plant species in North America for a single point in time (the fundamental niche) are described in the references reviewed herein. Invasive plant species analyzed to date include Hydrilla, Russian olive (*Elaeagnus angustifolia* L.), *Sericea lespedeza* (*Lespedeza cuneata* (Dum.-Cours.) G. Don), and garlic mustard (*Alliaria petiolata* (Bieb.) Cavara & Grande) (Peterson and others 2003a). To predict the spread of Asian long-horned beetle, the GARP approach has been combined with a spatial model of spread originally developed for forest fire. The GARP approach has several strengths for the regional risk analysis of invasive species, which are as follows:

1. It has been applied to a number of taxa of invasive and noninvasive species in the United States and elsewhere.
2. A freely available software tool has been developed that implements this approach.
3. Data requirements for this approach are modest.

Most weaknesses of the GARP approach are shared by all niche modeling approaches, which include:

1. Not all of the stages of the invasion process are modeled.
2. Only presence or absence of a species is predicted, not effects of invasive species.
3. Results may be biased, depending on thesource of data and the use of pseudo-absence data. Other approaches such as support vector machines and generalized additive model (GAM) approaches may be less biased and provide more optimal statistical solutions.

8.3.4 FHTET National Risk Mapping Approach

This approach is also a family of related approaches to predict tree mortality risk owing to an invasive insect or pathogen based on expert opinion, forest inventory data, and other GIS data (Michael Jeger *et.al.*, 2018). Specifically, predictions are made of the potential basal area loss of susceptible tree species owing to an invasive insect or pathogen. The location of suitable host species is interpolated using inverse-distance weighting based on forest inventory data. A multi-criteria risk ranking model is developed based on expert opinion about the factors that influence pest or pathogen establishment, spread, and tree mortality. An iterative process is used to develop risk maps, so the experts and analysts can alter the weighting of difference factors to adjust the maps to match expert opinion. This approach has been used to predict the potential effect of oak wilt in the North Central States and of wood wasp (Sirex noctilio Fabricius) throughout the conterminous United States.

Use of this approach requires one or more experts on the pest or pathogen, expertise in the use of FIA data, and expertise in GIS software. The spatial scope is the conterminous Unites States for a single time period. Required software includes ArcView 3.x, Spatial Analyst ModelBuilder (ESRI, Inc.), and IDRISI 32 (a raster GIS software package). Model output includes maps of predicted occurrence based on (1) hosts known to be susceptible and (2) hosts suspected to be susceptible. For regional and national risk analysis, the approach of mapping factors that influence a stressor and then combining these factors with weightings derived from expert opinion are intuitively appealing and fairly common. This flexible, iterative expert opinion-based approach can be used for virtually any pest or pathogen, and a risk map can be generated fairly quickly because the system is already in place. Other strengths of this approach include the use of national FIA data and the quantification of potential damage in terms of tree mortality. However, the flexible expert opinion-based approach is also a weakness because it is so open-ended, subjective, and difficult to validate. To date, it does not appear that an attempt has been made to determine which environmental factors were actually associated with pest presence, or to quantify uncertainties in GIS layers or predictions. In contrast, a statistical inference approach that made quantitative predictions of pest occurrence would be more useful because it could be better tested against validation data.

8.3.5 Meentemeyer Sudden Oak Death Approach

Meentemeyer and others (2004) used a rule-based function to predict spread of sudden oak death pathogen distributions in grid cells (30 by 30 m) throughout California. A prediction was made of the likelihood of presence of the disease based on rules derived from expert opinion and published data on plant species susceptibility, pathogen reproduction, and host climate (Michael Jeger *et.al.*, 2018). This method is focused on evaluating a single risk, the probability of oaks on a given site being infected by P. ramorum. More specifically, the method begins with mapping five predictor variables in a GIS and then using a set of rules to determine the risk of infection based on these predictor variables. The predictor variables are host species index, precipitation, maximum temperature, minimum temperature, and relative humidity. Host species index is weighted three times as strongly as precipitation and maximum temperature, which in turn are weighted twice as strongly as relative humidity and minimum temperature.

Input data for the model include host susceptibility, pathogen reproduction, and host climate suitability. Like many modeling approaches, this approach requires expertise in GIS and database analysis. The model output is in the form of a map with estimated risk of occurrence of the pathogen at a single time period – movement of the pathogen is not modeled. The spatial scope includes all of California, and the map unit is landscape cell (30 by 30 m). The approach uses the CALVEG database (USDA Forest Service RSL 2003) for vegetation alliance and presence of P. ramorum and the Parameter-elevation Regressions on Independent Slopes Model (PRISM) for elevation-based regression extrapolations from base weather stations for climate data, which are available for the conterminous United States.

The method meets the criterion of calculating the risk of detrimental environmental effect by mapping the probability of pathogen occurrence in each forest grid-cell and could be extended to predict the presence of pathogens in smaller regions or pixels. But the focus is assessment of effects over a region, specifically bioregions, rather than at all points within a region.

8.3.6 Nowak Host Range Approach

This approach predicts potential home range of an (invasive) insect or pathogen of trees by modeling the location of suitable host species based on forest inventory data (Michael Jeger *et.al.*, 2018). A model of urban forests (UFORE) is used to predict urban forest composition based on data from a limited number of cities in the United States. Predictions are also made of the amount of tree cover that could be lost owing to tree death and the costs of replacing killed trees. A simple model of spread (moving outward at a constant rate from one location) was used to predict the length of time required for invasion to occur in each major city. This approach has been used to predict the potential effect of Asian long-horned beetle throughout all urban areas in the United States and preliminary predictions have been made for nonurban areas. Model output is in the form of maps of predicted occurrence based on (1) hosts known to be susceptible and (2) hosts suspected to be susceptible. The model has been used at the scale of the conterminous United States for a single point in time.

One strength of this approach is the use of FIA data in conjunction with a model that has been used for many years. Another strength of this approach is the quantification of damage in terms of economic losses of urban trees. For urban trees, such economic losses are quite high, though for wildlands they will be much lower for an individual tree and much harder to estimate for a forested region. A limitation for regional risk assessment and management is that the focus of the model is urban areas. Another limitation, typical of most niche modeling efforts, is that not all steps in the process of invasive dispersion and reproduction are modeled, and that predictions are primarily of the potential host range of the pathogen, not of effects of the pathogen other than economic losses owing to the death of urban trees.

8.3.7 USGS and NASA Invasive Species MODIS Regression

In this approach, a logistic regression is developed to predict the suitability of each 1-km pixel as habitat for tamarisk throughout the conterminous United States (Morisette and others 2006). Various ground surveys of tamarisk occurrence were integrated into a single database as presence or absence of tamarisk (Michael Jeger *et.al.*, 2018). Land cover, normalized difference vegetation index (NDVI), and enhanced vegetation index (EVI) were derived from MODIS data products. A discrete Fourier transform was used to model a constant amplitude yearly sine wave to each pixel, and the mean, amplitude, and phase of both NDVI and EVI were used as potential predictor variables along with a fitted parameter for each land cover class in a logistic regression model to predict the likelihood of habitat suitable for tamarisk. The ground data were split into a training set to fit the model (67 percent of data) and a validation set (33 percent of data). The best model included land cover, and seasonal variability in NDVI and EVI. The proportion of correctly predicted observations using a threshold of 0.5 was 0.90. The main model inputs are MODIS data and surveys of tamarisk presence. Because it is a regression procedure, many other input data could be used, such as human population density, trail networks, air temperature, etc. The main model output is a relative ranking of the likelihood of suitable habitat for an invasive species.

This general approach would be useful for regional assessments because it uses remotely sensed data that cover the entire conterminous United States. However, for each invasive species, a large database of ground survey data is required. If FIA or other systematic survey data could be used for this purpose, that would make the approach useful for many more invasive species. A limitation of this approach is that it uses statistical correlation to make predictions, thus it cannot readily predict the effect of future environmental conditions such as changes owing to development, changes in hydrology, or changes in regional or global climate. Other examples of logistic regression to analyze invasive species include multiple species in South Africa and Russian knapweed (Acroptilon repens (L.) DC.) in Colorado

8.3.8 Dark Invasive Species Spatial Autoregressive Approach

This approach uses spatial statistical analysis to predict the distribution of invasive and noninvasive alien plants throughout all bioregions in California (Michael Jeger *et.al.*, 2018). Spatial autoregressive (SAR) models were used to assess the relationship between alien plant species distribution and native plant species richness, road density, population density, elevation, area of sample unit, and precipitation. Three predictors were found to be statistically significant for both invasive and noninvasive plants: elevation, road density, and native plant species richness. The best model (with all predictors) explained about 80 percent of the variance in the number of alien species in each bioregion. Additionally, there was significant spatial correlation for both invasive and noninvasive alien plants. Both invasive and noninvasive alien plants are found in regions with low elevation, high road density, and high native-plant species richness. Spatial data input requirements include a digital elevation model, precipitation (a modeled surface), road networks, native species richness, and occurrence of alien species. Because it is a regression procedure, many other input data could be used, such as population density, trail networks, air temperature, traffic volume, etc. The model has been applied to all of California for a single time, with bioregions as the map unit. Model outputs include maps of the number of invasive and noninvasive alien species by bioregion. The method could be extended to predict the presence of invasive species in smaller regions or pixels.

This general approach would be useful for regional probabilistic risk assessments because it uses widely available data in conjunction with a flexible spatial statistical approach. The approach could be improved by using more detailed data on vegetation types rather than bioregions. A limitation of this approach is that it uses statistical correlation to make predictions, thus it cannot readily predict the effect of future environmental conditions, such as changes owing to development, changes in hydrology, or changes in regional or global climate. However, it might be feasible to develop statistically based extrapolations from existing data. For example, if the number of nonindigenous species in a region can be predicted based on some measure of the transportation network, or other environmental factor, one could extrapolate to future conditions with more roads or a higher traffic volume. A future scenario of new road development or greater traffic or both on existing transportation networks could be developed based on planned State and Federal transportation projects. This scenario could be used to predict the subsequent increase in occurrence of nonindigenous species and invasive species.

8.3.9 Guo Support Vector Machine Approach

This method uses a type of machine learning algorithm called support vector machine (SVM) in a niche modeling approach to predict risk of occurrence of sudden oak death throughout California (Michael Jeger *et.al.*, 2018). A useful comparison is made of presence-only (one class SVM) versus presence with pseudo-absence data (2-class SVM). Based on their results, the use of pseudo-absence data does not appear to be a good choice for modeling invasive species— they inherently lead to bias because they conflate environmentally determined absence with absence on account of infestation not having occurred yet in a particular location. Input data include 14 environmental variables including mean annual temperature, mean annual precipitation, distance to roads, distance to patches of hosts, and presence of susceptible species. The use of this approach currently requires an analyst with not only GIS skills, but also substantial programming skill. Also, assistance may be needed from algorithm developers to modify code. Model output is a map of the potential location of the invasive species. The spatial scope includes all of California, and the map unit is a 1-km grid cell for a single time.

This approach would be useful for regional probabilistic risk assessments because it is a generic machine learning technique applied to niche modeling. Thus, it could be used for invasive plants, insects, diseases, and possibly other stressors. One-class SVMs appear particularly attractive because they are statistically based and unbiased and theoretically optimum, unlike some other machine learning methods and don't require a lot of model tuning. A weakness of the approach, at least for many potential users, is dependence on a library of computer code functions rather than a more mature and user-friendly software package, and assistance may be required from the library developers to apply the functions in an analysis. This approach also does not account for time, nor does it incorporate spatial processes such as dispersion. It may be difficult to specify weights for each variable. Like all niche models, it is dependent on data quality, and there will likely be issues of spatial support and spatial scaling.

8.3.10 Model selection-Final suggestions

The following suggestions to be considered when selecting modeling approaches for probabilistic risk assessment for invasive species at the regional scale:

1. Define management options and formulate the risk problem definition at

the same time so that predictions will be useful for making management decisions.

2. Ecosystems are spatially explicit, so use spatially explicit data, such as vegetation type, topography, stream networks, and elevation.
3. Use both socioeconomic and ecological information.
4. Do not assume that the initial conditions of a landscape can all be captured by a few regionalized variables because of the large role that site history often plays in shaping future dynamics.
5. Whenever possible, make quantitative predictions of risks rather than using ranks (such as low, medium, and high). Ranked values can lead to erroneous interpretations because it may not be clear what is meant by a high risk and also because of uncertainty about what happens at the boundaries of the rank categories.
6. Quantify important spatial and nonspatial sources of data uncertainty and address these uncertainties in the analysis.
7. Quantify important sources of uncertainty in model equations, including aggregation and scaling issues, and address these uncertainties in the analysis.
8. Whenever feasible, use multiple models based on different approaches and data.

Phytosanitary measures shown to be cost-effective and feasible : The benefit from the use of phytosanitary measures is that the pest will not be introduced and the PRA area will, consequently, not be subjected to the potential economic consequences. The cost-benefit analysis for each of the minimum measures found to provide acceptable security may be estimated. Those measures with an acceptable benefit-to-cost ratio should be considered.

Principle of 'minimal impact" : Measures should not be more trade restrictive than necessary. Measures should be applied to the minimum area necessary for the effective protection of the endangered area.

Reassessment of previous requirements : No additional measures should be imposed if existing measures are effective.

Principle of "equivalence" : If different phytosanitary measures with the same effect are identified, they should be accepted as alternatives.

8.4 Documentation of Pest Risk Analysis

The IPPC and the principle of "transparency" (ISPM Pub. No. 1: Principles of plant quarantine as related to international trade) require that countries should, on request, make available the rationale for phytosanitary requirements. The whole process from initiation to pest risk management should be sufficiently documented so that when a review or a dispute arises, the sources of information and rationale used in reaching the management decision can be clearly demonstrated. The main elements of documentation are: purpose for the PRA, pest, pest list, pathways, PRA area, endangered area, - sources of information, categorized pest list, conclusions of risk assessment, probability, consequences, risk management, options identified and options selected.

Uncertainty: Estimating the likelihood of pest introduction and of the consequences that could result involves many uncertainties. Uncertainty is always part of pest risk analysis; very often there is a lack of data necessary to reach secure conclusions. The subjective nature of pest risk analysis is also a source of uncertainty. ISPM 11 recognises that pest risk analysis involves many uncertainties, largely since estimates and extrapolations are made from real situations where the pest occurs to a hypothetical situation in the pest risk analysis area. In most cases analyses performed during pest risk analysis use historical data to forecast potential future events. It is important to document the areas of uncertainty and the degree of uncertainty in the assessment, and to indicate where expert judgement has been used. This is necessary for transparency and may also be useful for identifying and prioritizing research needs (Sansford , 1999).

Risk management : Risk management is where we decide what to do about the hazard. Because we aim to change the risk, this aspect of the analysis is necessarily linked to the risk assessment phase, which allows us to understand the degree to which the hazard is mitigated by applying one or another measure. In other words, risk assessment can be done in isolation, but risk management requires a link to risk assessment for the iterative process of evaluating the change in risk due to the application of mitigations.

Risk management is concerned with what can be done, and what are the options? what are the trade-offs (e.g. cost/benefit, etc.)? , what are the impacts of the risk management decisions?

'Measures' : Another group of terms that are important to understand, particularly when discussing pest risk management, are related to 'measures'. The broadest term that can be used is a 'phytosanitary measure', which is 'any legislation, regulation or official procedure having the purpose to prevent the introduction and/or spread of quarantine pests, or to limit the economic impact of regulated non-quarantine pests'. 'Phytosanitary measure' is therefore an umbrella term that includes several concepts related to managing pest risk. These concepts are each defined in their own right and includethe following.

Phytosanitary action: An official operation, such as inspection, testing, surveillance or treatment, undertaken to implement phytosanitary measures.

Phytosanitary legislation: Basic laws granting legal authority to an NPPO from which phytosanitary regulations may be drafted.

Phytosanitary procedure: Any official method for implementing phytosanitary measures including the performance of inspections, tests, surveillance or treatments in connection with regulated pests.

Phytosanitary requirement: Official rule to prevent the introduction and/or spread of quarantine pests, or to limit the economic impact of regulated nonquarantine pests, including establishment of procedures for phytosanitary certification.

Provisional measure: A phytosanitary regulation or procedure established without full technical justification owing to the current lack of adequate information. A provisional measure is subjected to periodic review and full technical justification as soon as possible.

Emergency measure: A phytosanitary measure established as a matter of urgency in a new or unexpected phytosanitary situation. An emergency measure may or may not be a provisional measure

Types of information are needed for PRA : The type of pest risk analysis will usually dictate what types of information are needed – and the information can be qualitative or quantitative. The most basic unit of most pest risk analyses is usually a specific pest. It is therefore not surprising that the most common types of information we need to do pest risk analysis cover basic biological information about pests. Information needed can include how to identify the pest (or diagnostic information), life history, ecology, host range, climatic conditions the pest requires, its global distribution and any other information about how the pest lives, feeds, reproduces or disperses.

Information is also needed on what type and level of damage the pest causes (both to crops and wild flora) within its native range, as well as in areas where it has been introduced. Depending on the type of pest, we may also want to gather information on pathways the pest is known to follow including interception records from ports of entry, if available (FAO, 1997; Devorshak and Griffin, 2002). Additional guidance on information for pest risk analysis has been considered by EPPO – they have a regional standard Checklist of Information Required for Pest Risk Analysis (PRA) (EPPO, 1998).

If a pest risk analysis is being conducted for the import or export of a commodity, we also need information on how the commodity is produced, processed, handled and shipped. In some cases, a country will consider the unmitigated risk (or the risk without any pre- or post-harvest practices) for a commodity and then compare the unmitigated risk to the level of risk associated with the commodity taking into account any of the handling practices (IPPC, 2004). This is because many pre- and post-harvest practices will have an effect on pest prevalence in the commodity, even though the processing is not directly intended to manage pest risk and the effects may be difficult to measure precisely. For instance, it is a common industry practice to wash mango fruits after harvest, primarily to remove sap (which may cause staining) from the skin of the mango (Christina Devorshak , 2012). Although the purpose of washing the fruit is not done to reduce pest risk, it will likely dislodge most external pests and this practice may be taken into account in the pest risk analysis.

8.5 Economic Analysis in PRA

Guidance found in the SPS Agreement and the ISPMs describe a broad range of approaches and techniques for assessing economic impacts in pest risk assessment. Most countries use a qualitative approach to estimating economic impacts in risk assessment. However a number of quantitative approaches can be taken. Time, resources for and complexity of these approaches vary, as do the implications of including or excluding gains from trade from the analysis. There is no clarifying WTO case law or jurisprudence relating to assessment of economic impacts. Soliman *et al.* (2010) argue that despite its limitations, partial budgeting would be the preferred quantitative approach for basic economic analysis because it is not data- or resource intensive and is relatively easy to understand and explain to decision makers. However, this approach is not without shortcomings. Other approaches, like partial equilibrium analysis, which expand the frame of analysis beyond impacts on producers, can give a more complete perspective on the economic effects of adopting a quarantine policy, but may not be feasible in all cases. Given the time and resource costs of developing even the simplest quantitative models and the time and resource constraints facing most risk assessment practitioners, it would appear that collaboration(s) between economic modellers and risk assessment practitioners to develop quantitative models with general applicability to host commodities of phytosanitary priority could be of value. Pest risk analysis is most commonly applied for four different purposes (IPPC, 2004, 2007), namely: Analysing risks for organisms, Analysing risks for pathways, Analysing risks for commodities, Supporting new policies or changes to existing policies and Prioritizing resources.

Commodity PRA : Pest risk analyses for commodities are a special group of pathway pest risk analyses that are focused on analysing the risk of pests entering an area or a country by moving with a specific commodity. Pest risk analyses conducted for imported commodities are often referred to as 'commodity pest risk analyses' (or CRAs), 'import risk analyses' (or IRAs), or 'pathway initiated pest risk analyses'. Most often, the technical justification is provided in the form of a pest risk analysis. When an exporting country makes a request to an importing country for market access for a commodity, the importing country should conduct a pest risk analysis to determine whether and to what extent phytosanitary measures should be applied to that commodity. The purpose of a commodity pest risk analysis is to analyse the risk of pests associated with a specific commodity moving from one area to another (Christina Devorshak , 2012).

Commodities may be imported into a country from many places and the types of pests associated with a given commodity are dependent on the origin of that commodity. If it is determined that the commodity poses risks for moving pests, then appropriate measures can be put in place to manage those risks. A commodity can be anything moving from one place to another, particularly in the context of international trade. Common commodities that are analysed for pest risk include fresh fruits and vegetables for consumption, nursery stock and plants for planting, seeds and germplasm, various types of wood products (e.g. wood packing material, raw lumber, logs, wood chips, etc.) and handicrafts made from plant materials (e.g. potpourri, baskets, decorative items). These types of commodities are well documented as providing pathways for pests to move great distances.

However, other types of commodities may also serve to transport pests even if they are not of agricultural origin. Snails are documented to adhere tightly to ceramic and marble tiles (as described in Section 8.4.1), attracted to the calcium in the tiles, and travelling great distances wherever they are shipped. Similarly, cars, vehicles and other types of machinery (e.g. farm equip equipment, military equipment) may move pests, pests may crawl into hidden spaces and/or lay eggs on the vehicles, or soil containing pathogens may be carried with vehicles from one place to another. To combine speed and safety of global movement implies compromise and acceptance of certain risks. It is here that "good governance" is essential for individual countries in order to determine those pests that could cause unacceptable damage to plants or plant products, the means and likelihood of their introduction and the options (measures) available to prevent such introduction. Together these three steps constitute the process known as pest risk analysis. The end result of the application of the pest risk analysis process is usually an advice to a national authority favouring or disfavouring phytosanitary measures (Hopper, 1993).

Pest risk analysis is one of the core elements for protecting plant health today, as reflected by the International Plant Protection Convention and its international standards for phytosanitary measures. The first international pest risk analysis standard became available in February 1996. Seven years later, in April 2003, the international plant protection community adopted a standard (supplementary to another pest risk analysis standard of 2001) to allow for specific focus on environmental risk.

8.6 Options of International Standards PRA

Pest risk analysis is a key element of all three major international agreements relating to plant protection, namely the International Plant Protection Convention, the World Trade Organization's Agreement on the Application of Sanitary and Phytosanitary Measures and the Convention on Biological Diversity. The IPPC framework, including the national plant protection organizations of all its contracting parties, could well provide an institutional framework that serves the objectives of the CBD and WTO as well.

Pest risk analysis is a framework for organizing biological and other scientific and economic information to calculate risk (essentially the probability of an unwanted event or hazard occurring multiplied by the magnitude of the consequences if it does happen). This assessment of risk is used to identify appropriate measures to reduce risk to an acceptable level. To comply with international agreements, countries imposing new phytosanitary regulations must either support them through a pest risk analysis or base them on relevant international standards. More specific international standards would be useful in harmonizing management options for pest risks. However, the wide variation in pest risks faced by individual countries in relation to introduction and spread of a pest (or pests) associated with a particular commodity usually makes it difficult to develop an agreed international standard. To date, only one specific commodity standard has been accepted under the IPPC as an international standard for phytosanitary measures, namely ISPM 15: Guidelines for regulating wood packaging material in international trade. This standard's acceptance was helped in that many countries were in agreement over the risks associated with wood packaging material as a pathway for pests. If the appropriate phytosanitary standard is not available, countries have the alternative of risk analysis to support restrictive measures.

Dispute settlement: International harmonization of phytosanitary measures will be successful only if contracting parties implement standards. Control systems are needed to verify whether this is done. The WTO has a dispute settlement mechanism in which the outcome of a dispute settlement is binding on the parties concerned. Phytosanitary disputes may be resolved within the IPPC framework, in a dispute settlement process which relies on the voluntary cooperation of both disputing parties, who are not bound by the findings. The use of the IPPC dispute settlement mechanism could provide valuable jurisprudence for consideration in phytosanitary disputes carried out through the WTO mechanism. To date, no dispute settlement has taken place within the IPPC framework.

Addressing environmental risks associated with plant pests: The contracting parties to the IPPC have developed a reference standard specifically to target the analysis of environmental risks posed by plant pests. This was adopted in April 2003 as a supplement (Analysis of environmental risks) to ISPM 11 [2001]: Pest risk analysis for quarantine pests. The standard has been further amended and supplemented to ISPM 11 [2004]: Pest risk analysis for quarantine pests, including analysis of environmental risks and living modified organisms. The process of revising ISPM 11 [2001] took into account the CBD's guiding principles for the prevention, introduction and mitigation of impacts of invasive alien species. ISPM 11 and other relevant ISPMs provide specific guidance to countries on how to use the information available to make timely decisions about potential threats to plants and plant communities from international trade. The international glossary of phytosanitary terms, ISPM 5, has been supplemented in 2001 with Guidelines on the interpretation and application of the concept of official control for regulated pests and in 2003 with Guidelines on the understanding of potential economic importance and related terms including reference to environmental considerations. These supplements have helped clarify what should be covered by pest risk analysis. Several other recent new or revised standards support pest risk analysis. Systems for the classification of pests of potential economic significance, generally arthropods and pathogens affecting agriculture or forestry, are well established; identifying species or pathways that pose an environmental risk is a relatively new challenge for NPPOs.

8.7 Overall PRA Strategy

Three different approaches for developing an overall PRA strategy to the management of invasive alien species are as follows:

i. **To expand monitoring and regulation to cover taxa and pathways that have not been covered previously:** This is particularly pertinent in a country that has not had legislation covering weeds. An NPPO could set up a quick screening process or a partial pest risk analysis on all plant species imported for planting. It would reduce the risk of introducing species of plants that "escape" their intended use and become countryside weeds or that become more invasive than in their native environment. If a species is found to be suspicious, a full pest risk analysis may be required. Although the focus is on plants for planting, any organism intentionally introduced that will establish a breeding population may warrant a pest risk analysis.

ii. **To identify the most important natural systems and/or native plants and review all potential threats to these:** NPPOs work with a broad range of stakeholders, including environmental non-governmental organizations, to decide which species or ecosystems that are not already the focus for plant protection effort are the most important to start protecting.

iii. **To focus efforts on particular geographic areas or areas with specific levels of protected status:** Limiting an NPPO's efforts to particular geographic areas, such as protected areas or parks, may be a good starting point if resources are limited. Information on such an area will exist, including ecological data; other agencies may have the mandate of protecting the area and add their authority to the outcome of analysis. The constant threat of re-entry of a pest to a protected area from the surrounding area could reduce the cost effectiveness of this approach. However, it could suit countries with natural barriers that will reduce the likelihood of internal spread.

Pests causing indirect injury to plants: An annex to the revision of ISPM 11 [2001] states that the scope of the IPPC also extends to organisms which are pests because they ... indirectly affect plants through effects on other organisms. In addition to pests that directly affect host plants, there are those, like most weeds/invasive plants, which affect plants primarily by other processes such as competition ... Such "secondary pests" are a difficult category for risk analysis. NPPOs could find themselves asked to regulate new pests and pathways for pests that were outside the experience, knowledge and training of their officials. Some of the organisms in this category of pests may be predicted (e.g. parasites of biological control agents or pollinators, species of exotic ants); for others, ecologists may have to sound the warning of the potential threat before an NPPO will know to act.

Resources for pest risk assessments: The principal tools and resources available for the pest risk assessment component of pest risk analysis include:

i. **National data sources:** National data will be needed from the PRA area and, to assess entry potential, from the country of origin.

ii. **International data sources: Examples include:** CAB International's Crop Protection Compendium, which can be used to select species that might be present in a commodity based on the pest distribution and host range in the exporting country; climatological information; abstracts of scientific literature; books; pest alerts from regional plant protection organizations; Web search engines.

iii. **Data sources for assessing entry potential:** Data on trade pathways and interceptions (detections in consignments) are most important but usually require specific enquiries for unpublished data. It includes data sources for assessing establishment potential ; if the area of origin and the host plants are known, climates in the area of origin can be compared with climates in the area under threat and the distribution of host plants determined and data sources for assessing economic, environmental and social impacts.

Assembling comprehensive data on a pest's impacts in its current range, sufficient biological data to predict its spread and population dynamics in the PRA area coupled with financial, economic, environmental and social data for the enterprises, ecosystems and people likely to be affected may prove difficult. However, assessments can still be made even if data are lacking.

8.8 Weed Risk Assessment

The movement of trade goods and aid of one sort or another throughout the world is essential for the well-being of all peoples. Not all these goods and gifts are benign, however, and some come with unwelcome surprises. Invasive species affect agricultural and other systems, and their impacts are second only to habitat destruction in terms of loss of biodiversity. These concerns have generated growing international interest in weed-risk assessment systems to prevent the introduction of new pests and to prioritise existing pests for control. Weed-risk assessment is a new discipline, and the first international symposium on the topic was held only recently in Australia (Peter A. Williams, 2003). This country, along with New Zealand, is at the forefront of developing and implementing strong quarantine protocols. Both countries are relatively isolated from the rest of the world, agriculture is important to their economies, and their citizens value natural landscapes and their ancient indigenous biodiversity. This report introduces the topic of weed-risk assessment and provides guidelines for countries wishing to strengthen their own quarantine protocols and to use scarce resources efficiently for prioritising existing pests for control. Fortunately, this task is becoming easier because the Internet allows rapid exchange of information and access to detailed databases on pests, example, the global compendium of weed.

International framework of weed-risk assessment : The actions taken to exclude a plant species from a country because of its weed potential must be consistent with the international standards regulating the movement of trade goods. These obligations are defined under the Agreement on the Application of Sanitary and Phytosanitary Measures (SPS agreement) of the World Trade Organisation (WTO 1994), and the International. Plant Protection Convention (IPPC) (1997 revised edition) deposited with the United Nations Food and Agricultural Organization (FAO, 1996). These two international agreements, whilst allowing countries to specify requirements for the entry of plant material, describe the obligations of countries so that import requirements are not unjustified trade barriers. A further international convention involving weeds concerns the need to conserve biodiversity. Article 8 (h) of the Convention on Biological Diversity states that: "Each Contracting Party shall, as far as possible and appropriate, prevent the introduction, control or eradicate those alien species which threaten ecosystems, habitats, or species." Not all countries are signatories to this convention.

Weed-risk assessment is concerned primarily with the first two stages of the pest risk assessment involving pest categorization, that is, the process for determining whether a pest has, or has not, the characteristics of a quarantine pest or those of a regulated non-quarantine pest . The minimum requirement of any weed-risk assessment system is that it satisfies the international agreements . To do this it must be built on explicit assumptions and must use scientific data. Weed-risk assessment systems designed for use only within a single sovereign state, and which do not have the potential to limit trade, need not comply with international agreements. But to be effective they must be based on similar sound principles.

8.9 Insect Pest Assessment-Case studies

The transport of adventive arthropods associated with rapidly expanding global trade has led to an ever increasing list of quarantine pests establishing beyond their native ranges. A significant number of these taxa have become serious forest pests, and some are directly threatening the viability of native tree species across their introduced ranges (Humble, 2010). The global movement and establishment of bark- and wood-borers such as *Anoplophora glabripennis* (Motschulsky), *Tetropium fuscum* (F.) and *Tomicus piniperda* (L.) in multiple international jurisdictions led to the recognition of the importance of solid wood packing (e.g. crating, pallets) as an introduction pathway. Regulatory inspections and rearing studies targeting wood packing pathways have identified both the diversity of taxa and the potential magnitude of pest movements associated with this route.

Concurrently, surveillance programmes initiated to detect invasive bark- and wood-borers in Canada and the United States have identified previously undetected establishment of multiple species of ambrosia- and bark-beetles (Curculionidae: Scolytinae), and woodborers (Cerambycidae) across North America. This paper reviews the lines of evidence that were used to support the development of the first pathway-based international standard for phytosanitary measures (ISPM), that for wood packing (ISPM 15). This standard requires mandatory treatment of wood used as dunnage, packaging, crating or pallets in international trade in order to mitigate populations of bark- and wood-borers potentially present in the raw wood.

The unmitigated pest risk potential for the importation of Pinus and Abies logs from all states of Mexico into the United States was assessed by estimating the probability and consequences of establishment of representative insects and pathogens of concern (Tkacz Borys M *et.al.*, 1998). Twenty-two individual pest risk assessments were prepared for Pinus logs, twelve dealing with insects and ten with pathogens. Six individual assessments were prepared for Abies logs. The selected organisms were representative examples of insects and pathogens found on the bark, in the bark, and in the wood of Pinus or Abies logs. Among the insects and pathogens assessed for Mexican pines, eight (*Dendroctonus mexicanus, Coptotermes crassus, Pterophylla beltrani, Ips bonanseai, Gnathotrichus perniciosus, Gnathotrichus nitidifrons, Fusarium subglutinan*s f. sp. *pini*, and *Ophiostoma* spp.) were rated a high risk potential. A moderate pest risk potential was assigned to nine other organisms or groups of organisms including *Pineus* spp., *Lophocampa alternata, Hylesia frigida, Hypoderma* spp., *Lophodermella* spp., *Synanthedon cardinalis, Heterobasidion annosum, Sphaeropsis sapinea, Bursaphelenchus xylophilus, Cronartium* spp., and *Peridermium* spp. The pests of concern with a moderate or high pest risk potential for *Abies* logs include *Lophocampa alternata, Scolytus mundus, S. aztecus, Pseudohylesinus variegatus, P. magnus, Ophiostoma abietinum,* and *Heterobasidion annosum.* For those organisms of concern that are associated with Mexican Pinus and *Abies* logs, specific phytosanitary measures may be required to ensure the quarantine safety of proposed importations.

Wheat is the most Important Imported food cereal m China. Zhang Congzhong and Xu Yanl (2001) analysed the risk of pests associated with imported wheat. Each potential pest was given a comprehensive value accordmg to the index system m the prelmunary quantitative assessment. The result shows that 6 quarantine stored pests and 5 field pests are potential quarantine pests associated with Imported wheat, which are *Trogoderrna granarium, Troxxlerma versicolor, Prostephanus truncatus, Trogoderma uiclusni, Triboluum. destructor, Caulophilus Iatuuisus, Cephus cinctus, Mayetiola destructor, Eurygaster mtemcep, Nysius hndtoni, and Cephus pygrnaeus*. It was suggested that the pests listed above should be the pnmary pests to be inspected when import wheat. Some options for nsk management were suggested.

8.10 Pest Risk Management Considerations

The IPPC has been identified as the international phytosanitary standard-setting authority by the WTO. The IPPC Standard for Phytosanitary Measures (ISPM) No. 19, Guidelines on Lists of Regulated Pests (2003), requires that pests regulated by National Plant Protection Organizations meet the criteria for either quarantine pests or regulated non-quarantine pests. To be considered a quarantine pest according to the IPPC's definition, an organism must be "a pest of potential economic importance to the area endangered thereby and not yet present there, or present but not widely distributed and being officially controlled". In 2001, the Interim Commission on Phytosanitary Measures approved Guidelines on the Interpretation and Application of the Concept of Official Control for Regulated Pests (see Supplement No. 1 in ISPM No.5). These guidelines include, among other elements, that measures be mandatory and that domestic and import requirements should have equivalent effect. Thus, for Canada to comply with WTO and IPPC guidelines and continue to regulate *H. glycines* as a quarantine pest on imports, it would have to strengthen significantly the enforcement of domestic movement restriction from any infested area in Canada (Anonymous, 2003).

Domestic movement regulations: The limitation in enforcing domestic regulation along with lack of movement compliance, may have been a contributing factor to the spread of the pest to new areas in Ontario. The recent case of *H. glycines* in Manitoba is a good example which reiterates the challenges in enforcing domestic movement regulations over a wider area. The single cyst detection was in a field well outside the flood zone of the Red River, and on a site where there was no record of previous soybean production. Thus, the likelihood of the pathway of introduction of *H. glycines* to this site may have been the seed from an infested field or spread of infested soil using agricultural equipment from other fields where soybeans are produced. However, it is noted that further cysts were not detected during a follow-up intensive survey of that particular field in Manitoba.

Spread by natural pathways: Natural pathways, such as wind, water and birds, also contribute to the spread of the nematode to new areas. These natural pathways and the spread cannot be controlled by regulations and enforcements. The continuous spread of the pest in Ontario to new production areas though cryptic, could be attributed to these natural pathways. The Manitoba soybean production area is in the Red River valley, which is contiguous with the production areas of Minnesota and North Dakota, where *H. glycines* has been reported within 300 km of the border. Considering the potential for annual flooding in this area it is very likely that *H. glycines* has spread into the soybean production areas of Manitoba along the Red River. Similarly, *H. glycines* has been reported in Eastern Ontario in 2008 and is likely present in the neighbouring soybean production area extending down the St. Lawrence valley.

Soil surveys: In order to effectively regulate *H. glycines* in Canada, additional extensive surveys would be required in order to determine the current distribution of the pest in soybean production areas. Soil surveys, which are highly resource-intensive exercises, would never fully describe the true distribution of this pest, given its biology and pathways of spread. It is reported that the population of the pest has to build to a detectable level before it can be confirmed by soil surveys.

Enforcement requirements on infested fields: To maintain the regulated status of *H. glycines*, containment measures would then have to be implemented by the CFIA that would parallel the current measures in place to contain the two species of potato cyst nematode that are present in Canada, *Globodera pallida* and *G. rostochiensis*, as per the Golden Nematode Order, SOR/80-260 (Saanich, British Columbia) and Golden Nematode Infested Places Order (Quebec) (Anonymous, 2003).

Risk management decision and option: The initial risk management discussion document, proposing the recommended option of deregulating *H. glycines* all across Canada, was circulated for stakeholder consultation in February 2011 and stakeholder feedback was received (Anonymous, 2003). While stakeholders from other provinces expressed support for deregulation of *H. glycines*, stakeholders from Quebec and Manitoba preferred continued regulation of *H. glycines*. Further, in December 2011, a Question and Answer document was distributed, which addressed specific concerns raised by stakeholders. Further discussions with concerned Quebec and Manitoba stakeholders were conducted in summer of 2012 to better understand their position, address specific concerns and provide timeline for the implementation of the deregulation of *H. glycines*. There was a concern from Manitoba and Quebec that the proposed deregulation will introduce *H. glycinesin* their soybean growing areas, where this pest has not been reported yet. During the consultation, theCFIA emphasized that all imports of seed and grain are required to be free of soil, as soil is the primary pathway for introduction of *H. glycines*.

The mandate of the CFIA is to prevent the introduction of pests/pathogens and to protect the agricultural resource base of Canada. The current status of *H. glycines*, widespread presence in Ontario, is such that it needs to move from being a quarantine pest to a management pest. It was also noted that despite the presence of *H. glycines*, the soybean acreage in Ontario has been increasing over the last two decades, which is largely attributed to the use of best management practices involving the development and use of *H. glycines* resistant soybean varieties. The discrepancy between Canada's strict import requirements and inconsistent enforcement of domestic regulations for *H. glycines* can be perceived as unjustified discrimination or a disguised restriction on international trade of some plant commodities. This makes Canada vulnerable to WTO challenges from trading partners. The decision was Heterodera glycines no longer fits the definition of a quarantine pest.

By way of this risk management document, the CFIA is announcing the deregulation of *H. glycines* in Canada. As of November 25, 2013, the CFIA will not enforce the import regulations related to *H. glycines*. The CFIA will not be enforcing the Schedule II of the Plant Protection Regulations (PPR) pertaining to the domestic movement restriction of *H. glycines* infested material. The risk-based decision was made considering several factors, including the low risk of the pest, the utilization of *H. glycines* best management practices, and discrepancy in domestic and import regulations. Soil is a prohibited substance and all imports are required to be free of soil. The implementation of the de-regulation of *H. glycines* would require that relevant plant health import directives be amended to remove reference to requirements for freedom from *H. glycines* or revoked. The Automated Import Reference Systems (AIRS) will also be updated to reflect the deregulation of *H. glycines*. As part of the overall regulatory review process, the CFIA will work towards the removal of *H. glycines*from the Schedule II of the Plant Protection Regulations and the List of Pests Regulated by Canada.

The following table (Table 16) discusses three pest risk management options considered, with the advantages and disadvantages listed for each (Anonymous, 2003).

Table 16: Pest Risk Management Options

Options	Advantages	Disadvantages
Status quo: Regulated Pest Status of Heterodera glycines Retain *H. glycines* in the List of Pests Regulated by Canada and Schedule II of Plant Protection Regulations on Restricted Movement within Canada. and Implement official control measures if H.glycines is found in a new area.	Control over import of seeds, potatoes and associated soil. Control over Domestic Movement of seed, and soil associated with seed, potato tubers, horticultural root crops, plant material, farm machinery and equipment etc. Authority to respond to positive finds and implementing official control measures. Authority to contain the pest by enforcing mitigation measures at point of origin, if from an infested area.	Already present in large areas and numerous soybean production counties of Ontario. No control over natural pathways of pest introduction, such as: Wind, Floods, Birds, etc. More resources needed to enforce domestic movement restrictions. Potential additional costs to the growers to follow quarantine measures. Additional costs in establishing and maintaining compliance agreements to handle regulated plant and plant parts from regulated areas Potential additional costs to equipment operators (e.g.cleaning costs). Potential for negative impact on trading partners and trading relationships due to continued oversight of imported commodities. Potential for WTO challenges from trading partners over discrepancy in domestic and import regulations, if domestic regulations are not enforced
Declare Ontario as "infested area" and continue regulation of *H. glycines* in rest of Canada	Authority to respond to positive finds at a location outside Ontario and implement official quarantine measures. Quarantine measures would include: 1. Declaring Ontario as a "soybean cyst nematode-infested	Domestic movement restrictions apply from Ontario to rest of Canada. No control over natural pathways of pest introduction, such as:

Options	Advantages	Disadvantages
Retain *H. glycines* in the List of PestsRegulated by Canada AND Schedule II of Plant Protection Regulationson Restricted Movement Within Canada. AND Make changes toSchedule II of Plant Protection Regulationson Restricted Movement from Ontario to the rest of Canada. AND Implement official control measures if H.glycines is found in a new area.	area". 2. Immediate issuance of notice of quarantine or equivalent regulatory order. 3. Categorization of lands in the infested zone, as infested, exposed and adjacent. All these categorized lands will be under quarantine. 4. Restriction on production of all host and minor hosts of *H. glycines*. 5. Restriction on movement of soil and equipment associated with infested soil. 6. Restriction on movement of plant and plant parts with infested soil. 7. Restriction on the movement of equipment and plant material from other category fields in the quarantine area. 8. Restriction on movement of grains intended for other end uses. Control over import of seeds, potatoes, horticultural root crops and associated soil into Canada except Ontario. Requirement on procurement of seed and seed potatoes intended for planting outside Ontario, from a field or area free of *H. glycines* Control over Domestic Movement of seed, and soil associated with seed, potato tubers, horticultural root crops, plant material, farm machinery and equipment etc., from Ontario to the rest of Canada. Better preparedness through official surveys, in soybean production areas outside Ontario.	Wind, Floods, Birds, etc. Potential impact on trade from Ontario More resources, potential partnerships with provinces and service providers needed to: 1. Apply potential quarantine measures. 2. Enforce domestic movement restrictions. 3. To conduct extensive annual surveys. Potential additional costs to the grower/regulated field owner when quarantine measures implemented: restriction on sale and movement of equipment, cleaning costs etc. Potential impact on soybean and potato seed producers in Ontario. Additional costs in establishing and maintaining compliance agreements to handle regulated plant and plant parts from regulated areas. Potential additional costs to Ontario equipment operators, like cleaning costs. Potential for negative impact on trading partners and trading relationships due to more stringent oversight of imported commodities. Potential for WTO challenges from trading partners over discrepancy in domestic and import regulations, if domestic regulations are not enforced.

Options	Advantages	Disadvantages
De-regulate *H. glycines* across Canada Remove *H. glycines*from the List of Pests Regulated by Canada AND Remove *H. glycines*from Schedule II of the Plant Protection Regulations on Domestic Movement Within Canada	Meet international obligations to IPPC,WTO-SPS, by adopting a harmonized approach to imports and domestic policies. No additional resources and costs for enforcement of regulations pertaining to *H. glycines*. No additional requirements and costs, related to *H. glycines* certification, for exporters to Canada. No domestic movement restrictions for commodities and equipment associated with *H. glycines*. Unrestricted movement between *H. glycines* infested counties/provinces, states in the USA and the rest of Canada. Soil – The most significant pathway for movement of *H. glycines* is regulated in imports.	No authority to regulate *H. glycines* on imports.

8.11 Pest Risk Analysis Request Form

India National Standard for PRA

DPPQ&S, Ministry of Agriculture, Government of India

Client details:

Name/ Organisation: ....

Address ....

....Postcode ....

Phone............................Fax....................E-mail.................................

PRA general parameters:

Activity (circle one): Import Export

Common/ Product name ....

Scientific/ botanical name (genus & species) ....

Scientific/ botanical name (Strain/ variety/ cultivar) ....

Country/ countries of origin ....

Quantity/ Volume ....

Product Type (circle one or more)

Processed/ Non-processed Living/ non-living

Plant/ Animal Genetically modified/ non-genetically modified

Seed/ plant/ soil Culture / non-culture

Other ....

Product processing (if applicable):

If seed: Ground/ kibbled/ whole/ preserved

If plant: Fresh/ dried/ freeze dried/ preserved

Processing refinement: Cooked/ frozen/ pulped/ steamed

Specify treatment details ....

....

Product origins (please state if question not relevant):

Source location (by country, origin & locality) ....

Production method, Certification scheme and / or accreditation type? ....

....

India National Standard for PRA

DPPQ&S, Ministry of Agriculture, Government of India

End use (circle one or more):

Human consumption / Processing/ Stock feed/ Pet food/ Fish food/ Seeds for sowing/ Nursery stock/ Multiplication/ Post-entry Quarantine/ Therapeutic/ Fertilisers/ In-vivo / In-vitro

Other ....

End destination (circle &/or specify):

Rural/ urban Multiple locations/ single

Specify Country, State & / or region (PRA defined area) ....

....

Entry (circle one or more)

Ship/ Air/ Ground transport/ Rail

Other ....

General comment:

(any further general comment or notes that need to be made, please make here).

8.12 Pest Information System

Pest surveillance forms an integral part of integrated pest management. Management decisions depend on pest information which are obtained from surveillance systems. Such systems involve data collection, retrieval, and analyses of field data information delivery to the various decision-makers. Data processing is facilitated and made more efficient with the use of micro-computers and mainframes. Data gathered from pest surveillance systems can assist gathering historical information that may provide further information into risk zones and risk periods. Knowledge and information are key to correct pest management decisions. Integrated pest management (IPM), a system that emphasizes appropriate decision making, is information intensive and depends heavily on accurate and timely information for field implementation by practitioners (Waheed *et.al.*, 2021).

The transfer of research and extension information to farmers plays the key role in the adoption of IPM. Electronic communication provides an effective multidirectional exchange of information. Electronic extension systems provide 24-hour access to an inquirer of specific information to be used in planning and decision support. In fact, it is rapidly changing the way individuals exchange information and make decisions. Now it is possible for extension services and applied researchers to deliver and receive information to and from much larger audiences via fax (both Internet- and telephone-based document delivery systems), multimedia programs, email, and the web. The emphasis is, however, beginning to shift from traditional one-way flow of information from research, then to extension, and finally to end-users of information, to the more egalitarian process where the pool of total experience and knowledge available in the community, from growers, industry, research and extension, is readily exchanged through electronic means, focused learning workshops, and increased on-farm applied research.

Information systems and actionable knowledge creation in rice-farming systems: Rice farmers in the Kumbungu District in Northern Ghana interact with information systems. Of interest here is the degree to which knowledge derived from such interaction is actionable. The paper addresses the overall question: what information systems are currently providing agricultural information to rice farmers, and to what extent does this result in actionable knowledge creation? Findings revealed that Farmer-to-Farmer systems contribute most to actionable knowledge creation (Andy Bonaventure Nyamekye *et.al.*, 2020). It was concluded that systems integration and local actor participation are essential for actionable knowledge creation in information systems.

National Agricultural Pest Information System (NAPIS): NAPIS is the database for Cooperative Agricultural Pest Survey and related pest detection surveys.About 5.58 million records summarize survey results for 6,417 insects, pathogens, weeds, mollusks and biological control organisms. Pest detection survey observations recorded in NAPIS emphasize exotic pests that may impact exports of U.S. agricultural products or damage agricultural production and natural resources. The NAPIS database is associated with the Cooperative Agricultural Pest Survey (CAPS), a program sponsored by USDA APHIS.

The impact of modern electronic technology for extension information communication has been anticipated over a decade ago; however, no one may have predicted the opportunities that have been opened with advent of the Internet and the Web. An Internet-based network provides one virtual platform for a decentralized organization like the cooperative extension service with its personnel and operation spread all over its jurisdiction, by bridging the distance gap and operating as a unit. There are a few barriers such as lack of time on the part of extension personnel, funds, training, and experience for efficiently using this technology for information delivery and analysis. Nevertheless, email and other Internet tools are widely used and preferred by Extension agents for exchanging time-sensitive information and networking with researchers and subject specialists. Extension agents view the web as an information resource with great potential for just-in-time communications.

An Online Pest Management Information System (PMISNET) was developed on major agricultural crops containing information on various aspects of IPM viz., general information, insect/pest, disease and weeds on major crops, IPM strategies etc. PMISNET has a three layer client-server architecture. The software runs on HTTP server and serves the request of the client that may on any computer connected with internet and having a graphic web browser. There is provision to browse, insert, update and delete the information through user friendly interface it has in built help and indexing of information to facilitate smooth navigation. For demonstration purpose two crops viz., sugarcane and cucurbits are used. More crops can be added in PMISNET for which information on IPM is available. Integrated pest management (IPM) is a sustainable approach to manage different pest related problems in the field of agriculture by combining biological, cultural, physical and chemical tools in a way that minimize economic, health and environmental risk. However, information on IPM is not readily available to the end user, a farmer or an extension worker. With the rapid growth of Internet, it is now possible to put the information in electronic format so that users can excess it anywhere anytime.

The Internet enables collaboration and information sharing on an unprecedented scale. It is becoming a prime medium for research and extension communication. The World Wide Web, the Internet's hypertext, multimedia publishing protocol, makes it possible to combine information from many different sites in a seamless fashion. The potential for using the web to integrate all types of static and interactive (dynamic) information is unique and unprecedented. The web provides excellent interfaces for all kinds of interactive network databases, and many kinds of online analyses and data processing. Web-based models and decision support systems (DSS) are becoming popular because little or no client software is required, thus reducing software management and distribution costs. No other medium offers such ability as simultaneous real-time weather information, multimedia, analytical processing and multi-way discussion and feedback.

Bibliography

Andy Bonaventure Nyamekye, Art Dewulf, Erik Van Slobbe and Katrien Termeer, 2020. Information systems and actionable knowledge creation in rice-farming systems in Northern Ghana. *African Geographical Rev.*, 39: 144-161

Isard, S. A., Russo, J.M. and DeWolf, E. D. 2006. The establishment of a national Pest Information Platform for Extension and Education. Online Plant Health Progress doi:10.1094/PHP-2006-0915-01-RV.

Michael Jeger, Claude Bragard,David Caffier, Thierry Candresse, Elisavet Chatzivassiliou, Katharina Dehnen-Schmutz, Jean-Claude Grégoire,Josep Anton Jaques Miret,Alan MacLeod,Maria Navajas Navarro,Björn Niere,Stephen Parnell,Roel Potting,Trond Rafoss, Vittorio Rossi, Gregor Urek, Ariena Van Bruggen, Wopke Van Der Werf, Jonathan West,Stephan Winter, Andy Hart, Jan Schans, Gritta Schrader, Muriel Suffert, Virag Kertész, Svetla Kozelska, Maria Rosaria Mannino, Olaf Mosbach-Schulz, Marco Pautasso, Giuseppe Stancanelli, Sara Tramontini and Sybren Vos, Gianni Gilioli, 2018. Guidance on quantitative pest risk assessment. *EFSA J.*, Vol.16, Issue 8, e05350

Robinet, C., Kehlenbeck, H., Kriticos, D.J., Baker, R.H.A., Battisti, A. and Brunel, S., 2012. A Suite of Models to Support the Quantitative Assessment of Spread in Pest Risk Analysis. *PLoS ONE* , 7(10): e43366

Waheed, I., Bajwa and Marcos Kogan, 2021. Internet-based IPM Informatics and Decision Support. Integrated Plant Protection Center (IPPC) , Oregon State University, Corvallis, OR 97331

9

Global Positioning System and Geographic Information System for Plant Biosecurity

9.1 Geospatial Technology

It is an organized collection of computer hardware, software, geographic data, and personnel. It is designed to efficiently capture, store, update, manipulate, analyze, and display geographically referenced information. It includes typical Geographic Information System (GIS) software packages (ArcGIS, ArcExplorer, MapPoint, Google Earth), Remote Sensing (Collecting and interpreting information about the environment from a distance using satellite imagery, radar, or aerial photography), Global Positioning Systems (GPS) (A system of radio-emitting and receiving satellites used for determining positions on the earth) and many others (Lisa Kennaway, 2009). Geospatial technology can be used in pest monitoring and detection, data visualization/query, survey data collection, management and analysis, risk and pathway analysis, change detection and many others.

The geospatial technology has been used from the early tasks of surveying the status of crop health until forecasting when the disease likely to be occurred. Although many big challenges are facing by the global and local agricultures to produce good outputs and to secure the world food population, however the rich of geospatial data and advancement of technologies have playing certain roles, particularly assisting the decision makers in forming strategies for combating various pests and diseases that affecting plant health and food crops (Sabtu *et.al.*, 2018). Geospatial technology has been used, from the early tasks of surveying the status of crop health to managing the collected data. This technology has been applied to complement the ground crew during field survey to automatically record the location of affected plants for further action. Distribution of pest and disease incidence on the affected crops has been mapped to visualise the events from a large-scale view. The power of spatial analysis has been used in predicting the possibilities of incidence likely to be occurred in a few years time. Global trade liberalisation and climate changes have presented major challenges for the local crops to produce good products and to secure food for the world population generally. Nevertheless, the advancement of geospatial technology has made the activities of combating various pests and diseases affecting plant health much easier than before.

International and national biosecurity policies consider risk assessment a critical component of overall plant health risk analysis. The Agreement on the Application of Sanitary and Phytosanitary Measures, the International Plant Protection Convention, and the Convention on Biological Diversity all provide guidelines and recommendations on how to use risk assessment. Lindgren Cory (2011) discussed how these instruments address risk assessment, and makes recommendations on how the risk assessment process needs to incorporate current geospatial predictive science and geographic information systems into the plant health biosecurity risk analysis toolbox. Geospatial technology is a rapidly growing and changing field. The evolution of Geographic Information System, the Global Navigation Satellite System (GNSS) and Remote Sensing (RS) technologies has enabled the collection and analysis of spatial and non-spatial agriculture data in a more accurate and timely manner.

Geospatial technologies like Global Positioning System integrated with Geographic Information System are being used in precision agriculture for integrated pest management (IPM). Remote sensing and spatial analysis and other tools like Global Navigation Satellite System (GNSS) are of additional value in planning crop management practices. On other hand, these technologies have some limitations; firstly, high spatial resolution imagery is not easily available for all areas, especially rural. Whereas as hyper-spectral imagery is required; secondly, there is a lack of technical knowledge about geospatial technology by consultants and end-users / farmers; finally this technology depends on time sensitive mapping and near real time image acquisition and product delivery. However, facing these challenges, Geospatial Technology can provide valuable information in an IMP context, allowing for a complete understanding (via remote mapping or spatial modeling) of the spatial complexity of the biotic and abiotic characteristic of a field and its crop, and providing information about the disease and pests populations that are present or likely to occur. The technology is highly effective for forestry where increasing and large-scale pest and disease attacks are increasingly reported. Geospatial technologies are making forestry management more precise and spatially comprehensive to better yield, insect pests and diseases control dynamics (Md. Arshad Anwer and Garima Singh, 2019). New access to data and technology will likely promote the transition of these tools from a research to an applied domain across both sectors. In this way geospatial data and technology plays a significant role in improving the production of yield and overcoming food security issues.

Global Positioning System is a satellite system that projects information to GPS receivers on the ground, enabling users to determine latitude and longitude coordinates. Global Information System refers to a software program that enables users to store and manipulate large amounts of data from GPS and other sources. The GIS created by computing background makes possible to generate complex view about fields and to make valid agro technological decisions. With the advent of the satellite-based GPS, farmers can gain the potential to take account of spatial variability. GPS is a system of satellites while, GIS software programs can merge data from GPS and numerous other sources. GPS and GIS technologies can assist producers' mitigation and preparedness efforts by identifying potential impact areas and areas vulnerable to intentional and unintentional biosecurity problems. These are a great tool for supporting plant health programs like Data visualization/ query, survey data collection, management and risk and pathway analysis etc (Mehta Vasu and Koranga Radha, 2020). They provide a method to understand economic implications, a proactive approach to safeguard agriculture and assist with quality control.

Global Positioning System: An agricultural producer may use a handheld GPS receiver to determine the latitude and longitude coordinates of a water source next to a field or vineyard.

Global Information System: Following a chemical spill, maps obtained from a GIS system can reveal environmentally-sensitive areas that should be protected during response and recovery phases.

The Global Navigation System Satellites (GNSS) has been used for various types of applications, such as location, navigation, tracking, mapping and timing.

9.2 Global Positioning System (GPS)

The GPS, originally Navstar GPS, is a satellite-based radio-navigation system owned by the United States government and operated by the United States Space Force. It is one of the global navigation satellite systems (GNSS) that provides geolocation and time information to a GPS receiver anywhere on or near the Earth where there is an unobstructed line of sight to four or more GPS satellites. The GPS does not require the user to transmit any data, and it operates independently of any telephonic or internet reception, though these technologies can enhance the usefulness of the GPS positioning information. The GPS provides critical positioning capabilities to military, civil, and commercial users around the world. The United States government created the system, maintains it, and makes it freely accessible to anyone with a GPS receiver (Anonymous, 2021).

GPS is a navigational system made up of a network of 24 satellites, placed into orbit by the Department of Defense. These satellites circle the globe twice a day in a very precise orbit and transmit signal information back to earth. GPS receivers take this information and use triangulation to calculate the user s exact location. GIS combines mapping software with database management tools to collect, organize and share many different types of information. GIS can store, display, and help your utility interpret virtually all the important attributes of your system. Mapping Services through TRWA Through the implementation of GPS and GIS technology, TRWA can effectively produce hardcopy and digital maps to be utilized by system personnel for water and/or wastewater facilities in the state.

The GPS project, originally Navstar GPS was started by the U.S. Department of Defense in 1973, with the first prototype spacecraft launched in 1978 and the full constellation of 24 satellites operational in 1993. Originally limited to use by the United States military, civilian use was allowed from the 1980s following an executive order from President Ronald Reagan after the Korean Air Lines Flight 007 incident. Advances in technology and new demands on the existing system have now led to efforts to modernize the GPS and implement the next generation of GPS Block IIIA satellites and Next Generation Operational Control System (OCX). The GPS does not require the user to transmit any data, and it operates independently of any telephonic or Internet reception, though these technologies can enhance the usefulness of the GPS positioning information. The GPS provides critical positioning capabilities to military, civil, and commercial users around the world. The United States government created the system, maintains and controls it, and makes it freely accessible to anyone with a GPS receiver.

9.2.1 How Does the GPS System work?

Precisely locating positions on Earth is not a new phenomenon. Navigators, sailors, explorers, and surveyors have done this for centuries as they traveled about the world. Most maps and globes display longitude and latitude or some other coordinate projection information. Points on Earth are given unique addresses on maps using specific coordinate systems (John Nowatzki *et.al.*, 2021). Agriculturists commonly use either a geographic system of latitude and longitude measured in degrees or a Universal Transverse Mercator coordinate system that locates positions in meters measured from a specific point. The GPS system uses measured distances to the precisely located GPS satellites to locate positions on Earth. Radio receivers in GPS units monitor radio signals broadcast from the GPS satellites. A GPS position is determined by simultaneously measuring the distance to at least three satellites. The distance to a satellite is measured by the time it takes a radio signal to travel from the satellite to the GPS receiver.

Computers in GPS units use information from the radio signals, including broadcast time and unique satellite information, to calculate positions. Information from at least four satellites is needed to calculate elevation. Signal reception from more satellites increases position accuracy. The global positioning system includes a constellation of 24 systematically arranged satellites orbiting Earth in six orbital planes with four satellites in each plane. The satellite orbits are approximately 12,500 miles above Earth. The constellation is arranged to guarantee radio reception from at least four satellites from any location anytime, anywhere on Earth (Figs. 15 and 16). GPS receivers normally receive signals from eight to nine satellites in location without obstructions such as buildings or trees.

In short, GPS is a system of 30+ navigation satellites circling Earth. We know where they are because they constantly send out signals. A GPS receiver in your phone listens for these signals. Once the receiver calculates its distance from four or more GPS satellites, it can figure out where you are.

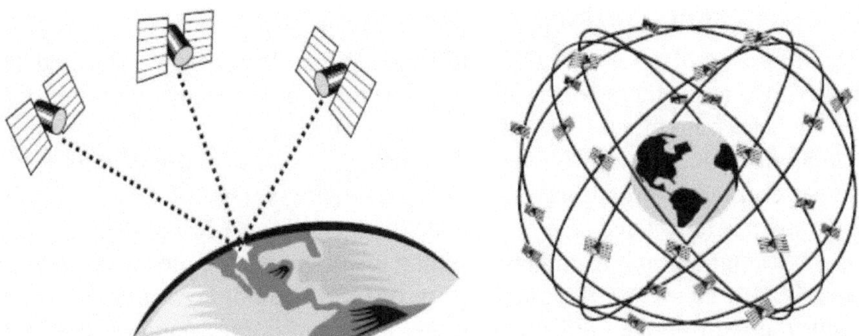

Fig. 15: Satellite System Representation **Fig. 16:** Satellite Constellation & Orbital Planes

Importance of GPS systems : GPS is important as it helps you to figure out where you are and where you are going when you are traveling from one place to another. Navigation and positioning are important but cumbersome activities, which GPS makes it easier. Once GPS locates your position, and then it starts tracing other factors like speed, bearing, tracks, trip distance, sunrise/sunset time, distance to destination and several other details. GPS uses 'man-made' stars as reference points to calculate positions accurate to a matter of meters. However, with recent forms of GPS you can make measurements much better than centimeter readings. So it is with the aid of GPS that you can give a unique and specified address to every square meter on the planet. So these days GPS finds its way into cars, planes, boats, construction equipments, smartphones, laptop computers and shoes and belts. In addition, GPS tracking system installed in the phone can greatly help an individual to get automated GPS information through their cell phones.

How accurate is GPS?: GPS device accuracy depends on many variables, such as the number of satellites available, the ionosphere, the urban environment and more.

Some factors that can hinder GPS accuracy include

- **Physical obstructions:** Arrival time measurements can be skewed by large masses like mountains, buildings, trees and more.
- **Atmospheric effects:** Ionospheric delays, heavy storm cover and solar storms can all affect GPS devices.
- **Ephemeris:** The orbital model within a satellite could be incorrect or out-of-date, although this is becoming increasingly rare.
- **Numerical miscalculations:** This might be a factor when the device hardware is not designed to specifications.
- **Artificial interference:** These include GPS jamming devices or spoofs.

9.2.2 Mobile Phone Tracking

The development of communications technology has long since surpassed the sole ability to access others when they are mobile. Today, mobile communication devices are becoming much more advanced and offer more than the ability to just carry on a conversation. Cell phone GPS tracking is one of those advances. All cell phones constantly broadcast a radio signal, even when not on a call. The cell phone companies have been able to estimate the location of a cell phone for many years using triangulation information from the towers receiving the signal (Patrick Bertagna, 2010). However, the introduction of GPS technology into cell phones has meant that cell phone GPS tracking now makes this information a lot more accurate.

9.2.3 Localization-based Systems and Errors

GSM localization is the use of multilateration to determine the location of GSM mobile phones, usually with the intent to locate the user. Localization-based systems can be broadly divided into Network based, Handset based and Hybrid.

i. **Network based:** Network-based techniques utilize the service provider's network infrastructure to identify the location of the handset. The advantage of network-based techniques (from mobile operator's point of view) is that they can be implemented non-intrusively, without affecting the handsets. The accuracy of network-based techniques varies, with cell identification as the least accurate and triangulation as the most accurate. The accuracy of network-based techniques is closely dependent on the concentration of base station cells, with urban environments achieving the highest possible accuracy.

ii. **Handset based:** Handset-based technology requires the installation of client software on the handset to determine its location for E-911 purposes. This technique determines the location of the handset by computing its location by cell identification, signal strengths of the home and neighboring cells, which is continuously sent to the carrier. In addition, if the handset is also equipped with GPS then significantly more precise location information is then sent from the handset to the carrier.

iii. **Hybrid:** Hybrid positioning systems use a combination of network-based and handset-based technologies for location determination. One example would be Assisted GPS, which uses both GPS and network information to compute the location. Hybrid-based techniques give the best accuracy of the three but inherit the limitations and challenges of network-based and handset-based technologies.

GPS errors : The quality of GPS units and operational errors associated with the GPS system determine the accuracy of GPS-located positions. There are several sources of GPS errors. GPS radio signals can "bounce off" objects such as buildings and trees prior to acquisition by the GPS receiver, resulting in lower accuracy. This is called multipath error. (Figs. 17 and 18) (John Nowatzki *et.al.*, 2021). The satellites use very accurate atomic clocks to generate the timing data received by the GPS receivers. However, even small errors in timing from clocks in the satellites and GPS units cause errors in GPS positions. Signal delay errors can be caused by atmospheric interference such as electrically charged particles in the ionosphere. A layer of water vapor located below the troposphere can also alter the speed of travel of radio signals. Errors from GPS satellites' orbit and location are also significant. Pressures from solar radiation and gravitational forces of the sun and moon can alter satellite locations. GPS receiver quality also affects GPS accuracy. More costly GPS units generally provide more accurate GPS positions than less expensive units.

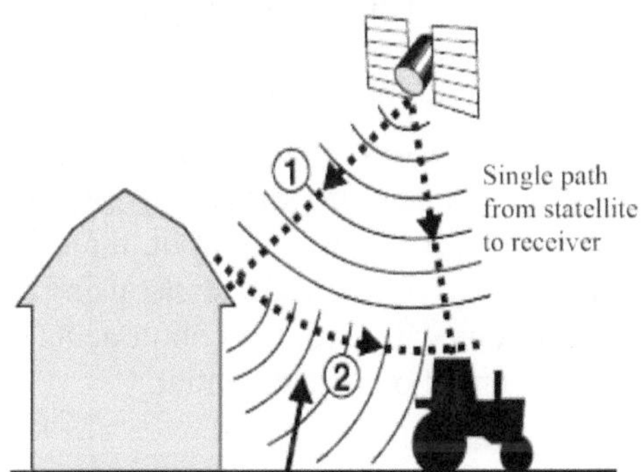

Fig. 17 : Multipath signal errors

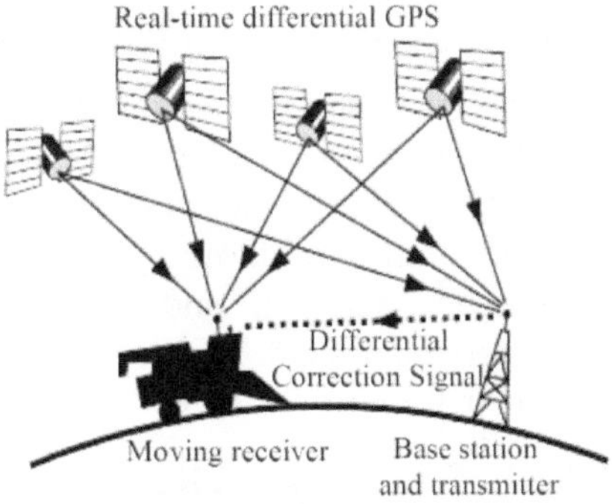

Fig. 18: Base station DGPS representation

9.3 Geographic Information System (GIS)

It is a conceptualized framework that provides the ability to capture and analyse spatial and geographic data. GIS applications (or GIS apps) are computer-based tools that allow the user to create interactive queries (user-created searches), store and edit spatial and non-spatial data, analyze spatial information output, and visually share the results of these operations by presenting them as maps. Simply, it is the scientific study of geographic concepts, applications, and systems.

Geographic information systems are utilized in multiple technologies, processes, techniques and methods. They are attached to various operations and numerous applications that relate to: engineering, planning, management, transport/logistics, insurance, telecommunications, and business.[2] For this reason, GIS and location intelligence applications are at the foundation of location-enabled services, that rely on geographic analysis and visualization. GIS provides the capability to relate previously unrelated information, through the use of location as the "key index variable". Locations and extents that are found in the Earth's space time, are able to be recorded through the date and time of occurrence, along with x, y, and z coordinates; representing, longitude (x), latitude (y), and elevation (z). All

Earth-based, spatial–temporal, location and extent references, should be relatable to one another, and ultimately, to a "real" physical location or extent. This key characteristic of GIS, has begun to open new avenues of scientific inquiry and studies.

The term and technology of GIS has existed for more than 25 years, and GIS has taken on several meanings. The acronym originally meant "Geographic Information Systems" and referred to the integrated software that enabled the collection, manipulation, analysis, and presentation of spatial data. Subsequently, as the numbers of users and applications grew, GIS was defined by some practitioners as "Geographic Information Science" stressing the analysis and rigor behind many of the applications used today. They point to areas of GIS such as remote sensing, modeling, and spatial statistics to show where GIS, under either definition, has matured from creating simple, static maps to a system that can produce dynamic, interactive applications (Kalaris Tom *et.al.*, 2014).

Applications and components of GIS: There are several applications of geoinformatics, viz., crop yield prediction, crop health monitoring, livestock monitoring, insect and pest control, irrigation control, flooding, erosion and drought control and farming automation. GIS enables the user to input, manage, manipulate, analyze, and display geographically referenced data using a computerized system. To perform various operations with GIS, the components of GIS such as software, hardware, data, people and methods are essential.

9.3.1 GIS in Biosecurity

The biosecurity system use GIS in several following ways.

- As tool for problem solving and decision making ;
- Visualisation of data in a spatial environment. ;
- Helping biosecurity teams to determine where things are
  i. The location of pests and diseases and the relationships to other pests and diseases
  ii. The quantities- where the most and/or least number of some pests or diseases are located
  iii. Densities
  iv. The density in any given space
  v. What is happening inside that area - what is happening nearby
  vi. Mapping the change - how a specific area has changed over time (and in what way)

9.3.1.1 GIS to Safeguard New and Emerging Diseases

The development of a real-time GIS-based (geographic information system) reporting system for new and emerging agricultural pathogens and pests is extremely relevant in the era of agricultural bioterrorism. The goal was to establish a real-time, GIS database network to detect, diagnose, report, monitor, map (temporally and spatially), and predict the spread of new and emerging plant diseases and pests. Such networks could also be used to geospatially and temporally monitor endemic pathogens/pests (Nutter Forrest, 2010).

Approach: The goal of this project was to enhance the value of the National Plant Diagnostic Network Center (NPDN). The NPDN has tremendous potential to facilitate the rapid exchange of critical diagnostic information among the five Regional Diagnostic Centers. During crises, the capability to exchange real-time information is paramount to effectively mitigate the potential impacts of new and emerging diseases and pests. This will require the rapid detection of new and emerging diseases and pests ; the documented geographical distribution (by using GIS to map disease/pest prevalence), the predicted geographic distribution and establishment of new and emerging diseases and pests ; the predicted risk (by generating real-time GIS risk maps) of pathogen/pest establishment beyond the initial point(s) of detection and the rapid dissemination of disease/pest management mitigation tactics. This goal can be accomplished only through a coordinated effort among the Regional Diagnostic Centers to develop compatible hardware and software platforms.

Rather than having each Regional Diagnostic Center develop its own GIS capabilities, it is more cost effective for our research team to develop compatible GIS, web-based capabilities for all five Regional Diagnostic Centers. Therefore, a primary goal of this project is to develop GIS, web-based products that will significantly enhance the value of NPDN and the Plant Diagnostic Information System (PDIS) database. Since ESRI GIS software is now the official USDA software package for all GIS applications, a dedicated ESRI GIS server will be used to allow institutional personnel to view specific geospatial data layers, such as historical/real time weather and confirmed disease/pest reports, etc. The Project Director and Principal Investigators have extensive experience in using ESRI GIS software to map disease/pest intensity (prevalence, incidence, severity) temporally and spatially at all spatial scales (within-field, among fields, county, state, region).

Vertical series of linked modules were developed, including: (i) a real-time, GIS disease/pest mapping capability that can generate real-time GIS maps (i.e. monitor the occurrence and spatial spread of new and emerging pathogens and pests in real-time), (ii) modified atmospheric transport models (such as HYSPLIT and MM5) to predict and geospatially depict pathogen (pest) short, meso, and long-distance dissemination, and, (iii) disease forecasting (warning) models to predict pathogen (pest) infection (establishment). These modules will be linked and delivered as a web-based system.

GIS to manage the spread of Plum pox virus: Viral diseases are among the main causes for declines in the productivity of stone-fruit plantations and their adaptability to abiotic factors. Virus-infected plants are often more susceptible to adverse ambient conditions and more sensitive to fungal and bacterial diseases (Pavlova *et.al.*, 2017). To date, Plum pox virus (PPV) is subject to quarantine list A2 of the European Plant Protection Organization (EPPO). The virus is on the list of regulated pests common in Ukraine, and is subject to internal quarantine. In Ukraine, the areas of PPV spread amount to 40,132,764 ha. In Odessa region, districts totalling 18.5 ha are in PPV quarantine. Six hotbeds of PPV infection totalling 28 ha were found in Odessa region. For the first time in Odessa region, PPV was found on cherry trees. PPV-D was the only strain of the virus detected in every hotbed in Odessa region. To conduct phytosanitary zoning and determine potential areas for the progress of the disease in Odessa region, we have used the modern method of geographical information system (GIS) technology. GIS-aided maps were constructed showing PPV spread in Odessa region. This is probably the first attempt to employ GIS technology to control plant pests in Ukraine. Provided it is properly maintained, the geospatial data, and the ability to generate detailed maps with it, is key to the success of PPV containment.

9.3.1.2 GIS in Dealing with Climate Change and Disaster Risks

The Land Resources Division of SPC is conducting a GIS consultation and training exercise in collaboration with Fiji's Ministry of Primary Industries (MPI), seeking broad inputs on the spatial information needed to improve decision making, and help identify and monitor climate change and disaster risks in the agriculture and forestry sectors. This collaborative activity is supported by the SPC/GIZ programme: Coping with Climate Change in the Pacific Island Region (CCCPIR), and is taking place in SPC Fiji from 28 May to 1 June 2012 (Anju Mangal, 2012). The consultation is intended to give LRD staff, and personnel from the Agriculture and Forestry Departments of the MPI an opportunity to learn about the potential uses of GIS in their climate change activities. This critical technology can provide useful information to help identify current and future risks and vulnerabilities, and can be used in designing and implementing adaptation programmes.

The consultation covers a vast pool of knowledge, information and best practices, in applying GIS to climate change. A number of datasets within LRD (such as those for pests, plant genetic resources and crop production) have been identified as suitable for mapping. These datasets will then be combined with other environmental datasets, such as climate and land cover, to measure their current distribution and potential future distribution under various climate change scenarios. For example, the LRD plant health team saw that mapping the results of plant pest and disease surveys on GIS can help to better understand current distributions, and the presence of specific plant pests and diseases, and future risks associated with the spread of these pests and diseases as a result of climate change and other factors.LRD is delivering a GIS training package in consultation with MPI, A second training workshop and, follow-up consultations, are scheduled for LRD staff and MPI personnel from the 18 June to the 6 July, 2012. A crucial outcome of this training will be improved GIS capability for both LRD and MPI, and long-term action plans, that identifies further needs for both organisations. It will also identify key climate change applications that would be useful in the context of climate change adaptation planning. This will be used to support training and the implementation of activities at the national level.

9.3.1.3 GIS as a Project Management Tool

All biosecurity data has a spatial component, hence it can be represented on a map. As such, a GIS is critical to any control/management/survey programme conducted by a biosecurity agency. A good illustration of a GIS as a management tool is the Citrus Canker Eradication Programme that occurred in Miami, Florida, during 2003–2006. Critical information that the programme needed to track for each individual property included, viz., citrus on the premises? When was a particular property last surveyed? Were suspect trees found? Have the canker finds been confirmed by a pathologist? Have homeowners given permission for tree removal? Were trees removed and destroyed? Are there other properties within 200 m with citrus trees?

This programme encompassed more than 100,000 properties and a million people. The ability of a manager to visualize the data on a map is invaluable. A GIS linked to the programme's database can assist with answering these kinds of questions, *viz*., Where have we surveyed and where will we need to survey next? How has a new positive detection changed our surveys? In which areas will we need more personnel? Which areas are ahead of or behind schedule? How is the pest spreading? Where are our management procedures not working? What areas are of greater risk? GIS can be an effective tool for data management and project oversight. GIS can help collect field data with tools like ArcPad, it can be used for QA/QC, and it can assist in managing personnel and equipment. A GIS technician is given a small project to carry out and manage, such as overseeing the update of a layer or performing a specific analysis and presenting the results. Managing simple projects like these may sound easy at first, since the new project manager has undoubtedly been part of project teams in the past. GIS will allow project managers and different people involved in project with different backgrounds to get the information about the progress of the project and support Decision Making. GIS will provide a common basis of understanding and communication among these people. GIS software lets you produce maps and other geographic displays to analyze and present information. The displays typically include points, lines, areas or raster images (from photos or scanned images).

9.3.1.4 GIS As An Analysis Tool

GIS provides tools that allow a more inclusive and complex analysis to support managers and programmes. Data may come from many sources and may include: Point data (survey data); line data (rivers and roads); polygon data (quarantine areas); or remotely sensed data (aerial photographs and satellite images). GIS allows an analyst to incorporate various kinds of data into a model to help predict introduction of a pest, the spread of a pest, the best quarantine boundaries, etc. One example of this kind of work is an analysis of Asian Gypsy Moth (AGM) trap locations in Washington and Oregon states. The goal of this analysis was to support trap placement by the two states' Departments of Agriculture. The analyst created a map showing areas of likely introduction. This "model" was based on many layers and combines data on population, trade, roads; ports, waterways, hosts, and intermodal facilities to estimate the risk on introduction of Asian Gypsy Moth into Washington and Oregon. The red areas are of higher risk.

The map from the model was compared to the current (previous year's) trap locations. For the project managers in Washington and Oregon, this was a valuable comparison to see where their map agreed with their trapping and where the two differed. It helped them evaluate their trap locations and trap densities. Note that the GIS model does not dictate to the manager where to place their traps. It is a tool to assist managers in the allocation of their resources.

9.3.1.5 GIS As A Communication Tool

The third area of focus for this brief overview of GIS involves using maps as a communication tool. Most people can read and interpret maps, which make maps ideal reporting, and summarizing tools. The legal description of a quarantine area is ponderous and difficult to visualize. In contrast, a map can clearly show the area. Maps can also show a manager or supervisor areas that have been surveyed, the locations of positives, and the locations that the programme will survey next. A map can summarize a risk or models into an easily understood package. Most people would rather look at a map than a table of figures. A map can't replace a detailed tabulation, but it can make the numbers more easily understood. The key is to produce maps that present their message in a clear, concise, and understandable manner.

The integration of remote sensing, Global Positioning Systems (GPS), and Geographic Information Systems (GIS) technologies, coupled with geostatistical analyses, can provide valid, science-based evidence concerning the presence and geospatial, distribution of invasive plant pathogens. Using the soybean rust pathosystem as a model, Nutter (2013) successfully extracted pathogen-specific spatial and temporal patterns from aerial, satellite, and ground-based sensors that can be used to detect, and accurately differentiate soybean rust, from other soybean diseases (with close to 100% accuracy). Using a GIS script that was developed called Gradient Finder, pathogen-specific spatial and temporal patterns can now be used to detect, identify, and map within-field anomalies caused by plant diseases. Using this approach, soybean rust disease foci can now be easily distinguished from disease patches caused by sudden death syndrome. Geospatial analyses can then be used to determine if spatial patterns are indicative of a natural or a deliberate introduction (i.e., a crime scene). The integration of remote sensing,

GPS, and GIS technologies can also be used to deliver precise GPS coordinates as to where investigators on the ground should obtain pathogen isolates (and other evidence) for genetic analyses concerning the population structure of pathogen isolates.

9.3.1.6 GIS in Plant Disease Epidemiology

GIS, being a computer system that can be installed on any recent model desktop computer and is capable of assembling, storing, manipulating, and displaying data that are referenced by geographic coordinates. GIS is adaptable to operations of any size, and data can be used at any scale from a single field to an agricultural region (George N. Agrios,2005). It is used to better understand and manage the environment, including the understanding and management of plant disease epidemics. GIS techniques allow one to make connections between events based on geographic proximity, connections that are essential to the understanding and management of epidemics but which often go unrecognized without GIS. GIS techniques can even incorporate disease forecasting systems, although the time and cost for it may be prohibitive. However, as high-resolution weather forecast data are often available, the development of plant disease epidemics can be predicted by knowing their dependency on some critical weather variable and from estimated geographic distribution of the pathogen inoculum within a GIS framework. GIS is often used for the spatial and temporal analysis of disease development over relatively large geographic areas and helps determine the role and relative importance of various parts of these areas in the initiation and development of the epidemic.

9.3.1.7 GIS in Disease Monitoring

Pests and diseases of bananas (Musa spp.) threaten the livelihoods of over 20 million people in the Great Lakes region. GIS provide valuable tools in monitoring, predicting, managing and fighting the spread of pests and diseases. The tools offer opportunities for cost-effective and efficient targeting of control interventions. In monitoring, GIS can be used to determine the spatial extent of a disease, to identify spatial patterns of the disease and to link the disease to auxiliary spatial data (Bouwmeester, *et.al.*, 2010). GIS can also be used to predict the projected spread of diseases, to provide input for risk assessment models in pest control and in quantifying changing thresholds of pests and diseases due to climate change. In order to use GIS techniques at a larger scale, a protocol for data collection and management is essential. This paper illustrates the use of GIS tools on data collected to identify critical intervention areas to combat the spread of Banana *Xanthomonas* wilt (BXW). In a survey covering the Great Lakes region, on-farm incidence of the disease was monitored and precise GPS coordinates of each sampled field were recorded. This enabled accurate mapping of the disease and performing the various spatial analyses, permitting an understanding of the geographical distribution of BXW infection. Data on food security and dependency on banana to rural populations was linked to the BXW severity to target priority areas of interventions and maximize impact.

9.4 Integrating GPS, GIS, and Geostatistics

GPS, GIS, and geostatistics can be used to advantage at different levels of management and expertise. For example, an extension agent with a handheld GPS unit could relay the coordinates of a serious plant disease situation to a specialist at a land grant college, who could use GIS to place the observation in a broader context with similar observations coming in from elsewhere. One of the most helpful applications of this technology is simple, to store and query spatially referenced point data (Merritt R. Nelson *et.al.*, 1999). The benefits of integration come from relating this simple format data to other data (roads, county boundaries, digital elevation models, hydrography, land ownership boundaries, and most importantly, other agricultural data) that are increasingly available from public and private sources via the World Wide Web. Where appropriate, point data can be interpolated to produce surface maps using geostatistics.

GIS maps of plant diseases and risk of plant disease can be created in a number of ways. Experts, including farmers, can synthesize personal observations from memory and generate maps simply by sketching polygons over appropriately referenced backgrounds. For example, a risk map at the continental scale of gray leaf spot of corn based on observations of the NCR-25 technical committee on corn and sorghum diseases was presented as part of APSnet Feature, 1 through 31 May 1998. Spatially located field observations of incidence combined with weather data can be used effectively for certain aerially dispersed fungal diseases. Continental-scale risk maps for blue mold of tobacco have been published on the World Wide Web 3 days a week during the tobacco growing season since 1996 by the North American Blue Mold Forecast Center at North Carolina State University. GIS risk assessment maps based on kriging of field observations were part of a tomato virus management program at the regional scale in the Del Fuerte Valley, Sinaloa, Mexico, discussed below. It is predicted that plant disease risk maps at a variety of scales will become increasingly available and helpful. Through an iterative process of comparing spatial data on risk with observed incidence, our understanding of the spatial and temporal aspects of disease processes should improve along with the ability to produce maps useful to farmers.

Research is needed on cost/benefit analyses of GIS applications in agriculture. GIS is of little practical value when a spatial pattern is uniform except perhaps in contrast to nonuniform patterns. On the other extreme, if the change in pattern is faster than data can be acquired, spatially referenced data might be skewed and misleading. There are theoretical limitations to the use of geostatistics, because the geostatistical model in which spatial analyses lead to surface maps does not always apply. Also, certain assumptions may not be appropriate in some cases (Nelson *et.al.*, 1999). A geostatistical analysis does not need to be a part of every spatial analysis or GIS. There is a personal subjective component to the suitability of this technology in plant disease management. Some people are more inclined than others to frame problems spatially. It is suspected that benefits from the communication possibilities of the technology will vary with individuals. The current most successful applications of this technology involve teamwork. A good combination is a team that includes an experienced field person, an experienced computer user with some background in statistics, and a patient data entry person. Such a combination is not always available and is rare in a single individual. Larger farms, younger farmers, and better-funded extension programs are more likely to apply these tools. A characteristic of agricultural GIS is that it strongly favors operations that have computer resources and skills. On the other hand, it may not be practical to avoid a resource that is already in wide use in government and industry, including the agricultural industry.

It is believed that the integration of GIS, GPS, and geostatistics provides a tool for the refined analysis of traditional and contemporary biological/ecological information on plant diseases. It will aid practitioners in the design of disease management in IPM programs, particularly on a regional scale. It will also provide a way of analyzing and communicating results of regional programs on a continuing basis. The availability of software capable of producing attractive maps provides an opportunity to visually communicate plant disease situations to a variety of audiences. A major benefit of this technology is derived from a combination of many images, not just a single image. Such output can be used by decisionmakers to stimulate coordinated action allowing available resources to be focused on the most significant problems. GIS and geostatistics are particularly useful in identifying recurring patterns of plant disease, as well as other problems such as insect and weed infestations. The association of environmental factors, landscape features and cropping patterns with the recurrence of disease or other problems can be readily communicated to key managers and decisionmakers.

Bibliographhy

Akagbuo Nwauzoma, 2016. A review of geographic information systems and digital imaging in plant pathology application. *African Journal of Agricultural Research.* 11: 4172-4180. 10.5897/AJAR2016.11641.

Anju Mangal, 2012. The role of geographical information systems in dealing with climate change and disaster risks.SPS Land Resources Division, SPC Campus, Nabua.

Anonymous, 2021. "What is GPS?". February 22, 2021. Archived from the original on May 6, 2021. Retrieved May 5, 2021.

Bouwmeester, H., Abele, S., Manyong, V.M., Legg, C., Mwangi, M., Nakato, V., Coyne, D. and Sonder, K.,2010. The Potential Benefits of GIS Techniques in Disease And Pest Control: An Example Based on a Regional Project in Central Africa. *Acta Hortic.*, 879: 333-340

George N. Agrios, 2005. Plant disease epidemiology, George N. Agrios (Ed.), Plant Pathology (Fifth Edition),Academic Press, 265-291

John Nowatzki, Vern Hofman, Lowell Disrud and Kraig Nelson, 2021. GPS Applications in Crop Production. National institute of Food and Agriculture, USDA

Kalaris Tom, Fieselmann Daniel, Magarey Roger, Colunga-Garcia Manuel, Roda Amy, Hardie Darryl, Cogger Naomi, Hammond Nichole, Martin P.A. Tony and Whittle Peter, 2014. "Chapter 11 The Role of Surveillance Methods and Technologies in Plant Biosecurity" . Other Publications in Zoonotics and Wildlife Disease. 172.

Lindgren Cory, 2011. Biosecurity Policy and the Use of Geospatial Predictive Tools to Address Invasive Plants: Updating the Risk Analysis Toolbox. Risk analysis: An Official Publication of the Society for Risk Analysis. 32. 9-15. 10.1111/j.1539-6924.2011.01642.x.

Lisa Kennaway, 2009. Using Geospatial Technology for Pest Monitoring and Detection. USDA APHIS Plant Protection and Quarantine (PPQ), Center for Plant Health Science & Technology (CPHST), Fort Collins, Colorado National Plant Diagnostic Network Meeting

Md. Arshad Anwer and Garima Singh, 2019. Geo-spatial Technology for Plant Disease and Insect Pest Management. *Bull. Env. Pharmacol. Life Sci.*, 8: 01-12

Mehta Vasu & Koranga Radha, 2020. Remote Sensing, GPS, GIS and Geostatistics in Agriculture: An Overview. In book: Advances in Agricultural Extension (pp.19-30), Publisher: AkiNik Publications

Merritt R. Nelson, Thomas V. Orum, Ramon Jaime-Garcia and Athar Nadeem, 1999. Applications of Geographic Information Systems and Geostatistics in Plant Disease Epidemiology and Management. *Plant Disease,* 83: 308-320

Nelson, M. R., Orum, T. V., Jaime-Garcia, R., & Nadeem, A. (1999). Applications of geographic information systems and geostatistics in plant disease epidemiology and management. *Plant disease*, 83(4), 308-319

Nutter , F., 2013. Forensic epidemiology: New sensor-based plant pathogen detection: Where to look for evidence in a 300-acre crop. The American Phytopathological Society, Iowa State University, Ames, IA, U.S.A.

Nutter Forrest, 2010. Safeguarding American Agriculture from New and Emerging Diseases and Pests: A GIS and Web-Based Disease Monitoring, Forecasting, and Information Delivery System. Iowa State University,USDA.

Nwauzoma, A. B., 2016. A review of geographic inofrmaiton systems and digital imaging in plant pathology application. *African J. Agri. Res.*, 11:4172-4180

Pavlova, S., Stakhurska, O., Budzanivska, I., Shevchenko, O. and Polischuk, V., 2017. Using geographical information system (GIS) technology to control the spread of Plum pox virus in Odessa region, Ukraine. *Acta Hortic.*, 1163: 147-152

Sabtu , N. M., Idris, N. H. and Ishak, M. H. I., 2018. The role of geospatial in plant pests and diseases: an overview. IOP Conf. Series: *Earth and Environ. Sci.*, 169: 012013

10

Pest, Disease and Epidemic Management

The term "plant pest", mainly applied to insect micropredators of plants, has a specific definition in terms of the International Plant Protection Convention and phytosanitary measures worldwide. A pest is any species, strain or biotype of plant, animal, or pathogenic agent injurious to plants or plant products. Worldwide, agricultural pest impacts are increased by higher degrees of interconnectedness. This is due to the increased risk that any particular pest problem anywhere in the world (as a system) will propagate across the entire system.

10.1 Pest

A pest is any animal or plant harmful to humans or human concerns. The term is particularly used for creatures that damage crops, livestock, and forestry or cause a nuisance to people, especially in their homes. Humans have modified the environment for their own purposes and are intolerant of other creatures occupying the same space when their activities impact adversely on human objectives. Thus, an elephant is unobjectionable in its natural habitat but a pest when it tramples crops.

'Pests' are defined by the Food and Agriculture Organization (FAO) as any species, strain or biotype of plant, animal or pathogenic agent injurious to plants or plant products, i.e. insects, mites, molluscs, nematodes, diseases and weeds. Indigenous pests may be chronic or occur in outbreaks, whereas introduced pests usually occur in an initial outbreak followed by continuous chronic damage. Both types of pest can cause severe losses. The damage they do results both from the direct injury they cause to the plants and from the indirect consequences of the fungal, bacterial or viral infections they transmit. Plants have their own defences against these attacks but these may be overwhelmed, especially in habitats where the plants are already stressed, or where the pests have been accidentally introduced and may have no natural enemies. The pests affecting trees are predominantly insects, and many of these have also been introduced inadvertently and lack natural enemies, and some have transmitted novel fungal diseases with devastating results.

A pest is any living thing, whether animal, plant, or fungus, which humans consider troublesome to themselves, their possessions, or the environment. It is a loose concept, as an organism can be a pest in one setting but beneficial, domesticated, or acceptable in another. Microorganisms, whether bacteria, microscopic fungi, protists, or viruses that cause trouble, on the other hand, are generally thought of as causes of disease (pathogens) rather than as pests (Julius Kühn-Institut, 2020). An older usage of the word "pest" is of a deadly epidemic disease, specifically plague. In its broadest sense, a pest is a competitor to humanity.

i. Animals as pests : Animals are considered pests or vermin when they injure people or damage crops, forestry, or buildings. Elephants are regarded as pests by the farmers whose crops they raid and trample. Mosquitoes and ticks are vectors that can transmit ailments but are also pests because of the distress caused by their bites. Grasshoppers are usually solitary herbivores of little economic importance until the conditions are met for them to enter a swarming phase, become locusts and cause enormous damage. Many people appreciate birds in the countryside and their gardens, but when these accumulate in large masses, they can be a nuisance. Flocks of starlings can consist of hundreds of thousands of individual birds, their roosts can be noisy and their droppings voluminous; the droppings are acidic and can cause corrosion of metals, stonework, and brickwork as well as being unsightly. Pigeons in urban settings may be a health hazard, and gulls near the coast can become a nuisance, especially if they become bold enough to snatch food from passers-by. All birds are a risk at airfields where they can be sucked into aircraft engines. Woodpeckers sometimes excavate holes in buildings, fencing and utility poles, causing structural damage; they also drum on various reverberatory structures on buildings such as gutters, down-spouts, chimneys, vents and aluminium sheeting. Jellyfish can form vast swarms which may be responsible for damage to fishing gear, and sometimes clog the cooling systems of power and desalination plants which draw their water from the sea.

Many of the animals that we regard as pests live in our homes. Before humans built dwellings, these creatures lived in the wider environment, but co-evolved with humans, adapting to the warm, sheltered conditions that a house provides, the wooden timbers, the furnishings, the food supplies and the rubbish dumps. Many no longer exist as free-living organisms in the outside world, and can therefore be considered to be domesticated (Jones Richard, 2015).

ii. Plants as pests: Plants may be considered pests, for example, if they are invasive species or weeds. There is no universal definition of what makes a plant a pest. Some governments, such as that of Western Australia, permit their authorities to prescribe as a pest plant "any plant that, in the local government authority's opinion, is likely to adversely affect the environment of the district, the value of property in the district, or the health, comfort or convenience of the district's inhabitants." An example of such a plant prescribed under this regulation is caltrop, Tribulus terrestris, which can cause poisoning in sheep and goats, but is mainly a nuisance around buildings, roadsides and recreation areas because of its uncomfortably sharp spiny burrs.

iii. Other organisms as pests: Some definitions encompass any hazardous or problematic organism, and so often include fungi, oomycetes, bacteria, nematodes and viruses.

10.2 Plant Disease and Pathogens

It is defined as the state of local or systemic abnormal physiological functioning of a plant, resulting from the continuous, prolonged 'irritation' caused by phytopathogenic organisms (infectious or biotic disease agents). Organisms causing a plant disease are known as pathogens.

Plant pathogens : In most pathosystems, virulence is dependent on hydrolases and the wider class of cell wall degrading proteins that degrade the cell wall. The vast majority of CWDPs are pathogen-produced and pectin-targeted (for example, pectinesterase, pectate lyase, and pectinases). For microbes the cell wall polysaccharides are themselves a food source, but mostly just a barrier to be overcome. Many pathogens also grow opportunistically when the host breaks down its own cell walls, most often during fruit ripening.

i. Fungi : Most phytopathogenic fungi belong to the Ascomycetes and the Basidiomycetes. The fungi reproduce both sexually and asexually via the production of spores and other structures. Spores may be spread long distances by air or water, or they may be soil borne. Many soil inhabiting fungi are capable of living saprotrophically, carrying out the part of their life cycle in the soil. These are facultative saprotrophs. Fungal diseases may be controlled through the use of fungicides and other agriculture practices. However, new races of fungi often evolve that are resistant to various fungicides. Biotrophic fungal pathogens colonize living plant tissue and obtain nutrients from living host cells. Necrotrophic fungal pathogens infect and kill host tissue and extract nutrients from the dead host cells. Significant fungal plant pathogens include the following.

Ascomycetes : Fusarium spp. (Fusarium wilt disease), Thielaviopsis spp. (canker rot, black root rot, Thielaviopsis root rot), Verticillium spp., Magnaporthe grisea (rice blast), Sclerotinia sclerotiorum (cottony rot).

Basidiomycetes: Ustilago spp. (smuts) smut of barley, Rhizoctonia spp., Phakospora pachyrhizi (soybean rust), Puccinia spp. (severe rusts of cereals and grasses), Armillaria spp. (honey fungus species, virulent pathogens of trees).

ii. Fungus-like organisms

Oomycetes : The oomycetes are fungus-like organisms. They include some of the most destructive plant pathogens including the genus Phytophthora, which includes the causal agents of potato late blight[4] and sudden oak death. Particular species of oomycetes are responsible for root rot. Despite not being closely related to the fungi, the oomycetes have developed similar infection strategies (Davis, 2009). Oomycetes are capable of using effector proteins to turn off a plant's defenses in its infection process. Plant pathologists commonly group them with fungal pathogens. Significant oomycete plant pathogens include Pythium spp.and Phytophthora spp., including the potato blight of the Great Irish Famine (1845–1849).

iii. Phytomyxea : Some slime molds in Phytomyxea cause important diseases, including club root in cabbage and its relatives and powdery scab in potatoes. These are caused by species of *Plasmodiophora* and *Spongospora*, respectively.

iv. Bacteria : Most bacteria that are associated with plants are actually saprotrophic and do no harm to the plant itself. However, a small number, around 100 known species, are able to cause disease.[8] Bacterial diseases are much more prevalent in subtropical and tropical regions of the world. Most plant pathogenic bacteria are rod-shaped (bacilli). In order to be able to colonize the plant they have specific pathogenicity factors. Five main types of bacterial pathogenicity factors are known: uses of cell wall–degrading enzymes, toxins, effector proteins, phytohormones and exopolysaccharides.

Pathogens such as *Erwinia* species use cell wall–degrading enzymes to cause soft rot. *Agrobacterium* species change the level of auxins to cause tumours with phytohormones. Exopolysaccharides are produced by bacteria and block xylem vessels, often leading to the death of the plant. Bacteria control the production of pathogenicity factors via quorum sensing.

Significant bacterial plant pathogens:

- Burkholderia
- Proteobacteria
  - *Xanthomonas* spp.
  - *Pseudomonas* spp.
- *Pseudomonas* syringae pv. tomato causes tomato plants to produce less fruit, and it "continues to adapt to the tomato by minimizing its recognition by the tomato immune system."

v. Phytoplasmas and Spiroplasmas: Phytoplasma and Spiroplasma are genera of bacteria that lack cell walls and are related to the mycoplasmas, which are human pathogens. Together they are referred to as the mollicutes. They also tend to have smaller genomes than most other bacteria. They are normally transmitted by sap-sucking insects, being transferred into the plant's phloem where it reproduces.

vi. Viruses, viroids and virus-like organisms: There are many types of plant viruses, and some are asymptomatic or latent. Under normal circumstances, plant viruses cause only a loss of crop yield. Therefore, it is not economically viable to try to control them, the exception being when they infect perennial species, such as fruit trees. Most plant viruses have small, single-stranded RNA genomes. However, some plant viruses also have double stranded RNA or single or double stranded DNA genomes. These genomes may encode only three or four proteins: a replicase, a coat protein, a movement protein, in order to allow cell to cell movement through plasmodesmata, and sometimes a protein that allows transmission by a vector. Plant viruses can have several more proteins and employ many different molecular translation methods.

Plant viruses are generally transmitted from plant to plant by a vector, but mechanical and seed transmission also occur. Vector transmission is often by an insect (for example, aphids), but some fungi, nematodes, and protozoa have been shown to be viral vectors. In many cases, the insect and virus are specific for virus transmission such as the beet leafhopper that transmits the curly top virus causing disease in several crop plants (Creamer *et.al.*, 2005). One example is mosaic disease of tobacco where leaves are dwarfed and the chlorophyll of the leaves is destroyed. Another example is Bunchy top of banana, where the plant is dwarfed, and the upper leaves form a tight rosette.

vii. Nematodes: Nematodes are small, multicellular, wormlike animals. Many live freely in the soil, but there are some species that parasitize plant roots. They are a problem in tropical and subtropical regions of the world, where they may infect crops. Potato cyst nematodes (*Globodera pallida* and *G. rostochiensis*) are widely distributed in Europe and North and South America and cause $300 million worth of damage in Europe every year. Root knot nematodes have quite a large host range, they parasitize plant root systems and thus directly affect the uptake of water and nutrients needed for normal plant growth and reproduction, whereas cyst nematodes tend to be able to infect only a few species (Huynh *et.al.*, 2016). Nematodes are able to cause radical changes in root cells in order to facilitate their lifestyle.

viii. Protozoa and algae: There are a few examples of plant diseases caused by protozoa (e.g., *Phytomonas*, a kinetoplastid). They are transmitted as durable zoospores that may be able to survive in a resting state in the soil for many years. Further, they can transmit plant viruses. When the motile zoospores come into contact with a root hair they produce a plasmodium which invades the roots. Some colourless parasitic algae (e.g., *Cephaleuros*) also cause plant diseases.

ix. Parasitic plants: Parasitic plants such as broomrape, mistletoe and dodder are included in the study of phytopathology. Dodder, for example, can be a conduit for the transmission of viruses or virus-like agents from a host plant to a plant that is not typically a host, or for an agent that is not graft-transmissible.

10.3 Categories of Pests

i. Based on occurrence following are pest categories

- **Regular pest:** Frequently occurs on crop - Close association e.g. Rice slem borer, Brinjal fruit borer

- **Occasional pest:** Infrequently occurs, no close association e.g. Caseworm on rice, Mango stem borer
- **Seasonal pest:** Occurs during a particular season every year e.g. Red hairy caterpillar on groundnut, Mango hoppers
- **Persistent pests:** Occurs on the crop throughout the year and is difficult to control e.g. Chilli thrips, mealy bug on guava
- **Sporadic pests:** Pest occurs in isolated localities during some period. e.g. Coconut slug caterpillar

ii. Based on level of infestation

- **Pest epidemic:** Sudden outbreak of a pest in a severe form in a region at a particular time e.g. Brown plant hopper in Tanjore, Tamil Nadu, India
- **Endemic pest:** Occurrence of the pest in a low level in few pockets, regularly and confined to particular area e.g. Rice gall midge in Madurai, Mango hoppers in Periyakulam, Tamil Nadu, India

Pest categories

i. **Key pest :** Most severe and damaging pests; GEP lies above EIL always ; Spray temporarily bring population below EIL ; These are persistent pests; The environment must be changed to bring GEP below EIL e.g. Cotton bollworm, Diamond backmoth.

ii. **Major pest:** GEP lies very close to EIL or coincides with EIL ; Economic damage can be prevented by timely and repeated sprays e.g. Cotton jassid, Rice stem borer

iii. **Minor pest/Occasional pest :** GEP is below the EIL usually ; Rarely they cross EIL ; Can be controlled by spraying e.g. Cotton stainers, Rice hispa, Ash weevils

iv. **Sporadic pests :** GEP generally below EIL ; Sometimes it crosses EIL and cause severe loss in some places/periods e.g. Sugarcane pyrilla, White grub, Hairy caterpillar

v. **Potential pests :** They are not pests at present ; GEP always less than EIL ; If environment changed may cause economic loss e.g. S. litura is potential pest in North India

Causes of pest outbreak: Activity of human beings which upsets the biotic balance of ecosystem is the prime cause for pest outbreak. The following are some human interventions , reason for outbreak

i. **Deforestation :** Pest feeding on forest trees are forced to feed on cropped ; Biomass/unit area more in forests than agricultural land ; Weather factors also altered ; Affects insect development.

ii. **Destruction of natural enemies :** Due to excess use of insecticides, natural enemies are killed ; This affects the natural control mechanism and pest outbreak occurs, e.g. Synthetic pyrethroid insecticides kill NE.

iii. **Intensive and extensive cultivation:** Monoculture (Intensive) leads to multiplication of pests ; Extensive cultivation of susceptible variety in large area ; No competition for food ; multiplication increases e.g. Stem borers in rice and sugarcane.

iv. **Introduction of new varieties and crops:** Varieties with favourable physiological and morphological factors cause multiplication of insects. e.g. Succulent, dwarf rice varieties favour leaf folder Combodia cotton favours stem weevil and spotted bollworm Hybrid sorghum (CSH 1), cumbu (HB1) favour shoot flies and gall midges.

v. **Improved agronomic practices :** Increased N fertilizer - High leaf folder incidence on rice ; Closer planting - BPH and leaf folder increases ; Granular insecticides - Possess phytotonic effect on rice

vi. **Introduction of new pest in new environment :** Pest multiplies due to absence of natural enemies in new area Apple wooly aphid Eriosoma lanigerum multiplied fast due to absence of Aphelinus mali (Parasite)

vii. **Accidental introduction of pests from foreign countries (through air/ sea ports):**
   - Diamondback moth on cauliflower (Plutella xylostella)
   - Potato tuber moth (Phthorimaea operculella)
   - Cottony cushion scale (Icerya purchase) on wattle tree
   - Wooly aphid (Eriosoma lanigerum) on apple
   - Psyllid (Heteropsylla cubana on subabul
   - Spiralling whitefly (Adeyrodichus dispersus) on most of horticultural crops

viii. **Large scale storage of food grains :** Serve as reservoir for stored grain pests urbanization - changes ecological balance.

10.4 Pest Epidemic

An epidemic (from Greek epi "upon or above" and demos "people") is the rapid spread of disease to a large number of people in a given population within a short period of time. For example, in meningococcal infections, an attack rate in excess of 15 cases per 100,000 people for two consecutive weeks is considered an epidemic (Green *et.al.*, 2002).

Epidemics of infectious disease are generally caused by several factors including a change in the ecology of the host population (e.g., increased stress or increase in the density of a vector species), a genetic change in the pathogen reservoir or the introduction of an emerging pathogen to a host population (by movement of pathogen or host). Generally, an epidemic occurs when host immunity to either an established pathogen or newly emerging novel pathogen is suddenly reduced below that found in the endemic equilibrium and the transmission threshold is exceeded.Simply, epidemic refers to the 'sudden outbreak of a pest in a severe form in a region at a particular time.

Examples; Lepidopterous fruit borers are generally the most important pests affecting production. Other important species include various leaf- and flower-eating caterpillars and beetles, bark borers, scales, leaf mites, fruit-sucking bugs, fruit-piercing moths and fruit flies etc., Under diseases, blue mold of tobacco (*Peronospora tabacina*), downy mildews of vine crops (*Pseudoperonospora cubensis*) and lima beans (*Phytophthora phaseoli*), late blight of potato and tomato (*Phytophthora infestans*), leaf spot of sugar beets (*Cercospora beticola*), and leaf rust of wheat (*Puccinia recondita tritici*) etc.,

The spread of transboundary plant pests and diseases has increased dramatically in recent years. Globalization, trade and climate change, as well as reduced resilience in production systems due to decades of agricultural intensification, have all played a part. Transboundary plant pests and diseases can easily spread to several countries and reach epidemic proportions. Outbreaks and upsurges can cause huge losses to crops and pastures, threatening the livelihoods of vulnerable farmers and the food and nutrition security of millions at a time. Locusts, armyworm, fruit flies, banana diseases, cassava diseases and wheat rusts are among the most destructive transboundary plant pests and diseases.

Plant pests and diseases spread in three principal ways, viz., trade or other human-migrated movement , environmental forces – weather and windborne and insect or other vector-borne – pathogens.

i. Cassava virus diseases : Cassava Mosaic and Brown Streak virus diseases continue to affect the main food crop – cassava – throughout the Great Lakes region of Eastern and Southern Africa. In Africa, an estimated 70 million people are dependent on cassava as a primary source of food contributing over 500 kcal per day per person. Cassava is produced mostly by smallholders on marginal and sub-marginal lands in the humid and semi-humid tropics. It is efficient in carbohydrate production, adapted to a wide range of environments and tolerant to drought and acidic soils.

The FAO strategic programme framework "Cassava diseases in central, eastern and southern Africa" (CaCESA) assists cassava-dependent vulnerable populations through better control and management of pests and diseases in central, eastern and southern regions of Africa. FAO provides technical assistance to national institutions to establish effective surveillance approaches, integrated management procedures, farmer training and capacity building. FAO initiatives promote integrated approaches, strengthening linkages among the stakeholders and promoting regional collaboration.

ii. Desert locust : The Desert Locust migrates in swarms across continents and is a potential threat to the livelihood of one-tenth of the world's population. This pest is a serious menace to agricultural production in Africa, the Near East and Southwest Asia. A locust can eat its own weight (about 2 grams) in plants every day. That means one million locusts can eat about one tonne of food each day, and the largest swarms can consume over 100 000 tonnes each day, or enough to feed tens of thousands of people for one year.

In 2012, the Desert Locust threat in the Sahel was controlled thanks to timely contributions of USD 8.2 million and a decade of strengthening national capacities and regional coordination in the framework of the FAO Emergency Prevention System (EMPRES), through the FAO Commission for Controlling the Desert Locust in the Western Region (CLCPRO). FAO's Desert Locust Information Service monitors the locust situation and provides early warning to countries and donors on an on-going basis. Through the FAO Emergency Prevention System (EMPRES) and the three regional locust commissions, national capacities in early warning, early reaction and contingency planning are constantly being strengthened so locust emergencies can be better managed and the frequency and duration of Desert Locust plagues can be reduced.

iii. Grape Phylloxera: Phylloxera is a serious insect pest of commercial grape production worldwide. The tiny aphid-like insect forms galls on leaves and roots of grapevines. The insect is believed to have originated from the Eastern U.S., where damage from the pest is most prevalent on the leaves of French-American hybrid grapevines. Premature defoliation and reduced shoot growth, yield, and quality of the crop are damage symptoms left by high populations of the insect pest. However, the major damage caused by this insect is due to root infestation. In Europe, phylloxera has great economic importance and has led to research and development programs for genetic resistance as a pest management tactic. European grape varieties should be grafted on American or hybrid grape root stocks. The leaf form of phylloxera doesn't cause much damage to the vine, so foliar insecticide sprays are of little value to control phylloxera during their wandering stage and are of limited value.

iv. Wheat rusts : Wheat rust diseases with continuous evolution of new pathotypes and airborne nature pose a serious threat to wheat production worldwide. Their impact is more pronounced across the major wheat growing regions including East Africa, North Africa, Middle East and Asia. It is estimated that 37% of world's wheat is under risk of potential epidemics of yellow, stem or leaf rust diseases. FAO promotes and supports global efforts for monitoring and management of wheat rust diseases, as a member of the Borlaug Global Rust Initiative. FAO provides technical support to countries at risk of rust epidemics. FAO assistance includes capacity building, surveillance and monitoring, seed systems, contingency planning, strengthening linkages among institutions and stakeholders, enhancing research–extension–farmer links, training of officers and farmers and emergency responses where necessary.

v. Potato blight: One of the most devastating and historically significant plant diseases is potato late blight. Epidemics of late blight destroyed the potato crops in Europe in the 1840s leading to mass starvation. One of the most significant effects of the disease on the population of the U.S. was the Great Irish Potato Famine from 1845 to 1847, where up to one million people died from the loss of their staple food crop, and nearly the same number of people emigrated to the rest of Europe and the U.S. to prevent starvation and death. Several factors contributed to the starvation, including the land-tenure system in Ireland at the time and the near total dependence of potato as food source for the poorer working population. Late blight is caused by an oomycete or fungus-like microorganism, *Phytophthora infestans*, which is a specialized pathogen of potato, tomato and other members of the solanaceous plant family. However, for many centuries, the causes of crop failures were a mystery, even for the 19th Century's potato famine in Ireland. Scientific explorations of potato late blight led to the discovery that plant diseases were caused by microorganisms. This led to the birth of plant pathology as a science. Late blight continues to be a major pest of potatoes, but the disease is managed through the uses of resistant potato cultivars, proper sanitation practices and fungicides.

vi. Downy mildew of grapes: Downy mildew is a common and serious disease of grapes that is partly responsible for the introduction of today's widely used fungicides. This highly destructive disease of grapevines is caused by the oomycete, or water mold, Plasmopara viticola and is a serious disease in all grape-growing areas of the world where there is spring and summer rainfall. It can be traced back to the 1850s in the Bordeaux region of France, where a vineyard producer was having problems with thieves pilfering from his vines. In hopes of making the grapes unattractive to the thieves, he applied a mixture of copper and lime to parts of his vineyard. Not only did it deter the thieves, it deterred the downy mildew disease incidence. This copper-lime mixture came to be known as the Bordeaux mixture, a commonly used fungicide that is still used today. This discovery was the beginning of modern fungicide use.

10.5 Transboundary Plant Pests

Transboundary pests are a serious threat to food security and environment, a condition exacerbated in recent decades by the globalized movement of people and commodities. India witnessed an upsurge of desert locust in 2020 with their swarms attaining epidemic proportions during COVID-19 pandemic. Rajasthan was on high alert with swarms entering Madhya Pradesh, Uttar Pradesh, Punjab, Haryana Gujarat, and Telangana between May and June. As on July first week, FAO had placed high alert with the possibility of more swarms of locusts likely to migrate from Somalia to along India-Pakistan border and it remains to be seen whether a locust plague is in the offing. Cassava mealybug (CMB) is the latest invasive insect in 2020 first observed in Thrissur, Kerala and has spread to Tamil Nadu causing 9 to 46 per cent infestation. Prevention of spread to unaffected areas and action for eradication and importation of CMB-specific parasitoid, *Anagyrus lopezi* is currently underway. The fall armyworm (FAW) invaded India on maize during May 2018, and spread across all maize growing states. Recently it was also reported from Bangladesh (Krishna Kumar and Vennila, 2020).

India recommended eight insecticides with conservation and augmentative biocontrol-cumcultural control interventions given prime importance. Rugose spiralling whitefly (RSW), first noticed on coconut from Tamil Nadu and Kerala in 2016, later spread to Andhra Pradesh, Karnataka, Goa and Assam, through infested seedlings and transportation of plant materials. Banana, mango, sapota, guava, cashew, maize, ramphal, oil palm, Indian almond, water apple, jack fruit and many ornamental plants are host crops of RSW. Natural build of the parasitoid, Encarsia in RSW endemic areas and enhancing its niche survival are given focus at present. South American tomato moth (SATM), an invasive insect on tomato both under greenhouse and field conditions, was reported in 2014 with its spread to several states, has established as a regular pest. While natural incidence of Metarhizium anisopliae on larval SATM was up to 35 per cent, resistance breeding through screening of wild and cultivated tomato genotypes is underway as a long-term management strategy.

Papaya mealybug (PMB) caused significant damage to agricultural and horticultural crops since its documentation in 2007 at Coimbatore, Tamil Nadu. Mulberry crop over 1,500 ha in Tamil Nadu too got destroyed. However, classical biological control using *Acerophagus papayae* from Puerto Rico is a success story that reduced incidence of PMB from 49 to 3 per cent. Cotton mealybug (CMB) first recorded in Gujarat in 2005 caused yield loss of 30-40 per cent in Punjab amounting INR 1 590 million during 2007 and 40-50 per cent in Gujarat. Infestation of CMB was reported amongst 71, 141, 124 and 194 species of plants belonging to 27, 45, 43 and 50 families, respectively, 5 across cotton growing zones in India and the parasitoid, *Aenasius bambawalei* offered fortuitous biological control. Invasive eucalyptus gall wasp (EGW) of 2001, spread across south, central, and northern states threatened the productivity of paper and pulp industry in 2007, however, is being kept under check presently by the native parasitoids.

Other established invasive plant pests in India include silver leaf whitefly, coconut eriopyhid mite), spiralling whitefly and coffee berry borer that are managed on need basis. Occurrence of Fusarium wilt (race 1) infecting Cavendish in 2010 and tropical race 4 (TR4) reported from Uttar Pradesh in 2017 was also recorded from Bihar, Madhya Pradesh, Gujarat, and Maharashtra (Krishna Kumar and Vennila, 2020). Productivity of banana, especially Cavendish varieties is highly reduced by TR4 in several parts of the country and the poor man's source of nutrition was at stake. Infected areas in Bihar and Uttar Pradesh saw a remarkable control of TR4 on account of microbial consortium developed by Indian Council of Agricultural Research. Citrus greening disease is destructive in major citrus belts of Maharashtra, Punjab, Southern and North-East India with its transmission through grafts and psyllid vector necessitating supply of disease-free citrus seedlings to reduce its incidence and damage.

10.6 Pests, Pandemics and Biosecurity

Most of the zoonotic viruses have significant possibilities in bioterrorism and have potential to wipe out humans and animals. Advances in molecular biology such as gene editing can make profound changes in genetic manipulations of organisms, and implications of such technologies in occurrence of pandemic and its mitigation need serious thinking. Institution of appropriate and timely biosecurity measures is an important instrument for protection and improvement of animal health. Breach in biosecurity due to ignorance and lapses in adoption of timely biosecurity measures in management of livestock, poultry and fish are salient reasons for the high incidence emerging and transboundary infectious diseases (Krishna Kumar and Vennila, 2020). India's stance, like most of the nations across the globe, to the ongoing COVID-19 biosecurity crisis is largely responsive and reactive than being proactive from a biosecurity perspective, exposing low level preparedness towards pandemics. In India, biosecurity has remained next to biosafety even after four decades of legislation. The proposed Agricultural Biosecurity Bill and the National Biotechnology Regulatory Authority aim to establish an integrated national biosecurity system covering plant, animal and marine issues. Under the Integrated disease surveillance programme, a network of public laboratories with biosafety practices and infrastructure was established although upgradations are needed to be continuous considering technological advancements. In India, about 30 bio-safety level (BSL) laboratories of the level of BSL-III or BSL-II+ are currently under operation with only two BSL-IV facilities. Prevention of transmission of pathogens across intra- and inter-country borders warrants devising of biosecurity measures at par with international standards.

International guidelines are developed by WHO, FAO and OIE in respect of human, plant and animal pests and pandemics. For handling the most dangerous transboundary pests more of BSL- III and BSL-IV laboratories in the country are required to ensure biosafety, biosecurity and biocontainment. Biosecurity needs to be observed from farm to national to regional and international levels in a bottom-up approach. Farm level biosecurity practices are available for crops, cattle, sheep, pig, poultry and fish production systems with best designs in terms of phyto/zoo sanitary measures such as quarantine, rodent and vector control, disinfection of animal sheds and premises, proper disposal of dung, urine, feed and fodder wastes and proper carcass disposal for effective management of infectious diseases although ground level adherence is still wanting. India needs to take a look into its biosecurity preparedness and plug all the big gaps to prevent being blindsided to dangerous biological agents- either man-made or natural. A number of biosecurity preparedness measures applicable for zoonotic and human diseases, have implications for plant quarantine which is lagging behind leading to a cascade of invasive pests affecting field and horticultural crops.

Epidemic management: In agroecosystems, crop yield is reduced by epidemics. At the field scale, epidemiology succeeded knowledge percolation across theory, empirical studies, and recommendations to actors. Achieving similar success at the landscape level requires understanding of ecosystems. Under the light of epidemiological knowledge, articulating the contributions of scientific disciplines is needed because the development of epidemics depends on the interaction in space and in time between host plants, pathogens, environment, and human action (Lydia Bousset *et.al.*, 2012).

At the field scale, epidemiology succeeded knowledge percolation across theory, empirical studies, and recommendations to actors. This success in rooted on a clear theoretical formalization, disease foci, and their growth in time and in space that allowed modelling prompted empirical studies and allowed links to other disciplines. The results of epidemiological studies on the modulation of quantitative growth of foci depending on the climate on the differentiation of pathogens into pathotypes and on quantitative resistance have been translated into guidelines for action in decision support models, into surveys of pathogen populations, into recommendations for varietal mixtures and into breeding strategies.

Cyclic epidemics, in which development depends on interaction in space and in time between host plants, pathogens, environment, and human actions was formulated (Lydia Bousset *et.al,* 2012). In agroecosystems, human actions exacerbate homogeneities alternating with sharp discontinuities on scales of time and space. The dynamics of cyclic epidemics takes discontinuities into account. This allows decomposing control at the field and at the landscape scales into goals to reach, corresponding to the components of the pluriannual dynamics of epidemics. Articulating disciplinary concepts open the prospect of optimization by identification of one's potential contributions. It was proposed that cyclic epidemics could be controlled by looking for a local solution, in a decentralized manner.

Formalization of the dynamics of cyclic epidemics: Formalization is well established in the continuous space consisting of the scale of one season in one field. Population numerical dynamics is described by initial inoculum and multiplication rate. Population spatial dynamics is described by the theory of disease foci. In the context of agroecosystems, the existence of sharp ruptures hampers the extension of these concepts over wider scales of time and space. Releasing the hypothesis of continuity has been achieved for time, by taking into account seasonality; for space, by taking into account heterogeneous landscape; for the continuity of epidemics, by taking into account human actions, but these aspects have not yet been formalized simultaneously (Lydia Bousset *et.al,* 2012). The alternation of continuities and ruptures on scales of time and space allows describing the pluriannual dynamics by three phases: the epidemics on one field, the inoculum production on the field, and the inoculum transmission between fields. Within each phase, the processes affecting population size or structure at lower scales of space (the field, the plant, and the vegetal organ) and time (the season and the duration of one pathogen generation) scales can be aligned.

The within field epidemic dynamic, that is, the quantitative fluctuations of size and the qualitative changes in genetic structure of the pathogen population, can be described by three time-chained or overlapping components within the season. The local onset of the epidemic in the field requires mobilization of propagules from a more or less distant inoculum pool connected to the field by migration; The pathogen population increases in size during the epidemic by secondary multiplication cycles; The pathogen population causes damages to plants, defined as the qualitative or quantitative yield loss caused by the disease, with subsequent economical consequences for farmers.

The production of inoculum during discontinuities in each field occurs either on the same plants or requires the infection of volunteer plants (Cook and Yarham, 2006) or necessitate migration to alternate hosts. This production of inoculum depends on the within field epidemics. Offspring release and contribution to the propagule pool in subsequent seasons depend on the pathogen's survival abilities. The inoculum transmission between fields coincides with intercrop discontinuities and depends on characteristics of both "source" and "target" fields, and on landscape spatial and temporal structures. Field characteristics for the source are the quantity and characteristics of the inoculum released; and for the targets, the host crop status and variety. Structure of the landscape includes for its spatial aspect, the location of new fields and the selection pressure they impose; for its temporal aspect, the time at which fields are sown and are susceptible to infection. The inoculum transmission phase results in strong quantitative (size) and qualitative (genetic structure) variations in pathogen populations.

In agroecosystems, the prospect is offered to maximise the efficiency of disease control by using human actions to interfere with the cyclic epidemic, both during the epidemic phases and during the survival phases.A shared description of processes pertinent to any kind of epidemics highlights the need for specific researches. Cyclic epidemics are different from polyetic epidemics and could therefore be managed in a different way, even if it is not excluded that decentralised strategies could also be appropriate for polyetic epidemics.

The role of initial inoculum (y_0), rate (r) of pathogen or disease development (infection), and period of time (t) that the pathogen and host populations interact during the cropping period is revisited in modeling plant disease epidemics. The importance of quantitative informations and the relationship between initial inoculum and the rate of disease development represent key elements for identification of the most useful disease models to be used (Nutte, 2007). For effective Integrated Disease Management of monocyclic or polycyclic epidemics, temporal population growth models of plant disease epidemics (monomolecular, exponential, logistic and Gompertz population models) were presented. Sanitation and disease management principles (exclusion, avoidance, eradication, protection, resistance and therapy) were described.

10.7 Farm Biosecurity and Management Practices

Farm Biosecurity and integrated pest management (IPM) are consistent in their goals of protecting crop health, often using a whole farm approach to good crop management. Many biosecurity risk mitigation strategies are synonymous with IPM preventative strategies and both rely on planning. This is particularly true for residential pests where there may be few chemical solutions available and their permanency within the farming system requires multiple long-term strategies to limit their movement, spread and subsequent potential damage. Farm biosecurity refers to a set of management practices and activities designed to prevent, minimise and control the introduction and spread of plant pests onto a property. These pests may be exotic to Australia, may be established in limited areas, or may be widely distributed and can spread from farm to farm. Biosecurity has a greater focus on practices that ideally exclude a pest from a property or limit its spread and establishment. Integrated pest management primarily involves strategically using the different practices to control a pest that is already present in a cropping system or is an imminent threat

Best management practices for field biosecurity: It is important to have a plan to minimize the spread of pests on your farm. Take time to develop a biosecurity plan for your farm. To reduce the likelihood of introducing new pests to the farm, it is recommended that producers develop a biosecurity plan for their operation (Anonymous, 2020). Those that have an animal enterprise on the farm are likely already familiar with biosecurity procedures implemented to reduce the spread of animal diseases and pests – the same basic principles can also be applied to our crop and forage fields. Implementation of biosecurity procedures may reduce the likelihood of pest introductions and can limit their potential spread among multiple crops or fields in operation.

Movement of pests : Transmission methods of pests includes movement via fomites, vectors, other living organisms, air currents, and water movement/ irrigation. Fomite transmission involves the movement of plant diseases or pests by inanimate objects. Planting, tillage, and harvesting equipment would be an example of potential fomites responsible for field to field movement of pests. In soybeans, the pathogen responsible for white mold forms small, hard lumps of fungus that can be moved from field to field in tire treads, on tillage implements, and even in combines.

Vectors are living organisms (such as insects) that can carry and transmit pathogens to our crops and forages, usually during feeding activity. Other living organisms can also carry pests on their physical form, and these may be introduced incidentally (for example introduction of weed seeds by birds or other wildlife) (Anonymous, 2020). One of the diseases that affect barley and wheat, barley yellow dwarf, is caused by a virus that is transmitted from infected to healthy plants by aphids when they feed. Airborne transmission may include the transportation of insects and diseases traveling via air currents. Waterborne transmission can occur due to impacted irrigation sources/equipment or by incidental surface runoff from contaminated fields. Several of the rust diseases that can be found in corn, soybean, and small grain crops move in weather systems from southern regions where the pathogens overwinter to our fields in Pennsylvania.

10.8 Pest Management Epidemic Models

A wide range of new infectious diseases, fatal and seemingly convoluted composition, are coming into existence along with the time, and are needed to be controlled by different tools. Moreover, such tools needed to be developed to check such diseases efficaciously. Hence apart from medical field specialists pioneers, experts and researchers from diversified fields are now partaking in to find an optimal way to check widespread of such diseases and if possible exterminate them as well. Fortunately, a valuable tool Mathematical modelling which can come in handy to analyze the nature of infectious diseases and can help to develop control methodology to deal with them, Daniel Bernoulli (1760) formulated the first epidemiological model for smallpox (Monika Badole *et.al.*, 2020).

But then came a cessation in the field of epidemiological modelling, which got its breakthrough when pioneer put forward the outcome of his research in the early twentieth century. Down the line, in 1927, Kermack and Mckendrick presented a preeminent deterministic compartment model in order to analyze the outbreak of Black Death in London in (1665-1666) and Mumbai plague havoc (1906). The results of that model were precise in predicting the nature of outbreaks and were in unison with recorded data of epidemics. Recently, many mathematical model has been proposed to control the effect of infectious diseases.

A compartmental epidemic model has been analyzed by considering susceptible, infected, and recovered through treatment control method. Moreover, available treatment control and infected individuals, both are considered as independent variables for treatment function. The complex-dynamics of the systemic model has been discussed through different facets, local and global asymptotic stability of the system has also been studied. The stability of the equilibrium states has also been carried out using some of the tested parameters from literature reviewed. Diseases, which are transmissible through human contact and which could be cured through effective medical treatment were the subject of studies. From analysis, this could be maintained that by reasonably higher treatment control, the endemic equilibrium turns unsteady. Hence, a stable endemic equilibrium could lose stability when treatment controls value is increased.

i. **Eco-epidemiological type predator–prey model:** An agricultural pest control system was proposed and analyzed (Anjana Das and Pal, 2018.). For this purpose, an eco-epidemiological type predator–prey model has been proposed with the consideration of a sound predator population and two classes of pest populations namely susceptible pest and infected pest. Further to consider uncertainty, model was modified and transformed it into a fuzzy system with incorporation of imprecise parameters. The dynamical behavior of the proposed model has been investigated by examining the existence and stability criteria of all feasible equilibria. An optimal control problem is formed by considering the pesticide control as the control parameter and then the problem is solved both theoretically and numerically with the help of some computer simulation works.

ii. **SIR Epidemic Model:** A SIR epidemic model with bilinear incidence rate has been proposed and the existing threshold requirements of all classifications of equilibrium points are obtained. The global and local stability of the disease-free and endemic equilibriums of the model was studied (Monika Badole *et.al.*, 2020). An optimal control problem was formed and solved. Some numerical simulations works were carried out to demonstrate our results. In this process, results generalized and improved any results in existing literature.

iii. **Daily epidemiological model of rice blast:** Early warning services for crop diseases are valuable when they provide timely forecasts that farmers can utilize to inform their disease management decisions. In South Korea, collaborative disease controls that utilize unmanned aerial vehicles are commonly performed for most rice paddies..

iv. **Age-dependent compartmental pest-pathogen model :** A compartmental pest management model was investigated, which divides the pest population into a susceptible and an infective class, while also including a class of pathogenic viruses. The model is age-structured, in the sense that it accounts for the differences in the amounts of pathogenic viruses released by infective pests at various infection ages (Hong Zhang and Paul Georgescu, 2009). First, the asymptotic behavior of the system is established via a monotonicity analysis which makes use of several integral inequalities. The linearized stability of the equilibria for the system is then discussed by means of Michailov criterion. As an outcome of our analysis, it is observed that the maximal length of the infective period plays an important role in the dynamics of the system. Several pest control strategies are further investigated by means of numerical simulations, it was concluded with a discussion on the biological significance of the mathematical findings.

vi. **Cassava pest protection model:** Modeling studies of the most economically important cassava diseases and arthropods were summarized, highlighting research gaps where modelling can contribute to the better management of these in the areas of surveillance, control, and host-pest dynamics understanding the effects of climate change and future challenges in modeling (Alonso Chavez *et.al.*, 2021).

Conceptual models of system dynamics were focused rather than statistical methods. Through the analysis areas were identified where modelling has contributed and areas where modelling can improve and further contribute. Research challenges were identified in the modelling developed for the surveillance, detection and control of cassava pests, and propose approaches to overcome these. Later, at the contributions that modelling has accomplished in the understanding of the interaction and dynamics of cassava and its' pests, highlighting success stories and areas where improvement is needed. The possibility that novel modelling applications can achieve to provide insights into the impacts and uncertainties of climate change was looked into. Research gaps, challenges, and opportunities were identified, where modelling can develop and contribute for the management of cassava pests, highlighting the recent advances in understanding molecular mechanisms of plant defence.

vii. **Simulation of chilli leaf curl disease dynamics :** Leaf curl, a whitefly-borne begomovirus disease, is the cause of frequent epidemic in chili. In the present study, transmission parameters involved in tripartite interaction were estimated to simulate disease dynamics in a population dynamics model framework (Roy *et.al.*, 2021). Epidemic was characterized by a rapid conversion rate of healthy host population into infectious type. Infection rate as basic reproduction number, $R_0 = 13.54$, has indicated a high rate of virus transmission. Equilibrium population of infectious host and viruliferous vector were observed to be sensitive to the immigration parameter. A small increase in immigration rate of viruliferous vector increased the population of both infectious host and viruliferous vector. Migrant iruliferous vectors, acquisition, and transmission rates as major parameters in the model indicate leaf curl epidemic is predominantly a vector -mediated process. Based on underlying principles of temperature influence on vector population abundance and transmission parameters, spatio-temporal pattern of disease risk predicted is noted to correspond with leaf curl distribution pattern in India. Temperature in the range of 15–35 °C plays an important role in epidemic as both vector population and virus transmission are influenced by temperature. Assessment of leaf curl dynamics would be a useful guide to crop planning and evolution of efficient management strategies.

10.8.1 The Dynamics of An Epidemic Model

Dynamical behavior of pest management models : A model was proposed which described the interaction between pest and its natural predator (Agus Suryanto *et.al.*, 2019). It was assumed that pest can be infected with diseases or pathogens such as bacteria, fungi, and viruses. The model was constructed by combining the Leslie-Gower model and S-I epidemic model. It also considered the effects of pest harvesting. Harvesting in this case was intended to take a number of pests as one of the pest population control strategies. The proposed model was analyzed dynamically to study its qualitative behaviour. The dynamical analysis included the determination of all possible equilibrium points and their stability properties. The implementation of pesticide control was discussed, where its optimal strategy was determined by Pontryagin's maximum principle. To support the analytical studies, some numerical simulations and their interpretation were also performed. Nonetheless, the presented dynamical behaviour, the optimal control approach, and numerical simulations can provide description of the possible outcomes of the model. Appropriate mathematical model which is suitable for a specific case of natural phenomena can be obtained by estimating real parameters. Such estimation can be performed by fitting the real world data with the proposed model.

Dynamical behaviors of an *SI* model with impulsive transmitting infected pests and spraying pesticides at fixed moment was discussed.

The epidemics of Flavescence dorée phytoplasma (FD) in grapevine: Insect-borne plant diseases recur commonly in wild plants and in agricultural crops, and are responsible for severe losses in terms of produce yield and monetary return. Mathematical models of insect-borne plant diseases are therefore an essential tool to help predicting the progression of an epidemic disease and aid in decision making when control strategies are to be implemented in the field (Federico Maggi *et.al.*, 2014). While retaining a generalized applicability of the proposed model to plant epidemics vectored by insects, the epidemics of Flavescence dorée phytoplasma (FD) was specifically investigated in grapevine plant Vitis vinifera specifically transmitted by the leafhopper *Scaphoideus titanus*. The epidemiological model accounted for life-cycle stage of *S. titanus*, FD pathogen cycle within *S. titanus* and V. vinifera, vineyard setting, and agronomic practices. The model was comprehensively tested against biological *S. titanus* life cycle and FD epidemics data collected in various research sites in Piemonte, Italy, over multiple years. The work represented a unique suite of governing equations tested on existing independent data and sets the basis for further modelling advances and possible applications to investigate effectiveness of real-case epidemics control strategies and scenarios.

Simulation of leaf curl disease dynamics in chili: Leaf curl, a whitefy-borne begomovirus disease, is the cause of frequent epidemic in chili. Transmission parameters involved in tripartite interaction were estimated to simulate disease dynamics in a population dynamics model framework (Buddhadeb Roy *et.al.*, 2021). Epidemic was characterized by a rapid conversion rate of healthy host population into infectious type. Infection rate as basic reproduction number, R0 = 13.54, had indicated a high rate of virus transmission. Equilibrium population of infectious host and viruliferous vector were observed to be sensitive to the immigration parameter. A small increase in immigration rate of viruliferous vector increased the population of both infectious host and viruliferous vector. Migrant viruliferous vectors, acquisition, and transmission rates as major parameters in the model indicate leaf curl epidemic was predominantly a vector -mediated process. Based on underlying principles of temperature infuence on vector population abundance and transmission parameters, spatio-temporal pattern of disease risk predicted was noted to correspond with leaf curl distribution pattern in India. Temperature in the range of 15–35 °C played an important role in epidemic as both vector population and virus transmission were influenced by temperature. Assessment of leaf curl dynamics would be a useful guide to crop planning and evolution of efficient management strategies.

Current finding provides an important clue for emphasizing prevention of migrant vectors rather than the vector control normally thought for. Based on underlying principles of temperature infuence on vector abundance and virus transmission spatio-temporal pattern of disease risk has been predicted. Spatio-temporal patterns for disease risk mapped across wide variants of agro-ecosystems may be used for crop planning if possible to avoid the period of high vector population. Modeling framework assumes susceptible chili cultivar and the viruliferous vectors are present in the agroecosystem. Therefore, parameter estimates are based on susceptible hosts. Adjustment of parameter values for host resistance may be required for general application of the finding. Epidemic analysis framework can be followed for testing and evolution of management strategies. A model framework of tripartite interaction necessary for evaluation of chili germplasms against leaf curl disease is now available. For the prediction of potential disease risk based on temperature thresholds, characteristic behavior of unimodal developmental response to temperature has been captured by three-parameter beta functions41. Te beta function of three parameters is more robust to capture the developmental rate as parameters have meaningful biological interpretation than the two-parameter models.

10.8.2 Quantification of Pest Dynamics

Quantitative estimates of crop yield loss from pests and diseases are commonly considered to be highly uncertain or imprecise, as a result of strong reliance to expert assessments or of observations which do not make use of standardized and uniform protocols. Yet, International standards and procedures exist to conduct field experiments that are specifically designed to quantify crop losses; quantify disease and pest injuries and measure crop losses (Anonymous, 2018). In practice, the quantification of yield losses involves four necessary elements, a quantification of injury(ies) caused by diseases, pests, and/or weeds; a quantification of the attainable yield (i.e., the yield level achieved in absence of injuries); a damage function translating injury into yield loss; the quantification of actual (harvested) yields. Yield loss data standards – "gold" and "silver" standards – may therefore be defined. The gold and silver standards would include the levels of disease or pest injuries; measurements of crop growth; successive development stages (phenology) of the crop; crop yield. The gold standard would additionally include information on crop history; details of crop management; geographical location and weather variables.

The interest to shift pest management strategies from the intensive use of agrochemicals to more sustainable and ecologically friendly practices has increased in recent years. One alternative to conventional farming systems is the implementation of diversification practices that increase diversity in- and around-the field to increase the incidence of natural enemies, reduce pest pressure and enhance crop production (Katja Poveda *et.al.*, 2008), who illustrated the theoretical framework on which diversification practices are based and contrast it with the empirical evidence. The detailed review of 62 original studies published in the last ten years, showed that diversification practices enhance natural enemies in 52%, reduce pest pressure in 53% and increase yield in only 32% of the cases where this was examined. These results were discussed on the basis of the reviewed studies providing key elements that should be taken into account to design diversification practices that can be implemented as competitive pest management strategies that cover the farmers' needs, reducing the intensive use of agrochemicals.

10.9 Plan Development and General Recommendations

Biosecurity plan development should involve all farm partners and employees. The plan should be available in writing, understandable to all, and should also be reviewed and updated annually to remain current. Effective plans are utilized by everyone, everyday and enforced to mitigate the potential movement of pests as described above.

General recommendations

- Require all employees and staff to read your biosecurity plan. Make the plan readily available by filing a copy in a known location accessible by all employees and staff.
- Diversify the operation to avoid uniform susceptibility by planting multiple crop species and/or varieties.
- Purchase certified seed free of diseases and weeds.
- Become aware of potential diseases and pests in your region, and scout fields regularly.
- Report any unusual plant diseases or pests to your local Extension educator or to the Pennsylvania Department of Agriculture.

- Post signs advising visitors of biosecure areas, and who to contact for entry permission.
- Limit access of property to certain areas through fencing or gated entry sites.
- Lock gates and buildings when not in use.
- Have designated (preferably paved) areas for vehicles to park that have visited other farms or agricultural sites.
- Vehicles arriving from other agricultural sites should be cleaned before allowing movement into fields. This includes mud and plant debris on vehicles and equipment.
- When traveling from properties with known disease or pest issues you may consider washing your hands, fully showering, and changing clothes. Disposable boot covers or multiple pairs of shoes can limit transmission.
- Consider performing field operations (tillage, spraying, harvesting) in areas with known pest issues only after performing those operations in "clean" fields first, to avoid transmitting any organisms to new fields.
- Properly contain plant material showing unusual symptoms or signs during transportation or mailing.

A number of research strategies have been initiated over the last decade to enhance plant biosecurity capacity at the pre-border, border and post-border frontiers. In preparation for emerging plant virus epidemics, diagnostic manuals for economically important plant viruses that threaten local industries have been developed and validated under local conditions (Rodoni, 2009). Contingency plans have also been prepared that provide guidelines to stakeholders on diagnostics, surveillance, survey strategies, epidemiology and pest risk analysis. Reference collections containing validated positive virus controls have been expanded to support a wide range of biosecurity sciences. Research has been conducted to introduce high throughput diagnostic capabilities and the design and development of advanced molecular techniques to detect virus genera. These diagnostic tools can be used by post entry quarantine agencies to detect known and unknown plant viral agents. Pre-emptive breeding strategies have also been initiated to protect plant industries if and when key exotic viruses become established in localized areas. With the emergence of free trade agreements between trading partners there is a requirement for quality assurance measures for pathogens, including viruses, which may occur in both the exporting and importing countries. These measures are required to ensure market access for the exporting country and also to minimize the risk of the establishment of a damaging virus epidemic in the importing country.

10.10 Future Thrusts and Conclusions

The experiencing of COVID-19 pandemic taught the humanity at large its vulnerability to life and living, direct and indirect impacts on nation's biosecurity and socio economy of. Despite all the interlinked challenges of security and safety of food, environment, health and biodiversity, India's focus is towards sustainably increasing agricultural productivity, farm incomes, food security and sectoral development by building resilience at multiple levels. The diversified Indian agro-ecosystems and sectors are replete with history of pests and pandemics. Research cum-developmental organizational set up and industries dealing with health system of human, livestock, poultry and fish have all the paraphernalia needed for an effective preparedness and management of pests and pandemics. But their operational success is fraught with shortcomings of lack of coordination and collaborations. Individual excellence should translate to collective management (Krishnakumar and Vennila, 2020). Slighter adjustments and reorientation in functioning of stakeholders interlinked through a common hub under 'one health' wheel would uplift the standards of diagnostics, preparedness and pest management.

Pre-import and post-entry quarantine require need based international cooperation and collaboration. Redressal for new and emerging pests need inter departmental coordination. National diagnostic laboratories equipped with tools and trained human resources, exclusive electronic pest surveillance system using standard protocols, and field workers functioning together by convergence of public and private organizations/institutions/departments/industries should be mandatory. Surveillance must be aided compulsorily by geo reference based mobile apps developed using protocols adhering to international phytosanitary standards for invasive pests and national sampling procedures for management emerging pests. An e-reporting system involving artificial intelligence for diagnosis and data analytics integrated at server level with edaphic and weather factors represented in a geo platform would help in geo spatial early warning and subsequent pest management preparedness. Updated scientific information system linked to real time pest scenario derived from e-surveillance would facilitate digital pest management advisories automated for dissemination to end users. Forging a self-reliant integrated one-health management system require partnership of public and private stakeholders for needful production and supply of demand driven human/veterinary vaccines and quality pest protection products for plant/animal/poultry/fish.

Vertical integration of agricultural education is the key to improve quality of human resource in the country and many more post graduate students must be encouraged to address zoonosis, with focus on epidemiology, ecology, biodiversity, molecular characterization with expertise in big data analytics. Empowering agri-graduates and diploma holders to take up contractual system of plant-animal-poultry-fish protection in identified areas such as field pest/epidemic monitoring, bioagent mass production, manufacture of sensor-based gadgets, co-ordination of input supply and delivery system of agrichemicals/vaccines at farm level would be prudent to generate employment and to serve as pathway for securing a better health for all. Developing an entrepreneurial capacity for mass production of macrobials (parasitoids and predators), microbials (growth promoters, antimicrobials/ antagonists), plant-based products and mechanical traps at cottage level would aid in sustaining natural farming systems.

Registration of biological control agents that is quick, scientific and with quality assurance will provide impetus to commercialize the technologies. Policy framework facilitating execution of proactive strategies of plant protection by governmental departments with hand holding of growers and input industries is essential. No other time is better than the present for use of digital tools and mass media to execute a unified 'one health' system. Enhancing production and income of farmers through reduction of yield losses caused by pests should be the motto of plant pest management towards fostering national food, and environmental security. Human public health services aregiven priority over veterinary and plant health in India including the poor insurance schemes for agriculture and allied sectors (Krishnakumar and Vennila, 2020). However, human and environmental health could be simultaneously improved by the same policy or management actions provided agriculture and animal husbandry expansion and intensification, and other modifications of natural landscapes are implemented in a way that minimizes biodiversity losses. Media should play a very big role in educating the do's and don'ts during pandemics and hence, media management must go hand in hand with strong scientific research outputs and information reaching public in a simplified way.

Political environment of India with its neighbours and the rest of the world shall contribute towards sustainable development through mitigation of pests and pandemics. 'We are healthy if our neighbour is healthy' should be the slogan. India has a framework of 'environment' governed under Ministry of Environment and Forest with Ministry of Agriculture and Farmer's Welfare that functions for agriculture, animal husbandry and related sectors. Since environment is beyond forest and wildlife, there needs to be a separate Ministry of Environment that could address issues such as climate change, depletion of corals, loss of biodiversity, water, air and ecosystem services holistically and the wider scale of land agriculture, forests, seas, oceans, and mountains put together. Environment is global and cannot be confined to forests. Indian agriculture and forestry are two sides of the same coin and both must be accounted together to tackle problems of shrinking forests, agriculture and allied sectors.

India should align closely with the global community on aspects of carbon emission or footprints and global warming pertaining to climate change both at regional and global level. The melting of ice especially in Himalayan regions and associated soil erosion down the plains need a preparedness (Krishnakumar and Vennila, 2020). Accurate monitoring of natural disasters of cyclones, drought hailstorms and floods and their forecasts strengthen the preparedness at local and macro level. Health management approach at local level for different sectors must be based on each of agroecological region of different agroclimatic zones of the country with revisits made once in five years for suitable calibrated changes. Sustainable production system begins with natural resource management.

Soil health and its enrichment come with enhancement of soil carbon and biodiversity through vegetation and other means. Rivers and water reservoirs are lifeline of entire population and the water consumed and utilized for various purposes need safety guards for which stringent policy decision for each river basin must be promulgated. Considering that agriculture and allied sectors are state subject in India, investment into improvement of organizational set up such as setting up of plant/animal/fish health clinics equipped with infrastructure for training and advisory and linking the supply chain of agrochemical/antimicrobial/vaccine marketing through such clinics are needed. Nevertheless, a governmental or contractual system of field pest management using standard operating procedures is a plausible strategy. Innovative institutional models, pro-agricultural policies and regulatory mechanisms would accelerate innovations, ensure food security, enhance livelihood opportunities of smallholders, and conserve natural resources.

Conclusion

Invasion by new pests and pandemics may continue but science-led global cooperation and collaboration should be able to mitigate their impact more effectively. Precise and quick diagnostics and immunizations with quality vaccines are a must considering the loss of lives and health of human, crop, livestock and fish due to viral pandemics. Improvements and utilization of resistant genetic stocks of crops/animals and fish with focus on increasing immunity in humans as producers and consumers, understanding the epidemiology and environmental interactions of pests and pandemics through advanced analytics, exploitation of biotechnological tools and food processing techniques of therapeutics with inbuilt biosecurity cum-biosafety supplemented with quarantine legislations associated with trade formulated by stakeholders and policy makers together need positive transformations towards sustaining food and health systems (Krishnakumar and Vennila, 2020). Like delayed justice, delayed mitigation of pandemic is a denied mitigation and there cannot be any policy paralysis in management of pandemic irrespective of region, religion, country, race or political system, political will, global collaboration and cooperation, science-led policy decisions, meta-analysis and data management supported by effective communication. Transparency and honesty across nations in sharing information and human resource development shall contribute to better preparedness leading to better mitigation and management of future pests and pandemics across globe. The time has come to appreciate biodiversity and ecosystem services better so as to answer many of our problems including pests and pandemics.

Bibliography

Adalberto, C., Café-Filho, Carlos Alberto Lopes and Mauricio Rossato, 2018. Management of plant disease epidemics with irrigation practices. *IntechOpen book series,* DOI:10.5772/ intechopen.78253

Agus Suryanto and Isnani Darti, 2019. Dynamics of Leslie-Gower pest-predator model with disease in pest including pest harvesting and optimal implememntation of pesticide. *Intl. J. Maths & Mathematical Sci.*, vol.2019, Article ID 5079171, 9 pages

Alonso Chavez, V., Milne, A.E. and van den Bosch, F., 2021. Modelling cassava production and pest management under biotic and abiotic constraints. *Pl. Mol. Biol.*, https://doi.org/10.1007/s11103-021-01170-8

Anjana Das and Pal, M., 2018. An Imprecise Eco-Epidemic Model with Pesticide in Relevance to Agricultural Pest Control. *Biophys. Rev. and Lett.,* 13 : 109-131

Anonymous, 2018. Improving the Quantification of Pests, Disease, and Weeds Impact on Crops, Jul 18, 2018, AgMIP, Center for Climate Systems Research, The Earth Institute, Columbia University 2880 Broadway, New York, NY, 10025

Anonymous, 2020. Best Management Practices for Field Biosecurity. PennState Extension. The Pennsylvania State University, P.A.16802

Buddhadeb Roy, Shailja Dubey, Amalendu Ghosh, Shalu Misra Shukla, Bikash Mandal and Parimal Sinha, 2021. Simulation of leaf curl disease dynamics in chili for strategic management options. *Scientifc Rep.,* 11:1010

Cook, R. J. and Yarham, D. J., 2006. Epidemiology in sustainable systems. In: The Epidemiology of Plant Diseases, B. M. Cooke, D. G. Jones, and B. Kaye (Eds), pp. 309–334, Springer, Dordrecht, The Netherlands, 2nd edition, 2006

Creamer, R., Hubble, H. and Lewis, A., 2005. Curtovirus Infection of Chile Pepper in New Mexico. *Pl. Dis.,* 89: 480–486

Davis, N., 2009. "Genome of Irish potato famine pathogen decoded". Haas *et al.* Broad Institute of MIT and Harvard. Retrieved 24 July 2012.

Federico Maggi , Domenico Bosco and Cristina Marzachì, 2014. Conceptual and mathematical modeling of insect-borne plant diseases: theory and application to flavescence dorée in grapevine. Research report R947, School of Civil Engineering, The University of Sydney

Green, M.S., Swartz, T., Mayshar, E., Lev, B., Leventhal, A., Slater, P.E. and Shemer, J., 2002. "When is an epidemic an epidemic?" (PDF). *The Israel Med. Assoc., J.*, 4 : 3–6.

Hong Zhang and Paul Georgescu, 2009. The Dynamics of an Age-Structured Pest Management Model. National Natural Science Foundation of China and the high-level talent project of Jiangsu University.

Huynh, B.L., Matthews, W.C., Ehlers, J.D., Lucas, M.R., Santos, J.R., Ndeve, A., Close, T.J. and Roberts, P.A., 2016. "A major QTL corresponding to the Rk locus for resistance to root-knot nematodes in cowpea (Vigna unguiculata L. Walp.)". TAG. Theoretical and Applied Genetics. *Theoretische und Angewandte Geneti,*. 129 : 87–95

Jones Richard, 2015. House Guests, House Pests: A Natural History of Animals in the Home. Bloomsbury Publishing. p. Preface. ISBN 978-1-4729-0624-3

Julius Kühn-Institut, 2020. Pests and pathogens". Federal Research Centre for Cultivated Plants.

Katja Poveda, María Isabel Gómez And Eliana Martínez, 2008 . Diversification practices: their effect on pest regulation and production. *Rev. Colomb. Entomol.,* 34, (2) Bogotá July/ Dec. 2008

Kim, K.H. and Jung, I., 2020. Development of a Daily Epidemiological Model of Rice Blast Tailored for Seasonal Disease Early Warning in South Korea. *Plant Pathol J.,* 36:406-417.

Krishna Kumar, N.K. and Vennila, S., 2020. Pests, Pandemics, Preparedness and Bio-Security. In: National Dialogue Indian Agriculture Towards 2030 Pathways for Enhancing Farmers' Income, Nutritional Security and Sustainable Food Systems. NITI Ayog, India-FAO.

Kropff, M. J., Tenga, P. S. and Rabbingec , R., 1995. The Challenge of Linking Pest and Crop Models. *Agri. Syst.*, 49 : 413-434

Lydia Bousset and Anne-Marie Chevre, 2012. Controlling cyclic epidemics on the crops of the agroecosystems: Articulate all the dimensions in the formulizations, but look for a local solution. *J. Bot.,* vol. 2012, Article ID 938218, 9 pages.

Monika Badole , Sandeep Kumar Tiwari and Aayush Sharma, 2020 . Stability Analysis of a Complex Dynamics of a SIR Epidemic Model with Bilinear Incidence Rate and Treatment. *J. Sci. Res.,*64, Issue 2.

Nutter, F.F.,2007. The Role of Plant Disease Epidemiology in Developing Successful Integrated Disease Management Programs. In: General Concepts in Integrated Pest and Disease Management. Ciancio A., Mukerji K.G. (Eds) Integrated Management of Plants Pests and Diseases, vol 1. Springer, Dordrecht.

Rodoni, B., 2009. The role of plant biosecurity in preventing and controlling emerging plant virus disease epidemics. *Virus Res*., 141:150-7

Roy, B., Dubey, S. and Ghosh, A, 2021. Simulation of leaf curl disease dynamics in chili for strategic management options. *Sci. Rep.,* 11: 1010

Tan, X., Tang, S. and Chen, X., 2017. A stochastic differential equation model for pest management. Adv. Differ. Equ., 197. https://doi.org/10.1186/s13662-017-1251-x

Wang Limin, Chen Lansun and Nieto Juan.,2010. The dynamics of an epidemic model for pest control with impulsive effect. *Nonlinear Anal-Real World App.,* 11. 1374-1386. 10.1016/j.nonrwa.2009.02.027.

Xia Wang Zhen and Guo Xinyu Song, 2016. Dynamical behavior of a pest management model with impulsive effect and nonlinear incidence rate. *Comput. Appl. Math.,* 30 : 381-398.

11

Agroterrorism and Biosecurity

11.1 Agroterrorism

'Agroterrorism', also known as 'agriterrorism and agricultural terrorism', is a malicious attempt to disrupt or destroy the agricultural industry and/or food supply system of a population through "the malicious use of plant or animal pathogens to cause devastating disease in the agricultural sectors". It is closely related to the concepts of biological warfare, chemical warfare and entomological warfare, except carried out by non-state parties. It refers to a hostile attack towards an agricultural environment including infrastructures and processes in order to significantly damage national or international political interests.

The terms 'agroterrorism, agroterror and agrosecurity', were coined by veterinarian pathologist Corrie Brown and writer Esmond Choueke in September 1999 as a means to spread the importance of this topic. The first public use of agroterrorism was in a report by Dr. Brown which was then reprinted in a front-page article of The New York Times on September 22, 1999, by reporter Judith Miller. Dr. Brown's article in 2,000 for Emerging Diseases of Animals (American Society for Microbiology) made these words a permanent fixture, and they soon ended up as part of everyday use. The Oxford Dictionary now recognizes the word 'agroterroism' and its derivatives. An initial debate by Dr. Brown and Mr. Choueke involved the spellings 'agriterror vs. agroterror'. The spelling with the "o" won, as it was closest to bioterrorism and thus would be easier to remember.

Agroterrorism has not been a serious problem in the period from 1945 to 2012. This might be the result of a lack of empiric data. In addition, the open-source information and the reliability of the references vary in quality. In two of the four agroterrorism cases described here, sufficient evidence to bring charges was lacking. Nevertheless, these cases demonstrate that incidents have taken place at various geographical places with different targets, using various biological agents. The attackers had various motives for the attacks, but all are related to political interests, including sabotage for economic gain. Atypical biological weapons or non–high-risk agents were used in these attacks (Haralampos Keremidis *et.al.*, 2013). The early period of biological warfare agent development in the UK was characterized primarily by developing antianimal and anticrop agents (Kohnen, 2000).

One of the many ways in which terror might be created is through the deliberate infection of animals with pathogenic microorganisms or contamination of foods of animal origin with toxic chemicals that could be introduced in the feed (agroterrorism). The effects of an act of agroterrorism might include, animal suffering, loss of valuable animals, cost of containment of outbreaks and disposal of carcasses, lost trade, and other economic effects involving suppliers, transporters, distributors, and restaurants (Anonymous, 2010). The $1 billion price tag on the dioxin contaminated animal feed in the Netherlands in 2006 and the $21 billion cost of the UK foot-and-mouth disease (FMD) outbreak in 2001 illustrate the potential economic impact of chemical contamination or infectious disease affecting animals. It has been suggested that strategic contamination or infection could create damage far more severe than that which occurs in accidental contamination or natural outbreaks.

There are several features of agroterrorism that are considered attractive for terrorists. Many of the infectious agents can be obtained quite easily and require little expertise to infect animals. Concentrated and intensive contemporary farming practices may facilitate the rapid spread of contagious agents. It has also been suggested that an emphasis on large herds at the expense of individual animals may delay recognition of signs of illness. Agroterrorism may be a means of using low-tech inexpensive methods to create havoc. It is estimated that a strong biological weapons arsenal could be developed at a cost of about $10 million, compared with $1 billion for a nuclear weapon. Although genetically altered insects that spread pathogens to infest crops is considered a more likely approach, use of infectious agents that attack animals is also a real possibility. Agents that have been used in threats or reported to have been deployed include *Bacillus anthracis* (anthrax), Burkholderia (*Pseudomonas*) *mallei* (glanders), fleas infected with the plague bacillus (*Yersinia pestis*), Newcastle disease virus, African swine fever virus, FMD virus. A number of the viral agents cause diseases that are foreign to North America.

Prevention is the best approach to being prepared; biosecurity is a major part of a good prevention strategy and has benefits that extend well beyond agroterrorism. Veterinarians are a central part of preparedness and it is well recognized that there is a need for an adequate number of veterinarians who can recognize and respond to foreign livestock diseases, educate producers, and promote adoption of biosecurity measures.Compared with the attention focused on such vital "nodes" as transportation and telecommunications, relatively little consideration has been paid to threats to the agriculture and food industries (Henry Parker, 2020).

11.2 Agriculture as a Target of Terrorism

The potential for terrorist attacks against agricultural targets (agroterrorism) is increasingly recognized as a national security threat, especially after the events of September 11, 2001. In this context, agroterrorism is defined as the deliberate introduction of an animal or plant disease with the goal of generating fear over the safety of food, causing economic losses, and/or undermining social stability (Anonymous, 2007). The response to the threat of agroterrorism has come to be called "food defense." An agroterrorist event would usually involve bioterrorism, since likely vectors include pathogens such as a viruses, bacteria, or fungi. People more generally associate bioterrorism with outbreaks of human illness (e.g., anthrax or smallpox), rather than diseases affecting animals or plants.

The goal of agroterrorism is not killing cows or plants. These are the means to the end of causing economic crises in the agricultural and food industries, social unrest, and loss of confidence in government. Human health could be at risk through contaminated food or if an animal pathogen is transmissible to humans (zoonotic). While agriculture may not be a terrorist's first choice because it lacks the "shock factor" of more traditional terrorist targets, an increasing number of terrorism analysts consider it a viable secondary target. Agroterrorism could be a low-cost but highly effective means toward an al-Qaeda goal of destroying the United States' economy. Evidence that agriculture and food are potential al Qaeda targets came in 2002 when terrorist hideouts in Afghanistan were found containing agricultural documents and manuals describing ways to make animal and plant poisons.

Possible pathogens in an agroterrorist attack: Of the hundreds of animal and plant pathogens and pests available to an agroterrorist, perhaps fewer than a couple of dozen represent significant economic threats. Determinants of this level of threat are the agent's contagiousness and potential for rapid spread, and its international status as a "reportable" pest or disease (i.e., subject to international quarantine) under rules of the World Organization for Animal Health (also commonly known as the OIE, the Office International des Epizooties).

A widely accepted view among scientists is that livestock are more susceptible to agroterrorism than cultivated plants. Much of this has to do with the success of efforts to systematically eliminate animals diseases from U.S. herds, which leaves current herds either unvaccinated or relatively unmonitored for such diseases by farmers and some local veterinarians. Once infected, livestock can often act as the vector for continuing to transmit the disease, facilitating an outbreak's spread, especially when live animals are transported. Certain animal diseases may be more attractive to terrorists because they can be zoonotic, or transmissible to humans. In contrast, a number of plant pathogens continue to exist in small areas of the U.S. and continue to infect limited areas of plants each year, making outbreaks and control efforts more routine. Moreover, plant pathogens generally are more difficult to manipulate from a technical perspective. Some plant pathogens require particular environmental conditions of humidity, temperature, or wind to take hold or spread. Other plant diseases may take a longer time than an animal disease to become established or achieve a level of destruction that a terrorist may desire.

11.2.1 Animal Pathogens

The Agricultural Bioterrorism Protection Act of 2002 (Subtitle B of P.L. 107-188, the Public Health Security and Bioterrorism Preparedness and Response Act) created the current, official list of animal pathogens that are of greatest concern for agroterrorism. The list is specified in the select agent rules implemented by USDA-APHIS and the Centers for Disease Control and Prevention (CDC) of the Department of Health and Human Services (HHS). The act requires that these lists (Table 17) be reviewed at least every two years. The select agent list for animal pathogens draws heavily from the enduring and highly respected OIE lists of high-concern pathogens. The select agent list is comprised of an APHIS-only list (of concern to animals) and an overlap list of agents selected both by APHIS and CDC (of concern to both animals and humans).

The 23 animal diseases listed exclusively by APHIS in 9 CFR 121.3, the left column of Table 17 include 20 of the OIE-listed diseases and three other disease agents (Akabane, Camel pox, and Menangle) considered to be emerging animal health risks for terrorism. The much larger OIE list includes other diseases that are not listed as "select agents." However, the select agent list was created to account for the additional risks perceived to be posed by terrorism.

Table 17: Livestock diseases in the select agent list:

Animal diseases and agents/toxins listed exclusively by APHIS 9 CFR 121.3		Overlap diseases and agents/toxins listed by both APHIS and CDC 9 CFR 121.4	
OIE class		OIE class	
African horse sickness	E	Anthrax (*Bacillus anthracis*)	M
African swine fever	S	*Botulinum neurotoxins*	
Akabane		Botulinum neurotoxin-producing species of *Clostridium*	
Avian influenza (highly pathogenic)	A		
Bluetongue (exotic)	M	Brucellosis of cattle (*Brucella abortus*)	B
Bovine spongiform encephalopathy	B	Brucellosis of sheep (*Brucella melitensis*)	C
Camel pox		Brucellosis of swine (*Brucella suis*)	S
Classical swine fever	S	Glanders (*Burkholderia mallei*)	E
Contagious caprine pleuropneumonia	C	Melioidosis (*Burkholderia pseudomallei*)	
Contagious bovine pleuropneumonia	B	*Clostridium* perfringens epsilon toxin	
Foot-and-mouth disease (FMD)	M	(Valley fever) *Coccidioides immitis*	
Goat pox	C	Q fever (*Coxiella burnetii*)	M
Heartwater (Cowdria ruminantium)	M	Eastern equine encephalitis	E
Japanese encephalitis	E	Tularemia (*Francisella tularensis*)	L
Lumpy skin disease	M	Hendra virus (of horses)	
Malignant catarrhal fever	B	Nipah virus (of pigs)	
Menangle virus		Rift Valley fever	M
Newcastle disease (exotic)	A	Shigatoxin	
Peste des petits ruminants	C	Staphylococcal enterotoxins	
Rinderpest	B	T-2 toxin	
Sheep pox	C	Venezuelan equine encephalitis	E
Swine vesicular disease	S		
Vesicular stomatitis	M		

11.2.2 Plant Pathogens

The Agricultural Bioterrorism Protection Act of 2002 (Subtitle B of P.L. 107-188) also instructed APHIS and CDC to create the current official list of potential plant pathogens. The Federal government lists biological agents and toxins for plants in 7 CFR 331.3 (Table 18). The act requires that these lists be reviewed at least every two years, and revised as necessary. Prior to the act, there was not a commonly recognized list of the most dangerous plant pathogens, although several diseases were usually mentioned and are now included in the APHIS select agent list.

The list of seven biological agents and toxins in 7 CFR 331.3 was compiled by the Plant Protection and Quarantine (PPQ) program in APHIS, in consultation with USDA's Agricultural Research Service; Forest Service; Cooperative State Research, Education, and Extension Service; and the American Phytopathological Society. The listed agents and toxins are viruses, bacteria, or fungi that can pose a severe threat to a number of important crops, including potatoes, rice, corn, and citrus. Because the pathogens can cause widespread crop losses and economic damage, they could potentially be used by terrorists. Other plant pathogens not included in the select agent list possibly could be used against certain crops or geographic regions. Examples include Karnal bunt, citrus canker, and soybean rust, all of which currently exist in the U.S. in regions quarantined or under surveillance by USDA. As with other agents, the effectiveness of an attack to spread such a disease may be dependent on environmental conditions and difficult to achieve.

Table 18: Plant diseases in the select agent list

Plant diseases caused by.	The select agents listed in 7 CFR 331.3
Citrus greening	*Liberobacter africanus, L. asiaticus*
Philippine downy mildew (of corn)	*Peronosclerospora philippinensis*
Bacterial wilt, brown rot (of potato)	*Ralstonia solanacearum,* race 3, biovar 2
Brown stripe downy mildew (of corn)	*Sclerophthora rayssiae* var. *zeae*
Potato wart or potato canker	*Synchytrium endobioticum*
Bacterial leaf streak (of rice)	*Xanthomonas oryzae* pv. *oryzicola*
Citrus variegated chlorosis	*Xylella fastidiosa*

11.3 Law Enforcement Against Agroterrorism

Recent events have shown that the United States is not impervious to acts of terrorism intended to inflict death, injury, and destruction of assets within our borders. Current information indicates that, regardless of location, American infrastructures and citizens will continue to be targets of terrorist activities. Terrorists have demonstrated their willingness to employ asymmetrical warfare to achieve their goals. Agroterrorism represents one such class of nontraditional warfare. Chemical, biological, and radiological agents pose new challenges to law enforcement, food and agriculture regulatory agencies, and public health officials in their efforts to minimize the effects of a terrorist attack and apprehend those responsible for the attack (Anonymous, 2008). In the past, law enforcement and food/agriculture regulatory agencies commonly conducted separate and independent investigations. An attack against the food or agriculture sector, however, requires a high level of cooperation between these disciplines to achieve their objectives of identifying the threat, preventing the spread of the disease or further contamination of a food product, preventing public panic, and apprehending those responsible. Lack of mutual awareness and understanding, as well as the absence of established communication procedures, could hinder the effectiveness of joint law enforcement investigations. Due to the continued likelihood of attacks against the U.S. food and agriculture sector, the effective use of all resources during an incident will be critical to ensure an efficient and appropriate response.

In a study, through interactive focus groups, input was obtained from law enforcement personnel, livestock producers, meat packers, truckers, feedlot managers, and animal health officials (Terry Knowles *et.al.*, 2005). The study also included two simulation exercises, field surveys, field interviews, and findings from prevention measures initiated in Kansas on a trial basis. Experts agree that the single greatest threat to the agricultural economy is foot-and-mouth disease (FMD), which has the potential to produce economic chaos. A FMD outbreak would likely require law enforcement officers to remain onsite for 60 days or more in order to enforce quarantines and stop-movement orders. Law enforcement's focus should be on prevention, which might include the identification of threats to the local agricultural industry, vulnerability assessment of potential agricultural targets, the development of new partnerships, the establishment of an effective intelligence network, and the development of local community policing programs for agriculture. Researchers concluded that law enforcement agencies currently lack the resources for an effective response to a FMD outbreak.

The recommendations include the development of a national law enforcement strategy; Department of Homeland Security coordination and funding of prevention measures implemented by local law enforcement agencies; additional funding for interdiction of illegal meat products being smuggled into the United States; a mandatory national animal identification system; improved local intelligence on criminal threats to agriculture; community policing programs for the agriculture industry; and agroterrorism awareness training for law enforcement personnel. 25 figures, 19 tables, 96 references, and appended definitions and relevant statues (Kansas), and study instruments.

11.4 Strategic Partnership Program Agroterrorism (SPPA) Initiative

The Department of Homeland Security (DHS), U.S. Department of Agriculture (USDA), Food and Drug Administration (FDA), and the Federal Bureau of Investigation (FBI) will collaborate with private industry and the States in a joint initiative, the Strategic Partnership Program Agroterrorism (SPPA) Initiative

(Anonymous, 2005). The SPPA Initiative will be a true partnership program, where an industry member or trade association or State may volunteer to participate. To volunteer, the industry or State member must submit a completed response form.

Program objectives: The federal government members in partnership with industry and State volunteers, plan to:

- Validate or identify sector-wide vulnerabilities by conducting critical infrastructure/key resources (CI/KR) assessments in order to:
  a. Identify gaps;
  b. Inform Centers of Excellence and Sector Specific Agencies (SSA) of identified research needs; and
  c. Catalog lessons-learned.
- Identify indicators and warnings that could signify planning for an attack.
- Develop mitigation strategies to reduce the threat/prevent an attack. Strategies may include actions that either industry or government may take to reduce vulnerabilities.
- Validate assessments conducted by the United States Government (USG) for food and agriculture sectors.
- Gather information to enhance existing tools that both USG and industry employ.
- Provide the USG and the industry with comprehensive reports including warnings and indicators, key vulnerabilities, and potential mitigation strategies.
- Provide sub-sector reports for the USG that combines assessment results to determine national critical infrastructure vulnerability points to support the National Infrastructure Protection Plan (NIPP) and national preparedness goals.
- Establish and/or strengthen relationships between Federal, State, and local law enforcement and the food and agriculture industry along with the critical food/agriculture sites visited.

Implementation: To facilitate this work, a series of site visits will be conducted at multiple food and agriculture and production facilities. Every Food and Agriculture Sector sub-sector will be studied (i.e. production, processing, retail, warehousing, and transportation) in order to assess the farm-to-table continuum. The primary purpose of the visit is to work with industry to validate or identify vulnerabilities at the specific site and the sector as a whole. These visits will be built upon the work done by the SSAs in order to assist in developing the National Infrastructure Protection Plan (NIPP), Federal Sector Specific Plans (SSP) and state Spp. All of the visits will be conducted on a volunteer basis.

The target start date for the SPPA program is September 1st, 2005. Two sites visits will be conducted each month - approximately one FDA and one USDA facility.

Teams comprised of knowledgeable personnel from the SSA, FBI, DHS, local and state officials, and industry will be formed to conduct the surveys.

Results: The desired results of the SPPA Initiative are as follows.

- Reports that details identified vulnerabilities, possible mitigation strategies, and warnings and indicators for each site. The reports will be distributed to all site participants.
- Reports that outline sector-wide vulnerabilities and lessons learned to effectively and appropriately prioritize national assets and resources. The reports will be distributed to DHS, USDA, FDA, and FBI.
- Each industry sub-sector will apply the CARVER assessment tool, and adapt, if necessary, to its unique production, processing, retrial, warehousing, and transportation system. Data sets will be set by GCC. Those data sets will be collected during the site visits and will be compiled by subsector (i.e. slaughterhouse, processing plant, etc). This data will be translated so outputs can be compared with other critical infrastructure sectors.
- CARVER + Shock templates
- Lessons learned
- Assessment templates for each 'system' by sub-sector that can be exported to other sites to identify vulnerabilities that incorporate existing tools.
- Sector-specific investigative templates and field guides for the food and agriculture/intelligence sector.
- Provide data to the NIPP working groups for further development of the NIPP and national preparedness plans.
- Increase awareness within industry and government needs regarding resources requirements and capabilities; current threats; and recognition of attack indicators.
- Identify and validate R&D initiatives related to the food and agriculture sector. Ensure that industry concerns and issues are carried forward to further R&D efforts.

11.5 Security Analysis for Agroterrorism in Developing Countries

Many developing countries are reliant on agricultural production for their wellbeing. The consequences of sharp declines in productivity may be famines, disruptions, and diversion of limited foreign aid to disaster management and away from developments and the loss of important sources of export earnings (Nicholas A. Linacre *et.al.*, 2005). Significantly countries with GDPs in the range of 250-5000 USD are typically heavily dependent on agricultural production for their economic prosperity and in this context agroterrorism has the potential to cause continued instability and slow growth, further destabilizing governments and creating favorable environments for insurgent activity, exacerbating the problems of underdevelopment. If it can be shown that (certain) developing countries are at risk of terrorist attacks on their food chains, it will be justified to spend resources to deal with this risk.

In some developing countries the potential exists for agroterrorism to cause widespread disruption through loss of sustenance, income and production. Defense of agriculture may also be problematic because of the lack stability and basic biosecurity infrastructure for the detection and prevention of diseases or invasive species. Currently new methodological approaches for terrorism risk assessments are being actively explored for resource prioritization. One such methodology for risk based allocation of resources is Threat, Vulnerability, and Consequence (TVC) Analysis. A qualitative application of the TVC framework is used to analyze the risk of agroterrorism in developing countries relative to industrialized countries. The analysis suggests that evidence exists to demonstrate general terrorist threats, vulnerability of agriculture and, depending on the country, potentially serious consequences arising from agroterrorism. Where specific threats emerge, action may be needed by the international community to strengthen biosecurity systems in developing countries through: increasing global cooperation, capacity building in monitoring, remediation and risk analysis technologies, and the dissemination of novel technologies for control of pests and diseases. The relative risk of agroterrorism between industrialized and developing countries was studied by applying Threat, Vulnerability, and Consequence (TVC) Analysis (Willis *et al.* 2004).

Application of the TVC analysis framework to the problem of agro-terrorism risk evaluation is motivated from an increasingly apparent need to provide national policy makers with risk assessment tools that can be used to help guide the allocation of security resources (Nicholas A. Linacre *et.al.*, 2005).. Broadly developed and developing countries share many characteristics that make may make them attractive targets for agroterrorism including:

1. The proliferation of terrorist groups who have grievances against both developed and developing countries;
2. The dependence of a significant portion of the economy on agricultural exports and imports; and
3. The large scale of agriculture.

Additionally developing countries suffer from:

1. A lack of capacity to monitor for potential agricultural pests and diseases;
2. A lack of expertise is in risk assessment practice and decision-making;
3. Poor existing security measures; and
4. Often fragile economic circumstances.

In any risk strategy there are three management options, viz., accept the risk, m anage the risk, or avoid the risk.

The default position of many developing countries is the acceptance of the risk of agroterrorism with very limited attempts at risk management. The presumption is that the risk is low. However, the previous analysis suggests that, while it is difficult to be clear about specific threats posed to agriculture in developing countries, it is conceivable that some developing countries will find that the general threat environment, vulnerability, and consequences are such that the risk is high. As the analysis has shown, developing countries are in general more vulnerable to agroterrorist attacks than industrialized countries and they have a lower capacity to deal with the consequences (Nicholas A. Linacre *et.al.*, 2005)

This does not mean that specific threats will materialize; however, it does mean that the potential exists for specific threats to develop as the security environment changes. Therefore more analysis is needed of specific emerging threats of agroterrorism in developing countries. This will help to identify situations in which spending scarce resources for preventing such threats is justified. There is, however, a problem with waiting for the emergence of such specific threats. When specific intelligence emerges it may be too late to take action on the development of biosecurity infrastructure. The potential threat of agroterrorism is an additional reason for the international community to invest more resources in activities that are already justified on more general grounds: contributing to the prevention of conflicts and to promoting security, including biosecurity, and assuring food safety and quality in developing countries.

11.6 Agroterrorism in Indian Context

Natural occurrence of pests and diseases and their control are integral components of crop and livestock farming systems. However, deliberate introduction of pests and diseases with a malicious intent is always possible (Shyama Ramani, 2018). Hence, every country must be prepared to thwart such incidences. Agents against crops, livestock, and other animals have been part of nearly every nation-sponsored offensive biowarfare program. Biological weapons can be used against agricultural targets as strategic economic weapons. This is important as global terrorism is rising. With increased globalization and connectedness via world trade, the threat from transboundary pests and pathogens arriving in countries in which they were previously absent is expected to increase. Countries that are large crop and animal producers are most at risk from invasive pests and pathogens in absolute terms. Even a minor outbreak of exotic pests or pathogens can lead to export restrictions on agriculture-dependent countries with severe economic consequences.

Major findings: Technical barriers to agroterrorism are lower than those to human-targeted bioterrorism, and the sector is unique as even a very small disease outbreak could prompt international export restrictions. Key vulnerabilities in the agriculture sector stem from, among others a variety of factors such as given below.

- **Insufficient monitoring, surveillance, and controls systems:** The surveillance system for agriculture related diseases in India is rather underdeveloped. For instance, the National Institute of Veterinary Epidemiology and Disease Informatics (NIVEDI) is the only institute in the country catering to the needs of surveillance and monitoring of livestock diseases and it has developed a web application which can forecast the probability of the occurrence of diseases in a particular district, two months in advance. Moreover, there is a need for a harmonized surveillance system at all levels – local, state and national.
- **Inefficient systems for reporting unusual occurrences and outbreaks of disease:** A vertically integrated plant disease diagnostic network of university and government diagnostic laboratories, research institutes, and field stations to address the problems of efficient and effective disease diagnosis and pathogen detection is lacking.
- **Porosity of borders:** The issue of quarantine is not dealt with seriously at airports. Customs official generally perform the task of quarantine and they are generally not fully aware of the technicalities of the issue. Unlike Australia which has county to county quarantine, in India we don't have any quarantine provisions for movement of agricultural materials and produce between states. Further, there is a lack of adequate human resources for quarantine purposes.
- **Need for institutional developments in terms of legislation and guidelines on crop and animal protection:** There is a need to rationalize regulatory functions by coordinating effort and synergy of expertise from various organizations responsible for research and development, food safety and distribution in the agricultural and food systems. Currently there are agencies under different ministries with intersecting and parallel discourses (e.g. Review Committee on Genetic Manipulation (RCGM) of the Department of Biotechnology, Ministry of Science and Technology, and the Genetic Engineering Approval Committee (GEAC) of the Ministry of Environment, Forest& Climate Change). This poses challenge in terms of meeting the core priorities and distribution of scarce resources among various agencies and strategic needs. Also, there is inadequate involvement of sub-national agencies as the state and local governments perform a number of important functions ranging from farm and food inspection, and collection of information on human and animal health.
- **Need for catch-up in early and accurate disease diagnostics and pathogen detection:** In India, traditional molecular diagnostic technologies are widely used to identify pathogenic agricultural organisms. Efforts are

being made to produce better diagnostic kits to detect pathogens. Most research in cutting-edge science and technology, such as nanotechnology for disease detection, diagnosis, and control and its application in agriculture, is at an early stage. Although establishment of private and public plant health clinics in India has increased the diagnostic capacity; however, effective coordination and networking of the clinics is yet to be accomplished.

- **Changes in legal framework:** The legal framework in India to protect the agricultural system from pests and diseases is ill-equipped to tackle the modern challenges of biosecurity. India's Agriculture Biosecurity Bill (2013) was expected to replace the Destructive Insects and Pest Act (DIPA) of 1914 and the Livestock Importation Act of 1898. However, the bill that was pending in the parliament has now lapsed.

11.7 Security Analysis for Agroterrorism

In some developing countries the potential exists for agroterrorism to cause widespread disruption through loss of sustenance, income and production. Defense of agriculture may also be problematic because of the lack stability and basic biosecurity infrastructure for the detection and prevention of diseases or invasive species (Nicholas A. Linacre *et.al.*, 2005). Currently new methodological approaches for terrorism risk assessments are being actively explored for resource prioritization. One such methodology for risk based allocation of resources is Threat, Vulnerability, and Consequence (TVC) Analysis. A qualitative application of the TVC framework is used to analyze the risk of agroterrorism in developing countries relative to industrialized countries. The analysis suggests that evidence exists to demonstrate general terrorist threats, vulnerability of agriculture and, depending on the country, potentially serious consequences arising from argoterrorism. Where specific threats emerge, action may be needed by the international community to strengthen biosecurity systems in developing countries through: increasing global cooperation, capacity building in monitoring, remediation and risk analysis technologies, and the dissemination of novel technologies for control of pests and diseases.

Broadly developed and developing countries share many characteristics that make may make them attractive targets for agroterrorism including:

1. The proliferation of terrorist groups who have grievances against both developed and developing countries;
2. The dependence of a significant portion of the economy on agricultural exports and imports; and
3. The large scale of agriculture.

Additionally developing countries suffer from

1. A lack of capacity to monitor for potential agricultural pests and diseases;
2. A lack of expertise is in risk assessment practice and decision-making;

3. Poor existing security measures; and
4. Often fragile economic circumstances.

By organizing and discussing these issues within the TVC framework we hope to demonstrate, at least qualitatively, the utility and applicability of the framework for the emerging issue of agroterrorism.

11.8 The 2015 Outbreak of Cotton Leaf Curl Disease in Punjab (India)

Cotton is economically the most important non-food crop, significantly contributing to several economies in Indian and African subcontinents (Sibnarayan Datta *et.al.*, 2020). Cotton leaf curl virus associated cotton leaf curl disease (CLCuD, transmitted by whiteflies) is the most devastating disease of cotton and repeated outbreaks of CLCuD have severely damaged cotton crops in Pakistan, and in northwestern India. Interestingly, outbreak causing virulent strains, namely Cotton leaf curl Multan virus (CLCuMuV), 'resistance breaking' Cotton leaf curl Kokhran virus-Burewala (CLCuKoV-Bu), their interspecies recombinants originally evolved and caused outbreaks in Pakistan, followed by their trans-border spread and large scale damage to cotton in contiguous regions in India. During 2015, severe infestation of whitefly was reported from southern Punjab (India), followed by severe outbreak of CLCuD, causing complete destruction of 2/3rd cotton crop estimating to loss of 630–670 million US dollars. Apart from badly shattering the national economy; unable to withstand losses, at least 15 cotton farmers committed suicide, triggering violent protests and socio-political chaos in Punjab.

11.9 Policy Recommendations

A variety of steps should be taken to ensure that national biodefense capabilities provide sufficient protection from emerging threats in the agricultural sector (Shyama Ramani, 2018).

i. **Enable Rapid diagnosis and forensics of pathogenic agents and their outbreak**: There is a need for an integrated national database to enable quick diagnosis of an outbreak. Public laboratories need to be connected centrally so that there is a quick flow of information and materials for diagnosis.

ii. **Standardize and harmonize safety protocols**: The food-supply and agricultural safety measures have to be standardized and streamlined within the framework of a single, integrated strategy that cuts across the missions and capabilities of agencies at national, state, and local level. A harmonized surveillance system at all levels – local, state and national is crucial to address potential agroterrorism threats.

iii. **Strengthen capabilities of public laboratories**: Develop and strengthen the bio-safety measures (laboratories handling risk group III and IV pathogens) at research laboratories and modernize the laboratories engaged in production of veterinary biologicals, quality testing and diagnostic services with appropriate Good Manufacturing Practice (GMP)/ Good Laboratory Practice (GLP)/ bio-safety compliant facilities under the new sanitary and phytosanitary regimes. Harness the latest scientific and technological advancements for quick on-site detection of plant pathogens.

iv. **Educate farmers**: Train farmers to watch-out for disease dispersion patterns and counter the threat of pests/ diseases.

11.10 The Agricultural Biosecurity Bill, 2013

Highlights of the Bill

- The Bill establishes the Agricultural Biosecurity Authority of India (ABAI) to protect plants, animals and related products from pests and diseases to ensure agricultural biosecurity.
- ABAI will regulate imports and exports of plants and animals, as well as their inter-state movement. Imports and exports of plants and animals shall only be allowed if they are issued permits by the respective authority in the originating country or by ABAI, respectively.
- ABAI will also conduct surveillance of pests and diseases in the country, undertake pest risk analysis, and interact with research institutes and state governments on plant and animal protection.
- ABAI may notify quarantine pests. It may also declare an area as a controlled area if it is suspected of being infested with pests. It shall communicate the measures to be taken by the state government.
- If a state government fails to take the required measures, ABAI can take necessary steps to eradicate or contain a quarantine pest in a controlled area. The state government shall reimburse ABAI with the costs incurred for such purposes.
- The central government may declare a biosecurity emergency on the recommendation of ABAI in case of a pest or disease outbreak.
- ABAI shall discharge international obligations under various international trade, sanitary and phytosanitary agreements.

Key issues and analysis

- Currently, import, export, quarantine, and the inter-state spread of plant and animal diseases are regulated by various entities under different laws. These functions will be subsumed under the proposed ABAI.
- Other countries have established biosecurity systems similar to the one proposed under the Bill. Most of them have established national authorities that regulate imports, exports, quarantine and inter-state movement of plants and animals.

- The Standing Committee examining the Bill recommended a higher representation of states in ABAI. It also recommended removing the requirement for states to reimburse ABAI for measures taken by it to contain a quarantine pest or disease.

Financial memorandum: Clause 7 of the Bill provides that the Central Government shall constitute an Authority to be called the Agricultural Biosecurity Authority of India and the Head Office of the Authority shall be at Faridabad. The Authority will exercise powers, conferred on and perform functions assigned to it under the proposed legislation. Clause 8 of the Bill provides that the Central Government shall appoint a Director General, two Deputy Director Generals and two members other than ex-officio Members.

Sub-clause

1. of clause 11 provides that the conditions of service, fee and allowances payable to Director General, Deputy Director General and the other members (other than ex-officio Member shall be such as may be prescribed under the rules).
2. Clause16 of the Bill provides that the Authority may appoint officers and employees for efficient performance of its functions and the scale of pay, allowances and other conditions of service will be such as may be provided by the Authority by regulations.
3. Clause 22 of the Bill provides for transfer of existing employees of the Directorate of Plant Protection and Quarantine Storage and Animal Quarantine Stations under the Central Government to the Authority. Sub clause (3) provides that the existing employees shall be given an option to express their willingness or otherwise to become employees of the Authority. The expenditure related to the employees so transferred to the Authority shall be met with the existing budgetary allocation.
4. Clause 45 of the Bill provides that the Central Government after the appropriations made by Parliament, may make to the Authority grants and loans for utilisation for the purposes of the proposed legislation. Clause 46 of the Bill provides for constitution of the Agricultural Biosecurity Fund to which all grants, loans, fees, penalties and charges received by the authorities shall be credited.
5. It is estimated that there would be an expenditure of 726 crores of rupees on the establishment of the Agricultural Biosecurity Authority, to be borne by the Central Government, this would include non-recurring capital expenditure of 506 crores of rupees on equipment, land and building and a further recurring expenditure of 220 crores of rupees towards salaries and allowances, establishment, etc.

6. The Bill if enacted and brought into operation would involve expenditure from the Consolidated Fund of India as mentioned above and is not likely to involve any other recurring or non-recurring expenditure.

11.11 Agroterrorism-Challenges Faced by India

Infectious diseases such as COVID-19, the disease caused by the novel coronavirus; severe acute respiratory syndrome (SARS); Middle East respiratory syndrome (MERS); and the diseases caused by the Ebola, Nipah, and Zika viruses have exposed countries' susceptibility to naturally occurring biological threats (Shruti Sharma, 2020). Even though scientists from multiple countries concluded that the virus responsible for the coronavirus pandemic shifted naturally from an animal source to a human host, the international community should not ignore the possibility of pathogens escaping accidentally from research labs and threats of deliberate manipulation to create more dangerous bioweapons.

India is especially vulnerable to such infections because of its geographical position, large population, low healthcare spending, minimal expenditure on research that benefits public health, weak coordination between central and state health authorities, limited involvement of private actors, poor awareness of biosecurity, and the rickety state of public health infrastructure. Most recently, COVID-19 has revealed the deep fault lines in India's public health infrastructure, including a shortage of healthcare workers, lack of trained epidemiologists, scarcity of medical equipment, poor access to healthcare facilities in rural areas, and inefficient disease reporting and surveillance in most states. The pandemic should therefore be a wake-up call for India to assess gaps in its public health infrastructure and divert its resources toward the healthcare sector to prepare itself for both natural and man-made biological emergencies.

Major Recommendations: As the spread of infectious diseases is a long-term, continuous, and evolving threat, India may need an agency specifically responsible for preventing and managing biological threats. India could consider investing in an agency that can coordinate policy responses for any biological emergency (Shruti Sharma, 2020). A full-time Office of Biological Threats Preparedness and Response (BTPR) under the National Disaster Management Authority (NDMA) is being suggested as an alternative. This paper sketched out this idea to stimulate further dialogue among interested stakeholders. This office could focus on naturally occurring diseases, threats emerging from laboratory accidents, and deliberate weaponization of diseases. Because India has numerous organizations that sometimes perform overlapping roles with limited or no coordination with each other, the office could become a nodal agency that brings together experts from different ministries, representatives from the private sector, and experts from the academic and scientific community.

Whether or not a new office is set up, it is important for India to review domestic measures needed to predict, prevent, and respond to both natural and man-made biological threats. These measures include:

- Periodic training of healthcare workers on nursing practices, safe handling of samples, decontamination procedures, and proper disposal of biomedical waste;
- Strengthening cooperation between central and state health authorities;
- Introducing clearer and stronger incentives for personnel to identify and report disease outbreaks among plants, animals, and humans to strengthen the disease surveillance network;
- Aggregating data obtained from different disease surveillance programs that collect data on plant, animal, and human health to detect outbreaks in a timely manner;
- Developing common disease reporting standards to harmonize data collection from all organizations reporting disease outbreaks;
- Creating an epidemiological model for diseases through collaboration between government, scientists, academicians, industry, epidemiologists, and data scientists;
- Implementing capacity-building measures, such as engaging with local donors to mobilize resources needed to ramp up public health infrastructure, increasing funding to research bodies, introducing incentives to invest in biotechnology research, and fostering collaboration between the scientific and the policy community, which should all be encouraged to strengthen India's preparedness for biological threats;
- Conducting surprise on-site inspections by members of the government-led Review Committee on Genetic Manipulation (RCGM), the Genetic Engineering Appraisal Committee (GEAC), and state regulatory authorities to ensure rigorous monitoring of biotechnological research;
- Harmonization of application protocols and introduction of standard evaluation forms for researchers applying for approvals to commercialize biotechnology-derived products;
- Introducing mandatory certification and validation for BSL-2 labs that sometimes work with high-risk pathogens;
- Developing a formal biosecurity policy that encompasses threats emerging from different sectors such as plant health, animal health, and public health to avoid any overlaps or coordination issues;
- Conducting specific training sessions for customs officials to identify specific pests or pathogens that might pose a risk to India's national security;
- Introducing simulation exercises to develop standard operating protocols that can be implemented during the time of a crisis, like inexpensive tabletop exercises that can help generate awareness among relevant agencies and can be useful for monitoring, assessing, and strengthening the capabilities of emergency policies, plans, and procedures.

Bibliography

Anonymous, 2005. NIJ Research Report Defining Law Enforcement's Role in Protecting American Agriculture from Agroterrorism Prepared for: National Institute of Justice Washington, D.C.

Anonymous, 2005. Strategic Partnership Program Agroterrorism (SPPA) Initiative. U.S. Food and Drug Administration, Department of Homeland Security, U.S. Department of Agriculture, Federal Bureau of Investigation

Anonymous, 2007. Agroterrorism: Threats and Preparedness. EveryCRSReport.com, University of North Texas Libraries Government Documents Department, Raw Metadata: JSON

Anonymous, 2008. Criminal Investigation Handbook for Agroterrorism. The Food and Drug Administration and U.S. Department of Agriculture,

Anonymous, 2010. Agroterrorism. What is The Threat and What Can Be Done About It? Rand National Defense Research Institute; [Last accessed February 3, 2010]. Available at http://www.rand.org/pubs/research_briefs/RB7565/RB7565.pdf

Chalk Peter, 2004. Agroterrorism: What Is the Threat and What Can Be Done About It?. Santa Monica, CA: RAND Corporation, California

Haralampos Keremidis, Bernd Appel, Andrea Menrath, Katharina Tomuzia, Magnus Normark, Roger Roffey and Rickard Knutsson, 2013. Historical Perspective on Agroterrorism: Lessons Learned from 1945 to 2012. Biosecurity and Bioterrorism: Biodefense Strategy, Practice and Science, 11: S17-S24

Henry Parker, 2020. *Agricultural Bioterrorism: A Federal Strategy to Meet the Threat*, McNair Paper 65, Washington, D.C.: Institute for National Strategic Studies, National Defense University, pp. 40–41.

Kohnen A, 2000. Responding to the threat of agroterrorism: Specific recommendations for the USDA, Cambridge, MA John F. Kennedy School of Government, Harvard University 2020.22.

Manuel, F.Z., 2017. Agro Terrorism: A Global Perspective. 2017. *J. Pol. Sci. Pub. Aff.*, 5:2-11.

Nicholas A. Linacre, Bonwoo Koo, Mark W. Rosegrant, Siwa Msangi, Jose Falck-Zepeda, Joanne Gaskell, John Komen, Marc J. Cohen, and Regina Birner , 2005. EPT Discussion Paper 138 Security Analysis for Agroterrorism: Applying the Threat, Vulnerability, Consequence Framework to Developing Countries. 2033 K Street, NW, Washington, DC 20006-1002 USA

Nicholas A. Linacre, Bonwoo Koo, Mark W. Rosegrant, Siwa Msangi, Jose Falck-Zepeda, Joanne Gaskell, John Komen, Marc J. Cohen, and Regina Birner, 2005. Security Analysis for Agroterrorism: Applying the Threat, Vulnerability, Consequence Framework to Developing Countries. Environment and Production Technology Division. International Food Policy Research Institute

Shruti Sharma, 2020. Biological Risks in India: Perspectives and Analysis. Carnegie Endowment for International Peace, Washington, DC 20036

Shyama Ramani, 2018. A bomb in your soup? Addressing possible biosecurity risks for Indian agriculture. SITE4society Brief No. 5-2018

Sibnarayan Datta, Hmuaka Vanlal and Dwivedi Sanjai. (2020). Agroterrorism in Indian Context. *Def. Life Sci. J.,* 5: 125-132

Terry Knowles, James Lane, Gary Bayens, Nevil Speer, Jerry Jaax, David Carter and Andra Bannister, 2005. Defining Law Enforcement's Role in Protecting American Agriculture from Agroterrorism. U S department of justice, Office of justice programs

Willis, H. H., Morral, A. R., Kelly, T. K. and Medby, J. J., 2004. Risk-based allocation of counterterrorism resources". RAND. Presented at the Society for Risk Analysis Annual Meeting Palm Springs 2004. www.sra.org.

12

Mitigation Planning and Integrated Approach for Biosecurity

The biosecurity risk assessment informs the biosecurity plan, which documents the mitigation measures put in place to address risks. Risks that fall outside of the risk tolerance threshold should be controlled through additional or enhanced mitigation measures. A cost-benefit analysis can assist in determining the mitigation measures in which to invest. Financial constraints and resource limitations may present challenges when looking to manage unacceptable risks. As a starting point, senior management may choose to initially focus mitigation measures on the most consequential risks and then control remaining risks as resources become available. In other instances, if the risks are determined to be too high or costly to mitigate, the organization's project or program may need to be modified or cancelled. Recommendations for mitigation measures should be documented in the final report of the biosecurity risk assessment.

It is recommended that the biosecurity risk assessment be reviewed routinely and updated when necessary to address changes that would affect the level of risk (e.g., threat environment, regulation and policy, after a biosecurity event occurs, program renewal, newly discovered vulnerabilities, construction of a new facility, and additions or subtractions to an organization's asset inventory. Once the assessment is complete, the findings and recommendations to senior management and decision makers should be presented in a comprehensive biosecurity risk assessment report. The decision to prepare a biosecurity risk assessment report is optional and left to the discretion of the assessment team since the biosecurity risk assessment is complete; however, summarizing the findings and placing emphasis on higher risks will facilitate communication and understanding of the biosecurity risk assessment. The report should summarize the biosecurity risk assessment and present the scenarios of highest risk as well as recommendations for reducing unacceptable risks. This report and the biosecurity risk assessment itself may contain sensitive information and are considered assets to be evaluated in the risk assessment process.

Ten sections are proposed for the biosecurity risk assessment report, viz., Executive Summary ; Purpose ; Scope ; Background ; Threat Environment ; Asset Inventory; Risk Assessment Results ;Risk Tolerance ;Recommendations ; Appendix (Asset Inventory, Biosecurity Event Table, Likelihood Table, Consequence Table, Risk Register, Schedule and Team members).

12.1 Risk Mitigation Plan Implementation and Action Items

Effective controls protect workers from workplace hazards; help avoid injuries, illnesses, and incidents; minimize or eliminate safety and health risks; and help employers provide workers with safe and healthful working conditions. The processes described in this section will help employers prevent and control hazards identified in the previous section (Anonymous, 2020).

12.1.1 Biosecurity Planning

The development of a biosecurity plan (BP) is an important step in improving preparedness for pest incursions. Biosecurity planning is undertaken at a national level through a partnership approach to identify the greatest biosecurity threats to Australia's plant industries. Through this process risk mitigation activities are identified to improve biosecurity practices and preparedness. Biosecurity planning is an obligation under the Emergency Plant Pest Response Deed (EPPRD) and BPs are generally reviewed every four to five years. Biosecurity plans are not designed to be on-farm manuals (Anonymous, 2021).

The biosecurity planning process : The first step in developing a BP is to identify the exotic pests (including insects, mites, molluscs (snail and slugs), pathogens (diseases) and nematodes) of the crop(s) produced commercially. The exotic pests are assessed by a Technical Expert Group for their potential to enter, establish and spread within a region and their economic impact to the crop(s) covered in the BP. Those that pose the highest risk are considered to be High Priority Pests (HPPs). The next step is to develop and agree upon effective biosecurity measures to protect against the identified the pests that pose the greatest risk to an industry. This involves the industry, governments, the relevant Research and Development Corporation (RDC) and PHA working in partnership with each other. Agreed risk mitigation activities are aligned to overarching strategies in the National Plant Biosecurity Strategy under:

- Capacity and Capability
- Plant Biosecurity Education and Awareness
- Preparedness and Response
- Surveillance
- Diagnostics
- Established Pests and Weeds
- Biosecurity Research, Development and Extension (RD&E)
- Legislative and Regulatory Issues of Importance

Once the BP has been developed, it is endorsed at the national level through the relevant peak industry body (or bodies) and by each of the state and territory governments and the Commonwealth government through the Plant Health Committee.

Responsibility for prevention and/or mitigation : As HMA, the Director General, DPIRD, is responsible for undertaking prevention and/or mitigation activities in relation to the hazard of animal or plant pests or diseases. Lead responsibility for prevention and mitigation of animal or plant biosecurity incursions rests with the HMA, in conjunction with the Commonwealth Department of Agriculture, Water and the Environment (DAWE). DAWE is responsible for biosecurity at the national border. State legislation administered by DPIRD contains responsibilities for border and biosecurity regulation within the State.

12.1.2 Prevention and/or Mitigation Strategies

State prevention strategies: Quarantine Services: DPIRD (through Biosecurity and Border Compliance) provides import and export inspection services under applicable legislation. Inspections are carried out on risk material, such as fresh fruit and vegetables, flowers, seeds, honey, vehicles and machinery and any other possible carriers of risk material, imported into WA. Activities include surveillance programs to detect and identify quarantine risk material at State border checkpoints (road, rail, sea and air), freight depots, post offices and other interstate entry points.

National prevention strategies: DAWE (Biosecurity and Compliance Group) provides quarantine inspection for international passengers, cargo, mail, animals, plants, and animal or plant products arriving in country. DAWE (Biosecurity and Compliance Group) manages risk through:

Border control: Passenger and cargo clearance at quarantine control entry points into Australia

Animal Quarantine: Quarantine controls apply to all animals and animal products (including insects, fish, and reptiles)

Plant quarantine: All plants or parts of plants (fruits, seeds, cuttings, bulbs, wood or bamboo items) must be examined and if necessary treated. Live plants must be kept at plant quarantine stations when they arrive in Australia to ensure they are not carrying pests or diseases.

Mitigation strategies

Strategy	Detail
Animal and Plant Disease Surveillance and Controls	DPIRD undertakes a number of active and passive surveillance programs to verify biosecurity status and to assist in early detection of significant incursions.
Diagnostics	DPIRD operates diagnostic and identification laboratories for animal and plant diseases and pest identification.
Livestock Identification and Brands	Western Australia has a comprehensive, mandatory livestock identification system. All livestock in Western Australia must be identified in accordance with the Biosecurity and Agriculture Management (Identification and Movement of Stock and Apiaries) Regulations 2013.

Livestock Movements	Control measures, requirements, procedures and protocols operate to control and record the movement of livestock within Western Australia. This also applies to import and export movements from other Australian States and Territories.
Emergency Animal Disease Watch hotline	A national hotline is available for reporting emergency animal pest and disease outbreaks or suspected cases – 1800 675 888.
Emergency Plant Pest (EPP) hotline	A national hotline is available for reporting emergency plant pest and disease outbreaks or suspected cases – 1800 084 881. The Pest and Disease Information Service (PaDIS) receives EPP calls for WA.

12.2 Tourism Biosecurity Risk Management and Planning

The state of national biosecurity planning was reviewed (Melly Domhnall and Hanrahan James, 2020). It was recognized tourist vectoring can increase the risk of invasive alien species (IAS) and disease; representing substantial biosecurity risk for tourism destinations worldwide. This research assessed the provision of biosecurity mitigation measures within national biosecurity plans and guidelines internationally. The author's position in this issue contends that a lack of sufficient biosecurity risk management and planning in place for tourism could have severe impacts on a destination's environment, society, and economy. The authors systematically reviewed national biosecurity planning through a mixed-method research approach. First, essential criteria identified from international literature allowed for content analysis to assess specific national biosecurity plans and strategies. Second, qualitative data was then gathered by conducting semi-structured interviews within national governing bodies and organizations. Planning for tourism biosecurity varies around the world with some destinations demonstrating highly evolved plans such as Hawaii, New Zealand, and Australia. However, this is not the case in Ireland where planning for biosecurity at a national level is severely limited. Biosecurity planning, pathway management, communication, quarantine and plans for tourism risk are inadequate to prevent the introduction and spread of IAS and disease in Ireland. Recommendations offer destinations globally and with "island status" an opportunity for biosecurity to be improved by using surveillance, communication, guidelines and specific capacities at the border stages within a specific national biosecurity plan. This is the state of national planning for biosecurity provides new knowledge specifically for tourism destinations worldwide, which can adopt the essential elements identified within this research for a national tourism biosecurity risk framework.

i. **Biosecurity- An integrated approach to manage risk to human, animal and plant life and health:** Biosecurity is a strategic and integrated approach to analysing and managing relevant risks to human, animal and plant life and health and associated risks for the environment. National stakeholders include relevant government agencies, agricultural producers and the food industry, scientific research institutes, specialist interest groups, nongovernmental organizations (NGOs) and the general public. International standard-setting organizations, international bodies and international legal instruments and agreements provide a governance framework for biosecurity (Anonymous, 2010). Benefits of biosecurity include early recognition of emerging pest and disease threats, ability to consider complete exposure pathways, integrated responses to threats, rationalization of controls, improved emergency preparedness and response, overall ensuring the more efficient use of available resources.

ii. **Harmonization and integration of approaches to biosecurity:** A traditional sector-based approach to biosecurity is increasingly under challenge, and many countries are revising the relevant legal and regulatory systems, institutional responsibilities and resources available for essential infrastructure in response. The aim is to ensure a more integrated approach, and ensure a faster and more effective response to biosecurity threats (Singh, 2008). Some countries have even made major changes to institutions to join all relevant responsibilities 'under one roof' – an example would be New Zealand; others have combined some responsibilities e.g. for plant and animal health inspection; while a larger group have instituted formal communication mechanisms (such as a national biosecurity committee or task force) to ensure regular and effective dialogue between the different stakeholders.

iii. **Requirements for a harmonized and integrated approach to biosecurity** The successful implementation of a harmonized and integrated biosecurity approach requires a clear policy and legal framework, an institutional framework that defines the roles and responsibilities of relevant stakeholders, adequate technical and scientific capability, including use of risk analysis, a well-functioning infrastructure for testing and control, and a system for communication and information exchange.

iv. **Requirements for a harmonized and integrated approach to biosecurity**: The successful implementation of a harmonized and integrated biosecurity approach requires a clear policy and legal framework, an institutional framework that defines the roles and responsibilities of relevant stakeholders, adequate technical and scientific capability, including use of risk analysis, a well-functioning infrastructure for testing and control, and a system for communication and information exchange. The Guide to Assess Biosecurity Capacity provides a process for assessing biosecurity capacity needs across all sectors and all sector organizations, which will help to identify requirements to pursue a harmonized and integrated biosecurity approach.

v. **Examples of enhanced outcomes:** In a modern biosecurity environment, considerable importance is placed on a holistic approach. Countries are encouraged to base their controls, as far as possible, on international standards where they exist. At the national level and internationally, there are likely to be significant benefits in integrating biosecurity activities to the extent practical . Examples of enhanced outcomes of biosecurity include better risk analysis, ability to consider complete exposure pathways, integrated responses to new and emerging diseases, rationalization of controls, improved emergency preparedness and response, integrated surveillance or traceability systems and more efficient use of available resources.

vi. **Policy framework:** A biosecurity policy framework sets out a broad course of action to address biosecurity risks in food and agriculture. It is based on appropriate public goals and a set of beliefs about the best way of achieving those goals. It provides a common basis for assessing biosecurity risks and priorities for action and gives direction and guidance to all the parties concerned.

vii. **Legal framework :** Sound biosecurity legislation (encompassing laws and regulations) is necessary to create an enabling environment of predictability and certainty through good governance and respect for the rule of law. Law clarifies the roles, responsibilities and rights of different stakeholders, including those parts of government with policy and delivery roles for biosecurity outcomes and programmes, in order to ensure consistency and accountability. It also defines appropriate powers to act, which is essential for enforcement.

viii. **Institutional framework :** A clear institutional framework within which to manage biosecurity is an important part of a more harmonized and integrated approach to biosecurity. The institutional framework identifies the competent authority or authorities responsible for establishing biosecurity controls and ensuring their implementation, as well as any other stakeholders involved. It also sets out the rules and procedures governing their roles and defines the mechanisms through which they work towards shared goals. The choice of institutional framework will be determined by factors which are specific to a country and biosecurity context (e.g. historical traditions, political orientation, financial and other resources).

ix. **Communication and information exchange:** The complexity inherent in managing biosecurity requires communication and information exchange among a wide range of national stakeholders including government agencies, the private sector (agricultural producers, processors, enterprises, importers/exporters, etc.), the scientific and research community, and the general public Transparency obligations under international agreements such as the SPS Agreement require governments to ensure transparency in the adoption of their sanitary and phytosanitary rules. This includes publishing proposed rules in advance and allowing time for comments from the public, as well as the establishment of enquiry points for consultations on rules and inspection and control procedures applicable to imports and exports. They also must open to scrutiny how they apply their food safety and animal and plant health regulations. National, regional and global networks all contribute to meeting the information needs of an integrated biosecurity system.

Competent authorities and new approaches: Competent authorities with adequate technical and scientific capability and infrastructure. Establishing biosecurity controls and ensuring their implementation is the core responsibility of competent authorities. They should have appropriate policies and regulations in place, as well as operational principles, procedures and capacity, and adequate resources. They should have, or have access to, adequate technical and scientific knowledge and skills, and should have adequate infrastructure.

Implementing national biosecurity mandates demands human resources with adequate technical capability. This includes personnel with specialized scientific knowledge and skills to carry out biosecurity functions (e.g. provision of scientific research and advice, inspection, verification and enforcement, diagnostic analysis, quarantine and certification, risk profiling and priority setting, standard setting and implementation, monitoring and surveillance, and emergency preparedness and response), based on a risk analysis approach wherever possible and practical. Technical resources in several of these areas may be shared across public agencies and the private sector. For instance, inspection activities may be carried out at any step in the hazard exposure pathway by the competent authority or by officially-recognized bodies. Similarly, diagnostic laboratories may be owned and operated by the public or private sector, or as a public-private partnership.

Emergency preparedness and response in the event of a disease outbreak are key elements of biosecurity systems and need for this capability is illustrated by recent disease outbreaks in many parts of the world. Emergency preparedness and response is a collective responsibility that requires partnerships between central government, competent authorities across all biosecurity sectors, industry and the public. Policy documents detailing joint roles and responsibilities, as well as decision-making and funding procedures in emergency situations are required, along with a series of standards and procedures governing monitoring and surveillance. Modern biosecurity concepts can only be applied if there is an effective infrastructure at the national level. Necessary infrastructure includes diagnostic laboratories with functioning equipment and supplies, facilities for storage and containment of samples and suspect consignments at checkpoints, as well as sanitation equipment, quarantine yards, inspection equipment, vehicles, and computers and communication equipment for the operation of monitoring, surveillance and emergency preparedness systems.

Willingness to explore new approaches : New approaches to biosecurity can be achieved in different ways depending on the particular circumstances and needs at the country level. There is not one single or best model. Generally, an integrated approach is pursued by merging services and functions. However, the extent of consolidation varies. For example, in New Zealand, policies and planning affecting different biosecurity sectors are more inclusive than in countries like Canada and Australia. In countries like France where there has been less consolidation, cooperation is pursued by means of formal and informal mechanisms of interaction, exchange and coordination among relevant bodies.

It is important to note that an integrated approach does not mean that all of the roles and responsibilities of the competent authorities involved should be harmonized. They often have distinct and sometimes separate roles, and contribute to biosecurity in different ways (e.g. a quarantine function presents a front line of defence against all hazards whereas a forestry management function may focus more on monitoring and remedial risk management of pests in either natural forests or plantations). Moreover, the situation is not static (e.g. rapid growth of aquaculture and technical breakthroughs in fish transgenics presents different biosecurity policy and functional needs compared with forestry). However, a common thread in all sectors is the increasing reliance on systematic risk analysis.

Conclusions: Improved health and well-being of human populations are the ultimate outcomes of functioning biosecurity systems. Biosecurity forms a bridge between agriculture and health. Poor practices in agriculture or food production can favour biosecurity threats, and directly have an impact on public health or threaten food security. The benefits of a more harmonized and integrated approach to biosecurity are already apparent in specific national situations. A more holistic approach to biosecurity will enable these benefits to be achieved in a manner that avoids inconsistencies, fills gaps, prevents the creation of unnecessary barriers to trade, and protect human health and consumer confidence in agricultural and food products.

12.3 National Agricultural Biosecurity System for Food Security Biotic Insecurity (NABS)

Establishing a National Agricultural Biosecurity System for Food Security Biotic insecurity (losses caused by pests, diseases, weeds, toxins, antinutritional factors and overall degenerating quality of food and agricultural products) impacts all the elements of food security, namely, production and availability, access, utilization and vulnerability. This relationship is getting increasingly intensified in the globalised world, as communications, transports, trade and travels are ever intensifying in the global village. Moreover, biosecurity concerns in the food and value chain may arise at any point in the production – post harvest handling – agroprocessing – value addition – retailing – distribution – product storage – consumption/utilization chain. The direct human health implications due to zoonoses, the health and environmental impacts of biotechnologically genetically modified unsafe organisms, biodiversity erosion and microbial imbalances and even bioterrorism notwithstanding, each of the nodes along the production-processing-marketing-consumption are variously impacted by the biological insecurity agents and their management systems. Therefore, as recommended by the National Commission on Farmers, it is essential to establish an effective national agricultural biosecurity system (NABS) which will minimise, if not completely eliminate, biorisks to production, trade, health and environment and lead to sustained and enhanced livelihood security. Functions, organizational structure and management of the proposed NABS are described below :

Functions : The NABS should determine the potential for synergies and harmonization within the national and sub-national regulatory frameworks that would result from a holistic and coordinated approach to biosecurity. Policy-makers should recognize the importance of biosecurity as a key element of sustainable development, and the benefits, including in trade, that can be gained from comprehensive approaches to biosecurity. They should also appreciate the cost of not fully recognising the role of biosecurity. Full awareness on part of all stakeholders is essential for sustaining and further strengthening this national movement. Strategy of the NABS should be to synergise linkages among science and technology, education and training and commercialization and utilization in the different subsectors capturing both commonalities and specialties for synergistically addressing the four main biosecurity components, namely, Preparedness and Prevention, Diagnostics, Surveillance and Input Management.

In a large country like India, NABS should recognize the efficiencies that may emanate from regional and sub-regional approaches to risk analysis, particularly in relation to animal and plant life and health and living modified organisms, and re-organise or establish agro-eco-regional facilities as per specific challenges and opportunities. The various quarantine, SPS and zoosanitary facilities should be updated and adequately staffed to be in an ever-ready condition. The Plant Protection staff should be duly rewarded with befitting incentives. Risk analysis and management frameworks are essential to achieve biosecurity. In the past, such frameworks have been mostly sectorial or used to address specific technical issues. In future, such frameworks should seek to improve collaboration among diverse interests and institutions (particularly agriculture, public health, environment, trade, and their associated stakeholders) to achieve biosecurity in a mutually supportive manner, thus avoiding duplication and possible inconsistencies. There are several such opportunities which should be grabbed. General principles for biological risk analysis in food and agriculture are the same, although procedures may differ depending on the hazards addressed. The IPPC, the *Codex Alimentarius*, the OIE, the CBD and its Cartagena Protocol, where appropriate, should apply coherent risk analysis methodologies in different sectors by jointly analyzing differences and commonalities in approaches, and use of terms in risk analysis. FAO may play an orchestrating role in this area and help develop tools, including tools to extend the Phytosanitary Capacity Evaluation to other sectors, to assist developing countries to analyse their capacity-building needs that take account of the full scope of biosecurity, including the communicational, legal, institutional, scientific and technical aspects.

The roles and responsibilities of both the public and private sectors should be considered in planning biosecurity capacity-building initiatives. Agriculture related industries should play greater positive role in strengthening the national biosecurity umbrella. The System should devise innovative measures to build partnerships involving all stakeholders. Appropriate linkages and coordination mechanisms among existing and planned biosecurity capacity-building initiatives should be established to enhance complementarity and avoid duplication of efforts, and to ensure that capacity building is directed at identified priorities. The System should give highest attention to obviate the serious shortcomings in quantity and quality of necessary databases. The need to share information and to ensure better understanding of the requirements for achieving biosecurity can hardly be over emphasized. The need for an Internet-based biosecurity portal to facilitate information exchange on biosecurity is a priority. The importance of information access and exchange in developing biosecurity capacity should also be recognized. India with its strength in bioinformatics can play a leading role in developing appropriate mechanisms for information exchange in biosecurity, and to participate in the development of information portals.

In order to lead from the front, the NABS should develop a specific methodology or adopt the ones already used by other national and international programmes for identification, establishment and maintenance of a given strategic area and render it pest free as per the international standards. For instance, in India such a project could be initiated involving State Governments, Farmers, Traders and other Stakeholders to "sensitize" and declare all areas under identified leading varieties of mangoes as pest free for export to USA under the recent Indo-USA agreement and also highlighted under the Indo-USA Knowledge Initiative. NABS may initiate projects in a few hot – spots in a highly scientific and professional manner-collection of ground facts and creation of database and benchmark information, undertaking detailed risk analysis and eradication of the risk (pathogen and pest) and monitoring the freedom of the area from the eradicated pest. It should also analyze impact of socio-economic, agro-ecological and climate change on overall biosecurity situation in the area.

Organizational structure: Necessary capacity must be put in place to establish and sustain the National Agricultural Biosecurity System and harmonized with international biosecurity standards for food and agriculture to take advantage of trade opportunities and technology sharing for enhanced and sustained agricultural production and farmers' income. Achieving biosecurity requires an understanding of, and the ability to analyse diverse and complex risks, and determine and apply measures in a coherent manner while respecting differences among sectors and organizations. Risk analysis and management, as mentioned earlier, is the most important unifying concept across different biosecurity sectors.

12.4 One Biosecurity Concept

Biosecurity is increasingly important in an ever more connected world that is exposed to multiple threats that impact human health, agriculture, and the environment sectors, yet policy is strongly sector specific. One Biosecurity bridges these sectors to allow greater foresight in the management of invasive alien plants, animals and pathogens that impact human health, animal health, plant health, and the environment (Hulme, 2020). Many invasive species impact multiple sectors but to date their overall threat to society and the economy are insufficiently well captured by current risk assessment tools resulting in unforeseen outcomes. The major future challenges to biosecurity such as urbanization, climate change, agricultural intensification, and increased human mobility all require the more holistic understanding provided by One Biosecurity to achieve a greater cross-fertilization of ideas and an improvement in approaches to deal with threats that impact multiple sectors.

In the wake of the SARS-CoV-2 pandemic, the world has woken up to the importance of biosecurity and the need to manage international borders. Yet strong sectorial identities exist within biosecurity that are associated with specific international standards, individual economic interests, specific research communities, and unique stakeholder involvement. Despite considerable research addressing human, animal, plant, and environmental health, the science connections between these sectors remain quite limited. One Biosecurity aims to address these limitations at global, national, and local scales. It is an interdisciplinary approach to biosecurity policy and research that builds on the interconnections between human, animal, plant, and environmental health to effectively prevent and mitigate the impacts of invasive alien species (Hulme, 2020). It provides an integrated perspective to address the many biosecurity risks that transcend the traditional boundaries of health, agriculture, and the environment. Individual invasive alien plant and animal species often have multiple impacts across sectors: as hosts of zoonotic parasites, vectors of pathogens, pests of agriculture or forestry, as well as threats to biodiversity and ecosystem function. It is time these risks were addressed in a systematic way.

One Biosecurity is essential to address several major sociological and environmental challenges to biosecurity: climate change, increasing urbanisation, agricultural intensification, human global mobility, loss of technical capability as well as public resistance to pesticides and vaccines. One Biosecurity will require the bringing together of taxonomists, population biologists, modellers, economists, chemists, engineers, and social scientists to engage in a new agenda that is shaped by politics, legislation, and public perceptions.

One Biosecurity provides a unified framework to address the many biosecurity risks that transcend the traditional boundaries of animal health, plant health, human health, and the environment (Fig.19). There are many examples where an alien species has impacts across multiple sectors including the environment and human health. Cross-sectorial impacts are not limited to vertebrates, invertebrates can also have impacts across human and livestock health. The giant African snail (*Achatina fulica*) has been introduced widely throughout the tropics where its feeding leads to considerable crop losses which are exacerbated by its role as a vector of plant pathogens (Phytophthora spp.), it outcompetes native gastropods and is also an intermediate host playing a role in the transmission of Angiostrongulus spp. the causative agents of eosinophilic meningoencephalitis in livestock and humans (Thiengo *et.al.*, 2007).

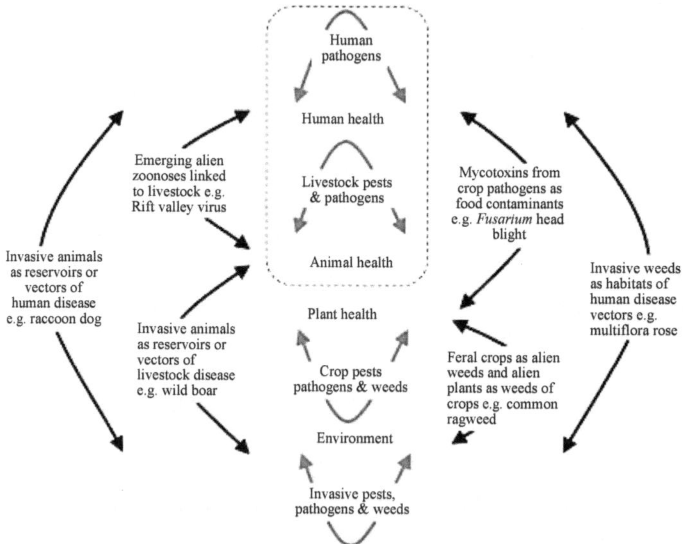

The dashed rectangle represents the sectors that are the focus of One Health and highlights the more comprehensive aspect of One Biosecurity.

Fig. 19 : Schematic representation of the One Biosecurity concept emphasising the links between human, animal, plant, and environmental health arising through the impacts of invasive alien plants, animals, and pathogens.

One Biosecurity -A better way forward : The earliest conceptualization of One Biosecurity stems from a review of Australia's quarantine and biosecurity arrangements undertaken in 2008 that sought to encourage a stronger partnership between the federal and state governments underpinned by a legal framework to support national responses to alien pests and diseases relevant to agriculture (Hulme, 2020). However, as has been shown in the previous sections, One Biosecurity must build on the interconnections among the health, agriculture, and environment sectors, have relevance at global, national and local scales and be interdisciplinary by embracing the natural and social sciences. By defining One Biosecurity in these broader and more enlightened terms the potential for transformative change in the management of invasive alien species is much more likely. A One Biosecurity perspective will require the bringing together of taxonomists, population biologists, modellers, economists, chemists, engineers, and social scientists to engage in an agenda that is shaped by politics, legislation, and public perceptions. However, One Biosecurity could easily suffer the same criticisms faced by One Health that it is no more than a buzzword and that in most cases the disciplinary divisions remain. How to avoid this ?

The first step is to ensure that the development of One Biosecurity sits front and centre of international and national policies. Under the SPS Agreement, World Trade Organization Members have the right to adopt sanitary and phytosanitary measures necessary for the protection of human, animal, and plant life or health. These measures must be science-based, not more trade restrictive than required and not arbitrarily or unjustifiably discriminatory against trading partners. This clearly argues for a holistic approach that embraces the health, agriculture, and environment sectors as captured by One Biosecurity. While this should mean that assessments need to capture the risks to human, animal, and plant life, at present the cross-sectorial nature of biological invasions is not captured effectively in the tools used for assessing biosecurity risks. For example, national and international animal health panels only examine the risks of introducing species that might directly or indirectly harm animal health, while risk assessments of plant pests (weeds, insects, and pathogens) focus on impacts on crop yields and ecosystem services] and those for invasive alien species examine impacts on biodiversity. Building interdisciplinary risk assessment tools should be a priority but is not without its challenges. It is likely that the environmental and social costs of biological invasions will be dwarfed by those of the human health and agriculture sectors, even where costs of control and eradication are similar and this will make the weighing up of risks more complex.

Developing a common risk assessment approach while essential, will not in itself drive the adoption of One Biosecurity. This first requires common understanding among scientists and policymakers of the shared threats invasive alien species pose to animal, plant, human, and environmental health. An initial mechanism may be to raise awareness of the benefits of One Biosecurity through the Intergovernmental Science-Policy Platform on Biodiversity and Ecosystem Services (IPBES) which has recently launched a specific assessment on the impacts of invasive alien species, including threats to human health and quality of life. With sufficient momentum, a specific International Convention addressing One Biosecurity could provide the essential governance oversight across existing multilateral agreements and conventions to ensure a more co-ordinated and synergistic approach to global biosecurity. However, such an option would need to overcome increasing resistance to multilateral initiatives (Welfens, 2020) which could be possible if the wider economic benefits as well as cost-effectiveness of an International Biosecurity Convention are made self-evident. However, the concept of One Health has gained momentum even in the absence of a dedicated International Convention. In the absence of a multilateral support, nation states such as New Zealand and Australia that already have strong biosecurity regulations could lead the way in developing national One Biosecurity frameworks which, if successful, could catalyse other nations to follow suit. Time will tell how feasible these options might be but hopefully it will not take another global pandemic for the logic of One Biosecurity to be realized.

12.5 Integrated Approach to Biosecurity

Biosecurity is a strategic and integrated approach to analyzing and managing relevant risks to human, animal and plant life and health and associated risks for the environment. National stakeholders include relevant government agencies, agricultural producers and the food industry, scientific research institutes, specialist interest groups, nongovernmental organizations (NGOs) and the general public. International standard-setting organizations, international bodies and international legal instruments and agreements provide a governance framework for biosecurity (Anonymous, 2010). Benefits of biosecurity include early recognition of emerging pest and disease threats, ability to consider complete exposure pathways, integrated responses to threats, rationalization of controls, improved emergency preparedness and response, overall ensuring the more efficient use of available resources.

Opportunity for integrated approaches to emerging cross-sectoral problems: There are a number of emerging biosecurity issues that are cross-sectoral in nature and that can benefit from increasingly integrated approaches, especially in terms of risk management. Antibiotic resistance arising from use of antimicrobials in agriculture and veterinary practice (including aquaculture) is a good example and it is recognized that a multidisciplinary and multiagency response is needed. New agricultural commodities derived from biotechnology (e.g. transgenic animals) presents another example where multi-sector experience will improve risk management. Harmonization of approaches to biosecurity is leading to new opportunities in terms of alignment of training of competent authority personnel. Common biosecurity concerns and methodologies mean that training materials and programmes can be shared and there is increasing cross-fertilization of ideas. Shared training opportunities also arise in technical exchanges between countries and capacity building; the latter being particularly important for developing countries.

Biosecurity- A new, holistic approach: It may be the time to consider changes in our biosecurity systems which make them more suitable for this new world (Waage and Mumford. 2008). Below three areas are suggested to start such a change.

12.5.1 Integrating Biosecurity Systems and Establishing a Common Toolkit

Governments are now dealing with a greater range of biosecurity threats, which extend well beyond traditional agricultural problems and priorities. It is inevitable that they will need to move away from established, sectoral biosecurity traditions focused on a narrow group of stakeholders, and develop a biosecurity system that allows comparison of threats across sectors and draws on a shared toolbox of best practices for measuring risk and evaluating the costs and benefits of prevention, eradication and control.

Besides established tools for risk analysis and cost–benefit analysis, a future toolbox must have means to evaluate:

- Indirect effects and externalities, including non-market effects of invasion and its control on the environment and health. This may require a mixture of quantitative and qualitative elements.
- Perception of risk, its elements and how it changes risk measures.
- Uncertainty and its measurement, whether formally through stochastic modelling or through adopting a more precautionary approach, or both.

Any assessment is limited by information, and we must do more to collect and analyse useful information on the patterns of movement and introduction of biosecurity threats; we have the tools for this, and increasingly the data, but we are not using these effectively. Advances in modelling will be critical to predicting biosecurity threats and the value of different prevention and control options. The result of this integrated, scientific approach should be a convergence of plant/animal and agricultural/environmental approaches to biosecurity, with a growing emphasis on proactive and preventative, rather than reactive measures for all. A common technology for detection, identification and monitoring is emerging for animal, plant and other biosecurity threats which will underpin this convergence.

It is not clear the extent to which this harmonization of approach can be achieved with common tools and procedures alone, or whether it requires an integration of different government bodies responsible for biosecurity (for both animal and plant health, or environmental and agricultural protection), as has been the pattern in Australia and New Zealand. Recent developments there, and in the USA with its National Invasive Species Council and Department of Homeland Security, should be treated as experiments and evaluated to examine whether fully integrated systems perform better than traditional ones.

12.5.2 Greater International Cooperation

Recent trends in biosecurity recommend a shift from a largely national approach to biosecurity towards greater international cooperation. International action against common threats moves biosecurity 'offshore' in a way which can benefit all parties through a shared approach to tackling new threats at their source. There are several specific opportunities for international action.

- Identification of the key pathways for the introduction of new threats and direction of international attention to these: the development and adoption of ballast water exchange conventions in response to marine invasions is an excellent example of a rapid, concerted international response to such a pathway.
- Improved international warning networks, like the new global early warning system of OIE, FAO and WHO which will assist in predicting and preventing zoonotic livestock disease problems through monitoring and epidemiological analysis.
- **International eradication programmes for animal diseases:** we have seen the benefit of global human disease eradication campaigns for smallpox, and the eradication of rinderpest may be imminent. What other animal or plant diseases and pests might be addressed in this manner?
- **Building biosecurity into all aspects of international cooperation:** where governments negotiate new activities which involve increased international movement, they should build into their planning the resources to address any increased biosecurity threat. New, cross-border roads, for instance, are notorious boosters of biosecurity threat, as are military campaigns and famine relief programmes.

Governments will be careful to weigh up the value to them of investing in national or international biosecurity. However, many are already driven towards the latter by political regionalization. Open trade within a borderless EU now means that a fish disease, for instance, is harder to stop at the UK border, and the UK self-interest shifts to stopping introduction at the outward borders of the EU. Establishment of common economic zones in Africa, Asia, Latin America and elsewhere will compound this effect.

Developing countries may have particularly strong views about international cooperation in biosecurity as a counter to these systems being used as non-tariff trade barriers. Better prevention systems in developed countries, based on risk analysis and new interception technology, disadvantage poorer countries where there is less capacity to measure, document and minimize biosecurity risks. If we accept that, where a country constitutes a 'weakest biosecurity link' in a trade network it is advantageous to all parties to improve its biosecurity, we will see investment in capacity building. New technology will come to be seen as 'trade enabling' rather than 'trade blocking'.

12.5.3 Building a Resilient System

Finally, we must accept that the biosecurity game, however well played, will have the results that more pests and diseases get to more countries. The perceived benefits of free trade may well prove greater than the justification for expending more and more national resources for pest and disease exclusion. Faced with this prospect, we should now begin to put serious consideration into learning to live with this problem, and making our agroecosystems more resilient to biological invasion. It would appear that much recent agricultural development has done the opposite. Opportunities for improving resilience are considerable and include, for example:

- Breeding of disease resistance into crops, assisted by new biotechnological tools for incorporating existing or new resistance mechanisms
- Development of vaccines for animal diseases which remove the need to exclude and stamp out diseases;
- Strategies of deployment of crops and livestock which reduce the risk of pest and disease outbreaks such as crop varietal mixtures which have proven effective in suppressing plant disease outbreaks
- Diversification of local production systems to be ecologically and economically resilient, reducing unnecessary movement of plants and animals.

This last point is particularly relevant to the UK. Intense movement of animals between farms and to distant abattoirs was a contributor to the scale of the 2001 FMD outbreak in the UK, while movement of horticultural material across Europe between planting, growth and sale may be driving the introduction of serious plant diseases into the UK. An analysis of the benefits and costs of indiscriminate movement of plants and animals may reveal strong economic arguments for reducing movement which carries a high public cost relative to a small private benefit.

Along with ecological resilience, there should be an improvement in 'economic resilience'. Spreading the financial costs of biosecurity more evenly across sectors, and public and private institutions will help to secure this, but it needs engagement of all stakeholders. Novel approaches may help the private sector to cope better with the pressures of biosecurity threats and responses. For example, a fair share of the benefits, and costs, of biosecurity measures should be shared among competing producers, importers and consumers. This may be achieved by broadening the group of stakeholders making decisions about biosecurity issues, to include consumers and overseas partners rather than allowing local producers to dominate decision-making.

A strategic shift from investment in exclusion to that in resilience would be a true challenge for governments playing today's biosecurity game, particularly given international agreements that could make investors in resilience 'trade pariahs'. But there are important drivers which might move this forward. In the case of animal biosecurity, for instance, animal welfare considerations may favour investment in vaccination as an alternative to culling strategies. More importantly, a resilience strategy could engage the private sector in picking up more of the cost of biosecurity. Most opportunities to increase resilience, including vaccines, pest and disease resistant crops and protective cropping or farming strategies will take the form of products or processes which producers will pay for themselves. By investing in resilience, governments distribute the burden of paying for biosecurity more evenly between the public and private sector. Some of this cost will inevitably be passed on to consumers, but perhaps this is more fair than sharing those costs across all tax payers in the form of government-funded programmes.

Conclusion

It is unlikely that today's biosecurity systems will change quickly. Many are presently 'locked in' by international agreements, such that change could carry severe trade penalties. But as the cost of running this system becomes greater with more frequent breaches and more expensive trade losses and eradication programmes, there will be mounting pressure to become more proactive and preventative to work together to stop new pests and diseases at their source, and ultimately to achieve freedom from introduced pests and diseases through building in resistance and resilience. New science which advances detection, monitoring and modelling of biosecurity threats and biotechnology for plant and animal resistance will be an important feature of this inevitable evolution of biosecurity systems.

Improved health and well-being of human populations are the ultimate outcomes of well-functioning biosecurity systems. These outcomes are strongly influenced by society and the environment and, in this context, agriculture and health are linked in many ways. Agriculture produces the world's food, fibre and materials for shelter, and is an important source of livelihoods. At the same time, agriculture can lead to poor health, especially in the form of infectious disease and malnutrition. The benefits of a more harmonized and integrated approach to biosecurity are already apparent in specific national situations. While the multi-sectoral character of biosecurity and the diverse range of interests involved make each national situation different, there are likely to be significant improvements in biosecurity systems and outputs if more coherent national and international approaches are applied. Benefits include improved regulatory and policy frameworks for human health (particularly food safety), improved animal and plant health, greater efficiencies in the use of human and financial resources, better understanding of potential risks (within and between sectors) and appropriate measures to manage them, and improved protection and sustainable use of the environment. Moreover, a more holistic approach to biosecurity will enable these benefits to be achieved in a manner that avoids inconsistencies, fills gaps, and prevents the creation of unnecessary barriers to trade.

Bibliography

Anonymous, 2010. Biosecurity: An integrated approach to manage risk to human, animal and plant life and health. INFOSAN (International Food Safety Authorities Network) Information Note No. 1/2010 – Biosecurity. WHO and FAO, United Nations

Anonymous, 2020. Biological Risk Assessment: General Considerations for Laboratories, OSHA, U.S. Department of Labor, Washington, D.C. 20210

Anonymous, 2021. State Hazard Plan-Animal and Plant Biosecurity. Department of Primary Industries and Regional Development (DPIRD) Pamela I'Anson, Director Incident & Emergency Management Sustainability and Biosecurity Directorate

Hulme, P.E., 2020. One Biosecurity: a unified concept to integrate human, animal, plant, and environmental health. *Emerg. Top. Life Sci.*, 4: 539-549.

Melly Domhnall and Hanrahan James, 2020. Tourism biosecurity risk management and planning: an international comparative analysis and implications for Ireland. *Tourism Rev.*, 76: 88-102.

Perry Suzy and Rebecca Laws, 2019. "Collaborative Planning and Shared Decision Making in Biosecurity Emergency Management" *Proceedings* 361:60

Singh, R.B., 2008. Biosecurity for Food Security. Paper presented at the National Commission on Farmers, 2006, for the South Asian Conference, 2008 and a Memorial Lecture, 2008

Thiengo S.C., Faraco F.A., Salgado N.C., Cowie R.H. and Fernandez M.A.,2007. Rapid spread of an invasive snail in South America: the giant African snail, *Achatina fulica*, in Brasil. *Biol. Invasions*, 9: 693–702

Waage, J K, and J D Mumford. 2008. "Agricultural biosecurity." *Philosophical transactions of the Royal Society of London. Series B, Biological Sciences*, 363:863-876.

Welfens, P.J.J.,2020. Trump's trade policy, Brexit, corona dynamics, EU crisis and declining multilateralism. *Int. Econ. Econ. Policy*, 17: 563–634

Bibliography

13

Biosafety, History Policies and Regulatory Mechanism

13.1 Biosafety

Biosafety is the prevention of large-scale loss of biological integrity, focusing both on ecology and human health. These prevention mechanisms include conduction of regular reviews of the biosafety in laboratory settings, as well as strict guidelines to follow. Biosafety is used to protect from harmful incidents. Many laboratories handling biohazards employ an ongoing risk management assessment and enforcement process for biosafety. Failures to follow such protocols can lead to increased risk of exposure to biohazards or pathogens. Human error and poor technique contribute to unnecessary exposure and compromise the best safeguards set into place for protection. The international Cartagena Protocol on Biosafety deals primarily with the agricultural definition but many advocacy groups seek to expand it to include post-genetic threats: new molecules, artificial life forms, and even robots which may compete directly in the natural food chain.

Biosafety in agriculture, chemistry, medicine, exobiology and beyond will likely require the application of the precautionary principle, and a new definition focused on the biological nature of the threatened organism rather than the nature of the threat. When biological warfare or new, currently hypothetical, threats (i.e., robots, new artificial bacteria) are considered, biosafety precautions are generally not sufficient (link to incident report, i.e. such as problems with CDC research labs in 2014). The new field of biosecurity addresses these complex threats. Biosafety level refers to the stringency of biocontainment precautions deemed necessary by the Centers for Disease Control and Prevention (CDC) for laboratory work with infectious materials.

Typically, institutions that experiment with or create potentially harmful biological material will have a committee or board of supervisors that is in charge of the institution's biosafety. They create and monitor the biosafety standards that must be met by labs in order to prevent the accidental release of potentially destructive biological material. Note that in the US, several groups are involved, and efforts are being made to improve processes for government run labs, but there is no unifying regulatory authority for all labs.

Biosafety is related to several fields, like

- In ecology (referring to imported life forms from beyond ecoregion borders),
- In agriculture (reducing the risk of alien viral or transgenic genes, genetic engineering or prions such as BSE/"MadCow", reducing the risk of food bacterial contamination)
- In medicine (referring to organs or tissues from biological origin, or genetic therapy products, virus; levels of lab containment protocols measured as 1, 2, 3, 4 in rising order of danger),
- In chemistry (i.e., nitrates in water, PCB levels affecting fertility)
- In exobiology (i.e., NASA's policy for containing alien microbes that may exist on space samples. See planetary protection and interplanetary contamination), and
- In synthetic biology (referring to the risks associated with this type of lab practice)

Hazards: Chemical hazards typically found in laboratory settings include carcinogens, toxins, irritants, corrosives, and sensitizers. Biological hazards include viruses, bacteria, fungi, prions, and biologically-derived toxins, which may be present in body fluids and tissue, cell culture specimens, and laboratory animals. Routes of exposure for chemical and biological hazards include inhalation, ingestion, skin contact, and eye contact (Anonymous, 2019). Physical hazards include ergonomic hazards, ionizing and non-ionizing radiation, and noise hazards. Additional safety hazards include burns and cuts from autoclaves, injuries from centrifuges, compressed gas leaks, cold burns from cryogens, electrical hazards, fires, injuries from machinery, and falls.

13.2 Biosafety Levels

A biosafety level (BSL), or pathogen/protection level, is a set of biocontainment precautions required to isolate dangerous biological agents in an enclosed laboratory facility. The levels of containment range from the lowest biosafety level 1 (BSL-1) to the highest at level 4 (BSL-4). In the United States, the Centers for Disease Control and Prevention (CDC) have specified these levels. In the European Union, the same biosafety levels are defined in a directive. In Canada the four levels are known as Containment Levels. Facilities with these designations are also sometimes given as P1 through P4 (for pathogen or protection level), as in the term P3 laboratory. At the lowest level of biosafety, precautions may consist of regular hand-washing and minimal protective equipment. At higher biosafety levels, precautions may include airflow systems, multiple containment rooms, sealed containers, positive pressure personnel suits, established protocols for all procedures, extensive personnel training, and high levels of security to control access to the facility. Health Canada reports that world-wide until 1999 there were recorded over 5,000 cases of accidental laboratory infections and 190 deaths.

The biosafety levels are thus designed to identify various protective measures that are to be taken in a laboratory setting to protect the researchers, the environment, and the microorganisms. These levels are defined by the Central for Disease Control and Prevention (CDC), where each of these levels is outlined with specific practices and safety requirements. Biosafety level designations are based on the combination of the design features, equipment, practices, and procedures required while working with agents from the various risk groups. The allocation of a pathogenic agent to a biosafety level for laboratory work must be based on the risk assessment. Such assessments take the risk group as well as other factors into consideration while establishing the appropriate biosafety level. The biosafety levels, thus, might differ from one region to another. As per the CDC, biosafety levels are of four types depending on the risk associated with the microorganism and the facilities available. The levels of containment range from the biosafety level 1 (BSL-1), which is the lowest to the level 4 (BSL-4), which is the highest.

13.2.1 Biosafety Level 1

Biosafety level 1 (BSL-1) is suitable for work with well-characterized agents which do not cause disease in healthy humans. In general, these agents should pose minimal potential hazard to laboratory personnel and the environment. At this level, precautions are limited relative to other levels. Laboratory personnel must wash their hands upon entering and exiting the lab. Research with these agents may be performed on standard open laboratory benches without the use of special containment equipment. However, eating and drinking are generally prohibited in laboratory areas.Potentially infectious material must be decontaminated before disposal, either by adding a chemical such as bleach or isopropanol or by packaging for decontamination elsewhere. Personal protective equipment is only required for circumstances where personnel might be exposed to hazardous material. BSL-1 laboratories must have a door which can be locked to limit access to the lab. However, it is not necessary for BSL-1 labs to be isolated from the general building.

This level of biosafety is appropriate for work with several kinds of microorganisms including non-pathogenic strains of Escherichia coli and Staphylococcus, Bacillus subtilis, Saccharomyces cerevisiae and other organisms not suspected to contribute to human disease.[13] Due to the relative ease and safety of maintaining a BSL-1 laboratory, these are the types of laboratories generally used as teaching spaces for high schools and colleges. Biosafety Level 1 is the level appropriate for work involving well-characterized agents not known to consistently cause disease in immune-competent adult humans and cause a minimal potential hazard to the laboratory personnel and the environment. Biosafety level 1 is the lowest safety level, and the precautions required for the level are thus limited and not as extensive. These laboratories provide general space in which work is done with viable agents that are not associated with disease in healthy adults.

Requirements

- The BSL-1 laboratories are not necessarily separated from the general traffic in the building.
- Most of the work is typically conducted on open bench tops using general microbiological practices.
- Unique laboratory design or containment equipment are not required but may be used depending on the risk assessment.
- Laboratory personnel must be provided with specific training in the procedures to be conducted in the laboratory, which is then supervised by a scientist with training in microbiology or related sciences.

The following are the standard practices, safety equipment, and facility requirements required in BSL-1.

Standard microbiological practices

- The laboratory supervisor should implement the policies regarding the access control to the laboratory.
- Laboratory personnel must wash their hands after working with potentially hazardous materials and before leaving the laboratory.
- Activities like eating, drinking, smoking, handling contact lenses, applying cosmetics, and storing food are not be permitted in laboratory areas.
- Mouth pipetting is prohibited; mechanical pipetting devices must be employed.
- All procedures to be conducted in the laboratory should be performed while avoiding the creation of splashes and aerosols.
- The work surface like the bench tops should be disinfected after work and after any spill of potentially hazardous biological material.
- The supervisor must ensure that all the laboratory personnel acquires appropriate training and necessary precautions while performing their tasks.

Safety practices: There are no safety specific safety practices required for BSL-1.

Safety equipment

- Special containment devices like the Bio-safety Cabinets are not required for BSL-1.
- In order to prevent the contamination of personal clothing, protective laboratory coats, gowns, or uniforms are recommended.
- While conducting tests with a high possibility of aerosol formation, protective eyewear can be used.

Uses

- Biosafety Level-1 is commonly used while performing tests on microbial agents that are not known to cause diseases in immune-compromised individuals.
- These laboratories include the laboratories used for teaching purposes in colleges and training centers

Organisms

- The common organisms that require Biosafety Level-1 containment include less hazardous organisms like *Agrobacterium* radiobacter, Aspergillus niger, *Bacillus thuringiensis*, Escherichia coli strain K12, Lactobacillus acidophilus, Micrococcus leuteus, Neurospora crassa, *Pseudomonas* fluorescens, Serratia marcescens.
- However, the requirement of the biosafety level might differ depending on the risk assessment of the pathogen.

13.2.2 Biosafety Level 2

At this level, all precautions used at Biosafety Level 1 are followed, and some additional precautions are taken. BSL-2 differs from BSL-1 in that

- Laboratory personnel have specific training in handling pathogenic agents and are directed by scientists with advanced training.
- Access to the laboratory is limited when work is being conducted.
- Extreme precautions are taken with contaminated sharp items.
- Certain procedures in which infectious aerosols or splashes may be created are conducted in biological safety cabinets or other physical containment equipment.

Biosafety level 2 is suitable for work involving agents of moderate potential hazard to personnel and the environment. This includes various microbes that cause mild disease to humans, or are difficult to contract via aerosol in a lab setting. Examples include hepatitis A, B, and C viruses, human immunodeficiency virus (HIV), pathogenic strains of Escherichia coli and Staphylococcus, Salmonella, Plasmodium falciparum, and Toxoplasma gondii. Prions, the infectious agents that transmit prion diseases such as vCJD, may be handled under Biosafety Level 2 or higher.

Biosafety level-2 laboratories are the laboratories that are used for the tasks involving microbial agents of moderate potential hazards to the laboratory personnel, the environment, and the agent. However, the infectious agents or the toxins might pose a moderate danger if accidentally inhaled, swallowed, or exposed to the skin. The precautions associated with biosafety level-2 are comparatively more extensive than BSL-1, but BSL-1 and BSL-2 laboratories are generally considered as basic laboratories.

Requirements

- BSL-2 laboratories like BSL-1 laboratories are not necessarily separated from the general traffic patterns in the building.
- However, access into the laboratory is limited while BSL-2 experiments are in progress.
- The annual inspection of the laboratories is also an important part of the BSL-2 requirements. These might include changing the filters or replacement of some devices.
- The work is mostly conducted on sterilized bench tops except for some processes that might form aerosols. The latter is conducted in safety cabinets.
- The precautions to be followed in BSL-2 include all the precautions of the BSL-1 and some additional precautions.

Standard microbiological practices

- All the laboratory personnel must wash their hands after using viable microorganisms and before leaving the laboratory.
- Eating, drinking, smoking, and handling contact lenses in the laboratory are strictly prohibited.
- Mechanical pipetting should be done instead of mouth pipetting.
- All contaminated cultures, glassware, plastic ware, and biologically contaminated waste must be treated as bio-hazards and thus, autoclaved.
- Work surfaces must be decontaminated with disinfectant at the end of the day or after any spills or splashes.
- Used hypodermic syringes and needles, Pasteur pipettes, razor blades, contaminated broken glass, and blood vials are treated as medical waste and discarded in puncture-resistant sharps disposal containers.

Safety practices

- People with increased risk of acquiring infections like the immune-compromised and pregnant individuals should not be allowed to enter the BSL-2 laboratories while the laboratories are at work.
- An annual review of the BSL-2 manual should be done to update the guidelines.
- Documented policies and procedures should be established that limit the entrance to individuals who know of the potential hazards and are appropriately trained.
- A biohazard symbol is placed on pieces of equipment where biohazardous materials are used or stored.

Safety equipment

- Protective coats are to be worn while entering the laboratory and then removed and kept in the laboratory post work.
- The laboratory design should be made such that it can be easily cleaned and decontaminated with minimum nooks and corners.
- The laboratory doors should be closed whenever work with hazardous biomaterials is conducted.
- An autoclave must be available.

Uses

- Biosafety level-2 laboratories are mostly used for routine analysis and culture of moderately hazardous agents.
- Besides, some of the laboratories used for teaching and training purposes are also BSL-2 laboratories.

Organisms

- The organisms that require BSL-2 laboratories include the pathogenic strains of E. coli, Staphylococcus, Salmonella, Plasmodium falciparum, Toxoplasma, and Herpes Simples Viruses.
- The allocation of organisms to the laboratories, however, might differ depending on the risk assessment.

13.2.3 Biosafety Level 3

Researcher at US Centers for Disease Control, Atlanta, Georgia, working with influenza virus under biosafety level 3 conditions, with respirator inside a biosafety cabinet (BSC).

Biosafety level 3 is appropriate for work involving microbes which can cause serious and potentially lethal disease via the inhalation route. This type of work can be done in clinical, diagnostic, teaching, research, or production facilities. Here, the precautions undertaken in BSL-1 and BSL-2 labs are followed, as well as additional measures including:

- All laboratory personnel are provided medical surveillance and offered relevant immunizations (where available) to reduce the risk of an accidental or unnoticed infection.
- All procedures involving infectious material must be done within a biological safety cabinet.
- Laboratory personnel must wear solid-front protective clothing (i.e. gowns that tie in the back). This cannot be worn outside of the laboratory and must be discarded or decontaminated after each use.

- A laboratory-specific biosafety manual must be drafted which details how the laboratory will operate in compliance with all safety requirements.

In addition, the facility which houses the BSL-3 laboratory must have certain features to ensure appropriate containment. The entrance to the laboratory must be separated from areas of the building with unrestricted traffic flow. Additionally, the laboratory must be behind two sets of self-closing doors (to reduce the risk of aerosols escaping). The construction of the laboratory is such that it can be easily cleaned. Carpets are not permitted, and any seams in the floors, walls, and ceilings are sealed to allow for easy cleaning and decontamination. Additionally, windows must be sealed, and a ventilation system installed which forces air to flow from the "clean" areas of the lab to the areas where infectious agents are handled. Air from the laboratory must be filtered before it can be recirculated.

A 2015 study by USA Today journalists identified more than 200 lab sites in the U.S. that were accredited biosafety levels 3 or 4. The Proceedings of a Workshop on "Developing Norms for the Provision of Biological Laboratories in Low-Resource Contexts" provides a list of BSL-3 laboratories in those countries.

Biosafety level 3 is commonly used for research and diagnostic work involving various microbes which can be transmitted by aerosols and/or cause severe disease. These include Francisella tularensis, Mycobacterium tuberculosis, Chlamydia psittaci, Venezuelan equine encephalitis virus, Eastern equine encephalitis virus, SARS-CoV-1, MERS-CoV, Coxiella burnetii, Rift Valley fever virus, Rickettsia rickettsii, several species of Brucella, chikungunya, yellow fever virus, West Nile virus, *Yersinia pestis*, and SARS-CoV-2.

- Biosafety level 3 (BSL-3) is the level where work is performed with agents that may cause severe or potentially lethal disease through inhalation or aerosol formation, to the personnel, and may even contaminate the environment.
- The tasks performed in the BSL-3 laboratories involve indigenous or exotic agents where the potential for infection by aerosols is high, and the disease may have lethal consequences.
- Autoinoculation and ingestion present primary hazards to personnel working with these agents at this level.
- Working in such laboratories require laboratory personnel with specific training in handling pathogenic and potentially lethal agents, along with supervisors scientists competent in handling infectious agents and associated procedures.

Requirements

- Biosafety Level 3 containment laboratories for animals and research are the most challenging containment level facilities to design and operate.

- These laboratories should be certified for use before initial operation and subsequently on an annual schedule or after a program change, renovation, or replacement of system components that may affect the operating environment of the laboratory.
- BSL-3 laboratories are also called the containment laboratory as they require containment equipment to protect the personnel, the microbial agent, and the environment.
- The requirements for BSL-3 include all the requirements of the BSL-1 and BSL-2 laboratories, along with some additional design features and special equipment.

Standard microbiological practices

- The entry to the BSL-3 laboratories is limited to individuals with appropriate training in handling BSL-3 organisms, all of whom are selected by the laboratory supervisor.
- Besides the general procedures and laboratory practices, the supervisor also formulates additional policies to limit the entry to the laboratory.
- All the procedures to be conducted in the BSL-3 must be conducted within a biosafety cabinet to prevent the exposure of the aerosols to the laboratory personnel.
- Personnel working in the laboratory must wear personal protective equipment before entering the laboratory and then remove them before leaving.
- The work surfaces and sinks should be decontaminated once every work shift or after any spills or splashes.
- The BSL-3 laboratories should be separated from the general traffic in a building to limit entry into the laboratories at all times.

Safety practices

- The doors of the BSL-3 laboratories are closed at all times with appropriate BSL-3 signs outside the suite, along with a universal biohazard sign and emergency contact information.
- Laboratory personnel must have medical surveillance and offered appropriate immunizations for agents handled or potentially present in the laboratory.
- Each institution should consider the collection and storage of serum samples from at-risk personnel.
- A laboratory-specific biosafety manual, which is available and accessible to all, must be prepared and adopted as a policy.

- The laboratory supervisor must check for the demonstration of proficiency in standard and special microbiological practices by all laboratory personnel before working with BSL-3 agents.
- Potentially hazardous materials must be placed in a durable, leak-proof container or vial during collection, processing, storage, or transport within a facility.
- All laboratory equipment should be routinely decontaminated after work or after any spills or splashes.
- The laboratory biosafety manual must define procedures t be adopted in the case of exposure to infectious materials, and these should be treated accordingly.
- No work in the BSL-3 laboratories should be conducted on an open bench or an open vessel. All the activities involving the infectious agents must be conducted within Biosafety cabinets or other physical containment devices.

Safety equipment

- Biosafety cabinets are to be used for the manipulation of all infectious agents.
- Individual protection gears like personal protective equipment, coats, gloves, and respiratory protection should be worn while entering the laboratories and then removed before leaving.
- The air flowing in the laboratory shouldn't be recirculated to any area of the laboratory and should be HEPA-filtered prior to being discharged to the outside.
- The filters, manuals, equipment, vacuum pipes, autoclaves, etc. should be revised and reviewed annually.

Uses

- BSL-3 laboratories are used for clinical, diagnostic, teaching, research, or production facilities.
- These laboratories are used for the handling and manipulation of highly infectious agents that prose direct severe effects on the health of the personnel.
- These are used for the studies regarding the effects of infectious agents and various toxins and their effects.

Organisms

- The pathogens that require BSL-3 laboratories include HIV, H1N1 flu, *Yersinia pestis*, Mycobacterium tuberculosis, SARS, Rabies Virus, West Nile Virus, Ricketts, etc.

- The placement of the organisms in different Biosafety levels, however, might defer and should also be determined after risk assessment.

13.2.4 Biosafety Level 4

Biosafety level 4 (BSL-4) is the highest level of biosafety precautions and is appropriate for work with agents that could easily be aerosol-transmitted within the laboratory and cause severe to fatal disease in humans for which there are no available vaccines or treatments. BSL-4 laboratories are generally set up to be either cabinet laboratories or protective-suit laboratories. In cabinet laboratories, all work must be done within a class III biosafety cabinet. Materials leaving the cabinet must be decontaminated by passing through an autoclave or a tank of disinfectant. The cabinets themselves are required to have seamless edges to allow for easy cleaning. Additionally the cabinet and all materials within must be free of sharp edges in order to reduce the risk of damage to the gloves. In a protective-suit laboratory, all work must be done in a class II biosafety cabinet by personnel wearing a positive pressure suit. In order to exit the BSL-4 laboratory, personnel must pass through a chemical shower for decontamination, then a room for removing the positive-pressure suit, followed by a personal shower. Entry into the BSL-4 laboratory is restricted to trained and authorized individuals, and all persons entering and exiting the laboratory must be recorded.

- Biosafety level 4 is the highest level that is employed while working with dangerous infectious agents that present a high individual as well as environmental risk in the form of life-threatening disease, aerosol transmission, or unknown risk of transmission.
- The BSL-4 laboratories are often used while handling and manipulating Risk Group 4 pathogens those are extremely dangerous, with no known vaccines or therapies, and require extreme precautions during work.
- The BSL-4 laboratories are of two types; cabinet laboratory where all the work is performed in a Class III biosafety cabinet or similar physical containment with very carefully formulated precautions and suit laboratory where all the laboratory personnel are required to wear full-body, air-supplied suits protective gears in the form of PPEs.

Requirements

- The requirements of BSL-4 laboratories are extensive with specific laboratory design, training procedures, and highly protective equipment and personal gears.
- These laboratories should be certified for use before initial operation and subsequently on an annual schedule or after a program change, renovation, or replacement of system components that may affect the operating environment of the laboratory.

- BSL-4 laboratories are also termed the maximum containment laboratories as they have secondary barriers to prevent hazardous materials from escaping into the environment.
- The BSL-4 laboratories should follow the requirements of all BSL-1, BSL-2, and BSL-3, along with additional specific precautions.

Standard microbiological practices

- No work conducted within the BSL-4 should be done on an open bench or an open vessel.
- The work stations, equipment, and sinks should be sterilized post work.
- The laboratory personnel should be in protective gear that might include full-body PPEs, gloves, masks, and coats.
- The doors of the laboratories should be closed at all times with the laboratory placed away from the general traffic in the building.
- Activities like drinking, eating, mouth pipetting should be avoided at all costs.
- Only people that are trained in handling the BSL-4 organisms and the equipment in the laboratory should be allowed into the laboratory.

Safety practices

- Viable or intact biological materials to be removed from the Class III cabinet in a BSL-4 are transferred in a nonbreakable, sealed primary container with a nonbreakable, sealed secondary container.
- No materials, except the biological materials that are to remain in a viable or intact state, are removed from the BSL-4 laboratory unless they have been autoclaved or decontaminated before they leave the facility.
- Only individuals whose presence in the facility is required for microbiological processes or support purposes are authorized to enter. Individuals that are at increased risk of acquiring an infection or for whom infection may be unusually hazardous are not allowed in the laboratory.
- Personnel can enter and leave the facility only after the clothing change and through the shower rooms.
- When the BSL-4 laboratory is at work or when infectious materials or infected animals are present in the laboratory, a hazard warning sign, along with the universal biohazard symbol, is placed on all access doors.
- A system is set up for reporting laboratory accidents, exposures, and the medical surveillance of potential laboratory-associated illnesses.

Safety equipment

- A Class III biological safety cabinet or Class I or II biological safety cabinets used in conjunction with one-piece personnel suits ventilated by a life support system are to be present in a BSL-4 while conducting all procedures within the facility.
- Walls, floors, and ceilings of the laboratories must form a sealed internal shell which facilitates fumigation and is animal and insect-proof.
- A double-doored autoclave is placed for decontaminating materials passing out of the facility.
- The exhaust air from the facility is filtered through HEPA filters before being discharged to the outside so as to prevent its entry into occupied buildings and air intakes.

Uses

- BSL-4 laboratories are used for diagnostic and research work on easily transmitted pathogens, causing fatal diseases.
- These laboratories are used for new and unknown pathogenic microbes, for which no vaccines or therapies are available.
- They are also used for clinical and production facilities that require highly sophisticated techniques and advanced processes.

Organisms: The BSL-4 level pathogens include the risk group IV organisms like Ebola virus, SARS-CoV-2, Central European Encephalitis virus, Hemorrhagic viruses, etc.

As with BSL-3 laboratories, BSL-4 laboratories must be separated from areas that receive unrestricted traffic. Additionally airflow is tightly controlled to ensure that air always flows from "clean" areas of the lab to areas where work with infectious agents is being performed. The entrance to the BSL-4 lab must also employ airlocks to minimize the possibility that aerosols from the lab could be removed from the lab. All laboratory waste, including filtered air, water, and trash must also be decontaminated before it can leave the facility.

Biosafety level 4 laboratories are used for diagnostic work and research on easily transmitted pathogens which can cause fatal disease. These include a number of viruses known to cause viral hemorrhagic fever such as Marburg virus, Ebola virus, Lassa virus, and Crimean-Congo hemorrhagic fever. Other pathogens handled at BSL-4 include Hendra virus, Nipah virus, and some flaviviruses. Additionally, poorly characterized pathogens which appear closely related to dangerous pathogens are often handled at this level until sufficient data are obtained either to confirm continued work at this level, or to permit working with them at a lower level. This level is also used for work with Variola virus, the causative agent of smallpox, though this work is only performed at the Centers for

Disease Control and Prevention in Atlanta, United States, and the State Research Center of Virology and Biotechnology in Koltsovo, Russia.

13.2.5 List of Biosafety Level 4 Organisms

BSL-4 category is specifically used for handling of Dangerous/exotic agents which pose high risk of life- threatening disease, aerosol- transmitted lab infections, or related agents with unknown risk of transmission.

Select agents

HHS human threats: Select agents and toxins

- *Crimean-Congo hemorrhagic fever orthonairovirus*
- *Ebolavirus*
- *Lassa mammarenavirus*
- *Lujo mammarenavirus*
- *Marburg virus*
- *Monkeypox virus*
- Reconstructed 1918 *influenza virus*
- *Chapare mammarenavirus*
- *Guanarito mammarenavirus*
- *Argentinian mammarenavirus* (formerly Junín virus)
- *Machupo mammarenavirus*
- *Brazilian mammarenavirus* (formerly Sabiá mammarenavirus)
- Far Eastern subtype *Flavivirus*
- Siberian subtype *Flavivirus*
- *Kyasanur Forest disease virus*
- *Omsk hemorrhagic fever virus*
- *Variola virus* (Smallpox virus)
- Variola virus minor (Alastrim)

HHS human or animal threats: Select agents and toxins

- *Hendra henipavirus*
- *Nipah henipavirus*
- Rift Valley fever phlebovirus
- Venezuelan equine encephalitis virus

USDA select agents and toxins

- *African horse sickness virus*
- *African swine fever virus*

- *Avian influenza virus*
- *Pestivirus C* (formerly *Classical swine fever virus*)
- *Foot-and-mouth disease virus*
- *Goatpox virus*
- *Lumpy skin disease virus*
- *Avian avulavirus* 1 (formerly Newcastle disease virus)
- *Small ruminant morbillivirus* (formerly *peste des petits ruminants virus*)
- *Rinderpest morbillivirus*
- *Sheeppox virus*
- Swine vesicular disease virus

Non-select agent

- Andes orthohantavirus

13.3 Historical Background of Biosafety

At first, a worthy milestone on biosafety was referred as "microbiological safety" dates back to 1908 where Winslow demonstrated a novel method of examination to enumerate bacteria present in the air. Additional study reviewed by Meyer and Eddie in 1941, described laboratory-acquired brucellosis which also revealed that similar infections could pose a threat to man has no relation to lab work . The principles of biosafety have developed together with the history of the American Biological Safety Association (ABSA) (Momtaz A. Shahein *et.al.*, 2021). Later in 1947, the NIH (national institutes of health) Building 7 had the first peacetime research laboratory especially tailored for microbiological safety. These historical landmarks and breakthroughs are just a few of the more studies which untied the importance and relevance of biosafety in healthcare and research institutions.

The principle and profession of biosafety have developed together with the history of the American Biological Safety Association (ABSA). As briefly described by the Federation of American Scientists, the first meeting was held in 1955 with the members of the military, as the focus addressed "The Role of Safety in the Biological Warfare Effort". Succeeding meetings attendees included the US Centers for Disease Control and Prevention (CDC) and the National Institutes of Health (NIH), universities, laboratories, hospitals and representatives from the industries. From then, written regulations covered the shipment of biological agents, safety training and programs, with the development of biological safety level classification. International issues on biosafety and studies on the individual or group of agents became the focus in the 1980s. At present, aside from studies focusing on specific biohazard level or pathogen, new strategies were developed to enhance risk assessment capacities, biosecurity, and biocontainment measures including the regulation of biosafety through national and international policies.

Other industries such as in agriculture and biotechnology are now considering biosafety application.

In the mid- to late 1800's, the science of microbiology had advanced to the point that the causative bacterial agent of common diseases such as tuberculosis, diphtheria and cholera were identified using Koch's postulates (Anonymous, 2019). Following close behind this initial work in the culture and purification of bacterial pathogens, LAIs were first reported. In the early- to mid-1900's, wooden and steel boxes were designed to prevent work-related LAIs, however it took many more years for the discipline of biosafety to develop. Biological Safety was pioneered at the U. S. Army Biological Research Laboratories in Fort Detrick Maryland led by the efforts of Arnold G. Wedum, Director of Industrial Health and Safety and the father of modern biological safety. Dr. Wedum was one of the original pioneers of the first Biological Safety Conference and was central in the formation of the American Biological Safety Association (ABSA). Today ABSA is an international organization (ABSA International) serving 37 countries with 1232 members. ChABSA (Chesapeake Area Biological Safety Association) is a local chapter of ABSA serving the Maryland, Virginia and Washington DC areas and dedicated to expanding biological safety awareness.

A Beginning

1854 - London cholera epidemic : Edwin Chadwick and Jack Snow ; Water pollution and disease disease transmission traced to Broad St. water pump ; Ignited "the sanitation revolution" (Emmett Barkley and Robert J. Hawley, 2012)

Lessons Learned

- Effective intervention does not always need accurate knowledge of disease causation
- Environmental measures may be more effective than changing individual behavior
- Always search for pragmatic solutions BMJ readers chose "the sanitation revolution" as the top medical milestone since its inaugural publication in 1840.

First Laboratory Infections

- 1885 Typhoid Typhoid lab infection, infection, unknown unknown cause
- 1893 Tetanus lab infection by syringe
- 1894 Cholera Cholera lab infection infection by pipette pipette
- 1897 Brucella lab infection by syringe
- 1898 Glanders lab infection by syringe
- 1899 Diphtheria lab infection by pipette

Contributions to Biosafety 1943 – 1969:

i. Occupational health program

- Health and safety of workers highest priority
- Treat every infection as a LAI until proven otherwise
- Reporting exposures was encouraged

ii. Risk assessment

- Number and severity of LAI
- Infectious dose for humans
- Availability of specific therapy or effective vaccine

iii. Applied research

- Pioneered risk assessment studies
- Developed and validated decontamination protocols
- Evaluated microbial hazards and protocols
- Evaluated efficiency of HEPA filters for capturing viral particles

iv. Created the Biological Safety Conference in 1956

- Provided guidance to federal agencies in support of the development of biosafety programs and guidelines
- Published papers on biosafety biosafety practices, risk assessments, and applied research projects
- Dr. W de um was a mentor to whomever asked for his guidance

v. Classification of etiologic agents on the basis of hazard

- Published by the CDC (DHHS) in 1969
- Four classes of hazard
- A fifth class of animal agents with USDA restrictions
- Scientific judgment of the PI (risk assessment)
- Competence of investigators
- Physical containment

Asilomar Conferences, Pacific Grove, CA: Asilomar Conferences, Pacific Grove, CA

1. Jan. 22-24, 1973– Experiments of concern

- SV40-lambda hybrid
- Non-defective adenovirus-SV40 hybrids

Led to publication of: NCI Safety Standards for Research Involving Oncogenic Viruses

- Three classes of potential hazard (Low Moderate High) Three classes of potential hazard (Low, Moderate, High)

- PI and individual responsibility
- Practices, safety cabinets, facilities
- Medical surveillance

2. February 24-27, 1975 – Experiments of Concern: Recombinant DNA Molecules

- Organized by Paul Berg; David Baltimore; Sydney Brenner; Richard Roblin III; Maxine Singer
- Conference agenda:

Review progress of rDNA research

- Consider potential biohazards Consider potential biohazards
- Consider ethical and legal concerns
- Draft a conference summary paper with recommendations

CDC/NIH Biosafety in Microbiological and Biomedical Laboratories (BMBL)

- First edition published in 1984
- Current (5th) edition published in 2009
- Advisory recommendations
- Voluntary co de o f practice
- Goal of upgrading operations
- Guide for laboratory construction or renovation
- Application to laboratories is based upon risk assessment risk assessment.

13.4 Biosafety Training

Safe practices in research involving oncogenic viruses and rDNA Safe practices in research involving oncogenic viruses and rDNA molecules (NIH 1972-1979)

- Train-the-trainer course on fundamentals for safe microbiological research (NIH/ASM 1977 research (NIH/ASM 1977 -1978)
- Instructors' Guide for Biosafety Training (NIH, 1983)
- American Biological Safety Association
  - Offering training since 1984 Offering training since 1984
  - Training primarily directed at biosafety professionals
- Midwest Regional Center of Excellence for Biodefense and Emerging Infectious Diseases Research
  - Among several RCE-based programs to offer training intended for investigators; first course offered in Feb, 2004
  - First biosafety fellowship program in the world
- NIH National Biosafety and Biocontainment Training Program

Interstate Shipment of Etiological Agents

- DOT 42 CFR P t 72 (1957) DOT 42 CFR Par t 72 (1957)
    - Occupational Exposure to Bloodborne Pathogens
- OSHA 29 CFR Part 1910.1030 (1991)
    - Possession Use and Transfer of Select Agents Possession, Use and Transfer of Select Agents – CDC 43 CFR Part 73 (2005)
- APHIS 9 CFR Part 121, and 7 CFR 331 (2005)

Public Health Security & Bioterrorism Preparedness and Response Act of 2002 Public Health Security & Bioterrorism Preparedness and Response Act of 2002

- Regulations for the transfer, possession and use of select agents
- Risk assessment (including children and vulnerable populations)
- Ensure appropriate training and skill in handling select agents
- Containment laboratories
- Security measures commensurate with the risk such agent or toxin poses to public health and safety (including the risk of use in domestic or international terrorism
- Availability of select agents for research, education and other legitimate purposes

As briefly described by the Federation of American Scientists, the first meeting was held in 1955 with the members of the military, as the focus addressed "The Role of Safety in the Biological Warfare Effort". The next meetings attendees included the US Centers for Disease Control and Prevention (CDC) and the National Institutes of Health (NIH), universities, laboratories, hospitals and representatives from the industries. From then, written regulations covered the shipment of biological agents, safety training and programs, with the development of biological safety level classification (Bayot , 2018). Biosafety studies on the individual or group of agents became the focus in the 1980s. Some from studies focusing on specific biohazard levels of pathogens and other new strategies were developed to enhance biorisk assessment capacities, biosecurity, and biocontainment measures including the regulation of biosafety through national and international policies. Other activities such as in agriculture and biotechnology are now considering biosafety applications.

13.5 Laboratory-Acquired Infections (LAIs)

Laboratory-acquired infections (LAIs) were considered significant because of the high risk in the laboratory workforce relative to the public, although the exposure to infectious agents can be higher in other groups of healthcare workers. Sulkin and Pike in 1949 studied several works of literature and mail surveys with an attempt to evaluate the risk of infection associated with employment in a clinical or research laboratory. New studies and reviews led to the identification and description of hazards unique to the laboratories, which later formed a basis for the development of approaches to prevent the emergence of LAIs. The incidence of laboratory-acquired infections varies among institutions conducting surveys to a specific of laboratories and facilities. Monitoring and evaluation of LAIs are still absent for some institutions which could be caused by the difficulties in the reporting schemes and lack of accurate data interpretation.

The reporting of LAI is not similar to that of notifiable diseases which is highly regulated for each healthcare institution across countries as implemented by their ministries of health. An example of Laboratory-acquired infections would be a person infected with tuberculosis, who could have an infection with TB bacilli but with no signs and symptoms, thus, cannot be considered as TB disease. The need for data collection for current LAIs should highlight the importance of improving biosafety, then LAI databases were created to contain all recently published studies and to verify its relevant findings. In 2018, Siengsanan-Lamont and Blacksell presented the results of a rapid review of LAI studies within the Asia-Pacific. Regarding potential biorisks for zoonotic diseases, viruses predominate, followed by bacteria and parasites. The importance of biorisk assessment and management was also emphasized, including preventive practices.Strict biosafety measures are a must for these working environments to protectthemselves as well as the community.

The first prototype Class III (maximum containment) biosafety cabinet was fashioned in 1943 by Hubert Kaempf Jr., then a U.S. Army soldier, under the direction of Arnold G. Wedum, Director (1944–69) of Industrial Health and Safety at the United States Army Biological Warfare Laboratories, Camp Detrick, Maryland. Kaempf was tired of his MP duties at Detrick and was able to transfer to the sheet metal department working with the contractor, the H.K. Ferguson Co. On 18 April 1955, fourteen representatives met at Camp Detrick in Frederick, Maryland. The meeting was to share knowledge and experiences regarding biosafety, chemical, radiological, and industrial safety issues that were common to the operations at the three principal biological warfare (BW) laboratories of the U.S. Army. Because of the potential implication of the work conducted at biological warfare laboratories, the conferences were restricted to top level security clearances. Beginning in 1957, these conferences were planned to include non-classified sessions as well as classified sessions to enable broader sharing of biological safety information. It was not until 1964, however, that conferences were held in a government installation not associated with a biological warfare program.

Over the next ten years, the biological safety conferences grew to include representatives from all federal agencies that sponsored or conducted research with pathogenic microorganisms. By 1966, it began to include representatives from universities, private laboratories, hospitals, and industrial complexes. Throughout the 1970s, participation in the conferences continued to expand and by 1983 discussions began regarding the creation of a formal organization. The American Biological Safety Association (ABSA) was officially established in 1984 and a constitution and bylaws were drafted the same year. As of 2008, ABSA includes some 1,600 members in its professional association.

In 1977 Jim Peacock of the Australian Academy of Science asked Bill Snowdon, then Chief CSIRO, AAHL if he could have the newly released USA NIH and the British equivalent requirements for the development of infrastructure for bio-containment reviewed by AAHL personnel with a view to recommending the adoption of one of them by Australian authorities. The review was carried out by CSIRO AAHL Project Manager Bill Curnow and CSIRO Engineer Arthur Jenkins. They drafted outcomes for each of the levels of security. AAHL was notionally classified as "substantially beyond P4". These were adopted by the Australian Academy of Science and became the basis for Australian Legislation. It opened in 1985 costing $185 million, built on Corio Oval. The Australian Animal Health Laboratory is a Class 4/ P4 Laboratory.

13.6 The Emergence of the Concept of Biosafety

"Biosafety" has several accepted definitions depending on the discipline involved (veterinary, food, medical or environmental), its linguistic origin or even the country in which it is being used. In Belgium, biosafety is defined as "The safety for human health and the environment, including the protection of biodiversity, during the use of genetically modified organisms or micro-organisms, and during the contained use of pathogenic organisms for humans" (Source: The cooperation agreement of 25 April 1997 between the Federal State and the Regions relating to administrative and scientific coordination in terms of biosafety) (Anonymous, 2021).

Biosafety therefore makes reference to safety for human health and the environment, genetically modified organisms (or micro-organisms) and pathogenic organisms (a pathogenic organism is a biological agent that can cause disease in immunocompetent humans and poses a risk for individuals directly exposed to it).

The exact moment of origin of biosafety cannot be clearly identified. This discipline has taken shape through different periods in recent history and through different fields (microbiology, molecular biology, veterinary science, guidelines relating to safety, etc). The first steps to this discipline appeared at the time of Pasteur and Koch (around 1890). It was indeed at that time that it appeared necessary to put in place some safety measures in response to the potential risk associated with the exposure to pathogenic micro-organisms. The first infectious diseases acquired in a laboratory were reported at that time. It was another few decades before the notion of a risk to human health linked to the handling of pathogenic micro-organisms was clearly defined. Pioneers actively contributed to the implementation of protective measures against biological risks following meticulous investigations carried out in microbiology laboratories. Safety measures in laboratories where pathogenic micro-organisms were handled were firstly implemented in North America and the United Kingdom at the beginning of the 1970s. They included working practices, personnel protection measures and physical containment measures aimed at limiting the spread of biological agents. Safety measures later applied in laboratories handling genetically modified organisms and micro-organisms (GMOs and GMMs) were largely inspired by these guidelines established in microbiology.

In the beginning, biosafety was considered as a sub-discipline of personnel safety (linked to legislation aimed at protecting workers against different types of risks such as chemical or radioactive). However, biological hazard is distinct from other sources of hazard (chemical, radioactive) by the fact that micro-organisms can multiply in vivo (in a host organism) as well as in vitro (in liquid or solid medium).

'Biosafety is based on scientific roots but it is not a science in itself': Over time biosafety became progressively an entirely separate discipline, partly as a result of two parallel developments: the implementation of an internationally recognized biological risk classification system, and the consequences from the Gordon conference on nucleic acids (1973) and two Asilomar conference (1973 and 1975).

Innovation and development of biosafety in the United States is reflected accurately in the history and pre-history of the American Biological Safety Association (ABSA). The first unofficial meeting was held on April 18, 1955 at Camp Detrick (now Fort Detrick) and involved members of the military representing Camp Detrick, Pine Bluff Arsenal, Arkansas (PBA), and Dugway Proving Grounds, Utah (DPG). In those days, the offensive BW program of the United States was in full swing: the opening keynote address was "The Role of Safety in the Biological Warfare Effort." Beginning in 1957, the yearly meetings began to include non-classified sessions to broaden the reach of the Association; representatives of the USDA were regular attendees through this "transition period." There were striking changes in the meetings in 1964-1965: the NIH and CDC joined for the first time, along with a number of other relevant federal agencies. All classified information was removed accompanied by a concerted effort to declassify safety studies and release them for public knowledge and advantage. By 1966, the attendees included universities, private laboratories, hospitals, and industry. Gradually, federal regulations began to appear. In 1973, the impact of new OSHA regulations was analyzed and debated at the ASBA meeting; interestingly, there was a range of responses to the new regulations.

In 1974, the United States Postal Service and Department of Transportation introduced regulations for shipping of etiologic agents (microorganisms and toxins that cause disease in humans). New safety programs and trainings were introduced. The designation of 4 levels of biosafety originated in the mid-1970s and the safety requirements for research with recombinant DNA were hotly debated. A survey of the ABSA meetings in the 1980s reveals increased focus on individual agents or groups of agents and coordination of international safety issues. ABSA now represents biosafety professionals in 20 countries, and reflects the organic nature of the topic: biosafety is a fast-moving field with constant research into and reevaluation of its tenets as threat perception change and technologies advance.

13.7 Biosafety-Indian Scenario

Biosafety in India is primarily focused on genetically modified (GM) agricultural research and ensuring environmental safety. This is evidenced by the Indian definition of biosafety as "the need to protect the environment including human and animal health from the possible adverse effects of the Genetically Modified Organisms (GMOs) and products thereof derived from the use of modern biotechnology." The Environment Protection Act (EPA) of 1986 created room for the development of India's first biosafety regulations,61 the 1989 Rules for the Manufacture/Use/Import/Export and Storage of Hazardous Microorganisms, Genetically Engineered Organisms or Cells (Anonymous, 2016). Since the publishing of the "Rules 1989," several other key guidelines have been released to provide revised legislation and guidance for studies involving recombinant DNA and transgenic plants. 3862,64 Each research institution working with rDNA or GMOs is required to have an Institutional Biosafety Committee (IBSC)62,65,66 that reports to the Ministry of Environment, Forests, and Climate Control (MoEFCC) and Department of Biotechnology (DBT), who together provide oversight for biological research institutions.

Pathogen characterization: According to the Recombinant DNA Safety Guidelines, 1990, microorganisms are categorized in to four risk groups based upon pathogenicity, transmissibility, host range, the availability of preventative or curative treatment, prevalence in the country, and its ability to cause disease in humans, animals and plants.36 Risk Group IV contains agents of the greatest risk to human safety, and Risk Group I contains those who pose the smallest risk. India uses the same BSL 1-4 containment facility categorization as the WHO and the United States.

Relevant regulations and legislation: Listed below are important biosafety guidelines and legislation in chronological order:

- The Environment Protection Act (EPA) of 198661 provides regulations for the manufacturing, use, export, import, and storage of GMOs, GMO products, and hazardous agents. This piece of legislation provided the foundations for the development of the "Rules 1989" and India's biosafety regulating framework.
- Drugs and Cosmetics Rules, 1988 (8th Amendment) designates the Ministry of Health and Family Welfare (DoH) as the regulatory body overseeing the import or manufacturing of biological and biotechnological products.
- Rules for the Manufacture/Use/Import/Export and Storage of Hazardous Microorganisms, Genetically Engineered Organisms or Cells, commonly

referred to as "Rules 1989," is India's official biosafety document. This document provides guidelines for the use, handling, transport, and production of microorganisms as well as GMOS, genetically engineered crops, and products made with engineered crops. Notified under the Environment Protection Act of 1986, this document establishes regulatory agencies and categorizes pathogens.

- The Revised Guidelines for Research in Transgenic Plants & Guidelines for Toxicity and Allergenicity Evaluation of Transgenic Seeds, Plants and Plant Parts (1998) provides rules for recombinant DNA research on plants including molecular analysis and field evaluation. These guidelines also address importing and exporting of GM plants for research use
- The Recombinant DNA Safety Guidelines (1990, 1994) includes agent classification and risk categorization, facility specifications, and laboratory protocol for working with recombinant DNA.
- The Guidelines for Generating Preclinical and Clinical Data for rDNA Vaccines, Diagnostics and Other Biologicals (1999) cover preclinical and clinical testing of rDNA vaccines, diagnostics and other biologicals. The guidelines are specific to ensuring the safety, quality, potency and effectiveness of the product.
- The Guidelines and Handbook for Institutional Biosafety Committees (2011) provides details on the necessary composition of the biosafety committees and each member's role, procedures for registration, the role of IBSCs in project approval, training requirements and committee meeting specifics, among other details.
- The Biotechnology Regulatory Authority of India Act (BRAI Act) (2013) mandates the establishment of the Biotechnology Regulatory Authority of India to "regulate the research, transport, import, manufacture and use of organisms and products of modern biotechnology and for matters connected therewith or incidental thereto." It was initially prepared by the Department of Biotechnology (DBT) in 2008, the bill went through several revisions before it was passed in 2013

Regulatory and Oversight Agencies: The Ministry of Environment, Forests, and Climate Change (MoEFCC) along with the Ministry of Science & Technology's Department of Biotechnology are responsible for enforcing the policies of the 1986 Environmental Protection Act. The Genetic Engineering Approval Committee (GEAC), the Review Committee on Genetic Manipulation, and Institutional Biosafety Committees (IBSCs) review and approve research projects. Both the State Biotechnology Coordination Committee (SBCC) and the District Level committee (DLC) monitor ongoing studies. The recombinants DNA Advisory Committee (RCAC) advises committees and government agencies when necessary.

Biosafety Officers and Institutional Biosafety Committees: The Guidelines and Handbook for Institutional Biosafety Committees identifies the need for a Biosafety Officer (BSO) "...if research is conducted on organisms that require special containment conditions (Biosafety Level 3 or 4)...or if large-scale rDNA research is conducted." According to these guidelines, the BSO must "also a member of the IBSC and act as a technical liaison between researchers and the IBSC (Anonymous, 2016). The Biosafety Officer should be adequately trained and be able to offer advice on specialized containment requirements." Specifics about the level of training are not provided, but the BSO must know enough to ensure that rDNA safety guidelines and good laboratory practices are followed.

Institutional Biosafety Committees (IBSC) are required for every facility working with rDNA and GMOs. These committees are authorized to approve laboratory studies, excluding field trials and hazardous genetic experiments or methods. Each committee is comprised of academics and researchers from the institute, including the head of the institution and a medical expert, plus a member of the DBT. There are currently about 500 IBSCs within India.

Accident and Incident reporting: The 1990 Recombinant DNA Safety Guidelines specify that an emergency plan should be in place in every facility for incidents such as breakage and spillage; accidental exposure to a chemical, toxin or agent; and natural disasters. Specific instructions are given for cleanup, sharps containers and disposal of contaminated materials. The Guidelines and Handbook for Institutional Biosafety Committees states that all accidents, illnesses, or breaches in protocol must be reported to the Principle Investigator (PI) who then must report the incident to the IBSC Chairperson within 24 hours and to the RCGM within 48 hours of the incident. In anticipation of an accidental GMO release, the PI of a study is required to have emergency protocol in place and distributed to appropriate regulatory committees, as is outlined in the "Rules, 1989.

Laboratory numbers: In 2008, it was reported that India had 14 BSL-3 facilities with 6 planned for construction and 1 BSL-4 facility with 2 planned for construction. In conflict with this report are several 2013 sources which reported that in March of 2013, India's first BSL-4 facility was completed at the National Institute of Virology (NIV) in Pune. At this time, the NIV had already constructed a BSL-3 facility at its Microbial Containment Complex campus. The new BSL-4 facility allows for extended research capabilities with agents like pandemic influenza, SARS, Nipah virus and Crimean-Congo hemorrhagic fever virus.

Research Status: India was listed 51st on the 2015 Scientific American Worldview Overall Scores for biotechnology innovation. Its highest scores were in productivity and intensity, and it lowest scores were in IP protection and enterprise support. India's Department of Biotechnology has dedicated significant effort increasing the nation's global contribution to biotechnology and scientific research. In the past few years, the DBT has worked to create an "Open Access

Policy," developing a database where funded researchers can share and learn from each other's work. Much of Indian investment and effort in biotechnology is directed towards genetically modified agriculture. India's first transgenic crop, *Bt* cotton, was introduced in 2002. Now, India produces 11.6 million hectares of *Bt* cotton and is the 4th highest producer of biotech crops in the world. Currently, India has developed 23 biotech crops with 67 biotech traits in different stages of development. These crops are developed by both the public and private sectors (39 public sector, 20 private, 8 autonomous institutes).

Biopharmaceuticals, bioinformatics, and bioservices are also major parts of India's biotechnology industry. India's biopharmaceutical industry accounts for 62% of the total biotechnology sector, with bioservices at 18%. The Bioservices sector continues to grow in capacity for contract research, clinical trials, and manufacturing engagements.

13.8 Biological Risk Classification

Originally, analysis of diseases acquired in the laboratory showed that certain pathogens were responsible for infectious diseases that were more severe than others. These observations led to a classification system for pathogenic micro-organisms.

From 1969, the Public Health Service in the United States worked on the definition of four risk groups for pathogenic micro-organisms for humans. The work lasted for five years and the classification criteria were adopted in 1974 (CDC (Centers for Disease Control). In 1979 the World Health Organization (WHO) set up a working group on good microbiological practices. The work of this group led to the publication of a manual suggesting measures aimed at the protection of workers, the population, animal breeding and the environment. As in the United States, the WHO adopted a classification system for pathogenic micro-organisms consisting of four risk groups. The work of the WHO would afterwards serve as a basis for a large number of national reference documents. In the UK four "Hazard Groups" were also adopted in 1984 (ACDP: Advisory Committee on Dangerous Pathogens. 1984. Categorization of biological agents on the basis of hazard and categories of containment. Health and Safety Executive), after a tentative trial of three categories (A, B and C) between 1975 and 1978 which was not finally implemented.

In Europe, Directive 90/679/EEC on the protection of workers against biological risk was adopted in 1990. It contains a non-exhaustive list pathogenic micro-organisms for humans, also distributed into four risk groups.

Belgium drew on all of this work in order to adopt in 1993 three lists assigning risk categories to several hundreds of micro-organisms pathogenic to humans, animals and plants (bacteria, fungi, parasites, viruses, including prion proteins linked to transmissible spongiform encephalopathies). These lists have been regularly updated since that time and can be consulted from this website. The setting up of risk groups of pathogenic organisms led, at the same time, to the implementation of good practices and containment measures aimed at ensuring individual and collective safety. As was done for organisms, four levels of containment were also defined to which increasingly strict safety measures were associated. These four levels were adopted on an international level, though not necessarily with unanimous classification and containment criteria. The different constitutive elements of what would be later known by the term "biosafety" were in place: risk classification of organisms on the one side and containment levels on the other. Between them, a key element would develop over time, the assessment of biological risks.

13.9 Recombinant DNA Techniques

In parallel to the use of pathogenic micro-organisms in the laboratory, 1970 marks the emergence of a new discipline: molecular biology. This discipline arose following a series of remarkable discoveries in fundamental research, starting just before the middle of the last century when Avery established that deoxyribonucleic acid (DNA) is the universal support of hereditary properties and contains the genetic information of living beings. This discovery was followed by the publication in 1953 of the work of Crick, Watson, Wilkins and Franklin identifying the double helix molecular structure of DNA, then in 1965 by the first description of restriction enzymes (by Linn and Arber), proteins capable of cutting DNA at specific sites.

This was the starting point for laboratory applications of molecular biology. The techniques (also referred to by the term "genetic engineering") allowing the insertion of a fragment of DNA (containing one or more genes) into another DNA will progressively refine with the aim of precisely and efficiently modifying the genome and the hereditary characteristics of living organisms. The aim of researchers was sometimes fundamental (better understanding of the functioning of the genome) but numerous teams of scientists also aimed to generate organisms with new properties through the manipulation of DNA. This is how the first genetically modified organisms (GMOs) were developed, and the so-called "modern biotechnologies" came into being.

The awareness of the potential risks associated with the use (still in its infancy) of these molecular biology techniques, and the products derived from them, arose rather rapidly. The very first debates on genetic engineering took shape towards the end of the 1960s within the scientific community, particularly in North America. The question was raised whether the combination of DNA sequences from different species, even unrelated (commonly known as "recombinant DNA"), could not result in new types of pathogenic organisms. This questioning culminated in 1972 with the work of the team of Paul Berg (David A. Jackson *et.al.*, 1972). The American biochemist, one of the pioneers of recombinant DNA, successfully carried out the cloning of a fragment of the oncogenic SV40 virus in a bacterial plasmid. However, his work led him, as well as some of his colleagues, to question the risks with which researchers handling this type of DNA were being confronted (they were particularly concerned about possible health consequences of the deliberate or accidental transfer of SV40 tumoral genes into Escherichia coli, a bacteria commonly used in the laboratory but also naturally present in the human digestive system). These fears were increased by the use of these techniques by a growing number of researchers and by the fact that the scientists working at that time on recombinant DNA were largely biochemists, less respectful of or less accustomed to the application of safety measures than microbiologists.

On Berg's initiative, North American scientists met in 1973, first at Asilomar, then during the "Gordon Conference on Nucleic Acids" (see text box). The scientific community committed itself to considering the potential risks linked to recombinant DNA techniques. Already at that time, the implementation of specific containment and personal protective measures was being considered. They were primarily aimed at guaranteeinging the safety of those exposed, essentially the scientists themselves, and to avoid any release into the environment. They were the first steps towards the concepts of "biosafety" and the assessment of risks associated with activities involving recombinant DNA in the laboratory.

However, the major outcome of these first conferences was the appeal, launched by Berg and some other scientists (including Watson), to the scientific community to impose a voluntary moratorium on experiments involving recombinant DNA until the holding of an international conference aimed at assessing the potential risks of this type of research. Despite the protests of certain scientists (who wished to continue this type of experiment without restrictions), this appeal was upheld, at first only by North American researchers but later also by some European and Japanese researchers. The second Asilomar conference ("Asilomar Conference on Recombinant DNA Molecules"), organised by Berg, was held in February 1975. It brought together 150 scientists, but also some legal experts and journalists. The participants decided (non-unanimously) to lift the moratorium imposed one year earlier. They particularly concluded to the necessity of managing research work involving recombinant DNA with strict guidelines.

GMOs from the laboratory to the field: With the development of recombinant DNA techniques and their adoption by a growing number of researchers across the world, potential applications of this technology also expanded. It rapidly became apparent that GMOs offered considerable possibilities in different applied fields such as medicine or the agro-food industry. The term "modern biotechnology" was used to distinguish applications arising from recombinant DNA techniques from those known as "traditional" that have been used within our societies, sometimes for centuries. In the agro-food applications of modern biotechnology, Belgium played a pioneering role in terms of research and development. Indeed, at the end of the 1970s, the work by the team of Professors Marc Van Montagu and Jozef Schell at the University of Ghent made a significant contribution to the development of genetically modified plants. By exploiting the DNA transfer capacity of the bacterium *Agrobacterium* tumefaciens into certain plants, these researchers showed that it is possible to express "foreign" genes into a plant and its offspring. This discovery paved the way for commercial exploitation of transgenic plants and the birth of numerous biotechnology companies (such as "Plant Genetic Systems" in Belgium).

This development also resulted in scientific products leaving the laboratory and coming into direct contact with the environment. As long as the genetic engineering developments were taking place in the laboratory, the assessment of potential risks focused on the impact to human health, essentially that of the laboratory staff. The deliberate release into the environment of genetically modified organisms (firstly for experimental purposes and then commercial) rapidly led to new questions about how to assess and manage potential risks linked specifically to this type of application. It was within this context that the Organization developed a series of scientific principles and recommendations specifically aimed at the assessment and management of risks linked to applications of recombinant DNA techniques in the environment.

Biosafety policies & procedures : This is a list of current policies and procedures approved by the Institutional Biosafety Committee regarding the use of biohazards in research, teaching, and/or diagnostic testing.

13.10 Biosafety Program Policies

i. General Biosafety Program Policy: This policy applies to research, teaching, diagnostic testing and other activities conducted at, sponsored by, or on behalf of UTK-A, and involving:

- Recombinant DNA molecules or synthetic nucleic acids as defined in the NIH Guidelines for Research Involving Recombinant or Synthetic Nucleic Acid Molecules (hereafter NIH Guidelines), including transgenic plants and animals;
- Biological agents (bacteria, viruses, fungi, protozoa, parasites, and prions) which may cause disease in humans, animals or plants and require containment and safeguards at Biosafety Level-2 (BSL-2) or higher (note: biological agents deemed low-risk or those not known to cause disease in humans, animals or plants are exempt from this policy per documented risk assessment);
- Acute biological toxins having an LD50 < 100 ng/kg in mammals and/or those defined as Select Agent Toxins;
- Human or nonhuman primate blood, blood products, tissues, secretions, excretions, or cell lines unless documented to be free of bloodborne pathogens or otherwise deemed low risk per documented risk assessment;
- Venomous animals manipulated and/or housed in laboratories or other indoor facilities;

- Poisonous plants posing a risk to humans via dermatological contact, inhalation, or other route of exposure;
- Novel nanoparticles conjugated to biologically active or cell-modifying molecules;
- Diagnostic specimens or environmental samples deemed likely to contain any of the above and posing a significant risk to humans, animals, plants or the environment per documented risk assessment.

ii. Laboratory Security policy: This document shall provide standards for laboratories with respect to the security of equipment, property, and hazardous materials.

Scope and Applicability: This standard addresses security for work with hazardous substances (including biological agents) and export control. Personal safety and security is not specifically covered in this document. This document shall apply to students, staff, and faculty in laboratories that contain hazardous materials or for labs that fall under the campus Export Control Program.

Policy Statement (Roles and Responsibilities)- Environmental Health and Safety shall

- Review and revise this policy and guidance when necessary.
- Disseminate information regarding this document through appropriate channels of campus communication.
- Consult on the topic of laboratory security upon request. This role may be delegated by EHS to the UT Police Department or appropriate responsible unit.

Responsible unit shall

- Ensure that appropriate facilities and equipment are available to meet the security requirements.
- Ensure that any security concerns or occurrences are communicated to the appropriate campus entities.
- Ensure that principal investigators (PIs) and/or laboratory supervisors carry out their roles and responsibilities.

Laboratory Supervisors and Principal Investigators shall

- Ensure students and employees are oriented to the relevant components of this guide and how they are applied in assigned lab spaces.
- Ensure students and employees follow this guide.
- Report any incident involving theft, unauthorized entry, vandalism and similar events to the UT Police Department.
- Be responsible for establishing security procedures for the laboratory

Employees that work in laboratories shall

- Participate in any orientation or training related to lab security
- Be familiar with the basic security requirements for the lab in which they work
- Report any deficiencies or breaches in security to the lab manager or PI

iii. Biosafety and Conservation of Plant Genetic Resource Regulatory Policies: Issues and Considerations for Developing Countries : Conservation and safe use of plant genetic resources (PGR) are the foundation of food and drug production, and the biological basis for food security, livelihoods and economic development. In the last three decades, the increasing appearance and commercialization of products from recombining DNA of plants constructed in the North, has created a need for access to PGR of the South and raised a flux of biosafety concerns of unintended consequences on human and the environment (Wagih *et.al.*, 2012). On one hand, the conservation of PGR in world"s Centers is a complex initiative and often contentious. The declared objective of the North"s initiatives in the conservation of PGR is perceived with suspicion by the South due to the emerging disputes over the sovereignty of new food; resulting from the current modus of access to this resource and inability to trace its use under the patent law (5). On the surface, it appears to be a simple issue of protecting plants from forces beyond the control of the owning nations in the South. Closer inspection reveals a complex tangle of conflicting issues concerning the South states" control over its PGR that is profoundly the base for biotechnology development by the North. On the other hand, the current world economic and legislative climate, competitive entry of the South into the biotechnology industries is fraught with many difficulties; such as lacking of technological and legal capacities for developing innovative biotechnology R&D, and relevant biosafety regulatory policy, inability to access new technology tools attached to intellectual property rights and missing the share of benefits regenerated from existing patents and infringements of owned PGR. In view of the continuing coalescence of the industry under the control of limited multinational companies, these difficulties seem to intensify. This paper discusses the conservation and regulations for a fare access to PGR and safe use of biotechnology in a sustainable and efficient manner, devoid of potential non-tariff/technical trade barrier as central to the principles of free trade under General Agreement on Tariffs and Trade, GATT.

13.11 Biosafety Issue and Constraints

The Cartagena Protocol on Biosafety (CPB) that followed the Convention on Biological Diversity (CBD) is an international treaty governing the movements of living modified organisms (LMOs) resulting from modern biotechnology from one country to another. It was adopted on 29 January 2000 as a supplementary agreement to the Convention on Biological Diversity and entered into force on 11 September 2003 (Wagih Mohamed and Al-Kiey, 2012).

The Protocol was ratified by the 171 member countries of the CBD, and became a legally binding instrument and its implementation became mandatory. This has become pivotal to trade and a potential trade barrier and therefore, required countries to develop the necessary legal and technical capacities in biotechnology for safe transfer, handling, use and identification of GMOs and their derivatives; particularly by protecting against unwanted human health and environmental outcomes. Environmental and ood safety risk assessments were recommended to be contracted out, where appropriate, on a cost recovery basis from applicants, as not to over burden the national regulatory agencies. While considering the need for developing relevant National Biosafety Guidelines to safeguard the environment and public health and to protect national interests and concerns, the North faces a number of major constraints, among them;

- understanding the potential impact of GMOs/ transgenic organisms and their products, including pharmaceuticals and genetically modified food (GMF), upon the environment, biodiversity and human health;
- dealing with public perception issues related to GMF, and the environmental impact of GMOs;
- assessing possible social and economic implications of the import of GMOs and their products in light of availability of safer substitutes;
- coping with the workload regulators are likely to face in preparing new regulatory policy and the cost of implementing relevant legislation; and
- relying on domestic public sector with limited support resources as opposed to strong private-sector industry that are in support of legislation in developed countries.

Considerations in the South's National Biosafety Guidelines : The development of National Biosafety Guidelines in biotechnology is based on certain global, regional and bilateral considerations that may intervene with or pressurize national interests in the course of optimizing a national framework for regulatory policy. Some of these considerations include the implications of World Trade Agreements (GATT), specially provisions concerning non-tariff trade barriers. The principles of free trade set at the final round of the GATT that was completed in Marrakech in 1993 is administered by the World Trade Organization (WTO), which became responsible for undertaking GATT obligations in the discrimination between justified non-tariff trade barriers, i.e. for reasons of environmental prudence and/ or adverse effect on human health, and restrictions that are unjustified under the principles of GATT. In this respect, the World Trade Agreements (WTAs) serve as a de facto means for harmonizing biosafety legislation and decisions through curtailing national sovereignty to rule unilaterally on new GMOs and their products under the principles of free trade (Wagih Mohamed and Al-Kiey, 2012).

Evidently, the implication of GATT compromises the sovereignty of individual nations, where its biosafety legislation might be construed as a spurious non-tariff trade barrier. In other words, the requirement is that national biosafety measures adopted for reasons of environmental protection are legitimate and scientifically justifiable. In case where national provisions deviate from international guidelines, a country must, if challenged, produce scientific evidence justifying such deviation. Under GATT, the potential Technical Barriers to Trade (TBT) were addressed in two agreements reflecting environmental biosafety issues, relying upon a range of international standards and guidelines. These agreements are: Agreement on Technical Barriers to International Trade (TBT); and the Agreement on the Application of Sanitary and Phytosanitary Measures (SPS).

Measures adopted for reasons of environmental protection are legitimate and scientifically justifiable. In case where the TBT Agreement requires that "Members shall ensure that technical regulations are not prepared, adopted or applied with a view to, or with the effect of creating unnecessary obstacles to international trade". However, the Agreement does itemize particular „legitimate objectives, according to which trade restrictions may be permitted. These objectives include exceptions, which were also made in Article XX of the GATT. Obstacles to trade based on the National Biosafety Guidelines may not be considered necessary or legitimate under TBT. This is obviously a potential of conflict of interest between principal exporters of biotechnology products (mainly developed countries), and developing countries. Therefore, many developing countries consistently supported the need for a protocol to cover not just the transboundary transfer of GMOs, but also their post entry safe handling and use. In the course of biosafety risk assessment and risk management, developing countries anticipate commitment from developed countries for more modest capacity building in biosafety (Wagih Mohamed and Al-Kiey, 2012).

The SPS Agreement requires members to base their sanitary or phytosanitary measures on international standards, guidelines or recommendations, where they exist. But, an option is available under Article 5.7, which allows adoption of provisional measures. "Members may introduce measures, which result in a higher level of protection than would be achieved by SPS measures, if there is a scientific justification." Substantial evidence is not a condition for applying restriction on imported GMOs or their products. Least-developing member countries were given till this year, 2000, to implement SPS obligation. Otherwise, a member country might apply the restriction with some evidence, and within a reasonable time, the member must provide additional evidence to justify further restriction. However, providing adequate scientific justification in a reasonable duration would be a problem case when risk is uncertain, impossible, or cannot be identified. This could mean that an embargo on an import of GMO or its products, based on provisions of the Biosafety Protocol, by one country is likely to bring trade disputes to the WTO. For this reason, the appropriateness of the „Biosafety Protocol has been challenged on the basis of its possible conflict with the GATT principle of free trade. Biosafety Protocol has been challenged on the basis of its possible conflict with the GATT principle of free trade.

13.12 Biosafety Regulatory Framework – Indian Scenario

In India, all activities related to Genetically engineered organisms (GE organisms) or cells and non-GE hazardous microorganisms and products thereof are regulated as per the Rules for the Manufacture, Use, Import, Export and Storage of Hazardous Microorganisms/Genetically Engineered Organisms or Cells 1989 (known as 'Rules, 1989') notified by the Ministry of Environment, Forest and Climate Change (MoEF&CC), Government of India, under the Environment (Protection) Act, 1986 (Anonymous, 2020).

The mandate and function of Competent Authorities entrusted with implementation of biosafety regulations under Rules 1989 are as follows.

Competent Authorities	**Mandate & Functions**
Recombinant DNA Advisory Committee (RDAC)	The RDAC functions in Department of Biotechnology (DBT) and shall review developments in biotechnology at national and international levels and shall recommend suitable and appropriate safety regulations for India in recombinant research, use and applications from time to time
Institutional Biosafety Committee (IBSC)	The IBSC is constituted by an occupier or any person including research institutions, handling hazardous microorganisms/ genetic engineered organisms {at Research & Development (R&D) level}. The occupier or any person including research institutions shall prepare, with the assistance of the IBSC, an up-to-date on site emergency plan according to the manuals /guidelines of the RCGM and make available copies to the DLC/SBCC and RCGM/GEAC.
Review Committee on Genetic Manipulation (RCGM)	RCGM functions in the DBT to monitor the safety related aspects in respect of on-going research projects and activities involving genetically engineered organisms/hazardous microorganisms. It shall bring out Manuals of guidelines specifying procedure for regulatory process with respect to activities involving genetically engineered organisms in research use and applications including industry with a view to ensure environmental safety. All ongoing projects involving high risk category and controlled field experiments shall be reviewed to ensure that adequate precautions and containment conditions are followed as per the guidelines. RCGM shall lay down procedures restricting or prohibiting production, sale, importation and use of such genetically engineered organisms of cells as are mentioned in the Schedule of Rules, 1989.
Genetic Engineering Appraisal Committee (GEAC)	The GEAC functions in the MOEF &CC and is responsible for approval of (i) activities involving large scale use of hazardous microorganisms and recombinants in research and industrial production from environmental angle (ii) proposals relating to release of genetically engineered organisms and products into the environment including experimental field trials. The committee or any person/s authorized by it shall have powers to take punitive actions under the Environment (Protection) Act, 1986.

Competent Authorities	Mandate & Functions
State Biotechnology Coordination Committee (SBCC)	SBCC reviews periodically the safety and control measures in the various installations/institutions handling genetically engineered organisms/hazardous microorganisms. SBCC have powers to inspect, investigate and take punitive action in case of violations of statutory provisions through the Nodal Department and the State Pollution Control Board/Directorate of Health &/ Medical Services
District Level Committee (DLC)	DLC monitors the safety regulations in installations/institutions engaged in the use of genetically modified organisms/ hazardous microorganisms and its applications in the environment. DLC/or any other person/s authorized in this behalf shall visit the installation engaged in activity involving genetically engineered organisms, hazardous microorganisms, formulate information chart, find out hazards and risks associated with each of these installations and coordinate activities with a view to meet any emergency. DLC shall also prepare an off-site emergency plan for field trials. The District Level Committee shall regularly submit its report to the State Biotechnology Coordination Committee/Genetic Engineering Approval Committee.

While the RDAC is advisory in functions, IBSC, RCGM and GEAC are involved in regulatory and approval functions. SBCC and DLC are responsible for monitoring the activities related to GMOs at state and district levels respectively.

13.13 Online Submission of Forms and Approval Process

IBSCs are responsible for assessing and monitoring the research involving in GMOs/LMOs/GE and hazardous organisms and ensure that the proposed risk assessment, risk management and emergency plans are sufficient. To achieve this objective, IBSCs need to ensure proper training is provided to all the IBSC members and researchers of the organisation. IBSCs at the institutional level are responsible for the assessing the research being conducted and providing approval to the submissions and recommending all the research applications to the RCGM for its consideration.

Submission of application forms at the IBKP portal : For conducting research in the organization, the Principal Investigators (PIs) must apply online by filling appropriate application form before initiating research. The application forms are available at the IBKP portal. The online application forms are to be filled and submitted for consideration by institute's IBSC and then by RCGM. The forms would be accessible to all the registered IBSCs on the portal. IBKP hosts application forms for import, export, transfer and R&D at contained (laboratory) and confined (field trial) activities specific for healthcare, agriculture and environmental use. All forms are interactive in nature where applicant can submit the application required information along with required enclosures in relevant application proforma.

Providing Sub-User access through the IBKP portal In view of multiple investigators from an organization being involved in work related to genetic engineering and/or hazardous microorganism that requires approval of IBSC and/ or RCGM, the portal has a provision where the Chairperson, IBSC can issue sub-user access to investigators to enable them to access the application forms. The salient features of sub-user access to portal are:

- The Chairman/ Member Secretary of the IBSCs having the login ID of the IBSC may provide access of the portal to the PIs for accessing and filling the application forms.
- Information submitted by a Sub-user (PI) will not be visible to other sub-users. However, all applications submitted by sub-user shall be visible to the authority holding the permanent IBSC log in ID and password.
- No other functions like change of IBSC information, compliance submission, etc. would be possible through sub-user access.

Process for online submission of applications at IBKP

- Using sub-user access credentials, applicant needs to access RCGM dashboard and initiate the online application filling process through 'submit new application' tab.
- From time to time, the forms will be auto-saved preventing loss of information filled in the process or else the applicant may save the document periodically.
- Periodically, Member Secretary, IBSC shall prepare agenda by including all theØ applications filled by sub-users and circulate agenda along with all applications one week in advance to be considered for convening the meeting of IBSC.
- Once IBSC minutes are approved, Member Secretary will fill the IBSC approvalØ details & upload IBSC minutes (preferably within a week of holding the meeting but no later than 15 days) and submit the application for consideration by RCGM.
- Only a completely filled application form will be accepted for consideration. TheØ applicant shall ensure correctness of the information provided in the form. Once submitted, the information cannot be changed.
- On successful submission, the applicant shall receive intimation throughØ registered email id and the same will be reflected in the notification panel of the portal.
- Each application will get a Unique Authorization Code (UAC code) which may beØ used by the applicant to track the status of application at IBKP.
- After verification of details in the submitted Application form, the application may be considered and decision taken be communicated to IBSC

so as to provide RCGM recommendation to PI. In case of any deficiencies/ clarifications, the same would be communicated to the IBSC through the portal to intimate PI for necessary actions. The information would be noted down in subsequent RCGM meeting.

13.14 Indian Biosafety Guidance

i. **Monitoring confined field trials of regulated genetically engineered plants:** Monitoring Manual: Before new genetically engineered (GE) plant varieties can be grown as commercial crops, researchers must test these plants in outdoor field trials to fully evaluate their agronomic utility and environmental impacts. Confined field trials (CFTs) are small-scale experimental trials of GE plant varieties, maintained under carefully controlled conditions, that are performed for the purpose of evaluating the plants before their introduction into agriculture (Anonymous, 2016). They represent a greater degree of environmental exposure than work performed in contained facilities.

ii. **Guidelines for the safety assessment of foods derived from genetically engineered plants**

To address the human health safety of foods derived from GE plants, there is a need to adopt a systematic and structured approach to their risk analysis. Risk analysis is a science based process comprised of risk assessment, risk management and risk communication and is an analytical tool to systematically evaluate safety concerns addressing human health safety of GM foods within a framework for decision making. It also provides further basis for reviewing the safety evaluation parameters as and when further information becomes available.

iii. **Protocols for food and feed safety assessment of GE crops**

A series of protocols have been developed by the Department of Biotechnology (DBT) as guidance to applicants seeking approval for the environmental release of genetically engineered (GE) plants in India under "Rules for the Manufacture, Use, Import, Export and Storage of Hazardous Microorganisms/Genetically Engineered Organisms or Cells 1989" (Rules, 1989) notified under the Environment (Protection) Act, 1986. Specifically, these protocols address key elements of the safety assessment of foods and/ or livestock feeds that may be derived from GE crops. The results of these studies are be submitted to the appropriate regulatory bodies (i.e., RCGM and GEAC) as required.

iv. **Recombinant DNA safety guidelines, 1990**

The guidelines cover areas of research involving genetically engineered organism. It also deals with genetic transformation of green plants, rDNA technology in vaccine development and on large scale production and dekliberate/ accidental release of organisms, plants, animals and products derived by rDNA technology into the environment. The issues relating to Genetic Engineering of human embryos, use of embryos and foetuses in research and human germ line gene therapy are excluded from the scope of the guidelines.

v. **Revised guidelines for safety in Biotechnology**

The current guidelines cover areas of research involving genetically engineered organisms, genetic transformation of plants, rDNA technology in vaccine and diagnostics development and on large scale production and deliberate/accidental release of organisms, plants, animals and products derived by rDNA technology into the environment and also import and shipment of GMOs/transgenics for laboratory research and large scale use.

vi. **Revised guidelines for research in transgenic plants, 1998**

The revised present document is meant for the researchers in the country who are involved in recombinant DNA research on plants. Earlier the Department of Biotechnology in January 1990 issued a compendium of guidelines under the title "Recombinant DNA Safety Guidelines". A revision was made in 1994 under the title "Revised Guidelines for Safety in Biotechnology". The current guidelines have been developed in the light of enormous progress that has been made in recombinant DNA research and its widespread use in developing improved microbial strains, cell lines and transgenic plants for commercial exploitation.

vii. **Guidelines for the conduct of confined field trials of regulated, GE plants:**

The "Guidelines for the conduct of confined field trials of regulated, GE plants in India" have been prepared to provide instructions to help applicants meet requirements for the application and authorization/approval of confined field trials of regulated, GE plants under Rules for Manufacture, Use, Import, Export and Storage of Hazardous Micro-Organisms, Genetically Engineered Organisms or Cells Rules, 1989 of the Environment (Protection) Act, 1986. These guidelines summarize the information requirements and procedures used by the two regulatory committees, the Review Committee on Genetic Manipulation (RCGM) and the Genetic Engineering Approval Committee (GEAC), that are responsible for evaluating and approving applications for confined field trials. The information provided in this document does not preclude additional regulatory requirements on case to case basis either from RCGM or GEAC or any other Ministries/regulatory bodies.

viii. **Recording formats for confined field trials**

This document contains the forms required for recording information associated with confined field trials.

ix. Guidelines for the monitoring of confined field trials of regulated, GE plants

The purpose of these guidelines is to provide clear and concise information to help those individuals designated as members of monitoring teams in the aforementioned monitoring agencies.

13.15 Biosafety Policy-Indian Scenario

Biological safety has not been, historically, a strong public policy thrust in Indian agriculture. Rural biological systems have long been threatened by low resource farmers who clear new lands and destroy natural habitat in order to graze animals or plant crops. Mismanagement of irrigation water has left once fertile croplands poisoned with salt. Careless application of chemical nutrients has contaminated rivers, streams and ponds, making water unsafe for human consumption and poisoning fish and amphibian species. Improper use of insecticides has killed non-target species while making target pests increasingly resistant to the poisons. Even after the enactment of India's powerful 1986 Environmental Protection Act (EPA), such conventional biohazards from farming were seldom put under effective regulation. It is thus somewhat surprising to see Indian officials paying so much attention to the biological safety of GM crops (Anonymous, 2011). India's official interest in genetic resource IPRs would have been strong even without GM crops as an issue, but the nation's current official interest in rural biological safety is to a large extent a product of the GM crop revolution itself.

India's first embrace of a formal national biosafety policy dates from the first emergence of rDNA agricultural and human health technologies in the 1980s. We might expect, given India's strong development aspirations and its prior history of low attention to the rural biohazards, that the nation's biosafety policies toward GM crops would have been permissive or even promotional, rather than precautionary or preventive. Instead, the Indian Government adopted from the start a set of procedures that ensured significant precaution, and as implemented - in the context of a growing global controversy over GM crops - these procedures sufficed to slow the progress of the technology considerably. As of 2000 farmers in India were not yet permitted to grow any GM crops, because none had yet received final approval from the nation's biosafety regulators.

The Indian Government first issued rules and procedures for handling GM organisms in December 1989. The Department of Biotechnology, inside the Ministry of Science and Technology, published these rules and procedures in January 1990 . These "Recombinant DNA Safety Guidelines" were revised slightly and republished as "Revised Guidelines for Safety in Biotechnology" in 1994. They describe the biosafety measures that must be undertaken in India both for contained research activities and also for large scale open environmental release of genetically altered agricultural and pharmaceutical materials. A subsequent 1998 revision of the Guidelines elaborated procedures for screening transgenic plants and seeds for toxicity and allergenicity.

India's crop biosafety Guidelines were written to require a screening of GM crop technologies for scientifically demonstrated risks, borrowing heavily from the demonstrated risk approach employed by USDA's Animal and Plant Health Inspection Service (APHIS) in the United States. This has given the Guidelines considerable permissive content on paper. Yet the implementation of the Guidelines is ultimately shared with India's Ministry of Environment, a procedure which inclines India's policy somewhat more toward precaution. The Guidelines create two separate review committees: a Review Committee on Genetic Manipulation (RCGM) which is empowered to approve (or not approve) applications for all small scale research activities in India designed to generate information on transgenic organisms, and a Genetic Engineering Approval Committee (GEAC) empowered to approve (or not) large scale research activities, plus actual industrial use or environmental release of all GM organisms.

Because the RCGM mostly screens research activities for biosafety, it is composed of representatives from India's leading public universities and research Institutes, including the Indian Council of Agricultural Research (ICAR), the Indian Council of Medical Research (ICMR), the Council of Scientific & Industrial Research (CSIR), and others. It is constituted by the Department of Biotechnology (DBT), and its member-secretary is adviser to DBT. It has thirteen members in all, and operates by majority vote. In contrast the GEAC is a statutory body under the Ministry of Environment and Forests (MoEF) and is empowered to approve or disapprove all large-scale use and environmental release of GM organisms. GEAC is thus India's most powerful biosafety policy gatekeeper. It is chaired by the Additional Secretary of MoEF, co-chaired by an expert nominee from DBT, and it includes representatives from DBT, from the Ministry of Industrial Development, the Ministry of Science and Technology, and the Department of Ocean Development. GEAC is designed to perform more than a technical function. It is the lead interministerial body empowered to shape - by consensus - the Government's final disposition toward large-scale use and environmental release of GM organisms. The GEAC is empowered to authorize or prohibit, conditionally or unconditionally, the import, export, transport, manufacture, processing, use, or sale of any GM organism.

This Indian biosafety system was intended to be permissive at the research stage, where RCGM was given the final word. Since RCGM is constituted by DBT - the agency explicitly tasked with promoting biotechnology in India - its membership comes from the very research institutes (in some cases the very individuals) who conduct biotech research and who rely on DBT for funding. When RCGM hands the regulatory process over to GEAC, after the research stage, a precautionary bias takes over, since GEAC is chaired by MoEF. This built-in tension between permissive research versus precautionary release might have some advantages in theory. By allowing research to go forward, India's process is capable of generating the necessary empirical data (e.g., from field trials) needed to make sound final release decisions. India's system may also be potentially useful for avoiding last minute jurisdictional disputes between biotechnology commissions and environmental ministries, of the kind seen in Brazil between CTNBio and IBAMA. In India, biosafety regulation stays legally under the EPA and within the jurisdiction of GEAC from start to finish. The public sector dominated Indian system also excludes representatives of private industry from direct participation, another contrast to Brazil, and it requires that RCGM send a Monitoring Committee to visit the experimental sites that are being employed by applicants to generate biosafety data.

India's Guidelines are quite thorough regarding the biosafety questions that must be answered prior to commercial release of GM technologies. In the area of transgenic plants, basic information must be provided on the characteristics of the donor organisms, the vectors used, the transgenic inserts, and then on the resulting transgenic plants themselves. Through lab, growth chamber, greenhouse and field trials, evaluations must be generated by applicants regarding toxicity and pathogenicity, possibility and extent of transgenic pollen escape and transfer to wild relatives, and consequences for both the environment and human and animal health. The data may be generated in India or anywhere else. Guidance is provided on proper (biosafe) procedures for conducting these experiments, along with the proper format for presenting this information as a formal registration document. Some of the data required by India's procedures go well beyond biosafety. RCGM is required in the latest version of the Guidelines to solicit from applicants and assemble information "on the comparative agronomic advantages of the transgenic plants" to assure GEAC that the technologies under review are "economically viable" as well as environmentally safe. While this might seem an added barrier to approval, it could also favor approvals in some cases by allowing evidence of social benefits to be presented alongside evidence of possible environmental risks.

This being the structure of India's biosafety review system for GM organisms, how has it operated in practice? In the area of health care, a number of GM products have been reviewed successfully with little controversy. As of March 2000, a total of ten different transgenic health care products in India (two locally produced, 8 imported) had moved all the way through the system and had been cleared for biosafety deregulation by GEAC. Social confidence regarding the biosafety of GM health care products has been consistently high in India. In the area of transgenic plants, however, the system has not moving so smoothly. Under criticism from NGOs, even RCGM has been forced into a highly cautious posture, and as of 2000 GEAC had yet to grant a commercial release for any GM crops.

India's biosafety review system for GM plants was pushed into this cautious posture in 1998, when RCGM was accused by NGOs in India of having exceeded its mandate by giving approval to limited field trials for Mahyco/Monsanto's Bt cotton plants. The original Guidelines had been ambiguous as to where RCGM's authority to manage small-scale research activities would end and where GEAC's authority to manage large-scale activities and "environmental release" would begin. When Mahyco applied to conduct contained field trials of its Bt cotton in 1998, RCGM assumed it had the authority to approve such trials, in part because it had earlier approved contained field trials of transgenic mustard in 1995 without incident. Yet the Bt cotton field trials approved for 40 locations in nine different states India in 1998-99 soon became a focus of intense national controversy, as noted above. NGOs in India, led by RFSTE, claimed that only GEAC had authority to approve field trials of GM crops, since even contained and limited trials constituted an environmental release. In February 1999, RFSTE then filed a PIL against the DBT for the manner in which it had authorized the trials (and also against Mahyco plus several other ministries), asking - without success - that further field trials be blocked.

RFSTE's challenge to the Bt cotton field trials was motivated not so much by any specific biosafety worry. The "Bollgard" Bt cotton seeds that Monsanto had brought in to cross with local Indian varieties had previously undergone extensive testing and monitoring in the United States to meet the biosafety standards set by APHIS and EPA. Bt cotton had been grown successfully in the United States on a commercial scale since 1996, and by 1998 it covered 40 percent of total U.S. cotton area. By 1998 it was also being grown widely and successfully in Australia, South Africa, and China. Bt cotton was potentially attractive in India as a solution to the bollworm infestations that farmers could no longer control with chemical sprays due to resistance in the pest population. This emergence of a real pest resistance problem from non-GM technologies made it harder to invoke hypothetical pest resistance problems as a reason not to plant GM cotton. resistant pest population in India in response to non-GM cotton production techniques made it harder to invoke hypothetical "superbugs" as a reason not to use GM cotton. Food safety risks were also hard to invoke for Bt cotton, since cotton is an industrial crop rather than a food. RFSTE was using biosafety rules and procedures as its means to try to block Bt cotton, but the core of its argument against this GM crop was that it was being introduced by the Monsanto company from abroad, the company which owned rights to a dreaded "terminator" technology.

Yet RFSTE also sought to make its case by pointing to procedural irregularities in RCGM's approval of the field trials. RCGM was faulted for having given field trial authority directly to a private company applicant, and for then allowing that applicant to launch the trials directly on leased farmers' fields, rather than on government research farms under direct public sector management. While this would seem a logical way to handle an application filed by a private company, RCGM might have known it would be seen in India as trusting the private sector too much. Second, RCGM had not secured advance approval from government authorities in several of the states where the trials were located, and in at least one of those states the trials even began before a local biosafety committee had been fully constituted. These several irregularities were exploited by RFSTE, and left

RCGM and DBT politically isolated when fears regarding the terminator gene subsequently broke. Nor did it help that RCGM's procedures in this case had been largely non-transparent. RCGM could point out that there was no formal obligation to publish deliberations or invite public comment prior to approving the limited trials, but when the trials became controversial RCGM and DBT were criticized all the more heavily for having operated behind closed doors. In the midst of this controversy in 1999, DBT's biosafety Guidelines were amended to clarify that GEAC's approval would be required for all field trials larger than one acre per location, or larger than 20 acres per year nationally. DBT also directed that the 1999 field trials for Mahyco's Bt cotton would take place either on university research farms or under closer national institute supervision. This, plus a DBT-led public awareness campaign in 1999, plus ministerial promises that the terminator gene would not be allowed into India, helped calm the public mood considerably.

Although India's Bt cotton trials went forward under protest and criticism, they did manage to generate important evidence of the GM plant's effectiveness against bollworm infestations. These pest attacks on cotton in India had emerged as a visible social issue due to the growing resistance of the pests to chemicals, and the resulting desperation of some small farmers. Excessive spraying of diluted and adulterated chemical insecticides - in some parts of India as many as 14 sprayings a season - had brought on the pest resistance problem, and some poorer cotton farmers in India who had borrowed heavily at a high rate of interest to buy chemicals that no longer worked were falling heavily into debt. Their plight came to public attention when bad weather worsened the pest crisis in 1998, and as many as 500 farmers resorted to suicide as their only way out. The first 1998 field trials of Bt cotton in India had seemed to confirm, meantime, that Bt varieties could provide at least a short-term solution to the pest control problem. DBT reported in 1998 that on average (in controlled field trials planted in eight different states in India) Bt cotton was able to reduce insect damage dramatically: yields for Bt cotton were 40 percent higher than for the non-Bt controls, and with an average of five fewer chemical sprays. For an Indian farmer with 5 ha of cotton, such a reduction in chemical use from switching to Bt cotton might by itself represent a savings of roughly Rs 2000 per crop (about $50).

RCGM was pleased and reassured by the 1998 field trial data, but nonetheless asked Mahyco in 1999 for ten more field trials, seeking more data on commercial advantages to farmers and also on a hypothetical geneflow concern, the possibility that insects could take Bt cotton pollen far enough away from the plant to result in unwanted gene transfer to other plants. Only after these additional trials had been completed did RCGM, in April 2000, formally express its technical confidence in Bt cotton. RCGM noted that its latest evidence showed farmers growing Bt cotton obtained 25-75 percent higher yields, while using six fewer chemical sprays, and with no evidence of harm to crops in adjacent plots.

Having received this approval at the research stage from RCGM, Mahyco promptly applied to GEAC for permission to begin large scale trials, and in response to this application, GEAC did approve large scale field trials for Bt cotton (up to 85 total hectares) in July 2000. Additional acreage was also permitted for seed production, in anticipation of a possible commercial release as early as 2001. To appease protesters GEAC made large scale field trials conditional on obtaining an independent Indian laboratory certification that the cotton plants did not contain the "terminator" gene. Even so, RFSTE was antagonized and submitted an additional petition against the trials. GEAC's prompt field trial approval increases somewhat the likelihood of an eventual commercial release for Bt cotton in India, although the first release may be only within a tight acreage restriction, once again on grounds of biosafety caution.

Has the Indian system been precautionary in this case only because the Bt cotton is a Monsanto variety? A second application working its way through India's system, from another company (ProAgro-PGS) to develop transgenic hybrid mustard, also met a cautious response from RCGM. Contained field trials of GM mustard began in 15 different locations in 1995, followed by open field trials, and none of these field trials was ever disrupted by an anti-GM activist, yet the process of bioscreening was highly precautionary just the same. In 1999, India's regulators asked ProAgro-PGS for one more year of field trials to screen for effects on soil micronutrients, an issue that had never been highlighted before. ProAgro-PGS hoped for eventual deregulation from GEAC, but recognized that it may be a conditional release only, perhaps with a requirement (hard to enforce) that farmers planting the GM mustard seeds separate themselves at least 40-50 meters from non-GM fields, as an extra precaution against unwanted geneflow. While India's anti-GM NGOs may be focused almost exclusively on keeping Monsanto's products out of the country, biosafety regulators in India have thus been highly precautionary toward all GM crop applications, whether from Monsanto or not.

13.16 Biosafety and Biotechnology

In agriculture, (including animal husbandry, fishery and forestry), the concept of biosafety involves assessing and monitoring the effects of possible gene flow, competitiveness and the effects on other organisms, as well as possible deleterious effects of the products on health of animals and humans. Policy decisions taken in regard to biosafety may have long-term implications for the sustainability of agriculture and food security. Within the WTO, biosafety in relation to GMOs appears to fall chiefly under the Agreement on the Application of Sanitary and Phytosanitary Measures (SPS Agreement). There are also implications for GMOs in standard-setting under the Agreement on Technical Barriers to Trade (TBT Agreement). This Agreement covers a large number of technical measures that seek to protect consumers from economic fraud and deception and measures concerning human, animal and plant life and health not covered by the SPS Agreement, and the environment.

International standard setting within the framework of FAO- Food standards: the *Codex Alimentarius* Commission: The purpose of the *Codex Alimentarius* Commission is to protect the health of consumers, to ensure fair practices in food trade, and to promote coordination of all food standards work undertaken by international governmental and non-governmental organizations. The Commission's Medium-term Objectives include inter alia "consideration of standards, guidelines or other recommendations as appropriate for foods derived from biotechnology or traits introduced into foods by biotechnology on the basis of scientific evidence and risk-analysis and having regard, where appropriate, to other legitimate factors relevant for the health protection of consumers and promotion of fair practices in food trade".

The 23rd Session of the *Codex Alimentarius* Commission (June/July 1999) established an Ad Hoc Intergovernmental Task Force on Foods derived from Biotechnology to fulfil these objectives, which will take full account of the work of national authorities, FAO, WHO, other international organizations and other relevant international fora. The Task Force will submit a preliminary report to the *Codex Alimentarius* Commission in 2001, and a full report in 2003. The Codex Committee on Food Labelling is also working on the development of recommendations for the labelling of foods obtained through biotechnology.It should be noted that Codex standards apply to all types of foods, and, for this reason, Codex will need to deal with foods of plant, animal and fish origin. The impact of feeding GMO plants to animals, and the nature of the resulting foods from these animals will also need to be addressed.

Phytosanitary Standards: The International Plant Protection Convention (IPPC): The IPPC's purpose is common and effective action to prevent the introduction and spread of pests of plants and plant products, and the promotion of appropriate control measures. It covers both cultivated and wild plants; the direct and indirect effects of pests; and the prevention of the introduction and spread of weeds, and their control. The IPPC also covers the movement of biological control agents, and other organisms of phytosanitary concern claimed to be beneficial. The IPPC provides the global standard setting mechanism for phytosanitary measures. It may be concerned with evaluating the potential "pest" characteristics (including weediness) of GMOs, that is, whether a GMO may be detrimental to plant life or health.

At the second meeting of the Interim Commission on Phytosanitary Measures (ICPM) in October 1999, a number of members gave high priority to standard setting in relation to GMOs in particular to risk assessment and testing and release of GMOs. They indicated that this could be addressed within the framework of the IPPC. Others advocated a more cautious approach while some indicated the need to give sufficient priority to development of standards for plant quarantine. The ICPM decided that an exploratory working group would address the issues of biosafety in relation to GMOs and of invasive species and report back to the 3rd meeting of the ICPM in April 2001.

Commission on Genetic Resources for Food and Agriculture (CGRFA): Since 1989, the CGRFA has regularly considered reports on technical and policy issues regarding biosafety, within the context of biotechnology as it relates to genetic resources for food and agriculture. In 1991, it requested the preparation of a draft Code of Conduct on Biotechnology, with the aim of maximizing the positive effects, and minimizing the possible negative effects of biotechnology. The draft Code of Conduct, drawn up following a survey of over 400 international experts, contained four modules, one of which was on biosafety. In 1993, noting that the CBD was considering the development of a biosafety protocol, the CGRFA recommended that FAO participate in this work, in order to ensure that the aspects of biosafety relevant to genetic resources for food and agriculture were appropriately covered. The biosafety component of the draft Code of Conduct was forwarded to the Executive Secretary of the CBD, at the request of the CGRFA, as an input to the biosafety protocol.

The CGRFA has suspended work on the draft Code of Conduct pending the completion of the negotiations for the revision of the International Undertaking on Plant Genetic Resources. The Eighth Session of the CGRFA, in April 1999, noting that the negotiations for the revision of the International Undertaking were expected to be completed during 2000, requested a report on the status of the draft Code of Conduct, at its Ninth Session, in 2001.

13.16.1 GMO's in Fisheries

The fishery sector has recognized that GMOs are a diverse class of organisms that share many common features with introduced or alien species. FAO's Regional Fisheries bodies have adopted, in principle, codes of practice on the use of introduced species and GMOs, produced by FAO's European Inland Fishery Advisory Commission (EIFAC) and the International Council for the Exploration of the Sea (ICES). The general principles in such codes of practice, which include general principles for environmental assessment, contained use, advanced notification and the application of the Precautionary Approach, have been incorporated into the FAO Code of Conduct for Responsible Fisheries. FAO continues to work with regional bodies, professional fishery associations and national governments in the harmonization and refinement of these codes, and in methods for appropriate risk assessment. A recent international meeting 1 convened by FAO and the International Centre for Living Aquatic Resources Management (ICLARM) on developing policies for aquatic genetic resources recognized "that in the formulation of biosafety policy and regulations for living modified organisms, the characteristics of the organisms and of potentially accessible environments are more important considerations than the processes used to produce those organisms". In terms of aquatic animal health and quarantine, FAO and OIE are working together. According to the SPS agreement, the accepted international standards governing the movement of aquatic animals are those of OIE. FAO closely collaborates with OIE in providing assistance to developing countries to improve their capacities in the effective application of these standards.

13.16.2 Biotechnology and Biosafety Policy at OECD- Future Trends

The Organization for Economic Co-operation and Development (OECD) Council Recommendation on Recombinant DNA Safety Considerations is a legal instrument which has been in force since 1986. It outlines the safety assessment practices that countries should have in place for agricultural and environmental biotechnology. This article suggests possible updates to make it suitable for the modern era (Peter *et.al.*, 2021) (Fig. 20, Table 19). The Organization for Economic Co-operation and Development (OECD) is an intergovernmental organization with 37 member countries. In 1986, the OECD Council adopted a Recommendation entitled Recombinant DNA Safety Considerations: Safety Considerations for Industrial, Agricultural and Environmental Applications of Organisms Derived by Recombinant DNA Techniques. This OECD legal instrument was intended to promote international understanding of safety issues raised by recombinant DNA (rDNA) techniques so as to take steps towards an international consensus on the protection of health and the environment as well as a reduction of non-tariff barriers to trade. It also marked the start of OECD's engagement with modern biotechnology, which continues today with a track record of seminal events and publications that have facilitated the international harmonization of risk assessment principles.

In view of the importance of harmonizing regulations at the regional and sub-regional levels related to the testing and release of GMOs, FAO will continue to strengthen its normative and advisory work, in coordination and cooperation with other relevant organizations. Recent advances make it likely that a diverse set of GMO-based technologies, and transgenic animals, will be brought into agricultural production environments (including for modified milk) in the near future. This will require more systematic consideration of the biosafety questions involved. At the international level, there are as yet no immediately relevant instruments. There is an evident need for harmonization over a wider range of biosafety issues involved within animal agriculture, beginning with systematic consultation among the relevant international organizations. In terms of technical advice and capacity-building, FAO will: advise member governments on regulatory issues (including in the context of implementing the Biosafety Protocol, if adopted); advise on harmonization at regional and international levels; offer legal advice for the establishment of any regulatory bodies required; assist in establishment of the capacity for risk assessment; and seek to mobilize extra-budgetary funds and to cooperate with other relevant organizations. Member countries have also requested FAO's assistance in establishing and implementing regulations regarding the quality and safety of foods derived from biotechnology; in this context, existing regulatory instruments elaborated by national food safety authorities should also be considered.

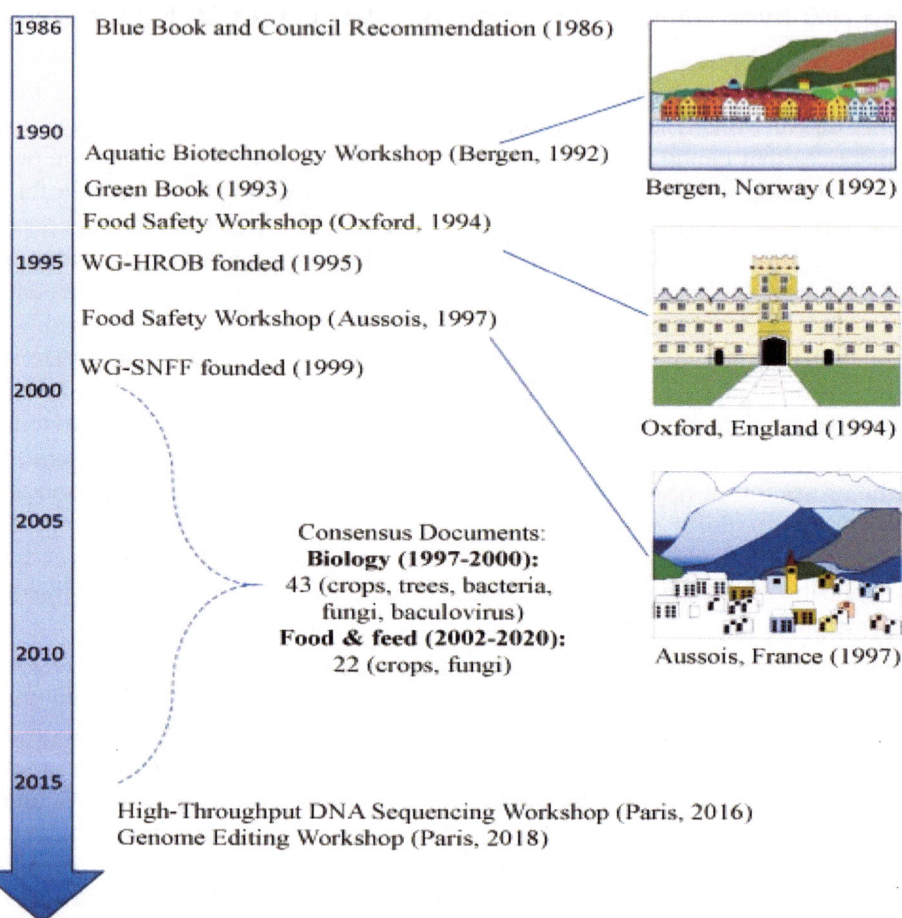

Fig. 20: Chronology of Seminal OECD Events and Publications on the Safety of Modern Biotechnology (1986–Present).

Abbreviations: WG-HROB, OECD Working Group on Harmonisation of Regulatory Oversight in Biotechnology; WG-SNFF, OECD Working Group on the Safety of Novel Foods and Feeds.

Table 19: Summary of Relevant OECD Principles and Guidance (Consensus Documents)

Principle or guidance	Source
GMO development follows the principle of a stepwise approach, with risk/safety assessment being performed at each stage, from the laboratory to the growth chamber and greenhouse, to limited field testing, and finally to large-scale field testing.	OECD, 1986 (Blue Book) OECD, 1986 (Council Recommendation)

Principle or guidance	Source
Principle of case-by-case assessment. This principle implies that the risk/safety assessment of each case is performed individually using assessment criteria relevant to the particular case.	OECD, 1986 (Blue Book). OECD, 1986 (Council Recommendation)
Principle of risk/safety assessment and risk management. This process includes hazard identification and, if a hazard is identified, risk assessment. The process is based on the characteristics of the organism, the trait introduced, and the environment into which the organism is introduced, as well as the intended application.	OECD, 1993 Published under the auspices of the former OECD Group of National Experts on Safety in Biotechnology
The principle of familiarity is used to address the environmental safety of GMOs. It is based on the understanding that most GMOs are developed from organisms such as crop plants whose biological characteristics are well understood. It allows risk assessors to use previous knowledge and experience with the introduction of GMOs into the environment, and this can indicate management measures.	OECD, 1993 Published under the auspices of the former OECD Group of National Experts on Safety in Biotechnology
The OECD regularly publishes environmental safety Consensus Documents. These are in line with the principle of familiarity in that they contain information that member countries have agreed is relevant to the risk/safety assessment of GM plants and other GMOs. The subject of each document is a crop plant or another organism where GM varieties have been commercialised or are expected in the future.	OECD, 2017
The principle of 'substantial equivalence', otherwise known as the comparative approach, is used in the food and feed safety assessment of products derived from GMOs. It involves the comparison of a new food or feed derived from a GMO with a similar product having a history of safe use, such that the safety evaluation can subsequently focus on any differences identified.	OECD, 1993 *Codex Alimentarius*, 2009
The OECD regularly publishes food/feed safety Consensus Documents which support the comparative approach. Each document has a crop plant as its subject, and includes data on key compositional parameters (nutrients, anti-nutrients, toxins, allergens) for the analysis of new variants of food- or feed-producing organisms including GM crops and several fungal species.	OECD, 2019

The Recommendation should take into account the accomplishments since 1986, and how the intentions of the Recommendation have largely come true, stating the importance of the principles of familiarity and the comparative approach as well as the case-by-case nature of the pre-market safety assessment of products of modern biotechnology. The revised Recommendation could look forward to the likely and preferred future developments within the purview of modern biotechnology, reflecting present-day discussions similarly to the original Recommendation.

13.16.3 Regulations of Living Modified Organisms (LMOs) in India

The Cartagena Protocol on Biosafety (CPB) to the Convention on Biological Diversity (CBD) is an international treaty governing the movements of living modified organisms (LMOs) resulting from modern biotechnology from one country to another. It was adopted on 29 January, 2000 as a supplementary agreement to the CBD and entered into force on 11 September, 2003. The CPB has been developed in response to advancements in the area of modern biotechnology and associated concerns that LMOs resulting from modern biotechnology may have negative effects on biodiversity and human health (Anonymous, 2019). As on January 2019, 171 countries have ratified or acceded to the Protocol. India is a party to CBD and CPB.

LMOs are usually considered to be the same as genetically modified organism (GMO), genetically engineered (GE) organism but definitions and interpretations of the term vary widely. there are increasing debates about safety concerns among various stakeholders related to their impact on health, environment and biodiversity. Many countries with active biotechnology research programmes including India have put in place biosafety regulations to deal with these safety issues. The Ministry of Environment, Forest and Climate Change (MoEFCC) is the nodal Ministry for implementation of biosafety regulations in India and implementation of CPB. MoEFCC has taken several initiatives for capacity building of multiple stakeholders for creating awareness about CPB and national regulations. Resource documents and outreach material have been prepared as part of UNEP/GEF supported "Phase II Capacity Building Project on Biosafety" being implemented by MoEFCC for regulators, scientists, enforcement agencies, students etc. Safety of LMOs is evaluated in a comprehensive process that involves several steps. Systematic safety assessment methodologies have been developed at national and international level that give conclusion on whether or not the LMO is as safe as its conventional counterpart. LMOs are permitted to be grown and used only after they have passed safety assessment and are not likely to pose risks for human health and environment.

The use of LMOs in India is governed by the "Rules for the Manufacture/Use/Import/Export and Storage of Hazardous Microorganisms, Genetically Engineered Organisms or Cells, 1989" (Rules 1989) notified under the Environment (Protection) Act, 1986. Rules, 1989 essentially cover entire spectrum of activities involving GMOs and products thereof including the sale, storage, exportation, importation, production, manufacturing, packaging etc. The Rules, 1989 are implemented by the MoEFCC, the Department of Biotechnology (DBT) of Ministry of Science & Technology and State Governments through six competent authorities viz. (i) Recombinant DNA Advisory Committee (RDAC), (ii) Institutional Biosafety Committee (IBSCs), (iii) Review Committee on Genetic Manipulation (RCGM) and (iv) Genetic Engineering Appraisal Committee (GEAC) (v) State Biotechnology Coordination committee (SBCC) and (vi) District Level Committee

(DLC) (Anonymous, 2019). While the RDAC is advisory in function, the IBSC, RCGM, and GEAC are of regulatory function, SBCC and DLC are for monitoring purposes.

Biosafety guidelines in India

Contained use : Recombinant DNA Safety Guidelines, 1990 (Updated, 2017) ; Revised Guidelines for Research in Transgenic Plants, 1998

Confined field trials : Guidelines for Conduct of Confined Field Trials (CFTs) of Regulated GE Plants, 2008 ; Standard Operating Procedures (SOPs) for CFTs of Regulated GE Plants, 2008; Guidelines for Monitoring of CFTs of Regulated GE Plants, 2008

Food safety assessment: Guidelines for the Safety Assessment of Foods Derived from GE Plants, 2008 (Updated in 2012) ; Protocols for Food and Feed Safety Assessment of GE Crops, 2008

Environmental safety assessment : Guidelines for Environmental Risk Assessment (ERA) of GE Plants, 2016 ; Risk Analysis Framework, 2016 ; ERA of GE Plants: A Guide for Stakeholders, 2016

Others : Guidelines for generating preclinical and clinical data for rDNA vaccines, diagnostics and other biologicals, 1999 ;Guidelines and Handbook for Institutional Biosafety Committees (IBSCs), 2011; Guidelines on Similar biologics: Regulatory requirements for Marketing Authorization in India, 2012 (updated in 2016).

In India, no person can import, export, transport, manufacture, store, process, use or sell any LMOs, substances or cells except with the approval of GEAC. RCGM is authorized to permit imports only for research purpose. Deliberate or unintentional release of LMOs is not be allowed. Production in which LMOs are generated or used cannot be commenced except with the approval of GEAC. All approvals are for limited period as per Rules, 1989. GEAC has powers to revoke approvals in case of any new information on harmful effects of LMOs, any damage to the environment that could not be envisaged when approval was given or non-compliance of any conditions stipulated by GEAC).

Major policies: In addition to the Rules, 1989, provisions in other acts, rules and policies are also applicable to GMOs/LMOs (Anonymous, 2019). Relevant details concerning import/export of LMOs are as under.

Plant Quarantine Order, 2003 : It covers regulation of import of germplasm/ GMOs/transgenic plant material for research purpose (Figure 3). The National Bureau of Plant Genetic Resources (NBPGR) has been designated as the Competent Authority to issue the import permits1 for import of seeds for research purposes after getting permission under Rules, 1989 and to receive import material from customs authorities for quarantine inspection. A phytosanitary certificate issued from exporting country is needed during the imports of transgenic material.

Food Safety and Standards Act, 2006: This Act regulates manufacture, storage, distribution, sale and import of food which includes GM food.

DGFT Notification Relating to Inclusion of GM Policy in Foreign Trade Policy (2006-09): At time of imports of a consignment, declaration stating that it contains genetically modified material is needed or importer is liable to penal action under the Foreign Trade (Development and Regulation) Act of 1992.

13.16.4 Mandate of Ministries

Pursuant to these acts, rules, policies and guidelines, the mandate of various ministries/ departments is summarized in table 20.

Table 20: Mandate of Ministries/Departments

Ministry / Department	Mandate
Ministry of Environment, Forest and Climate Change	• Primarily responsible for conservation and protection of environment, ensuring environmental and human health safety before release of GMOs / LMOs • Nodal agency for implementing Rules, 1989 and the Cartagena Protocol on Biosafety
Department of Biotechnology (Ministry of Science & Technology)	• Nodal department for promoting biotechnology programs • Provides scientific support in implementation of biosafety regulations
Ministry of Agriculture and farmers welfare	• Policies aimed at agriculture growth • Indian Council of Agricultural Research (ICAR) responsible for monitoring agronomic benefits of GM technology • Monitoring post-release performance of GM crops
Ministry of Health and Family Welfare	• Policies aimed at protecting and monitoring human health. • Food Safety and Standards Authority of India responsible for regulating GE foods
Ministry of Commerce and Industries	• Enhance trade with other countries through export/import policies • Nodal agency for implementing DGFT notification on GMOs
Central Board of Excise and Customs, Department of Revenue, Ministry of Finance	• Enforcement of regulation pertaining to transboundary movement of GMOs/LMOs at point of entry

13.17 Cartagena Protocol on Biosafety

In addition to national regulations, international agreements also impact the activities involving LMOs. A supplementary agreement under the ageis of Convention on Biological Diversity (CBD) was adopted on 29 January, 2000, the Cartagena Protocol on Biosafety (CPB). The CPB entered into force on 11 September, 2003 and has 171 countries Party to it as on March, 2019 (Anonymous, 2019). India ratified the CPB on January 23, 2003. Ministry of Environment, Forest and Climate Change (MoEF&CC), Govt. of India is the nodal ministry. The CPB applies to transboundary movement, transit, handling and use of all LMOs that may have adverse effects on the conservation and sustainable use of biological diversity, taking also into account risks to human health.

LMOs covered under the CPB are categorized as under

- LMOs for intentional introduction into the environment (seedlings, trees, animals for breeding, live fish, bacteria or other microorganisms)
- LMOs intended for direct use as food or feed, or for processing (e.g. agricultural commoditiescorn, canola, cotton)
- LMOs for contained use (e.g. bacteria for laboratory scientific experiment)

Exemptions under the Protocol include LMOs that are pharmaceutical for humans if they are covered by other international agreements or arrangements and products derived from LMOs such as processed food (e.g. soybean oil, corn flour). The CPB promotes biosafety by establishing practical rules and procedures for the safe transfer, handling and use of LMOs, with specific focus on regulating the transboundary movement of LMOs. There are 40 Articles in the CPB, which could be categorized into key elements and supporting tools and mechanisms. The four key elements include procedures for transboundary movement of LMOs, risk assessment and risk management, handling, transport, packaging and identification and information sharing.

Specific procedures have been defined for LMOs for intentional introduction into the environment, that are subjected to advanced informed agreement procedures vs. LMOs for direct use as FFP that may be subjected to a separate procedure. Biosafety protocol requires decisions on import of LMOs for intentional introduction into the environment in accordance with scientifically sound risk assessments. These assessments aim at identifying and evaluating the potential adverse effects of LMOs. The Protocol sets out principles and methodologies on how to conduct a risk assessment. The Protocol also requires Parties to adopt measures and strategies for preventing adverse effects and for managing and controlling risks identified by risk assessments. The methodology of the risk assessment follows the conventional risk assessment paradigm, beginning with identification of a potential hazard, such as characteristics of an LMO, which may have an adverse effect on biodiversity. Risks are then characterized based on combined evaluation of the likelihood of adverse effects, and the consequences should those effects be realized (Anonymous, 2019).

Article 15, 16 and Annex III of the CPB outlines requirements that relate to risk assessment and risk management of LMOs as indicated below:

Article 15 on Risk Assessment establishes the basic requirements for risk assessment under the CPB and refers to Annex Ill for further guidance.

Annex Ill sets forth the objectives of the risk assessment, what the risk assessment will be used for, general principles that the risk assessment must follow, the methodology of the risk assessment and particular points to consider when assessing the potential risks of LMO. The general principles include:

- Risk assessment should be carried out in a scientifically sound and transparent manner
- Lack of scientific knowledge of scientific consensus should not necessarily be interpreted as indicating a particular level of risk, an absence of risk, or an acceptable risk
- Risk associated with LMOs or products thereof, should be considered in the context of risks posed by the non-modified recipient or parental organisms in the likely potential receiving environment.
- Risk assessment should be carried out on a case by case basis

Article 16 on Risk Management deals with the management of risks of those organisms that fall within the scope of the CPB. The CPB requires each Party to manage and control any risks that may be identified by a risk assessment. Parties are required to do the following:

- adopt measures and strategies for preventing adverse effects and for managing and controlling the risks identified by risk assessments
- take measures to prevent unintentional transboundary movements
- ensure that LMOs undergo appropriate periods of observation prior to use
- cooperate in identifying LMOs or traits that may pose risks to biodiversity and take appropriate management measures.

LMO Unique Identifiers : Documentation requirements have been defined for various categories of LMOs viz., LMOs for contained use, LMOs-FFP and LMOs for intentional introduction into the environment. However, all categories require reference to a unique identifier code(Anonymous, 2019).

- Documentation requirements for all categories of LMOs require reference to a unique identifier code. To date, only one unique identification system exists: OECD Unique Identifiers for Transgenic Plants
- OECD Unique Identifier is a simple alpha numeric code that is given to each living modified plant that is approved for commercial use
- Developers of transgenic plants are the ones to assign the unique identifier
- 9- digit code composed of 3 elements separated by dashes
    - 2 or 3 alpha numeric digits to designate the applicant;
    - 5 or 6 alpha numeric digits to design at the transformation event; and
    - 1 numerical digit for verification Example: MON-00810-6M on santo's YieldGardMaize
- Unique identifier codes can be used to search BCH for information about specific LMOs.

LMO Quick-links are a tool developed to assist in the identification of LMOs in documentation accompanying their transboundary movement. LMO Quick-links are small image files, which can be easily copied and pasted in documentation accompanying LMOs for the purpose of providing information on a specific living modified organism. LMO Quick-links identify a LMO through the organism's unique identifier (for plants), trade name and a link to the BCH where more information on the LMO is available.

Biosafety Clearing House (BCH): It is a website set up under the CPB to facilitate exchange of information on GMOs/LMOs by Parties. It is a repository of up-to-date information on LMOs and biosafety including information about the National laws, regulations, guidelines, competent national authorities and final decisions taken by countries that is Party. Information available on BCH is organized into National Records that are submitted by Parties and Reference Records that are submitted by general BCH users. To facilitate easier understanding about results of queries involving decisions on LMOs, different icons have been used in the BCH.

13.18 Biosafety Regulatory Mechanisms

Under Biosafety Research programme main emphasis is given to facilitate the implementation of biosafety procedures, rules and guidelines under Environment (Protection) Act 1986 and Rules 1989 to ensure safety from the use of Genetically Modified Organisms (GMOs) and products thereof in research and application to the users as well as to the environment. A three tier mechanism comprising Institutional Biosafety Committees (IBSC) at the Institute/ company; the Review Committee on Genetic Manipulation (RCGM) in the Department of Biotechnology; and the Genetic Engineering Approval Committee (GEAC) in the Ministry of Environment & Forests (MoE&F) for granting approval for research and development activities on recombinant DNA products, environmental release of genetically engineered (GE)crops and monitoring and evaluation of research activities involving recombinant DNA technology has been established. Applications in pharma / agriculture sectors for import/ export/ transfer / exchange of GE materials including GE seeds, conduct of pre-clinical toxicity studies, evaluation of pre-clinical study reports and recommendations to DCGI for appropriate phase of clinical trials of new drug(s) or similar biologics, confined field trials on GE crops etc., were examined by the RCGM and appropriate decisions were taken. RCGM has taken several policy decisions on biosafety research on agricultural/ bio-pharmaceuticals / industrial products (Anonymous, 2019).

13.18.1 DNA Profiling Bill

"The DNA Technology (Use and Application) Regulation Bill - 2019"

Recognizing the need for regulation of the use and application of Deoxyribonucleic Acid (DNA) technology in the country, "The DNA Technology (Use and

Application) Regulation Bill - 2019", has been formulated for establishing identity of missing persons, victims, offenders, under trials and unknown deceased persons. The Bill has the provision to establish a DNA Regulatory Board (DRB) to carry out the functions and exercise the power assigned to it under the proposed Bill. It also has the provisions to establish National and Regional DNA Data Banks for the maintenance of the national forensic DNA database for the purposes of identification of the aforementioned category of persons. This will aid in scientific up-gradation and streamlining of the DNA testing activities in the country. The Bill is under consideration by the Government-National Guidelines for Stem Cell Research – 2017.

The Department of Biotechnology formulates guidelines to help conduct research in the different areas of biosciences and promote its use in the industry and utility among the people.

Mandatory registration of the Institutional Committee for Stem Cell Research (IC-SCR)

As per the National Guidelines for Stem Cell Research-2013 all institutions carrying out research on human stem cells must constitute an Institutional Committee for Stem Cell Research (IC-SCR) and register with the National Apex Committee for Stem Cell Research and Therapy (NAC-SCRT). The registration of IC-SCR is mandatory and all the institutions working in the field are required to comply with the guidelines.

DBT statement on Research Misconduct: The Department of Biotechnology formulates guidelines to help conduct research in the different areas of biosciences and promote its use in the industry and utility among the people.

Mandatory registration of the Institutional Committee for Stem Cell Research (IC-SCR): As per the National Guidelines for Stem Cell Research-2013 all institutions carrying out research on human stem cells must constitute an Institutional Committee for Stem Cell Research (IC-SCR) and register with the National Apex Committee for Stem Cell Research and Therapy (NAC-SCRT). The registration of IC-SCR is mandatory and all the institutions working in the field are required to comply with the guidelines.

13.18.2 Guidelines for evaluation of Nanopharmaceuticals in India

The 'Guidelines for Evaluation of Nanopharmaceuticals in India' are compiled with an aim of evaluation of the pharmaceutical preparations containing nanomaterials where the novel application of nanotechnology imparts the significant advantages over the existing active pharmaceutical ingredients (API) in terms of targeted delivery to disease site, higher efficacy and lower toxicity. The nanometer dimension of a material provides unique properties with a marked difference from its bulk counterpart. These unique properties of the nanoformulation can be utilized for targeted drug delivery and other applications. The properties of any nanomaterial depend on its physical and chemical characteristics. Variation of size and surface charge of the same material can show different functionalities. Thus, these nanopharmaceuticals need additional tests for ensuring uniform safety and efficacy(Anonymous, 2019). The general principle of regulatory guideline for API as mentioned in the New Drugs and Clinical Trial Rules 2019 issued by CDSCO has been followed. These guidelines are also aligned with ICH guidelines and international norms followed by other countries commercializing such products. The goal of these guidelines is to help the regulators to assess the quality, safety and efficacy of the nanoformulation/nanopharmaceuticals in a systematic manner. It will also guide the innovators and industries to generate the data according to the nanomaterial specific requirements. The current guidelines are applicable to nanopharmaceuticals and not for cosmetics and nanoenabled devices or implants.

The nanotechnology intervention has opened a new horizon for targeted delivery of approved drugs and repurposing of drugs. Every year several new nanopharmaceuticals/ nanomedicine are being introduced into the market globally. The rapid progress in this emerging field is expected to change the current therapeutic practice in near future. These guidelines will encourage the Indian innovators and industries to develop and commercialize new nanopharmaceuticals which will make our country a global leader in this area. With rapid advances in basic sciences, our understanding about the safety parameters of nanomaterials is continuously updated. The novel multifunctional nanomaterials may need additional new tests for safety assessment in future. So these first guidelines may need modification with new edition from time to time.

Scope of the guidelines: The guidelines apply to the nanopharmaceuticals in the form of finished formulation as well as API of a new molecule or an already approved molecule with altered nano- scale dimensions, properties or phenomenon associated with the application of nanotechnology intended to be used for treatment, in vivo diagnosis, mitigation, cure or prevention of diseases and disorders in humans. These guidelines do not apply to the conventional drug with incidental presence of nanoparticles or drug products containing microorganisms or proteins, which are naturally present in the nanoscale range. Theseguidelines are also not applicable to medical devices, in vitro diagnostics, tissue engineered products using nanotechnology and nanoparticle modified cell based therapies. These guidelines may serve as a useful document for manufacturers, importers of nanopharmaceuticals and other stakeholders involved in research and development of nanopharmaceuticals.

General considerations of the guidelines : Safety studies should be conducted as per general guidelines specified in Second Schedule of New Drugs and Clinical Trials Rules, 20193 . However, in case any specific study is not included in the general requirements of the said Schedule, the studies may be planned, designed and conducted as per the principles of USFDA4 , ICH guidelines for pharmaceuticals5 or OECD guidelines for chemicals6 . These guidelines are in conformity with the provisions of Drugs and Cosmetics Act, 1940 and New Drugs and Clinical Trials Rules, 2019, with certain specific aspects of quality, safety and efficacy applicable to nanopharmaceuticals (Anonymous, 2019).

These guidelines have evolved with considerations of the following documents.

- Schedule Y of D & C Rules, 1945 prevalent before 19 March of 2019 as well as the New Drugs and Clinical Trials Rules, 2019.
- Drug Products, Including Biological Products that Contain Nanomaterials Guidance for Industry .
- Second Regulatory Review on Nanopharmaceuticals, European Union, 2012, Nanomaterials: Commission proposes 'case by case approach' to assessment .
- Identification of regulatory needs for nanomedicines: 1st EU-NCL survey with the "Nanomedicine" working group of the international pharmaceutical regulators.
- Regulatory Aspects of the Nanopharmaceutical in the EU, 20179.

In these guidelines, the nanopharmaceuticals (nanomedicines) have been classified according to their degradability, organicity, function and status of approval. Accordingly, the safety and efficacy data requirements have been described for different categories of nanopharmaceuticals. Specific scientific evidence required for approval of any nanopharmaceutical and the strategies for pharmacovigilance of such products have been incorporated in these guidelines. Each application should be considered on its own merit based on data submitted, using scientific judgment and logical argument. For new generation of nanomaterials, development of methods for risk assessment, safety testing and availability of quality data on nanomaterials for regulatory assessment are essential.

Information required for evaluation of Nanopharmaceuticals : As already mentioned, the information required for nanopharmaceuticals should be decided on 'case-bycase approach' basis. However, in general, the following data should be submitted to the regulatory authority along with the application to conduct clinical trials and manufacture of nanopharmaceuticals for marketing in India.

1. Introduction

- A brief description of the nanopharmaceutical
- Indication for which it is intended to be used
- Category to which it belongs
- Justification for developing nanopharmaceutical

2. Chemical and Pharmaceutical Information

i. Information on the ingredients

- Drug information (Chemical Name, Generic Name, International non-proprietary name)

- Information on nanomaterial used, excipient/ inactive ingredients
- V Brief description and rationality of the nanopharmaceutical

ii. Physiochemical characterization data of nanopharmaceuticals

- Complete description of Individual component(s) [e.g. API, Nanocarrier material (single/multiphase) surface functional groups, contrast agents (theranostic), excipients, targeting ligands, etc]
- Chemical name, structure, crystal structure of drug and nanomaterial(s) » Empirical formula (drug and nanomaterials)
- Molecular weight (drug and nanomaterials)
- Description of the product with
  - Nano-size range by number and/ or intensity distribution, average size and polydispersion index, percentage of particle under each distribution with standard deviation
  - Shape, surface texture information
  - Surface charge with standard deviation (zeta potential)
  - Percentage of drug loading with standard deviation
  - Encapsulation efficiency, loaded versus free drug content, with standard deviation
  - Osmolality (wherever applicable)
  - Solubility / dispersion information (for injectable product)
  - Colloidal stability information for injections (06 batches)
  - State of drug(API) in nanomaterial (chemically conjugated/loaded/ complexed with the nanocarrier)
  - Scalable GMP process description of the nanopharmaceutical preparation
  - Average pH (wherever applicable)
  - Viscosity (wherever applicable)
  - Mechanical integrity/Properties (as applicable)
  - Endotoxin/microbial load level for parental nanoformulations
  - Residual solvent content as per ICH guidelines
  - Sterilization protocols/methods/stability post sterilization (as applicable)
  - Waste disposal method

Note: From the full list of the product's physico-chemical parameters, some of them need to be identified as critical quality attributes. They should be listed along with the product specifications to ensure quality and reproducibility from batchto-batch. In addition, a detailed description of the manufacturing process, process controls and waste disposal methods should be provided.

iii. Analytical data (nanocarrier/ API/ nanopharmaceutical)

Size (DLS, TEM, AFM, SEM, etc.)

Quantitative Elemental analysis (AAS, ICP, XRF, etc.)

Mass spectrum analysis

NMR spectra (state of the API as encapsulated or chemically bonded, or intercalated may be identified and specified)

FT-IR spectra » Fluorescence spectra Ramanspectra

UV-VIS spectra

XRD, XPS

Polymorphic changes identification (during the shelf-life of the product)

Note: Data to be generated using appropriate scientifically validated analytical methods not limited to the ones mentioned above for a particular nanopharmaceutical depending upon the claim of novelty

iv. Complete monograph specification for the nanopharmaceutical

Defined criteria for unique identification of nanopharmaceutical

Identity and quantification of impurities

In vitro/ in vivo release kinetics of the drug/active ingredient (as applicable)

In vitro/ in vivo degradation kinetics of nanopharmaceutical in various simulated media

Stability data

v. Analytical method validations for nanopharmaceutical

Assay method » Impurity estimation method

Residual solvent/other volatile impurities (OVI) estimation method

Dissolution test method

vi. Stability studies of nanopharmaceuticals

vii. Data on nanopharmaceutical formulation

Dosage form

Route of administration

Composition

Details about loading process, chemical bonding/ conjugation between active ingredient and carrier, surface coating/ modification and functionalization

In process quality control check

Finished product specification » Excipient compatibility study

Validation of the analytical method

viii. Comparative evaluation of innovator product or approved Indian product, if applicable

Container and closure system

Content uniformity

Impurities

pH

ix. Stability evaluation in market intended pack at proposed storage conditions

x. Packing specifications

xi. Process validation

Pharmacovigilance of Nanopharmaceuticals: Pharmacovigilance must be carried out throughout the lifecycle of the nanopharmaceutical. A detailedpharmacovigilance plan alongwith marketing authorization application must be submitted by the product developer.

Pharmacovigilance plan must mention:

Safety data from clinical development

All the potential risks of the nanopharmaceutical

Summary of anticipated risks

Population at risk

Situations not adequately studied

All the potential drug-drug and drug-food interactions of the nanopharmaceutical either as a separate document with pharmacovigilance plan or pharmacovigilance strategies or in the section referring to safety specifications of the document.

For nanopharmaceuticals of antimicrobials, monitoring of patterns of resistance will be an important component of pharmacovigilance. Hence, strategies for monitoring and prevention of the resistance should be mentioned in a separate section of the document. For nanopharmaceuticals of antimicrobial agents, a signal will be generated if there is an alarming rise in the incidence of resistance to the particular claim proposed. If any significant safety concerns arise during clinical trials which warrant studies in special population such as children, elderly, pregnant women or in hepatic or renal failure patients, the protocol of such studies should be submitted along with the pharmacovigilance plan (Anonymous, 2019). Protocols for comparative observational studies (cross sectional/case control/ cohort), drug utilization study or any targeted clinical evaluation to be conducted as a part of pharmacovigilance plan should be the part of the document.

Conclusion: General requirements and guidelines specified for approval of manufacture/import of any new drug or to undertake clinical trial as specified in the New Drugs and Clinical Trials Rules, 2019 especially in Second Schedule of the said Rules and other applicable regulations apply to nanopharmaceuticals also.

However, the requirement of special or additional tests for safety and efficacy evaluation of a particular nanopharmaceutical should be decided on a 'case-bycase approach' basis which will depend upon various factors such as physiochemical and biological nature, and other aspects including the back ground data available on the API or nanocarrier, the regulatory status in other countries, etc., as enumerated in this document. Successful translation of nanopharmaceuticals from nonclinical proof of concept to clinics is challenging. Like development of any new drug, it requires effective integration of nanotechnology with chemistry, life sciences and medicine. However, because of complexity in nanotechnology, the system necessitates a 'case-bycase approach' with involvement of varied expertise for successful development of nanopharmaceuticals.

13.18.3 Understanding Biosafety Regulation

There is complex chain of regulatory Framework in India for purpose of biosafety. DBT, Department of Science and Technology and Ministry of Environment and Forest (MOef) are the two main regulatory body. DBT has been setup under Ministry of science and technology while GEAC has been setup under MOeF. There are various committee setup under the said ministry. Their main work is to regulate the various activities from research activity to storage or trials of GMOs in India. These committees have six statutory authority also. DBT has right to appoints the members of the committees (Sheelendra, 2015). The GEAC implements any decision based on report of the State Biotechnology Coordination Committees (SBCC) and District Level Committees (DLC) after any events of post release. Other committees are (1) The Institutional Biosafety Committees (IBSC), (2) Review Committee on Genetic Manipulations (RCGM) and both involve in implementation of guidelines. Recombinant DNA Advisory Committee (RDAC) issue guidelines for working over R-DNA technology. The main objective of National biosafety framework is to (1) Regulate all the process related GMOs in all state to Ensures safety of GMO

Rules for the Manufacture, Use, Import, Export and Storage of Hazardous Microorganisms/ Genetically Engineered Organisms or Cells, 1989 (Rules, 1989)" notified under the Environment Protection Act, (EPA) 1986 covers broad array of rules related to biosafety of GMO. RDAC is advisory while RCGM, IBSC, and GEAC has role of approval of R-DNA products while under state government SBCC and DLC works as monitoring role. RDAC was setup in 1990, and mainly concern with biotech policy framework regarding r-DNA work. Review committee on genetic manipulation (RCGM) has members from various scientific community which has to conduct meeting time to time in every months. Genetic engineering appraisal (approval) committee GEAC, is the Apex committee to permit/authorize the use of GMO and products thereof for large scale field trials and commercial applications. State biotechnology coordination committee (SBCC) is related with monitoring the activity of GMO and in case of violation has role in punitive action in every state. District level committees (DLC), has role in post monitoring of biosafety related issue. Commercial release is also monitored on the basis of invasiveness, gene persistence, gene flow in unrelated organism, adaptation for resistance property. The check is performed for gene contamination via pollen grains or via other mode in ecosystem and for imported items phytosanitary certification is essential along with quarantine procedures.

13.19 Regulatory Mechanism in India

Guidelines for safety have been issued by the Department of Biotechnology (DBT) in 1990 covering research in biotechnology, field trials and commercial applications. DBT had also brought out separate guideline for research in transgenic plats in 1998 and for clinical products in 1999 (Kannaiyan, 2006). Activities involving GMOs are also covered under other policies such as the Drugs and Cosmetics Act (8th Amendment), the Drug Policy, 2002, and the National Seed Policy, 2002. Presently, there are six competent authorities for implementation of regulations and guidelines in the country:

1. Recombinant DNA Advisory Committee (RDAC)
2. Review Committee on Genetic Manipulation (RCGM)
3. Genetic Engineering Approval Committee (GEAC),
4. Institutional Biosafety Committee (IBSC) attached to every organization engaged in rDNA research
5. State Biosafety Coordination Committees (SBCC) and
6. District Level Committee (DLC)

Of the above committees, the IBSC is constituted by organizations involved in research with GMOs with the approval of DBT. The IBSC is the nodal point for interaction within the institution for implementation of the guidelines. (Every research project using GMOs has an identified investigator who is required to get the research project approved from safety angle and inform the IBSC about the status and results of the experiments being conducted). The functions of IBSC include to:

- Reviewing and giving clearance to project proposals falling under restricted category as per DBT guidelines
- Recommending Category III risk or above experiments to RCGM approval
- Tailoring biosafety programme to the level of risk assessment
- Training of personnel on biosafety
- Adopting emergency plans

The role of IBSC assumes major importance since it is the only Statutory committee, which operate from the premises of institution and hence is in a position to conduct onsite evaluation, assessment and monitoring of adherence to the biosafety guidelines. The decisions taken by the next higher committee i.e

Review Committee on Genetic Manipulation (RCGM), which operates from DBT are based on the applications submitted by the investigators with the approval of IBSC on the status of the project and its conformity with the regulatory guidelines.

The IBSC comprises head of the institution / nominee, three or more scientists engaged in DNA work or molecular biology, an outside expert in the relevant discipline, a member with medical qualifications and nominee of DBT. The DBT nominee to oversee the adherence to biosafety guideline and server as the link between the Department of Biotechnology and the respective IBSC. There are approximately 225 IBSCs in various institutions and laboratories in industries in India overseeing a range of research projects including development of genetically modified microorganisms, transgenic plants, transgenic animal and recombinant products. DBT has substantially funded research and development in biotechnology in the country, s also agencies like CSIR, ICMR and ICSR, Several private companies have also initiated research either independently or in collaboration with Indian and foreign institutions/companies. As a result of these efforts, a few products are in advanced stages of development and are likely to e commercialized shortly. 18 recombinant therapeutic products are already in the market, out of which five therapeutic are being manufactured indigenously. In the area of agriculture, four varieties of Bt cotton have been approved for commercial cultivation in the country and several crops are in the approval pipeline.

13.19.1 Biosafety and Bioregulatory Mechanisms for Transgenic Crops

The biotechnological tools promise the delivery of genes or genomic fragments across the species, leading to the development of transgenic/ genetically modified organisms with unique and economically important traits. Since the achievements from technological advances run parallel with the intrinsic risks and with exponential growth of the technological power, as evidenced from the transgenesis strategies such as CRISPR/Cas9, the responsibilities of the scientific community also raise. The initial part of the chapter by Deepu Mathew (2018) details the various risks associated with genetic modification, its assessment and management. Subsequently, the biosafety and bioregulatory mechanisms with respect to genetic modified (GM) crops in India, including the Acts and Rules, agencies involved, roles played by the statutory bodies, laboratory, greenhouse and field trials for GM crops and their Standard Operating Procedures and conditions for large scale seed production and environmental release of the crops, were discussed, which also examines the bioregulatory mechanisms and biosafety protocols followed by the major global players in this area, the United States of America, European Union and Canada.

13.19.2 Risk Assessment

Assessing Environmental Risks- Gene Escape: Gene escape could happen through vertical gene transfer or horizontal gene transfer. Vertical gene transfer refers to the transfer of genes from parents to offspring, through the pollen fertilizing the female plant belonging to the same or related genera. Horizontal gene transfer represents nonsexual gene transfer, more frequently between the microorganisms or rarely between plants and microbes. Since microbes have the capability to integrate these genes from plants into their genome, though the event may be extremely rare, this has to be viewed seriously in the context of the antibiotic resistance genes used in the transformation protocol. As of now there are many strategies that avoid the use of these marker genes through recombinase enzymes and specialized vectors. Furthermore, most breeders adopt large-scale crossing of the transgenic with the parent non-GM plants to select the recombinant plants with transgenes devoid of antibiotic resistance genes. However, the presence of the marker has to be confirmed through polymerase chain reaction (PCR)-based techniques using the flanking primers for these genes. Gene escape by vertical transfer could be easily assessed by random sampling of seeds from the vicinity of the lineage-specific transcripts (LSTs) and subjecting the seeds to PCR analysis using the gene-specific primers.

Transgenic crops-Biosafety concerns and regulations in India

The first-generation GM crops have improved traits like Herbicide-resistant crops (soybeans and maize, Pest resistance (Cotton and corn). Second-generation GM crops involve enhanced quality traits, such as higher nutrient content. "Golden Rice," one of the very first GM crops, is biofortified to address vitamin A deficiency. Other biofortification projects include corn, sorghum, cassava, and banana plants, with enhanced minerals and vitamins. Crops can also be modified to ward off plant viruses or fungi. Even though the seed is more expensive, these GM crops lower the costs of production by reducing inputs of machinery, fuel, and chemical pesticides. Important environmental benefits, such as controlling farm runoff that otherwise pollutes water systems, are associated with reduced spraying of chemical insecticides and highly toxic herbicides.

Now-a-days with the rapid advance research and development in agricultural biotechnology, countries are approving many genetically modified crops for commercial release and agricultural production. ISAA reported in 2017, the 21st year of commercialization of biotech crops, 189.8 million hectares of biotech crops were planted by up to 17 million farmers in 24 countries. From the initial planting of 1.7 million hectares in 1996 when the first biotech crop was commercialized, the 189.8 million hectares planted in 2017 indicates ~112-fold increase.

In India, *Bt* cotton was approved by Government of India in March 2002 as the first transgenic crop for commercial cultivation for a period of three years. Apart from cotton, there are more than 20 crops under research and development in about 50 public and private sector organizations in India. Out of these, 13 crops have been approved for contained limited field trials in India. Though, it is widely claimed that transgenic crops offers dramatic promise for meeting some of greatest challenges but like all new technologies, it also poses certain risks, because of the fact that transgenic crops can bring together new gene combinations which are not found in nature having possible harmful effects on health, environmental and non-target species. However, as more and more transgenic crops are being released for field-testing and commercialization, concerns have been expressed about the potential risks associated with their impact to human health, environment and biological diversity.

13.20 Biosafety Concerns

As more and more transgenic crops are released for field-testing and commercialization, concerns have been expressed regarding potential risks to both human health and environment. Biosafety describes the principles, procedures and policies to be adopted to ensure the environmental and personal safety. Recognizing the need of biosafety in GE research and development activities, an international multilateral agreement on biosafety "the Cartagena Protocol on Biosafety (CPB)" has been adopted by 167 parties, including 165 United Nations countries, Niue, and the European Union. The Protocol entered into force on 11 September 2003, and its main objectives are:

1. To set up the procedures for safe trans-boundary movement of living modified organisms
2. Harmonize principles and methodology for risk assessment and establish a mechanism for information sharing through the Biosafety Clearing House (BCH).

Research work in the area of GE and GMOs requires prior approval from the appropriate regulatory authorities of the country. Following guidelines provided for minimizing biosafety issues is mandatory. The primary regulatory body at research institute level is the Institutional Biosafety Committee (IBSC) or its equivalent body consisting of experts from different relevant disciplines. The IBSC ensures existence of the basic biosafety equipment required as per the safety level of the experiments to be conducted. There has been increasing awareness among the researchers, producers and users of GMOs, administrators, policy makers, environmentalists and general public about biosafety all over the world. Transgenic crops are not toxic nor are likely to proliferate in the environment. However, specific crops may be harmful by virtue of novel combinations of traits they possess. This means that the concerns associated with use of GMOs can differ greatly depending on the particular gene organism combination and therefore a case-by-case approach is required for risk assessment and management.

The major biosafety concerns falls into these categories:

Bio-safety of human and animal health

- Risk of toxicity, due to the nature of the product or the changes in the metabolism and the composition of the organisms resulting from gene transfer.

- Newer proteins in transgenic crops from the organisms, which have not been consumed as foods, sometimes has the risk of these proteins becoming allergens.
- Genes used for antibiotic resistance as selectable markers have also raised concerns regarding the transfer of such genes to microorganisms and thereby aggravate the health problems due to antibiotic resistance in the disease causing organisms.

Ecological concerns

- Gene flow due to cross pollination for the traits involving resistance can result in development of tolerant or resistant weeds that are difficult to eradicate.
- GM crops could lead to erosion of biodiversity and pollute gene pools of endangered plant species.
- Genetic erosion has occurred as the farmers have replaces the use of traditional varieties with monocultures.

Environmental concerns

- Effect of transgenic plants on population dynamics of target and non-target pests, secondary pest problems, insect sensitivity, evolution of new insect biotypes, environmental influence on gene expression, development of resistance in insect population, development of resistance to herbicide
- Gene escape into the environment- accidental cross breeding GMP plants and traditional varieties through pollen transfer can contaminate the traditional local varities with GMO genes resulting in the loss of traditional varieties of the farmers.

Public attitude: Consumer response depends on perceptions about risks and benefits of genetically modified foods. The media, individuals, scientists and administrator, politicians and NGO have the responsibility to educate the people about the benefits of GM foods.

Socio economic and ethical consideration

- Potential benefits to the consumers and farmers. Due to increasing seed market, the developing countries may get dependent on few suppliers.
- **GURT:** Gene Use Restriction Technologies. India has totally banned the use of GURT in plant variety for registration under PPV & FRA, 2001.

Bibliography

Anonymous, 2011. AgBioWorld, P.O. Box 85, Tuskegee Insitute, AL, 36087-0085, USA

Anonymous, 2016. Indian Council of Agricultural Research Biosafety Portal. Assistant Director General (Plant Protection & Biosafety), Indian Council of Agricultural Research, New Delhi, India

Anonymous, 2016. National Biosafety Systems -Case studies to analyze current biosafety approaches and regulations for Brazil, China, India, Israel, Pakistan, Kenya, Russia, Singapore, the United Kingdom, and the United States. UPMC Center for Health Security

Anonymous, 2019. Chesapeake Area Biological Safety Association, American Biological Safety Association (ABSA)

Anonymous, 2019. Guidelines for Evaluation of Nanopharmaceuticals in India Government of India New Delhi October 2019. Published by: Department of Biotechnology Government of India, New Delhi- 110003

Anonymous, 2019. Laboratory Safety Guidance" (PDF). U.S. Occupational Safety and Health Administration. 2011. pp. 9, 15, 21, 24–28. Retrieved 2019-01-17.

Anonymous, 2019. Lmos, Biodiversity and Biosafety National and International Regulations. Prepared by Ministry of Environment, Forest and Climate Change (MoEFCC) and Biotech Consortium India Limited, New Delhi under the UNEP/GEF supported Phase II Capacity Building Project on Biosafety, New Delhi

Anonymous, 2020. HANDBOOK for Institutional Biosafety Committees (IBSCs), Department of Biotechnology Ministry of Science & Technology, Government of India

Anonymous, 2021, Belgian Biosafety server

Anonymous, 2021. Microbe Notes, Kathmandu, Nepal

Chaturvedi, S., 1997, "Biosafety Policy and Implications in India." Biotechnology and Development Monitor, No. 30, p. 1013.

David A. Jackson and Robert, H., Symonst and Paul Berg, 1972. Biochemical Method for Inserting New Genetic Information into DNA of Simian Virus 40: Circular SV40 DNA Molecules Containing Lambda Phage Genes and the Galactose Operon of Escherichia coli (molecular hybrids/DNA joining/viral transformation/genetic transfer). *Proc. Nat. Acad. Sci.*, USA, 69: 2904-2909

Deepu Mathew, 2018. Biosafety and Bioregulatory Mechanisms for Transgenic Crops. In book: Genetic Engineering of Horticultural Crops (pp.273-314), Edition: 1, Chapter: 13, Publisher: Academic Press, Elsevier Inc., Cambridge, Editors: Rout G., Peter K.V.10.1016/B978-0-12-810439-2.00013-1.

Emmett Barkley and Robert J. Hawley, 2012. History of Biosafety and U.S. Biodefense Programs Bioterrorism, Biological Warfare, and Defenses. Level 3 Biosafety Training Course Midwest Regional Center of Excellence Midwest Regional Center of Excellence for Biodefense and Emerging Infectious Disease Research Sept. 24, 2012 Washington University in St. Louis St Louis MO

Kannaiyan, S., 2006. Biotechnology And Biosafety.- National Biodiversity Authority, Government of India

Momtaz A. Shahein, Amany N. Dapgh, Essam M. Ibrahem Essam E. Kamel, Heba H. Hassan and Dalia M.A. Elmasry, 2021. Biosafety in laboratories: History, Application and Expected community outcomes. *Egyptian J. Ani. Health, Article* **9,** 1: 96-102

Peter W.E. Kearns, Gijs A. Kleter, Hans E.N. Bergmans and Harry A. Kuiper, 2021. Biotechnology and Biosafety Policy at OECD: FutureTr. in Biotechnol., 39: 965-969,

Sheelendra, B.M., 2015. GMO's Foods: Regulatory Mechanism and Challenges in India. *J. Bioremed. Biodeg.*, 6:e165. doi:10.4172/2155-6199.1000e165

Usha Kiran Betha, 2019. ICAR-Indian Institute of Oilseeds Research, Rajendranagar, Hyderabad, India

Wagih Mohamed and Al-Kiey, T.M., 2012. Biosafety and Conservation of Plant Genetic Resource Regulatory Policies: Issues and Considerations for Developing Countries.

Wagih, M. E., Wagih, E. E. and Al-Kiey, T.M., 2012. Biosafety and Conservation of Plant Genetic Resource Regulatory Policies: Issues and Considerations for Developing Countries. First International Conference on Applications of Biotechnology in Agriculture, Benha University, Egypt, Feb 18-22, 2012.

14

Cartagena Protocol on Biosafety and Its Implications

The Cartagena Protocol on Biosafety entered into force on 11 September 2003. It is a legally binding international agreement under the United Nations' Convention on Biological Diversity (CBD).

What is a protocol ?: A protocol is a binding international instrument, separate from, but related to, another treaty. It is a separate instrument, that must be individually negotiated, signed and eventually ratified. It is only binding on States that become parties to it. It thus has its own parties, and creates separate rights and obligations for them, as any other treaty.

The unique characteristic of a protocol is that it is related to a 'parent' treaty, through substantive, procedural, and institutional links. Most importantly, a protocol under a specific treaty must comply with the parent treaty's provisions authorizing and regulating the adoption of protocols under its auspices. Any protocol adopted as a result of these 'enabling' provisions in the parent treaty must comply with them. In particular, it may not deal with subjects which are beyond the purview of these provisions, or if these provisions are not restrictive in this regard, with subjects which are beyond the purview of the parent instrument (Mackenzie Ruth *et.al.*, 2013). Such enabling provisions usually restrict (as is the case for the Cartagena Protocol) participation in a protocol to parties to the parent treaty. In addition, the parent treaty usually defines basic institutional and procedural links between the two instruments, for example it may indicate that provisions in the treaty itself (e.g. related to dispute settlement) will also apply to any protocol adopted under it.

14.1 Cartagena Protocol on Biosafety (CPB)

The protocol itself may, however, add further links to the parent treaty, for example by designating mechanisms existing under the treaty (e.g. the Conference of the Parties) also to serve the protocol. This is the case for the Cartagena Protocol. The CPB is a legally binding (to those countries that have ratified it), international agreement, supplemental to the Convention on Biological Diversity (CBD), with the objective to protect the world's biological diversity from the potential risks associated with the transfer, handling and use of living modified organisms (LMOs), that result from modern biotechnology. It therefore focusses on potential environmental impacts and does not address food safety issues directly - although potential risks to human health may be taken into account. Currently, 170 countries are party to the Protocol.

The Cartagena Protocol on Biosafety is the first international law to specifically regulate genetic engineering, and this largely reflects the global climate of concern about the safety, health and environmental risks of genetically modified organisms (GMOs), along with the wider political and socio-economic implications of this technology. For the first time in international law, there is an implicit recognition that GMOs are inherently different from naturally occurring organisms, and carry special risks and hazards, hence the need to have a legally binding international instrument. The Protocol recognizes that GMOs may have biodiversity, human health and socio-economic impacts, and that these impacts should be risk assessed or taken into account when making decisions on GMOs. Precaution is the basis for the Protocol itself, and is operationalized in decision-making and risk assessment.

The entry into force of the Protocol was an important defining moment in global biosafety regulation. It followed years of negotiations, from when the need for a biosafety protocol to address the risks of genetic engineering was first articulated in Article 19(3) of the CBD in 1992, to its adoption by more than 130 countries in the year 2000 in Montreal. The Protocol's entry into force means that it is legally binding in the international legal system and in the legal systems of countries that have ratified, approved, accepted, or acceded to it (depending on a country's legal system) (Traavik and Lim, 2007). As of March 2007, there are 140 Parties to the Protocol. The Protocol enters into force in a country 90 days after it deposits its instrument of ratification, approval, acceptance, or accession with the United Nations Secretary General.

The Biosafety Protocol seeks to protect biological diversity from the potential risks posed by genetically modified organisms resulting from modern biotechnology. The Biosafety Protocol makes clear that products from new technologies must be based on the precautionary principle and allow developing nations to balance public health against economic benefits. It will for example let countries ban imports of genetically modified organisms if they feel there is not enough scientific evidence that the product is safe and requires exporters to label shipments containing genetically altered commodities such as corn or cotton. The required number of 50 instruments of ratification/accession/approval/acceptance by countries was reached in May 2003. In accordance with the provisions of its Article 37, the Protocol entered into force on 11 September 2003. As of July 2020, the Protocol had 173 parties, which includes 170 United Nations member states, the State of Palestine, Niue, and the European Union (Anonymous, 2020).

Background: The Cartagena Protocol on Biosafety, also known as the Biosafety Protocol, was adopted in January 2000, after a CBD Open-ended Ad Hoc Working Group on Biosafety had met six times between July 1996 and February 1999. The Working Group submitted a draft text of the Protocol, for consideration by Conference of the Parties at its first extraordinary meeting, which was convened for the express purpose of adopting a protocol on biosafety to the CBD. After a few delays, the Cartagena Protocol was eventually adopted on 29 January 2000. The Biosafety Protocol seeks to protect biological diversity from the potential risks posed by living modified organisms resulting from modern biotechnology.

The first Conference of the Parties serving as the Meeting of the Parties to the Protocol (COP-MOP 1) was held in Kuala Lumpur, Malaysia, 23–27 February 2004. COP-MOP 2 was held in Montreal, Canada, 30 May–3 June 2005, and COP-MOP 3 was held in Curitiba, Brazil 13–17 March 2006. The COP-MOP is the Protocol's supreme decisionmaking body, which negotiates and adopts decisions that take forward the development, interpretation and implementation of the Protocol. COP-MOP decisions are binding on the Parties. Subsequent COP-MOPs will be held every two years, back to back with the Conferences of the Parties (COPs) of the CBD. The next COP-MOP will be held in Bonn, Germany in May 2008. Prior to the Protocol's coming into force, the Intergovernmental Committee of the Cartagena Protocol (ICCP) met three times to move forward the work of the Protocol in the interim.

14.1.1 Entry into Force

The Protocol entered into force on 11 September 2003, ninety days after the deposit of the fiftieth instrument of ratification. In accordance with Article 29, paragraph 1, of the Protocol, the COP to the Convention serves as the meeting of the Parties to the Protocol (COP-MOP), the governing body of the Protocol. An overview of the COP-MOP activity is provided in the Meetings of the COP-MOP page.

The road to the Cartagena protocol on biosafety (and beyond):

Phase 1: 1970s and 1980s (problem identification)

Phase 2: late 1980s and beginning of 1990s (framework development)

Phase 3: 1989–1992 (Biodiversity Convention negotiating process)

Phase 4: 1992–1995 (issue definition)

Phase 5: 1996–2000 (negotiation)

Phase 6: 2000–entry into force (interim period)

1996–2000. The negotiation phase 1.

i. Elements definition phase

1996 (July, 5 days) First meeting of the BSWG (Biosafety Working Group)

1997 (May, 5 days) Second meeting of the BSWG

ii. Drafting and negotiation phase

1997 (October, 5 days) Third meeting of the BSWG

1998 (February, 7 days) Fourth meeting of the BSWG

1998 (August, 14 days) Fifth meeting of the BSWG

1999 (14-24 February, 9 days) Sixth meeting of the BSWG and ExCOP (Extraordinary Conference of the Parties)

iii Final negotiation phase

1999 (July) Informal consultations - to consider whether to resume negotiations

1999 (September) Informal consultation meeting - paving the way for resolving differences

2000 (January) Informal consultations followed by the resumed ExCOP (24-29 Jan) to resolve remaining core issues and adopt the Protocol.

14.1.2 Examples of Genetic Modification

i. **GM bacteria:** Possibly the most important area of genetic modification, albeit in containment, is that of single-cell organisms modified to act as chemical factories for the production of food additives (including flavour enhancers) and fine chemicals. In 1997, the U.S. Environmental Protection Agency approved the first genetically engineered bacteria for agricultural use. The bacterium, a strain of Rhizobium meliloti, contained genes from five different species and was genetically altered to enhance its ability to provide nitrogen to alfalfa plants on farmland.

ii. **GM agricultural crops:** One of the most prominent developments of genetic modification technology has been the creation of transgenic agricultural crop varieties. Many millions of hectares of commercial transgenic crops are grown annually, although it is impossible to obtain exact figures as official data are not always available. In 2001 alone there were 35.7 million hectares of GM crops grown in the United States, 3.2 million hectares in Canada, 11.8 million hectares in Argentina and at least 1.5 million hectares in China. From the two traits currently used, herbicide-tolerant crops are grown on 77% of the area, crops producing the Bt-toxin on 15%, "stacked" varieties producing Bt-toxin and showing herbicide-tolerance on 8%. Most of the harvest is used as animal feed. Many other traits have been inserted into agricultural crops but are grown on a small scale or have not yet been commercialized. Papaya has been modified to provide resistance to papaya ring spot virus. Rice yellow mottle virus attacks rice in Africa – modern biotechnology has produced a rice resistant to the virus. Vaccines against diseases of the gastro-intestinal tract have been produced in bananas and potatoes.

iii. **GM Trees:** Biotechnology companies have linked up with key players in the industrial forest sector to support research that will increase tree growth rates, modify wood structure, alter trees' reproductive cycles, improve tolerance to certain herbicides and even store more of the gases that are responsible for global warming. While forest-related biotech research is still in its infancy compared with agriculture, field trials of GM trees have proliferated around the world. Recent research shows that, since 1988, there have been 184 GM tree field trials globally. More trials have been conducted with poplar than any other species due to its popularity as a pulp and paper species. The U.S. has released the largest number of GM trees via field trials, with 74% of the world-wide total.

iv. **GM Animals :** The first GM animal was a mouse, which was developed in early 1988, when the Harvard Oncomouse was patented in the USA. The technology has been applied during the 1990s to some mammals, including cattle, pigs, sheep and mice. It has also been applied to poultry. The creation and use of GM animals continues to increase. In Great Britain in 2000 there were 581,740 procedures in which GM animals were used or bred, 14% more than in 1999. Around 99% of these involved mice.

v. **GM Fish :** Commercial aquaculture has made use of GM technology and there is also specialist interest for aquarium species. The Atlantic and Pacific salmon has received most media attention, particularly those that contain an additional gene for growth hormone production and an anti-freeze gene. These fish have shown three-fold growth rate increases and potential to exploit colder waters. Reports indicate that transgenic salmon have also displayed severe deformities.

vi. **GM Insects:** The fruitfly Drosophila melanogaster was one of the first organisms to be genetically engineered over 20 years ago, and has been regularly used in medical and scientific research. The genetic modification of other insects has begun only recently. For instance, researchers are trying to create mosquitoes engineered not to host the malaria virus.

14.2 Timeline and Scope of the Cartagena Protocol on Biosafety

1993	The Convention on Biological Diversity enters into force on 29 December 1993		
1995	COP2	Second meeting of the Conference of the Parties - Consideration of the need for and modalities of a protocol for the safe transfer, handling and use of living modified organisms. Jakarta, Indonesia, 6 - 17 November 1995	Decision II/5
1996	COP3	Third meeting of the Conference of the Parties - Issues related to biosafety. Buenos Aires, Argentina, 4 - 15 November 1996	Decision III/20
1996	BSWG1	First meeting of the Open-Ended Ad Hoc working Group on Biosafety. Aarhus, Denmark, 22 - 26 July 1996	Meeting Documents
1997	BSWG2	Second meeting of the Open-Ended Ad Hoc working Group on Biosafety. Montreal, Canada, 12 - 16 May 1997	Meeting Documents
1997	BSWG3	Third meeting of the Open-Ended Ad Hoc working Group on Biosafety. Montreal, Canada, 13 - 17 October 1997	Meeting Documents

1993	The Convention on Biological Diversity enters into force on 29 December 1993		
1998	BSWG4	Fourth meeting of the Open-Ended Ad Hoc working Group on Biosafety. Montreal, Canada, 5 - 13 February 1998	Meeting Documents
1998	COP4	Fourth meeting of the Conference of the Parties - Issues related to biosafety. Bratislava, Slovakia, 4 - 15 May 1998	Decision IV/3
1998	BSWG5	Fifth meeting of the Open-Ended Ad Hoc working Group on Biosafety. Montreal, Canada, 17 - 28 August 1998	Meeting Documents
1999	BSWG6	Sixth meeting of the Open-Ended Ad Hoc working Group on Biosafety. Cartagena, Colombia, 14 - 19 February 1999	Meeting Documents
1999	BSIC1	Informal Consultation on the process to resume the Extraordinary Meeting of COP to adopt a protocol on Biosafety. Montreal, Canada, 1 July 1999	Meeting Documents
1999	BSIC2	Second Informal Consultation on the process to resume the Extraordinary Meeting of COP to adopt a protocol on Biosafety. Vienna, Austria, 15 - 19 September 1999	Meeting Documents
1999 - 2000	EXCOP1	First Extraordinary Meeting of the Conference of the Parties - Decisions on the continuation of the first extraordinary meeting of the Conference of the Parties to the Convention on Biological Diversity, adoption of the Cartagena Protocol and interim arrangements. Cartagena, Colombia 22 - 23 February 1999 and Montreal, Canada, 24 - 28 January 2000	Decisions EM-I/1-3
2000	COP5	The Cartagena Protocol on Biosafety is opened for signature. Fifth meeting of the Conference of the Parties - Work plan of the Intergovernmental Committee for the Cartagena Protocol on Biosafety. Nairobi, Kenya, 15 - 26 May 2000	Decision V/1
2000	ICCP1	First meeting of the Intergovernmental Committee for the Cartagena Protocol on Biosafety. Montpellier, France, 11 - 15 December 2000	Meeting Documents
2001	ICCP2	Second meeting of the Intergovernmental Committee for the Cartagena Protocol on Biosafety. Nairobi, Kenya, 1 - 5 October 2001	Meeting Documents
2002	COP6	Sixth meeting of the Conference of the Parties - Intergovernmental Committee for the Cartagena Protocol on Biosafety. The Hague, Netherlands, 7 - 19 April 2002	Decision VI/1
2002	ICCP3	Third meeting of the Intergovernmental Committee for the Cartagena Protocol on Biosafety. The Hague, The Netherlands, 22 - 26 April 2002	Meeting Documents
2003	**The Cartagena Protocol on Biosafety enters into force on 11 September 2003**		

Objective: In accordance with the precautionary approach, contained in Principle 15 of the Rio Declaration on Environment and Development, the objective of the Protocol is to contribute to ensuring an adequate level of protection in the field of the safe transfer, handling and use of 'living modified organisms resulting from modern biotechnology' that may have adverse effects on the conservation and sustainable use of biological diversity, taking also into account risks to human health, and specifically focusing on transboundary movements.

A number of points can be made with regard to the objective of the Protocol.

First, the precautionary approach as contained in the Rio Declaration is clearly identified to be the basis of the Protocol, and the objective of the Protocol is taken to be in accordance with the precautionary approach in Principle 15. (The preamble of the Protocol also reaffirms the precautionary approach contained in Principle 15.). The idea is to ensure that there is an adequate level of protection in the undertaking of all activities, in particular, the transboundary movement of living modified organisms (LMOs). Protection against adverse effects on biological diversity, 'taking also into account risks to human health', is the objective of the Protocol. Clearly, protection from risks to human health is part of the objective of the Protocol (Traavik and Lim, 2007). The Protocol always uses this language formulation whenever making reference to impacts on human health. This reflects the compromise that was reached on this issue, between the majority of developing countries that wanted the protection of human health to be included as an objective of the Protocol and those that only wanted the Protocol to ensure protection of biological diversity. It is clear from this formulation that impacts on human health as a result of adverse effects on biological diversity are captured, while direct impacts on human health (e.g. from consuming a GMO) may also arguably be captured.

Living Modified Organisms (LMOs): The protocol defines a 'living modified organism' as any living organism that possesses a novel combination of genetic material obtained through the use of modern biotechnology, and 'living organism' means any biological entity capable of transferring or replicating genetic material, including sterile organisms, viruses and viroids. 'Modern biotechnology' is defined in the Protocol to mean the application of in vitro nucleic acid techniques, or fusion of cells beyond the taxonomic family, that overcome natural physiological reproductive or recombination barriers and are not techniques used in traditional breeding and selection. 'Living modified organism (LMO) Products' are defined as processed material that are of living modified organism origin, containing detectable novel combinations of replicable genetic material obtained through the use of modern biotechnology (Anonymous, 2017). Common LMOs include agricultural crops that have been genetically modified for greater productivity or for resistance to pests or diseases. Examples of modified crops include tomatoes, cassava, corn, cotton and soybeans. 'Living modified organism intended for direct use as food or feed, or for processing (LMO-FFP)' are agricultural commodities from GM crops. Overall the term 'living modified organisms' is equivalent to genetically modified organism – the Protocol did not make any distinction between these terms and did not use the term 'genetically modified organism.

Many countries use the terms 'LMO' and 'GMO' interchangeably, and consider that the terms refer to the same thing. A number of countries use the term 'GMO' in their national laws, and interpret the definition of LMO in the Protocol to be consistent with the definition of GMO in their national laws. Laws in the European Union and a number of other countries use the term 'GMO' to refer to LMOs covered by the Protocol. Malaysia, for example, made a written declaration on signing the CBD that the term 'LMO' would be understood as meaning 'GMO'. The definition of LMO in the Protocol is instructive on this point. What are clearly excluded from the definition of LMO, and by using the term 'living organism', are products from LMOs, which are not living, and which are therefore not covered by the scope of the Protocol. This includes, for example, oil produced from genetically modified (GM) canola or meat from GM animals.

General scope of the protocol: A key fight during the course of the Protocol negotiations was for the inclusion of 'products thereof' in the general scope of the Protocol. This was strongly advocated by the Like-Minded Group of Developing Countries. 'Products thereof' include products derived from LMOs such as processed foods containing GM soya, cotton clothing made from GM cotton, etc.

However, 'products thereof' are excluded from the scope of the Protocol and, as such, remain largely unregulated internationally. However, there are two references to 'products thereof' in the Protocol. First, in the Risk Assessment Annex (Annex III) of the Protocol 'Risks associated with LMOs or products thereof, namely, processed materials that are of LMO origin, containing detectable novel combinations of replicable genetic material obtained through the use of modern biotechnology, should be considered in the context of the risks posed by the nonmodified recipients or parental organisms in the likely potential receiving environment', and in Article 20.3 (c) which requires relevant information on 'products thereof' in the context of risk assessments or environmental reviews to be made available to the Biosafety Clearing House, where appropriate (Traavik and Lim, 2007). The Protocol's scope applies to the 'transboundary movement, transit, handling, and use of all living modified organisms that may have adverse effects on the conservation and sustainable use of biological diversity, taking also into account risks to human health' (Article 4). The terms 'transboundary movement, transit, handling and use' are wide enough to include all activities related to LMOs under the scope of the Protocol. The general scope of the Protocol in Article 4 provides for a comprehensive scope, covering all LMOs, and does not specifically exclude any category of LMOs.

14.2.1 Main Principles and Provisions

A number of key principles underpin the Protocol. The principle of prior informed consent applies and there should be no transboundary movement without the prior knowledge and authorization of the importing Party. The onus is on the exporters and exporting Parties to notify and furnish relevant information to the importing Party before an LMO crosses national boundaries. An importing Party makes its own decision based on risk assessment and applying precaution, and national sovereignty in decision making is therefore one of the principles that the Protocol establishes. The right to say 'no' is also clearly established.

Advance informed agreement (AIA) procedure: The Protocol has a special focus on transboundary movements. The procedure by which transboundary movements of LMOs are regulated is known as the advance informed agreement (AIA) procedure which involves a few steps. Firstly, the Party of export notifies or requires its exporters to notify the Party of import if there is an intention to export a LMO. The notification must include at least the information required in Annex I (Information Required in Notifications under Articles 8, 10 and 13) of the Protocol. The notification is then acknowledged by the Party of import. The Party of import may make a decision on the notification according to its domestic regulatory framework, which must be consistent with the Protocol or proceed according to the decision procedure in the Protocol.

The decision by the Party of import is based on risk assessment and precaution, and the Party of import may take into account socio-economic considerations when making its decision. A Party is obliged to consult its public in the decision-making process, and must make the results of such decisions available to the public. A Party may make the following decisions: unconditional approval, approval with conditions, prohibition of the import, request for additional relevant information, or extension of the time period for making a decision (Traavik and Lim, 2007). There is time periods specified in the AIA procedure. The Party of import is required to acknowledge receipt of the notification within 90 days, and has a total of 270 days from the time it receives the notification to communicate its decision on the transboundary movement.

However, the issue of time frames was very contentious during the Protocol negotiations and a number of flexibilities have been built into the provisions. The Party of import may make its decision according to its domestic regulatory framework, which may not necessarily strictly adhere to the time periods specified in the Protocol, but which must be 'consistent with' the Protocol. Moreover, the clock stops (i.e. the time keeping is suspended) once the Party of import has requested for additional relevant information. The decision by the Party of import may also be to extend the time period. A failure to acknowledge receipt of the notification does not imply the consent of the Party of import. Neither does a failure by the Party of import to communicate its decision within the specified time period imply that it has consented to the transboundary movement.

14.2.2 Risk Assessment

1. The Protocol requires that decisions on proposed imports be based on formal risk assessments.
2. Risk assessments must be undertaken in a scientific manner based on recognized risk assessment techniques, taking into account advice and guidelines developed by relevant international organizations.
3. Lack of scientific knowledge or scientific consensus must not necessarily be interpreted as indicating a particular level of risk, an absence or risk, or an acceptable risk.
4. Risks associated with LMOs should be considered in the context of the risks posed by the non-modified recipients or parental organisms in the likely potential receiving environment.
5. Risk assessment should be carried out on a case by case basis.

Other key provisions in the Protocol

- Risk assessment and risk management
- Socio-economic considerations
- Public awareness, education and participation
- Review of decisions
- Unintentional transboundary movements and emergency measures
- Illegal transboundary movement of GMOs
- Handling, transport, packaging, and identification of transboundary shipments
- Identification of LMO-FFPs
- Identification of LMOs destined for contained use
- Identification of LMOs for deliberate release and other LMOs within the scope of the Protocol
- Information sharing and the Biosafety Clearing-House
- Confidential information
- Capacity building
- Liability and redress
- Non-Parties and bilateral, regional, multilateral agreements and arrangements
- Compliance with the Protocol
- Relationship with other international agreements
- Review -The Protocol will be reviewed five years after its entry into force, i.e. at COP-MOP 4 in 2008, and at least every five years thereafter to

evaluate its effectiveness, including an assessment of its procedures and annexes.

14.3 Biosafety Clearing House (BCH)

The Protocol established a Biosafety Clearing-House, in order to facilitate the exchange of scientific, technical, environmental and legal information on, and experience with, living modified organisms; and to assist parties to implement the Protocol (Article 20 of the Protocol, SCBD 2000). It was established in a phased manner, and the first meeting of the Parties approved the transition from the pilot phase to the fully operational phase, and adopted modalities for its operations. Information is not limited to existing laws, regulations, or guidelines related to the implementation of the Protocol, but also include summaries of risk assessments environmental reviews of LMOs and final decisions regarding the importation or release of LMOs.

The Clearing-House Mechanism of the CBD (Article 18(3) CBD): The Conference of the Parties, at its first meeting, shall determine how to establish a clearing-house mechanism to promote and facilitate technical and scientific cooperation (Ruth Mackenzie *et.al.*, 2003).

14.4 Article 18(3) of the Convention on Biological Diversity (CBD)

CBD created a mechanism to translate the goal of partnerships and cooperation into action – the Clearing-House Mechanism (CHM). The CHM was created to "promote and facilitate technical and scientific cooperation between the Parties to the CBD" and is a key to achieving the CBD's three principal objectives. It also facilitates access to and the exchange of information on biodiversity around the world. It is a network of Parties and partners working together to facilitate implementation of the CBD. The Parties directed the CBD Secretariat to take a leadership role in facilitating the implementation of the CHM, and also created an Informal Advisory Committee (IAC) to provide the Secretariat with feedback and advice through the CHM development process. The activities of the CHM are directed by the CBD COP as well as by the advice of the Subsidiary Body on Scientific, Technical and Technological Advice. The CBD COP designated 1996-1998 as the Pilot Phase of CHM operations, during which activities and services would evolve in response to the needs of countries and partners working to implement the CBD. The Parties also made a commitment to commissioning an Independent Review of the CHM after completion of the Pilot Phase. This report was published in September 1999.

The CHM depends on a decentralized process to gather and organize the information that its users need. Driving this process are networks of focal points, international centres and institutions with expertise that co-ordinate initiatives among themselves on topics of common interest. Each focal point also contributes to the Clearing- House information system, which is then made accessible to all users. In this way, focal points encourage networking among government agencies, expert groups, non-governmental organizations and private enterprise at all levels.

Capacity-Building: The Parties shall cooperate in the development and/or strengthening of human resources and institutional capacities in biosafety, including biotechnology to the extent that it is required for biosafety, for the purpose of the effective implementation of this Protocol, in developing country Parties, in particular the least developed and small island developing States among them, and in Parties with economies in transition, including through existing global, regional, subregional and national institutions and organizations and, as appropriate, through facilitating private sector involvement. For the purposes of implementing paragraph above in relation to cooperation, the needs of developing country Parties, in particular the least developed and small island developing States among them, for financial resources and access to and transfer of technology and know-how in accordance with the relevant provisions of the Convention, shall be taken fully into account for capacity-building in biosafety. Cooperation in capacity building shall, subject to the different situation, capabilities and requirements of each Party, include scientific and technical training in the proper and safe management of biotechnology, and in the use of risk assessment and risk management for biosafety, and the enhancement of technological and institutional capacities in biosafety. The needs of Parties with economies in transition shall also be taken fully into account for such capacity-building in biosafety.

Precautionary approach: One of the outcomes of the United Nations Conference on Environment and Development (also known as the Earth Summit) held in Rio de Janeiro, Brazil, in June 1992, was the adoption of the Rio Declaration on Environment and Development, which contains 27 principles to underpin sustainable development. Commonly known as the precautionary principle, Principle 15 states that "In order to protect the environment, the precautionary approach shall be widely applied by States according to their capabilities. Where there are threats of serious or irreversible damage, lack of full scientific certainty shall not be used as a reason for postponing cost-effective measures to prevent environmental degradation."

Elements of the precautionary approach are reflected in a number of the provisions of the Protocol, such as:

- The preamble, reaffirming "the precautionary approach contained in Principle 15 of the Rio Declaration on environment and Development";
- Article 1, indicating that the objective of the Protocol is "in accordance with the precautionary approach contained in Principle 15 of the Rio Declaration on Environment and Development";
- Article 10.6 and 11.8, which states "Lack of scientific certainty due to insufficient relevant scientific information and knowledge regarding the extent of the potential adverse effects of an LMO on biodiversity, taking

into account risks to human health, shall not prevent a Party of import from taking a decision, as appropriate, with regard to the import of the LMO in question, in order to avoid or minimize such potential adverse effects."; and

- Annex III on risk assessment, which notes that "Lack of scientific knowledge or scientific consensus should not necessarily be interpreted as indicating a particular level of risk, an absence of risk, or an acceptable risk."

Application: The Protocol applies to the transboundary movement, transit, handling and use of all living modified organisms that may have adverse effects on the conservation and sustainable use of biological diversity, taking also into account risks to human health (Article 4 of the Protocol, SCBD 2000).

Parties and Non-parties: The governing body of the Protocol is called the Conference of the Parties to the Convention serving as the meeting of the Parties to the Protocol (also the COP-MOP). The main function of this body is to review the implementation of the Protocol and make decisions necessary to promote its effective operation. Decisions under the Protocol can only be taken by Parties to the Protocol. Parties to the Convention that are not Parties to the Protocol may only participate as observers in the proceedings of meetings of the COP-MOP. The Protocol addresses the obligations of Parties in relation to the transboundary movements of LMOs to and from non-Parties to the Protocol. The transboundary movements between Parties and non-Parties must be carried out in a manner that is consistent with the objective of the Protocol. Parties are required to encourage non-Parties to adhere to the Protocol and to contribute information to the Biosafety Clearing-House.

Relationship with the WTO: A number of agreements under the World Trade Organization (WTO), such as the Agreement on the Application of Sanitary and Phytosanitary Measures (SPS Agreement) and the Agreement on Technical Barriers to Trade (TBT Agreement), and the Agreement on Trade-Related Aspects of Intellectual Property Rights (TRIPs), contain provisions that are relevant to the Protocol. This Protocol states in its preamble that parties:

- Recognize that trade and environment agreements should be mutually supportive;
- Emphasize that the Protocol is not interpreted as implying a change in the rights and obligations under any existing agreements; and
- Understand that the above recital is not intended to subordinate the Protocol to other international agreements.

Main features: The Protocol promotes biosafety by establishing rules and procedures for the safe transfer, handling, and use of LMOs, with specific focus on transboundary movements of LMOs. It features a set of procedures including one for LMOs that are to be intentionally introduced into the environment called the advance informed agreement procedure, and one for LMOs that are intended to be used directly as food or feed or for processing. Parties to the Protocol must ensure that LMOs are handled, packaged and transported under conditions of safety. Furthermore, the shipment of LMOs subject to transboundary movement must be accompanied by appropriate documentation specifying, among other things, identity of LMOs and contact point for further information. These procedures and requirements are designed to provide importing Parties with the necessary information needed for making informed decisions about whether or not to accept LMO imports and for handling them in a safe manner.

The Party of import makes its decisions in accordance with scientifically sound risk assessments. The Protocol sets out principles and methodologies on how to conduct a risk assessment. In case of insufficient relevant scientific information and knowledge, the Party of import may use precaution in making their decisions on import. Parties may also take into account, consistent with their international obligations, socio-economic considerations in reaching decisions on import of LMOs.Parties must also adopt measures for managing any risks identified by the risk assessment, and they must take necessary steps in the event of accidental release of LMOs. To facilitate its implementation, the Protocol establishes a Biosafety Clearing-House for Parties to exchange information, and contains a number of important provisions, including capacity-building, a financial mechanism, compliance procedures, and requirements for public awareness and participation.

14.5 Procedures for Moving LMOs Across Borders

i. Advance Informed Agreement: The "Advance Informed Agreement" (AIA) procedure applies to the first intentional transboundary movement of LMOs for intentional introduction into the environment of the Party of import. It includes four components: notification by the Party of export or the exporter, acknowledgment of receipt of notification by the Party of import, the decision procedure, and opportunity for review of decisions. The purpose of this procedure is to ensure that importing countries have both the opportunity and the capacity to assess risks that may be associated with the LMO before agreeing to its import. The Party of import must indicate the reasons on which its decisions are based (unless consent is unconditional). A Party of import may, at any time, in light of new scientific information, review and change a decision. A Party of export or a notifier may also request the Party of import to review its decisions.

However, the Protocol's AIA procedure does not apply to certain categories of LMOs, viz., LMOs in transit, LMOs destined for contained use and LMOs intended for direct use as food or feed or for processing. While the Protocol's AIA procedure does not apply to certain categories of LMOs, Parties have the right to regulate the importation on the basis of domestic legislation. There are also allowances in the Protocol to declare certain LMOs exempt from application of the AIA procedure.

ii. LMOs intended for food or feed, or for processing: LMOs intended for direct use as food or feed, or processing (LMOs-FFP) represent a large category of agricultural commodities. The Protocol, instead of using the AIA procedure, establishes a more simplified procedure for the transboundary movement of LMOs-FFP. Under this procedure, A Party must inform other Parties through the Biosafety Clearing-House, within 15 days, of its decision regarding domestic use of LMOs that may be subject to transboundary movement. Decisions by the Party of import on whether or not to accept the import of LMOs-FFP are taken under its domestic regulatory framework that is consistent with the objective of the Protocol. A developing country Party or a Party with an economy in transition may, in the absence of a domestic regulatory framework, declare through the Biosafety Clearing-House that its decisions on the first import of LMOs-FFP will be taken in accordance with risk assessment as set out in the Protocol and time frame for decision-making.

iii. Handling, transport, packaging and identification: The Protocol provides for practical requirements that are deemed to contribute to the safe movement of LMOs. Parties are required to take measures for the safe handling, packaging and transportation of LMOs that are subject to transboundary movement. The Protocol specifies requirements on identification by setting out what information must be provided in documentation that should accompany transboundary shipments of LMOs. It also leaves room for possible future development of standards for handling, packaging, transport and identification of LMOs by the meeting of the Parties to the Protocol.

Each Party is required to take measures ensuring that LMOs subject to intentional transboundary movement are accompanied by documentation identifying the LMOs and providing contact details of persons responsible for such movement. The details of these requirements vary according to the intended use of the LMOs, and, in the case of LMOs for food, feed or for processing, they should be further addressed by the governing body of the Protocol. The first meeting of the Parties adopted decisions outlining identification requirements for different categories of LMOs (Decision BS-I/6, SCBD 2004). However, the second meeting of the Parties failed to reach agreement on the detailed requirements to identify LMOs intended for direct use as food, feed or for processing and will need to reconsider this issue at its third meeting in March 2006.

14.6 The Protocol and Human Health Issues

- The appropriate treatment of human health issues in the Protocol was contentious from the outset of the negotiations. Article 19(3) of the CBD makes no reference to human health. In the discussion of the mandate to negotiate a Protocol, the subject of human health was nevertheless considered – and proved controversial. To some, an instrument on biosafety which failed to cover human health issues was not a viable proposition; for others, however, human health should not be covered at all in the context of a Protocol to the CBD (Ruth Mackenzie *et.al.*, 2003).

- Ultimately, the negotiators compromised, and the final version of the Protocol recognizes throughout that risks to human health are to be "taken also into account". Thus, the Protocol specifically mentions human health in various provisions, including in Article 4 on Scope: This Protocol shall apply to… [LMOs] that may have adverse effects on the conservation and sustainable use of biological diversity, taking also into account risks to human health.
- The wording "taking also into account risks to human health" has its origin in Article 8(g) of the CBD. Independently of any other instrument in this field, including the Protocol, Article 8(g) requires Parties to the CBD to "regulate, manage or control the risks associated with use and release of living modified organisms resulting from biotechnology which are likely to have adverse environmental impacts, that could affect the conservation and sustainable use of biological diversity, taking also into account the risks to human health".
- In the absence of any additional explanatory provision in either the CBD or the Protocol, however, the meaning of the phrase "taking also into account the risks to human health" remains somewhat unclear. This is all the more so as not much of the debate on this subject during the negotiations has been recorded. The wording was introduced by the European Union at an early stage. Several delegations considered that direct impacts of LMOs on human health should not be covered by the Protocol – as they were dealt with in other contexts. However, many others – including particularly those from developing countries – wished to give the same weight to impacts of LMOs on human health and on biological diversity.
- The first approach leads to the conclusion that risks posed by a LMO to human health are taken into account under the Protocol only if they result from the potential adverse effects of the same LMO on biological diversity.
- The other approach leads to the conclusion that risks posed by a LMO to human health are to be taken into account under the Protocol also in the absence of, or separately from, potential adverse effects of the LMO considered on biological diversity. This would be the case, for instance, for any change in allergenic properties of pollen as a result of genetic modification, or the consumption of GM food.
- Both interpretations can be supported by the phrase "taking also into account risks to human health". The practical effect of the absence of unambiguous guidance in the Protocol itself on this issue, along with the lack of consensus on one or the other above mentioned interpretations, appears to be that, under the Protocol at least, Parties will have a certain

latitude and flexibility in deciding which human health aspects to cover in their implementation of the Protocol – unless and until they decide upon an authoritative interpretation collectively in the meeting of the Parties to the Protocol.

Possible elements of national biosafety regulations: In the elaboration of a national biosafety legal framework, some elements which States have considered include the following:

- Define the objectives of the framework
- Define the scope of the framework – what activities and organisms are covered
- Place responsibility for implementation of the framework on a Minister or Ministers and on particular government department(s) or agency
- Establish or designate advisory body(ies) to advise on technical aspects of regulatory decisions
- Establish a general prohibition on activities involving LMOs unless an authorization / licence or other approval has been obtained in accordance with regulations
- Establish a system of permits or authorizations for activities involving LMOs
- Allow for exemptions or “fast-track” or simplified procedures for certain LMOs with which there is extensive experience under the regulations, or which have been deemed to be “low-risk”
- Provide for public information and consultation on permit applications and policy issues Set out information required in an application for a permit (information required may vary according to the type of LMO and/or the intended activity)
- Address the protection of commercial confidential information
- Establish a risk assessment procedure, whereby risks associated with the release or other activity are identified, in accordance with risk assessment criteria
- Allow for risk management conditions to be attached to permits, including any applicable labeling or marking requirements
- Set out procedures for monitoring and review of activities subject to permit, including compliance with conditions
- Set out penalties and sanctions for non-compliance
- Make provisions for liability for any damage arising out of activities involving LMOs
- Address unintentional releases and emergency measures

- Make certain transitional arrangements in respect of pre-existing activities or applications

14.7 The *Codex Alimentarius*

This is a non-binding Code developed by the *Codex Alimentarius* Commission, a body of FAO/World Health Organization which elaborates standards, general principles, guidelines, and recommended codes of practice in relation to food safety and related issues (Ruth Mackenzie *et.al.*, 2003). The Codex is significant in relation to LMOs because standards may be adopted in future on safety of foods derived from biotechnology (for example, addressing issues of potential allergenicity; possible gene transfer from LMOs; pathogenicity deriving from the organism used; nutritional considerations; risk assessment and authorization procedures; and appropriate labelling). The Codex has underway at least three processes of relevance to LMOs. The Task Force on Foods Derived from Biotechnology is working, inter alia, on Principles for Risk Analysis of Foods Derived from Modern Biotechnology. The Committee on General Principles is elaborating Draft Working Principles for Risk Analysis. The Committee on food labelling is preparing recommendations for the labelling of food obtained through biotechnology.

i. *Codex Alimentarius* and Genetically Modified Foods : The *Codex Alimentarius* Commission is a FAO/WHO body which elaborates standards, general principles, guidelines and recommended codes of practice in relation to food safety. As of 2002, there are at least three processes underway in the *Codex Alimentarius* Commission of relevance to the safety assessment and labelling of foods derived from modern biotechnology. In 1999, the *Codex Alimentarius* Commission decided to establish a Task Force on Foods derived from Biotechnology. The Task Force met for the first time in March 2000 and decided to elaborate (a) a set of broad general principles for risk analysis of foods derived from biotechnology; (b) specific guidance on the risk assessment of foods derived from biotechnology; and (c) a list of available analytical methods including those for the detection or identification of foods or food ingredients derived from biotechnology. The Task Force is due to present its final report to the Commission in 2003. At its third meeting in March 2002 the Task Force agreed to advance Draft Principles for the Risk Analysis of Foods Derived from Modern Biotechnology and a Draft Guideline for the Conduct of Food Safety Assessment of Foods Derived from Recombinant DNA Plants for consideration by the *Codex Alimentarius* Commission in 2003.

In the meantime, the Codex Committee on General Principles is undertaking work on Draft Working Principles for Risk Analysis. These may address, among other things, the role of precaution in risk management. The Committee on Food Labelling is in the process of considering draft recommendations for the Labelling of Foods Obtained through Certain Techniques of Genetic Modification/Genetic Engineering. Other Codex Committees undertaking work relevant to foods derived from modern biotechnology include the Committee on Food Import and Export Inspection and Certification Systems and the Committee on Methods of Analysis and Sampling. Domestic food safety measures that are in conformity with standards, guidelines and recommendations of the *Codex Alimentarius* are (rebuttably) presumed to be consistent with the WTO Agreement on Sanitary and Phytosanitary Measures.

Sierra Leone became the latest country to ratify the Cartagena Protocol on 15th June 2020. Uzbekistan had ratified the protocol on 25th October 2019. The total parties to the Cartagena Protocol as of June 2021 are 173. The Cartagena Protocol completed its 20th year of adoption in the year 2020.The dates of the meetings of COP-15 to the CBD, COP-10 to the Cartagena Protocol on Biosafety, COP-4 to the Nagoya Protocol have been revised to 11th-24th October 2021, which were planned to happen in May 2021. It is expected to be held in Kunming, Yunnan Province, China. The upcoming meeting will revolve around discussing and agreeing on the two key documents:

1. The recently published fifth Global Biodiversity Outlook, and
2. The updated zero draft of the post-2020 global biodiversity framework.

ii. National implementation of the Protocol: The Protocol sets minimum standards for the regulation of LMOs – Parties may take action that is more protective of the conservation and sustainable use of biological diversity than that called for in the Protocol. However, the action must be consistent with the objectives and provisions of the Protocol and be in accordance with the Parties' other obligations under international law. Parties are also under an obligation to take the necessary and appropriate legal, administrative and other measures to implement their obligations under the Protocol (Traavik and Lim, 2007). This means that national measures such as a national biosafety law should be put in place to implement Protocol obligations.

The Protocol contains many important principles, which are now established in international law. However, it is a negotiated text with deficiencies for biosafety. While strengthening the Protocol and rectifying its deficiencies should be the long-term goal, it is critical that national governments, and developing countries in particular, formulate domestic biosafety laws that improve on the scope and standards set by the Protocol, and which also comprehensively regulate the domestic development and use of GMOs. As an international law that is binding on countries that are Party to it, the Protocol presents obligations on and opportunities for sovereign countries. As a negotiated text, many flexibilities for interpretation and implementation are available for countries to utilize, putting real biosafety at the heart of national regulation. In conclusion, the Protocol is just the start of the long and difficult road to effective international regulation of genetic engineering. Much more needs to be done, and countries must act to ensure that real biosafety becomes a reality.

14.8 Global Scenario of Cartagena Protocol on Biosafety

Although data about the spread of genetically modified (GM) crops worldwide is difªcult to come by, Figures compiled by the International Service for the Acquisition of Agri-biotech Applications (ISAAA) claim that, from 1996, when genetically modified varieties were first grown commercially, the global area planted to such crops has increased over 50-fold, from 1.7 million hectares to 90.0 million hectares in 2005. Biotech crops are now grown in 21 countries. Of these, the leader is the United States, with 49.8 million hectares, followed by Argentina (17.1 million ha), Brazil (9.4 million ha), Canada (5.8 million ha), China (3.3 million ha), Paraguay (1.8 million ha), India (1.3 million ha) and South Africa (0.5 million ha) (Aarti Gupta and Robert Falkner, 2006). Mexico and twelve other countries make up the rest, with less than 0.3 million hectares each.8 It is important to note that, although both developed and developing countries are growing transgenic crops, the United States alone accounts for over half of the total area devoted to such crops.

Apart from the few developing countries growing transgenic crops in commercial quantities, others are still carrying out field testing and experimental research, if they participate in the process at all. However, irrespective of whether countries grow transgenic crops, most have to contend with an increasingly global trade in agricultural commodities and food containing genetically modified material. The growth of a globalized biotechnology industry and of trade in biotech crops requires countries to develop regulatory systems, forcing them to consider the impact that the spread of GM seed and crops to their countries might have on the sustainability of their agricultural systems, on the prospects for biosafety and food security, and on their current and future position in global agricultural trade.

The need for global governance of genetic engineering has thus been recognized since the late 1980s, when the first calls were made for an international biosafety treaty. It took until January 2000 for an agreement to be reached on the Cartagena Protocol on Biosafety, after nearly four years of increasingly contentious negotiations. The Protocol is the center-piece of the emerging global governance architecture for genetic engineering in agriculture. Other key elements of this governance architecture include the WTO's Agreement on the Application of Sanitary and Phytosanitary Measures (SPS Agreement) and Agreement on Technical Barriers to Trade (TBT Agreement). Furthermore, the *Codex Alimentarius* Commission, a global food safety standard-setting body, is debating global safety standards for food produced via use of genetic engineering. This emerging governance framework has to contend with a wide range of concerns (including ecological, human health, social and ethical) associated with genetic engineering in agriculture. The governance challenge is made more complex by the fact that the existence, nature and manageability of risks remain deeply contested. Moreover, the emerging system of rules and institutions is far from coherent or consistent. Instead, it remains unclear how components of this rapidly expanding set of global rules interact with and influence one another. This is partly because these regimes are still evolving, and their obligations are still being interpreted or expanded within global fora, as well as via national interpretation and implementation. It is also because of the potential for conflict between the Cartagena Protocol's obligations and WTO agreement.

This transatlantic GMO conflict underlines the fact that no shared global approach to biosafety regulation currently exists, which might be diffused to different domestic contexts via a global governance regime. Instead, two dominant regulatory approaches persist, one serving as a model for comprehensive and precautionary biosafety regulation (the EU model), the other emphasizing a "sound science" approach to biosafety, whereby restrictive regulatory action is justified only in the face of scientific evidence of harm (the US model).15 In the absence of such evidence, the US model assumes the "substantial equivalence" of GM and non-GM varieties of seed and food crops. The two regulatory approaches also differ with regard to labeling, segregation, traceability and threshold requirements for domestically authorized GMOs (all of which are either mandatory or more stringent in the EU, as compared to the US). There are no signs of these two models converging towards one consensual regulatory model. Instead, negotiation of the Cartagena Protocol has provided one more site where these conflicts have played out, and its influence must therefore be considered within this larger context.

14.9 Indian Scenario of Cartagena Protocol on Biosafety

India has opted to join the Cartegena Protocol on Biosafety. This makes India responsible for not only ensuring biosafety for her own people, but also for the world community. International Union for the Conservation of Nature (IUCN) is responsible for getting this protocol come into force. Under the UN Convention on Biological Diversity, adoption of the Cartagena Protocol on Biosafety had been a significant development with a record number of signatories on the opening day of the fifth conference of Parties (CoP-5). Using precautionary principle is one of the options under Article 10, 11 and 26 of the Cartagena Protocol (Anonymous, 2002). Articles 15 and 16 deal with the risk assessments in respect of any new technology product, particularly the genetically modified organisms (GMOs). These principles embodied in the Cartagena Protocol are soft and quite flexible in nature and yet many countries have kept themselves away from signing and ratifying this protocol. The difficulty of the protocol coming into force so far has been the issues of capacities countries have to develop to implement the protocol. In fact, many developing countries including India do not have adequate capacity to assess risks and ensure safety.

India will become the 36th country to ratify the protocol which aims to provide a safe mechanism for the transfer, handling and use of living modified organisms (lmos). The pact will come into effect only after 50 countries have ratified it. Negotiated under the aegis of the Convention on Biological Diversity, it was adopted in 2000 after five years of talks and two aborted attempts. Experts say that the weak link is that the protocol does not override provisions of other international agreements such as the World Trade Organization treaty. The main flaw in the pact pertains to implementation of the 'precautionary principle'. Though according to the rule a country can ban genetically modified products even in the absence of scientific data about their harmful effects, WTO regulations mandate that an import can be barred only on the basis of scientific evidence.

The protocol would be ineffective for India as the country lacks scientific know-how at the local level. Further, while the WTO has a strong dispute settlement system that has presided over several environment-related disputes, the CBD (under which the cpb operates) does not have one. Experts feel that this could make the provisions of CPB toothless. Sri Lanka was dragged into settling the dispute through the WTO clause when it banned the import of GM foods. "How to deal with such potential conflicts?" However proponents of the pact argue that as India would be playing a major role in biotechnology-based trade in the future, the decision to ratify will prove beneficial. "India can lead developing countries in the biotechnology sector". As for the threat of the protocol getting subverted by the WTO regime, ministry officials say that it is too early to arrive at any conclusion. " WTO is a decade-old international body. CP B would also mature and define its own dispute settlement system,"

Some of the perceived risks associated with GMOs are that transformed organisms could displace existing species and change the ecosystem, gene flow among plants could transfer the novel gene into related species making super weeds and novel genes could have unintended harmful effects on non-target organisms. Some new products that were intentionally altered for nutritional profiles can be challenging. For example, the genetically engineered low-glutelin rice produced for low glutelin levels was associated with unexpected increase in levels of prolamines. Increased levels of prolamines may not affect industrial use of rice but may affect nutritional quality of rice and may increase the allergic potential of modified rice. It was engineered to express high levels of beta-carotene, a precursor to vitamin A. But it was found that this genetic modification was associated with increased levels of xanthophylls.

Such unexpected biochemical changes can occur in modifications. However, it cannot be said that genetic engineering always caused such surprising unintended expressions. To nullify all possibilities of such situations it is important to take a precautionary approach and develop better methods to assess the risks. Examples of research data on impact of genetically modified cereals like maize becoming a weed, genetic contamination of GM crops with native population in Mexico are revealing to ensure that biological assessments are needed more thoroughly. Any classic case of adverse impact of GM crop originates from the fact that the GM crops have more biological takeover compared to native or unmodified crops/

organisms. However, the ecological and environmental impacts are trickier to assess and might take anything between few months to more than a decade. The other concern is the issue of outcrossing between GM organisms and pathogens. Another adverse impact of modified organisms can be the negative effect on population of non-target organism. Apart from these, other impacts include unintended effects of modified organisms on target organisms like changes in soil biota, especially introduction of modified soil microbes and plants like nitogen fixers with modified rhizobia.

Sixth National Report to the Convention of Biological Diversity: India submitted its sixth national report (NR6) to the Convention on Biological Diversity (CBD) during the inaugural session of the meeting of the State Biodiversity Boards (SBBs) organized by the National Biodiversity Authority (NBA) (Anonymous, 2018). The NR6 highlights the progress India has made in achieving the 12 National Biodiversity Targets (NBT) set under the convention process. With this India is among the first five countries in the world, the first in Asia and the first among the biodiversity-rich mega diverse nations, to have submitted Sixth National Report (NR6) to the Convention on Biological Diversity (CBD).The submission of national reports is a mandatory obligation on parties to international treaties, including the CBD. India developed 12 National Biodiversity Targets in line with 20 global Aichi biodiversity targets.

Highlights of the Report : India is one of the few countries where forest cover is on the rise, according to the 15th India State of Forest Report (ISFR) 2017. While India has exceeded/overachieved two NBTs, it is on track to achieve eight NBTs and in respect of the remaining two NBTs also, India is striving to meet the targets by the stipulated time of 2020. More than 20% of India's total geographical area is under biodiversity conservation, India has exceeded the terrestrial component of 17% of Aichi target 11. India published the first internationally recognized certificate of compliance (IRCC) under the Protocol in 2015, and since then published nearly 75% of the IRCCs. Thereby, achieving target relating to access and benefit sharing (ABS) by operationalizing the Nagoya Protocol on ABS. The population of Lion has risen to over 520 in 2015, and elephants to 30,000 in 2015. One-horned Indian Rhino which was on the brink of extinction during the early 20th century, now number 2400. Further, while globally over 0.3 % of total recorded species are critically endangered, in India only 0.08% of the species recorded are in this category.

Measures have been adopted for sustainable management of agriculture, fisheries and forests, with a view to provide food and nutritional security to all without destroying the natural resource base while ensuring intergenerational environmental equity. Programmes are in place to maintain genetic diversity of cultivated plants, farms livestock and their wild relatives, towards minimising genetic erosion and safeguarding their genetic diversity. Mechanisms and enabling environment are being created for recognising and protecting the vast heritage of coded and oral traditional knowledge relating to biodiversity.

Emerging global and national governance regimes for biosafety or the safe use of biotechnology in agriculture were examined (Gupta Aarti, 2000). The central concern is with examining the nature of the transnationalnational interface in biosafety governance, i.e. the relationship between multilaterally negotiated rules and national-level biosafety decision-making. The relevance of the recently concluded Cartagena Protocol on Biosafety was examined, dealing with the transboundary movement of genetically modified organisms (GMOs), for biosafety governance in India. In its call for "informed consent" prior to transfer of certain GMOs, the Cartagena Protocol validates the need for national-level choice in biosafety decision-making. However, a competing imperative is standardization of rules governing such choice, in order to enhance predictability and reduce national differences in biosafety decision-making. These potentially contradictory goals are reconciled in the Cartagena Protocol through reliance upon a minimalist scope and ambiguous decision-criteria for informed consent. In light of this, it is argued that the Cartagena Protocol can be relevant to national biosafety governance in three ways: first, it legitimizes the existence of domestic biosafety regulations; second, notwithstanding a minimalist scope, it shifts the burden for information sharing to producers of GMOs; and third, given ambiguous decision-criteria, it leaves unchanged national discretion in GMO decision-making. The likely relevance of such impacts for national-level biosafety governance in India was evaluated, through examining the processes of biosafety decision-making and information sharing currently in place. The transnational-national interface and the role for multilateral rule-making, in facilitating governance of contested decision-areas such as biosafety were also considered.

Legitimizing Indian Biosafety Regulations: At the very least, the Cartagena Protocol embodies within it a norm that LMOs require special regulatory attention. This was contested through much of the early stages of the negotiation, given divergent views on the uniqueness of risk posed by LMOs (see Gupta 1999 for a detailed discussion of this initial history). Furthermore, the protocol also legitimizes the restriction of trade in LMOs under certain conditions. Until the process of negotiating a protocol commenced in the mid-1990s, there was no explicit international norm on the need to regulate such trade. Thus, attempts in this period to restrict LMO transfers pursuant to domestic regulations could be seen as hostile to a de facto prevailing international norm of the unhindered transfer of such entities (Gupta Aarti, 2000). The relevance of a completed protocol for the transnational-national biosafety interface is then that it serves first and foremost to legitimize the existence of domestic biosafety regulations. Clearly, however, while the protocol lends legitimacy to domestic regulation of LMOs, can it simultaneously legitimize contradictory approaches to LMO regulation, embodied within different domestic frameworks? As seen in the previous section, conflicts between existing domestic approaches, in particular the intra-OECD conflict over decision-criteria lay at the heart of contestation in this negotiation.

This highlights the importance of examining the domestic sources of international obligations, an area of research recently gaining currency in the international relations literature. Previous writings in this literature had tended to assume that international agreements mandate new rules and obligations, which are then to be implemented at the national level.

However, in an increasing number of international negotiations, including biosafety, the source of conflict centers less around devising new obligations and more on whose already existing national approaches will be internationalized. How such conflicts are resolved will be crucial to the relevance of the final agreement for distinct national contexts. Although internationalizing their domestic approaches was most characteristic of the intra-OECD conflict in protocol negotiations, it was also an important concern for India. Thus, Indian negotiators sought to ensure that a completed protocol would not limit the already existing scope or decision-criteria for biosafety governance in domestic regulations (interviews). The extent to which the protocol's obligations go beyond or fall short of scope and criteria for biosafety decision-making in India is examined next.

14.10 Implications for India

i. Implications for India of "sound scientific" decision-making: The protocol mandates that consent decisions for LMO transfers have to be based upon a scientifically sound risk assessment. Risk assessments are also the center point of domestic biosafety evaluations in India. Yet, what constitutes a scientifically sound risk assessment in the Indian context? Information to be generated in such assessments includes the minimum "points to consider" mandated by the Cartagena Protocol. Equally important, however, do Indian information requirements go beyond what is mandated by the protocol, to cover ecologically or socioeconomically specific concerns in the Indian context? One example is often highlighted by biosafety regulators in claiming that risk assessment in India is "even stricter than the best models elsewhere" (Gupta Aarti, 2000). This is the requirement to assess allergenicity and toxicity potential from consuming transgenic plants for ruminants such as goats, seen as relevant to the Indian context. Biosafety regulators argue that generation of such data reflects the stringency of domestic risk assessment, since such requirements are not an integral part of risk assessments elsewhere. This is seen, however, as unscientific by some producers of LMOs required to generate this data (at present mainly the private sector). According to them, this is scientifically untenable, given that adverse impact testing is usually undertaken for smaller animals, with extrapolations about those higher up in the food chain. Such requirements are thus perceived by LMO producers as reflecting the regulators' need for the appearance of stringency rather than being a scientifically sound judgment that such data are necessary. This dispute reveals not only the differential interpretations of what is considered a scientifically sound risk assessment, but also that the protocol's injunctions cannot aid in mediating such conflicts. One outcome, then, is to leave national discretion with regard to sound scientific decision-making unchanged.

Differing "science-based" understandings of risks and benefits posed by LMOs are also evident from submissions to the Supreme Court by the plaintiff and defendants in the public interest litigation pending against field testing of transgenic cotton in India. The submissions by NGOs, government biosafety regulators and private sector producers of the transgenic crop reflect both the sources of information considered scientifically sound and authoritative, and the networks that these groups seek to align themselves with.

ii. Implications for India of precautionary decision-making: While the "precautionary principle" is not explicitly mentioned in Indian biosafety regulations, the 1986 Environmental Protection Act under which LMOs are regulated calls for regulation of substances that "may be or tend to be" injurious to the environment. This suggests that unambiguous scientific proof of injury to the environment is not a prerequisite for regulation. Thus, regulating LMOs under this Act can itself be construed as precautionary. Furthermore, if precautionary decision-making is understood as going beyond scientific uncertainties about quantifiable ecological and health impacts, then the clearest precautionary actions on LMOs in India have been related to socioeconomic concerns. As seen earlier, these "socially precautionary" actions have included restrictions on imports of transgenic commodities following the soybean controversy, and restrictions on future import of terminator technology following the controversy about its alleged field-testing in India. In both these cases, there were no environmental or human health rationales offered for restrictions on imports. Instead, attention to the issue in the media rather than assessment of potential hazards to human health from transgenic soybean fueled the decision not to import such commodities, whilst the terminator controversy was centrally concerned with socioeconomic issues of dependence on foreign technologies and farmer's rights.

iii. The implications for India of socioeconomic considerations: In fact, a striking characteristic of decision-criteria for LMO use in India is the fuzziness between sound scientific, precautionary and socio-economic considerations in such decision-making. This is evident, for example, from the merging of scientific risk assessment with socioeconomic considerations in the requirement to generate economic benefit data during the "technical" biosafety evaluation of a transgenic crop. The requirement to demonstrate that a transgenic crop is both "environmentally safe and economically viable" is the clearest evidence of these interlinkages (Gupta Aarti, 2000). This requirement can be seen as both "scientifically sound" (if scientific assessments include a broader conception of assessable impacts to include the socioeconomic) as well as "socially precautionary" thus effectively merging all three decision-criteria for consent. While debate continues about whether socioeconomic concerns can be included in a risk assessment or whether they belong to a political risk management calculus, they have been included in both phases of decision-making in India, highlighting the problematic nature of this dichotomy in this context.

Furthermore, the data to be generated is about socioeconomic benefits and not risks, muddying the categories still further. As regulators argue, a biosafety evaluation in the Indian context must, of necessity, include not just ecological and human health considerations, but also responsiveness to the needs, constraints and priorities of Indian agriculture (interviews), since these cannot be divorced from the biosafety evaluation. The Cartagena Protocol, on the other hand, maintains the demarcation between the technical and the political, with provision on risk assessment calling only for data to be generated on adverse ecological and human health impacts.

The compatibility of taking socioeconomic considerations into account in biosafety decision-making with obligations under the WTO were, as noted earlier, a central concern in protocol negotiations, and thus merit comment here. The compatibility with WTO of taking economic benefit data into account before approving an LMO for domestic use has not yet been put to the test in India, in part because such evaluations have to date been of domestically produced and not imported transgenic crops. However, resolution of the terminator controversy in India reveals that conflicts with the WTO over taking socioeconomic considerations into account are likely to be preempted by unilateral actions such as Monsanto's categorical "promise" that it "will only bring to India technologies that are thoroughly tested and approved by the Indian government". Monsanto's statement was made notwithstanding whether such testing and approvals by the Indian government were in keeping with India's WTO obligations. Clearly, in highly contested areas such as biosafety, public acceptance or the "court of public opinion" is as critical a mediator of conflict as determinations of legal rights under the WTO. Notwithstanding the narrow inclusion of socioeconomic considerations in the Cartagena Protocol, then, national discretion to take such factors into account appears to be little affected.

Bibliography

Aarti Gupta and Robert Falkner, 2006. The Influence of the Cartagena Protocol on Biosafety: Comparing Mexico, China and South Africa. Global Environ. *Politics*, 6:23-55

Anonymous, 2002. India And The Cartagena Protocol On Biosafety. Dec.2, 2002, The Financial Express

Anonymous, 2017. Frequently Asked Questions (FAQs) on the Cartagena Protocol". Convention on Biological Diversity. United Nations Environment Programme.

Anonymous, 2018. sixth National Report to the Convention of Biological Diversity. Daily Updates-Biodiversity and Environment.

Anonymous, 2020. "About the Protocol". *Convention on Biological Diversity*. Retrieved 17 September 2020.

Gupta Aarti, 2000. "Governing Biosafety in India: The Relevance of the Cartagena Protocol." Belfer Center for Science and International Affairs (BCSIA) Discussion Paper 2000-

24, Environment and Natural Resources Program, Kennedy School of Government, Harvard University, 2000.

Mackenzie Ruth, Burhenne-Guilmin Françoise, La Viña Antonio G.M. and Werksman Jacob D., Ascencio, Alfonso, Kinderlerer, Julian, Kummer, Katharina and Tapper, Richard, 2003. An Explanatory Guide to the Cartagena Protocol on Biosafety. IUCN, Gland, Switzerland and Cambridge, UK. xvi + 295pp.

Ruth Mackenzie, Françoise Burhenne-Guilmin, Antonio G.M. La Viña and Jacob D. Werksman in cooperation with Alfonso Ascencio, Julian Kinderlerer, Katharina Kummer and Richard Tapper Ruth Mackenzie, Françoise Burhenne-Guilmin, Antonio G.M. La Viña and Jacob D. Werksman in cooperation with Alfonso Ascencio, Julian Kinderlerer, Katharina Kummer and Richard Tapper, 2003. An Explanatory Guide to the Cartagena Protocol on Biosafety- An Explanatory Guide to the Cartagena Protocol on Biosafety. IUCN Environmental Policy and Law Paper No. 46 Environmental Policy and Law Paper No. 46.

Traavik, T. and Lim, L.C. , 2007. Biosafety First, Tapir Academic Publishers.

15

Issues Related to the Genetically Modified Crops

Genetically engineered crops refer to alterations in the genetic makeup of the crop by introgression new traits such as herbicide tolerance, virus resistance, drought, flood and frost resistance, delay in maturation time of the crop and increased crop yield. They can be made resistant to pests and diseases which can significantly reduce the consumption of insecticide. Biodiversity is the feedstock for biotechnology industries. The issues relating to the Genetically Modified (GM) Foods have generated intense public debate in many parts of the world. Even though the issues under debate include the costs and benefits of the GM crops and the inherent safety concerns, the outcome of the debate differs from country to country, depending on its geographical location, strength and resilience of the farm sector, attitudes of people towards food, and so on (Anonymous, 2009). In India also, this debate has engaged the attention not only of the Government but also of the farming community and the civil society.

15.1 Genetically Modified Crops / Food—Relevant Issues

GM foods are developed and marketed because there is some perceived advantage either to the producer or consumer of these foods. This is meant to translate into a product with a lower price, greater benefit (in terms of durability or nutritional value) or both. Initially GM seed developers wanted their products to be accepted by producers and have concentrated on innovations that bring direct benefit to farmers (and the food industry generally). One of the objectives for developing plants based on GM organisms is to improve crop protection. The GM crops currently on the market are mainly aimed at an increased level of crop protection through the introduction of resistance against plant diseases caused by insects or viruses or through increased tolerance towards herbicides.

Resistance against insects is achieved by incorporating into the food plant the gene for toxin production from the bacterium *Bacillus thuringiensis* (Bt). This toxin is currently used as a conventional insecticide in agriculture and is safe for human consumption. GM crops that inherently produce this toxin have been shown to require lower quantities of insecticides in specific situations, e.g. where pest pressure is high. Virus resistance is achieved through the introduction of a gene from certain viruses which cause disease in plants. Virus resistance makes plants less susceptible to diseases caused by such viruses, resulting in higher crop yields. Herbicide tolerance is achieved through the introduction of a gene from a bacterium conveying resistance to some herbicides. In situations where weed pressure is high, the use of such crops has resulted in a reduction in the quantity of the herbicides used.

15.1.1 Is the Safety of GM foods Assessed Differently from Conventional Foods?

Generally consumers consider that conventional foods (that have an established record of safe consumption over the history) are safe. Whenever novel varieties of organisms for food use are developed using the traditional breeding methods that had existed before the introduction of gene technology, some of the characteristics of organisms may be altered, either in a positive or a negative way. National food authorities may be called upon to examine the safety of such conventional foods obtained from novel varieties of organisms, but this is not always the case. In contrast, most national authorities consider that specific assessments are necessary for GM foods. Specific systems have been set up for the rigorous evaluation of GM organisms and GM foods relative to both human health and the environment. Similar evaluations are generally not performed for conventional foods. Hence there currently exists a significant difference in the evaluation process prior to marketing for these two groups of food.

The WHO Department of Food Safety and Zoonoses aims at assisting national authorities in the identification of foods that should be subject to risk assessment and to recommend appropriate approaches to safety assessment. Should national authorities decide to conduct safety assessment of GM organisms, WHO recommends the use of *Codex Alimentarius* guidelines.

WHO has been taking an active role in relation to GM foods, primarily for two reasons:

- On the grounds that public health could benefit from the potential of biotechnology, for example, from an increase in the nutrient content of foods, decreased allergenicity and more efficient and/or sustainable food production; and
- Based on the need to examine the potential negative effects on human health of the consumption of food produced through genetic modification in order to protect public health. Modern technologies should be thoroughly evaluated if they are to constitute a true improvement in the way food is produced.

WHO, together with FAO, has convened several expert consultations on the evaluation of GM foods and provided technical advice for the *Codex Alimentarius* Commission which was fed into the Codex Guidelines on safety assessment of GM foods. WHO will keep paying due attention to the safety of GM foods from the view of public health protection, in close collaboration with FAO and other international bodies. The safety assessment of GM foods generally focuses on (a) direct health effects (toxicity), (b) potential to provoke allergic reaction (allergenicity); (c) specific components thought to have nutritional or toxic properties; (d) the stability of the inserted gene; (e) nutritional effects associated with genetic modification; and (f) any unintended effects which could result from the gene insertion.

Though, it is widely claimed that biotechnology, particularly genetically engineered food offers dramatic promise for meeting some of the 21st century's greatest challenges; like all new technologies, it also poses certain apprehensions and risks, both known and unknown. It is, therefore, paramount in this context, to know the basic processes involved in genetic modification for proper appreciation of the related issues and challenges.

15.1.2 The Basic Processes Involved in Genetic Modification

i. **Commercial venture :** The first commercially grown GM food crop was Tomato (called Flavr Savr), modified to ripen without softening by a Californian company Calgene, which took the initiative to obtain approval for its release in 1994. Currently, a number of food crops such as soyabean, corn, cotton, tomatoes, Hawaiian papaya, potatoes, rapeseed (canola), sugarcane, sugar beet, field corn as well as sweet corn and rice have been genetically modified to enhance either their yield, or size,or durability, etc. Scientists are also working on crops which they hope will be useful for industry, such as plants that produce oil for the cosmetics industry, crops with altered nutritional value, and even crops that produce pharmaceutical drugs (Anonymous, 2009). Major producers of transgenic crops include USA, Argentina, Brazil, India, Canada, China, Paraguay, South Africa, among others.

ii. **Issues of crop protection:** The initial objective for developing GM plants was to improve crop protection. The GM crops currently in the market are mainly aimed at an increased level of crop protection through the use of one of the three basic traits: resistance to insect damage; resistance to viral infections; and tolerance towards herbicides. All the genes used to modify crops so far are derived from micro-organisms.

   - Insect resistance is achieved by incorporating into the food plant the gene for toxin production from the bacterium *Bacillus thuringiensis* . This toxin is used as a conventional insecticide in agriculture and is safe for human consumption. GM crops that permanently produce this toxin have been shown to require lower quantities of insecticides;
   - Virus resistance is achieved through the introduction of a gene from certain viruses which cause disease in plants. Virus resistance makes plants less susceptible to diseases caused by such viruses, resulting in higher crop yields;

- Herbicide tolerance is achieved through the introduction of a gene from a bacterium conveying resistance to some herbicides. In situations where weed pressure is high, the use of such crops has resulted in a reduction in the quantity of the herbicides used.

iii. **Understanding risks and benefits:** The risk-benefit analysis of the GM crops can be summarized as below:

Benefits	• Improved resistance to diseases, pests and herbicides • Improved tolerance to cold/heat • Improved tolerance to drought/ salinity • Reduced maturation time • Increased nutrients, yields, quality and stress tolerance • Increased food security for growing population • Food with greater shelf life or food with medicinal benefits, such as edible vaccines—for example, bananas with bacterial or rotavirus antigens
Issues of concern (Human health risks and environmental safety concerns)	• Potential impact on human health including allergens, transfer of antibiotic resistance markers and 'out crossing'. The movement of genes from GM plants into conventional crops or related species in the wild (referred to as 'out crossing'), as well as the mixing of crops derived from conventional seeds with those grown using GM seeds, may have an indirect effect on food safety and food security. It has been found that genes inserted into GM food survive digestive processes and are transferred into the human gut. • Potential impact on environment, including transfer of transgenes through cross-pollination, unknown effects on other organisms (e.g., soil microbes), and loss of flora and fauna biodiversity.

iv. **Some other concerns:** Critics of genetically modified food have also pointed out certain other aspects apart from human health risks and environmental safety concerns as mentioned below (Anonymous, 2009).

Access and Intellectual Property Rights	• Critics claim that patent laws give developers of the GM crops a dangerous degree of control over the food supply • Domination of world food production by a few companies • Increasing dependence of developing countries on industrialized nations • Biopiracy, or foreign exploitation of natural resources
Ethical concerns	• Violation of natural organisms' intrinsic values by mixing among species • Objections to consuming animal genes in plants

v. **Safety assessment:** The starting point for the safety assessment of genetically engineered food products is to assess if the food is 'substantially equivalent' to its natural counterpart. In deciding whether a modified product is substantially equivalent, the product is tested by the manufacturer for unexpected changes in a limited set of components such as toxins, nutrients or allergens that are present in the unmodified food. The data is then assessed by an independent regulatory body. If these tests show no significant difference between the modified and the unmodified products, then no further food safety testing is required. However, if the product has no natural equivalent, or shows significant differences from the unmodified food, then further safety testing is carried out. This method has, however, been severely criticized by some scientists since it is not clear what level of similarity makes something 'substantially equivalent'.

Although the benefits of transgenic technology are clear, the potential risks have created public concerns about the wisdom of releasing and consuming genetically modified (GM) crops. Biotechnological tool such as recombinant DNA technology has come a long way in solving the problem of food security. Genetic modification can help humankind to face new challenges as a result of high population growth, biodiversity loss and climate change (Mishra and Kumari, 2018). Therefore, it is imperative to have robust biosafety protocols/ procedures for India. While designing GM crops, the native species and gene in question needs to be taken into account. GM crops might become agricultural weeds or invade natural habitats if proper risk assessment (RA) is not performed prior to their release.

15.2 Impacts of GM Crops and Mitigation Techniques

The possible impacts of GM crops are as follows:

i. Weediness and invasiveness: One of the potential concerns about genetically modified organisms (GMOs) is that they will become agricultural weeds or invade natural habitats, as the traits introduced by GMOs might increase the reproductive success or fitness of the crop, thereby increasing its competitive ability. One conjectural risk is that GMOs will either cause the host species to become invasive or will escape from the original host species or cause other species to become invasive. New combinations can create genotypes with different and surprising ecological behaviors. Researchers have shown that the gene flow from transgenic crop is easy to escape to the weedy relative Brassica campestris. Canola is also capable of cross pollinating with several other weed species including wild radish (*Raphanus raphanistrum*) and buchan weed (*Alternanthera philoxeroides*).

ii. Gene flow from GM crops: The transgene escape to weedy relatives through pollen is one of the potential risks of GM crops. Gene flow indicates the movement of genes or genetic materials from one population into another. There are three avenues for gene flow to occur: pollen-mediated, seed-mediated and vegetative-propagule-mediated gene flow. To minimize the possibility of transgene flow, a number of strategies have been developed or proposed, applying physical and biological approaches include confined field trial, transgenic mitigation, maternal inheritance, male sterility, cleistogamy, apomixis, incompatible genomes, temporal control via inducible primers and seed sterility. These are called "genetic use restriction technologies" or GURTs.

Mitigation techniques: Some of the mitigation techniques are as follows:

a. Confined field trial strategy: One of the ways to understand the gene flow is to conduct confined field trial (CFT). CFT is a small-scale experiment, done in the open field, with the intention of confining plant genes and plant material to trial site. CFTs are needed for breeding trials to incorporate traits into locally adapted varieties or to create populations for genetic study, to collect safety data to inform regulatory decisions on GM crops commercial release, to scale up experimental crops so that sufficient seed or other plant material is available for animal-feeding studies, or to study possible environmental impacts such as plant characteristics, potential for weediness, changes in pollen production, or gene flow.

b. Transgenic mitigation strategy : A transgenic mitigation (TM) strategy is also available for reducing the potential risks of escaped transgene(s) to the weedy or wild populations by co-introducing "mitigator" genes that are tandemly linked to the target transgene(s) to deliberately reduce the fitness of any hybrid and its progene i.e. the individuals carrying those traits would be eliminated in natural populations through competition with other more highly fit native individuals (Mishra and Kumari, 2018). Some of the deleterious traits that have been proposed are abolition of secondary dormancy, dwarfing and inhibition of shattering of seeds. A mitigator dwarfing Δgai (gibberelic acid-insensitive) gene, when transformed into tobacco, reduced fitness by 17% and was predicted to slow escape from a few generations to many thousands, depending on rates of gene flow and levels of recombination. Thus, TM would limit transgene escape through pollen and seed flow.

c. Chloroplast transformation: To prevent gene flow via pollen, transgenes can be targeted to chloroplast genomes, which are generally transmitted only through ovules of the female parent. Numerous transgenes have been successfully integrated into chloroplasts in wide variety of plant species (Daniell and Kumar, 2005) and this approach has been shown to block pollen flow of the transgene in tobacco and tomato. Although targeting transgenes to the chloroplast will not completely limit all the gene flow, as it does not restrict transgene movement via seed dispersal.

iii. Male sterility: Inserting transgenes into male sterile lines is another means of preventing transgene escape via pollen flow. Either naturally derived male sterile lines can be used or male-sterility mutants can be engineered. One approach is to use tapetum-specific promoters to derive expression of a recombinant RNase gene. Plant Genetics Systems (Ghent, Belgium) has engineered male-sterile and male-restorer lines of GM rapeseed utilizing two genes from Bacillus amyloliquefaciens-barnase, which cleaves RNA and barstar, a protein that binds to barnase and prevents its function. The central pitfall of using male-sterile lines, is it can only be used for vegetative crops (Mishra and Kumari, 2018).

iv. Cleistogamy : Cleistogamy or automatic self-pollination is a process involving plant propagation by self-pollinating flowers. A deemed superwoman1-

cleistogamy has been isolated from rice.[9] The drawbacks of this approach are that transgenic seed movement is not restricted; the crop must be self-fertile, which may suppress the creation of genetically superior plants and characteristic must be stably expressed (Fargue *et.al.*, 2006).

v. Apomixis: Apomixis relates to asexual reproduction by plants without fertilization. In some of the species, seed is derived from apomictic origin and not the product of sexual reproduction. Apomictic embryos can be produced from the integument or nucellus or from megaspore mother cells or nucellar cells. An apomictic Maize/ Tripsicum hybrid has been patented (Kindiger *et.al.*, 1998). Apomixis can prevent transgene escape through pollen flow but cannot restrict movement of the transgene via seeds.

vi. Inducible promoter/gene: If transgenes could be removed before flowering by temporal control of inducible promoters, escape through both pollen and seed flow can be prevented.[12] suggested that this could be done by placing a chemically induced or fruit-specific promoter in front of a construct with a site-specific recombinase gene such as Cre that recognized lox sites flanking the transgene. When Cre expression is induced, the transgene would be removed. The disadvantage of this approach is that it is not applicable to traits required throughout the plants' life, the inducer must completely penetrate all the relevant plant tissues and somebody must induce the system. A related approach might be to use a chemically induced promoter that allows expression of the transgene only when a chemical is applied. If the elicitor is not found outside of the agronomic system, then the transgene would not be expressed in nature. One other mode of restricting transgene escape into relatives is to incorporate inducible genes that make seeds sterile.

A system was developed where an inducible promoter is used to induce a site-specific recombinase (Cre) that removes DNA sequences flanked by lox sites, allowing for expression of a lethal gene proposed a system where Cre is placed adjacent to the transgene and removes both when induced (Hill *et.al.*, 2007). This technique requires induction of all the cells which is cumbersome. Any failure could result in some viable seeds being produced. To counter this problem approaches have been developed where the inducible seed termination mechanism is "on" until the inducer turns it "off", rather than the reverse. In this system, a seed promoter is linked to the barnase gene and the barstar gene is linked to an inducible promoter, while this method has great potential in limiting gene flow, considerable resistance has been expressed to it because these systems would also prevent farmers from saving seed.

15.3 Impacts of GMOs on Non-target Organisms

The negative impact of GM crops on non-target organism is also a serious threat. Transgenic plants that produce insecticidal substances should continue to be subject to careful testing to ensure safety and minimizing the environmental risks.

A tiered approach to an ecological risk assessment is recommended, particularly in a regulatory context, as a screening tool to determine potential risks to non-target organisms from releasing a GE crop in the environment. Non-target organisms include invertebrates beneficial to maintaining a healthy ecosystem in an agricultural setting such as pollinators, biological control organisms and decomposers. Other wild lives that are not intended to be harmed by a GE crop such as birds, mammals and fishes are also considered as non-target organisms. Many direct and indirect effects could result from release of a GE crop in the environment, such as fitness costs to an organism, a reduction in the abundance or diversity of non-target organism or a reduction in functional responses. A tiered approach allows for a common testing framework and a standardized sequence of tests to be used as screening tools to evaluate and focus risk considerations.

Effects of insect resistant GM plants on non-target organisms: Insect resistance, conferred via expression of a variety of *Bacillus thuringiensis* delta-endotoxins, is the second most commonly used trait, after herbicide resistance, in commercial GM crops. Bt is a ubiquitous gram positive, spore forming bacteria found in soil, insects, stored-product dust, and deciduous and coniferous leaves. The insecticidal activity starts by the formation of parasporal crystal of the bacterium during its stationary phase. The crystals are made of protoxins, referred to as cry toxins or endotoxins, which, when ingested by an insect, are activated by proteases in the insect mid-gut. Cry toxins readily bind to receptors on the apical brush border of the mid gut microvillae of susceptible insects and insert into the membrane. This insertion leads to the formation of pores causing lysis of cells, leading to starvation, paralysis, septicemia and death of the insect. The potential hazard of transgenic crops engineered with plant-incorporated-protectant (PIP) trait is toxicity to non-target beneficial organism (Hill *et.al.*, 2006). The exposure assessment predicts the likelihood that non-target organism will have the dietary exposure to the expressed PIP at or above the hazard threshold level. An evaluation requires studies that address acute and chronic mortality effects, changes in reproductive processes and trophic level disruptions of the ecosystem.

A reduction in host or feeding directly plant parts may have adverse effects on natural enemies. Beneficial insects if allowed to persist in Bt fields may aid in controlling secondary pests. GM crops can affect non-target organisms such as avian including bobwhite quail, mallard duck, broiler chicken and migratory birds; wild mammals such as rodents, deer, and cats; aquatic animals like catfish, freshwater invertebrate such as Daphnia magna; aquatic insects like caddis flies. One of the major concerns of PIP GM crops is the toxic effect on the pollinators, (bees, flies) and parasitoids (Nasonia vitripennis). The adult and larval predators feed primarily on arthropods. Indirect exposure may occur when predatory species feed on herbivores that have fed on Bt crops e.g. lady beetles, green lays wing etc. Another consent is the potential impact of PIP GM crops on beneficial non-target invertebrates, therefore the monarch butterfly has been used as bio indicators species to test the relative risk of Bt corn pollen. Ecological risk assessment for PIP GM crops should have a tiered approach. Lower tier test which includes dietary laboratory test, potential toxicity or fitness cost as well as trophic level effect that include function or changes in the structure and species diversity of the invertebrate community. Since pest resistant plant varieties utilizes significantly lesser amount of insecticides and because of their compatibility with biological and cultural practices, such as evaluation of impact on non-target organism, would go along with it. Therefore, scientific analysis of risks and benefits should be conducted before commercial use.

15.4 Ecological Risk Assessment

Risk assessment is a process that uses scientific data to determine potential hazards, exposure and the likelihood of adverse impacts on populations of organisms in the environment. This is usually implemented through the integration of hazard identification and characterization of all of the elements of risk associated with a new GM crop or derived product. Typical categories of hazards arising from the introduction of transgenic crops include: possible unintended negative health effects in a susceptible subgroup of the target population; the evolution of resistance in the targeted pest/ pathogen population when the transgene confers resistance to a pest or pathogen; non-target hazards associated directly or indirectly with the transgenic plant or transgene product outside the plant; and those associated with the integration and subsequent expression of the transgene in a different organism or species following gene flow.

The hazardous effects can either have a direct or indirect, immediate or delayed impact on the environment or human and animal health (European Parliament 2001). An RA is also conducted with a view to determine if there is a need for risk management and if so, the most appropriate methods to be used. The results of RA can then be used by the relevant competent authority to make an informed decision regarding whether or not to give approval for import or cultivation.

The potential risk of GE crop could be of following types:

i. **Transgenic organism persists without cultivation :** The trait expressed in transgenic crop might release it from constraints, thus allowing it to persist, become invasive in non-agricultural situation. This would lead to decreased biodiversity.

ii. **Transgenic organisms interbreed with related taxa:** If a GM crop crosses out with a wild relative then transgene might get transferred to progene or population of wild relative. That introgressed trait then might cause the recipient population to release from a natural constraint on population size.

iii. **Evolution of resistance by the pest to the insect or disease resistance:** The transgene can shorten the useful life of GM product. This is of concern if the GM product is designed to replace chemical pesticides. The evolution of resistance is also a concern to the developer of the GM crop because it needs a short period of time during which the developer can market the crop.

Development of pest resistance to pesticides and GM crops: Pesticide resistance is a genetically based phenomenon. Resistance occurs when a pest population is exposed to a pesticide. When this happens, not all insects are killed. Those individuals that survive frequently have done so because they are genetically predisposed to be resistant to that pesticide. Repeated applications and higher rates of the insecticide will kill increasing numbers of individuals but some resistant insects will survive. The offspring of these survivors will carry the genetic makeup of their parents. These offspring, many of which will inherit the ability to survive the exposure to the insecticide, will become a greater proportion with each succeeding generation of the population. Insects developing resistance to GM crops is also a major concern. All of the current commercially available insect-resistant crops use genes derived from *Bacillus thuringiensis* (Bt). Diamondback moth (Plutela xylostella), Indian meal moth (Plodia interpunctella), Colorado potato beetle (Leptinotarsa decemlineata) and several other insects have shown the ability to adapt to resist BT foliar sprays in the field or laboratory (Whalon *et.al.*, 2010).

15.5 Pesticide and Herbicide Resistance Management

Resistance is genetically based, controlled by one or more genes and passed from parents to offspring. Resistance is preadaptive i.e. the application of a pesticide does not "create" resistant individuals; resistant individuals are already present in the pest population, although usually at very low frequency. Application of the pesticide introduces selection pressure that results in increased frequency of the resistant genotypes. Resistance involves genetic variation, differential survival and reproduction by survivors. For example, in a field experiment, potatoes containing Colorado potato beetle larvae were sprayed with carbofuran at the commercially labeled rate. Control was outstanding - 99.98% of the larvae were killed and only 96 larvae survived. These larvae carried a dominant gene for acetylcholinesterase that made them insensitive to carbofuran. (Table 21) (Difonzo and Collen, 2012).

Table 21: Considerations for designing a refuge

Pest biology/ behavior	Cropping system
Larval movement	Farm/ field size
Adult dispersal	Rotational crops
Mating time	Neighboring host and non-host crops (including others with Bt)
Mating location	Other pests in the crop
Mating frequency	Type, frequency, timing of pesticide applications in the Bt and refuge crops
Generations/ year	
Alternate hosts	

i. Insecticide resistance management: The key to managing insecticide resistance is to reduce selection pressure. Under the principles of IPM, insecticide applications are made based on scouting and threshold levels of pests. Several tactics could be employed in order to reduce or minimize insecticidal resistance like:

- Alternate or mix insecticides with different modes of action, for instance apply a carbamate followed by a pyrethroid or tank-mix a pyrethroid with an organophosphate. In both the cases, if an insect is resistant to one type of insecticide and survives its application, the second insecticide will still kill it.
- Use a high dose of active ingredient, i.e., use the highest label rate coupled with excellent coverage.
- Leave an unsprayed refuge. A refuge is an area that is not sprayed, leaving a population of unexposed individuals. This area provides susceptible insects to mate with resistant insects generated in sprayed fields.

Management of resistant BT crops: *Bacillus thuringiensis* (Bt), is a naturally occurring bacterium that produces spores and protein crystals that kill insects. Based on resistant colonies and field populations, several resistance mechanisms are proposed for Bt.[23] These include modification of the alkaline pH, changes in enzymes that deactivate toxin, enhanced repair mechanisms in the gut and most importantly, a reduction or loss of binding sites in the gut.

Management of BT crops could be done by applying "alternate Bt toxins". In this system, plants expressing different Bt toxins may be grown from field to field or from year to year. In this case, if an insect is able to feed on plants expressing some toxin say toxin1, it is still susceptible to the other toxin say toxin2 and is killed by plants expressing that Bt as insects carrying genes for resistance to both toxins are extremely rare. Another management system may involve using "combine Bt toxins" which performs "gene pyramiding". For instance, Bollgard II cotton expresses both Cry 1Ac and Cry 2Ab to control cotton bollworm and tobacco budworm (Helicoverpa virescens). Gene pyramiding is designed to improve efficacy of the crop and to reduce the formation of resistance.

In the absence of genes to pyramid, the main strategy for resistance management is the high-dose/ refuge strategy, implemented for BT cotton and Bt corn modified to kill pest caterpillars.

1. High dose: The high dose is expected to kill both homozygous susceptible (SS) and heterozygous resistant (RS) individuals.
2. Untreated refuge: The refuge area maintains a population of unexposed, susceptible (SS) individuals to mate with the rare resistant (RR) survivors of the high-dose Bt The resulting offspring from this mating are heterozygous resistant (RS) yet still susceptible and thus killed by the high Bt crop.

ii. Herbicide resistance management: Several biological factors affect the risk of producing herbicide-resistant weeds, these factors include: frequency of resistant plants, fitness/ competitiveness of resistant plants, seed production, dissemination potential and means of inheritance.

Frequency of resistant plants: The initial frequency of resistant plants in the environment is the benchmark by which the frequency will change because of selection pressure. If the initial frequency is very low, a greater amount of selection pressure will be required for the resistant plants to become dominant. If the initial frequency is high, the resistant plants will become dominant with much less selection pressure.

Plant fitness/competitiveness: Plants that are well adapted and competitive in their environment will increase in frequency more rapidly than those that are less competitive than their non-resistant cohort.

Seed production and dissemination potential: The strength of selection pressure is based on how greatly seed production is reduced in non-resistant plants compared to resistant ones. If the resistant plant species have high seed production capacity, the number of resistant plants will increase rapidly. In contrast, if the resistant plant species produce low numbers of seeds, the rate of increase in the population of resistant plants will be slower.

Mode of inheritance : If a resistant trait is related to a single dominant gene, resistance is likely to spread more rapidly, than with a single recessive gene, since the resistance will be expressed in the heterozygous individuals.

Management of herbicide resistance in weeds: Mechanism involved: There are primarily three mechanisms that are involved in herbicide resistance in weeds, viz., modified site of action; metabolism and compartmentalization.

- It is the most common and most irresistible mechanism for herbicide resistance. Herbicide sites of action are often proteins or enzymes. The herbicide will bind to the site of action and inactivate the protein or enzyme, resulting in the disruption of a biochemical or physiological process. Therefore, the herbicide is unable to disrupt the biochemical or physiological process and the plant is not affected by exposure to the herbicide.
- In herbicide metabolism, herbicide-resistant plants are capable of rapidly detoxifying the herbicide. The rapid detoxification of the herbicide protects the plant from significant herbicide injury.
- In compartmentalization, plants make herbicide molecules in compartments of plant cell vacuoles which results in the herbicide molecules prevention from interacting with the herbicide site of action and therefore the plant is not affected.

Practices to reduce risk of herbicide-resistant weeds:

- Herbicide rotation: It is important to avoid consecutive applications of herbicides with the same site of action against the same weed species unless other effective control practices are also included in the management system.
- Herbicide combinations: Here, two or more herbicides are applied together that have different sites of action.
- Crop rotation: Rotating crops within a field can be an effective practice to minimize selection pressure for resistant weeds.

15.6 Indian Scenario

Bt cotton was approved by Government of India in March 2002 as the first transgenic crop for commercial cultivation for a period of three years. Bt cotton incorporates a gene from a bacterium *Bacillus thuringiensis*, which is effective against the American bollworm, the major pest on cotton. *Bt* cotton has since been grown in six states, i.e., Madhya Pradesh, Gujarat, Maharashtra, Andhra Pradesh, Karnataka and Tamil Nadu. The area under Bt cotton from 72,000 acres in 2002 increased to 2,30,000 acres in 2003 and 13,00,000 acres in 2004. Initially, three hybrids produced by M/s Maharashtra Hybrid Seeds Company Ltd. (MAHYCO) were approved in 2002. Other seed companies such as M/s Rasi Seeds, M/s Ankur Seeds, M/s. Krishidhan Seeds and M/s Ajeet Seeds have initiated development of Bt cotton hybrids in association with MAHYCO. On April 1, 2004, one hybrid of M/s Rasi Seeds has been accorded approval for commercial cultivation and 12 more varieties are under large scale field trials (Anonymous, 2005). Companies such as Nath Seeds, Syngenta and J.K. Agrigenetics are developing transgenic cotton using different genes/events.

Apart from cotton, there are more than 20 crops under research and development in about 50 public and private sector organizations in India. These include rice, potato, brinjal, cabbage, cauliflower, groundnut and pigeon peas. The target traits include insect resistance, herbicide tolerance, viral and fungal disease resistance and stress tolerance. Out of these, 13 crops have been approved for contained limited field trials in India. These are mostly related to insect resistance using Cry genes. However, as more and more transgenic crops are being released for field-testing and commercialization, concerns have been expressed about the potential risks associated with their impact to human health, environment and biological diversity. Questions centre around increased toxicity and allergenicity, impact of introduced traits introgressing into other related species throughout crossing, the potential buildup of resistance in insect populations to engineered insecticidal traits, unintended secondary effects on non-target organisms, and potential effects on biodiversity.

Biosafety legislation and regulatory institutions to implement them have been put in place by many countries including India, engaged in transgenic research and commercialization. There are elaborate steps to manage these risks and it is the responsibility of the scientists, industry, and the government to assure the public of the safety of the novel products commercialized (Anonymous, 2005).

India has a well-defined regulatory mechanism for development and evaluation of GMOs including transgenic crops and the products thereof. Rules notified in 1989 under Environmental Protection Act, 1986 (EPA) define the competent authorities and composition of such authorities for handling regulation of GMOs and products thereof.

Presently, there are six competent authorities:

1. Recombinant DNA Advisory Committee (RDAC)
2. Review Committee on Genetic Manipulation (RCGM)
3. Genetic Engineering Approval Committee (GEAC)
4. Institutional Biosafety Committees (IBSC) attached to every organization engaged in rDNA research
5. State Biosafety Coordination Committees (SBCC)
6. District Level Committees (DLC).

Guidelines for safety in biotechnology have been issued by the Department of Biotechnology (DBT) in 1990 covering research, field trials and commercial applications. DBT also brought out separate guidelines for research in transgenic plants in 1998. The National Seed Policy, 2002 also has a separate section on transgenic plant varieties.

As the first commercial transgenic crop in the country, *Bt* cotton has been subject to extensive monitoring by both at the central and the state levels during the last three years of commercial cultivation. Simultaneously, there have been various initiatives towards capacity building of various stakeholders. A series of eight workshops on "Biosafety issues related to GMOs" was organized in 2002 by Biotech Consortium India Limited in association with Ministry of Environment & Forests (MoEF), Government of India and Department of Biotechnology, Govt. of India. MoEF sponsored another series of workshops on "Biosafety issues related to transgenic crops" in 2003 in the six Bt cotton growing states covering various stakeholders such as government officials, research institutions, agricultural universities, NGOs and farmers for sensitization and creation of awareness.

15.7 Recombinant DNA Guidelines, 1990

With the advancement of research in biotechnology initiated by various Indian institutions and industry, Department of Biotechnology had formulated Recombinant DNA Guidelines in 1990. These guidelines were further revised in 1994 to cover R&D activities on GMOs, transgenic crops, large-scale production and deliberate release of GMOs, plants, animals and products into the environment, shipment and importation of GMOs for laboratory research (Anonymous, 2005). For research, the guidelines have been classified into three categories, based on the level of the associated risk and requirement for the approval of competent authority.

- Category I activities include those experiments involving self cloning using strains and also inter-species cloning belonging to organism in the same exchanger group which are exempt for the purpose of intimation and approval of competent authority.
- Category II activities which require prior intimation of competent authority and include experiments falling under containment levels II, III and IV (details of each containment level provided separately in the guidelines).
- Category III activities that require review and approval of competent authority before commencement include experiments involving toxin gene cloning, cloning of genes for vaccine production, and other experiments as mentioned in the guidelines.

The levels of risk and classification of the organisms within these categories have been defined in these guidelines. Appropriate practices, equipment and facilities necessary for safeguards in handling organisms, plants and animals in various risk groups have been recommended. The guidelines employ the concept of physical and biological containment and the principle of good laboratory practices. For containment facilities and biosafety practices, recommendations in the WHO laboratory safety manual on genetic engineering techniques involving microorganisms of different risk groups have been incorporated therein.

The guidelines categorize experiments beyond 20 liters capacity for research and industrial purposes as large-scale. In case of cultivation of plants, this limits is 20 acres area. The guideline gives principles of occupational safety and hygiene for large-scale practice and containment. Safety criteria have also been defined in the guidelines. Physical containment conditions that should be ensured for large-scale experiments and production have been specified in the guidelines. For release to the environment the guidelines specify appropriate containment facilities depending on the type of organisms handled and potential risks involved. The guidelines require the interested party to evaluate rDNA modified organism for potential risk prior to application in agriculture and environment like properties of the organism, possible interaction with other disease causing agents and the infected wild plant species. An independent review of potential risks should be conducted on a case-to-case basis.

15.8 Guidelines for Research in Transgenic Plants, 1998

In 1998, DBT brought out separate guidelines for carrying out research in transgenic plants called the Revised Guidelines for Research in Transgenic Plants. These also include the guidelines for toxicity and allergenicity of transgenic seeds, plants and plant parts. These guidelines cover areas of recombinant DNA research on plants including the development of transgenic plants and their growth in soil for molecular and field evaluation (Anonymous, 2005). The guidelines also deal with import and shipment of genetically modified plants of research purposes. Genetic engineering experiments on plants have been grouped under three categories.

- Category I include routine cloning of defined genes, defined non-coding stretches of DNA and open reading frames in defined genes in E. coli or other bacterial/fungal hosts which are generally considered as safe to human, animals and plants.
- Category II experiments include experiments carried out in lab and green house/net house using defined DNA fragments nonpathogenic to human and animals for genetic transformation of plants, both model species and crop species.
- Category III includes experiments having high risk where the escape of transgenic traits into the open environment could cause significant alterations in the biosphere, the ecosystem, plants and animals by dispersing new genetic traits the effects of which cannot be judged precisely. This also includes experiments having risks mentioned above conducted in green houses and open field conditions.

To monitor the impact of transgenic plants on the environment over a period of time, a special Monitoring cum Evaluation Committee (MEC) has been set up by the RCGM. The committee undertakes field visits at the experimental sites and suggests remedial measures to adjust the trial design, if required, based on the on-the-spot situation. This committee also collects and reviews information on the comparative agronomic advantages of the transgenic plants and advises the RCGM on the risks and benefits from the use of transgenic plants under evaluation. The guidelines include complete design of a contained green house suitable for conducting research with transgenic plants. Besides, it provides the basis for generating food safety information on transgenic plants and plant parts.

15.9 Seed Policy 2002

The Seed Policy 2002 contains a separate section (No. 6) on transgenic plant varieties. It has been stated that all genetically engineered crops/varieties will be tested for environment and biosafety before their commercial release as per the regulations on guidelines of the EPA, 1986. Seeds of transgenic plant varieties for research purposes will be imported only through the National Bureau of Plant Genetic Resources (NBPGR) as per the EPA, 1986. Transgenic crops/varieties will be tested to determine their agronomic value for at least two seasons under the All India Coordinated Project Trials of ICAR, in coordination with the tests for environment and bio-safety clearance as per the EPA before any variety is commercially released in the market. After the transgenic plant variety is commercially released, its seed will be registered and marketed in the country as per the provisions of the Seeds Act (Anonymous, 2005). After commercial release of a transgenic plant variety, its performance in the field, will be monitored for at least 3 to 5 years by the Ministry of Agriculture and State Departments of Agriculture.

It has also been mentioned that transgenic varieties can be protected under the PVP legislation in the same manner as nontransgenic varieties after their release for commercial cultivation. A copy of seed policy is placed in Annex-2. In addition the above, Ministry of Agriculture has issued a notification on November 12, 2003 nominating the Central Institute of Cotton Research (CICR) to act as a referral laboratory for Bt cotton seeds.

The Government of India trusts that the National Seeds Policy will receive the fullest support of State Governments/Union Territory Administrations, State Agricultural Universities, plant breeders, seed producers, the seed industry and all other stakeholders, so that it may serve as a catalyst to meet the objectives of sustainable development of agriculture, food and nutritional security for the population, and improved standards of living for farming communities. The National Seeds Policy will be a vital instrument in attaining the objectives of doubling food production and making India hunger free. It is expected to provide the impetus for a new revolution in Indian agriculture, based on an efficient system for supply of seeds of the best quality to the cultivator. The National Seeds Policy will lay the foundation for comprehensive reforms in the seed sector. Significant changes in the existing legislative framework will be effected accompanied by programmatic interventions. The Policy will also provide the parameters for the development of the seed sector in the Tenth and subsequent Plans. The progress of implementation of the Policy will be monitored by a High Level Review Committee.

RULES, 1989: The Ministry of Environment & Forests, Government of India notified the rules and procedures for the manufacture, import, use, research and release of GMOs as well as products made by the use of such organisms on December 5, 1989 under the Environmental Protection Act 1986 (EPA). These rules and regulations, commonly referred as Rules 1989 cover the areas of research as well as large scale applications of GMOs and products made there from throughout India. The Rules, 1989 order compliance of the safeguards through voluntary as well as regulatory approach and any violation and non-compliance including non-reporting of the activity in this area would attract punitive action provided under the EPA.

The two main agencies identified for implementation of the rules are the Ministry of Environment & Forests and the Department of Biotechnology, Government of India. The rules have also defined competent authorities and the composition of such authorities for handling of various aspects of the rules. There are six competent authorities as per the rules, viz., Recombinant DNA Advisory Committee (RDAC),

Review Committee on Genetic Manipulation (RCGM), Genetic Engineering Approval Committee (GEAC), Institutional Biosafety Committees (IBSC, State Biosafety Coordination Committees (SBCC) and District Level Committees (DLC).

Out of these, the three agencies that are involved in approval of new transgenic crops are:

1. IBSC set-up at each institution for monitoring institute level research in genetically modified organisms.
2. RCGM set-up at DBT to monitor ongoing research activities in GMOs and small scale field trials.
3. GEAC in the Ministry of Environment and Forests set-up to authorize large-scale trials and environmental release of genetically modified organisms.

The Recombinant DNA Advisory Committee (RDAC) constituted by DBT takes note of developments in biotechnology at national and international level and prepares suitable recommendations. The State Biotechnology Coordination Committees (SBCCs) set up in each state where research and application of GMOs are contemplated, coordinate the activities related to GMOs in the state with the central ministry (Anonymous, 2005). SBCCs have monitoring functions and therefore have got powers to inspect, investigate and to take punitive action in case of violations. Similarly, District Level Committees (DLCs) are constituted at district level to monitor the safety regulations in installations engaged in the use of GMOs in research and application.

The approvals and prohibitions under Rules 1989 are summarized below:

- No person shall import, export, transport, manufacture, process, use or sell any GMOs, substances or cells except with the approval of the GEAC.
- Use of pathogenic organisms or GMOs or cells for research purpose shall be allowed under the Notification, 1989 of the EPA, 1986.
- Any person operating or using GMOs for scale up or pilot operations shall have to obtain permission from GEAC.
- For purpose of education, experiments on GMOs IBSC can look after, as per the guidelines of the Government of India.
- Deliberate or unintentional release of GMOs not allowed.
- Production in which GMOs are generated or used shall not be commenced except with the approval of GEAC
- GEAC supervises the implementation of rules and guidelines.
- GEAC carries out supervision through SBCC, DLC or any authorized person.
- If orders are not complied, SBCC/DLC may take suitable measures at the expenses of the person who is responsible.

- In case of immediate interventions to prevent any damage, SBCC and DLC can take suitable measures and the expenses incurred will be recovered from the person responsible.
- All approvals shall be for a period of 4 years at first instance renewable for 2 years at a time.
- GEAC shall have powers to revoke approvals in case of:
  i. Any new information on harmful effects of GMOs.
  ii. GMOs cause such damage to the environment as could not be envisaged when approval was given.
  iii. Non-compliance of any conditions stipulated by GEAC.

The Recombinant DNA Advisory Committee (RDAC) constituted by DBT takes note of developments in biotechnology at national and international level and prepares suitable recommendations. The State Biotechnology Coordination Committees (SBCCs) set up in each state where research and application of GMOs are contemplated, coordinate the activities related to GMOs in the state with the central ministry. SBCCs have monitoring functions and therefore have got powers to inspect, investigate and to take punitive action in case of violations. Similarly, District Level Committees (DLCs) are constituted at district level to monitor the safety regulations in installations engaged in the use of GMOs in research and application.

15.10 Controversies and Moratoriums Associated with GM Crops in India – Timeline (Alex Andrews George, 2020)

- 2002 : Bt cotton introduced in India.
- 2006 : Activists filed a PIL against GM crops in the Supreme Court.
- 2010 : The then environmental minister Jairam Ramesh blocked the release of *Bt* Brinjal until further notice owing to a lack of consensus among scientists and opposition from brinjal-growing states. No objection certificates from states were made mandatory for field trials.
- 2012 : Parliamentary standing committee on agriculture, in its 37th report asked for an end to all GM field trials in the country.
- 2013 July : New crop trials have been effectively on hold since late 2012, after a supreme court-appointed expert panel recommended suspension for 10 years until regulatory and monitoring systems could be strengthened. Though the SC panel suggested moratorium on GM trails, there was no official verdict from the Supreme Court on this issue.
- 2013 July :Environment minister Jayanthi Natarajan put on hold all trials following SC panel suggestions.
- 2014 : Her successor, Veerappa Moili cleared the way for trails. (NB: Two of Manmohan Singh's own environment ministers had stalled GM trials

earlier, but Veerappa Moily took an opposite stand and the process of approving the one-acre field trials restarted.)

- 2014 March : GEAC (UPA government) approved field trials for 11 crops, including maize, rice, sorghum, wheat, groundnut and cotton.
- 2014 July : 21 new varities of genetically modified (GM) crops such as rice, wheat, maize and cotton have been approved for field trials by the NDA government in July 2014. The Genetic Engineering Appraisal Committee (GEAC) — consisting mostly of bio-technology supporters — rejected just one out of the 28 proposals up for consideration. Six proposals were rejected for want of more information.
- 2016: GEAC gave green signal to GM Mustard for field trial, but SC stayed the order and sought public opinion on the same.
- There are as many as 20 GM crops already undergoing trails at various stages.

15.11 Challenges and Way Forward

Rapid research, development and commercialization of GM crops in past few years has also generated considerable concerns with regard to their potential impacts on the environment, biodiversity and consequently on the human and animal health. Biosafety concerns, impact on environment and ethical issues are some of the major challenges for GM crops research and deregulation in entire world including India. Several news articles, reports, and documentaries have been published by media on safety of GMOs on regular basis. Ecological risk assessment of transgenic crops, issue of gene flow, development of secondary pest resistance and ecological risks involved with pollen flow are some of the issues which should be addressed before release of any GM crop for open field trials and commercialization (Manish Shukla *et.al.*, 2018). Biosafety issues should be addressed at all the stages of development and release of GM crops on a case-to-case basis. It is essential to have robust biosafety guidelines to evaluate the GM crops and need to be addressed at all the stages of development and release of transgenic crops. Since, many transgenic crops are being developed and released for open field trials and commercial cultivation regularly; concerns have been raised about the potential risks associated with their impact on environment, biological diversity and human health.

Assessments of environmental risk of GMOs have been performed extensively worldwide during the last decade and extensive risk assessment framework has been formulated. There are many important traits introduced into plants and urgent need to address their impact on agriculture and agricultural practices in several countries. Gene flow from GM crops, factors affecting gene flow, consequences of gene flow and transfer of traits to wild weedy relatives from genetically engineered crops must be explored extensively before release of any crop for commercialization. Many environmental issues related to genetically modified crops have been studied and discussed by scientists and policy makers thoroughly since the release of first transgenic crop for commercial cultivation in 1980s. These include effects of transgenic crops on biodiversity, gene flow, out-crossing, invasiveness, weediness, horizontal gene transfer, effect on non-target organisms, and evolution of virulent strains of pests and impact on soil microorganisms. However, many issues are still being debated because the potential long term cumulative effects are difficult to predict. Horizontal gene transfer is one the biggest concerns among all the ethical issues related to GMOs. Novel gene in an organism could be a source of potential harm to the human health or the environment. Introduced gene from GM crops may confer a novel trait in another organism such as the transfer of antibiotic resistance genes to a pathogen has the potential to compromise human or animal therapy (Bennett *et al.*,2014).

There is relatively very little research has been conducted on the long-term effects on human health and environment because GMO technology has been available for such a short period of time. The greatest danger lies not in the effects that we have studied, but in those which we cannot anticipate at this point. Some new proteins which have never been ingested before by humans are now part of the foods that people consume every day and their potential effects on the human health are as of yet unknown. The use of transgenic plants or genetically modified organisms is a practice still in its infancy. The long-term effects of this technology are yet to be seen, and thus we must proceed with caution as we develop our practices and guidelines. Well planned, systematic research, development, exchange and commercialization of transgenic crops after addressing all the concerns pertaining to the environment followed by regular post-release monitoring to evaluate the long-term impacts would be required to sustain the biodiversity and to harness the benefits of genetically modified crops (Manish Shukla *et.al.*, 2018).

Current research and development in crop biotechnology in India is focused on the development of biotech food, feed and fiber crops that can contribute to higher and more stable yields and also enhanced nutrition. Rice being the major staple food, the research on genomics of rice is being pursued aggressively for conferring biotic and abiotic stress. Field trials with Bt rice are already underway. Nutritional quality improvement and reduction of post harvest losses particularly in fruits and vegetables through delayed ripening gene is also a major research area. Several transgenic crops developed by public sector are already in field trials and 85 different plant species for various traits are under different stages of experiments for the development of genetically engineered crops in India ((Manish Shukla *et.al.*, 2018).

After commercial release of *Bt* Cotton in 2002, there was rapid increase in research on *Bt* technology and as a result different foreign genes were deployed in Brinjal, Cabbage, Cauliflower, Tomato, Lady's finger, Rice, Corn and Groundnut for many beneficial traits. These crops are being evaluated under confined field trials except Bt brinjal which is facing moratorium as of now. Political leaders, scientists as well as technocrats in India have noticed these opportunities, and they are now routinely endorsing the potential contributions that biotechnology - including transgenic crops - might make to agricultural productivity, growth and poverty reduction in the years ahead.

Many top leaders in India have endorsed the value of agribiotechnology in general, and treasury resources have even been allocated to promote GM crop research within India's national agricultural research system. But there are also few expert opinions and a report on GM crops by parliamentary standing committee recommends that introducing transgenics in agriculture is not a sustainable way forward to achieve a food security in coming years in India. Critics of GM crops were able to work within India's open and democratic political system to push for a precautious or even a preventive approach toward GM crops instead, especially in the area of biosafety policy. Urbanization and industrial growth diminishes the available arable land rapidly. Increasing the agricultural productivity by second green revolution can only solve the problem of food and nutrition security in India.

Requirement of NOCs from State Governments to conduct field trials of GM crops, illegal cultivation of HT cotton and resistance developed by pink boll worm to Bt cotton are important challenges for GM crops during and post deregulation process. GEAC has introduced the requirement of NOC (No Objection Certificate) from the State Governments to conduct field trials since agriculture is state subject. Many State Governments have declined to issue NOC despite of approval of GEAC. Government is working to establish the Notified Field Trial Sites (NFTS) in all agro-climatic regions to resolve this issue of NOC requirements. Meanwhile, GEAC in its 130th meeting in August 2016 has exempted the NOC requirements from State Governments for Event Selection Trials (ESTs) (Manish Shukla *et.al.*, 2018). Few reports emerged about illegal cultivation of herbicide-tolerant (HT) cotton in the States of Telangana and Andhra Pradesh. GEAC has not approved any HT cotton event for commercial use. DBT constituted a Field Inspection and Scientific Evaluation Committee (FISEC) to examine the extent of illegal cultivation of HT cotton in India.

Though respective State Governments have been advised to take required action but the necessary guidelines and precautionary measures will be issued after FISEC report submission. Development of resistance by pink boll worm to BGI and BGII cotton pose a serious challenge to researchers and regulators about application of GM technology in agriculture. Several incidence of damage to bolls and yield loss in Bt cotton have been reported from Gujarat in last few years. *Bt* cotton Bollgard-II contains cry1Ac & cry2Ab gene for effective control of pest. Planting a refuge crop around the cotton crop could delay the resistance as suggested by regulators and scientists. It is imperative to develop new product for farmers affected with pink boll worm resistance and also remind us that GM technology is not the very effective solution for all agricultural problems.

Despite several challenges, the futures of genetically modified crops look promising in India. Many developing countries, especially the Asian countries are expected to plant GM crops next 2–3 years. Currently, many genetically engineered crops are in final stages of development in India and scientists are looking towards government anxiously for policy change and approval for commercialization. GM crops is not a panacea but they have the potential for achieving food security targets and make a substantial contribution for poverty alleviation in developing world by increasing crop productivity on limited resources. The world will have to produce 70% more food, feed and biomass on about the same area of land by 2050(Manish Shukla *et.al.*, 2018). Increasing population, urbanization, income and dietary preferences in developing countries requires extraordinary steps to intensify agriculture production to meet the growing demand by substantially increasing crop yield and optimizing the use of input resources).

Conclusion: The wide advantages of transgenic crops for a society to solve food security or nutrition security issues have been well established. Many added benefits such as higher nutritional value, herbicide tolerance, virus resistance, tolerance to various abiotic stresses, increase the shelf life of a fruit and thus, can account for a good market for farmers. There is an urgent need for India to carry on its GM crop research program to sustain its food and nutrition security targets. Debate on GM crops about safe or unsafe will never end although there is hardly any substantial scientific evidence against safety of GM foods. Surprisingly, few public sector intuitions also showed their concerns for GM foods. Intuitions funded by GOI should follow same broad policies of Indian Government and must show their strength with the government to fight poverty and malnutrition. This valid point is raised due to opposite views of the members of Technical Expert Committee appointed by the Supreme Court of India for Safety and Guidelines for GM crop research and make recommendations for future of GM crop research in India(Manish Shukla *et.al.*, 2018).

Although, this is a fact that India does not have basic infrastructure and stringent guidelines for GM crop research and risk assessment but taking in to account India's urgent need, halt this program cannot be halted. Ideally, India must continue on its research on GM crop and its deregulation along with building basic infrastructure facilities and preparing stringent biosafety and marketing guidelines. Although portals like GEAC, IGMORIS (Indian GMO Research Information System), Biosafety Clearing House are doing their role for assessing biosafety and their regulation of GM crops but there is an urgent need to build a single window system and online portal for assessment, control, regulations and approval of GM crops. It should become mandatory for every company and public sector institutions to register themselves with this portal whenever they develop any event for transgenic development and starting their field trials before submitting for approval. Every new transgenic event under development must show their registry number and date of registry with a website or online portal specifically designed to obtain approval in any country for marketing. This portal should also contain publication list associated with the development of any specific genetically modified crop so that any individual who have an interest in any GM crop development event can get all the details at single place along with its current status. This type of portal will be very useful and public friendly for making positive effects among people at large for GM food research, safety of GM foods and its present status.

According to International Services for the Acquisition of AgriBiotech Applications (ISAAA) India has the fourth largest area planted under transgenic crops, a total 11.6 million hectares (mh) was under transgenic plantation in year 2014, only after, Brazil (42.2 mh), Argentina (24.3 mh) and USA (73.1 mh) (Sonam Sneha, 2018). Currently across the world approximately 181.1 mh area was used for the transgenic crop plantation. In year 2009, India's Maharashtra Hybrid Seeds Company (Mahyco) and US based company Monsanto with has developed *Bt* eggplant (Solanum melongena) by inserting a crystal gene (Cry1Ac) from *B. thuringiensis*. But still the Indian Government has imposed a moratorium on its release after its commercialization due to the public resentment. Nearly 96% (nearly ~ 11.57 mh) of the country's cotton area is now covered by Bt hybrids.

15.12 Pros and Cons of Transgenic Crops

Despite of various advantages the use of transgenic plants for human welfare has been restricted owing to various concerns raised by the public and the critics. These concerns are divided into different categories, namely, environmental, health, nutritional, ecological, socioeconomic, and ethical concerns. Certain groups of public, including religious bodies, find it very unethical or inhumane to introduce human or animal genes into plants (Whiteman, 2000). For example, the transfer of animal genes such as α-interferon gene into plants is objectionable to the vegetarians. Such concern was one of the reasons due to which the concept of "edible vaccines" did not gain much impetus.

There is a potential risk that the GM plants may hybridize (or cross-breed) with sexually compatible wild-type species. This genetic exchange is possible due to wind pollination, biotic pollination or seed dispersal. This may have an impact on the environment through the production of hybrids and their progeny. It is also speculated that the nutritional composition of GM products may be affected in GM plants. Another concern is that the transgenes from animals (obtained from fishes, mouse, human, and microbes) introduced into GM plant for molecular farming may pose a risk of changing the fundamental nature of vegetables. The public is worried about the risk that the GM plants can spread through nature and interbreed with natural organisms, thereby contaminating "non-GM" environments. Non target effect, that is, undesirable effect of a novel gene (usually conferring pest or disease resistance) on "friendly" organisms in the environment, is another concern related to GM plants (Sonam Sneha, 2018). The public has long been worried about the loss of plant biodiversity due to global industrialization, urbanization, and the popularity of conventionally-bred high-yielding varieties. It is speculated that the biodiversity will be further threatened due to the encouraging use of GM plants. This is because development of GM plants may favor monocultures, that is, plants of a single kind, which are best suitable for one or other conditions or produce one product . Further, the transformation of more natural ecosystems into agricultural lands for planting GM plants is adding to this ecological instability.

It is speculated that the random gene insertion, transgene instability, and genomic disruption due to gene transfer may result in unpredictable gene expression. Such a risk is, however, unlikely to be unique to GM plants or of any significance considering our current knowledge of genomic flux in plants. Plants adapt to the fluctuations in the environment through changing their genes and developing better races called "evolved races." These mutations, however, occur at a very low frequency (i.e., one in about 109/gene/generation). It is hypothesised that the cultivation of GM plants by the farmers at an increasing rate throughout the world may change the evolutionary pattern drastically. Another concern is the evolution of non-GM plants through hybridisation with GM plants (Sonam Sneha, 2018). Public is also concerned about the potential risks associated with gene transfer from plants to microbes. The third health risk is related to the ability of GM plants to create new toxic organisms. It is speculated that some non-pest microbial strains may acquire pathogenic trait by gene flow from GM plants.

15.13 Regulatory Options for Genetically Modified Crops in India

The introduction of semi-dwarfing, high-yielding and nutrients-responsive crop varieties in the 1960s and 1970s alleviated the suffering of low crop yield, food shortages and epidemics of famine in India and other parts of the Asian continent. Two semi-dwarfing genes, Rht in wheat and Sd-1 in rice heralded the green revolution for which Dr. Norman Borlaug was awarded the Nobel Peace Prize in 1970. In contrast, the revolutionary new genetics of crop improvement shamble over formidable obstacles of regulatory delays, political interferences and public misconceptions. India benefited immensely from the green revolution and is now grappling to deal with the nuances of GM crops. The development of GM mustard discontinued prematurely in 2001 and insect-resistant Bt cotton varieties were successfully approved for commercial cultivation in 2002 in an evolving nature of regulatory system. However, the moratorium on Bt brinjal by MOEF in 2010 meant a considerable detour from an objective, science-based, rigorous institutional process of regulatory approval to a more subjective, nonscience-driven, political decision-making process. A study by Bhagirath Choudhary *et.al.*, (2014) examined what ails the regulatory system of GM crops in India and the steps that led to the regulatory logjam. Responding to the growing challenges and impediments of existing biosafety regulation, it suggested options that are critical for GM crops to take roots for a multiplier harvest.

Regulatory beginning: The National Biotechnology Board, which was constituted in 1982 issued a set of biotechnology safety guidelines in 1983 to undertake biotech research in laboratory and contained use settings. In 1986, the National Biotechnology Board was promoted to a full fledge Department of Biotechnology (DBT) under the Ministry of Science and Technology (MOST) (Bhagirath Choudhary *et.al.*, 2014) In the early years, DBT monitored developments in the biotech field globally, developed safety guidelines and made efforts to promote large-scale use of indigenously relevant biotechnologies in the country. Realizing the importance of adequately assessing biosafety, biodiversity and environmental risks (Ghosh, 1997), the research, product development and commercial release involving GMOs, hazardous microorganisms and trans-boundary movement of the living modified organisms (LMOs) by default were reallocated to MOEF. The Government of India (Allocation of Business) Rules 1961 assigned the responsibilities of 'biodiversity conservation' and 'environment protection' to MOEF in 1961 (Government of India, 1961).

Thereafter, MOEF began regulating genetically modified organisms and products thereof under the existing Environmental Protection Act 1986, commonly referred as EPA 1986, which was enacted by the Parliament of India in 1986. Whereas the EPA 1986 does not describe GMOs and GM crops in the law per se, it lays down the legislative provisions to regulate 'hazardous substances' *(Hazardous substance of the EPA 1986 section 3, 2 (iv) means any substance or preparation which, by reason of its chemical or physicochemical properties or handling, is liable to cause harm to human beings other living creatures, plants, microorganism, property or the environment) and to make administrative rules to regulate environmental pollution caused by hazardous substances. Henceforth, MOEF drafted and notified 'the rules for the manufacture, use, import, export and storage of hazardous microorganisms, genetically engineered organisms or cells in 1989' referred as the EPA Rules 1989 under 'hazardous substances' section of the EPA 1986. As a result, GM crops, GMOs and the products of genetic engineering were de facto categorized as 'inherently harmful' in the same manner as hazardous substances that cause harm to human beings or other living creatures, property or the environment. Notably, the EPA Rules 1989 to regulate GMOs and GM crops were issued by an 'administrative order' through publication in the Gazette of India vide notification GSR 1037(E) dated 5 December 1989 and came into force vide notification S.O.677(E) dated 13 September 1993 (Gazette of India, 1989, 1993).

The EPA Rules 1989 cover a range of activities involving manufacture, use, import, export, storage and research of all genetically engineered organisms including microorganisms, plants and animals and products thereof. The Rules also apply to hazardous microorganisms that are pathogenic to human beings, animals or plants, regardless whether they are genetically modified. The Rules 1989 not only regulate research, development and large-scale commercialization of GM crops but also order compliance of the safeguard through regulatory approach, post-approval monitoring of violation and non-compliance . The Rules define competent authorities and composition of such authorities for handling of various aspects of GMOs. There are six competent authorities that function into a three-tier system; the first tier includes the 'Policy Advisory Committee' such as the Recombinant DNA Advisory Committee (RDAC); the second tier consists of 'Regulating and Approval Committees' such as the Institutional Biosafety Committee (IBSC), the Review Committee on Genetic Manipulation (RCGM) and the Genetic Engineering Approval Committee (GEAC), and finally, the third tier includes the 'Post Monitoring Committee' comprising of the State Biotechnology Coordination Committee (SBCC) and the District Level Committee (DLC).

The functions of each of the committees have been articulated in the Rules 1989. In the spirit of interministerial coordination, the implementation of the Rules 1989 was fast tracked with the subject matter experience and expertise of DBT (Bhagirath Choudhary *et.al.*, 2014). As a matter of fact, the EPA Rules 1989 assigned the biosafety, risk assessment and risk management related aspects of GM crops to DBT. Notably, DBT was made an integral part of the EPA Rules 1989, which was a unique feature allowing both ministries – MOEF and DBT of MOST – to regulate and safeguard from any foreseeable harm and weigh risks and benefits of GM crops and hazardous microorganisms under the Rules 1989. However, in case of post-monitoring of GM crops, the EPA Rules assigned the responsibility to the respective State (s). It established a regulatory framework involving multiple government departments and delineated the administrative structure, authority, procedure and requirements for the regulation of GM crops at Union and States level. However, the Rules 1989 were unclear on the role and responsibility of the Ministry of Health and the Ministry of Agriculture – the important ministries that are empowered to regulate seed and human health and matters related to regulation of GM crops. Over the years, the biosafety regulatory system has evolved into a dynamic and comprehensive regulatory framework that involves different ministries; first, the ministries authorized under the EPA Rules 1989 and second, the ministries that indirectly deal with GM crops.

The regulation of GM crops from development, environmental release to commercial approval has been covered by three legislative Acts enacted by the Parliament of India and administered by different ministries. These included the Environment Protection Act 1986 implemented by MOEF, the Seed Act 1966 & the Seeds (Control) Order by Ministry of Agriculture (MOA) and the Food Safety and Standard Act 2006 (subsumed the Prevention of Food Adulteration Act 1954) by the Ministry of Health and Family Welfare.

As per the EPA Rules 1989, the Recombinant DNA Advisory Committee (RDAC) set up by DBT brought out a first set of 'Recombinant DNA Safety Guidelines' in 1990 to regulate rDNA technology in medicine and agriculture. These guidelines were revised in 1994 as 'Revised Guidelines for Safety of Biotechnology'.

Realizing the need for comprehensive guidelines for transgenic plants in the mid-nineties, DBT framed and released a comprehensive guide for GM crops in 1998 referred to as 'Revised Guidelines for Research in Transgenic Plants and Guidelines for Toxicity and Allergenicity Evaluation of Transgenic Seeds, Plants and Plant Parts' to regulate GM crops and products. With regard to the application of GM crops, the guidelines 1990, 1994 and 1998 outline safety procedures, testing and use of genetically modified organisms and products. Considering the ecological consequences and the potential risks associated with the environmental release of GM crops, the guidelines prescribe the biosafety evaluation and risk assessment of the environmental aspects and agronomic performance on a case-by-case basis taking into consideration specific crop, trait and agro-ecological system. These guidelines also call for regulatory measures to ensure safety of imported GM materials in the country (Randhawa and Chhabra, 2009).

The regulatory system evolved along with the import of transgenic crops, GM mustard by Proagro and *Bt* cotton by Mahyco for R&D purpose in the mid-nineties. The development of GM mustard *Brassica juncea* was discontinued in 2001 by Bayer CropScience that acquired Proagro at the penultimate stage of commercial approval. The insect-resistant *Bt* cotton varieties primarily Gossypium hirsutum developed by Mahyco in collaboration with Monsanto received approval for commercial cultivation in 2002. The approval process witnessed an intense debate and protest as a result of the evolving nature of regulatory system responding to scientific, technological, policy and social challenges.

According to the EPA Rules 1989, GEAC is the statuary committee with a sole mandate to regulate GM crops and accords approval for contained, confined and environmental release of GM crops in the country. GEAC is also entrusted with various provisions of the EPA Rules 1989 that require approval including:

- Approval for the import of genetically modified microorganism for research purpose (section 7(1)).
- Approval for the use of hazardous microorganisms and recombinants in research and industrial production (section 4(3) (i)).
- Approval for all kind of experimental field trials of GM crops (section 4(3) (i)).
- Approval for measure concerning discharge of hazardous microorganisms (section 7(3)).
- Approval for licenses for scaling up pilot project involving genetically engineered microorganism (section 7(4)).
- Approval for the deliberate release of genetically engineered organism (section 9(2)).
- Approval for certain substances containing genetically engineered oraganisms (section 10).

- Approval for food stuffs containing GMOs (section 11) and finally.
- Hold power to revoke any approval (section 13(2)).

Changing the name of GEAC from approval to appraisal committee contradicts the aforesaid provisions of the EPA Rules 1989. It also precludes the committee to exercise the authority to grant permission for approval of import, field trials and environmental release of GM crops in the country (Bhagirath Choudhary *et.al.*, 2014).

Conclusion: Three fundamental flaws were identified in the current biosafety regulatory framework in the form of the EPA Rules 1989 that need to be rectified for the Indian regulatory system to function in a cost-effective and time-bound manner. Firstly, GM crops are categorized as 'inherently harmful' under the 'hazardous substance' provision of the Environmental Protection Act 1986, which is scientifically incorrect and gives rise to misperceptions about the safety and potential risk of GM crops to health and environment. Secondly, the EPA Rules 1989 to regulate GM crops were issued not by a 'legislative act' but by an 'administrative order' that remains untenable and liable to change with the desire of MOEF, which affects the predictability of the regulations and ignores the need to take into account the views and policies of other concerned ministries and the Parliament of India. Finally, the Union environment ministry administers the regulation of GM crops in India whereas agriculture falls under the respective State(s) (Bhagirath Choudhary *et.al.*, 2014). This often confronts approvals posing a 'Union Vs State' conflict in decision-making on GM crops.

It is an imperative for the country to rely and guide the agricultural growth and food production with the help of scientific expertise and institutional capacity that have been built with painstaking efforts and enormous investment over a period of time. At no point in time should it endanger the institutional capacity of R&D, innovation and product deployment that are required to reverse the declining availability of food production. The new generation biotech traits, input traits in the short run and output traits in the long run have the potential to contribute to sustainable food, feed and fibre production. The development and deployment of novel traits would require the pooling of expertise and resources of both public and private sector institutions. Similarly, the regulatory processes have also to solicit the expertise of professionals and accredited laboratories from within and outside India to perform the regulatory oversight, stewardship and monitoring. Therefore, the current and future regulatory systems have to continue to harness the collective strength of interdisciplinary institutions to ensure the safety and efficacy of biotech crops. The current regulatory system of RCGM and GEAC has effectively utilized the existing expertise and capacity to build a system that regulates GM crops. However, it remains untenable due to the fundamental flaws in the EPA Rules 1989. The impact of the draft law on the Biotechnology Regulatory Authority of India (BRAI), which was recently introduced in the Parliament of India to create a new biotech regulator will largely depend on the extent to which it clarifies the Union's and States' jurisdiction and consolidates the decision-making responsibility in a non-politicized fashion, by adhering to the legislative parameters of purposefulness, clarity, transparency, predictability and efficiency.

As the Food and Agriculture Organisation (FAO) has rightly pointed out in 2004, "Science cannot declare any technology completely risk free. Genetically engineered crops can reduce some environmental risks associated with conventional agriculture, but will also introduce new challenges that must be addressed. Society will have to decide when and where genetic engineering is safe enough". Arguments both for and against the cultivation and use of the GM crops are varied and there is a wide consensus that assessment should take place on a case-by-case basis before genetically modified food is brought to the market. These assessments should be done by Government or an independent credible regulatory authority or private agencies and these should not be driven by any commercial interests. Moreover, educating public opinion is also very important as food is always a sensitive cultural issue. Merely indicating whether a product is genetically modified or not, without providing any additional vital information, would not serve any purpose; rather information on its content and possible risks or benefits should be provided.

To sum up in the words of M.S. Swaminathan (2009), "GM foods have the potential to solve many of the world's hunger and malnutrition problems, and to help protect and preserve the environment by increasing yield and reducing reliance upon chemical pesticides. Yet there are many challenges ahead for governments, especially in the areas of safety testing, regulation, industrial policy and food labelling."

15.14 Public Concern-Global Scenario

In the late 1980s, there was a major controversy associated with GM foods even when the GMOs were not in the market. But the industrial applications of gene technology were developed to the production and marketing status. After words, the European Commission harmonized the national regulations across Europe. Concerns from the community side on GMOs in particular about its authorization have taken place since 1990s and the regulatory frame work on the marketing aspects underwent refining. Issues specifically on the use of GMOs for human consumption were introduced in 1997, in the Regulation on Novel Foods Ingredients (258/97/EC of 27 January 1997) (Bawa and Anilakumar, 2013). This Regulations deals with rules for authorization and labelling of novel foods including food products made from GMOs, recognizing for the first time the consumer's right to information and labelling as a tool for making an informed choice. The labelling of GM maize varieties and GM soy varieties that did not fall under this Regulation are covered by Regulation (EC 1139/98). Further legislative initiatives concern the traceability and labelling of GMOs and the authorization of GMOs in food and feed.

The initial outcome of the implementation of the first European directive seemed to be a settlement of the conflicts over technologies related to gene applications. By 1996, the second international level controversy over gene technology came up and triggered the arrival of GM soybeans at European harbours. The GM soy beans by Monsanto to resist the herbicide represented the first large scale marketing of GM foods in Europe. Events such as commercialisation of GM maize and other GM modified commodities focused the public attention on the emerging biosciences, as did other gene technology applications such as animal and human cloning. The public debate on the issues associated with the GM foods resulted in the formation of many non-governmental organizations with explicit interest. At the same time there is a great demand for public participation in the issues about regulation and scientific strategy who expresses acceptance or rejection of GM products through purchase decisions or consumer boycotts.

Most research effort has been devoted to assessing people's attitudes towards GM foods as a technology. Numerous "opinion poll"—type surveys have been conducted on national and cross-national levels. Ethical concerns are also important, that a particular technology is in some way "tampering with nature", or that unintended effects are unpredictable and thus unknown to science.

Consumers attitude towards GM crops: Consumer acceptance is conditioned by the risk that they perceive from introducing food into their consumption habits processed through technology that they hardly understand. In a study conducted in Spain, the main conclusion was that the introduction of GM food into agro-food markets should be accompanied by adequate policies to guarantee consumer safety. These actions would allow a decrease in consumer-perceived risk by taking special care of the information provided, concretely relating to health. For, the most influential factor in consumer-perceived risk from these foods is concern about health (Bawa and Anilakumar, 2013).

A study was conducted and aimed to identify the factors that affect consumers purchasing behaviour towards food products that are free from GMO (GM Free) in a European region and more precisely in the Prefecture of Drama-Kavala-Xanthi. Field interviews conducted in a random selected sample consisted of 337 consumers in the cities of Drama, Kavala, Xanthi in 2009. Principal components analysis (PCA) was conducted in order to identify the factors that affect people in preferring consuming products that are GM Free. The factors that influence people in the study area to buy GM Free products are: (a) products' certification as GM Free or organic products, (b) interest about the protection of the environment and nutrition value, (c) marketing issues and (d) price and quality. Furthermore, cluster and discriminant analysis identified two groups of consumers: (a) those influenced by the product price, quality and marketing aspects and (b) those interested in product's certification and environmental protection.

Twelve long-term studies (of more than 90 days, up to 2 years in duration) and 12 multigenerational studies (from 2 to 5 generations) were conducted on the effects of diets containing GM maize, potato, soybean, rice, or triticale on animal health. They referenced the 90-day studies on GM feed for which long-term or multigenerational study data were available. Many parameters have been examined using biochemical analyses, histological examination of specific organs, hematology and the detection of transgenic DNA. Results from all the 24 studies do not suggest any health hazards and, in general, there were no statistically significant differences within parameters observed. They observed some small differences, though these fell within the normal variation range of the considered parameter and thus had no biological or toxicological significance. The studies reviewed present evidence to show that GM plants are nutritionally equivalent to their non-GM counterparts and can be safely used in food and feed.

15.15 Issues with Respect to India

n a major setback to the proponents of GM technology in farm crops, the Parliamentary Committee on Agriculture in 2012 asked Indian government to stop all field trials and sought a bar on GM food crops such as Bt. Brinjal (Bawa and Anilakumar, 2013). Raising the "ethical dimensions" of transgenics in agricultural crops, as well as studies of a long-term environmental and chronic toxicology impact, the panel noted that there were no significant socio-economic benefits to farmers.

Countries like India have great security concerns at the same time specific problems exist for small and marginal farmers. India could use a toxin free variety of the Lathyrus sativus grown on marginal lands and consumed by the very poor. GM mustard is a variety using the barnase-barstar-bar gene complex, an unstable gene construct with possible undesirable effects, to achieve male sterile lines that are used to make hybrid mustard varieties. In India we have good non-GM alternatives for making male sterile lines for hybrid production so the Proagro variety is of little use. Being a food crop, GM mustard will have to be examined very carefully. Even if there were to be benefits, they have to be weighed against the risks posed to human health and the environment. Apart from this, mustard is a cross-pollinating crop and pollen with their foreign genes is bound to reach non-GM mustard and wild relatives. We do not know what impact this will have. If GM technology is to be used in India, it should be directed at the real needs of Indian farmers, on crops like legumes, oilseeds and fodder and traits like drought tolerance and salinity tolerance.

Basmati rice and Darjeeling tea are perhaps India's most easily identifiable premium products in the area of food. Basmati is highly prized rice, its markets are growing and it is a high end, expensive product in the international market. Like Champagne wine and truffles from France, international consumers treat it as a special, luxury food. Since rice is nutritionally a poor cereal, it is thought that addition of iron and vitamin A by genetic modification would increase the nutritional quality. So does it make any sense at all to breed a GM Basmati, along the lines of Bt Cotton? However, premium wine makers have outright rejected the notion of GM doctored wines that were designed to cut out the hangover and were supposed to be 'healthier'. Premium products like special wines, truffles and Basmati rice need to be handled in a special, premium way (Sahai, 2003).

15.16 Risks, Precautions and Ethical Concerns of Genetically Modified Organisms

Commercial potential of biotechnology is immense since the scope of its activity covers the entire spectrum of human life. The most potent biotechnological approach is the transfer of specifically constructed gene assemblies through various techniques. However, this deliberate modification and the resulting entities thereof have become the bone of contention all over the world (Dhan Prakash *et.al.*, 2011). Benefits aside, genetically modified organisms (GMOs) have always been considered a threat to environment and human health. In view of this, it has been considered necessary by biosafety regulations of individual countries to test the feasibility of GMOs in contained and controlled environments for any potential risks they may pose. Various aspects of risk, its assessment, and management which are imperative in decision making regarding the safe use of GMOs were discussed. Efficient efforts are necessary for implementation of regulations. Importance of the risk assessment, management, and precautionary approach in environmental agreements and activism is also discussed.

Risks related to the use of GMO's: The application of genetic modification allows genetic material to be transferred from any species into plants or other organisms. The introduction of a gene into different cells can result in different outcomes, and the overall pattern of gene expression can be altered by the introduction of a single gene. The sequence of the gene and its role in the donor organism may have a relatively well-characterized function in the organism from which it is isolated. However, this apparent "precision" in the understanding of a gene does not mean that the consequences of the transfer are known or can be predicted (Dhan Prakash *et.al.*, 2011). Copies of a gene may be integrated, additional fragments inserted, and gene sequences rearranged and deleted—which may result in lack of operation of the genes instability or interference with other gene functions possibly cause some potential risks. Therefore, there could be a number of predictable and unpredictable risks related to release of GMOs in the open environment. The report prepared by the Law Centre of IUCN, the World Conservation Union (2004), enlists numerous environmental risks likely to occur by the use of GMOs in the field. These risks are as follows.

Each gene may control several different traits in a single organism. Even the insertion of a single gene can impact the entire genome of the host resulting in unintended side effects, all of which may not be recognizable at the same time. It is difficult to predict this type of risk.

Genetic Contamination/Interbreeding: Introduced GMOs may interbreed with the wild-type or sexually compatible relatives. The novel trait may disappear in wild types unless it confers a selective advantage to the recipient. However, tolerance abilities of wild types may also develop, thus altering the native species' ecological relationship and behaviour.

Competition with natural species: Faster growth of GMOs can enable them to have a competitive advantage over the native organisms. This may allow them to become invasive, to spread into new habitats, and cause ecological and economic damage.

Increased selection pressure on target and nontarget organisms : Pressure may increase on target and nontarget species to adapt to the introduced changes as if to a geological change or a natural selection pressure causing them to evolve distinct resistant populations.

Ecosystem impacts: The effects of changes in a single species may extend well beyond to the ecosystem. Single impacts are always joined by the risk of ecosystem damage and destruction.

Impossibility of followup: Once the GMOs have been introduced into the environment and some problems arise, it is impossible to eliminate them. Many of these risks are identical to those incurred with regards to the introduction of naturally or conventionally bred species. But still this does not suggest that GMOs are safe or beneficial, nor that they should be less scrutinized.

Horizontal transfer of recombinant genes to other microorganisms: One risk of particular concern relating to GMOs is the risk of horizontal gene transfer (HGT). HGT is the acquisition of foreign genes (via transformation, transduction, and conjugation) by organisms in a variety of environmental situations. It occurs especially in response to changing environments and provides organisms, especially prokaryotes, with access to genes other than those that can be inherited.

HGT of an introduced gene from a GMO may confer a novel trait in another organism, which could be a source of potential harm to the health of people or the environment. For example, the transfer of antibiotic resistance genes to a pathogen has the potential to compromise human or animal therapy. HGT has been observed for many different bacteria, for many genes, and in many different environments. It would therefore be a mistake to suppose that recombinant genes would not spread to other bacteria, unless precautions are taken. Recent evidence from the HGT technology confirms that transgenic DNA in GM crops and products can spread by being taken up directly by viruses and bacteria as well as plant and animals cells. It was reported that HGT also moved from a nuclear monocot gene into the genome of the eudicot parasite witchweed, which infects many grass species in Africa.

Some of the important potential impacts of HGT from GMOs include the following.

Adverse effects on the health of people or the environment: These include enhanced pathogenicity, emergence of a new disease, pest or weed, increased disease burden if the recipient organism is a pathogenic microorganism or virus, increased weed or pest burden if the recipient organism is a plant or invertebrate, and adverse effects on species, communities, or ecosystems.

Unpredictable and unintended effects: HGT may transfer the introduced genes from a GMO to potential pests or pathogens and many yet to be identified organisms. This may alter the ecological niche or ecological potential of the recipient organism and even bring about unexpected changes in structure or function. Furthermore, the gene transferred may insert at variable sites of the recipient gene, not only introducing a novel gene but also disrupting an endogenous gene, causing unpredictable and unintended effects.

Loss of management / control measures: Regulatory approvals for field trials of GMOs often require measures to limit and control the release in space and time. With the spread of the introduced gene(s) to another species by HGT, a new GMO is created. This new GMO may give rise to adverse effects which are not controlled by management measures imposed by the original license or permit.

Long-term effects : Sometimes the impact of HGT may be more severe in the long term. Even under relatively strong selection pressure, it may take thousands of generations for a recipient organism to become the dominant form in the population. In addition, other factors such as timing of appropriate biotic or abiotic environmental conditions and additional changes in the recipient organism could delay adverse effects.

Ethical concerns: Various ethical issues associated with HGT from GMOs have been raised including perceived threats to the integrity and intrinsic value of the organisms involved, to the concept of natural order and integrity of species, and to the integrity of the ecosystems in which the genetically modified organism occurs. Several scientific evidence that has emerged on GMOs over the last couple of years shows that there are several clear risks to human health and the environment. When genetic engineers create GMO or transgenic plants, they have no means of inserting the gene in a particular position. The gene ends up in a random location in the genetic material, and its position is not usually identified. There are already several examples of such undesired effects being identified in the US after approval (e.g., GM cotton with deformed cotton bolls; increased lignin in GM soya etc.). Releasing genetically modified plants or crop into the environment may have direct effects, including gene transfer to wild relatives or conventional crops, weediness, trait effects on nontarget species, and other unintended effects.

15.17 Risk Management

Once a risk is assessed, it must be managed. The management of risk is an exclusively political action, resulting in a decision regarding whether to accept or not the risk previously estimated. It can take additional aspects (e.g., socioeconomic or ethical) into consideration and concerns methods used to reduce the scientifically identified risk (Dhan Prakash *et.al.*, 2011). Many frameworks of risk assessment methodology separate risk assessment from risk management. Some frameworks, however, consider only certain aspects of risk management (e.g., monitoring) as separate from risk assessment but other aspects of risk management (e.g., consideration of risk mitigation options) to be part of risk assessment methodology, since a final characterization of risks must take into account the effects of any mitigation options that reduce risks. The important aspect is, of course, the iterative and interlinked relationship between risk assessment and risk management.Often decisions are made with incomplete information, and this leads to uncertainty. This uncertainty needs to be handled to assess the impact it might have on a decision. Biosafety regulatory frameworks should serve as mechanisms for ensuring the safe use of biotechnology products without imposing unintended constraints to technology transfer.

The Article 16 of the Cartagena Protocol of Biosafety is purely relevant to risk management of GMO. The protocol establishes and maintains appropriate mechanisms and measures strategies to regulate, manage, and control risks identified in the provision of risk assessment.

- The potential to harmonize national regulatory frameworks thus ensures appropriate biosafety decision making based on scientific risk assessment. If properly implemented, the protocol has the potential to encourage innovation, development, technology transfer, and capacity building in relation to biotechnology, while also achieving the goals of conservation, sustainable agriculture, and equitable sharing of the technology's benefits.
- To realize its potential, however, decisions concerning protocol implementation must be carefully considered and should not place undue burdens on a technology that possesses such great potential to contribute positively to sustainable agriculture and development throughout the world.
- A first-things-first approach where initial efforts focus on bringing all parties to the protocol into compliance with it as quickly as possible. Developing further requirements or fine-tuning obligations at this stage only worsens the degree of noncompliance already in existence.
- Therefore, capacity building should remain the primary area of focus under the Biosafety Protocol to ensure the safe adoption of this technology. In this regard, material exists to help national governments.
- The users and developers of agricultural biotechnology embrace their share of the duty in the protocol implementation process and will continue to campaign for fair, science-based regulations and assist with and contribute to effective capacity building.

Risk management process also forms a second focus of the economic/political component of the GMO biosafety issue. Whereas a risk/benefit analysis concludes that risks exist with regard to a GMO introduction or other activity, but are sufficiently outweighed by the benefits of that action, it will probably still be required both practically and legally to take steps to manage the risk and to ensure that damage will be minimized. Elements of currently used and proposed risk management process include a variety of different kinds of activities. To a large extent, the specific protective measures imposed on the GMO user will be determined based on scientific factors linked to specific details of the GMO and the proposed use. These issues, too, turn on the ability of the decision maker to rely on unbiased scientific experts who are able to analyze each proposal or application and determine what controls are needed, and what the best available technologies and practices are.

The three important components was design for risk management. These components are impact assessment, public awareness/participation, and the design of regulatory systems. These concepts, all very important in this field, are critically important for GMO-related governance. It is not possible to overstate the importance of the public's contribution to effective decision making, as well as the importance of public awareness, within the context of government decisions on matters and activities affecting the environment.

15.18 Current Debate

The genetic modification of crops has been a controversial issue since the first commercial production of GMF. The proponents of such technologies claim that bio-engineering of food is absolutely safe and it is similar to what has been happening through traditional agriculture for thousands of years (Maghari and Ardekani, 2011). However, in selective breeding when two parental plants are crossed to obtain a desirable trait, it is likely that other unpleasant characteristics are transferred as well. Therefore, taking out the undesirable traits is a slow process and requires trial and errors through several generations of plants breeding. In this context, modern biotechnology has allowed us to go beyond natural physiological reproductive barriers in a manner that gene transfer among evolutionarily divergent organisms is now possible and therefore, individual genes expressing certain traits in animals or microorganisms can be precisely incorporated to the plant genome.

GM advocates believe that conventional breeding can achieve similar results using transferred gene but only within related species and in a lengthy and imprecise process. However, GMF opponents explain that genetic engineering bears no resemblance to natural breeding as it forcibly combines genes from unrelated species together; species that were perfectly separated over billions of years of evolution. They believe that the genetic engineering is not an alternative to traditional breeding as natural crossing of plants contributes thousands of genes to the offspring through the elegant dance of life. Agri-biotech companies claim that recombinant DNA techniques can bring advantages for consumers such as nutritional enhancement as well as improving the quality and yield of food and non-food plants such as cotton and pharmaceuticals. Most of the claims about the benefits of GMF have been proposed by the seed industry. However, independent scientists warn that the publications on the success of the GM in offering more nutritious and safe food is not based on expected scientific standards.

Drug studies funded by pharmaceutical companies are more likely to report positive result in favor of the sponsor than independently funded studies. The biased results might be achieved by the type of experiment design, selection of data and briefing the actual findings to what is expected. The same might be happening with researches conducted by the seed industry. The majority of research experiments on transgenic plants are being performed by the private sector and those carried out in universities are funded by the industry. Therefore, independent scientists should urgently follow strict precautionary approach in designing experiments on GMF. GM plants have to meet the criteria of the guidelines in order to get approval for entering the market. However, the regulatory and scientific capacities to implement such guidelines need to be built up worldwide specifically in developing countries.

Intellectual Property Rights (IPR) is one of the important factors in the current debate on GMF. The GM crops are patented by Agri-business companies leading to monopolization of the global agricultural food and controlling distribution of the world food supply. Social activists believe that the hidden reason why biotech companies are eager to produce GM crops is because they can be privatized, unlike ordinary crops which are the natural property of all humanity. It is argued for example that to achieve this monopoly, the large Agri-biotech company, Monsanto, has taken over small seed companies in the past 10 years and has become the biggest Agri-biotech Corporation in the world. The patent right for vegetable forms of life also affect the livelihoods of family farmers as they are required to sign a contract preventing them from saving and re-planting the seeds, thus they have to pay for seeds each year.

To sum up, taking everything into consideration, GM crops are alive; they can migrate and spread worldwide. In this regard, clear signals should be sent to biotech companies to proceed with caution and avoid causing unintended harm to human health and the environment. It is widely believed that it is the right of consumers to demand mandatory labeling of GM food products, independent testing for safety and environmental impacts, and liability for any damage associated with GM crops. We are aware that many regulatory laws already exist for risk assessments which are performed on three levels of impacts on Agriculture (gene flow, reducing biodiversity), Food and Food safety (allergenicity, toxicity), and Environment (including non target organism); And at the same time, in recent years Cartagena protocol has created laws and guidelines and has obliged countries and companies to obey them for production, handling and consumption of GM materials.

15.19 Future Development

The GM foods have the potential to solve many of the world's hunger and malnutrition problems, and to help protect and preserve the environment by increasing yield and reducing reliance upon synthetic pesticides and herbicides. Challenges ahead lie in many areas viz. safety testing, regulation, policies and food labelling (Bawa and Anilakumar, 2013). Many people feel that genetic engineering is the inevitable wave of the future and that we cannot afford to ignore a technology that has such enormous potential benefits.

Future also envisages that applications of GMOs are diverse and include drugs in food, bananas that produce human vaccines against infectious diseases such as Hepatitis B, metabolically engineered fish that mature more quickly, fruit and nut trees that yield years earlier, foods no longer containing properties associated with common intolerances, and plants that produce new biodegradable plastics with unique properties. While their practicality or efficacy in commercial production has yet to be fully tested, the next decade may see exponential increases in GM product development as researchers gain increasing access to genomic resources that are applicable to organisms beyond the scope of individual projects.

One has to agree that there are many opinions about scarce data on the potential health risks of GM food crops, even though these should have been tested for and eliminated before their introduction. Although it is argued that small differences between GM and non-GM crops have little biological meaning, it is opined that most GM and parental line crops fall short of the definition of substantial equivalence. In any case, we need novel methods and concepts to probe into the compositional, nutritional, toxicological and metabolic differences between GM and conventional crops and into the safety of the genetic techniques used in developing GM crops if we want to put this technology on a proper scientific foundation and allay the fears of the general public. Considerable effort need to be directed towards understanding people's attitudes towards this gene technology. At the same time it is imperative to note the lack of trust in institutions and institutional activities regarding GMOs and the public perceive that institutions have failed to take account of the actual concerns of the public as part of their risk management activities.

Bibliography

Alex Andrews George, 2020 . Genetically Modified Crops and Regulations in India. ClearIAS-Current Affair Notes

Anonymous, 2005. Biosafety issues related to transgenic crops (with focus on Bt cotton). Prepared by biotech consortium India limited, New Delhi in association with ministry of environment and forests government of India. 94 pp.

Anonymous, 2009. Genetically genetically modified crops issues and challenges in the context of India the context of India. Research unit (LARRDIS) research unit (LARRDIS) Rajya Sabha Secretariat, New Delhi, India

Bawa, A.S. and Anilakumar, K.R., 2013. Genetically modified foods: safety, risks and public concerns-a review. *J. Food Sci. Technol.*, 50:1035-1046.

Bennett,P.M., Livesey, C.T., Nathwani, D., Reeves, D.S., Saunders, J.R.and Wise, R., 2004. An assessment of the risks associated with the use of antibiotic resistance genes in genetically modified plants: report of the working party of the british society for antimicrobial chemotherapy. *J. Antimicrob. Chemother.*, 53:418–431.

Bhagirath Choudhary, Godelieve Gheysen, Jeroen Buysse , Piet van der Meer and Sylvia Burssens, 2014. Regulatory options for genetically modified crops in India. Review article. *Pl. Biotechnol. J.*, 12: 135–146

Daniell, H. S. and Kumar, N.,2005. Dufourmantel. Breakthrough in chloroplast genetic engineering of agronomically important crops. *Trends Biotechnol.*, 23:238–245.

Dhan Prakash, Sonika Verma, Ranjana Bhatia, B.N. abd Tiwary, 2011. Risks and precautions of Genetically Modified Organisms. Intl. Schol.Res. Notices, vol. 2011. Article ID 369573, 13 pages

Difonzo, C. and Collen, E., 2012. Handy Bt trait table.

Fargue, A., Colback, N. and Pere, J., 2006. Predictive study of the advantages of cleistogamy in oilseed rape for limiting unwanted gene flow. *Euphytica*, 151:1–13.

Gazette of India,1993. Ministry of environment and forest – notification, S.O.677 (E). The Gazette of India –Extraordinary, Govt of India.

Gazette of India,1989. Ministry of environment and forest – notification. GSR 1037(E), The Gazette of India –Extraordinary, Govt of India.

Government of India,1961. The Government of India (Allocation of Business) Rules. New Delhi, India: Government of India, Rashtrapati Bhawan

Hill, M.J., Hall, L. and Arnison, P.G.,2007. Genetic use restriction technologies (GURTs): Strategies to impedentransgene movement. *Trends Plant Sci.*,12:177–183.

Kindiger, B.K. and Sokolov, V., 1998. Apomictic maize. U.S. Patent.5710367.

Maghari, B.M. AND Ardekani, A.M., 2011. Genetically modified foods and social concerns. *Avicenna J Med Biotechnol.*, 3:109-117.

Manish Shukla, Khair Tuwair Al-Busaidi, Mala Trivedi and Rajesh K. Tiwari ,2018. Status of research, regulations and challenges for genetically modified crops in India, GM Crops & Food, 9:173-188,

Mishra, M. and Kumari, S., 2018. Biosafety issues related to genetically engineered crops. *MOJ Res. Rev.*,1:272-276

Randhawa, G.J. and Chhabra, R.,2009. Import and commercialization of transgenic crops: an Indian perspective. *Asian Biotechnol. Dev. Rev.*, 11: 115–130

Sahai, S., 2003. Genetically modified crops: issues for India. *Fin Agric.*, 35:7–11.

Sonam Sneha, 2018. Current scenario of transgenic crops in India. Insights in Aquaculture and Biotechnology, Vol.2, Number 2.

Swaminathan, M.S., 2009. 'GM: Food for thought', The Asian Age, 26 August 2009.

Usha Kiran Betha, 2019. ICAR-Indian Institute of Oilseeds Research, Rajendranagar, Hyderabad, India

Whalon, M., Mota-Sanchez, D. and Hollingworth, R., 2010. Arthropod pesticide resistance database.

Whitman, D.B., 2000. "Genetically modified foods: harmful or helpful?" CSA, 2000, http://www.csa.com/.

16

Operational Biosafety Practices and Procedures

A standard operating procedure (SOP) is a set of step-by-step instructions compiled by an organization to help workers carry out routine operations. SOPs aim to achieve efficiency, quality output and uniformity of performance, while reducing miscommunication and failure to comply with industry regulations. The military (e.g. in the U.S. and UK) sometimes uses the term standing (rather than standard) operating procedure because a military SOP refers to a unit's unique procedures, which are not necessarily standard to another unit. The word "standard" can imply that only one (standard) procedure is to be used across all units. The term can also be used facetiously to refer to practices that are unconstructive, yet the norm. In the Philippines, for instance, "SOP" is the term for pervasive corruption within the government and its institutions.

A Safe Operating Procedure is a written plan of a project, test, or experiment that assesses its potential hazards and. explains how the hazards have been eliminated or minimized.

Procedures are extensively employed to assist with working safely. They are sometimes called safe work methods statements (SWMS, pronounced as 'Swims'). They are usually preceded by various methods of analysing tasks or jobs to be performed in a workplace, including an approach called job safety analysis, in which hazards are identified and their control methods described. Procedures must be suited to the literacy levels of the user, and as part of this, the readability of procedures is important (Taylor, 2012).

How to write a standard operating procedure document?

1. Step 1: Begin with the end in mind. ...
2. Step 2: Choose a format. ...
3. Step 3: Ask for input. ...

4. Step 4: Define the scope. ...
5. Step 5: Identify your audience. ...
6. Step 6: Write the SOP. ...
7. Step 7: Review, test, edit, repeat.

Sops should include In general, administrative/programmatic SOPs will consist of five elements: Title page, Table of Contents, Purpose, Procedures, Quality Assurance/Quality Control, and References

Biosafety Standard Operating Procedures (SOPs) List

SOP BIO-001 Lab Glassware Use and Disposal ;
SOP BIO-002 Sharps Usage and Disposal ;
SOP BIO-003 Disposal of Solid Biohazardous Waste ;
SOP BIO-004 Decontamination of Liquid Biohazardous Waste
SOP BIO-005 Decontamination of Reusable Lab ware, Work Surfaces and Equipment
SOP BIO-006 Use of Autoclave for Sterilization of Materials and Biological Waste ; SOP BIO-007 Cleaning Biological Spill Inside the Centrifuge
SOP BIO-008 Cleaning and Decontamination of Small Spills in the Lab or BSC
SOP BIO-009 Cleaning Instruments and Materials Used for Handling Potentially Prion Infected Neural Tissue
SOP BIO-010 Use and Cleaning of the Biosafety Cabinet
SOP BIO-011 For Cleaning Blood and Body Fluid Spills;
SOP BIO-012 Biosafety Level 2 Practices

16.1 Biosafety Level (BSL) Practices Chart

Biosafety Levels (BSL): Research and teaching activities involving infectious agents requires prior approval by the Institutional Biosafety Committee (IBC) via the Biohazard Use Authorization (BUA) review process. Each laboratory space where biohazardous materials are used is assigned one of 3 internationally recognized biosafety levels or BSL. The biosafety level is commensurate with the:

- Degree of risk posed by the biohazardous materials
- Activities carried out with those materials

The IBC uses the biosafety levels recommended by the CDC and NIH as the usual standards of containment to be set for work with a given biohazardous material. Containment requirements are subject to modification by the IBC at its discretion, depending on the circumstances presented by a specific project (Anonymous, 2020).

*Containment is the term used to describe methods, practices, procedures, facilities, and equipment used to safely manage biohazardous materials in the laboratory. The purpose of containment is to reduce or eliminate exposure to people or the environment to potentially hazardous agents.

Biosafety levels

BSL	Agents	Practices Safety	Equipment (Primary Barriers)	Facilities (Secondary Barriers)
	Not known to consistently cause disease in healthy adults	Standard Microbiological Practices	None required	Open bench top, sink required
	Associated with human disease, hazard = percutaneous injury, ingestion, mucous membrane exposure	BSL-1 practice plus: Limited access • Biohazard warning sign • "Sharps" precautions • Biosafety manual defining • any needed waste decontamination or medical surveillance policies	Primary barriers = Class I or II BSCs or other physical containment devices used for all manipulations of agents that cause splashes or aerosols of infectious materials; PPEs: laboratory coats; gloves; face protection as needed	BSL-1 plus: Autoclave available
	Indigenous or exotic agents with potential for aerosol transmission; disease may have serious or lethal consequences.	BSL-2 practice plus: • Controlled access • Decontamination of all • waste Decontamination of lab • clothing before laundering • Baseline serum	Primary barriers = Class I or II BCSs or other physical containment devices used for all open manipulations of agents; PPEs: protective lab clothing; gloves; respiratory protection as needed	BSL-2 plus: Physical separation • from access corridors Self-closing, doubledoor access • Exhausted air not • recirculated Negative airflow into • laboratory

Guidelines for good laboratory practices at BSL-2: BSL-2 applies to work with a broad spectrum of moderate-risk agents that are generally present in the environment at large and are associated with human disease of varying severity (Anonymous, 2012). Microorganisms assigned to this containment level include Salmonella spp., Toxoplasma spp., and Hepatitis B. With the use of good microbiological techniques, much of this work can be done on open bench tops as long as there is limited potential for splashes and aerosol creation.

1. Laboratory employees must immediately notify the laboratory supervisor or PI; and they should in turn notify RU EHS, OEP, and RUPD, in case of an accident, injury, illness, or overt exposure associated with laboratory activities. When deemed necessary, proceed to the RU Student Health Center for appropriate medical surveillance and/or treatment, after proper decontamination.
2. Access to the laboratory must be limited or restricted by the PI when experiments or work with cultures or specimens is in progress. In addition:
    - Only personnel advised of the special hazards and meeting any specific entry requirements, i.e., FBI clearance for work with select agents, appropriate immunizations, or serum sampling, are permitted in the laboratory. All bio-hazardous procedures, provided by the PI, must be understood and followed by laboratory occupants.
    - Lockable, self-closing doors and other security measures must be provided for facilities maintaining select agents and toxins to control access.
    - Ensure that when bio-hazardous agents and toxins are in use or stored in the laboratory, a biohazard sign is posted on the lab access door. This sign identifies the agent(s) in use, the biosafety level, any required immunizations, the PI's name and telephone number, and any PPE that must be worn in the laboratory.
    - The PI must ensure that all laboratory personnel receive appropriate training on hazards associated with the agents/toxins involved, the necessary precautions to prevent exposures, and exposure evaluation procedures. Personnel must receive training before starting work with the agents/toxins, as well as annual updates and additional training as necessary for procedural or policy changes. The laboratory director is responsible for ensuring that, before working with organisms at Biosafety Level 2, all personnel demonstrate proficiency in standard microbiological practices and techniques.
3. Laboratory employees must wash hands frequently and always after handling viable material or animals, after removing gloves, and before leaving the laboratory. A sink for hand washing must be present in each

laboratory. Foot, knee, or automatically operated hand washing sinks should be considered. An eyewash and safety shower must be readily accessible.

4. Eating, drinking, smoking, chewing gum, handling contact lenses, or applying cosmetics is not allowed in the laboratory. Persons wearing contact lenses in the laboratory should also wear goggles or a face shield.
5. Food, medications, or cosmetics should not be brought into the laboratory for storage or later use. Food must be stored outside the work area in cabinets or refrigerators designated specifically for that purpose.
6. Mouth pipetting is forbidden; only mechanical pipetting devices are permitted. 7. All procedures must be performed carefully to minimize the creation of splashes or aerosols.
8. Establish and follow policies for safe handling of sharps, including:
   - Use a high degree of caution when handling any contaminated sharp item, such as needles and syringes, slides, pipettes, capillary tubes, and scalpels;
   - Restrict needles and syringes or other sharp instruments in the laboratory for use only when there is no alternative, such as for parenteral injection, phlebotomy, or aspiration of fluids from laboratory animals and diaphragm bottles.
   - Substitute plastic-ware for glass whenever possible.
   - Handle broken glassware with a brush and dustpan, tongs or forceps, not directly with hands;
   - Use only needle-locking syringes or disposable syringe-needle units (i.e., needle is integral to the syringe) for injection or aspiration of infectious material. Syringes which re-sheathe the needle, needle-less systems and other safety devices should be used whenever possible and appropriate;
   - Do not bend, shear, break, recap, or remove used needles from disposable syringes or otherwise manipulate such units by hand before disposal. Dispose of needles and syringes in the puncture resistant sharps container provided in the laboratory for this purpose. Place full sharps containers in an autoclave bag and sterilize before disposal in biohazard containers.
9. Lab coats, gowns, smocks, or other provided protective garments must be worn while in the lab. When exiting the lab, remove lab coats and other protective clothing in the lab for disposal or laundering.
10. Gloves must be worn when manipulating bio-hazardous agents or when hands must otherwise contact contaminated surfaces. Remove and change gloves when overtly contaminated or when torn or punctured. Do not wear contaminated gloves outside the lab. Do not wash or reuse disposable gloves. Consider alternatives to latex gloves to prevent allergic response.

11. Appropriate face protection (goggles, mask, face shield or other splatter guard) must be worn for anticipated splashes or sprays of bio-hazardous agents to the face when agents must be handled outside the BSC.
12. Equipment and work surfaces must be decontaminated at completion of work, at the end of the day, and following spills of viable materials.
13. Bench tops must be impervious to water and resistant to solvents, acids, alkalis, and chemicals used for surface decontamination. Laboratory surfaces and spaces between fixtures must be designed to be easily cleaned; no carpets or rugs are allowed in the laboratory.
14. Work in the open laboratory is permitted, except for those instances where a properly maintained biological safety cabinet is required, such as:
    - Procedures with a potential for creating infectious aerosols or splashes are conducted. These may include centrifuging, grinding, blending, vigorous shaking or mixing, sonic disruption, opening containers of infectious materials whose internal pressures may be different from ambient pressures, inoculating animals intranasally, and harvesting infected tissues from animals or embryonated eggs; and
    - High concentrations or large volumes of bio-hazardous agents or toxins are used. Such materials may be centrifuged in open laboratory if sealed rotor heads or centrifuge safety cups are used, and if these rotors or safety cups are opened only in a biological safety cabinet.
    - Air sampling studies have shown that most of the common manipulations of bacterial and viral cultures in research laboratories release aerosols of viable organisms. This must be considered when evaluating need for use of the biological safety cabinet or other physical containment device.
15. Dispose of all bio-hazardous wastes and associated wastes as outlined by RU EHS Policy.
16. All containers of cultures, tissues, specimens of body fluids, or other potentially infectious waste must be covered to prevent leakage during collection, handling, processing, storage, transport, or shipping.
17. An insect and rodent control program is in place at RU. Occupants should routinely ensure screens are fitted on exterior windows that open into the lab and contact Facilities Management when pest control is needed.
18. Illumination must be adequate for all activities, avoiding reflections and glare that could impede vision.

16.2 Biological Safety Cabinets (BSCs)

BSCs are classified as Class I, Class II, or Class III cabinets. When properly maintained and operated, they effectively contain and capture microbial contaminants and infectious agents using HEPA (High Efficiency Particulate Air) filters. Biosafety cabinets should not be confused with clean benches which only protect the material being worked with and are not suitable for work with infectious or toxic material. (Although clean benches, like BSCs, have HEPAfiltered air, in clean benches the air flows over the experimental material toward the user rather than being drawn away) (Anonymous, 2012). BSCs should also not be confused with conventional fume hoods that do not filter microorganisms.

Guidelines for Working in a BSC

- Turn off the ultraviolet lamp if one is in use. Turn on the fluorescent lamp.
- Inspect the air intake grilles for obstructions and foreign material and remove if necessary.
- Adjust view screen to proper height.
- Turn the cabinet on for at least 10 minutes prior to use, if the cabinet is not left running.
- Prepare a written checklist of materials necessary for the particular activity.
- Wash hands and arms with mild soap. Put on a rear-fastening, long-sleeved gown with tight-fitting cuffs. Put on safety glasses and a pair (or two pairs) of high quality nitrile gloves.
- Disinfect work surface with a suitable disinfectant.
- Place items into the cabinet so that they can be worked with efficiently without unnecessary disruption of the airflow, working with materials from the clean to the dirty side.
- Adjust the working height of the stool so that the worker's face is above the front opening.
- Delay manipulation of materials for approximately one minute after placing hands/arms inside the cabinet.
- Minimize the frequency of moving hands in and out of the cabinet.
- Do not disturb the airflow by covering any portion of the grillwork with materials.
- Work at a moderate pace to prevent the air flow disruption that occurs with rapid movements.
- Wipe the bottom and side of the hood surfaces with disinfectant when work is completed.

Note: Be very careful when using small pieces of materials such as chemwipes in the hood. These can be blown into the hood and disrupt the motor operations.

Certification of the BSC: Certification is a series of performance tests on the BSC to confirm that it will provide the user and experimental material the protection for which it is designed. The airflow, filters, and cabinet integrity are checked to ensure that the cabinet meets minimum performance standards. Certification is arranged through each department and provided by an outside vendor. BSCs intended for user protection and/or BSL2 work must be certified:

- After they are received and installed (before use with infectious materials);
- After filter changes;
- After being moved (even a few feet); and
- Annually

BSC decontamination (using the formaldehyde gas production process) must be provided by an outside vendor and needs to be done:

- before any maintenance work requiring disassembly of the air plenum, including filter replacement;
- prior to cabinet recertification;
- before moving the cabinet to a new laboratory; and
- before discarding or salvaging.

The production of formaldehyde gas is a health concern; therefore, extreme caution should be used when having the procedure performed.

Laboratory emergency postings

- Names of responsible individuals to be contacted in case of emergencies must be posted outside of entrance doors leading into each laboratory.
- A list of the significant hazards found within the laboratory must to be posted for notification of staff and emergency response personnel. The list of hazards that must be identified by signage posted at entrances to laboratories includes (but is not limited to):
- Use/storage of bio-hazardous agents, acute carcinogens and toxic chemicals, radiological agents, and flammable materials.
- Presence of strong magnetic equipment.
- Emission of X-rays.
- Required PPE.
- A listing of all alarms in the laboratory and whom to contact if an alarm is sounding

Evacuation routes: Principal investigators must ensure that staff receives adequate training regarding emergency evacuation procedures that includes the following elements:

- Familiarization with primary and secondary (alternate) evacuation routes.
- Awareness of alarm method(s) used to signal a building evacuation.
- Designation of post evacuation meeting areas for laboratory staff.

16.3 Biohazard, Levels and Declaration

Biohazard: A biological hazard or biohazard, is a biological substance that poses a threat to the health of living organisms, primarily humans. This could include a sample of a microorganism, virus or toxin that can adversely affect human health. A biohazard could also be a substance harmful to other animals. The term and its associated symbol are generally used as a warning, so that those potentially exposed to the substances will know to take precautions. The biohazard symbol was developed in 1966 by Charles Baldwin, an environmental-health engineer working for the Dow Chemical Company on the containment products.It is used in the labeling of biological materials that carry a significant health risk, including viral samples and used hypodermic needles.

In Unicode, the biohazard symbol is U+2623 (Fig.21).

Fig. 21: The Biohazard Symbol

Hazard Clearance and Declaration: The purpose of the Hazard Clearance and Declaration Form (a.k.a. Clean Sheet) is to make Physical Facilities (PF) personnel aware of the potential hazards that may exist in areas where they are requested to perform work. PF personnel are not expected to handle equipment such as fume hoods, refrigerators, freezers, furniture, or other laboratory accessories that is contaminated with hazardous materials. It is the responsibility of the person making the work request (customer) to ensure that all surfaces have been properly decontaminated before PF personnel arrive to perform work in the area. It is also the customer's responsibility to declare any unique hazards (e.g., electrical, physical) that may still remain in the area. Even though decontamination of surfaces potentially contaminated with hazardous materials is required, all personnel should still wear appropriate personal protective equipment (PPE) when handling the equipment as a precautionary measure (e.g., chemical-resistant gloves).

The Hazard Clearance and Declaration Form facilitates the necessary communication between the customer and PF personnel. The Hazard Clearance and Declaration Form will be initiated by PF personnel if there is a need to request more information about the potential hazards of an area, or it can be initiated by the customer at the time work is requested to lessen the potential for work delays. However, it is always the customer's responsibility to complete Section 1 of this form when asked to do so.

REM's Role: If radioisotopes have been used the form must be completed and Radiological and Environmental Management (REM) must be contacted (49-46371): Radioisotope use is the only circumstance where it is necessary to contact REM. If the customer properly decontaminates an area and declares the hazards that may still remain on the form, then it is not necessary for REM to get involved. However, if the customer or PF personnel need REM to provide a professional opinion about possible needs for more cleaning, analytical testing, or additional PPE requirements they can do so by calling 49-46371.

Decommissioning and Cleaning: The following procedures must be completed by the customer before PF personnel perform services:

- If radioisotopes were used, decontaminate all accessible surfaces and contact REM (49-46371) to verify that no detectable surface contamination exists.
- If biological agents were used, decontaminate all accessible surfaces with an appropriate disinfectant (e.g., 10% bleach solution). Indicate the type of decontamination procedures that were used in Section 1 of the Hazard Clearance and Declaration Form. Explain any stains or other evidences of spills that may still remain after cleaning has been performed.
- When requesting PF personnel to dispose of equipment, ensure that all surfaces have been decontaminated if necessary. Explain any stains or other evidences of spills that may still remain after cleaning has been performed. Ensure that all hazard labels (e.g., Biohazard, Carcinogen) have been removed or defaced. The exception to this rule is for radioactive labels; REM will remove these labels once the equipment is confirmed to have no contamination.
- Remove all chemical containers, beakers, test tubes, sharps, and all other potentially hazardous materials from the immediate area (e.g., remove chemicals from the hood before PF personnel arrive to work on the hood). Any additional hazards that may be present should be abated or clearly stated in Section 1 of this form if abatement is not possible.
- **Communicate all potential hazards and recommend appropriate PPE in Section 1 of this form.** PF personnel will be able to perform their job more safely if they are aware of all hazards that are present.

Regulation: Biohazardous safety issues are identified with specified labels, signs and paragraphs established by the American National Standards Institute (ANSI). Today, ANSI Z535 standards for biohazards are used worldwide and should always be used appropriately within ANSI Z535 Hazardous Communications (HazCom) signage, labeling and paragraphs. The goal is to help workers rapidly identify the severity of a biohazard from a distance and through colour and design standardization.

Biological hazard symbol design

- A red or white-coloured background is used behind a black biohazard symbol when integrated with a DANGER sign, label or paragraph.
- An orange or white-coloured background is used behind a black biohazard symbol when integrated with a WARNING sign, label or paragraph.
- A yellow or white-coloured background is used behind a black biohazard symbol when integrated with a CAUTION sign, label or paragraph.
- A green or white-coloured background is used behind a black biohazard symbol when integrated with a NOTICE sign, label or paragraph.

DANGER is used to identify a biohazard that will cause death. WARNING is used to identify a biohazard that may cause death. CAUTION is used to identify a biohazard that will cause injury, but not death. NOTICE is used to identify a non-injury biohazard message (e.g. hygiene, cleanup or general lab policies).

OSHA requires the use of proper ANSI HazCom where applicable in American workplaces. States and local governments also use these standards as codes and laws within their own jurisdictions. Proper use of ANSI Z535 signs, labels and paragraphs are written into many of OSHA's standards for HazCom and crafted to integrate with ISO symbols.

Biohazard Classification

Biohazardous agents are classified for transportation by UN number:

- Category A, UN 2814 : Infectious substance, affecting humans: An infectious substance in a form capable of causing permanent disability or life-threatening or fatal disease in otherwise healthy humans or animals when exposure to it occurs.
- Category A, UN 2900 : Infectious substance, affecting animals (only): An infectious substance that is not in a form generally capable of causing permanent disability or life-threatening or fatal disease in otherwise healthy humans and animals when exposure to themselves occurs.
- Category B, UN 3373 : Biological substance transported for diagnostic or investigative purposes.
- Regulated Medical Waste, UN 3291 – Waste or reusable material derived from medical treatment of an animal or human, or from biomedical research, which includes the production and testing.

Levels of Biohazard: The United States Centers for Disease Control and Prevention (CDC) categorizes various diseases in levels of biohazard, Level 1 being minimum risk and Level 4 being extreme risk. Laboratories and other facilities are categorized as BSL (Biosafety Level) 1–4 or as P1 through P4 for short (Pathogen or Protection Level).

- **Biohazard Level 1:** Bacteria and viruses including Bacillus subtilis, canine hepatitis, Escherichia coli, and varicella (chickenpox), as well as some cell cultures and non-infectious bacteria. At this level precautions against the biohazardous materials in question are minimal, most likely involving gloves and some sort of facial protection.
- **Biohazard Level 2:** Bacteria and viruses that cause only mild disease to humans, or are difficult to contract via aerosol in a lab setting, such as hepatitis A, B, and C, some influenza A strains, Human respiratory syncytial virus, Lyme disease salmonella, mumps, measles, scrapie, dengue fever, and HIV. Routine diagnostic work with clinical specimens can be done safely at Biosafety Level 2, using Biosafety Level 2 practices and procedures. Research work (including co-cultivation, virus replication studies, or manipulations involving concentrated virus) can be done in a BSL-2 (P2) facility, using BSL-3 practices and procedures.
- **Biohazard Level 3:** Bacteria and viruses that can cause severe to fatal disease in humans, but for which vaccines or other treatments exist, such as anthrax, West Nile virus, Venezuelan equine encephalitis, SARS coronavirus, MERS coronavirus, SARS-CoV-2, Influenza A H5N1, hantaviruses, tuberculosis, typhus, Rift Valley fever, Rocky Mountain spotted fever, yellow fever, and malaria.
- **Biohazard Level 4:** Viruses that cause severe to fatal disease in humans, and for which vaccines or other treatments are not available, such as Bolivian hemorrhagic fever, Marburg virus, Ebola virus, Lassa fever virus, Crimean–Congo hemorrhagic fever, and other hemorrhagic diseases, as well as Nipah virus (Tigabu Bersabeh *et.al.*, 2014). Variola virus (smallpox) is an agent that is worked with at BSL-4 despite the existence of a vaccine, as it has been eradicated and thus the general population is no longer routinely vaccinated. When dealing with biological hazards at this level, the use of a positive pressure personnel suit with a segregated air supply is mandatory. The entrance and exit of a Level Four biolab will contain multiple showers, a vacuum room, an ultraviolet light room, autonomous detection system, and other safety precautions designed to destroy all traces of the biohazard. Multiple airlocks are employed and are electronically secured to prevent doors from both opening at the same time. All air and water service going to and coming from a Biosafety Level 4 (P4) lab will undergo similar decontamination procedures to eliminate the possibility of an accidental release. Currently there are no bacteria classified at this level.

Biohazards are infectious agents or biologically-derived infectious materials that present a potential risk to the health of humans or animals, either directly through infection or indirectly through environmental contamination (Anonymous, 2012). Infectious agents have the ability to replicate and give rise to potentially large populations when small numbers are released in nature from a controlled situation. Principal investigators should indicate below any of the hazard categories which are stored or in use within laboratories under their charge. If boxes are checked, identity and biosafety level of agents meeting classification should be listed in the space provided below each category, additional spaces may be added if required:

- Pathogens: human, animal, and plant pathogens, including bacteria, prions, rickettsia, fungi, viruses, and parasites.
- Human Blood and Other Potentially Infectious Materials (OPIM): All human blood, blood products, unfixed clinical tissues, and certain body fluids
- Cells/Cell Lines: Cultured cells from humans, non-human primates, and other mammalian species and the potentially infectious agents these cells may contain
- Allergens: (adjuvants, animal's dander, latex, etc.):
- Toxins: (bacterial, fungal, plant, etc.):
- Recombinant DNA and related products:
- Clinical Specimens:
- Infected animals: including live animals, animal tissues, animal bedding/ waste materials, and other materials derived from known or potentially infected animals.
- Select agents: including and of the select agent materials.

16.4 Record Keeping Requirements

Standard Operating Procedures (SOPs) Manual: Laboratory SOPs for procedure involving bio-hazardous and rDNA materials should be attached in Appendix A of this manual. Standard Operating Procedures (SOPs) must be developed for all laboratory procedures that involve known or potentially bio-hazardous agents (Anonymous, 2012). Laboratory staff must be trained annually in regards to each SOP or whenever new procedures are added to the work regimen. Required elements of an SOP include,

1. Descriptive title defining purpose of operation.
2. Preparation and revision dates.
3. Identification of department/laboratory for which SOP is applicable.
4. Brief statement of purpose. 5. Indication of potential undesirable outcomes.
6. Identification of regulatory standards that apply to procedures.
7. Listing by category of all materials, tools, and equipment required to complete SOP. It is critical that all safety equipment be identified (required PPE, BSCs, etc.).

8. Listing of environmental conditions, time constraints, or other factors which may have a negative impact on the execution of the SOP.
9. An overview of the sequence of the SOP describing major functions and anticipated/potential health and safety and environmental impact.
10. Definitions of terms.
11. Prominent display of warnings and cautions prior to description of each task with potential danger involved.
12. Listing of all tasks included within SOP in sequential order.

Biological Material Safety Data Sheets: In addition to the inventories of MSDSs compiled for hazardous chemicals, all laboratories performing research with known or potential bio-hazardous agents must maintain a comprehensive collection of MSDSs for each bio-hazardous agent. Questions or concerns regarding acquisition of MSDSs of bio-hazardous materials should be directed to the Biosafety Officer.

Job Safety Analysis (JSA): OSHA requires that supervisors prepare assessments (JSAs) of safety issues relating to the workplace and work activities identifying the range of hazards and determining whether existing precautions are adequate. If existing safety precautions are not adequate, appropriate corrections must be identified and implemented. Supervisors are further required to explain and discuss the completed JSA with employees and to maintain the JSA within the laboratory.

Exposure Control Plan (ECP): In accordance with the requirements of the OSHA Blood borne Pathogen Standard, laboratories performing research where there is a potential occupational exposure to bloodborne pathogens are required to develop and maintain an ECP. The Office of Environmental Health and Safety has developed a program which provides a uniform policy for protection of university personnel who, as part of their job function, face reasonably anticipated exposure to bloodborne pathogens.

Biosafety Levels 1, 2, 3, & 4: Biological safety levels are ranked from one to four and are selected based on the agents or organisms on which the research or work is being conducted. Each level up builds on the previous level, adding constraints and barriers (Fig.22)

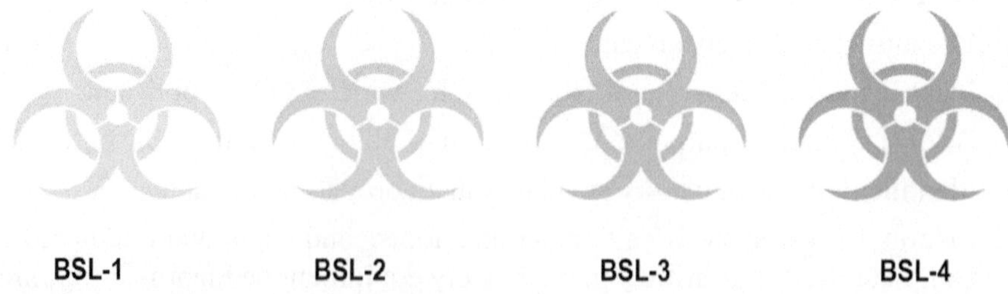

Fig. 22: Biosafety Levels

16.5 Requirements and Practices for Your Assigned BSL

Follow requirements and practices for your assigned BSL: (Anonymous, 2020)

- A. Hazard levels
- B. Standard microbiological practices
- C. Safety equipment
- D. Laboratory facilities

Biosafety levels (BSL)	BSL–1	BSL–2	BSL-2+	BSL–3
A. Hazard levels				
1. Degree of hazard	Low risk: Well characterized agents not known to cause disease in healthy adult humans	Moderate: Agents that cause human disease of moderate hazard	High: Agents involved in laboratory-acquired infections or where disease can have serious or potentially lethal consequences.	High: Agents that cause disease of moderate to high hazard that have serious or potentially lethal consequences
2. Examples	Escherichia coli (laboratory strain)	Streptococcus pyogenes	Listeria monocytogenes	Mycobacterium tuberculosis
B. Standard microbiological practices				
Biosafety levels (BSL)	BSL–1	BSL–2	BSL–2+	BSL–3
1. Access to the laboratory	Access does not have to be restricted – however, doors cannot be propped open (in violation of fire code).	Doors to the laboratory are closed when BSL-2 work is being conducted to prevent public access.	Doors to the laboratory are closed when BSL-2 work is being conducted to prevent public access.	Doors to the laboratory are closed and locked to prevent untrained personnel access.
2. Biohazard signage	Biohazard sign must be posted.	Biohazard sign must be posted.	Biohazard sign must be posted.	Biohazard sign must be posted.
3. Biohazard solid waste decontamination	Biomedical waste vendor.	Biomedical waste vendor or steam sterilize with EH&S approval.	Biomedical waste vendor or steam sterilize with EH&S approval. Pathological waste or infected animals must be incinerated.	Steam sterilize in laboratory – EH&S may grant exceptions in extenuating circumstances.

Biosafety levels (BSL)	BSL–1	BSL–2	BSL-2+	BSL–3
4. Biohazardous liquid culture decontamination	10% bleach/water made fresh daily with bleach having an EPA registration number (e.g., Chlorox) for 30 minutes.	10% bleach/water made fresh daily with bleach having an EPA registration number (e.g., Chlorox) for 30 minutes or steam sterilize with EH&S approval.	10% bleach/water made fresh daily with bleach having an EPA registration number (e.g., Chlorox) for 30 minutes or steam sterilize with EH&S approval. Prion liquid waste must be added to 1 Normal NaOH and collected as hazardous waste.	Steam sterilize in laboratory – EH&S may grant exceptions in extenuating circumstances.
5. Eating, drinking, application of cosmetics or contact lenses	Permitted only in designated clean areas.	Permitted only in designated clean areas. Not permitted if Aerosol Transmissible Disease Pathogens are used.	Not permitted at any time.	Not permitted at any time.
6. Contaminated sharps (e.g., needles, blades, glass)	Safe handling practices must be developed and implemented. Substitute plastic ware for glassware whenever possible.			
7. Decontamination of work surfaces	Daily, after finishing work and following spills.			
8. Pipetting	Mechanical device – no mouth pipetting.			
9. Storage of biohazardous waste material	Double red bags held in rigid, leak proof containers with biohazard labels on the top and side. Biohazardous waste must be under direct control of the responsible laboratory until it is placed in an EH&S approved storage area.			
10. Hand washing	Required after working with potentially hazardous materials and before leaving the laboratory.			
11. Training	The laboratory supervisor must ensure that laboratory personnel receive appropriate training regarding their duties, the necessary precautions to prevent exposures, and exposure evaluation procedures.			

Biosafety levels (BSL)	BSL–1	BSL–2	BSL-2+	BSL–3
12. Medical surveillance	Recommended where personal health status may result in increased susceptibility to infection or inability to receive vaccinations or prophylactic interventions.			Required. Laboratory personnel must be provided with medical surveillance and offered appropriate immunizations.
13. Equipment decontamination	Equipment must be cleaned of residues and green tagged by EH&S before repair, maintenance, or removal from laboratory.			Equipment must be decontaminated and green tagged by EH&S before repair, maintenance, or removal from laboratory.
14. Animals and plants not associated with the work	Allowed if approved by laboratory director and university policy.			Not allowed in the laboratory.
15. Institutional Biosafety Committee approval required for use of glassware	Not applicable.	Recommended	Required for HBV, HCV, and HIV	Required.
C. Safety equipment				
1. Class II Biological safety cabinet (with annual certification)	Not required.	Required for all aerosol generating processes**	Required for all work.	Required for all work.

Biosafety levels (BSL)	**BSL–1**	**BSL–2**	**BSL-2+**	**BSL–3**
2. Sealed rotors or safety cups for centrifuging	Not required.	Required for high concentrations or large volumes of infectious agents. Exception: Centrifuges without secondary containment can be operated inside a certified biosafety cabinet.	Required for all work.	Required for all work.
3. Laboratory coats	Required.	Required.	Required.	Required (solid front disposable gown).
4. Gloves (alternatives to latex gloves should be available)	Required.	Required.	Required.	Required.
5. Eye protection (safety glasses, goggles)	Required. This includes work in the biosafety cabinet.	Required. This includes work in the biosafety cabinet.	Required. This includes work in the biosafety cabinet.	Required. This includes work in the biosafety cabinet.
6. Sleeve protectors	Not required.	Recommended.	Recommended	
D. Laboratory facilities				

Biosafety levels (BSL)	**BSL–1**	**BSL–2**	**BSL-2+**	**BSL–3**
1. Ventilation	Negative pressure is required if adjacent area is a lower biosafety level or non-laboratory space. Single-pass air is required. Reference: Institutional Biosafety Committee			
2. Hand washing facilities	Required.	Required (foot/elbow/ electronic operation)		
3. Autoclave	Not required.	Required in laboratory.		
4. Eye wash station	Recommended. However, use of hazardous chemicals may change this to a requirement.	Required.	Required	Required in laboratory.
5. Doors	Required.	Required. Doors should be self-closing and have locks.	Required. Doors should be self-closing and have locks.	Required. A series of 2 self-closing doors is the basic requirement for entry. The space between the 2 doors is called the anteroom. Palm scanners are used to restrict access.
6. Chairs	Chairs used in laboratory work must be covered with a non-porous material that can be easily cleaned and decontaminated with appropriate disinfectant.			
7. Cleaning and decontamination	Laboratory design should allow the facility to be easily cleaned and decontaminated. Carpets and rugs are not appropriate.	Laboratory design should allow the facility to be easily cleaned and decontaminated. Carpets and rugs are not appropriate.	Laboratory design should allow the facility to be easily cleaned and decontaminated. Carpets and rugs are not appropriate. Seams, floors, walls, and ceiling surfaces should be sealed. Spaces around doors and ventilation openings should be capable of being sealed to allow for space fumigation.	

16.6 Agents that Require Enhanced Precautions (BSL-2+)

The Institutional Biosafety Committee (IBC) has approved specific Exposure Control Plans (ECP) that require additional precautions and safety equipment for certain agents (see the list below). Handling macaque monkeys or their blood, tissue, or body fluids requires using the macaque ECP.

Agents that require enhanced precautions (BSL-2+)	
• Escherichia coli (shiga toxin producing strains) • Hepatitis B Virus (HBV) • Hepatitis C Virus (HCV) • Human Immunodeficiency Virus (HIV) • Influenza virus • Infectious Prions • Listeria monocytogenes • Lymphocytic Choriomeningitis Virus (LCMV)	• Macaque tissue • Oncogenes used in viral vectors • Rabies Virus or vectors • Salmonella typhi • Shigella dysenteriae (Type 1) • Simian Immunodeficiency Virus (SIV) • Toxoplasma gondii • Vaccinia virus or vector • Zika virus

Aerosol generating processes

- Centrifuging
- Grinding
- Blending
- Vigorous shaking or mixing
- Sonic disruption
- Opening containers with high internal pressures
- Inoculating animals intranasally
- Harvesting tissues from animals or embryonated eggs

Indian Scenario

Guidelines for the Establishment of Containment Facilities: Biosafety Level 2 (BSL-2) & 3 (BSL-3) and Certification of BSL-3 Facility:

In India, all activities related to Genetically Engineered organisms (GE organisms) or cells and hazardous microorganisms and products thereof are regulated as per the "Manufacture, Use/Import/Export and Storage of Hazardous Microorganisms/ Genetically Engineered Organisms or Cells, Rules, 1989" (Rules, 1989) notified by the Ministry of Environment, Forest and Climate Change (MoEF&CC), Government of India under the Environment (Protection) Act, 1986 (EPA 1986). The Regulations & Guidelines for Recombinant DNA Research and Biocontainment, 2017 notified by the Department of Biotechnology provide operational guidance on containment of biohazards and levels of biosafety that are required to be complied with by all the IBSCs and the host institutions involved in research, development and handling of the genetically engineered (GE) microorganisms and non-GE hazardous microorganisms (Anonymous, 2020).

With increasing number of BSL 2 and BSL3 facilitates across the country, it was felt necessary to bring out a document on establishment of BSL 2 and BSL 3 containment facility and certification procedure. The "Guidelines for the Establishment of Containment Facilities: Biosafety Level 2 (BSL-2) & 3 (BSL-3) and Certification of BSL-3 facility" notified here is an addition to the Regulations & Guidelines for Recombinant DNA Research and Biocontainment, 2017 with reference to the containment principle and procedures, containment levels, BSL3 facility operation and certification.

The "Guidelines for the Establishment of Containment Facilities: Biosafety Level 2 (BSL-2) & 3 (BSL-3) and Certification of BSL-3 facility" was drafted in accordance with the national and international reference, guidance and regulations substantiated by other inter-ministerial committees, recommendations from engineering experts, and detailed deliberations by the Expert Committee Chaired by Prof. Rakesh Bhatnagar, Vice Chancellor, Banaras Hindu University, Varanasi and RCGM experts. The guidelines notified here provide the definition of containment, procedures and special practices, safety equipment and facility designs associated with various levels of containment in the Indian scenario. The establishment of the BSL-2 facility described in Chapter 2 addresses the requirements to be followed in the facility's design and related standards and practices for biosafety. Establishment of BSL-3 facility described in Chapter 3 addresses the additional design features and containment practices, recommendations for technical standards for the engineering controls for BSL-3 laboratory, essential tests during installation of BSL-3 facility , procedures for validation in BSL-3 facility, Standard Operating Procedures (SOPs) for specific considerations for working in the BSL-3 facility and mechanism of Certification for the operation of BSL-3 facility including the application forms, and checklist featuring set standards and procedures for BSL-3 operation for annual inspection (Anonymous, 2020).

These guidelines are applicable to diagnostics, research and development organizations, stakeholders as well as academic institutions. This guidance document "Guidelines for the Establishment of Containment Facilities: Biosafety Level 2 (BSL-2) & 3 (BSL-3) and Certification of BSL-3 facility" with subsequent additional information on technical standards for engineering controls, major tests to be conducted during installation, and methodology for validation of BSL-3 facility, after detailed deliberations from the experts was approved in the 184th and 191st meetings of the Review Committee on Genetic Manipulation (RCGM) held on 04th June 2020 and 15th October 2020, respectively. It is important to mention that while the guidelines notified here provide specifications for BSL-2 facility, those for the BSL-3 facility are essential components of compliance required for the Certification of the BSL-3 facility.

Bibliography

Anonymous, 2012. Biosafety Manual & Standard Operating Procedures, Radford University Office of Environmental Health & Safety Current Revision, November 2012

Anonymous, 2020. Biosafety Level (BSL) Practices Chart, UC San Diego, 9500 Gilman DR.La Jolla, CA 92093 (858)534-2230

Anonymous, 2020. Guidelines for the Establishment of Containment Facilities: Biosafety Level 2 (Bsl-2) & 3 (Bsl-3) and Certification of Bsl-3 Facility. Department of Biotechnology, Ministry of Science and Technology ,Government of India. 84 pp.

Taylor, G.A.,2012. Readability of OHS documents - A comparison of surface characteristics of OHS text between some languages. *Safety Sci.*, 50: 1627-1635.

Tigabu Bersabeh, Rasmussen Lynn, White E. Lucile, Tower Nichole, Saeed Mohammad, Bukreyev Alexander, Rockx Barry, LeDuc James W and Noah James W., 2014. "A BSL-4 High-Throughput Screen Identifies Sulfonamide Inhibitors of Nipah Virus". *Assay and Drug Development Technologies*, 12 :155–161.

Glossary of Terms

Allele: A variant form of a gene.

Allergen: A foreign substance that causes an allergic reaction. Allergens are all around us and are often proteins or chemical compounds presented to the body through touch, ingestion and/or inhalation

Allergenicity: Refers to the potential of a substance to be an allergen. The two most important determinants of allergenicity are identity and exposure .

Allergy: A damaging immune response by the body to a substance; often associated with a particular food, pollen, fur, or dust, to which it has become hypersensitive.

Bacillus thuringiensis: It is a soil bacterium, which produces a crystalline protein (*Bt* Cry protein) that - when ingested by certain types of insects – binds to gut cell receptors and in doing so kill the insect. *Bt* toxin is considered an effective insecticide to a wide variety of pest insects.

Biodiversity: The wide diversity and complex interactions interrelatedness of earth organisms based on genetic and environmental factors.

Biosecurity: The protection of a country, region, location's or firm's from economic, environmental and/or human health from harmful organisms.

Biosecurity risk analysis: The process of risk analysis that follows three main stages, namely the likelihood of hazard identification, assessment of its impact in the event of an incursion or threat and risk management.

Biosecurity threats: Those matters or activities which, individually or collectively, may constitute a biological risk to the ecological welfare or to the well-being of humans, animals or plants of a country.

Biosensor technology: The use of cells or biological molecules in an electronic system to detect specific substances. It consists of a biological sensing agent coupled with a microelectronic circuit.

Biosynthesis: Production of a chemical by a living organism.

Biotechnology: (1) The application of science and technology to living organisms as well as parts, products and models thereof, to alter living or non-living materials for the production of knowledge, goods and services; (2) the scientific manipulation of living organism, especially at the molecular genetic level, to produce useful product. Gene splicing and use of recombinant DNA (rDNA) are major techniques used.

Bioterrorism: (1) The intentional use of biological and chemical agents for the purpose of causing harm; (2) the malicious use by terrorists of pathogens, parts of them, or their toxins in direct or indirect acts against humans, livestock or crops; (3) terrorism by intentional release or dissemination of biological agents (bacteria, viruses, or toxins); these may be in a naturally-occurring or in a human-modified form.

Biotic stress: Living organisms which can harm plants, such as viruses, fungi, and bacteria, and harmful insects.

Bonding social capital: (1) Exclusive social capital refers to the cohesive bonds (strong ties) that facilitate social relations within relatively homogenous social groups (e.g. families, ethnic groups, religious cliques); (2) ties within subgroups of villagers which may maintain a high level of intimacy and trust; (3) characterized by high level of close, closed and density-knit networks and associated high levels of familiar/personalized trust and reciprocity.

Brainstorming: A facilitated group discussion whereby members are encouraged to share their ideas about a particular topic, the main purpose of which is to get participants to react to the topic and express ideas in a creative fashion and provide opportunity to gather diverse opinions and generate new ideas, and to learn.

Bridging social capital: (1) Inclusive social capital refers to the weaker horizontal ties that exist between distant friends, associate or colleagues (e.g. civil rights movement, ecumenical religious organization); (2) relations between members of different such subgroups which involves more sparse ties with people/organizations that are diverse, or heterogeneous. These types of connections relate to generalized trust, beyond trusting relationship with people who are familiar or known.

Brokering social capital: A quality of individual or special sub-groups, who are able to extract extra social benefit by forming strategic bridges across areas of reduced relational density and thereby mobilize useful resources, access novel information and develop innovative solutions.

Brokers: (1) Individual or organizational actors who carry many exclusive links, that is, links to groups that would otherwise not be in direct contact with each other; (2) facilitator of information exchange

Buffer zone: An area surrounding or adjacent to an area officially delimited for phytosanitary purposes in order to minimize the probability of spread of the target pest into or out of delimited area, and subject to phytosanitary or other control measures, if appropriate.

Build capital: (1) Any engineered work, for example irrigation canal; it includes the infrastructure that supports the other capitals; (2) local infrastructure. Comprises housing and other buildings; roads and bridges; energy supplies; communications; markets; and transportation.

Bulbs and tubers: A commodity class for dormant underground parts of plants intended for planting (includes corms and rhizomes).

Bioethics: The study of the ethical and moral implications of new biological discoveries, biomedical advances, and their applications as in the fields of genetic engineering and drug research. It considers all living organisms and the environment, from the level of the individual to the biosphere.

Biosafety: Biosafety refers to the safe handling practices, procedures and proper use of living organisms to prevent accidental harm either directly or indirectly to human health or to the environment.

Biotechnology: Biotechnology integrates scientific and engineering principles with utility considerations to develop and improve the products and processes of living organisms. Biotech applications can roughly be divided into three different groups based on the level of conceptual complexity.

Cartagena Protocol on Biosafety: The Cartagena Protocol on Biosafety is a supplement to the Convention on Biological Diversity and was adopted by the parties in early 2000, and is open for ratification by those parties. It came into effect on 11th September 2003. It aims to protect biological diversity and human health from the potential risks arising from the import and export of genetically modified organisms (GMO's) developed using modern biotechnology.

Cas: A CRISPR associated protein.

Codex Alimentarius: The *Codex Alimentarius* is a programme established by the Food and Agriculture Organisation (FAO) and the World Health Organisation (WHO). It provides an international set of standards, best practices codes, guidelines and recommendations relating to food quality and safety, including codes governing hygienic processing practices, recommendations relating to compliance with standards, limits for pesticide residues, and guidelines for contaminants, food additives and veterinary drugs.

Co-existence: Co-existence is defined as the ability of farmers being to be able to make practical choices between conventional, organic and genetically modified crop production, in compliance with the relevant legislation on labelling rules and purity standards.

Containment: Refers to the techniques and systems used to limit the environmental exposure of hazardous or potentially hazardous biological agents or their products.

Chromosome: A thread-like structure of nucleic acids and protein found in the nucleus of most living cells, carrying genetic information in the form of genes.

Cisgene: An intact, functional gene sequence that occurs naturally in a target organism or a sexually compatible species, including its natural regulatory cis-sequences such as its promoter and terminator.

Decision under Article 11(1): a decision regarding domestic use, including placing on the market, of an LMO that may be subject to transboundary movement for direct use as food or feed, or for processing.

Decision under Article 11(4): a decision on the import of living modified organisms intended for direct use as food or feed, or for processing, under its domestic regulatory framework.

Decision under Article 11(6): a decision taken by a developing country Party or a Party with an economy in transition, in the absence of a domestic regulatory framework, prior to the first import of an LMO intended for direct use as food or feed, or for processing.

Declaration under Article 11(6): a declaration that a decision prior to the first import of an LMO intended for direct use as food or feed, or for processing, on which information has been provided under Article 11(1).

Diagnostic Procedure: A technical procedure or method for diagnosis or detection of a specific plant pest or group of plant pests.

Diagnostic Protocol: Any document that contains detailed information about a specific plant pest or group of plant pests relevant to its diagnosis. A diagnostic protocol will include diagnostic procedure/s and data on: the pest, its biology and taxonomy, detection, identification, acknowledgements, references, and contact information for expert/s.

Emergency Diagnostic Procedure (EDP) : A technical procedure developed for the diagnosis or detection of an EPP for use in an emergency response event, where a National Diagnostic Procedure is not available or found to be not suitable. The EDP will be submitted to SPHD for endorsement.

Emergency Response Event : The sequence of events put into place if an EPP is discovered.

Environmental Impact: Impacts on human beings, ecosystems and man-made capital resulting from changes in environmental quality.

Environmental impact assessment (EIA): It is a process of evaluating the likely environmental impacts of a proposed activity taking into account inter-related socio-economic, cultural and human-health impacts, both beneficial and adverse

Expert Review Panel: A group of Plant Health Experts selected by SPHD to review an evaluation report submitted by an Independent Laboratory for the approval of a diagnostic procedure/protocol in accordance to SPHD Reference Standard No. 3: Guidelines to the Development and Approval Processes for National Diagnostic Protocols

Field Trials (Confined field trials or CFTs): When referring to GMOs it involves testing, trying, or putting to proof a new technique or crop variety outside a laboratory but with specific containment requirements, e.g. limited locations, limited plot size, strict operating procedures, preventative measures so that they may not enter the food and feed supply, etc.

Gene: A hereditary, functional unit consisting of a sequence of DNA that occupies a specific location on a chromosome and encodes a specific functional product (i.e. a RNA molecule and/or protein) which determines a particular characteristic in an organism.

Gene drive: Refers to a genetic element that promotes the inheritance of a particular allele/gene to increase its prevalence in a population (also used to describe the practice of inducing such non-Mendelian inheritance).

Gene Flow (Pollen-mediated gene flow in plants): The movement of genes from one individual or population to another genetically compatible individual or population.

Gene therapy: A technique for delivering functional genes (to replace aberrant ones) into living cells by means of a genetically modified vector or by physical means in order to genetically alter the living cell (from GMO Act).

Genetic Engineering(GE) & Genetic Modification(GM): Refers to the direct modification of an organism's genome by introducing, eliminating or rearranging specific genes using the methods of modern molecular biology, particularly those techniques referred to as recombinant DNA techniques.

Genetically Modified Organism (GMO): An organism whose genome has been altered through genetic engineering and as a result may have one or more novel genetic traits not normally associated with the organism. Similar to Living Modified Organism (LMO).

Genome (& Genotype): The full complement of an organism's genetic material, which is unique for every individual except clones, e.g. plants grown from cuttings and identical twins. The human genome contains an estimated 20 to 25 thousand protein coding genes.

Genome editing: The precise modification of the nucleotide sequence of a genome.

GM event (& GM line): Refers to a unique GMO – generated from a unique DNA recombination event that took place in a single cell during transformation, which was subsequently regenerated into a complete GMO.

Guidelines: Are documents that accompany regulations and acts and are produced by regulatory authorities. They provide the steps an individual should follow with respect to a given act or regulation

Hazard: A hazard is any potential source of harm.

Herbicide: Herbicides are chemicals that are toxic to a particular group of plants and are used to manage weeds. The molecular mechanism of the toxicity is highly specific for every group of herbicides, e.g inhibition of protein synthesis, and allows a basis for the engineering of herbicide tolerance

Herbicide Tolerance: Herbicide tolerance refers to a plant's ability to survive the activity of a particular herbicide and is based on a molecular mechanism that somehow neutralises the activity of a herbicide; e.g. via deactivation or substitution reactions.

Identity Preservation: Identity preservation refers to a system of production, handling and marketing practices that preserves the identity of the source or the nature of the materials for the separate trade of grains and seeds in order to meet customer demands. Identity preservation is an important feature in markets where GM and non-GM products co-exist

Indigenous Population: Refers to communities of natural inhabitants in a defined geographical area. Indigenous people are typically a minority population that has maintained distinct cultural, language and social characteristics

Insect Resistance: The development or selection of heritable traits (genes) in an insect population that allow individuals expressing the trait to survive in the presence of levels of an insecticide (biological or chemical control agent) that would otherwise debilitate or kill this species of insect.

Insect Resistant: Usually refers to a crop plant that contains substances (e.g. bio-active proteins or hard fibres) that kill or deter plant-eating insects. Insect resistance is the second-most common commercial GM trait for crop plants after herbicide tolerance and is predominantly obtained through the use of *Bt* toxin proteins originating from the soil bacterium *Bacillus thuringiensis*.

Intellectual Property: Intellectual property often refers to the exclusive rights that developers of GM technology have to their intellectual or creative contribution. Copyrights, patents and trademarks are common types of intellectual property

Living Modified Organism : Living modified organism (LMO) refers to any living organism that possesses a novel combination of genetic material obtained through the use of genetic engineering.

Meiosis: Cell division that produces reproductive cells (gametes) in sexually reproducing organisms ; the nucleus divides into four nuclei each containing half the chromosome number (n), as opposed to diploid (2n), somatic cells with two sets (one each from each parent).

National Biosafety Framework (NBF): It is a national legislative instrument established in a manner appropriate and practical for a country meeting all the national priorities and global obligations.

National Development Plan: The National Development Plan is a long-term South African development plan, developed by the National Planning Commission in collaboration with South Africans from all walks of life. It serves as an action plan for securing the future of South Africans as charted in the Constitution.

National System of Innovation: It is a system of interconnected institutions to create, store and transfer the knowledge, skills and artefacts which define new technologies.

New breeding techniques: A non-specific collective name for a wide and evolving range of techniques aimed at modifying genomes and/or gene expression.

Next Industrial Revolution: The Next Industrial Revolution has been defined as technological developments that blur the lines between the physical, digital and biological spheres.

Novel Foods: Novel foods are products that have never been used as a food; foods which result from a process that has not previously been used for food; or, foods that have been modified by genetic manipulation. This last category of foods has been described as genetically modified foods.

Novel Trait: Novel traits are genetic characteristics in an organism that have been developed either through the use of traditional or modern biotechnology.

Null segregant: Offspring in which the GM element used to engineer the parent organism has been removed through conventional breeding (GM traits are only used transiently).

Open Innovation: The basic premise of open innovation is to introduce more actors into the innovation process so that knowledge can circulate more freely and be transformed into products and services that create new markets, fostering a stronger culture of entrepreneurship

Open Reading Frame: Part of a gene that has the potential to code for a protein or peptide

Open Science: Open Science refers to an approach to research based on greater access to public research data enabled by information and communications technology tools and platforms, broader collaboration in science, including the participation of non-scientists, and the use of alternative copyright tools for diffusing research results.

Organism: A biological entity, cellular or non-cellular, capable of metabolism, replication, reproduction or of transferring genetic material and includes micro-organisms.

Peer Review: A process by which an Independent Expert approved by SPHD, reviews the accuracy and currency of the introduction, taxonomic information (4.2 and 4.3 from SPHD RS No.2) and any other information requested by SPHD provided in the submitted diagnostic protocol.

Performance: The efficiency with which a test or procedure reacts.

Phenotype: The set of, or a specific, observable characteristic(s) of an individual organism resulting from the interaction of its genotype with the environment.

Plant Health Committee (PHC): A Standing Committee of the National Biosecurity Committee.

Plant Health Expert/ Expert: A plant health professional recognised by SPHD, PHC, CCEPP and/or NMG who has expertise in Diagnostics;• Surveillance; Management; and/or Response to EPP(s).

Plant Pest Detection: The process of finding an organism either in symptomatic or asymptomatic plant material.

Plant Pest Identification: A process of ascertaining the taxonomic identity of a plant pest.

Pleiotropic: When one gene influences two or more seemingly unrelated phenotypic traits.

Precautionary Principle: The precautionary principle is an approach to risk management that has been developed in circumstances of scientific uncertainty, emphasising the need for action in the face of potentially serious risk to humans or their environment.

Prokaryote: A unicellular organism that lacks a distinct cell nucleus and the DNA is not organised into chromosomes. It also lacks the internal structures bound by membranes called organelles, such as mitochondria.

Quadruple helix: The quadruple helix approach is grounded in the idea that innovation is the outcome of an interactive process involving government, academia/the research sector, the private sector and civil society, each contributing according to its "institutional" function in society. Inclusion of the fourth helix (i.e., civil society) is critical as science, technology and innovation is increasingly evaluated by its social robustness and inclusivity.

Recombinant DNA technology: A series of procedures that are used to join together (recombine) DNA segments. A recombinant DNA molecule is constructed from segments of two or more different DNA molecules.

Resistance Development: Usually refers to the development of resistance against some sort of treatment. Pathogenic bacteria can, for example, develop resistance against antibiotics, plants/weeds can develop resistance against herbicides and insects can develop resistance against insecticides

Resistance Development Management: Refers to the techniques and mechanisms used to prevent resistance development in targeted populations.

Reverse Breeding: A novel breeding technique designed to produce homologous parental lines for a superior heterozygous plant.

Risk: Risk is the probability of a harm occurring under defined circumstances. Risk is estimated by considering both the likelihood and consequence of a harm occurring (risk = likelihood x consequence).

Risk Analysis: Risk analysis integrates the contextualisation, assessment, management and communication of risk posed by, or as a result of activities with GMOs.

Risk Assessment: Risk assessment is defined as a formalised basis for the objective quantitative or qualitative evaluation of risk in a manner in which assumptions and uncertainties are clearly considered and presented.

Risk Perception: The subjective judgement that people make about the characteristics and severity of a risk.

Risk Management: The evaluation of whether the risks identified by the risk assessment process are acceptable and manageable, then selecting and implementing the control measures as appropriate to ensure that risks are minimised or controlled.

Science, Technology and Innovation (STI) systems: These are systems to spur economic growth, reduce poverty and increase its competitive in the global trading system.

Segregation: In a GMO context it is the physical separation of GM and non-GM components to allow for two distinct product offerings. Requires record-keeping, testing, separate value chains, etc.

Socio-economic Impacts: Socio-economic impacts refer to all social and economic impacts that result or potentially can result due to a change in amongst others, the environment, industry or policy when adopting a technology.

Species : It is one of the basic units of biological classification. They are comprised of reproductive communities and populations that are distinguished by their collective variation with respect to many different characteristics and qualities.

Stakeholder: An institution, organisation, or group that has some interest in a particular sector or system.

Substantial Equivalence: The concept of substantial equivalence is used as a guide in the safety assessment of genetically modified foods by comparing the novel food to its unmodified counterpart which has a history of safe use.

Sustainability/Sustainable Development: The ability to meet present needs without compromising those needs of future generations. It relates to the continuity of economic, social, institutional and environmental aspects of human society, as well as the non-human environment.

Synthetic biology: A modern biotechnological approach that combines science, technology and engineering to facilitate the understanding, redesign, and/or modification of existing biological systems and components or the design and manufacture of new biological systems and components, based on the selective and purposeful use of genetic material.

Thesaurus of the BCH: This point of contact is set out for the purposes of receiving notifications under Article 17 - regarding unintentional transboundary movements of living modified organisms and emergency measures.

Toxicity: The degree to which something is toxic or poisonous, or a substances potential to exert a harmful effect on humans, animals, or plants and a description of the effect and the conditions or concentration under which the effect takes place.

Toxin: A complex and poisonous organic substance, often a protein, that is produced by living cells or organisms and is capable of causing disease when introduced into the body tissues or cells of certain organisms at a specified dose.

Traceability: *Codex Alimentarius* defines traceability as "the ability to follow the movement of a food through specified stage(s) of production, processing and distribution".

Transboundary Movement: Transboundary movement refers to the movement of living modified organisms (LMOs) across boundaries of countries and includes intended as well as unintended movement.

Transdisciplinary: Transdisciplinary research is defined as research efforts conducted by investigators from different disciplines working jointly to create new conceptual, theoretical, methodological and translational innovations that integrate and move beyond discipline-specific approaches to address a common problem.

Transgene: Transgene refers to a gene(s) or genetic material that is isolated and transferred from one organism and incorporated by another organism, either naturally or through genetic engineering techniques. The process is known as transgenesis.

Unintended Effects: Unintended effects refer to the outcomes that are not (or not limited to) the results originally intended in a particular situation. The unintended results may be foreseen or unforeseen.

Unique Identification (UID): Unique identification systems are being established to identify living modified organisms. They are usually a combination of alphabets and numbers used to distinguish one organism from the other, and allow for attaching and retrieving specific information on the organism.

Validation: Validation is a process to determine the fitness of a procedure/protocol for a particular use and includes procedure optimisation and demonstration of performance characteristics and evaluation of sensitivity and specificity.

Vector : A carrier of an infectious agent.

Verification: A process by which an Independent Laboratory demonstrates that the diagnostic procedures can be followed and if necessary, identifies critical improvements in the methods described.

Variety: A subdivision of a species for taxonomic classification also referred to as a 'cultivar.' A variety is a group of individual plants that are uniform, stable, and distinguishable from other groups of individuals in the same species by distinct genetic and phenotypic characteristics.

Willingness to pay: The amount an individual is willing to pay to acquire a product or service. This may be obtained from stated or revealed preference approaches.